ACETOGENESIS

CHAPMAN & HALL MICROBIOLOGY SERIES

Physiology / Ecology / Molecular Biology / Biotechnology

Series Editors

C. A. Reddy
Department of Microbiology & Public Health
Michigan State University
East Lansing, MI 48824-1101

A. M. Chakrabarty
Department of Microbiology & Immunology
University of Illinois Medical Center
835 S. Wolcott Avenue
Chicago, IL 60612

Arnold L. Demain
Rm. 56-123
Massachusetts Institute of Technology
Cambridge, MA 02139

James M. Tiedje
Center for Microbiology Ecology
Department of Crop & Soil Sciences
Michigan State University
East Lansing, MI 48824

Previously published

James G. Ferry, ed.   *Methanogenesis*

ACETOGENESIS

EDITED BY

Harold L. Drake

CHAPMAN & HALL
New York • London

This edition published by
Chapman & Hall
One Penn Plaza
New York, NY 10119

Published in Great Britain by
Chapman & Hall
2-6 Boundary Row
London SE1 8HN

Printed in the United States of America

QR
92
.A3
A26
1994

Library of Congress Cataloging in Publication Data

Acetogenesis / Harold L. Drake, editor.
    p. cm.
    Includes bibliographical references.
    ISBN 0-412-03211-2
    1. Acetic acid—Metabolism. 2. Acetylcoenzyme A. 3. Microbial
metabolism.   I. Drake, Harold L., 1952– .
QR92.A3A26   1994
589.9—dc20   93-43138
                                                   CIP

British Library Cataloguing in Publication Data available

Please send your order for this or any Chapman & Hall book to **Chapman & Hall,
29 West 35th Street, New York, NY 10001, Attn: Customer Service Department.**
You may also call our Order Department at 1-212-244-3336 or fax your purchase
order to 1-800-248-4724.

For a complete listing of Chapman & Hall's titles, send your requests to **Chapman &
Hall, Dept. BC, One Penn Plaza, New York, NY 10019.**

Dedication

This book is dedicated to two people, Harland and Millie Wood. On one hand, this book is an extension of Harland's insight and hard work on CO_2 fixation over the course of nearly 6 decades. On the other hand, this book is an extension of Millie's continuous support and neverending belief in Harland.

Harland G. Wood

2 September 1907–12 September 1991

Contents

Preface

Originally thought of as a specialized, if not curious and obscure process, and initially attracting relatively little attention from the scientific community, the acetyl-CoA pathway is now recognized as a widespread anaerobic process that is fundamental to the global carbon cycle. Indeed, it has been postulated that this pathway may have constituted the first autotrophic process on our planet. Unfortunately, the acetyl-CoA pathway and acetogenic bacteria are often discussed in sandwich fashion with other related topics, such as anaerobic bacteria, methanogenesis, C_1 metabolism, and so on. Given the extent of information now available, it seems appropriate at the present time to bring this process and biological group into the limelight. Thus, the primary goals of this book are to provide (i) a comprehensive review of this topic and (ii) a forum for the exchange of thoughts that might be a stimulus for future work in this field. In both cases, I have encouraged the authors to tell it the way they see it.

It has been an honor to have worked together with numerous colleagues and friends in putting together the first book on acetogenic bacteria and acetogenesis. I am grateful for their excellent contributions and patience. I have not been the best editor, but no one else would take the job. I am also very grateful to Gregory Payne, Scientific Editor for Chapman and Hall Inc., for interest in this project, as well as tolerance for lateness. Also, a special thanks to Friederike Rothe for assistance with the index, and to Belle Kaluza for supplying coffee along the way.

I seldom read long prefaces, so I will close by saying I was one of many lucky individuals who learned about acetogenesis from Harland G. Wood. Although many individuals have contributed to the study of acetogenic bacteria, their ecological roles, and biochemical and physiological processes, Harland was, over the years, an essential and persistent driving force in resolution of the acetyl-CoA pathway. He was a never aging, never tiring scientist, and a strong supporter of basic science and young investigators. He passed away in 1991. The authors and I are collectively indebted to him.

Thanks, Harland. And thanks, Millie.

Harold L. Drake
Bayreuth, October 1993

Contributors

Jan R. Andreesen, Institut für Mikrobiologie, Martin-Luther-Universität Halle, Weinbergweg 16 a, D-06099 Halle, Germany.

John A. Breznak, Department of Microbiology and Center for Microbial Ecology, Michigan State University, East Lansing, Michigan, 48821-1101 USA

Marvin P. Bryant, Departments of Animal Sciences and Microbiology, University of Illinois, Urbana, Illinois 61801, USA

Steven L. Daniel, Lehrstuhl für Ökologische Mikrobiologie, BITÖK, Universität Bayreuth, D - 95440 Bayreuth, Germany

Gabriele Diekert, Institut für Mikrobiologie der Universität Stuttgart, Allmandring 31, D-70569 Stuttgart, Germany

Harold L. Drake, Lehrstuhl für Ökologische Mikrobiologie, BITÖK, Universität Bayreuth, D - 95440 Bayreuth, Germany

James G. Ferry, Department of Biochemistry and Anaerobic Microbiology, Virginia Polytechnic Institute and State University, Blacksburg, Virginia 24061-0305, USA

Anne Cornish Frazer, Center for Agricultural Molecular Biology, Cook College, Rutgers, The State University of New Jersey, New Brunswick, New Jersey 08903, USA

Georg Fuchs, Angewandte Mikrobiologie, Universität Ulm, D - 89069 Ulm, Germany

Gerhard Gottschalk, Institut für Mikrobiologie der Universität Göttingen, Grisebachstr. 8, D - 37077 Göttingen, Germany

Oleg R. Kotsyurbenko, Institute of Microbiology, Russian Academy of Sciences, Pr. 60-let Octyabrya 7, kor.2, 117811 Moscow, Russia

Kirsten Küsel, Lehrstuhl für Ökologische Mikrobiologie, BITÖK, Universität Bayreuth, D - 95440 Bayreuth, Germany

Lars G. Ljungdahl, Center for Biological Resource Recovery, Department of Biochemistry, The University of Georgia, Athens, Georgia 30602, USA

Charles R. Lovell, Department of Biological Sciences, University of South Carolina, Columbia, South Carolina 29208, USA

Roderick I. Mackie, Department of Animal Sciences, Unviersity of Illinois, Urbana, Illinois 61801, USA

Carola Mathies, Lehrstuhl für Ökologische Mikrobiologie, BITÖK, Universität Bayreuth, D - 95440 Bayreuth, Germany

Terry L. Miller, Wadsworth Center for Laboratories & Research, New York State Department of Health, Box 509, Albany, New York 12201-0509, USA

Volker Müller, Institut für Mikrobiologie der Universität Göttingen, Grisebachstr. 8, D - 37077 Göttingen, Germany

Alla N. Nozhevnikova, Institute of Microbiology, Russian Academy of Sciences, Pr. 60-let Octyabrya 7, kor.2, 117811 Moscow, Russia

Caroline M. Plugge, Department of Microbiology, Wageningen Agricultural University, Hesselink van Suchtelenweg 4, 6703 CT Wageningen, The Netherlands

Margarete A. Pusheva, Institute of Microbiology, Russian Academy of Sciences, Prospect 60-letiya Oktyabrya 7k.2, 117811 Moscow, Russia

Stephen W. Ragsdale, Department of Biochemistry, East Campus, The University of Nebraska, Lincoln, Nebraska 68583-0718, USA

Bernhard Schink, Fakultät für Biologie der Universität Konstanz, Postfach 5560, D - 78434 Konstanz, Germany

Marija V. Simankova, Institute of Microbiology, Russian Academy of Sciences, Pr. 60-let Octyabrya 7, kor.2, 117811 Moscow, Russia

Alfons J. M. Stams, Department of Microbiology, Wageningen Agricultural University, Hesselink van Suchtelenweg 4, 6703 CT Wageningen, The Netherlands

Erhard Stupperich, Angewandte Mikrobiologie, Universität Ulm, D-89069 Ulm, Germany

Ralph S. Tanner, Department of Botany and Microbiology, University of Oklahoma, Norman, Oklahoma 73019, USA

William B. Whitman, Department of Microbiology, University of Georgia, Athens, Georgia 30602, USA

Juergen Wiegel, Department of Microbiology, The University of Georgia, Athens, Georgia 30602, USA

Carl R. Woese, Department of Microbiology, University of Illinois, Urbana, Illinois 61801, USA

Gert Wohlfarth, Institut für Mikrobiologie der Universität Stuttgart, Allmandring 31, D-70569 Stuttgart, Germany

Meyer J. Wolin, Wadsworth Center for Laboratories & Research, New York State Department of Health, Box 509, Albany, New York 12201-0509, USA

George A. Zavarzin, Institute of Microbiology, Russian Academy of Sciences, Prospect 60-letiya Oktyabrya 7k.2, 117811 Moscow, Russia

Tatjana N. Zhilina, Institute of Microbiology, Russian Academy of Sciences, Prospect 60-letiya Oktyabrya 7k.2, 117811 Moscow, Russia

Stephen H. Zinder, Section of Microbiology, Wing Hall, Cornell University, Ithaca, New York 14853, USA

I

INTRODUCTION TO ACETOGENESIS

1

Acetogenesis, Acetogenic Bacteria, and the Acetyl-CoA "Wood/Ljungdahl" Pathway: Past and Current Perspectives

Harold L. Drake

1.1 Introduction: A Brief Reflection on Harland G. Wood

We will get to acetogens shortly. Before that, I must note that it was never my intent to deliver the introductory chapter for this book. That was Harland's job, and as editor of this book, I was elated when he agreed to take on that project. I knew full well no one but he was up to that task. "Time passes," he once noted to me as we were discussing a difficult problem that required an unacceptable amount of time to resolve. Unfortunately, time does pass, and Harland G. Wood passed away in September 1991 and was unable to complete his preparation of this chapter. He was 84.

Harland was an uncompromising investigator and worked with seemingly endless energy until the job was done. His capacity to put in overtime everyday held true until the time of his passing. He greatly enjoyed stalking problems in the laboratory, as well as deer in the wild. Perhaps he is best described as an enduring hunter. His annual hunting adventures are as legendary as his scientific exploits, and his "sticking-to-the-trail-no-matter-what" attitude was instrumental in bagging many great trophies, including the resolution of the acetyl-CoA pathway, the autotrophic process that constitutes the basis of this book and the careers of many of its authors, including its editor.

I first met Harland in 1977 when I was a doctoral student with James M. Akagi at the University of Kansas. Harland had been invited to KU to give a lecture on CO_2 fixation and, prior to his lecture, was making the rounds in the Microbiology Department. Harland was legendary to many of the professors in the department, and Professor Akagi had often told me of Professor Wood's importance in the scientific community. Holding his hand high above his head to indicate the height of Professor Wood, he warned me repeatedly that I would have to look up to this great scientist. By the time of his arrival in the department, I (and many other students) had conjured up the image of someone with an exceptionally tall, impressive stature. When he entered the laboratory, I was changing the oil of the lyophilizer vacuum pump and was bent over wrestling a greasy wrench. Rising up quickly to meet this giant, I was shocked we were the same height, meeting each other at eye level! He did not hesitate to shake my greasy hand and seized the moment to reflect on his repairing vacuum pumps, thermal diffusion columns, and mass spectrometers in the "good old days." With his rough, stern sort of voice (you had no trouble knowing who was yelling in the laboratory when it was Harland!), he told me not to trivialize the importance of doing a few dirty jobs in the laboratory. I was lucky that he later asked me to be a postdoctoral student in his group at "Western Reserve" (he admonished me and others soundly for using the word "Case" instead of "Western Reserve"), and as his postdoc, I saw Harland on more than one occasion with dirty hands in the laboratory. He showed by example. Although Jim played a good joke on me, it was not long before I nonetheless looked up to Harland.

The scientific community is extremely lucky that Harland remained active in the laboratory until the time of his passing because he produced some of his best work in recent years; he consequently left us with numerous publications and reviews that are still current. Two of his most recent reviews (Wood, 1991, delivered upon receiving the William C. Rose Award; Wood, 1989) are written with a personal touch and cover most of the historical moments relative to the activities of his and other research groups on resolving the acetyl-CoA pathway. Harland spent most of his career studying various aspects of CO_2 fixation, and additional historical and somewhat colorful moments on related topics are provided in additional reviews (Wood, 1972, 1976, 1982, 1985). I refer the reader to these references for Harland's statement.

To close this section, I would like to emphasize that Harland's work strongly influenced thinking in the scientific community. His breakthrough discovery in the mid-thirties on heterotrophic CO_2 fixation ultimately required reassessment of the metabolic differences between autotrophs (such as trees) and all other life forms on the planet. He is often referred to as a biochemist, but his discoveries on CO_2 fixation were basic to the whole of biology. Many realize that he was considered more than once for a Nobel Prize in recognition of his contributions; few know the rest of the story. Harland was certainly an admired colleague by many Nobelists. Although never the recipient, I include a photo that illustrates the fact that, for many of us, Harland was nonetheless a head above the rest (Fig. 1.1).

Figure 1.1. From left to right: Harland G. Wood, and Nobelists Carl F. Cori, Fritz Lipmann, and Severo Ochoa; taken in Cleveland, Ohio, on the occasion of Harland's 70th birthday celebration.

1.2 Historical Perspectives

1.2.1 Early Studies on CO_2 Fixation to Acetate

The Dutch microbiologist Wieringa discovered (reported) the first acetogenic bacterium in 1936. The organism, *Clostridium aceticum,* was a Gram-negative, spore-forming, mesophilic rod and grew at the expense of a reaction that had not been previously reported for a bacterium (Wieringa, 1936, 1939–1940, 1941):

$$4 H_2 + 2 CO_2 \rightarrow CH_3COOH + 2 H_2O \qquad (I)$$

However, 4 years earlier, the capacity of unknown organisms in sewage to reduce CO_2 to acetate at the expense of H_2 had already been observed by Fischer et al.

(1932). Although Wieringa recognized the uniqueness of this H_2-dependent process and noted that acetate was also formed from sugars, relatively little was done with *C. aceticum,* initially perhaps because of World War II but later because the culture was lost and unavailable for study until nearly four decades later (see Braun et al., 1981).

As with many fortuitous success stories in science, this loss was perhaps lucky because *Clostridium thermoaceticum* (Fig. 1.2), discovered a few years later by Fontaine and co-workers (Fontaine et al., 1942), became by default the only acetogen available for laboratory study. It was to become the model acetogen, the star of the show so to speak, on which all central studies for the elucidation of the chemolithoautotrophic acetyl-CoA pathway of acetogens were to be derived. In contrast to *C. aceticum, C. thermoaceticum* was thermophilic and also grew to very high cell densities. (With the old tomato-juice-extract medium, optical densities of 5–10 are obtainable.) These properties provided later biochemists with high quantities of relatively thermostable components with which to work. Resolution of the acetyl-CoA pathway may have been more difficult with *C. aceticum.*

Ironically, *C. thermoaceticum* was isolated as an obligate heterotroph. Nonetheless (and luckily for many of us!), it contained a new autotrophic CO_2-fixation process. Indeed, it was not until 1990, well after the enzymology of the "autotrophic" pathway for the organism was firmly established, that this obligate heterotroph was shown to be autotrophic (Daniel et al., 1990). Despite the fact that the autotrophic capacity of *C. thermoaceticum* was to remain silent for nearly 50 years, Fontaine et al. (1942) noted that this bacterium catalyzed an unprecedented growth-supportive reaction that hinted toward a new CO_2-fixing process:

$$C_6H_{12}O_6 \text{ (glucose)} \rightarrow 3 \text{ CH}_3\text{COOH} \qquad \text{(II)}$$

At first glance, this reaction may look quite dissimilar to reaction I for *C. aceticum.* However, it was recognized by Fontaine et al. (1942) that there was no known process that could account for the formation of more than 2 moles of acetate per mole of glucose consumed (via glycolytic 3-3 cleavage). It seemed likely that, as in the case of *C. aceticum,* CO_2 was being produced via oxidation and then subsequently reduced to acetate. This proposal was formulated by Fontaine et al. (1942) as follows:

> . . . it seems probable that either there is some primary cleavage of glucose other than the classical 3-3 split, or that a one-carbon compound is being reabsorbed. Of these two possibilities, the recent work on carbon dioxide uptake makes the latter seem more likely.

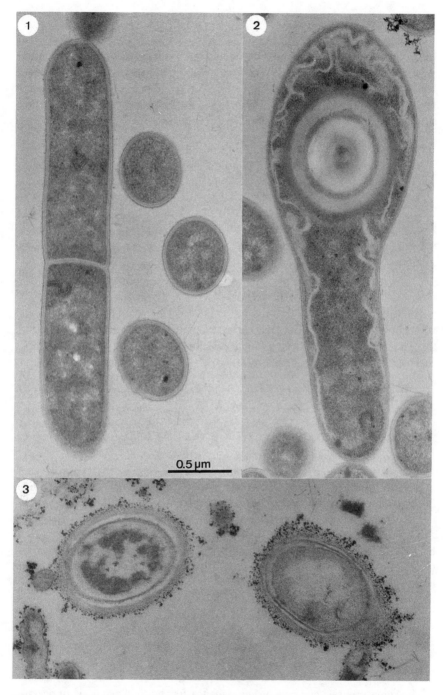

Figure 1.2.   Thin sections of chemolithoautotrophically cultivated *Clostridium thermo-aceticum*. Panel 1: vegetative cell; panel 2: sporulated cell; panel 3: spore.

In the latter statement, they were of course referring to Harland's classic work on heterotrophic CO_2 fixation (Wood and Werkman, 1936, 1938; Wood et al. 1941a, 1941b).

Subsequently, a direct correlation between the autotrophic and heterotrophic systems was formulated by Barker (1944):

Oxidation: $C_6H_{12}O_6 + 2 H_2O \rightarrow 2 CH_3COOH + 2 CO_2 + 8H$ (III)

Reduction: $8 H + 2 CO_2 \rightarrow CH_3COOH + 2 H_2O$ (IV)

Net reaction: $C_6H_{12}O_6$ (glucose) $\rightarrow 3 CH_3COOH$ (V)

Reaction IV was conceived as the equivalent of reaction I. Barker (1944) extended the problem by demonstrating that the growth of *C. thermoaceticum* was also supported by the following reaction:

$4 CH_3COCOOH$ (pyruvate) $+ 2 H_2O \rightarrow 5 CH_3COOH + 2 CO_2$ (VI)

With glucose out of the picture, this simple fermentation balance clearly suggested that CO_2 was being produced and subsequently utilized as a terminal electron acceptor:

Oxidation: $4 CH_3COCOOH + 4 H_2O \rightarrow 4 CH_3COOH + 4 CO_2 + 8 H$ (VII)

Reduction: $8 H + 2 CO_2 \rightarrow CH_3COOH + 2 H_2O$ (VIII)

Net reaction: $4 CH_3COCOOH + 2 H_2O \rightarrow 5 CH_3COOH + 2 CO_2$ (IX)

Again, we see reaction VIII as the equivalent of reaction I.

In 1945, after the wartime restrictions on isotope-related research were lifted, Barker teamed up with Kamen (Barker and Kamen, 1945) and demonstrated in the first published tracer experiments with [14]C (see Kamen, 1963, p. 241) that [14]CO_2 was indeed incorporated equally into both carbon atoms of acetate. This landmark experiment reinforced the similarity between the autotrophic and heterotrophic processes of *C. aceticum* and *C. thermoaceticum*, respectively, and uniformed future thinking. Barker and Kamen (1945) concluded

> . . . that the acetic acid fermentation of glucose by *C. thermoaceticum* involves a partial oxidation of the substrate to two moles each of acetic acid and carbon dioxide followed by a reduction and condensation of the carbon dioxide to a third mole of acetic acid.

The stage was set, and conclusive evidence for the total synthesis of acetate from CO_2 came in 1952 when Harland repeated the ^{14}C experiments of Barker and Kamen with $^{13}CO_2$ (Wood, 1952a). In this classic mass-balance experiment, Harland definitively proved that a molecule of acetate was synthesized from two molecules of CO_2. This exercise was critical because it differentiated between the random incorporation of CO_2 into either the methyl or carboxyl carbons of a population of acetate molecules, and the true, total synthesis of acetate from CO_2. The conclusion that the third molecule of acetate from glucose was a result of CO_2 fixation was supported by Harland's finding that carbons 3 and 4 of [3,4-^{14}C]-glucose gave rise to $^{14}CO_2$ (Wood, 1952b). Thus, glucose was, indeed, subject to the classic 3-3 split, and the CO_2 produced via decarboxylation of pyruvate fixed via an unknown autotrophic mechanism into acetate. Harland grasped the significance of this finding and, with *C. thermoaceticum*, set out to resolve the pathway responsible.

1.2.2 Naming of the "Wood/Ljungdahl" Pathway

Those early moments were back in the "good old days" when thousands of pure enzymes (not even one!) could not be ordered from catalogs, when the now textbook catabolic pathways were unresolved, when the Calvin cycle for autotrophic CO_2 fixation was not known, when labeled compounds were just emerging as biological tools, when paper and thin-layer chromatography were not available, and when most biologists thought DNA was too simple to be genetic information. The time required to resolve the pathway and the chemolithotrophic potentials responsible for the total synthesis of acetate and biomass from CO_2 would span the rest of Harland's career. This length of time is explained in part by the fact that most of the key enzymes involved are very oxygen labile and thus difficult to study using standard techniques. Milestones are highlighted in Table 1.1, and elsewhere throughout Chapters 1, 2, 3, and 5.

Many unexpected and important breakthroughs finally resulted in the resolution of the acetyl-CoA pathway. However, a key event occurred in 1958 when Lars G. Ljungdahl became Harland's graduate student. For the next 33 years, they were to remain focused on resolving how *C. thermoaceticum* synthesized acetate from CO_2. In their first review together in the late sixties (Ljungdahl and Wood, 1969), they outlined the work that demonstrated that reaction I constituted a new form of autotrophic CO_2 fixation, being dissimilar to Calvin-type fixation, reverse citric acid cycle fixation, and the type of CO_2 fixation occurring in the purinolytic clostridia. In this review, they proposed a model for the fixation of CO_2 by acetogens that was to become the basis of all subsequent studies on the pathway. Over the years that followed, Lars chipped away at the "methyl branch" and bioenergetics of the pathway (see Section 1.5.1), whereas Harland focused more on the mechanism by which two C_1 units were condensed in the formation of a C_2 compound. As outlined in their last review together (Wood and Ljungdahl, 1991), they collectively assembled most of the biochemical pieces of the puzzle.

Table 1.1   Events pertaining to the resolution of the acetyl-CoA pathway and chemolithoautotrophic potentials of *Clostridium thermoaceticum*.

Year	Event (Reference)
1932	H_2-dependent conversion of CO_2 to acetate with unknown organisms in sewage sludge (Fischer et al., 1932)
1936	Discovery of first acetogen, *Clostridium aceticum*, and the total synthesis of acetate from CO_2 + H_2 (Wieringa, 1936, 1939–1940) (Culture was subsequently lost)
1942	Discovery of second acetogen, *Clostridium thermoaceticum*, and the stoichiometric conversion of glucose to acetate (Fontaine et al., 1942)
1944	Acetogenic conversion of pyruvate, by Barker (1944)
1945	Synthesis of double-labeled acetate from $^{14}CO_2$ (Barker and Kamen, 1945)
1952	Synthesis of double-labeled acetate from $^{13}CO_2$ and demonstration of classic 3-3 split and decarboxylation of glucose (Wood, 1952a, 1952b)
1955	Differential labeling of acetate from [^{14}C]formate showing formate as a methyl-group precursor (Lentz and Wood, 1955)
1964–1966	Differential labeling of acetate from [^{14}C]methyl-B_{12} showing methyl-B_{12} as a methyl-group precursor (Poston et al., 1964, 1966)
1965–1966	Isolation of native corrinoids as methyl donors and proposal of pathway for the tetrahydrofolate/corrinoid-mediated synthesis of acetate from CO_2 (Ljungdahl and Wood, 1965; Ljungdahl et al., 1966)
1965	$^{14}CO_2$-labeling pattern showing potential for autotrophic synthesis of cell-carbon precursors from CO_2 (Ljungdahl and Wood, 1965)
1969	Proposal for a C_1 pathway for the autotrophic synthesis of acetate from CO_2 (Ljungdahl and Wood, 1969)
1973–1986	Resolution of formate dehydrogenase and tetrahydrofolate pathway and enzymes for conversion of CO_2 to methyltetrahydrofolate [Ljungdahl and co-workers, see Ljungdahl (1986) for review]
1978	Oxidation of CO to CO_2 and demonstration of CO dehydrogenase (Diekert and Thauer, 1978)
1981	Purification of a five-component enzyme system for synthesis of acetyl-CoA from pyruvate and methyltetrahydrofolate (Drake et al., 1981a)
1981–1982	Demonstration that CO (i) replaces pyruvate in the methyltetrahydrofolate-dependent synthesis of acetate and (ii) undergoes exchange with acetyl-CoA with purified enzyme system (Drake et al., 1981b; Hu et al., 1982)
1982	Demonstration of hydrogenase (Drake, 1982)
1983	Purification of CO dehydrogenase (Diekert and Ritter, 1983; Ragsdale et al., 1983)
1983	Usage of H_2 and CO as energy sources (Kerby and Zeikus, 1983)
1984	Resolution of minimal nutritional requirements (Lundie and Drake, 1984)
1984–1986	Demonstration that CO dehydrogenase is acetyl-CoA synthase, the central condensing enzyme in the autotrophic synthesis of acetyl-CoA (Pezacka and Wood, 1984a, 1984b; Ragsdale et al., 1985), and that "CO" is the carbonyl precursor in the pathway under growth conditions (Diekert et al., 1984; Martin et al., 1985).

(Continued)

10

Table 1.1   *Continued*

Year	Event (Reference)
1984	Hydrogenase-coupled enzyme system for H_2-dependent synthesis of acetyl-CoA (Pezacka and Wood, 1984b)
1985–1993	Enzymological and molecular studies to resolve the catalytic and regulatory mechanisms of acetyl-CoA synthase and other components of the acetyl-CoA pathway (Wood, Ragsdale, Ljungdahl, Diekert, Lindahl, and co-workers (see Ragsdale, 1991, Morton et al., 1992, and chapter 3 for reviews; see also Shin and Lindahl, 1992a, 1992b, 1993)
1986–1993	Resolution of electron transport-mediated system coupled to the synthesis of ATP from chemolithotrophic substrates (Hugenholtz and Ljungdahl, 1989, 1990; Das et al., 1989; Ivey and Ljungdahl, 1986; Das and Ljungdahl, 1993; see Chapter 2)
1990	Chemolithoautotrophic growth and utilization of H_2, CO_2, and CO (Daniel et al., 1990)
1991	Integrated model for the flow of carbon and energy via the acetyl-CoA pathway (Wood and Ljungdahl, 1991)

Lars was always the faithful student and was the first to propose that the acetyl-CoA pathway be termed the Wood pathway in honor of Harland (Ljungdahl, 1986). The numerous other investigators who made key discoveries along the way not withstanding, it was the over three decades of work by both Lars and Harland that were most essential in resolving the major framework of the acetyl-CoA pathway. Because of this, the pathway has also been referred to as the Ljungdahl/Wood pathway and Wood/Ljungdahl pathway (Ladapo and Whitman, 1990; Drake, 1993). However, I would propose that we settle on usage of the name "Wood/Ljungdahl" pathway, to not only honor both investigators but also to recognize the fundamental insight and leadership role of Harland in the study of CO_2 fixation, both of which were paramount to resolution of the pathway.

Independent of such honors, it is good to recall from where we started. We have certainly moved some distance from the pioneering works of Wieringa, Fontaine, Barker, Kamen, and others. As we get farther away from initial discoveries and deeper into biochemical mechanisms, it is easy to lose sight of the fact that acetogenesis as a process was discovered at the microbiological level, just as was heterotrophic fixation of CO_2. Although subsequent enzymological studies confirmed and resolved a new pathway, the process (i.e., target) was already identified: The stoichiometry for reaction I has not changed.

1.3   Definitions and Usefulness of the Terms Acetogen and Homoacetogen

The term "acetogenic" is somewhat nebulous and any organism making acetate or acetic acid could be termed an acetogen. However, the metabolic processes by which acetate is formed, and the role of these processes relative to growth, are not equivalent among the acetate-forming bacteria. If it is important to distinguish between acetate-forming bacteria, a differential nomenclature (i.e., a differential usage of the term acetogen) must be applied. This point should not be trivialized. For example, the term acetogen has been used to describe *Thermobacteroides proteolyticus* and the syntroph PA-1 because they form acetate from glucose, but the flow of reductant is to H_2, not the reduction of CO_2 to acetate (Ollivier et al., 1985; Brulla and Bryant, 1989). Protons are used as terminal electron acceptors, and there is no evidence that such organisms use the acetyl-CoA pathway for glucose fermentation and the reductive synthesis of acetate.

The term "acetogenesis" could also be used to describe the process by which any organism forms acetate. However, it has been suggested that its usage be restricted to processes by which two molecules of CO_2 are used to form one molecule of acetate (Wood and Ljungdahl, 1991). Unfortunately, such nomenclature again fails to adequately distinguish between the three known mechanisms by which acetate is formed from CO_2: (i) the acetyl-CoA pathway, (ii) the glycine synthase-dependent pathway used by purine and glycine fermenters, and (iii) the reductive citric acid cycle used by, e.g., certain sulfur-reducing autotrophs (Fuchs, 1986, 1989; Thauer, 1988; Wood and Ljungdahl, 1991). Although this book deals mostly with the acetyl-CoA pathway and associated organisms, the latter two processes are dealt with in Chapters 19 and 23.

To resolve such problems, the following definition has been applied in this book unless otherwise indicated or qualified (modified from Drake, 1992):

> **Acetogens** are obligately anaerobic bacteria that can use the acetyl-CoA pathway as their predominant (i) mechanism for the reductive synthesis of acetyl-CoA from CO_2, (ii) terminal electron-accepting, energy-conserving process, and (iii) mechanism for the synthesis of cell carbon from CO_2.

Although this definition is simply an extension of some used previously, note that the production of acetate is not an element of this definition. The fate of acetyl-CoA is less important than the process by which it is formed (see Section 1.4). For example, the acetogens *Eubacterium limosum, Butyribacterium methylotrophicum,* and *Clostridium pfennigii* can form butyrate from acetyl-CoA (Lynd and Zeikus, 1983; Zeikus, 1983; Krumholz and Bryant, 1985; Zeikus et al., 1985; Loubiere et al., 1992), whereas the acetogen *Acetobacterium woodii* can form ethanol from acetyl-CoA (Buschhorn et al., 1989). Note also that the definition *does not* assume acetogens *must* use the pathway (see below).

The advantage of this definition is that it distinguishes those bacteria that use the acetyl-CoA pathway as defined earlier from those bacteria that use the acetyl-CoA pathway for other purposes, and also from those bacteria that synthesize acetate by all other mechanisms. The disadvantage of this definition is that it excludes two other important groups of anaerobic bacteria that also synthesize acetate from CO_2 but via either the glycine synthase-dependent pathway (Chapter 23) or the reductive citric acid cycle (Chapter 19) rather that the acetyl-CoA pathway, both of which are often referred to as acetogenic processes (Wood and Ljungdahl, 1991). In this regard, the above definition may not satisfy the view of all microbiologists. Nonetheless, it is clear that loose usage of the terms "acetogen" or "acetogenesis" is often not constructive.

Homoacetogens are acetogens (per earlier definition) that form acetate as the sole reduced end product from certain substrates. *C. thermoaecticum* and *Acetobacterium woodii* are examples of organisms to which this term has been used, and it has been suggested that strict application of this term would avoid confusion between acetyl-CoA pathway utilizers and the other different acetate-forming "acetogens" such as symbiotic proton reducers or ethanol oxidizers (Schink and Bomar, 1992). However, to conclude that an organism is homoaceto-genic is dangerous because homoacetogenesis is dependent on growth conditions as much as on the acetogen. Many (if not all) so-called homoacetogens are not homoacetogenic, their potential to synthesize acetate being conditional rather than absolute. CO (Diekert, et al. 1986), H_2 (Martin et al., 1983; Lorowitz and Bryant, 1984; Savage et al., 1987), CH_4 (Savage et al., 1987; Buschhorn et al., 1989), butyrate (Lynd and Zeikus, 1983; Krumholz and Bryant, 1985; Worden et al., 1989; Grethlein et al., 1991), ethanol (Buschhorn et al. 1989), lactate (Lorowitz and Bryant, 1984; Drake, 1993; Misoph and Drake, unpublished results), succinate (Dorn et al., 1978; Lorowitz and Bryant, 1984, Matthies et al., 1993), reduced aromatic acrylates (Tschech and Pfenning, 1984; Parekh et al. 1992), reduced aromatic aldehydes (Lux et al., 1990), and even sulfide (Heijthuijsen and Hansen, 1989; Beaty and Ljungdahl, 1991) and nitrite/ammonia (Seifritz et al., 1993) are examples of minor or sole reduced end products of "homoacetogens," the production of which can constitute the sole energy-conserving, growth-supportive mechanism of the cell (see also Drake, 1993, and chapter 10). For example, if *C. thermoaceticum* had been originally isolated with a nitrate-enriched medium, a nitrate-dissimilating rather than CO_2-fixing bacterium may have been described by Fontaine and co-workers in 1942; this "homoacetogen" does not reduce CO_2 to acetate in the presence of 1–5-mM concentrations of nitrate but, instead, switches to nitrate dissimilation (Seifritz et al., 1993). Nonetheless, although the term homoacetogen may be inadequate, and although additional enzymological information is needed to definitively define an organism as an acetogen, the production of acetate as the sole or primary reduced end product from certain carbohydrates (e.g., glucose), H_2/CO_2, or CO strongly suggests that the organism in question utilizes the acetyl-CoA pathway per the above definition.

Lastly on this theme, it has often been discussed that an alternative term be coined for acetogens, i.e., those organisms that fall under the definition above. Stephen Zinder has recently mentioned the term "acetosyn" to me, being a spin-off of the word "acetosynthesis" (instead of acetogenesis). As pointed out by Stephen, use of this word might more properly emphasize that acetogens (aceto-syns) synthesize, rather than merely form, acetate. This reminds me of the fact that in the last edition of *Bergey's Manual of Systematic Bacteriology*, *Clostridium thermoaceticum* and *Clostridium formicoaceticum* were renamed *Clostridium thermaceticum* and *Clostridium formicaceticum*. Keying in on the logic applied to change the names of these two species, perhaps we could simply drop the "o" from "acetogen." Personally, I favor putting the "o" back in. This view likely makes little difference since rumor has it the names might change even more drastically next time around. In effort to more properly define things that were predefined at an earlier date, we change terms or invent new ones. Though this practice is usually well-intended, it is often a nightmare for many of us. Nonetheless, in keeping with this trend, perhaps we should determine if we need a new term for "acetogen." Alternatively, we could simply discontinue using this term indiscriminately.

1.4   Acetogenic Bacteria

1.4.1   Origins of Isolates to Date

Those bacteria isolated to date that might be considered acetogenic bacteria by the above definition are listed in Table 1.2. However, relatively few of these bacteria have been examined in detail; thus, a few risks have been taken in compiling this list. Although definitive information is often lacking, the apparent homoacetogenic tendencies of a particular organism has been taken into account. In this regard, three main metabolic features of these organisms are

(i)   the use of chemolithoautotrophic substrates (H_2/CO_2 or CO/CO_2) as sole sources of carbon and energy under strictly anaerobic conditions,

(ii)   the capacity to convert certain sugars stoichiometrically to acetate,

(iii)   the use of aromatic compounds under acetogenic conditions.

Most, but not all, acetogens display all three characteristics.

Various morphologies are observed, and staining properties are sometimes variable within a genus (Schink and Bomar, 1992). The thermophilic spore-forming species are very enduring relative to high temperatures. For example, *C. thermoaceticum* cultures survive boiling for 10 min, and spore suspensions remain viable after 8 h at 100°C and after 15 min at 120°C; spores do not remain active after 30 min at 120°C (Cato et al., 1986). This fact makes it essential to autoclave all glassware for a minimum of 30 minutes that have had contact with such acetogens; failing to do so has caused the demise of more than one "uninoculated control" in our laboratory.

The number of known acetogens has increased dramatically in recent years, a fact that illustrates the current interest level in this biological group. Acetogens can be found in almost all anaerobic environments, including some extreme habitats and, as indicated by the isolation of strain SS1 (Liu and Suflita, 1993), in deep subsurface sediments. The acetyl-CoA pathway was, in fact, resolved with a thermophile, and Chapters 15 and 16 in this book address the occurrence and activities of acetogens under extreme halophilic and low-temperature conditions. They are found in numerous gastrointestinal habitats, including that of humans, and can often be predominant in the terminating stage of carbon turnover under anaerobic conditions (see Chapter 10–13).

A few recent isolates are noteworthy, though their use of the acetyl-CoA pathway is presently less than certain based on available information. A *Clostridium* species isolated from Mullet (a fish) gut (Mountfort, 1992) catalyzes a homoacetogenic fermentation of alginic acid, a large seaweed-derived polymer (av. M_r = 240,000) composed of D-mannosyluronic and L-galosyluronic acids; per unit uronic acid monomer, the homoacetogenic capacity of this organism is equivalent to the homoacetogenic capacities of glucose-utilizing acetogens such as *C. thermoaceticum* (Fontaine et al., 1942) in that reductant is recovered stoichiometrically in acetate. Likewise, *Eubacterium pectinii,* also isolated from Mullet gut (Mountfort, 1992), ferments pectin (a polymer of D-galacturonic acid with M_r ranging from 20,000 to 400,000) stoichiometrically to acetate. If future studies confirm the use of the acetyl-CoA pathway for acetate formation by these isolates, they would constitute important examples of acetogens capable of degrading large molecular weight polymers. In this regard, the human isolate I52 is reported to degrade amorphous but not crystalline cellulose (Chapter 13). On the assumption that this isolate is acetogenic, it constitutes the first known cellulose-degrading acetogen.

It should be noted that a starch- and inulin- (a starch-like polysaccharide; av. M_r = 5000) degrading strain of *Clostridium thermoautotrophicum* (strain I-1) has been reported (Drent and Gottschal, 1991). However, the fermentation balance of fructose was 1 fructose → 0.42 formate + 0.68 acetate + 1.27 ethanol + 1.0 H_2 + 0.97 CO_2 + 0.60 cell carbon. As the above stoichiometry is clearly in contrast to that of the homoacetogenic type strain (JW 701/3) of *C. thermoautotrophicum* (Wiegel et al., 1981), in the absence of additional information it should be questioned whether strain I-1 actually is an acetogen. Although this bacterium also grew with H_2/CO_2, no details of this growth or products thereof were stated. Such problems further accentuate the difficulties outlined in Section 1.3 regarding application of the term "acetogen" in describing or classifying an organism.

Table 1.2   Acetogenic bacteria isolated to date[a]

Acetogen	Reference and Year Isolated	Source of Isolate	Gram Stain and Cell Morph.	Growth Temp.[b]	G + C (mol%)	Type Strain
Acetitomaculum ruminis	Greening and Leedle, 1989	Rumen fluid, steer	+ Rod	Mesophilic	34	ATCC 43876
Acetoanaerobium noterae	Sleat et al., 1985	Sediment	− Rod	Mesophilic	37	ATCC 35199
Acetoanaerobium romashkovii	Charakhch'yan et al., 1992	Oil field	− Rod	Mesophilic	40	
Acetobacterium carbinolicum	Eichler and Schink, 1984	Freshwater sediment	+ Rod	Mesophilic	38	DSM 2925
Acetobacterium malicum	Tanaka, Pfennig, 1988	Freshwater sediment	+ Rod	Mesophilic	44	DSM 4132
Acetobacterium wieringae	Braun and Gottschalk, 1982	Sewage digester	+ Rod	Mesophilic	43	DSM 1911
Acetobacterium woodii	Balch et al., 1977	Marine estuary	+ Rod	Mesophilic	39	ATCC 29683
Acetobacterium sp. KoB58	Wagener and Schink, 1988	Sewage sludge	+ Rod	Mesophilic	44	
Acetobacterium sp. HP4	Conrad et al., 1989	Lake sediment	+ Rod	Psychrotrophic	NR[c]	
Acetobacterium sp. MrTac1	Emde and Schink, 1987	Marine sediment	+ Rod	Mesophilic	NR	
Acetobacterium sp. OyTac1	Emde and Schink, 1987	Freshwater sediment	+ Rod	Mesophilic	NR	
Acetobacterium sp. AmMan1	Dörner and Schink, 1991	Freshwater sediment	+ Rod	Mesophilic	36	
Acetobacterium sp. 69	Inoue et al., 1992	Sea sediment	+ Rod	Mesophilic	48	
Acetobacterium sp. HA1	Schramm and Schink, 1991	Sewage sludge	+ Rod	Mesophilic	NR	
Acetobacterium sp. B10	Sembiring and Winter, 1989, 1990	Water-waste pond	+ Rod	Mesophilic	NR	
Acetogenium kivui	Leigh et al., 1981	Lake sediment	− Rod	Thermophilic	38	ATCC 33488
Acetohalobium arabaticum	Zhilina and Zavarzin, 1990	Saline lagoon	−[d] Rod	Mesophilic	34	DSM 5501
Acetonema longum	Kane, Breznak 1991	Wood-eating termite, gut	− Rod	Mesophilic	52	DSM 6540
Butyribacterium methylotrophicum	Zeikus et al., 1980	Sewage digestor	+ Rod	Mesophilic	49	ATCC 33266
Clostridium aceticum	Wieringa, 1936; Braun et al., 1981[e]	Soil	− Rod	Mesophilic	33	DSM 1496

16

Organism	Reference	Source	Gram/Shape		No.	Collection
Clostridium fervidus (?)	Patel et al., 1987	Hot spring	− Rod	Thermophilic	39	ATCC 43204
Clostridium formicoaceticum	Andreesen et al., 1970[f]	Sewage	− Rod	Mesophilic	34	DSM 92
Clostridium ljungdahlii	Tanner and Yang, 1990; Tanner et al., 1993	Animal waste	+ Rod	Mesophilic	22	ATCC 49587
Clostridium magnum	Schink, 1984	Freshwater sediment	− Rod	Mesophilic	29	DSM 2767
Clostridium mayombei	Kane et al., 1991	Soil-feeding termite, gut	+ Rod	Mesophilic	26	DSM 6539
Clostridium pfennigii	Krumholz and Bryant, 1985	Rumen fluid, steer	+ Rod	Mesophilic	38	DSM 3222
Clostridium thermoaceticum	Fontaine et al., 1942	Horse manure	+/− Rod	Thermophilic	54	ATCC 35608
Clostridium thermoautotrophicum	Wiegel et al., 1981	Hot spring	+/− Rod	Thermophilic	54	ATCC 33924
Clostridium sp. CV-AA1	Adamse and Velzeboer, 1982	Sewage sludge	− Rod	Mesophilic	42	
Clostridium sp. (?)	Mountfort, 1992	Marine fish, gut	NR Rod	NR	NR	
Eubacterium limosum[g]	Sharak-Genthner et al., 1981	Rumen fluid, sheep	+ Rod	Mesophilic	48	ATCC 8486
Eubacterium pectinii (?)	Mountfort, 1992	Marine fish, gut	NR Rod	NR	NR	
Peptostreptococcus productus U1[h]	Lorowitz and Bryant, 1984	Sewage digestor	+ Coccus	Mesophilic	45	ATCC 35244
Sporomusa acidovorans	Ollivier et al., 1985	Distillation waste water	− Rod	Mesophilic	42	DSM 3132
Sporomusa malonica	Dehning et al., 1989	Freshwater sediment	− Rod	Mesophilic	44	DSM 5090
Sporomusa ovata	Möller et al., 1984	Silage	− Rod	Mesophilic	42	DSM 2662
Sporomusa paucivorans	Hermann et al., 1987	Lake sediment	− Rod	Mesophilic	47	DSM 3637
Sporomusa sphaeroides	Möller et al., 1984	River mud	− Rod	Mesophilic	47	DSM 2875
Sporomusa termitida	Breznak et al., 1988	Wood-eating termite, gut	− Rod	Mesophilic	49	DSM 4440
Syntrophococcus sucromutans	Krumholz and Bryant, 1986	Rumen fluid, steer	− Coccus	Mesophilic	52	DSM 3224

Continued

Table 1.2 Continued

Acetogen	Reference and Year Isolated	Source of Isolate	Gram Stain and Cell Morph.	Growth Temp.[b]	G + C (mol%)	Type Strain
Unclassified						
AOR	Lee and Zinder, 1988	Thermophilic digestor	+ Rod	Thermophilic	47	
DMB	Braus-Stromyer et al., 1993[i]	Digestor	− Rod/vibrio	Mesophilic	NR	
MC	Traunecker et al., 1991	Sewage digester sludge	+ Coccus	Mesophilic	48	
RMMac1	Schuppert and Schink, 1990	Marine sediment	−/+ Rod	Mesophilic	48	
22	Ohwaki and Hungate, 1977	Sewage sludge	+ Rod	Mesophilic	NR	
X-8	Samain et al., 1982	Vegetable wastewater	− Rod	Mesophilic	NR	
TH-001	Frazer and Young, 1985	Sewage sludge	− Rod	Mesophilic	NR	
TMBS4	Bak et al., 1992	Freshwater ditch mud	− Rod	Mesophilic	NR	DSM 6591
ZM[j]	Kotsyurbenko et al., 1992, 1993[j]	Cattle manure	+ Rod	Psychrotrophic	NR	
ZS[j]	Kotsyurbenko et al., 1992, 1993[j]	Paper-mill waste	+ Rod	Psychrotrophic	NR	
ZB[j]	Kotsyurbenko et al., 1992, 1993[j]	Bog sediment	+ Rod	Psychrotrophic	NR	
ZT[j]	Kotsyurbenko et al., 1992, 1993[j]	Tundra soil	+ Rod	Psychrophilic[k]	NR	
SS1	Liu and Sufita, 1993	406-m deep sediment	+ Oval rod	Mesophilic	NR	
417/2	Charakhch'yan et al., 1992	Oil field	− Rod	Mesophilic	43	
417/5	Charakhch'yan et al., 1992	Oil field	− Rod	Mesophilic	43	
I52	Wollin and Miller, 1993[l]	Human feces	− Cocco/rod	Mesophilic	NR	
CS1Van	Wollin and Miller, 1993[l]	Human feces	+ Rod	Mesophilic	NR	
CS3Glu	Wollin and Miller, 1993[l]	Human feces	+ Cocco/rod	Mesophilic	NR	
CS7H	Wollin and Miller, 1993[l]	Human feces	+ Rod	Mesophilic	NR	

[a] Bacteria that appear to use the acetyl-CoA pathway for the synthesis of acetate and growth [modified from Drake (1992)]. See reference for complete description of cell morphology and staining properties. Some rod-shaped species (e.g., *Acetobacterium woodii*) are often oval shaped. Gram staining is often highly variable (Cato and Stackebrandt, 1989; Schink and Bomar, 1992). The actual acetogenic nature of organisms marked with a "?" (*Clostridium fervidus*, *Clostridium* sp., and *Eubacterium pectinii*) is not clear (see text). Type strains are available from American Type Culture Collection (ATCC), 12301 Parklawn Drive, Rockville, Maryland, U.S.A., and Deutsche Sammlung von Mikrooganismen und Zellkulturen GmbH (DSM), Mascheroder Weg 1 b, Braunschweig, Germany.

[b] General temperature preference: psychrophilic (5–10°C), psychrotrophic (16–18°C), mesophilic (31–34°C), and thermophilic (58–62°C).

[c] NR, not reported.

[d] Based on properties of membrane.

[e] See also Adamse (1980).

[f] See also El Ghazzawi (1967).

[g] See also Moore and Cato (1965).

[h] See also Geerligs et al. (1987); *P. productus* strain Marburg, ATCC 43917.

[i] Personal communication.

[j] May be a species of *Acetobacterium*, personal communication; see Chapter 15, including Addendum of that chapter.

[k] Appears to have equivalent growth potentials under both psychrophilic and psychrotrophic conditions.

[l] Personal communication; see Chapter 13.

1.4.2   Methods for the Isolation and Study of Acetogens and Their Activities

Acetogens are strict anaerobes. Consequently, anaerobic methods must be applied to isolate them and investigate their activities. Isolation and enrichment protocols are described in detail in Chapter 7. The basic Hungate technique (Hungate, 1969) or modifications thereof are widely used for cultivation purposes. Growth media must be anaerobic and have a reasonably low redox potential; sodium sulfide, cysteine, dithionite, or dithiothreitol are often used to maintain redox potentials of less than -200 mV. Alternatively, titanium (III) reducer (Zehnder and Wuhrman, 1976; Moench and Zeikus, 1983) can also be used. It has the potential advantage of not being as toxic as sulfide-based reducers. Cadmium chloride ($CdCl_2$) has also been used as reducer for acetogen media (Breznak and Switzer, 1986; Breznak et al., 1988). It has been the experience of the author that during growth, the redox potential of growth media is lowered concomitant to the growth of certain acetogens (as shown by the reduction of certain redox indicators). This suggest that some acetogens might be able to compensate for nonoptimal redox conditions. However, it should be assumed that acetogens have limited tolerance relative to positive redox potentials.

Table 1.3 outlines the basic medium used in our laboratory for the cultivation of numerous acetogens; it is similar to those used in many other laboratories. Essentially all acetogens studied to date have growth pH optima approximating either 6.7 or 7.7, and we have had excellent success in cultivating acetogens with this medium. Additional modifications relative to pH or carbonate content can be made quite easily by buffer modification. Something not widely appreciated is the original observations of Wieringa (1941) with C. aceticium that indicated enrichment of acetogens under alkaline conditions (pH 8–9) might favor their isolation. In this regard, although a true alkalinophilic acetogen has not been reported in the literature, their existence seems likely.

Because CO_2 is used as a terminal electron acceptor by acetogens, many acetogens are unable to grow in its absence under certain conditions. In particular, this is true when growth is at the expense of methyl-level reductant. When designing a medium, this fact must be remembered. Also important is the inclusion of trace metals, as acetogens are rich in metallo enzymes (Ljungdahl, 1986). Certain acetogens have trace vitamin requirements (Drake, 1992), and the inclusion of vitamins during initial enrichment is common practice. Some acetogens have not been successfully cultivated under defined conditions, and unknown nutritional factors are often required. For example, the protocol used to elucidate the nutritional requirements of C. thermoaceticum (Lundie and Drake, 1984) and C. thermoautotrophicum (Savage and Drake, 1986) was not successfully applied to Peptostreptococcus productus; inclusion of vitamins, amino

Table 1.3 A typical medium for the cultivation of acetogens[a]

Salts	mg/L	Vitamins	mg/L
NaCl	400	Nicotinic acid	0.25
NH$_4$Cl	400	Cyanocobalamin	0.25
MgCl$_2$ · 6H$_2$O	330	p-Aminobenzoic acid	0.25
CaCl$_2$ · 2H$_2$O	50	Calcium D-pantothenate	0.25
		Thiamine · HCl	0.25
Trace Elements	mg/L	Riboflavin	0.25
MnSO$_4$ · H$_2$O	2.5	Lipoic acid	0.30
FeSO$_4$ · 7H$_2$O	0.5	Folic acid	0.1
Co(NO$_3$)$_2$ · 6H$_2$O	0.5	Biotin	0.1
ZnCl$_2$	0.5	Pyridoxal · HCl	0.05
NiCl$_2$ · 6H$_2$O	0.25		
H$_2$SeO$_4$	0.25	Buffer for pH 6.7 Media	mg/L
CuSO$_4$ · 5H$_2$O	0.05	NaHCO$_3$	7,500
AlK(SO$_4$)$_2$ · 12H$_2$O	0.05	KH$_2$PO$_4$	500
H$_3$BO$_3$	0.05	Gas phase	100% CO$_2$
Na$_2$MoO$_4$ · 2H$_2$O	0.05		
Na$_2$WO$_4$ · 2H$_2$O	0.05	Buffer for pH 7.7 Media	mg/L
		NaHCO$_3$	6,000
Reducers	mg/L	Na$_2$CO$_3$ · 10H$_2$O	16,500
Cysteine · HCl · H$_2$O	250	KH$_2$PO$_4$	500
Sodium sulfide	250	Gas phase	100% N$_2$

[a] The water used in preparing the medium is made anaerobic by boiling and subsequent cooling under CO_2. Final gas used during dispensing is as indicated for the two buffer systems. Resazurin (1 mg/L) can be added as a redox indicator, and yeast extract (0.5–2 g/L), tryptone (0.5–2 g/L), clarified rumen fluid (50 ml/L), or casamino acids (0.5–2 g/L) can be added to create an enriched undefined medium. Deletion of the vitamins yields a basal medium. Crimp-sealed tubes or bottles are used for cultivation. See Daniel et al. (1990) and Matthies et al. (1993) for complete protocols.

acids, fatty acids, and other nutrients did not substitute for unknown growth factors in yeast extract (Drake and co-workers, unpublished data). Nonetheless, some chemolithoautotrophic acetogens grow without any trace organic nutrients [e.g., *Acetogenium kivui* (Leigh et al., 1981)].

Numerous substrates can be oxidized and deliver reductant to the acetyl-CoA pathway and the reductive synthesis of acetate. Consequently, the substrate used to isolate an acetogen is critical to the "catabolic type" selected for during enrichment. Utilization of certain substrates for isolation tends to favor the growth of acetogens. For example, H_2/CO_2 will quite often select for acetogens under either defined or undefined conditions; this approach has been used in the isolation of many acetogens, including *A. kivui* (Leigh et al., 1981), *Acetobacterium woodii* (Balch et al., 1977), and *Sporomusa termitida* (Breznak et al., 1988). Likewise, the capacity to utilize CO/CO_2 as a sole source of carbon and energy can also be selective; this approach was utilized in the enrichment of *P. productus* (Lorowitz and Bryant, 1984; Geerligs et al., 1987). The capacity to utilize methoxylated aromatic compounds, first observed for *A. woodii* (Bache and Pfennig, 1981), is another general metabolic potential of acetogens that appears not to be widespread in other anaerobic groups and may also be useful for selective enrichment of acetogens. Many recent isolates have been obtained by using substrates that are not typically identified as being appreciably metabolized by known bacteria under anaerobic conditions. Mandalate [yielding *Acetobacterium* strain AmMan1 (Dörner and Schink, 1991)], trimethylamine [yielding *Acetohalobium arabaticum* (Zhilina and Zavarzin, 1990)], methoxyacetate [yielding strain RMMac1 (Schuppert and Schink, 1990)], and methyl chloride [yielding strain MC (Traunecker et al., 1991)] are examples of such substrates. In all such cases, sequential transfer of the initial enrichment on the substrate selected yielded a highly enriched acetogenic population of one or more species. Subsequent isolation by the roll-tube method of Hungate (1969) or anaerobic streak plating on agar or Gelrite (preferred for thermophiles because of its increased capacity to remain solid at higher temperatures) will normally yield pure cultures of acetogens. On solid surfaces, acetogenic colonies can be differentiated from nonacetogenic colonies by their ability to form zones of clearing with calcium carbonate medium (Balch et al., 1977) or zones of color change with media supplemented with a pH indicator (e.g., bromocresol green) (Braun et al., 1979; Leedle and Greening, 1988).

Because of their diverse metabolic potentials, things are easily missed with acetogens when first isolated. As noted earlier, *C. thermoaceticum* was isolated as an obligate heterotroph, being incapable of using H_2 (Fontaine et al., 1942). When I left KU headed for Western Reserve, Jim told me, "I don't know what you'll find up there in Cleveland, but one thing is certain, you'll never find hydrogenase in *Clostridium thermoaceticum*." He and Lars had teamed up while both were at Western Reserve and had looked for it because it was clearly a key missing enzyme to the autotrophic capabilities of the organism (see Ljungdahl et al., 1981, p. 419). It surfaced some 20 years subsequent to their search (Drake, 1982), a finding that was in part responsible for resolution of the autotrophic capabilities of this "obligate heterotroph."

Thus, the initial growth potentials of acetogens are often deceiving. Although *C. thermoaceticum* was finally shown to be both chemolithotrophic (Kerby and Zeikus, 1983) and chemolithoautotrophic (Daniel et al., 1990), being competent in H_2 uptake as well as H_2 evolution (Martin et al., 1983), the chemolithotrophic potentials of *C. thermoautotrophicum* were used to classify it as a different species from *C. thermoaceticum* prior to such findings (Wiegel et al., 1981).

We have not been able to distinguish these "two" acetogens at the level of carbon-substrate utilization patterns, nor have others found significant differences in their electron transport/ATPase/energy-conserving systems (Hugenholz and Ljungdahl, 1989). *Clostridium magnum* was likewise originally isolated as an obligate heterotroph (Schink, 1984). However, as with *C. thermoaceticum,* it was later found to be capable of H_2-dependent lithotrophic growth, containing hydrogenase(s) also capable of both H_2 uptake and H_2 evolution (Bomar et al., 1991). [See also Lux and Drake (1992) for reassessment of *C. formicoaceticum.*] Thus, as the list of acetogens becomes longer, detailed evaluation of metabolic potentials should be emphasized.

Acetogens often share the same habits with methanogens, and bromoethanesulfonate, an analogue of coenzyme M and a metabolic inhibitor of methangens (Gunsalus et al., 1978; Zehnder et al., 1980; Smith and Mah, 1981), is often added in 10–50 mM concentrations to media for the selective enrichment of acetogens on substrates that could be used by competing methanogens (e.g., H_2/CO_2) (Greening and Leedle, 1989; Kane and Breznak, 1991). Lumazine (a pterin compound: 2,4-[1H, 3H]-pteridinedione) is a more potent inhibitor of methanogens [0.11 mM yields 100% inhibition of the growth of *Methanobacterium thermoautotrophicum* (Nagaranthal and Nagle, 1992)] and may be more effective than bromoethanesulfonate if shown to be selective during enrichment. Lumazine appears to inhibit methyl-S-CoM reductase, a key enzyme in methanogenesis (Nagaranthal and Nagle, 1992).

Many of the enzymes from acetogens are extremely susceptible to inactivation by oxidation, thus necessitating the application of anaerobic techniques for their purification and study. This fact was a real handicap to fast progress on resolving the pathway, and the first efforts to purify acetyl-CoA synthase and the multicomponent system responsible for the terminal synthesis of acetyl-CoA were largely thwarted because of oxygen lability. When Harland was in Germany on his last sabbatical with Feodor Lynen, the author was struggling with this problem and was fortunate to have the help of Warwick Sakami to explain "how we old timers had solved such problems decades ago!" He had come to terms with a similar problem during efforts to study oxygen-sensitive coenzymes and cofactors (Sakami, 1962). Warwick was very insightful, as well as an excellent glassblower, and the "real breakthrough" (as Harland would say) was not when anyone realized the problem but when someone saw the solution. That was Warwick. He had never seen an anaerobic chamber and did not worry about not having one! He designed an anaerobic protein chromatography system that finally provided initial access to a purified acetyl-CoA synthesizing system (Drake et al., 1980, 1981a, 1981b; Hu et al., 1982). Few know of the critical role played by Warwick in resolving the pathway. Anaerobic chambers (Fig. 1.3) have since become indispensible for the study of numerous oxygen-sensitive enzymes central to acetogenesis, as well as for the study of whole cell-mediated processes (Ragsdale

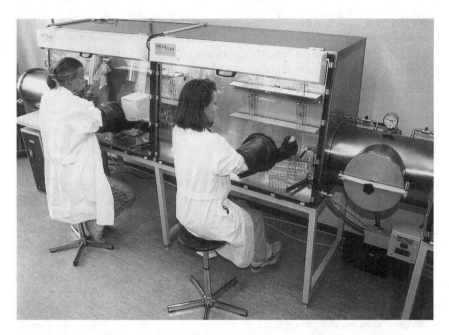

Figure 1.3.   A Mecaplex (Grenchen, Switzerland) anaerobic chamber used in the author's laboratory for the study of acetogens and other anaerobic bacteria.

et al., 1983; Lundie and Drake, 1984; Kellum and Drake, 1986; Ljungdahl, 1986; and Wood, 1991; Heise et al., 1989, 1991; Hsu et al., 1990). In more recent years, anaerobic UV/Vis and EPR spectrometric assay methods for the study of catalysis have also been implemented (Gorst and Ragsdale, 1991; Ragsdale, 1991; Shin et al., 1992; Shin and Lindahl 1992a, 1992b, 1993).

1.4.3   Classification and Identification of Acetogens

Acetogens are not a closely related group. Although utilization of the acetyl-CoA pathway unifies them, they are extremely diverse genetically; the guanine plus cytosine (G+C) content of genomes varies from 22 mol% with *Clostridium ljungdahlii* to 54 mol% with *C. thermoaceticum* (Table 1.2). However, all isolates to date are eubacteria, including the extreme halophilic isolate *A. arabaticum* (Zhilina and Zavarzin, 1990). New isolates are keyed out according to (i) morphology, (ii) presence, shape, and intracellular location of spores, (iii) staining properties, (iv) motility and type of flagella, (v) temperature, pH, and salinity requirements, (vi) G + C content, (vii) substrate/product profiles, (viii) comparative analysis of 16S ribosomal RNA, and (ix) DNA–DNA hybridization properties to known species. Not all of the acetogens listed in Table 1.2 have been characterized relative to their pylogeny (Tanner et al., 1981; Cato and Stackebrandt, 1989) (see Chapter 9).

The occurrence of acetyl-CoA synthase in a new isolate can be useful in determining if the organism is an acetogen. Acetyl-CoA synthase of acetogens is an important enzyme in the acetyl-CoA pathway and catalyzes two reactions, that of CO oxidation and that of acetyl-CoA synthesis (see Section 1.5, and Chapters 2, 3, and 5). Hence, two names are often applied to this catalyst:

CO dehydrogenase: $CO + H_2O \rightarrow CO_2 + 2H^+ + 2e^-$ X

acetyl-CoA synthase: $[CH_3^-] + [CO] + Coenzyme\text{-}A \rightarrow acetyl\text{-}CoA$ XI

(Note: $[CH_3^-]$ and $[CO]$ indicate enzyme-bound intermediates; see Section 1.5.1.) Although the occurrence of this enzyme is evidence of the pathway, the method used for its assay can yield misleading results because reaction X is catalyzed by nonacetogens that contain "CO dehydrogenase" activity (e.g., *Clostridium pasteurianum*). Indeed, it was this fact that preceded the discovery of "CO dehydrogenase" in acetogens, not vise versa (Fuchs et al., 1974; Thauer et al., 1974; Diekert and Thauer, 1978). It is, therefore, recommended that, prior to asserting that acetyl-CoA synthase is present and thus the acetyl-CoA pathway operative in a particular isolate, the capacity to form acetyl-CoA via the pyruvate/ CO-coupled methyltetrahydrofolate assay (Schulman et al., 1973; Drake et al., 1981a; Hu et al., 1982) be ascertained. Although more complicated than the CO dehydrogenase assay, the acetyl-CoA synthase assay can be performed with cell extracts and will confirm the occurrence of reaction XI, a reaction fundamentally more important for identification purposes. In addition, it has been shown with *C. thermoaceticum* that mediation of both reactions X and XI by acetyl-CoA synthase in cell extracts may not yield parallel results and, additionally, may vary with growth conditions (Kellum and Drake, 1986). Such potential differences emphasize the need to ascertain the validity of these reactions on a case-by-case basis.

1.5   The Acetyl-CoA Pathway: Nuts and Bolts of Acetogenesis

Given the definition applied for acetogens (Section 1.3), it stands to reason that the acetyl-CoA pathway serves three main functions for the cell:

   (i)   as a terminal electron-accepting process,

   (ii)   as an energy-conserving process,

   (iii) as a mechanism for the autotrophic assimilation of carbon.

Given these functions, it is clear that the main purpose of the pathway may vary with the growth conditions of the cell. In this section, these functions will be introduced; the reader is directed to Chapters 2–5 and 10 for further details of these functions and associated processes.

1.5.1   Acetyl-CoA Pathway as a Reductant Sink

Acetogens are just like all other anaerobes in that they require a terminal electron acceptor other than oxygen. CO_2 serves this purpose when the acetyl-CoA pathway is engaged. Thus, the pathway is primarily used as an electron sink under certain heterotrophic conditions. During glucose-dependent growth, for example, the critical chemolithoautotrophic functions of the pathway (i.e., the production of acetyl-CoA for biosynthesis and the conservation of energy) are arguably less significant because the cell already has access to ATP, acetyl-CoA, and other biosynthetic precursors via the breakdown of glucose. Under such conditions, its main function is the recycling of reduced electron carriers (NAD, ferredoxin, etc.) generated during the oxidation of glucose.

These points are conceptualized in Figure 1.4; in essence the acetyl-CoA pathway is one of three integrated processes during glucose utilization. Glucose is first oxidized to pyruvate via glycolysis (box A). Pyruvate is subsequently decarboxylated and oxidized to acetyl-CoA, which is then subject to phosphorylation and conversion to acetate via acetate kinase (box B). Processes conceptualized in boxes A and B collectively yield 2 moles of acetate per mole of glucose. CO_2 is subsequently used as the terminal sink for the eight reducing equivalents generated during glycolysis and the oxidation of pyruvate (box C). This last process, that of box C, is the acetyl-CoA pathway and is responsible for the formation of the third molecule of acetate.

When box C in Figure 1.4 is expanded, we see there are two branches of the acetyl-CoA pathway, both of which "fix" CO_2 (Fig. 1.5). On the methyl branch, formate dehydrogenase initiates the fixation of CO_2 via reduction to formate (Yamamoto et al., 1983) and makes possible the subsequent reduction of formate to the methyl level via the tetrahydrofolate pathway (Ljungdahl, 1986). Formate dehydrogenase from C. thermoaceticum is rich in metals (2 moles selenium, 2 moles tungsten, and approximately 36 moles iron per mole enzyme) and appears to contain a tungstopterin prosthetic group (Yamamoto et al., 1983; Ljungdahl, 1986). Formate dehydrogenase from C. thermoaceticum was the first enzyme in which tungsten was shown to be a biologically active trace metal. On the carbonyl branch, acetyl-CoA synthase (Chapter 3) fixes CO_2 via reduction to the carbonyl [CO] level and facilitates the carbonyl-dependent synthesis of acetyl-CoA, which subsequently is converted to acetate.

Figure 1.6 outlines the location of all reductant-generating and reductant-consuming processes during the stoichiometric conversion of glucose to three

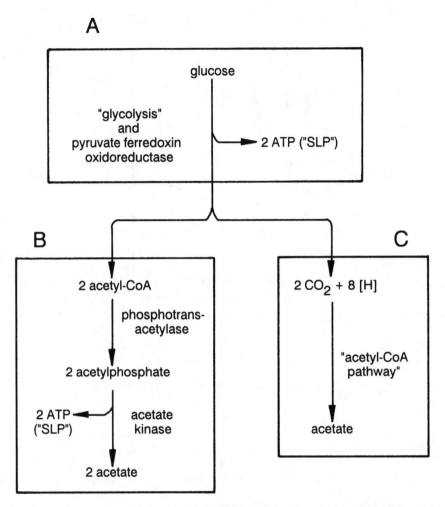

Figure 1.4. Overview of the three main processes of acetogens that collectively yield three molecules of acetate per molecule of glucose. SLP, substrate-level phosphorylation.

acetates, and Table 1.4 identifies the specific catalyst at each step. The large enclosed box in Figure 1.6 is the acetyl-CoA pathway and illustrates those reactions that collectively constitute box C in Figure 1.4. As illustrated, acetyl-CoA synthase is responsible for the synthesis of acetyl-CoA subsequent to the reductive conversion of CO_2 to the methyl and carbonyl levels. Recent evidence with acetyl-CoA synthase of *C. thermoaceticum* suggests that (i) an NiFe complex is essential to acetyl-CoA synthase activity (reaction XI) but not to the CO

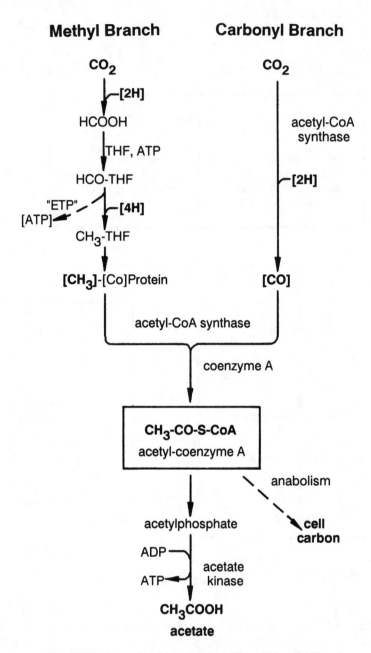

Figure 1.5. Simplified scheme for the acetyl-CoA "Wood/Ljungdahl" pathway. THF, tetrahydrofolate; [Co]protein, corrinoid enzyme. Broken line and bracketed ATP in the methyl path indicates the conservation of energy via chemiosmotic processes and electron transport phosphorylation (ETP); the exact amount of ATP formed via ETP is not known and likely depends on both growth conditions and acetogen.

Table 1.4 Enzymes of the acetyl-CoA pathway as outlined in Figure 1.6[a]

No. in Figure 1.5	Enzyme	M_r	Subunit Composition	Function in Pathway Illustrated in Figure 1.5	$\Delta G^{o'}$ (kJ per Rx)	Reference
1	Formate dehydrogenase	340,000	$\alpha_2\beta_2$ (96,000 and 76,000)	NADPH-dependent reduction of CO_2 to formate	+ 3.4	Yamamoto et al., 1983
2	Formyltetrahydrofolate (HCO-THF) synthetase	240,000	α_4	Conversion of formate to HCO-THF	−8.4	Mayer et al., 1982
3 and 4	Methenyltetrahydrofolate (CH-THF) cyclohydrolase and methylenetetrahydrofolate (CH_2-THF) dehydrogenase complex	55,000	α_2	HCO-THF converted to CH-THF and CH-THF reduced to CH_2-THF	−4.0 −23.0	O'Brien et al., 1973
5	Methylenetetrahydrofolate (CH_2-THF) reductase	289,000	α_8	Reduction of CH_2-THF to CH_3-THF	−57.3	Park et al., 1991
6	Methyltransferase	58,900	α_2	Transfer of methyl unit from CH_3-THF to corrinoid enzyme		Drake et al., 1981a, 1981b
7	Corrinoid enzyme	89,000	$\alpha\beta$ (34,000 and 55,000)	Methylation of acetyl-CoA synthase		Hu et al., 1984
8	Acetyl-CoA synthase	440,000	$\alpha_3\beta_3$ (78,000 and 71,000) (81,730 & 72,928)[b]	Reduction of CO_2 to the [CO] level and synthesis of acetyl-CoA from CH_3-THF, [CO], and CoASH	+ 20.1 + 21.8	Ragsdale et al., 1983, Diekert and Ritter, 1983
9	Acetyl-CoA synthase disulfide reductase	225,000	α_4	Reduction of disulfide of CoA binding site of acetyl-CoA synthase		Pezacka and Wood, 1986
10	Phosphotransacetylase	88,100	α_4	Conversion of acetyl-CoA to acetylphosphate	+ 9.0	Drake et al., 1981a
11	Acetate kinase	60,000	Not reported	Conversion of acetylphosphate and ADP to acetate and ATP	− 13.0	Schaupp and Ljungdahl, 1974
12	Pyruvate-Fd oxidoreductase	225,000	α_2	Oxidation/decarboxylation of pyruvate to acetyl-CoA	− 19.2	Drake et al., 1981a

[a] Enzymes indicated have been purified from C. thermoaceticum. Abbreviations: THF, tetrahydrofolate; CoA, coenzyme A; Fd, ferredoxin. Data for $\Delta G^{o'}$ were obtained from Thauer et al. (1977) and Fuchs (1986).
[b] As deducted from amino acid composition.

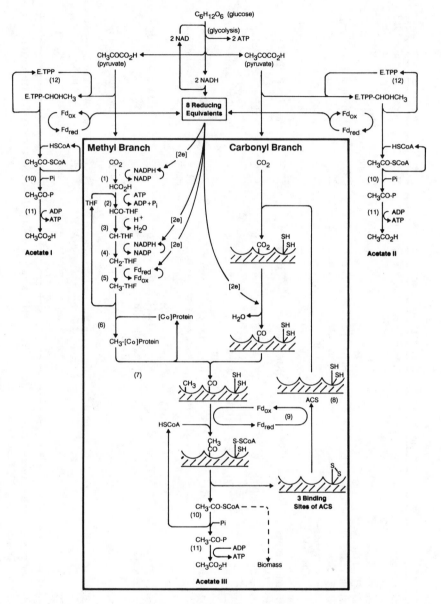

Figure 1.6. The acetyl-CoA "Wood/Ljungdahl" pathway (enclosed box) coupled to the oxidation of glucose (Drake, 1992). The overall scheme was elucidated from studies with *Clostridium thermoaceticum;* variations may occur with other acetogens. Pyruvate oxidoreductase (giving rise to acetates I and II) and glycolysis collectively yield the eight reducing equivalents required for the reductive synthesis of acetate III. Abbreviations: THF, tetrahydrofolate; [Co]protein, corrinoid enzyme; E.TPP, thiamine pyrophosphate; Fd, ferredoxin; ACS, acetyl-CoA synthase, illustrated with three potential binding sites (including the disulfide-CoA binding site); Pi, inorganic phosphate; HSCOA, coenzyme A; [2e], two reducing equivalents. See Table 1.4 for characteristics of enzymes and thermodynamics of reactions.

dehydrogenase activity also inherrent to the native enzyme (reaction X) (Shin and Lindahl 1992a, 1992b) and (ii) the nickel environment of the enzyme is heterogenous rather than uniform (Shin et al., 1993). See Chapter 3 for further details on the biochemistry of acetyl-CoA synthase from acetogens.

1.5.2 Acetate Formation and the Conservation of Energy

The second main function of the acetyl-CoA pathway is the conservation of energy. ATP is formed via substrate-level phosphorylation during the oxidation of glucose (boxes A and B, Fig. 1.4). However, there is no net production of ATP (i.e., conserved energy) via substrate-level phosphorylation during the reductive synthesis of acetate from CO_2 via the acetyl-CoA pathway (box C, Fig. 1.4). During the reductive synthesis of acetate, one ATP is consumed in the activation of formate and is later regained via acetate kinase, thus yielding a breakeven situation relative to ATP gain and substrate-level events (Fig. 1.6). In addition, the reduction of CO_2 to the CO level on the carbonyl branch is not thermodynamically favorable and may also require energy (estimated at one-third ATP equivalent) (Diekert et al., 1986; Wohlfarth and Diekert, 1991; Diekert, 1992).

However, in classic growth studies, acetogens were discovered to have cell yields in excess of that explainable from substrate-level events. In general, 1 mole of ATP yields approximately 10 g dry weight of biomass, and acetogens like *C. formicoaceticum, C. aceticum,* and *A. woodii* grown heterotrophically yield 50–70 g cell dry weight per mole glucose-equivalent carbohydrate (Andreesen et al., 1973; Tschech and Pfennig, 1984). Because the maximum ATP predicted from substrate-level events would be 4 mole ATP per mole glucose utilized (Fig. 1.4, boxes A and B), thus yielding a maximal cell yield of 40 g cell dry weight per mole glucose, these unexpected high cell yields led to the realization that acetogens conserved energy by chemiosmotic mechanisms. In this regard, they were among the first obligate anaerobes investigated for anaerobic oxidative phosphorylation. The capacity of acetogens to grow at the sole expense of H_2/CO_2 reinforced the belief that acetogens were capable of electron transport phosphorylation or chemiosmotic energy conservation (as such growth has no known potential for net ATP gains via substrate-level phoshorylation). Indeed, subsequent studies revealed that acetogens are rich in redox centers, electron carriers, oxidoreductases, ion pumps and antiporters, and membranous ATPases.

Thermodynamic factors govern the cell's potential for the chemiosmotic conservation of energy, and the relative changes in Gibbs free energy of each step in the acetyl-CoA pathway (Table 1.4) suggest that the methylenetetrahydrofolate reductase-mediated step (step 5 in Fig. 1.6) is associated with energy conservation. Although we are only beginning to understand the detailed mechanisms engaged during energy conservation, one fact is clear: different acetogens use different mechanisms for conserving energy by nonsubstrate-level events. This fact is illustrated by the growth behaviors of *C. thermoaceticum* and *A. kivui* under different conditions. Although ATPase inhibitors have similar effects on growth under organotrophic and chemolithoautotrophic conditions (thus suggesting both employ similar mechanisms for energy conservation), under these same conditions, these acetogens respond differently to sodium deprivation. Such observations reinforce the fact that acetogens are capable of either proton- or sodium-mediated chemiosmotic conservation of energy (Heise et al., 1989; Geerligs et al., 1989; Yang and Drake, 1990). Acetogens are rich in both membranous and cytoplasmic electron carriers (examples include ferredoxins, rubredoxins, quinones, and cytochromes), and emphasis in work to date has focused on characterizing membranous proton and sodium ATPases that are coupled to electron transport phosphorylation and electrochemical gradients (Das et al., 1989; Das and Ljungdahl, 1983; Hugenholtz and Ljungdahl, 1989; Heise et al., 1991, 1992; Gottschalk, 1989). The various chemiosmotic and electron transport-mediated mechanisms by which energy can be conserved during acetogenesis are reviewed in detail in Chapters 2, 4, and 5.

However, in addition to ATP, pyrophosphate can also be utilized as an energy source by certain anaerobic bacteria (Liu et al., 1982; Varma and Peck, 1983). *C. thermoaceticum* has been found to form and subsequently consume high levels of intracellular pyrophosphate during growth (Heinonen and Drake 1988). Although this implies a role for this high energy phosphate in cell energetics, no direct correlations to cell yields have been established. Interestingly, with *A. woodii*, pyruvate phosphate dikinase (reaction XII) and phosphoenolpyruvate carboxytransphosphorylase (reaction XIII), both of which yield pyrophosphate (PP_i), have been shown to facilitate the flow of carbon during autotrophic assimilation of CO_2 (Eden and Fuchs 1982, 1983) (see section 5.3, below):

$$\text{pyruvate} + \text{ATP} + P_i \rightarrow \text{phosphoenolpyruvate} + \text{AMP} + PP_i \qquad \text{XII}$$
$$\text{phosphoenolpyruvate} + P_i + CO_2 \rightarrow \text{oxaloacetate} + PP_i \qquad \text{XIII}$$

Nonetheless, it is important to point out that we know relatively little about high-energy storage compounds and alternative high-energy phosphates used by acetogens.

Energy conservation coupled to acetogenesis by either substrate-level phosphorylation or a chemiosmotic mechanism is dependent on the oxidation of acetogenic substrates (this assumes, for example, that decarboxylation reactions are not energy conserving [Dimroth 1987]). Comparative evaluation of reducing equivalents consumed or acetate formed might reveal correlations to cell energetics. However, the thermodynamics of acetogenesis varies with the substrate utilized, and the theoretical energy available for growth does not translate directly to cell yields (Table 1.5). For example, acetogenesis from either H_2 or glucose have very similar Gibbs free energies on a mole acetate basis, yet growth yields of *C. thermoaceticum* per reductant pair consumed are markedly dissimilar between these two acetogenic substrates. Such observations may be due in part to differences in activation energies in the formation and removal of intermediates, as well as differences between metabolic interfaces of anabolism, catabolism, and transport with different acetogenic substrates. In addition, different acetogens grow differently with identical substrates. For example, H_2- and glucose-dependent cell yields of *A. kivui* are approximately 2-fold higher than those of *C. thermoaceticum* (Daniel et al. 1990). Are such differences due to more efficient usage of H_2 or glucose, or the engagement of additional energy-conserving sites during reductant flow? Detailed chemostat studies would better resolve such problems.

Because of its role in acetogenesis, CO is an interesting growth-supportive substrate for most acetogens. Per acetate synthesized, CO has approximately twice the potential energy as that of H_2; likewise, at least in the case of *C. thermoaceticum,* CO yields approximately three times the amount of biomass per pair reducing equivalent as that of H_2 [Table 1.5; the same applies to other acetogens (Daniel et al., 1990)]. It is not clear what biochemical mechanism accounts for this large growth differential. It was proposed that CO might enter the tetrahydrofolate pathway at the formyl level and, thus, bypass the ATP-requiring formyltetrahydrofolate synthetase step (Fig. 1.6, reaction 2); however, synthesis of formyltetrahydrofolate from CO was shown to require both ATP and NADP, suggesting that CO nonetheless passes through CO_2 on the methyl branch (Kellum and Drake, 1986). On the carbonyl branch, CO likely enters directly at the carbonyl level and would thus save the theoretical one-third ATP needed to reduce CO_2 to the CO level (Diekert, 1992). However, this small

Table 1.5   Growth differentials of *Clostridium thermoaceticum* on different acetogenic substrates[a]

Substrate Consumed (mM)	Biomass Formed (mg/ml)	Acetate-to-Biomass Ratio	Biomass per 2[H] Ratio[b]	Standard Change in Gibbs Free Energy[c]
Glucose (4.7)	0.154	68	8.2	− 104
Oxalate (14.4)	0.071	54	4.9	− 166
Glyoxylate (12.9)	0.121	44	4.7	− 171
CO (53.6)	0.146	60	2.7	− 175
H_2 (80.5)	0.062	276	0.8	− 95

[a] In part from Daniel and Drake (1993).
[b] Cell biomass (mg dry weight) synthesized per pair (mM) reducing equivalents.
[c] Theoretical kJ per mole acetate synthesized from indicated substrate under standard conditions [obtained or calculated from Fuchs (1986) or Daniel and Drake (1993)].

amount of conserved energy does not adequately explain the large differential in growth between CO and H_2 cultures. With *Sporomusa,* different electron carriers appear to be used for H_2- and betaine-derived reductant; b- and/or c-type cytochromes are required for usage of betaine-derived reductant, and the bioenergetics of such reductant flow may, therefore, be dissimilar to that derived from H_2 (Kamlage et al., 1993; Kamlage and Blaut, 1993). If this difference can be extrapolated to other acetogens and other substrates, the bioenergetic differences between CO- and H_2-dependent chemolithoautotrophy might be explained in part at the level of different electron transport systems rather than metabolic bypasses that are energetically more favorable.

1.5.3 Chemolithoautotrophic CO_2 Fixation and Assimilation of Carbon

Although most research efforts have focused on resolving the steps of the acetyl-CoA pathway that, in the end, yield acetate, the acetyl-CoA formed via the pathway must also be utilized as the main assimilatory building block under autotrophic conditions. This third fundamental role of the pathway is essential to biosynthesis and the autotrophic growth of acetogens; however, it has only been explored in detail with *A. woodii* (Eden and Fuchs, 1982, 1983). For this acetogen, the Calvin cycle is not responsible for CO_2 assimilation, and ribulose-1,5-bisphosphate carboxylase is not present. Acetyl-CoA is reductively carboxylated to pyruvate by pyruvate : acceptor oxidoreductase; pyruvate is subsequently converted to phosphoenolpyruvate and further assimilated via an incomplete Krebs (tricarboxylic acid) cycle and gluconeogenesis (Eden and Fuchs, 1983). Earlier labeling patterns obtained with *C. thermoaceticum* (Ljungdahl and Wood, 1965; Ljungdahl, 1986) support these findings. It is not known if the assimilatory function of the acetyl-CoA pathway is tightly regulated during chemolithoautotrophic and heterotrophic growth.

1.6 The Concept of Critical Enzymes in CO_2 Fixation and Chemolithoautotrophy

Acetyl-CoA synthase is often highlighted as the most interesting and key enzyme in the acetyl-CoA pathway. The basis for such views stems from its apparent central role in the pathway and its relatively high concentration in the cell [2% of the soluble cell protein (Ragsdale et al., 1983)]. However, it seems shortsighted to single out one enzyme as most important because the lack of any one enzyme uncouples the pathway.

In this regard, it should also be noted that formate dehydrogenase (Yamamoto et al., 1983) and formyltetrahydrofolate synthetase (Lovell et al., 1990) are arguably of equal importance to that of acetyl-CoA synthase relative to chemolithoautotrophic CO_2 fixation:

(i)   both formate dehydrogenase and acetyl-CoA synthase reductively "fix" CO_2 in the acetyl-CoA pathway to formate and CO, respectively,

(ii) both formyltetrahydrofolate synthetase and acetyl-CoA synthase covalently "fix" these reduced forms of CO_2 in the acetyl-CoA pathway.

Additionally, under chemolithoautotrophic conditions, the pathway must not only fix carbon but also conserve energy. The reaction most likely associated with energy conservation at a thermodynamic level is that facilitated by methylenetetrahydrofolate reductase (Park et al., 1991; Clark and Ljungdahl, 1984; Wohlfarth and Diekert, 1991). Furthermore, all such considerations deeply undercut the essentiality of the other energy-conserving processes and associated enzymes (e.g., sodium-mediated chemiosmotic synthesis of ATP that is potentially coupled to oxidoreductase/transmethylation reactions between methylenetetrahydrofolate and corrinoids) of which we know very little but are nonetheless central to the real value of the acetyl-CoA pathway relative to the cell's capacity to synthesize ATP (see Chapters 2, 4, and 5; also Heise et al., 1993).

Furthermore, the fact that any one enzyme is present in a high concentration in the cell may merely indicate that enzyme and reaction catalyzed is thermodynamically less efficient (relative to preceding or subsequent enzymes in a pathway), thus necessitating the enzyme be present in a high concentration to compensate. In this regard, of all the reactions of the acetyl-CoA pathway, the reactions catalyzed by acetyl-CoA synthase are thermodynamically the least favorable (Table 1.4).

Lastly, H_2/CO_2-coupled chemolithotrophic growth is strictly dependent on CO_2 fixation. This process requires activation of reductant via hydrogenase. Because this enzyme likely catalyzes the first step in the overall process, it is difficult to highlight the importance of subsequent events by overshadowing the initiating, fundamental role of hydrogenase in chemolithoautotrophic CO_2 fixation.

1.7   Regulation and Genetics

This will be a short section: We know very little about the regulatory and genetic mechanisms employed by acetogens. Indeed, it has only been in recent years that we have started to appreciate that acetogens are likely quite normal regarding regulation, i.e., that their metabolic potentials, including the acetyl-CoA pathway, may, in fact, be subject to regulatory control. In this respect, far more is known regarding the molecular biology and genetics of methanogens (Reeve, 1992; Ferry, 1992). As mentioned earlier, the only focal point for many years was the biochemistry of the apparent constituitive capacity of acetogens to reduce CO_2 to acetate. Only when some of their additional metabolic potentials were realized did we also learn that regulatory switches engage these potentials as required by environmental conditions.

Although relatively few studies have actually addressed the regulation of the diverse metabolic potentials of acetogens, it is nonetheless quite easy to see significant differences in the expression of proteins and catalytic potentials under different metabolic states (Kellum and Drake, 1986; Savage et al., 1987; Wu et al., 1988; Hsu et al., 1990; Daniel and Drake, 1993). For example, hydrogenase and acetyl-CoA synthase/CO dehydrogenase levels may be significantly influenced by growth substrates (Braun and Gottschalk, 1981; Kellum and Drake, 1986; Daniel et al., 1990; Lux and Drake, 1992), and the capacity to utilize aromatic carboxyl and methoxyl groups appears to be an inducible, rather than constitutive, potential (Hsu et al., 1990; Wu et al., 1988; De Weerd et al., 1988) (see Chapter 17). We have recently shown that nitrate dissimilation (i.e., use of nitrate as an energy-conserving terminal electron acceptor) totally represses the function or engagement of the acetyl-CoA pathway of C. thermoaceticum (Seifritz et al., 1993), thus suggesting that acetogenesis is, in fact, not a constituitive metabolic process of this acetogen (see Chapter 10). The molecular details of how differential processes are regulated are basically unexplored.

Several thermophilic genes encoding enzymes of the acetyl-CoA pathway of C. thermoaceticum have been cloned and successfully expressed in Escherichia coli [for a review see Morton et al. (1992)]. Overexpression of these thermophilic enzymes in a mesophile has practical advantages relative to the production and study of thermostable catalysts. Formyltetrahydrofolate synthetase, for example, has been expressed in active form in E. coli, and genetic probes based on this enzyme have been developed for the identification of acetogens (see Chapter 8). In contrast, genes encoding for acetyl-CoA synthase of C. thermoaceticum have been cloned and expressed in E. coli, but not in active form (Roberts et al., 1989). These genes are located on a 10-kilobase segment of the C. thermoaceticum genome that also contains genes for the corrinoid protein and methyltransferase (Table 1.4); however, how this gene cluster is actually regulated is not resolved.

Little has been reported relative to the use of native plasmids or phages of acetogens as vectors for genetic studies. A 5.6-kilobase plasmid has been isolated and characterized by restriction enzyme mapping from C. aceticum (Lee et al., 1987); curing experiments showed that the occurrence of the plasmid was not correlated to autotrophic growth potentials. An 88-kilobase plasmid from C. thermoaceticum has been isolated, but its potential function in the cell is also unknown (Morton et al., 1992). No phages of acetogenic bacteria have been reported in the literature. This brief accounting, though superficial, illustrates that the regulation and genetics of acetogenic bacteria would be a very fruitful and important future area for study.

1.8 Ecosystems and the Acetyl-CoA Pathway

Another glance over Table 1.2 reminds us that acetogens appear to be community members in practically all anaerobic environments. However, the magnitude and importance of their role(s) in these habitats are not well understood. This is mainly due to (i) their system-level roles having not been rigorously evaluated and (ii) the fact that the main product acetogens form, i.e., acetate, is not as easily assessed in ecosystem-level studies as is anaerobically produced gaseous products such as CH_4. Consequently, our awareness of their activities is not has sharply honed as is our knowledge of, for example, methanogenesis.

Ecosystems vary in size and complexity. Evaluation of the role acetogens may play in the gut of the termite is certainly a different issue than is the evaluation of the potential role acetogens play in anaerobic compartments of soils. For the most part, enrichment, enumeration, and physiological studies have only shown that acetogens are widely distributed and harbor arguably the largest metabolic arsenal of any obligately anaerobic bacterial group characterized to date. This section only illustrates the potential importance of acetogens and the acetyl-CoA pathway at a more global level; future studies must clearly resolve this matter more precisely.

1.8.1 Diverse Habitats, Diverse Physiological Potentials

Just as diverse as their habitats are the metabolic potentials of acetogens. If the acetyl-CoA pathway explains their group identity at the biochemical level, it is likely that their other diverse metabolic processes explain their function at the habitat level. As emphasized throughout Part 3 of this book, their activities impact on a wide range of substrates and can have a profound consequence to the overall flow of matter and energy in their native habitats. Potential sources of reductant include saccharides, C_1 compounds including CO, formate, methyl chloride, H_2, aromatic methoxyl groups, aromatic aldehydes, C_2 compounds including oxalate and glyoxylate, and various alcohols. When reductant from such substrates flows toward CO_2 and the acetyl-CoA pathway, varying substrate : acetate ratios are obtained experimentally (Table 1.6).

However, many acetogens are not strictly dependent on the acetyl-CoA pathway for energy conservation and can make use of alternative nonacetogenic, terminal electron-accepting sinks, as illustrated in the following examples:

Table 1.6 Representative growth-supportive substrates and overall substrate-product stoichiometries of acetogenic bacteria that can form acetate as their predominant reduced end product[a]

Substrate	Overall Stoichiometry for Acetate Production	Representative Acetogen
Acetoin	$2CH_3COCHOHCH_3 + 2CO_2 + 2H_2O \rightarrow 5CH_3COOH$	A. carbinolicum
Alcoxyethanols	$4RO{-}CH_2CH_2OH + 2CO_2 + 2H_2O \rightarrow 4ROH + 5CH_3COOH$	A. malicum
e.g., 2-methoxyethanol	$4CH_3OCH_2CH_2OH + 2CO_2 + 2H_2O \rightarrow 4CH_3OH + 5 CH_3COOH$	A. malicum
e.g., 2-ethoxyethanol	$4C_2H_5OCH_2CH_2OH + 2CO_2 + 2H_2O \rightarrow 4C_2H_5OH + 5CH_3COOH$	A. malicum
Benzaldehydes	4 4-hydroxybenzaldehyde $+ 2CO_2 + 2H_2O \rightarrow CH_3COOH +$ 4-hydroxybenzoate	C. formicoaceticum
2,3-Butanediol	$4CH_3CHOHCHOHCH_3 + 6CO_2 + 2H_2O \rightarrow 11CH_3COOH$	C. magnum
Carbon monoxide	$4CO + 2H_2O \rightarrow CH_3COOH + 2CO_2$	C. thermoaceticum
Citrate	$4C_6H_8O_7 + 2H_2O \rightarrow 9CH_3COOH + 6CO_2$	C. magnum
Cellobiose	$C_{12}H_{22}O_{11} + H_2O \rightarrow 6CH_3COOH$	P. productus
Ethanol	$2CH_3CH_2OH + 2CO_2 \rightarrow 3CH_3COOH$	C. formicoaceticum
Formate	$4HCOOH \rightarrow CH_3COOH + 2CO_2 + 2H_2O$	C. thermoaceticum
Fructose	$C_6H_{12}O_6 \rightarrow 3CH_3COOH$	C. formicoaceticum
Glucose	$C_6H_{12}O_6 \rightarrow 3CH_3COOH$	C. thermoaceticum
Glycerol	$4HOCH_2CHOHCH_2OH + 2CO_2 \rightarrow 7CH_3COOH + 2H_2O$	A. carbinolicum
Glyoxylate	$2HOOCCHO \rightarrow CH_3COOH + 2CO_2$	C. thermoaceticum
Hydrogen + carbon dioxide	$4H_2 + 2CO_2 \rightarrow CH_3COOH + 2H_2O$	C. aceticum
Hydrogen + carbon monoxide	$2H_2 + 2CO \rightarrow CH_3COOH$	C. thermoaceticum
Hydrogen + formate	$2H_2 + 2HCOOH \rightarrow CH_3COOH + 2H_2O$	C. thermoaceticum
Malate	$2HOOCCHOHCH_2COOH \rightarrow 3CH_3COOH + 2CO_2$	C. magnum
Methanol	$4CH_3OH + 2CO_2 \rightarrow 3CH_3COOH + 2H_2O$	C. thermoaceticum
Methoxyacetate	$4CH_3OCH_2COOH + 2CO_2 + 2H_2O \rightarrow 3CH_3COOH + 4HOCH_2COOH$	Strain RMMac1
Methoxylated aromatics	4 aromatic-$[OCH_3] + 2CO_2 + 2H_2O \rightarrow$ 4 aromatic-$[OH] + 3CH_3COOH$	(Many acetogens)
e.g., syringate	2 syringate$[-OCH_3]_2 + 2CO_2 + 2H_2O \rightarrow$ 2 gallate$[-OH]_2 + 3CH_3COOH$	A. woodii
e.g., vanillate	4 vanillate$[-OCH_3] + 2CO_2 + 2H_2O \rightarrow$ 4 protocatechuate$[-OH] + 3CH_3COOH$	A. woodii
e.g., syringate + H_2/CO_2	syringate$[-OCH_3]_2 + 2CO_2 + 2H_2 \rightarrow$ gallate$[-OH]_2 + 2CH_3COOH$	Strain SS1
Methyl chloride	$4CH_3Cl + 2CO_2 + 2H_2O \rightarrow 3CH_3COOH + 4HCl$	Strain MC
Oxalate	$4HOOCCOOH \rightarrow CH_3COOH + 6CO_2 + 2H_2O$	C. thermoaceticum
Pyruvate	$4CH_3COCOOH + 2H_2O \rightarrow 5CH_3COOH + 2CO_2$	C. thermoaceticum
Xylose	$2C_5H_{10}O_5 \rightarrow 5CH_3COOH$	C. thermoaceticum

[a] No distinction is made (a) between acids and their dissociated salt forms or (b) between CO_2 and its carbonate or bicarbonate forms. Although only one representative acetogen is listed per substrate, many acetogens may be capable of utilizing each substrate. For example, almost all acetogens can utilize H_2/CO_2 or methoxylated aromatic compounds for acetate synthesis.

P. productus:	pyruvate $+ 2e^- \rightarrow$ lactate	(XII)
C. formicoaceticum:	fumarate $+ 2e^- \rightarrow$ succinate	(XIII)
A. woodii:	aromatic acrylates $+ 2e^- \rightarrow$ aromatic hydroacrylates	(XIV)
C. thermoaceticum:	dimethylsulfoxide $+ 2e^- \rightarrow$ dimethylsulfide	(XV)
C. thermoaceticum:	thiosulfate $+ 2e^- \rightarrow$ sulfide	(XVI)
C. thermoaceticum:	nitrate $+ 8e^- \rightarrow$ ammonium	(XVII)
C. thermoaceticum:	nitrate $+ 2e^- \rightarrow$ nitrite	(XIX)
P. productus:	benzaldehyde $+ 2e^- \rightarrow$ aromatic alcohols	(XX)
A. woodii:	$2H^+ + 2e^- \rightarrow H_2$	(XXI)

All of these sinks may be coupled to energy conservation, either directly or indirectly. For example, although the reduction of aromatic aldehydes has not been shown to be directly energy conserving, it might provide a mechanism for the generation of CO_2 under conditions of CO_2 deprivation; this CO_2 subsequently serves as an energy-conserving sink via acetogenesis (Drake, 1993). The oxidation of aromatic aldehydes is growth supportive for *C. formicoaceticum* and *C. aceticum* (Gößner et al., 1994). Under some conditions, the use of protons and the formation of H_2 as a terminal sink might also be coupled to energy flow, though this potential is largely unexplored; however, nearly all acetogens we have examined have the potential to form at least trace levels of H_2. Such activities might be linked to the syntrophic potentials of acetogens capable of interspecies H_2 transfer (Winter and Wolfe, 1980; Heijthuijsen and Hansen, 1986; Cord-Ruwisch and Ollivier, 1986) (see Chapters 12 and 14).

Thus, although the acetyl-CoA pathway has made acetogens a somewhat famous bacteriological group, they are far less specialized or restricted than is apparent if one evaluates them only from the standpoint of this pathway. Collectively, they have extensive, diverse catabolic potentials that run tangent to, or intersect with, the acetyl-CoA pathway. This fact differentiates them from other more metabolically restricted anaerobes, such as methanogens (Whitman, 1985; Jones et al., 1987; Whitman et al., 1992). However, in contrast to the well-studied enzymes of the acetyl-CoA pathway, little is known of the other metabolic features of acetogens at the enzymological level.

1.8.2 Impact of the Acetyl-CoA Pathway: Global and Evolutionary Perspectives

Although originally seen as just an interesting process in a few obscure organisms, an important discovery was made when acetyl-CoA synthase and the acetyl-CoA pathway were shown to be widely spread in nature, serving multiple functions for different bacteriological groups (Table 1.7; Schauder et al., 1986; Fuchs, 1986, 1989; Thauer et al., 1989). Major bacterial groups employing this

Table 1.7. Potential functions of acetyl-CoA synthase and the acetyl-CoA pathway in obligately anaerobic bacteria[a]

Type of Function	Representative Bacterial Group	Representative Organism	Typical Reduced End Product of Metabolism	Interfaced to Catabolic or Anabolic Processes	General Function of Acetyl-CoA Relative to Carbon Flow	Chapter in Book for Overview of Topic
A. Acetyl-CoA forming	Acetogens	Clostridium thermoaceticum	Acetate	Both	$CO_2 + [H] \rightarrow$ [acetyl-CoA] $\rightarrow$ acetate + biomass	1–5, 10
	Autotrophic methanogens	Methanobacterium thermoautotrophicum	Methane	Anabolic	$CO_2 + [H] \rightarrow$ [acetyl-CoA] $\rightarrow$ biomass	20
	Autotrophic S-reducing bacteria	Desulfobacterium autotrophicum	Sulfide	Anabolic	$CO_2 + [H] \rightarrow$ [acetyl-CoA] $\rightarrow$ biomass	19 (see also Fuchs, 1986, 1989)
B. Acetyl-CoA degrading	Acetate-oxidizing methanogens	Methanosarcina barkeri	Methane	Catabolic	acetate $\rightarrow$ [acetyl-CoA] $\rightarrow CO_2 + CH_4$	21
	Acetate-oxidizing S-reducing bacteria	Desulfotomaculum acetoxidans	Sulfide	Catabolic	acetate $\rightarrow$ [acetyl-CoA] $\rightarrow CO_2 + [H]$	19 (see also Fuchs, 1986, 1989)
Bidirectional	? (bidirectional acetogens?)	"Reversibacterium"[b] (strain AOR)	Acetate or hydrogen	Both	acetate $\leftrightarrows$ [acetyl-CoA] $\leftrightarrows CO_2 + H_2$	14

[a] Fermentative organisms (e.g., Clostridium pasteurianum) that have the capacity to oxidize CO via what appears to be acetyl-CoA synthase-dependent activity are not included as such organisms do not have a known physiological function for this cryptic activity (from Drake, 1992).
[b] An unclassified isolate capable of growth at the expense of either acetate or H_2. When growth is in the direction of H_2 synthesis, the partial pressure of H_2 must be kept very low; these conditions could occur in syntrophic association with H_2 consumers.

pathway in either the direction of acetate synthesis or acetate degradation include acetogens, methanogens, and sulfate-reducing bacteria. The occurrence of CO dehydrogenase in other anaerobes (e.g., *C. pasterianum*) implies an even more widespread background level of parts of this pathway. It is important to emphasize that the enzymes and cofactors involved in these different bacterial groups are different, but descriptively function in a relatively similar manner.

Although we now understand a great deal about the enzymology of the acetyl-CoA pathway, we know far less about the overall impact of this pathway at the global level. From the preceding comments, it can be anticipated that the acetyl-CoA pathway performs a major global function in the fixation of CO_2 and breakdown of acetate in anaerobic communities. Nature's capacity to form (and thus use) acetate is enormous. Acetate is an essential trophic-level metabolite in many ecosystems (see Chapters 10–12, 15, and 16). At the global level, it is estimated that approximately 10^{13} kg of acetate are formed/metabolized annually in anaerobic habitats such as soils and sediments; 10% of this acetate might be derived from CO_2 fixation via acetogenesis (Wood and Ljungdahl, 1991). It is also estimated that 3.3×10^6 tons (3.35×10^9 kg) of acetate per day are produced bacteriologically in the hindgut of termites alone, a number that exceeds by severalfold the amount of methane [6×10^5 tons (6.36×10^8 kg) per day] formed via biogenic reduction of CO_2 (Breznak and Kane, 1990, Seiler, 1984). Acetogens are also common to the human gut and may produce nearly 1.25×10^{10} kg of acetate per year from H_2/CO_2; additionally, it is estimated that homoacetate fermentation accounts for most of the carbohydrate turnover in 90% of the world's population (Lajoie, 1988; M. Wolin and T. Miller, personal communication; see Chapter 13). Even in beech forest soils of our region, as much as 25% of the total organic carbon of the soil may have the potential to be turned over through acetate under low-temperature, anaerobic conditions (Küsel and Drake, 1994; Chapter 10). The synthesis of 21 g and 38 g acetate per kg dry wt. soil has been obtained at 5°C and 20°C, respectively, from endogenous soil substrates, and this activity is concomitant with the H_2-dependent acetogenic potentials of these soils.

Of the three known mechanisms for complete autotrophic CO_2 fixation (the Calvin cycle, the reductive tricarboxylic acid cycle, and the acetyl-CoA pathway; these exclude at the present time the glycine synthase-coupled pathway because it is not yet known to be truly autotrophic), the acetyl-CoA pathway can be viewed as biochemically the most simple (Wood and Ljungdahl, 1991). Its occurrence and use in the Archea suggest that it was used as an early CO_2-fixing process (Brock, 1989). It has been proposed that the methanogens were the first autotrophs (Schopf et al., 1983), and, as noted earlier, these archaebacteria use the acetyl-CoA pathway for chemolithoautotrophic growth; neither the Calvin cycle nor the reductive tricarboxylic acid cycle have been detected in methanogens. It can, thus, be postulated that the acetyl-CoA pathway may have been the first process used for the autotrophic fixation of CO_2 (Wood and Ljungdahl, 1991; Ragsdale, 1991). Nonetheless, this point, along with the initial reason (catabolic versus anabolic) for evolution of the process, is open to debate (see Chapters 9 and 19).

Although this body of information falls short of determining the global importance of acetogenesis, collectively, it illustrates that the acetyl-CoA Wood/Ljungdahl pathway, like the Calvin cycle, likely plays a fundamental role in the carbon cycle of our planet. An important outcome of future studies will hopefully be a resolution of the magnitude of this process at the global level.

1.9   Potential Biotechnological Applications

The potential applications of acetogens in industry are addressed in detail in Chapter 18 and are also recently reviewed (Lowe et al., 1993). Acetogens make acetate, and this potential has been the main focal point relative to application. Acetic acid is an important organic compound to the petrochemical industry (see Chapter 18). It is produced synthetically from fossil feedstocks such as methanol, and production in the United States and worldwide approximated 4 and 8 billion pounds, respectively, in 1989 (Busche, 1991). The 1973 oil crisis fueled interest in the commercial application of acetogenesis for the production of acetic acid. Such interests yielded mainly laboratory-level investigations rather than process-level applications, mainly because oil prices and availability returned to levels that have not necessitated a shift away from synthetic processes. Nonetheless, acetic acid production from a renewable resource seems inevitable given the current emphasis on energy conservation and pollution.

Acetic acid can be produced biologically from glucose by industry by two main processes. One is the classic two-stage vinegar process, in which 1 mole of glucose is converted to 2 moles of acetic acid via the sequential activities of *Saccharomyces cerevisiae* (stage 1, initially yielding 2 ethanol/glucose) and *Acetobacter aceti* (stage 2, overall yielding 2 acetic acid/glucose). The other is a one-stage process employing acetogenic bacteria. These two process have been evaluated relative to potential production costs (Busche, 1991); at the present time, neither process competes with the production costs of the synthetic process (estimated at $0.30/lb acetic acid in 1991). Furthermore, the two-stage vinegar process (production costs estimated at $0.38/lb acetic acid in 1991) is viewed more economically feasible than is the one-stage acetogenic process (based on application of *C. thermoaceticum;* production costs estimated at $0.42/lb acetic acid in 1991) (Busche, 1991). This is in contrast to the often-stated theoretical advantage of acetogenesis for the commercial production of acetic acid: Acetogenesis conserves all of the carbon of glucose in the product acetic acid, thus increasing overall yields per glucose molecule by 50%.

However, a basic problem with the acetogenic process is that the acetogens evaluated are inhibited by high concentrations of acetate and low pH (Wang and Wang, 1984). Many studies have addressed this problem with batch, reactor, and immobilized cell systems (Wang and Wang, 1983; Sugaya et al., 1986; Ljungdahl et al., 1985, 1989), and acid tolerant mutants of *C. thermoaceticum* have been obtained and characterized (Schwartz and Keller, 1982; Brumm, 1988; Parekh and Cheryan, 1991; Cheryan and Parekh, 1992). Additional and similar work has been conducted with *A. kivui* (Klemps et al., 1987; Eysmondt et al., 1990; Ibba and Fynn, 1991). In this regard, it is postulated that production costs of the acetogenic process could outcompete the synthetic process should an acid tolerant, high acetic acid-producing acetogen (or mutant) be obtained (Brusche, 1991). The use of calcium-magnesium acetate as an environmentally safe road de-icer and in the potential control of sulfur emissions from the combustion of high-sulfur coal continues to focus attention on the future application of acetogenesis for the commercial production of acetic acid (Ljungdahl et al., 1985, 1989; Wiegel et al., 1990; Parekh and Cheryan, 1991; Cheryan and Parekh, 1992).

The bioconversion of synthesis gas (obtained from coal by indirect liquefaction and consisting mainly of H_2, CO, and CO_2) by acetogenic bacteria has also been addressed in exploritory studies (Grethlein and Jain, 1992). Both acetate and butyrate are possible products from synthesis gas by *B. methylotrophicum*, butyrate being formed mostly during stationary phase of growth at pH 6.8; butyrate becomes the favored product (on a per gram basis) when cells at the onset of stationary phase are shifted to pH 6 (Worden et al., 1989). In addition, *n*-butanol has recently been identified as the predominant product from CO of *B. methylotrophicum* at pH 5.5 (Grethlein et al., 1991). The potential to produce ethanol from synthesis gas by metabolically altered acetogens has also been addressed, though, to date, these studies do not indicate ethanol production by acetogens is a favored process (Buschhorn et al., 1989; Grethlein and Jain, 1992; Tanner et al., 1993). The potential use of acetogens for the commercial production of vitamin B_{12} and cysteine has also been addressed in laboratory studies (Inoue et al., 1992; Koesnandar et al., 1991).

Acetogens are cited as having bioremediation potentials. Such potentials have not been extensively examined. Acetogens have a remarkable capacity to use CO as an energy and carbon source in laboratory studies, and they are often thought of as being able to detoxify CO from the environment (Ragsdale, 1991). However, this potential may be unrealistic in nature, at least relative to the detoxification of CO introduced into the atmosphere by human activities. It seems more likely that abiotic processes in the atmosphere or aerobic CO-utilizing soil carboxydotrophs, such as *Pseudomonas carboxydovorans* or CO-oxidizing species of *Streptomyces,* serve as the primary sinks for CO (Meyer, 1988, Meyer et al., 1993). Some acetogens have been shown to be capable of dehalogenation, either in pure or coculture (Freedman and Gossett, 1991; Egli et al., 1988;

Traunecker et al., 1991). However, the ability to dehalogenate has not been shown to be a widespread or high-level potential of acetogens, and likewise the capacity to mineralize aromatic rings is not readily observed with acetogens.

1.10   Unresolved Problems

Throughout this chapter, unresolved problems have been mentioned that might be considered relevant questions for future studies:

(i)   How different are the enzymes and cofactors of the acetyl-CoA pathway of different acetogens? What is the nature of the hydrogenases of acetogens? What is the biochemistry of the numerous other catabolic processes? Most of our enzymological knowledge is restricted to one acetogen, *C. thermoaceticum,* and is, even for this organism, not complete. For example, recent studies with *Sporomusa ovata* on the "methyl-level contact points" that interface to the acetyl-CoA pathway indicate that such efforts would be productive as well as important (Stupperich and Konle, 1993).

(ii)   How are the acetyl-CoA pathway and other important metabolic potentials of acetogens regulated? How is the flow of reductant toward different terminal electron acceptors controlled? Answers will require physiological, biochemical, and molecular studies.

(iii)   Are acetogens in nature active in the degradation of cellulose and other polymers? In addition, are there conditions under which acetogens play an important role in the cleavage of aromatic rings? Although many new metabolic potentials have been discovered in recent years, the capacity of acetogens to utilize polymers and degrade aromatic rings does not appear to be a widespread potential in acetogens.

(iv)   What is the real impact of acetogens in complex habitats such as soils and sediments? How widespread is their influence in extreme environments? What is the overall global impact of the acetyl-CoA pathway? At present, we can only speculate on such basic questions.

(v)   Can the metabolic potentials of acetogens be harnessed in a commercial application? In a practical sense, this would require compatible market prices for the product.

If Harland would have been able to finish his preparation of this chapter, I am sure he would at this point say, "Now that this damn writing is finished, it's time to get back to the lab and get some real work done on these problems." I agree, Harland.

Acknowledgments

The author expresses deep appreciation to Harland G. Wood and Lars G. Ljungdahl for the many years of colleagueship and advice, as well as the shared wine and catfish at Taylor Grocery in Taylor, Mississippi. Your signatures can still be found there on the north and west walls. The author also thanks Jan R. Andressen and Steven L. Daniel for critical reading and editorial review of the manuscript. The authors of this book have kindly provided unpublished data that have been useful in preparing the scope of the text; my thanks to all of you. Figure 1.2 was kindly provided by Georg Acker. Current support for the author's laboratory is derived in part from funds from the Bundesministerium für Forschung und Technologie (0339476A0) and is gratefully acknowledged. Earlier support obtained from the National Institutes of Health, the National Science Foundation, and the State of Mississippi was also fundamental during the author's earlier investigations at the University of Mississippi.

References

Adamse, A. D. (1980). New isolation of *Clostridium aceticum* (Wieringa). *Antonie van Leeuwenhoek* **46:**523–531.

Adamse, A. D., and C. T. M. Velzeboer. 1982. Features of a *Clostridium*, strain CV-AA1, an obligatory anaerobic bacterium producing acetic acid from methanol. *Antonie van Leeuwenhoek* **48:**305–313.

Andreesen, J. R., G. Gottschalk, and H. G. Schlegel. 1970. *Clostridium formicoaceticum* nov. spec. isolation, description and distinction from *C. aceticum* and *C. thermoaceticum*. *Arch. Microbiol.* **72:**154–174.

Andressen, J. R., A. Schaupp, C. Neurauter, A. Brown, and L. G. Ljungdahl. 1973. Fermentation of glucose, fructose, and xylose by *Clostridium thermoaceticum*: Effect of metals on growth yield, enzymes, and the synthesis of acetate from CO_2. *J. Bacteriol.* **114:**743–751.

Bache, R., and N. Pfennig. 1981. Selective isolation of *Acetobacterium woodii* on methoxylated aromatic acids and determination of growth yields. *Arch. Microbiol.* **130:**255–261.

Balch, W. E., S. Schoberth, R. S. Tanner, and R. S. Wolfe. 1977. *Acetobacterium*, a new genus of hydrogen-oxidizing, carbon dioxide-reducing, anaerobic bacteria. *Int. J. Sys. Bacteriol.* **27:**355–361.

Bak, F., K. Finster, and F. Rothfuß. 1992. Formation of dimethylsulfide and methanethiol from methoxylated aromatic compounds and inorganic sulfide by newly isolated anaerobic bacteria. *Arch. Microbiol.* **157:**529–534.

Barker, H. A. 1944. On the role of carbon dioxide in the metabolism of *Clostridium thermoaceticum*. *Proc. Natl. Acad. Sci.* **30:**88–90.

Barker, H. A., and M. D. Kamen. 1945. Carbon dioxide utilization in the synthesis of acetic acid by *Clostridium thermoaceticum*. *Proc. Natl. Acad. Sci. USA* **31:**219–225.

Beaty, P. S., and L. G. Ljungdahl. 1991. Growth of *Clostridium thermoaceticum* on methanol, ethanol, propanol, and butanol in medium containing either thiosulfate or dimethylsulfoxide, Abstr. K-131, p. 236, *Abstr. Ann. Meet. Am. Soc. Microbiol.* 1991.

Bomar, M., H. Hippe, and B. Schink. 1991. Lithotrophic growth and hydrogen metabolism by *Clostridium magnum*. *FEMS Microbiol. Lett.* 83:347–350.

Braun, K., S. Schoberth, and G. Gottschalk. 1979. Enumeration of bacteria forming acetate from H_2 and CO_2 in anaerobic habitats. *Arch. Microbiol.* 120:201–204.

Braun, K., and G. Gottschalk. 1981. Effect of molecular hydrogen and carbon dioxide on chemo-organotrophic growth of *Acetobacterium woodii* and *Clostridium aceticum*. *Arch. Microbiol.* 128:294–298.

Braun, M., F. Mayer, and G. Gottschalk. 1981. *Clostridium aceticum* (Wieringa), a microorganism producing acetic acid from molecular hydrogen and carbon dioxide. *Arch. Microbiol.* 128:288–293.

Braun, M., and G. Gottschalk. 1982. *Acetobacterium wieringae* sp. nov., a new species producing acetic acid from molecular hydrogen and carbon dioxide. *Zbl. Bakt. Hyg.*, I. abt. Orig. C3, pp. 368–376.

Braus-Stromeyer, S. A., R. Hermann, A. M. Cook, and T. Leisinger. 1993. Dichloromethane as the sole carbon source for an acetogenic mixed culture and isolation of a fermentative, dichloromethane-degrading bacterium. *Appl. Environ. Microbiol.* 59:3790–3797.

Breznak, J. A., and J. M. Switzer. 1986. Acetate synthesis from H_2 plus CO_2 by termite gut microbes. *Appl. Environ. Microbiol.* 52:623–630.

Breznak, J. A., J. M. Switzer, and H.-J. Seitz. 1988. *Sporomusa termitida* sp. nov., an H_2/CO_2 utilizing acetogen isolated from termites. *Arch. Microbiol.* 150:282–288.

Breznak, J. A., and M. D. Kane. 1990. Microbiol H_2/CO_2 acetogenesis in animal guts: nature and nutritional significance. *FEMS Microbiol. Rev.* 87:309–314.

Brock, T. D. 1989. Evolutionary relationships of the autotrophic bacteria. *In: Autotrophic Bacteria*, H. G. Schlegel and B. Bowien (eds.), pp. 499–512. Science Tech Publishers, Madison, WI.

Brulla, W. J., and M. P. Bryant. 1989. Growth of the syntrophic anaerobic acetogen, strain PA-1, with glucose or succinate as energy source. *Appl. Environ. Microbiol.* 55:1289–1290.

Brumm, P. J. 1988. Fermentation of single and mixed substrates by the parent and an acid-tolerant, mutant strain of *Clostridium thermoaceticum*. *Biotechnol. Bioengineer.* 32:444–450.

Busche, R. M. (1991). Extractive fermentation of acetic acid: Economic tradeoff between yield of *Clostridium* and concentration of *Acetobacter*. *Appl. Biochem. Biotechnol.* 28/29:605–621.

Buschhorn, H., P. Dürre, and G. Gottschalk. 1989. Production and utilization of ethanol by the homoacetogen *Acetobacterium woodii*. *Appl. Environ. Microbiol.* 55:1835–1840.

Cato, E. P., W. L. George, and S. M. Finegold. 1986. Genus *Clostridium* Prazmowski

1880. *In: Bergey's Manual of Systematic Bacteriology*, P. H. A. Sneath (ed.), Vol. 2, pp. 1141–1200. Williams and Wilkins, Baltimore, MD.

Cato, E. S., and E. Stackebrandt. 1989. Taxonomy and phylogeny. *In: Clostridia*, N. P. Minton, and D. J. Clarke (eds.), pp. 1–26. Plenum Press, New York.

Charakhch'yan, D.-I. A., A. N. Mileeva, L. L. Mityushina, and S. S. Belyaev. 1992. Acetogenic bacteria from oil fields of Tataria and western Siberia. *Mikrobiologiya* **61**:306–315.

Cheryan, M., and S. Parekh. 1992. Acetate and calcium magnesium acetate (CMA) production with mutant strains of *Clostridium thermoaceticum* ATCC 49707. *Abstr. Ann. Meet. Am. Soc. Microbiol.*, p. 315, Abstr. 0-39.

Clark, J. E., and L. G. Ljungdahl. 1984. Purification and properties of 5,10-methylenetetrahydrofolate reductase, an iron-sulfur flavoprotein from *Clostridium formicoaceticum*. *J. Biol. Chem.* **259**:10845–10849.

Conrad, R., F. Bak, H. J. Seitz, B. Thebrath, H. P. Mayer, and H. Schütz. 1989. Hydrogen turnover by psychrotrophic homoacetogenic and mesophilic methanogenic bacteria in anoxic paddy soil and lake sediment. *FEMS Microbiol. Ecol.* **62**:285–294.

Cord-Ruwisch, R., and B. Ollivier. 1986. Interspecific hydrogen transfer during methanol degradation by *Sporomusa acidovorans* and hydrogenophilic anaerobes. *Arch. Microbiol.* **144**:163–165.

Daniel, S. L., T. Hsu, S. I. Dean, H. L. Drake. 1990. Characterization of the H_2- and CO-dependent chemolithotrophic potentials of the acetogens *Clostridium thermoaceticum* and *Acetogenium kivui*. *J. Bacteriol.* **172**:4464–4471.

Daniel, S. L., H. L. Drake. 1993. Oxalate- and glyoxylate-dependent growth and acetogenesis by *Clostridium thermoaceticum*. *Appl. Environ. Microbiol.* **59**:3062–3069.

Das, A., J. Hugenholtz, H. van Halbeek, and L. G. Ljungdahl. 1989. Structure and function of a menaquinone involved in electron transport in membranes of *Clostridium thermoautotrophicum* and *Clostridium thermoaceticum*. *J. Bacteriol.* **171**:5823–5829.

Das, A., L. G. Ljungdahl. 1993. F_0 and F_1 parts of ATP synthases from *Clostridium thermoautotrophicum* and *Escherichia coli* are not functionally compatible. *FEBS Lett.* **317**:17–21.

Dehning, I., M. Stieb, B. Schink. 1989. *Sporomusa malonica* sp. nov., a homoacetogenic bacterium growing by decarboxylation of malonate or succinate. *Arch. Microbiol.* **151**:421–426.

DeWeerd, K. A., A. Saxena, D. P. Nagle, Jr., J. M. Suflita. 1988. Metabolism of the ^{18}O-methoxy substituent of 3-methoxybenzoic acid and other unlabeled methoxybenzoic acids by anaerobic bacteria. *Appl. Environ. Microbiol.* **54**:1237–1242.

Diekert, G., and R. K. Thauer. 1978. Carbon monoxide oxidation by *Clostridium thermoaceticum* and *Clostridium formicoaceticum*. *J. Bacteriol.* **136**:597–606.

Diekert, G., and M. Ritter. 1983. Purification of the nickel protein carbon monoxide dehydrogenase of *Clostridium thermoaceticum*. *FEBS Lett.* **151**:41–44.

Diekert, G., M. Hansch, and R. Conrad. 1984. Acetate synthesis from $2 CO_2$ in acetogenic bacteria: is carbon monoxide an intermediate? *Arch. Microbiol.* **138**:224–228.

Diekert, G., E. Schrader, and W. Harder. 1986. Energetics of CO formation and CO oxidation in cell suspensions of *Acetobacterium woodii*. *Arch. Microbiol.* **144**:386–392.

Diekert, G. 1992. The acetogenic bacteria. *In:* A. Balows, H. G. Trüper, M. Dworkin, W. Harder, K.-H. Schleifer (eds.) *The Prokaryotes,* 2nd ed., pp. 517–533. Springer-Verlag, New York.

Dimroth, P. 1987. Sodium ion transport decarboxlases and other aspects of sodium ion cycling in bacteria. *Microbiol. Rev.* **51**:320–340.

Dorn, M., J. R. Andreesen, and G. Gottschalk. (1978). Fermentation of fumarate and L-malate by *Clostridium formicoaceticum*. *J. Bacteriol.* **133**:26–32.

Dörner, C., and B. Schink. 1991. Fermentation of mandelate to benzoate and acetate by a homoacetogenic bacterium. *Arch. Microbiol.* **156**:302–306.

Drake, H. L., S.-I. Hu, and H. G. Wood. 1980. Purification of carbon monoxide dehydrogenase, a nickel enzyme from *Clostridium thermoaceticum*. *J. Biol. Chem.* **255**:7174–7180.

Drake, H. L., S.-I. Hu, and H. G. Wood. 1981a. Purification of five components from *Clostridium thermoaceticum* which catalyze synthesis of acetate from pyruvate and methyltetrahydrofolate: properties of phosphotransacetylase. *J. Biol. Chem.* **255**:7174–7180.

Drake, H. L., S.-I. Hu, and H. G. Wood. 1981b. The synthesis of acetate from carbon monoxide plus methyltetrahydrofolate and the involvement of the nickel enzyme, CO dehydrogenase, Abstr. K42. p. 144. *Abstr. Ann. Meet. Am. Soc. Microbiol,* 1981.

Drake, H. L. 1982. Demonstration of hydrogenase in extracts of the homoacetate-fermenting bacterium *Clostridium thermoaceticum*. *J. Bacteriol.* **150**:702–709.

Drake, H. L. 1992. Acetogenesis and acetogenic bacteria. *In: Encyclopedia of Microbiology,* J. Lederberg (ed.), Vol. 1, pp. 1–15. Academic Press, San Diego, CA.

Drake, H. L. 1993. CO_2, reductant, and the autrophic acetyl-CoA pathway: alternative origins and destinations. *In: Microbial Growth on C_1 Compounds,* C. Murrell, and D. P. Kelly (eds.), pp. 493–507. Intercept Limited, Andover, England.

Drent, W. J., and J. C. Gottschal. 1991. Fermentation of inulin by a new strain of *Clostridium thermoautotrophicum* isolated from dahlia tubers. *FEMS Microbiol. Lett.* **78**:285–292.

Eden, G., and G. Fuchs. 1982. Total synthesis of acetyl coenzyme A involved in autotrophic CO_2 fixation in *Acetobacterium woodii*. *Arch. Microbiol.* **133**:66–74.

Eden, G., and G. Fuchs. 1983. Autotrophic CO_2 fixation in *Acetobacterium woodii* II. Demonstration of enzymes involved. *Arch. Microbiol.* **135**:68–73.

Eichler, B., and B. Schink. 1984. Oxidation of primary aliphatic alcohols by *Acetobacterium carbinolicum* sp. nov., a homoacetogenic anaerobe. *Arch. Microbiol.* **140**:147–152.

Egli, C., T. Tschan, R. Scholtz, A. M. Cook, and T. Leisinger. 1988. Transformation of tetrachloromethane to dichloromethane and carbon dioxide by *Acetobacterium woodii*. *Appl. Environ. Microbiol.* **54**:2819–2824.

El Ghazzawi, E. 1967. Neuisolierung von *Clostridium formicoaceticum* Wieringa und stoffwechselphysiologische Untersuchungen. *Arch. Mikrobiol.* **57**:1–19.

Emde, R., and B. Schink. 1987. Fermentation of triacetin and glycerol by *Acetobacterium* sp. No energy is conserved by acetate excretion. *Arch. Microbiol.* **149**:142–148.

von Eysmondt, J., Dj. Vasic-Racki, and Ch. Wandrey. 1990. Acetic acid production by *Acetogenium kivui* in continuous culture–kinetic studies and computer simulations. *Appl. Microbiol. Biotechnol.* **34**:344–349.

Ferry, J. G. (1992). Methane from acetate. *J. Bacteriol.* **174**:5489–5495.

Fischer, F., R. Lieske, and K. Winzer. 1932. Biologische Gasreaktionen. II. Über die Bildung von Essigsäure bei der biologischen Umsetzung von Kohlenoxyd und Kohlensäure mit Wasserstoff zu Methan. *Biochem. Z.* **245**:2–12.

Fontaine, F. E., W. H. Peterson, E. McCoy, M. J. Johnson, and G. J. Ritter. 1942. A new type of glucose fermentation by *Clostridium thermoaceticum* n. sp. *J. Bacteriol.* **43**:701–715.

Frazer, A. C., and L. Y. Young. 1985. A gram-negative anaerobic bacterium that utilizes O-methyl substituents of aromatic acids. *Appl. Environ. Microbiol.* **49**:1345–1347.

Freedman, D. L., and J. M. Gosset. 1991. Biodegradation of dichloromethane and its utilization as a growth substrate under methanogenic conditions. *Appl. Environ. Microbiol.* **57**:2847–2857.

Fuchs, G., U. Schnitker, and R. K. Thauer. 1974. Carbon monoxide oxidation by growing cultures of *Clostridium pasteurianum*. *Eur. J. Biochem.* **49**:111–115.

Fuchs, G. 1986. CO_2 fixation in acetogenic bacteria: variations on a theme. *FEMS Microbiol. Rev.* **39**:181–213.

Fuchs, G. 1989. Alternative pathways of autotrophic CO_2 fixation. *In: Autotrophic Bacteria*, H. G. Schlegel and B. Bowien (eds.), pp. 365–382, Science Tech, Madison, WI and Springer-Verlag, Berlin.

Geerligs, G., H. C. Aldrich, W. Harder, and G. Diekert. 1987. Isolation and characterization of a carbon monoxide utilizing strain of the acetogen *Peptostreptococcus productus*. *Arch. Microbiol.* **148**:305–313.

Geerligs, G., P. Schönheit, and G. Diekert. 1989. Sodium dependent acetate formation from CO_2 in *Peptostreptococcus productus* (strain Marburg). *FEMS Microbiol. Lett.* **57**: 253–258.

Gorst, C. M., and S. W. Ragsdale. 1991. Characterization of the NiFeCO complex of carbon monoxide dehydrogenase as a catalytically competent intermediate in the pathway of acetyl-coenzyme A synthesis. *J. Biol. Chem.* **266**:20687–20693.

Gößner, A., S. L. Daniel, and H. L. Drake. 1994. Acetogenesis coupled to the oxidation of aromatic aldehyde groups. *Arch. Microbiol.* **161**:126–131.

Gottschalk, G. 1989. Bioenergetics of methanogenic and acetogenic bacteria. *In* H. G.

Schlegel and B. Bowien (eds.), *Autotrophic Bacteria,* pp. 383–396. Science Tech, Madison, WI.

Greening, R. C., and J. A. Z. Leedle. 1989. Enrichment and isolation of *Acetitomaculum ruminis,* gen. nov., sp. nov.: acetogenic bacteria from the bovine rumen. *Arch. Microbiol.* **151**:399–406.

Grethlein, A. J., R. M. Worden, M. K. Jain, and R. Datta. 1991. Evidence for production of *n*-butanol from carbon monoxide by *Butyribacterium methylotrophicum. J. Ferment. Bioengineer.* **72**:58–60.

Grethlein, A. J., and M. K. Jain. 1992. Bioprocessing of coal-derived synthesis gases by anaerobic bacteria. *TIBTECH* **10**:418–423.

Gunsalus, R. P., J. A. Romesser, and R. S. Wolfe. 1978. Preparation of coenzyme M analogs and their activity in the methyl-coenzyme M reductase in *Methanobacterium thermoautotrophicum. Biochemistry* **17**:2374–2377.

Heijthuijsen, J. H. F. G., and T. A. Hansen. 1986. Interspecies hydrogen transfer in co-cultures of methanol-utilizing acidogens and sulfate-reducing or methanogenic bacteria. *FEMS Microbiol. Ecol.* **38**:57–64.

Heijthuijsen, J. H. F. G., and T. A. Hansen. 1989. Selection of sulphur sources for the growth of *Butyribacterium methylotrophicum* and *Acetobacterium woodii. Appl. Microbiol. Biotechnol.* **32**:186–192.

Heinonen, J. K., and H. L. Drake. 1988. Comparative assessment of inorganic pyrophosphate and pyrophosphatase levels of *Escherichia coli, Clostridium pasteurianum,* and *Clostridium thermoaceticum. FEMS Microbiol. Lett.* **52**: 205–208.

Heise, R., V. Müller, and G. Gottschalk. 1989. Sodium dependence of acetate formation by the acetogenic bacterium *Acetobacterium woodii. J. Bacteriol.* **171**:5473–5478.

Heise, R., J. Reidlinger, V. Müller, and G. Gottschalk. 1991. A sodium-stimulated ATP synthase in the acetogenic bacterium *Acetobacterium woodii. FEBS Lett.* **295**:119–122.

Heise, R., V. Müller, and G. Gottschalk. 1992. Presence of a sodium-translocating ATPase in membrane vesicles of the homoacetogenic bacterium *Acetobacterium woodii. Eur. J. Biochem.* **206**:553–557.

Heise, R., V. Müller, and G. Gottschalk. 1993. Acetogenesis and ATP synthesis in *Acetobacterium woodii* are coupled via a transmembrane primary sodium ion gradient. *FEMS Micrbiol. Lett.* **112**:261–268.

Hermann, M., M.-R. Popoff, and M. Sebald. 1987. *Sporomusa paucivorans* sp. nov., a methylotrophic bacterium that forms acetic acid from hydrogen and carbon dioxide. *Int. J. Syst. Bacteriol.* **37**:93–101.

Hsu, T., M. F. Lux, and H. L. Drake. 1990. Expression of an aromatic-dependent decarboxylase which provides growth-essential CO_2 equivalents for the acetogenic (Wood) pathway of *Clostridium thermoaceticum. J. Bacteriol.* **172**:5901–5907.

Hu, S.-I., H. L. Drake, and H. G. Wood. 1982. Synthesis of acetyl coenzyme A from carbon monoxide, methyltetrahydrofolate, and coenzyme A by enzymes from *Clostridium thermoaceticum. J. Bacteriol.* **149**:440–448.

Hu, S.-I., E. Pezacka, and H. G. Wood. 1984. Acetate synthesis from carbon monoxide by *Clostridium thermoaceticum*: purification of the corrinoid protein. *J. Biol. Chem.* **259**:8892–8897.

Hugenholtz, J., and L. G. Ljungdahl. 1989. Electron transport and electrochemical proton gradient in membrane vesicles of *Clostridium thermoautotrophicum*. *J. Bacteriol.* **171**:2873–2875.

Hugenholtz, J., and L. G. Ljungdahl. 1990. Amino acid transport in membrane vesicles of *Clostridium thermoautotrophicum*. *FEMS Microbiol. Lett.* **69**:117–122.

Hungate, R. E. 1969. A roll tube method for cultivation of strict anaerobes. *In: Methods in Microbiology*, J. R. Norris and D. W. Ribbons (eds.), Vol. 3B, pp. 117–132. Academic Press, New York.

Ibba, M., and G. H. Fynn. 1991. Two stage methanogenesis of glucose by *Acetogenium kivui* and acetoclastic methanogenic sp. *Biotechnol. Lett.* **13**:671–676.

Inoue, K., S. Kageyama, K. Miki, T. Morinaga, Y. Kamagata, K. Nakamura, and E. Mikami. 1992. Vitamin B_{12} Production by *Acetobacterium* sp. and its tetrachloromethane-resistant mutants. *J. Ferment. Bioengineer.* **73**:76–78.

Ivey, D. M., and L. G. Ljungdahl. 1986. Purification and characterization of the F_1-ATPase from *Clostridium thermoaceticum*. *J. Bacteriol.* **165**:252–257.

Jones, W. J., D. P. Nagle, Jr., and W. B. Whitman. 1987. Methanogens and the diversity of archaebacteria. *Microbiol. Rev.* **51**:135–177.

Kamen, M. D. 1963. The early history of carbon-14. *J. Chem. Educ.* **40**:234–242.

Kamlage, B., and M. Blaut. 1993. Isolation of a cytochrome-deficient mutant strain of *Sporomusa sphaeroides* not capable of oxidizing methyl groups. *J. Bacteriol.* **175**:3043–3050.

Kamlage, B., A. Boelter, and M. Blaut. 1993. Spectroscopic and potentiometric characterization of cytochromes in two *Sporomusa* species and their expression during growth on selected substrates. *Arch. Microbiol.* **159**:189–196.

Kane, M. D., and J. A. Breznak. 1991. *Acetonema longum* gen. nov. sp. nov., an H_2/CO_2 acetogenic bacterium from the termite, *Pterotermes occidentis*. *Arch. Microbiol.* **156**:91–98.

Kane, M. D., A. Brauman, and J. A. Breznak. 1991. *Clostridium mayombei* sp. nov., an H_2/CO_2 acetogenic bacterium from the gut of the African soil-feeding termite, *Cubitermes speciosus*. *Arch. Microbiol.* **156**:99–104.

Kellum, R., and H. L. Drake. 1984. Effects of cultivation gas phase on hydrogenase of the acetogen *Clostridium thermoaceticum*. *J. Bacteriol.* **160**:466–469.

Kellum, R., and H. L. Drake. 1986. Effects of carbon monoxide on one-carbon enzymes and energetics of *Clostridium thermoaceticum*. *FEMS Microbiol. Lett.* **34**:41–45.

Kerby, R., and J. G. Zeikus. 1983. Growth of *Clostridium thermoaceticum* on H_2/CO_2 or CO as energy source. *Curr. Microbiol.* **8**:27–30.

Klemps, R., S. M. Schoberth, and H. Sahm. 1987. Production of acetic acid by *Acetogenium kivui*. *Appl. Microbiol. Biotechnol.* **27**:229–234.

Koesnadar, N. Nishio, A. Yamamoto, and S. Nagi. 1991. Enzymatic reduction of cystine into cysteine by cell-free extract of *Clostridium thermoaceticum*. *J. Ferment. Bioengineer*. **72**:11–14.

Kotsyurbenko, O. R., M. V. Simankova, N. P. Bolotina, T. N. Zhilina, and A. N. Nozhevnikova. 1992. Psychrotrophic homoacetogenic bacteria from several environments. Abstr. C136. *7th Int. Symp. C₁-Compounds*. 1992.

Krumholz, L. R., and M. P. Bryant. 1985. *Clostridium pfennigii* sp. nov. uses methoxyl groups of monobenzenoids and produces butyrate. *Int. J. Syst. Bacteriol*. **35**:454–456.

Krumholz, L. R., and M. P. Bryant. 1986. *Syntrophococcus sucromutans* sp. nov. gen. nov. uses carbohydrates as electron donors and formate, methoxymonobenzenoids or *Methanobrevibacter* as electron acceptor systems. *Arch. Microbiol*. **143**:313–318.

Küsel, K., and H. L. Drake. 1994. Acetate synthesis by soil from a Bavarian beech forest. *Appl. Environ. Microbiol*. **60**:1370–1373.

Ladapo, J., and W. B. Whitman. 1990. Method for isolation of auxotrophs in the methanogenic archaebacteria: role of the acetyl-CoA pathway of autotrophic CO_2 fixation in *Methanococcus maripaludis*. *Proc. Natl. Acad. Sci. USA* **87**:5598–5602.

Lajoie, S. F., S. Bank, T. L. Miller, and M. J. Wolin. 1988. Acetate production from hydrogen and [^{13}C]carbon dioxide by the microflora of human feces. *Appl. Environ. Microbiol*. **54**:2723–2727.

Lee, C.-K., P. Dürre, H. Hippe, and G. Gottschalk. 1987. Screening for plasmids in the genus *Clostridium*. *Arch. Microbiol*. **148**: 107–114.

Lee, M. J., and S. H. Zinder. 1988. Isolation and characterization of a thermophilic bacterium which oxidizes acetate in syntrophic association with a methanogen and which grows acetogenically on H_2-CO_2. *Appl. Environ. Microbiol*. **54**:124–129.

Leedle, J. A. Z., and R. C. Greening. 1988. Postprandial changes in methanogenic and acidogenic bacteria in the rumens of steers fed high- or low-forage diets once daily. *Appl. Environ. Microbiol*. **54**:502–506.

Leigh, J. A., F. Mayer, and R. S. Wolfe. 1981. *Acetogenium kivui*, a new thermophilic hydrogen-oxidizing, acetogenic bacterium. *Arch. Microbiol*. **129**:275–280.

Lentz, K., and H. G. Wood. 1955. Synthesis of acetate from formate and carbon dioxide by *Clostridium thermoaceticum*. *J. Biol. Chem*. **215**:645–654.

Liu, C.-L., N. Hart, and H. D. Peck, Jr. 1982. Inorganic pyrophosphate: energy source for sulfate-reducing bacteria of the genus *Desulfotomaculum*. *Science* **217**:363–364.

Liu, S., and J. M. Suflita. 1993. H_2/CO_2-dependent anaerobic O-demethylation activity in subsurface sediments and by an isolated bacterium. *Appl. Environ. Microbiol*. **59**:1325–1331.

Ljungdahl, L., and H. G. Wood. 1965. Incorporation of C^{14} from carbon dioxide into sugar phosphates, carboxylic acids, and amino acids by *Clostridium thermoaceticum*. *J. Bacteriol*. **89**:1055–1064.

Ljungdahl, L., E. Irion, and H. G. Wood. 1966. Role of corrinoids in the total synthesis of acetate from CO_2 by *Clostridium thermoaceticum*. *Fed. Proc*. **25**:1642–1648.

Ljungdahl, L. G. and H. G. Wood. 1969. Total synthesis of acetate from CO_2 by heterotrophic bacteria. *Ann. Rev. Microbiol.* **23**:515–538.

Ljungdahl, L. G., F. Bryant, L. Carreira, T. Saiki, and J. Wiegel. 1981. Some aspects of thermophilic and extreme thermophilic anaerobic microorganisms. *In: Trends in the Biology of Fermentations,* A. Hollaender (ed.), pp. 397–419. Plenum Press, New York.

Ljungdahl, L. G., L. H. Carreira, R. J. Garrison, N. E. Rabek, and J. Wiegel. 1985. Comparison of three thermophilic acetogenic bacteria for production of calcium magnesium acetate. *Biotechnol. Bioengeer. Symp.* **15**:207–223.

Ljungdahl, L. G. 1986. The autotrophic pathway of acetate synthesis in acetogenic bacteria. *Ann. Rev. Microbiol.* **40**:415–450.

Ljungdahl, L. G., J. Hugenholtz, and J. Wiegel. 1989. Acetogenic and Acid-Producing Clostridia. *In: Clostridia,* N. P. Minton and D. J. Clarke (eds.), pp. 145–191. Plenum Press, New York.

Lorowitz, W. H., and M. P. Bryant. 1984. *Peptostreptococcus productus* strain that grows rapidly with CO as the energy source. *Appl. Environ. Microbiol.* **47**: 961–964.

Loubiere, P., E. Gros, V. Paquet, and N. D. Lindley. 1992. Kinetics and physiological implications of the growth behaviour of *Eubacterium limosum* on glucose/methanol mixtures. *J. Gen. Microbiol.* **138**:979–985.

Lovell, C. R., A. Przybyla, and L. G. Ljungdahl. 1990. Primary structure of the thermostable formyltetrahydrofolate synthetase from *Clostridium thermoaceticum. Biochemistry* **29**:5687–5694.

Lowe, A., M. K. Jain, and J. G. Zeikus. 1993. Biology, ecology, and biotechnological applications of anaerobic bacteria adapted to environmental stresses in temperature, pH, salinity, or substrates. *Microbiol. Rev.* **57**:451–509.

Lundie, Jr., L. L. and H. L. Drake. 1984. Development of a minimally defined medium for the acetogen *Clostridium thermoaceticum. J. Bacteriol.* **159**:700–703.

Lux, M. F., E. Keith, T. Hsu, and H. L. Drake. 1990. Biotransformation of aromatic aldehydes by acetogenic bacteria. *FEMS Microbiol. Lett.* **67**:73–78.

Lux, M. F., and H. L. Drake. 1992. Re-examination of the metabolic potentials of the acetogens *Clostridium aceticum* and *Clostridium formicoaceticum:* chemolithoautotrophic and aromatic-dependent growth. *FEMS Microbiol. Lett.* **95**:49–56.

Lynd, L. H., and J. G. Zeikus. 1983. Metabolism of H_2-CO_2, methanol, and glucose by *Butyribacterium methylotrophicum. J. Bacteriol.* **153**:1415–1423.

Martin, D. R., L. L. Lundie, R. Kellum, and H. L. Drake. 1983. Carbon monoxide-dependent evolution of hydrogen by the homoacetate-fermenting bacterium *Clostridium thermoaceticum. Curr. Microbiol.* **8**:337–340.

Martin, D. R., A. Misra, and H. L. Drake. 1985. Dissimilation of carbon monoxide to acetic acid by glucose-limited cultures of *Clostridium thermoaceticum. Appl. Environ. Microbiol.* **49**:1412–1417.

Matthies, C., A. Freiberger, and H. L. Drake. 1993. Fumarate dissimilation and differential reductant flow by *Clostridium formicoaceticum* and *Clostridium aceticum*. *Arch. Microbiol.* **160**:273–278.

Mayer, F., J. I. Elliott, D. Sherod, and L. G. Ljungdahl. 1982. Formyltetrahydrofolate synthetase from *Clostridium thermoaceticum*. *Eur. J. Biochem.* **124**:397–404.

Meyer, O. 1988. Biology and biotechnology of aerobic carbon monoxide-oxidising bacteria. *In: Biotechnology Focus 1*, M. Schlingmann, W. Crueger, K. Esser, R. Thauer, and F. Wagner (eds.), pp. 3–31. Hanser Publishers, Munich, Vienna.

Meyer, O., K. Frunzke, and G. Mörsdorf. 1993. Biochemistry of the aerobic utilization of carbon monoxide. *In: Microbial Growth on C_1 Compounds*, J. C. Murrell, and D. P. Kelly (eds.), pp. 433–459. Intercept Ltd., Andover, England.

Möller, B., R. Oßmer, B. H. Howard, G. Gottschalk, and H. Hippe. 1984. *Sporomusa*, a new genus of gram-negative anaerobic bacteria including *Sporomusa sphaeroides* spec. nov. and *Sporomusa ovata* spec. nov. *Arch. Microbiol.* **139**:388–396.

Moench, T. T., and J. G. Zeikus. 1983. An improved preparation method for a titanium (III) media reductant. *J. Microbiol. Methods* **1**:199–202.

Moore, W., and E. Cato. 1965. Synonymy of *Eubacterium limosum* and *Butyribacterium rettgeri*. *Int. Bull. Bacteriol. Nomen. Taxon.* **15**:69–80.

Morton, T. A., C-F. Chou, and L. G. Ljungdahl. 1992. Cloning, sequencing, and expressions of genes encoding enzymes of the autotrophic acetyl-CoA pathway in the acetogen *Clostridium thermoaceticum*. *In: Genetics and Molecular Biology of Anaerobic Bacteria*, M. Sebald (ed.), pp. 389–406. Springer-Verlag, New York.

Mountfort, D. O. 1992. Ecophysiological significance of anaerobes in the gastrointestinal tracts of marine fish. Abstr. C1-4-4, p. 91. *Sixth Internat. Symp. on Microbial Ecology (ISME-6)* 1992.

Nagaranthal, K. R., and D. P. Nagle, Jr. 1992. Inhibition of methanogenesis in *Methanobacterium thermoautotrophicum* by lumazine, Abstr. I-23, p. 240. *Ann. Meet. Am. Soc. Microbiol.* 1992.

O'Brien, W. E., J. M. Brewer, and L. G. Ljungdahl. 1973. Purification and characterization of thermostable 5,10-methylenetetrahydrofolate dehydrogenase from *Clostridium thermoaceticum*. *J. Biol. Chem.* **248**:403–408.

Ohwaki, K., and R. E. Hungate. 1977. Hydrogen utilization by clostridia in sewage sludge. *Appl. Environ. Microbiol.* **33**:1270–1274.

Ollivier, B. M., R. A. Mah, T. J. Ferguson, D. R. Boone, J. L. Garcia, and R. Robinson. 1985. Emendation of the genus *Thermobacteroides: Thermobacteriodes proteolyticus* sp. nov., a proteolytic acetogen from a methanogenic enrichment. *Int. J. Syst. Bacteriol.* **35**:425–428.

Ollivier, B., R. Cordruwisch, A. Lombardo, and J.-L. Garcia. 1985. Isolation and characterization of *Sporomusa acidovorans* sp. nov., a methylotrophic homoacetogenic bacterium. *Arch. Microbiol.* **142**:307–310.

Parekh, M., E. S. Keith, S. L. Daniel, and H. L. Drake. 1992. Comparative evaluation of the metabolic potentials of different strains of *Peptostreptococcus productus*: utilization and transformation of aromatic compounds. *FEMS Microbiol. Lett.* **94**:69–74.

Parekh, S. R., and M. Cheryan. 1991. Production of acetate by mutant strains of *Clostridium thermoaceticum*. *Appl. Microbiol. Biotechnol.* **36**:384–387.

Park, E. Y., J. E. Clark, D. V. DerVartanian, and L. G. Ljungdahl. 1991. 5,10-Methylenetetrahydrofolate reductases: iron-sulfur-zinc flavoproteins of two acetogenic clostridia. *In: Chemistry and Biochemistry of Flavoenzymes*, F. Müller (ed.), Vol. 1, pp. 389–400. CRC Press, Boca Raton, FL.

Patel, B. K. C., C. Monk, H. Littleworth, H. W. Morgan, and R. M. Daniel. 1987. *Clostridium fervidus* sp. nov., a new chemoorganotrophic acetogenic thermophile. *Int. J. Syst. Bacteriol.* **37**:123–126.

Pezacka, E., and H. G. Wood. 1984a. Role of carbon monoxide dehydrogenase in the autotrophic pathway used by acetogenic bacteria. *Proc. Natl. Acad. Sci. USA* **81**: 6261–6265.

Pezacka, E., and H. G. Wood. 1984b. The synthesis of acetyl-CoA by *Clostridium thermoaceticum* from carbon dioxide, hydrogen, coenzyme A and methyltetrahydrofolate. *Arch. Microbiol.* **137**:63–69.

Pezacka, E., and H. G. Wood. 1986. The autotrophic pathway of acetogenic bacteria. Role of CO dehydrogenase disulfide reductase. *J. Biol. Chem.* **261**:1609–1615.

Poston, J. M., K. Kuratomi, and E. R. Stadtman. 1964. Methyl-vitamin B_{12} as a source of methyl groups for the synthesis of acetate by cell-free extracts of *Clostridium thermoaceticum*. *Ann. N.Y. Acad. Sci.* **112**:804–806.

Poston, J. M., K. Kuratomi, and E. R. Stadtman. 1966. The conversion of carbon dioxide to acetate: I. The use of cobalt-methylcobalamin as a source of methyl groups for the synthesis of acetate by cell-free extracts of *Clostridium thermoaceticum*. *J. Biol. Chem.* **241**:4209–4216.

Ragsdale, S. W., J. E. Clark, L. G. Ljungdahl, L. L. Lundie, and H. L. Drake. 1983. Properties of purified carbon monoxide dehydrogenase from *Clostridium thermoaceticum*, a nickel, iron-sulfide protein. *J. Biol. Chem.* **258**:2364–2369.

Ragsdale, S. W., H. G. Wood, and W. E. Antholine. 1985. Evidence that an iron-nickel-carbon complex is formed by reaction of CO with the CO dehydrogenase from *Clostridium thermoaceticum*. *Proc. Natl. Acad. Sci. USA* **82**:6811–6814.

Ragsdale, S. W. 1991. Enzymology of the acetyl-CoA pathway of CO_2 fixation. *Crit. Rev. Biochem. Mol. Biol.* **26**:261–300.

Reeve, J. N. 1992. Molecular biology of methanogens. *Annu. Rev. Microbiol.* **46**:165–191.

Roberts, D. L., J. E. James-Hagstrom, D. K. Garvin, C. M. Gorst, J. A. Runquist, J. R. Baur, F. C. Haase, and S. W. Ragsdale. 1989. Cloning and expression of the gene cluster encoding key proteins involved in acetyl-CoA synthesis in *Clostridium thermoaceticum:* CO dehydrogenase, the corrinoid/Fe-S protein, and methyltransferase. *Proc. Natl. Acad. Sci. USA* **86**:32–36.

Sakami, W. 1962. Anaerobic gradient elution chromatography. *Anal. Biochem.* **3**:358–360.

Samain, E., G. Albangnac, H. C. Dubourguier, and J-P. Touzel. 1982. Characterization of a new propionic acid bacterium that ferments ethanol and displays a growth factor-dependent association with a gram-negative homoacetogen. *FEMS Microbiol. Lett.* **15**:69–74.

Savage, M. D., and H. L. Drake. 1986. Adaptation of the acetogen *Clostridium thermoautotrophicum* to minimal medium. *J. Bacteriol.* **165**:315–318.

Savage, M. D., Z. Wu, S. L. Daniel, L. L. Lundie, Jr., and H. L. Drake. 1987. Carbon monoxide-dependent chemolithotrophic growth of *Clostridium thermoautotrophicum*. *Appl. Environ. Microbiol.* **53**:1902–1906.

Schaupp, A. and L. G. Ljungdahl. 1974. Purification and properties of acetate kinase from *Clostridium thermoaceticum*. *Arch. Microbiol.* **100**:121–129.

Schauder, R., B. Eikmanns, R. K. Thauer, F. Widdel, and G. Fuchs. 1986. Acetate oxidation to CO_2 in anaerobic bacteria via a novel pathway not involving reactions of the critic acid cycle. *Arch. Microbiol.* **145**:162–172.

Schink, B. 1984. *Clostridium magnum* sp. nov., a non-autotrophic homoacetogenic bacterium. *Arch. Microbiol.* **137**:250–255.

Schink, B., and M. Bomar. 1992. The genera *Acetobacterium, Acetogenium, Acetoanaerobium,* and *Acetitomaculum. In: The Prokaryotes,* 2nd ed., A. Balows, H. G. Trüper, M. Dworkin, W. Harder, K.-H. Schleifer (eds.), pp. 1925–1936. Springer-Verlag, New York.

Schopf, J. W., J. M. Hayes, and M. R. Walter. 1983. Evolution of the earth's earliest ecosystems: recent progress and unsolved problems. *In: Earth's Earliest Biosphere,* J. W. Schopf (ed.), pp. 361–384. Princeton University Press, Princeton, NJ.

Schramm, E., and B. Schink. 1991. Ether-cleaving enzyme and diol dehydratase involved in anaerobic polyethylene glycol degradation by a new *Acetobacterium* sp. *Biodegradation* **2**:71–79.

Schulman, M., R. K. Ghambeer, L. G. Ljungdahl, and H. G. Wood. 1973. Total synthesis of acetate from CO_2. VII. Evidence with *Clostridium thermoaceticum* that the carboxyl of acetate is derived from the carboxyl of pyruvate by transcarboxylation and not by fixation of CO_2. *J. Biol. Chem.* **248**:6255–6261.

Schuppert, B., and B. Schink. 1990. Fermentation of methoxyacetate to glycolate and acetate by newly isolated strains of *Acetobacterium* sp. *Arch. Microbiol.* **153**:200–204.

Schwartz, R. D., and F. A. Keller, Jr. 1982. Isolation of a strain of *Clostridium thermoaceticum* capable of growth and acetic acid production at pH 4.5. *Appl. Environ. Microbiol.* **43**:117–123.

Seifritz, C., S. L. Daniel, A. Gößner, and H. L. Drake. 1993. Nitrate as a preferred electron sink for the acetogen *Clostridium thermoaceticum*. *J. Bacteriol.* **175**:8008–8013.

Seiler, W. 1984. Contribution of biological processes to the global budget of CH_4 in the atmosphere. *In: Current Perspectives in Microbiol Ecology,* M. J. Klug, and C. A. Reddy (eds), pp. 468–477. American Society of Microbiology, Washington, D.C.

Sembiring, T., and J. Winter. 1989. Anaerobic degradation of *o*-phenylphenol by mixed and pure cultures. *Appl. Microbiol. Biotech.* **31**:89–92.

Sembiring, T., and J. Winter. 1990. Demethylation of aromatic compounds by strain B10 and complete degradation of 3-methoxybenzoate in co-culture with *Desulfosarcina* strains. *Appl. Microbiol. Biotechnol.* **33**:233–238.

Sharak-Genthner, B.R., C. L. Davies, and M. P. Bryant. 1981. Features of rumen and sewage sludge strains of *Eubacterium limosum*, a methanol- and H_2-CO_2-utilizing species. *Appl. Environ. Microbiol.* **42**:12–19.

Shin, W., and P. A. Lindahl. 1992a. Function and CO binding properties of the NiFe complex in carbon monoxide dehydrogenase from *Clostridium thermoaceticum*. *Biochemistry* **31**:12870–12875.

Shin, W., and P. A. Lindahl. 1992b. Discovery of a labile nickel ion required for Co/acetyl-CoA exchange activity in the NiFe complex of carbon monoxide dehydrogenase from *Clostridium thermoaceticum*. *J. Am. Chem. Soc.* **114**:9718–9719.

Shin, W., and P. A. Lindahl. 1993. Low spin quantitation of NiFeC EPR signal from carbon monoxide dehydrogenase is not due to damage incurred during protein purification. *Biochim. Biophys. Acta* **1161**:317–322.

Shin, W., P. R. Stafford, and P. A. Lindahl. 1992. Redox titrations of carbon monoxide dehydrogenase from *Clostridium thermoaceticum*. *Biochemistry* **31**:6003–6011.

Shin, W., M. E. Anderson, and P. A. Lindahl. 1993. Heterogenous nickel environments in carbon monoxide dehydrogenase from *Clostridium thermoaceticum*. *J. Am. Chem. Soc.* **115**:5522–5526.

Sleat, R., R. A. Mah, and R. Robinson. 1985. *Acetoanaerobium noterae* gen. nov., sp. nov.: an anaerobic bacterium that forms acetate from H_2 and CO_2. *Int. J. Syst. Bacteriol.* **35**:10–15.

Smith, M. R., and R. A. Mah. 1981. 2-Bromoethanesulfonate: a selective agent for isolating resistant *Methanosarcina* mutants. *Curr. Microbiol.* **6**:321–326.

Stupperich, E., and R. Konle. 1993. Corrinoid-dependent methyl transfer reactions are involved in methanol and 3,4-dimethoxybenzoate metabolism by *Sporomusa ovata*. *Appl. Environ. Microbiol.* **59**:3110–3116.

Sugaya, K., D. Tusé, and J. L. Jones. 1986. Production of acetic acid by *Clostridium thermoaceticum* in batch and continuous fermentations. *Biotechnol. Bioengineer.* **28**:678–683.

Tanaka, K., and N. Pfennig. 1988. Fermentation of 2-methoxyethanol by *Acetobacterium malicum* sp. nov. and *Pelobacter venetianus*. *Arch. Microbiol.* **149**:181–187.

Tanner, R. S., E. Stakebrandt, G. E. Fox, and C. R. Woese. 1981. A phylogenetic analysis of *Acetobacterium woodii*, *Clostridium barkeri*, *Clostridium butyricum*, *Clostridium lituseburense*, *Eubacterium limosum*, and *Eubacterium tenue*. *Curr. Microbiol.* **5**:35–38.

Tanner, R. S., and D. Yang. 1990. *Clostridium ljungdahlii* PETC sp. nov., a new, acetogenic, gram-positive, anaerobic bacterium. Abstr. R-21, p. 249. *Abstr. Ann. Meet. Am. Soc. Microbiol.*, 1990.

Tanner, R. S., L. M. Miller, and D. Yang. 1993. *Clostridium ljungdahlii* sp. nov., and acetogenic species in clostridial rRNA homology group I. *Int. J. Syst. Bacteriol.* **43**:232–236.

Thauer, R. K. 1988. Citric acid cycle, 50 years on: modification and an alternative pathway in anaerobic bacteria. *Eur. J. Biochem.* **176**:497–508.

Thauer, R. K., G. Fuchs, B. Käufer, and U. Schnitker. 1974. Carbon-monoxide oxidation in cell-free extracts of *Clostridium pasteurianum*. *Eur. J. Biochem.* **45**:343–349.

Thauer, R. K., K. Jungermann, and K. Decker. 1977. Energy conservation in chemotrophic anaerobic bacteria. *Bacteriol. Rev.* **41**:100–180.

Thauer, R. K., D. Möller-Zinkhan, and A. M. Spormann. 1989. Biochemistry of acetate catabolism in anaerobic chemotrophic bacteria. *Annu. Rev. Microbiol.* **43**:43–67.

Traunecker, J., A. Preuß, and G. Diekert. 1991. Isolation and characterization of a methyl cloride utilizing, strictly anaerobic bacterium. *Arch. Microbiol.* **156**:416–421.

Tschech, A., and N. Pfennig. 1984. Growth yield increase linked to caffeate reduction in *Acetobacterium woodii*. *Arch. Microbiol.* **137**:163–167.

Varma, A. K., and H. D. Peck, Jr. 1983. Utilization of short and long-chain polyphosphates as energy sources for the anaerobic growth of bacteria. *FEMS Microbiol. Lett.* **16**:281–285.

Wagener, S., and B. Schink. 1988. Fermentative degradation of nonionic surfactants and polyethylene glycol by enrichment cultures and by pure cultures of homoacetogenic and propionate-forming bacteria. *Appl. Environ. Microbiol.* **54**:561–565.

Wang, G., and D. I. C. Wang. 1983. Production of acetic acid by immobilized whole cells of *Clostridium thermoaceticum*. *Appl. Biochem. Biotechnol.* **8**:491–503.

Wang, G., and D. I. C. Wang. 1984. Elucidation of growth inhibition and acetic acid production by *Clostridium thermoaceticum*. *Appl. Environ. Microbiol.* **47**:294–298.

Whitman, W. B. 1985. Methanogenic bacteria. *In:* C. R. Woese and R. S. Wolfe (eds.) *The Bacteria,* Vol. VIII, pp. 3–84. Academic Press, San Diego, CA.

Whitman, W. B., T. L. Bowen, and D. R. Boone. 1992. The methanogenic bacteria. *In:* A. Balows, H. G. Trüper, M. Dworkin, W. Harder, K.-H. Schleifer (eds.). *The Prokaryotes,* 2nd ed., pp. 719–767, Springer-Verlag, New York.

Wiegel, J., M. Braun, and G. Gottschalk. 1981. *Clostridium thermoautotrophicum* species novum, a thermophile producing acetate from molecular hydrogen and carbon dioxide. *Curr. Microbiol.* **5**:255–260.

Wiegel, J., L. H. Carreira, R. J. Garrison, N. E. Robek, and L. G. Ljungdahl. 1990. Calcium magnesium acetate (CMA) manufacture from glucose by fermentation with thermophilic homoacetogenic bacteria. *In: Calcium Magnesium Acetate,* D. L. Wise, Y. Levendis, and M. Metghalchi (eds.), pp. 359–416. Elsevier, Amsterdam.

Wieringa, K. T. (1936). Over het verdwijnen van waterstof en koolzuur onder anaerobe voorwaarden. *Antonie van Leeuwenhoek* **3**:263–273.

Wieringa, K. T. 1939–1940. The formation of acetic acid from carbon dioxide and hydrogen by anaerobic spore-forming bacteria. *Antonie van Leeuwenhoek J. Microbiol. Seriol* **6**:251–262.

Wieringa, K. T. 1941. Über die Bildung von Essigsäure aus Kohlensäure und Wasserstoff durch anaerobe Bazillen. *Brennstoff-Chemie* **22**:161–164.

Winter, J. U., and R. S. Wolfe. 1980. Methane formation from fructose by syntrophic associations of *Acetobacterium woodii* and different strains of methanogens. *Arch. Microbiol.* **124**:73–79.

Wohlfahrt, G., and G. Diekert. 1991. Thermodynamics of methylenetetrahydrofolate reduction to methyltetrahydrofolate and its implications for the energy metabolism of homoacetogenic bacteria. *Arch. Microbiol.* **155**:378–381.

Wood, H. G., and C. H. Werkman. 1936. Mechanism of glucose dissimilation by the propionic acid bacteria. *Biochem. J.* **30**:618–623.

Wood, G. H., and C. H. Werkman. 1938. The utilization of CO_2 by the propionic acid bacteria. *Biochem. J.* **32**:1262–1271.

Wood, H. G., C. H. Werkman, A. Hemingway, and A. O. Nier. 1941a. Heavy carbon as a tracer in heterotrophic carbon dioxide assimilation. *J. Biol. Chem.* **139**:365–376.

Wood, H. G., C. H. Werkman, A. Hemingway, and A. O. Nier. 1941b. The position of carbon dioxide carbon in succinic acid synthesized by heterotrophic bacteria. *J. Biol. Chem.* **139**:377–381.

Wood, H. G. 1952a. A study of carbon dioxide fixation by mass determination on the types of C^{13}-acetate. *J. Biol. Chem.* **194**:905–931.

Wood, H. G. 1952b. Fermentation of 3,4-C^{14}- and 1-C^{14}-labeled glucose by *Clostridium thermoaceticum*. *J. Biol. Chem.* **199**:579–583.

Wood, H. G. 1972. My life and carbon dioxide fixation. *In: The Molecular Basis of Biological Transport, Miami Winter Symposium Vol. 3*, J. F. Woessner, Jr., and F. Huijing (eds.), pp. 1–54. Academic Press, New York.

Wood, H. G. 1976. Trailing the propionic acid bacteria. *In: Reflections on Biochemistry*, A. Kornberg, B. L. Horecker, L. Cornudella, and J. Oro (eds.), pp. 105–115. Permagon Press, Oxford.

Wood, H. G. 1982. The discovery of the fixation of CO_2 by heterotrophic organisms and metabolism of the propionic bacteria. *In: Of Oxygen, Fuels, and Living Matter, Part 2*, G. Semenza (ed.), pp. 173–250, John Wiley and Sons, New York.

Wood, H. G. 1985. Then and now. *Annu. Rev. Biochem.* **54**:1–41.

Wood, H. G. 1989. Past and present of CO_2 utilization. *In: Autotrophic Bacteria*, H. G. Schlegel and B. Bowien (eds.), pp. 33–52. Science Tech. Madison and Springer-Verlag, Berlin.

Wood, H. G. 1991. Life with CO or CO_2 and H_2 as a source of carbon and energy. *FASEB J.* **5**:156–163.

Wood, H. G., and L. G. Ljungdahl. 1991. Autotrophic character of the acetogenic bacteria. *In: Variations in Autotrophic Life*, J. M. Shively, and L. L. Barton (eds.), pp. 201–250. Academic Press, San Diego, CA.

Worden, R. M., A. J. Grethlein, J. G. Zeikus, and R. Datta. 1989. Butyrate production from carbon monoxide by *Butyribacterium methylotrophicum*. *Appl. Biochem. Biotechnol.* **20/21**:687–698.

Wu, Z., S. L. Daniel, and H. L. Drake. 1988. Characterization of a CO-dependent O-demethylating enzyme system from the acetogen *Clostridium thermoaceticum*. *J. Bacteriol.* **170**:5747–5750.

Yamamoto, I., T. Saiki, S.-M. Liu, and L. G. Ljungdahl. 1983. Purification and properties of NADP-dependent formate dehydrogenase from *Clostridium thermoaceticum*, a tungsten-selenium-iron protein. *J. Biol. Chem.* **258**:1826–1832.

Yang, H., and H. L. Drake. 1990. Differential effects of sodium on hydrogen- and glucose-dependent growth of the acetogenic bacterium *Acetogenium kivui*. *Appl. Environ. Microbiol.* **56**:81–86.

Zehnder, A. J. B., and K. Wuhrmann. 1976. Titanium III citrate as a nontoxic oxidation-reduction buffering system for the culture of obligate anaerobes. *Science* **194**:1165–1166.

Zehnder, A. J. B., B. A. Huser, T. D. Brock, and K. Wuhrmann. 1980. Characterization of an acetate-decarboxylating non-hydrogen oxidizing methane bacterium. *Arch. Microbiol.* **124**:1–11.

Zeikus, J. G., L. H. Lynd, T. E. Thompson, J. A. Krzycki, P. J. Weimer, and P. W. Hegge. 1980. Isolation and characterization of a new methylotrophic, acidogenic anaerobe, the Marburg strain. *Curr. Microbiol.* **3**:381–386.

Zeikus, J. G. 1983. Metabolism of one-carbon compounds by chemotrophic anaerobes. *Adv. Microbial. Physiol.* **24**:215–299.

Zeikus, J. G., R. Kerby, and J. A. Krzycki. 1985. Single-carbon chemistry of acetogenic and methanogenic bacteria. *Science* **227**:1167–1173.

Zhilina, T. N., and G. A. Zavarzin. 1990. Extremely halophilic, methylotrophic, anaerobic bacteria. *FEMS Microbiol. Rev.* **87**:315–322.

II

BIOCHEMISTRY AND ENERGETICS OF ACETOGENESIS AND THE ACETYL-CoA PATHWAY

2

The Acetyl-CoA Pathway and the Chemiosmotic Generation of ATP during Acetogenesis

Lars G. Ljungdahl

2.1 Introduction and Historical Notes

Fischer et al. in 1932 reported that a culture obtained from sewage sludge growing either on sewage sludge or an inorganic nutrient solution converted carbon monoxide to carbon dioxide and hydrogen. They further noted that CO_2 in the presence of molecular hydrogen was reduced to form acetate, which then was metabolized, yielding CO_2 and methane. The report appears to be the first describing the formation of acetic acid by the reduction of CO_2. It demonstrated also the importance of acetate as a precursor of methane. Wieringa in 1936 isolated a spore-forming anaerobic bacterium that formed acetate from CO_2 and H_2. He later described this bacterium and named it *Clostridium aceticum* (Wieringa, 1940). To grow, it required a low redox potential. Karlsson et al. (1948) found that *C. aceticum* could grow on a synthetic medium of glucose, glutamic acid, biotin, pyridoxamine, and pantothenic acid. No further studies were done with *C. aceticum* until it was reisolated in 1980 (Adamse, 1980) and revived from an old culture (Braun et al., 1981).

This volume is dedicated in part to Harland G. Wood. He started graduate work in 1931 at Iowa State College and was assigned to study propionic acid bacteria. At that time, the dogma was that heterotrophic organisms were unable to fix CO_2, which clearly separated them from autotrophs. Therefore, when Wood and Werkman (1935) reported that propionic acid bacteria-fermenting glycerol fix CO_2, this discovery was greeted with much skepticism. This disbelief lingered on until the heavy carbon isotope ^{13}C became available. Wood became a pioneer using isotopes in studies of biochemical reactions and with $^{13}CO_2$ Wood et al. (1941) conclusively demonstrated that CO_2 is fixed by all forms of life.

The discovery of heterotrophic CO_2 fixation was undoubtedly Wood's most important contribution; it affects every living cell as we now know. Wood devoted much of his scientific career to the metabolism of CO_2 and his discoveries include several CO_2-fixation reactions, the role and mechanism of biotin, transcarboxylation, and, as evidenced in this book, the autotrophic acetyl-CoA pathway of acetogenic bacteria.

Clostridium thermoaceticum was discovered in 1942 by Fontaine et al. It was described as a heterotrophic bacterium-fermenting glucose, fructose, and xylose with acetic acid as the only product (reactions I and II).

$$C_6H_{12}O_6 \rightarrow 3\ CH_3COOH \qquad\qquad (I)$$
$$2\ C_5H_{10}O_5 \rightarrow 5\ CH_3COOH \qquad\qquad (II)$$

The formation of 3 mol of a two-carbon compound (acetic acid) from 1-mol of glucose or fructose was unusual and suggested a possible new type of cleavage of the hexose other than classical glycolysis involving a 3-3 split. However, considering the work by Wood, Fontaine et al. (1942) suggested a pathway involving the formation of 2 mol of acetic acid and 2 mol of a one-carbon compound (CO_2) that was reabsorbed and converted to acetic acid (reactions III and IV)

$$C_6H_{12}O_6 + 2H_2O \rightarrow 2CH_3COOH + 2CO_2 + 8H^+ + 8e^- \qquad (III)$$
$$2CO_2 + 8H^+ + 8e^- \rightarrow CH_3COOH + 2H_2O \qquad\qquad (IV)$$
$$\text{Sum} = C_6H_{12}O_6 \rightarrow 3CH_3COOH \qquad\qquad\qquad (I)$$

This latter suggestion was strongly supported when the fermentation of glucose was performed in the presence of $^{14}CO_2$ (Barker and Kamen, 1945). It was observed that about one-third of the acetate was formed from $^{14}CO_2$ and chemical degradation of the [^{14}C]-acetate showed that both carbons were labeled. Barker and Kamen (1945) noted also that a considerable part of the cell material was synthesized from CO_2. However, it was pointed out by Wood (1952a) that a chemical degradation of [^{14}C] acetate does not distinguish between acetate labeled in both carbon and a mixture of methyl- and carboxyl-labeled acetate. Therefore, Wood (1952a) evaluated the fermentation of glucose in the presence of a high concentration of $^{13}CO_2$. The acetate produced was converted to ethylene and the quantities of $H_2^{13}C{=}^{13}CH_2$ (mass = 30), $H_2^{12}C{=}^{13}CH_2$ (mass = 29), and $H_2^{12}C{=}^{12}CH_2$ (mass = 28) were determined by mass analysis. The experiment established that one-third of the acetate was totally synthesized from CO_2 as predicted in reactions I, III, and IV. In addition, some acetate was found labeled exclusively in the carboxyl group. This observation made also by Barker and Kamen (1945). Wood (1952a) demonstrated that this labeling occurs by an exchange reaction between the carboxyl group of acetate and $^{13}CO_2$. This exchange can now be explained and is catalyzed by carbon monoxide dehydrogenase/acetyl-CoA synthase (see Chapter 3). The experiment using $^{13}CO_2$ and mass analysis of produced acetate was later extended to other "acetogenic" bacteria and it was established that *Clostridium formicoaceticum, Clostridium acidiurici*, and *Clostridium cylindrosporum* totally synthesize acetate from CO_2 (Schulman et al., 1972).

That the fermentation of glucose proceeds by the Embden Meyerhof glycolytic pathway was established for *C. thermoaceticum* by Wood (1952b) and *C. formicoaceticum* by O'Brien and Ljungdahl (1972) using differently labeled [^{14}C]-glucose and [^{14}C]-fructose samples. Wood's work (1952a, 1952b) firmly established reactions III and IV and strongly suggested the concept that CO_2 served as an electron acceptor of electrons generated during glycolysis. However, it did not establish any mechanisms for the total synthesis of acetate from CO_2. Nevertheless, Wood and his colleagues (Utter and Wood, 1951; Wood and Stjernholm, 1962) suggested that this synthesis could constitute a unique autotrophic pathway. Clearly, Wood had this vision until he died, and on occasions he expressed the hope that he could stay in the field long enough to see the synthesis of acetate from CO_2 clarified. He did, and throughout, he was an inspiration to all who worked to elucidate the autotrophic acetyl-CoA (Wood) pathway.

2.2 Figuring out the Wood Acetyl-CoA Pathway

The only acetogenic bacterium known and available from 1952 until 1967 was *C. thermoaceticum* and, consequently, all of the early work to resolve the pathway of synthesis of acetate from CO_2 was carried out with it. *C. formicoaceticum* was discovered in 1967 (El Ghazzawi, 1967; Andreesen et al., 1970), and *Acetobacterium woodii* in 1977 (Balch et al., 1977). The isolation of *A. woodi* and the demonstration of its ability to grow autotrophically on CO_2 and H_2 put, for the first time, the concept of autotrophic acetogens on a firm footing. *C. aceticum* was reisolated in 1980 (Adamse, 1980; Braun et al., 1981), and *Clostridium thermoautotrophicum* was described in 1981 (Wiegel et al., 1981); both are capable of autotrophic growth. *C. thermoaceticum* was still considered an obligate heterotroph growing only on fructose, glucose, xylose, and pyruvate. In 1982, Drake showed the presence of hydrogenase in *C. thermoaceticum*. This discovery was soon followed by the demonstration that it can grow on a mixture of CO_2 and H_2 or on CO (Kerby and Zeikus, 1983). These experiments were conducted with a complex medium and did not establish the ability of *C. thermoaceticum* to grow autotrophically (chemolithoautotrophically). That capacity was established after the development of a minimally defined medium (Lundie and Drake, 1984) and a further evaluation of the chemolithoautotrophic potential of *C. thermoaceticum* (Daniel et al., 1990). Now, *C. thermoaceticum* has been shown to metabolize methanol and other alcohols, formate, oxalate, phenylmethylethers, and similar compounds (see Chapters 1, 7, 10, 17, and 18). It also uses a wide range of electron acceptors including sulfite, dimethylsulfoxide, and nitrate. It is actually somewhat astonishing that neither Harland G. Wood nor I, having worked with *C. thermoaceticum* for over 30 years, were part of the discovery of its metabolic capacity. We considered this from time to time. Perhaps an explanation is that we were too entranced and busy with the establishment of the pathway and the isolation and studies of the enzymes involved.

The first approach to find the pathway of acetate synthesis from CO_2 was with $^{14}CO_2$ in short-term experiments using resting cell suspensions. Lentz and Wood (1955) noted a very rapid isotope interchange between formate and CO_2 and that ^{14}C-formate was incorporated more rapidly than $^{14}CO_2$ into the methyl group of acetate. Labeled $^{14}CO_2$ was, on the other hand, a better precursor of the carboxyl group than was formate. This was the first indication that formate was an intermediate in the conversion of CO_2 to the methyl group of acetate. Subsequently, an NADP-dependent formate dehydrogenase was discovered (Li et al., 1966) which proved to be a tungsten-selenium-iron protein (Andreesen and Ljungdahl, 1973; Yamamoto et al. 1983). It was the first demonstration of tungsten as a biologically important element.

Sugar phosphates, carboxylic acids, and amino acids were isolated from cells of *C. thermoaceticum* exposed for 5 s to $^{14}CO_2$ (Ljungdahl and Wood, 1965). The distributions of ^{14}C in some of these compounds were determined. These experiments ruled out the Calvin photosynthetic cycle and that the acetate synthesis did not involve serine, aspartic acid, and dicarboxylic acids. Only formate and the carboxyl group of lactate (reflecting the carboxyl group of pyruvate) were found labeled higher than the acetate. However, the experiments clearly indicated that acetyl-CoA formed from $^{14}CO_2$ was a precursor for cell material. This has been elaborated on relative to autotrophic growth (Ljungdahl, 1986).

A very significant discovery was reported by Poston et al. (1964, 1966). They showed that the methyl group of synthetic Co-methylcobalamin was incorporated into the methyl group of acetate by cell-free extracts of *C. thermoaceticum*. This incorporation required pyruvate. The bond between the methyl group and cobalt of Co-methylcobalamin is very sensitive to light. Consequently, a Co-methylcobalamin type of compound naturally occurring in *C. thermoaceticum* would not have been observed in the puls experiment conducted by Ljungdahl and Wood (1965). Therefore, these experiments were repeated, performed in the darkroom, and the corrinoids were isolated also in the darkroom (Irion and Ljungdahl, 1965). It was found that *C. thermoaceticum* contains a large number of corrinoids (over 20 different ones were isolated and identified) and synthesizes them. Most importantly, Co-(methyl)-5-methoxybenzimidazolylcobamide and Co-methylcobyric acid were identified with their Co-methyl groups heavily labeled with ^{14}C

from $^{14}CO_2$. The methyl group of these corrinoids were in the presence of pyruvate incorporated into the methyl group of acetate by cell-free extracts of *C. thermoaceticum* (Ljungdahl et al., 1965). These experiments established that the methyl group of the Co-methylcorrinoids were formed from CO_2 and also served as a source of the methyl group of acetate, clearly indicating the methylcorrinoids as intermediates in the synthesis of acetate from CO_2 (see also Chapter 6).

The synthesis of methionine, which involves the transmethylation of the methyl group of 5-methyltetrahydrofolate (5-methyl-THF) to homocysteine, was known to be catalyzed by two different reactions, one of which is B_{12} dependent [for review, see Taylor (1982)]. The role of B_{12} in the latter is as a transmethylation agent. Formate was known as a precursor of the methyl group of 5-methyl-THF and it seemed logical to test this compound as a precursor for the methyl group of acetate. It was found that extracts of *C. thermoaceticum* convert the methyl group of N-5-$^{14}CH_3$-THF to acetate in a reaction requiring pyruvate (Ljungdahl et al., 1966). Based on these observations, a pathway of acetate synthesis from CO_2 and pyruvate was considered (Fig. 2.1). It, in large measure, is similar to the pathway now considered to be correct (Wood and Ljungdahl, 1991; Ragsdale, 1991; Drake, 1992). Clearly, the reduction of CO_2 to CO and the involvement of carbon monoxide dehydrogenase/acetyl-CoA synthase, which catalyzes key reactions in acetate synthesis, were not considered. Carbon monoxide dehydrogenase was discovered in 1978 (Diekert and Thauer, 1978). That it also catalyzes acetyl-CoA synthesis was found later in Wood's laboratory (Hu et al., 1982; Ragsdale and Wood, 1985).

In the scheme of Figure 2.1, a Co-carboxymethylcorrinoid was postulated as an intermediate. It was based on the isolation from *C. thermoaceticum* of a compound with properties resembling Co-carboxymethyl-B_{12} (Ljungdahl and Irion, 1966), but subsequent experiments did not confirm this isolation. It should be noted that the synthesis of the acetyl moiety was considered involving a metal, the cobalt of the corrinoid, a not too farfetched idea from the present concept involving the Ni-iron cluster of carbon monoxide dehydrogenase/acetyl-CoA synthase (see Chapter 3).

Pyruvate plays a central role in the scheme of Figure 2.1. The carboxyl group of pyruvate is rapidly labeled with ^{14}C in puls-labeling experiments with $^{14}CO_2$ (Ljungdahl and Wood, 1963, 1965), and it is possible that the carboxyl group of pyruvate is a better precursor of the carboxyl group of acetyl-CoA than CO_2. This, in fact, has been demonstrated (Schulman et al., 1973; Pezacka and Wood, 1984). We know now that CO_2 and CO directly react with carbon monoxide dehydrogenase/acetyl-CoA synthase and are precursors to the carboxy group of acetyl-CoA. We must also consider that the carboxyl group of pyruvate is a precursor to the carboxy group. However, the mechanism of transfer of the carboxyl group of pyruvate to carbon monoxide dehydrogenase/acetyl-CoA syn-

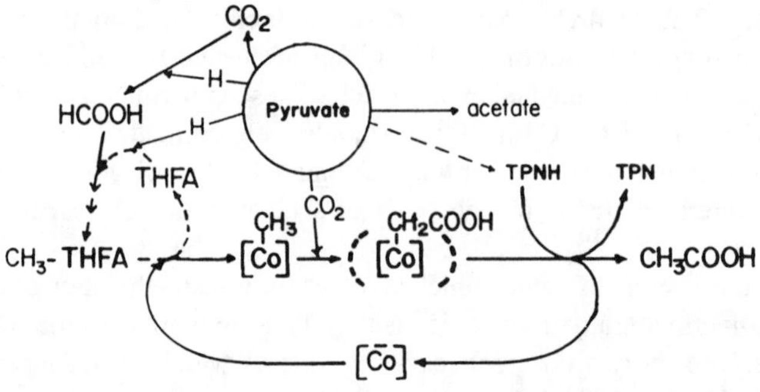

Figure 2.1.   A concept of the total synthesis of acetate from CO_2 in *C. thermoaceticum* as envisioned in 1966. CH_3-THFA is N-5-methyltetrahydrofolate, [Co] signifies a corrinoid, and TPNH is NADPH. Pyruvate was required for the conversion of CH_3-THFA and Co-methylcorrinoids to the methyl group of acetate. TPNH was required for the cleavage of the Co-carboxymethyl corrinoid to acetate. In the overall conversion, pyruvate was postulated to be oxidized to acetate and CO_2, the latter serving as the hydrogen acceptor (Ljungdahl et al., 1966).

thase, and its incorporation into acetyl-CoA is still not understood. Otherwise, the acetyl-CoA pathway is well established. The remainder of this chapter will deal with the coupling of the acetyl-CoA pathway to the generation of energy by chemiosmosis.

2.3   The Acetyl-CoA Pathway: Evolving of Autotrophy

When *C. thermoaceticum* and other homoacetogenic bacteria ferment sugars, energy is generated by substrate-level phosphorylation. Four mol of ATP are generated per mol of glucose converted to acetate and CO_2 (reaction III). When the acetogens grow autotrophically on CO_2/H_2, it should be noted that 1 mol of ATP is consumed in the formyltetrahydrofolate synthetase reaction and that 1 mol is generated when acetate is formed from acetyl-CoA by phosphotransacetylase and acetate kinase. A net gain of ATP is not obtained; in fact, a net loss must be considered because some of the acetyl-CoA is used for the synthesis of cell material. Thus, of necessity, acetogens growing autotrophically must generate energy by a mechanism of electron transport phosphorylation or chemiosmosis. How anaerobic microorganisms, like acetogens and methanogens, growing autotrophically generate energy when reducing CO_2 is not yet understood. It is evidenced in this volume that it occurs by several mechanisms.

C. thermoaceticum and *C. thermoautotrophicum* have H^+-ATPase (Ivey and Ljungdahl, 1986; Mayer et al., 1986), contain cytochromes (Hugenholtz et al. 1987), and grow independently of sodium (Yang and Drake, 1990). *A. woodii* (Heise et al., 1989), *Acetogenium kivui* (Yang and Drake, 1990), and *Peptostreptococcus productus* (Geerlings et al., 1989) are examples of acetogens using a Na^+-ATPase and they apparently lack cytochromes. The elucidation of the different mechanisms for the generation of energy and electron transport in anaerobic autotrophic microorganisms is a major task within the field of anaerobic microbiology. In Chapter 4, Müller and Gottschalk discuss sodium-dependent energy generation. This chapter reviews the proton-dependent process used by *C. thermoaceticum* and *C. thermoautotrophicum*. It should be noted that these two clostrida are genetically and physiologically very similar (Bateson et al., 1989).

With the discoveries of hydrogenase in *C. thermoaceticum* (Drake, 1982) and that it grows at the expense of H_2 as an energy/electron source (Kerby and Zeikus, 1983), it became obvious that the reduction of CO_2 to acetate was coupled to energy conservation. However, this was considered earlier as results of growth yield experiments (Andreesen et al., 1973). According to reactions III and IV, fermentation of sugars generates CO_2 and electrons. The CO_2 is then used as an electron acceptor. The transfer of electrons from the glycolytic pathway to the acetate synthesis pathway could involve electron transport phosphorylation. It should be noted that when 1 mol of glucose is fermented, 2 mol of NADH is generated in the glyceraldehyde-phosphate dehydrogenase reaction, and that 2 mol of reduced ferredoxin is formed in the pyruvate-ferredoxin oxidoreductase reaction (Thauer, 1972) (Fig. 2.2). The reduction of CO_2 to acetate requires 2 mol of NADPH [one each for the formate dehydrogenase (Yamamoto et al., 1983) and the methenyl-THF dehydrogenase reactions (Ljungdahl et al., 1980)] and 2 mol of reduced ferredoxin (or equivalent electron carriers) for the reductions of CO_2 to CO (Ragsdale et al., 1983) and methylene-THF to methyl-THF (Park et al., 1990). Ferredoxin appears to be a general donor of electrons to other carriers including $NADP^+$. Thus, two of the reductions can be explained. However, the transfer of electrons from NADH to any of the electrons carriers involved in CO_2 reduction and subsequent reductions of the intermediated of the acetyl-CoA pathway has not yet been fully explained. A search for a pyridine nucleotide transhydrogenase in *C. thermoaceticum* has not been successful. Instead, electrons from NADH are transferred to electron carriers including cytochromes and a flavoprotein located in the cytoplasmic membrane. An NADH dehydrogenase catalyzes this transfer (Ivey, 1987). Conceivably, the membrane-localized electrons are then used for the reductions of CO_2 to CO and methylene-THF to methyl-THF and perhaps also $NADP^+$. Consistent with this are the findings that carbon monoxide dehydrogenase/acetyl-CoA synthase and methylene-THF reductase are membrane associated (Hugenholtz et al., 1987). The above seems

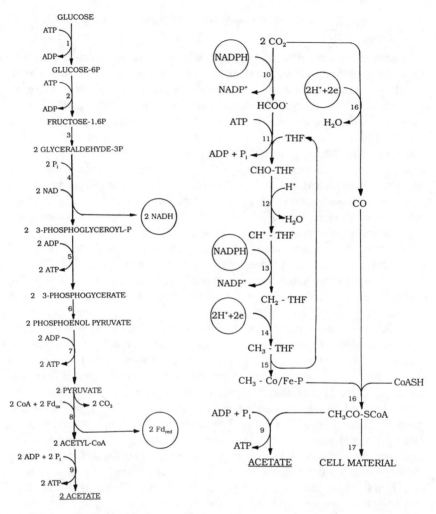

Figure 2.2. The fermentation of glucose to 3 mol of acetate by *C. thermoaceticum*. The left side shows the fermentation of glucose to 2 mol of acetate and 2 mol of CO_2 via the Embden–Meyerhof pathway. Two mol of NADH are formed in reaction 4 (glyceraldehyde-P dehydrogenase) and 2 mol of reduced ferredoxin (Fd) in reaction 8 (pyruvate-ferredoxin oxidoreductase). The right side shows the reduction of 2 mol of CO_2 to acetate via the acetyl-CoA pathway. The reductions require 2 mol of reduced NADPH in reactions 10 and 13 catalyzed by formate dehydrogenase and methylenetetrahydrofolate dehydrogenase, and 2 mol of a reduced electron carrier probably ferredoxin or Cyt b_{559} in reactions 14 and 16 catalyzed by methylenetetrahydrofolate reductase and carbon monoxide dehydrogenase, respectively. The transfer of electrons between electron carriers (marked with circles) of two pathways remains to be explained, but it is likely to involve the cytoplasmic membrane.

to explain, in a very sketchy manner, growth on hexose concomitant with synthesis of acetate from CO_2 and the electron transport required in this process, but it does not explain the generation of ATP by chemosmosis.

It should be noted that when acetogens grow on sugars, there is no requirement for a hydrogenase as net H_2 is neither produced nor taken up. As a historical note, it should be mentioned that Wood and I, on several occasions, looked for hydrogenase in *C. thermoaceticum* and never were able to demonstrate it (Ljungdahl and Wood, 1982). When Drake (1982) showed hydrogenase in this bacterium, it was very surprising to us. This discovery, together with the findings of carbon monoxide dehydrogenase in *C. thermoaceticum* (Diekert and Thauer, 1978) were important for the idea that it could grow chemolitoautotrophically (Daniel et al., 1990).

2.4 Electron Carriers, Membrane Components, and ATPase of *C. thermoaceticum* and *C. thermoautotrophicum*

All enzymes believed needed for the acetyl-CoA pathway have been purified and characterized. Their function and properties are summarized in Table 1.4 and in reviews by Wood and Ljungdahl (1991) and Ragsdale (1991). Like most clostridia, the acetogenic *C. thermoaceticum* and *C. thermoautotrophicum* contain several iron-sulfur proteins such as ferredoxins and rubredoxins as well as flavodoxins. Clearly, this was to be expected considering the redox processes in these bacteria. In addition, cytoplasmic membranes of these acetogens contain *b*-type cytochromes, menaquinone, a flavoprotein, and ATPase. Several enzymes are associated with the membranes. Already mentioned are carbon monoxide dehydrogenase/acetyl-CoA synthase, methylene-THF reductase, and NADH dehydrogenase. Other enzymes and activities demonstrated in membrane preparations are hydrogenase, formaldehyde dehydrogenase, and methanol dehydrogenase (Hugenholtz and Ljungdahl, 1990a), CO-dependent O-demethylation (Daniel et al., 1991), and dimethylsulfoxide reductase (Beaty, personal communication). Recently, it was found that several compounds other than CO_2 serve as ultimate electron acceptors and can sustain growth of *C. thermoaceticum* and *C. thermoautotrophicum*. Acetate may not then be a product. These acceptors include dimethylsulfoxide, thiosulfate (Beaty and Ljungdahl, 1991), and nitrate (Seifritz et al., 1993).

We noted previously that ferredoxin was an electron acceptor in the pyruvate ferredoxin oxidoreductase reaction and that it could be an electron donor in several of the reactions of the acetyl-CoA pathway. The presence of ferredoxin in *C. thermoaceticum* was first reported by Poston and Stadtman (1967), who found it to stimulate the conversion of Co-$^{14}CH_3$-cobalamin to acetate in the presence of pyruvate. It is now known that *C. thermoaceticum* contains two ferredoxins, named I and II (Yang et al., 1977; Elliott and Ljungdahl, 1982). Ferredoxin I being clostridial is unusual in that it contains only one [Fe$_4$S$_4$] cluster. It has been sequenced (Elliott et al., 1982), and its primary structure is more similar to the one-iron cluster ferredoxins of *Desulfovbrio gigas* (Bruschi, 1979) and *Bacillus stearothermophilus* (Hase et al., 1976) than to ferredoxins of many clostridia. Ferredoxin II has two [Fe$_4$S$_4$] clusters and appears to be a more like typical clostridial ferredoxins (Rabinowitz, 1972). It should be noted that ferredoxin II can be obtained simply by washing frozen–thawed cells, whereas ferredoxin I is not. The latter is obtained from the supernatant fraction of cell-free extracts. This indicates that the ferredoxins are in separate compartments of the cell. This and the ability of ferredoxin II to undergo two one-electron transfers per molecule rather than only one as for ferredoxin I suggest that the ferredoxins serve different cellular roles that remain to be clarified. The redox potentials of the ferredoxins are similar: −350 mV for ferredoxin I and −365 mV for ferredoxin II.

Two rubredoxins from *C. thermoaceticum* have been characterized (Yang et al., 1980). They differ slightly. Rubredoxin I has an M_r of 6200 and a redox potential of −27 mV. Corresponding properties for rubredoxin II are 5800 and 20 mV. The function of the rubredoxins are not known. However, an NAD(P)H-rubredoxin oxidoreductase has been isolated from *C. thermoaceticum* (Yang, 1977; Yang and Ljungdahl, 1977). It is a flavoprotein containing FMN, has an M_r of 32,000, and consists of two apparently identical subunits. It catalyzes the oxidation of NADH and NADPH using the rubredoxins, mammalian cytochrome C, and dichlorophenol-indophenol as electron acceptors. The reactions are practically not reversible and the purified enzyme does not catalyze a pyridine nucleotide transhydrogenase reaction. Electron paramagnetic resonance studies of membrane vesicles from *C. thermoautotrophicum* demonstrated the presence of signals in the g-1.91–2.00 region, which indicated the presence of [Fe$_4$-S$_4$] clusters. Other signals at g=4.28 and g=9.3 have been attributed to rubredoxin-type iron centers (Hugenholtz et al., 1989). These results are evidence for the presence of ferredoxin and rubredoxin in the membrane, but they do not prove it as other components of the membrane have iron-sulfur centers, e.g., methylenetetrahydrofolate reductase and carbon monoxide dehydrogenase/acetyl-CoA synthase. The presence of the latter in the membrane was indicated in the EPR studies, which showed strong signals also of g=2.00–2.11 attributed to Ni.

The presence of b-type cytochromes in *C. thermoaceticum* and *C. formicoaceticum* were reported by Gottwald et al. (1975). It has now been shown that *C. thermoaceticum* and *C. thermoautotrophicum* each contain two *b*-type cytochromes (Ivey, 1987; Hugenholtz et al., 1987). This was demonstrated by spectroscopic examination of membrane vesicles at −196°C. The cytochromes have been purified from such vesicles by extraction first with cholate which released a cytochrome, designated b_{559}, and by a second extraction of the residue from the cholate extraction with Zwittergent 3–14 which solubilized the second cytochrome, designated b_{554} (Ivey, 1987). Cytochrome b_{554} has absorption peaks at 554, 525, and 424 nm, a relatively high potential of -48 mV, and M_r of about 55,000. Cytochrome b_{559} has absorption peaks at 559, 530, and 428 nm, a rather low potential of -200 mV, and M_r of about 70,000.

In addition to the cytochromes, Ivey (1987) isolated a membrane-bound flavoprotein. It was obtained with the cytochrome b_{559} in the extraction with cholate. It has an apparent M_r of 27,300 and seems to be a dimer of identical subunits. The midpoint potential is -221 mV. This protein has many properties similar to the NAD(P)H-rubredoxin oxidoreductase purified by Yang (1977) and discussed above. However, Yang obtained his protein from the supernatant fraction of cell-free extracts, whereas Ivey obtained his flavoprotein from the membrane fraction recovered from cells broken with the French pressure cell. This seems to indicate that the two flavoproteins are different, as one comes from the cytoplasmic fraction and the other from the membrane fraction.

Menaquinone was found in *C. thermoaceticum* and *C. formicoaceticum* some time ago (Gottwald et al., 1975). It has now been demonstrated that *C. thermoaceticum* and *C. thermoautotrophicum* contain the same menaquinone identified as MK-7 (2 methyl-3-heptaprenyl-1.4-naphthoquinone) (Das et al., 1989). The menaquinone was almost exclusively found in the cytoplasmic membrane fraction of both bacteria. The amount was from 1.3 to 1.8 μmol/g of protein in the membranes.

The ATPases of *C. thermoaceticum* and *C. thermoautotrophicum* are very similar or identical (Ivey, 1987). In whole cells of *C. thermoaceticum*, ATP synthesis occurs in response to a pH jump (Ivey and Ljungdahl, 1986), and the isolated F_1F_0 ATPase reconstituted in membrane vesicles carriers out an ATP-dependent pumping of protons (Ivey, 1987). The structure of the F_1 portion on electron microscopic observations is very similar to ATPases of a number of sources including those from *Escherichia coli,* mitochondria, and chloroplast (Mayer et al., 1986; Mayer, 1993). However, the *C. thermoaceticum* F_1-ATPase lacks the ε subunit and has an $\alpha_3\beta_3\gamma\delta$ structure. Nevertheless, the *C. thermoaceticum* enzyme, like its counterparts of respiratory organisms, synthesizes ATP by using an electron chemical gradient of protons as the driving force. More recent studies (Ivey, 1987; Ivey, Das, and Ljungdahl, unpublished) have demonstrated that the F_0 moiety of the *C. thermoautotrophicum* contains only two subunits instead of three as found for F_0 of *E. coli*. The apparent M_r values of these subunits are 19,000 and 8000. The 8000 peptide has been identified as the DCCD-binding proteolipid. N^1N^1-dicyclohexyl-carbodiimide (DCCD) is an inhibitor of H^+-ATPases (Hermolin and Filligame, 1989). It should be noted that antibodies to the three subunits of the F_0 moiety of *E. coli* ATPase do not cross-react with the subunits of the F_0 ATPase of *C. thermoautotrophicum* (Das, unpublished). It has also been demonstrated that F_1 and F_0 moieties of *E. coli* and *C. thermoautotrophicum* do not functionally interact in cross-reconstitution experiments (Das and Ljungdahl, 1993). Apparently the ATPases of *C. thermoautotrophicum* and *E. coli* being similar in many aspects are nevertheless structurally and immunologically different.

Thauer et al. (1977), and Fuchs (1986) in their reviews on energy conservation in chemotropic bacteria discussed the different reaction steps from CO_2 to acetate of the acetyl-CoA pathway. They pointed out that the reduction of methylene-THF to methyl-THF yields enough energy needed for the production of ATP. The generation of a proton motive force in whole cells of *A. woodii* has been observed with the oxidation of CO (Diekert et al., 1986). This was demonstrated by an increase of ATP inside the cells when they were exposed to CO and the CO-dependent uptake of histidine by whole cells. These observations indicate that the reduction of methylene-THF to methyl-THF and the oxidation of CO to CO_2 could be involved in membrane-associated chemiosmosis (see Chapter 5). A requisite for this would be that the enzymes catalyzing these reactions are associated with the cytoplasmic membrane. Although all enzymes directly involved in the acetyl-CoA pathway have been isolated and characterized, they were all isolated from the soluble (supernatant) fraction of extracts prepared with the French pressure cell—an indication that they are not associated with the membrane fraction. Nevertheless, it was demonstrated by preparing membrane vesicles using lysozyme treatment that both methylene-THF reductase and carbon monoxide dehydrogenase/acetyl-CoA synthase are associated with these vesicles (Hugenholtz et al., 1987). Clearly, French pressure cell treatment disintegrates the cytoplasmic membrane too much and causes release of the membrane-associated enzymes, whereas the milder treatment with lysozyme allows the preparation of membrane vesicles with the associated enzyme still attached. It should be noted that membranes prepared by the French pressure cell have ATPase, the *b*-type cytochromes, and some other enzyme, which apparently are firmly embedded in the membrane.

Methylene-THF reductase has been isolated from *C. formicoaceticum* (Clark and Ljungdahl, 1984) and *C. thermoaceticum* (Park et al. 1990). The enzymes from the two clostridia are flavoproteins containing [Fe-S] centers, which have yet to be characterized. The *C. formicoaceticum* enzyme has an $\alpha_4\beta_4$ structure containing two different subunits of $M_r = 26,000$ and 46,000. The *C. thermoaceticum* enzyme appears to be an octamer of a single-type subunit of $M_r = 36,000$. Reduction of methylene-THF is achieved with $FADH_2$ and reduced ferredoxin but not by reduced rubredoxin and NADH or NADPH. It was proposed that reduced ferredoxin serves as the physiological electron donor. However, the redox potential of the methylene-THF/methyl-THF is -120 mV and membrane-bond electron carriers such as cytochrome b_{559} and the flavoprotein with redox potentials -215 and -221 mV, respectively, should also be considered as physiological electron donors.

That carbon monoxide dehydrogenase/acetyl-CoA synthase is membrane associated is clearly in accord with the finding that CO serves as the sole carbon and energy source. The oxidation of CO to CO_2 yields enough energy for ATP synthesis (Diekert et al., 1986). A Y_{CO} of 2.5 g of dry cells per mol of CO consumed has been reported for *C. thermoautotrophicum* grown on a defined minimal medium (Savage et al., 1987). The carbon monoxide dehydrogenase is an hexamer comprised of two different subunits, which have been sequenced (Morton et al., 1991). This work was a combined effort among the laboratories of Harland G. Wood, Steve Ragsdale, and mine. The evening before the day Harland passed away he was notified that the manuscript had been accepted. His acknowledgement was, "Great." Harland was indeed active to the last day of his life. How fortunate! The purification of the CO dehydrogenase was achieved by Ragsdale et al. (1983) and Diekert and Ritter (1983). Per dimer, the enzyme contains apparently 2 nickel, 1 zinc, 11 iron, and 14 acid-labile sulfur (Ragsdale et al., 1983). Intensive spectroscopic studies of the enzyme have revealed the presence of at least four different iron centers, including a nickel-iron-carbon complex with the proposed composition of $Ni_1Fe_{3-4}S_4C_1$ (Fan et al., 1991). This complex is likely the site of acetyl-CoA synthesis. A detailed discussion of CO dehydrogenases of acetogens and methanogens can be found in Chapters 3 and 21, respectively. Methyl viologen, flavodoxin, ferredoxin, and rubredoxin function as electron acceptors in the oxidation of CO catalyzed by the carbon monoxide dehydrogenase function. Of these, rubredoxin is by far the best. However, ferredoxin may be the natural electron carrier. This is based on the observations that ferredoxin stimulates the exchange of CO with $CH_3^{14}COSCoA$ catalyzed by carbon monoxide dehydrogenase (Ragsdale and Wood, 1985) and that ferredoxin interacts with the enzyme (Shanmugasundaram and Wood, 1992). It is interesting to note that carbon monoxide dehydrogenase of *C. thermoaceticum* is itself an electron acceptor for the oxidation of molecular hydrogen by hydrogenase of *Acetobacterium woodii* (Ragsdale and Ljungdahl, 1984). Thus, it appears as carbon monoxide dehydrogenase has the ability to interact with several different electron carriers. This property may depend on the different types of iron clusters of the enzyme. Purified carbon monoxide dehydrogenase has no hydrogenase or formate dehydrogenase activity.

2.5 Proton Motive Force and Electron Transport in Membranes of *C. thermoaceticum* and *C. thermoautotrophicum*

In the preceding section we learned about components of the electron transport system. ATPase and other enzymes of membrane vesicles of *C. thermoaceticum* and *C. thermoautotrophicum*. The redox potentials of these components and reactions catalyzed by the membranes are listed in Table 2.1. Membrane vesicles from the clostridia can be prepared using either the French pressure cell or by treatment with lysozyme and subsequent osmotic shock. The latter method yields membrane vesicles more complete and with the right side out (Hugenholtz, et al., 1987). As mentioned earlier, these vesicles contain carbon monoxide dehydrogenase and, apparently, a complete electron transport system including the proton-pumping component (Hugenholtz and Ljungdahl, 1990a).

A proton motive force (Δp) is generated by these membrane vesicles when they are exposed to CO (electron donor) and ferricyanide (electron acceptor) (Hugenholtz and Ljungdahl, 1989). The size of the Δp generated is dependent on the level of carbon monoxide dehydrogenase activity in the membrane vesicles. A minimum of 4 U (μmol of CO oxidized per min) per mg of membrane protein is required. With 15 U/mg, a $\Delta \Psi$ of 126 mV was generated at pH 7.5. This value is similar to those reported for whole cells of *C. thermoaceticum* (Baronofsky et

Table 2.1   Oxidation–reduction potentials for electron carriers isolated from cytoplasmic membranes of *C. thermoaceticum* and *C. thermoautotrophicum* and reaction couples important for acetogenesis

E Carrier	E'_0 (mV)
Ferredoxin I	-350
Ferredoxin II	-365
Rubredoxin I	-27
Rubredoxin II	$+20$
Flavoprotein	-221
Menaquinone	-74
Cytochrome b_{559}	-215
Cytochrome b_{554}	-57
FAD/FADH$_2$	-219
NADP$^+$/NADPH	-324
NAD$^+$/NADH	-320
CH$_2$-THF/CH$_3$-THF	-120
CH-THF$^+$/CH$_2$-THF	-295
CO$_2$/CO	-524
CO$_2$/HCOO$^-$	-420
CH$_2$O/CH$_3$OH	-182
HCOO$^-$/CH$_2$O	-535
H$^+$/H$_2$	-421
Benzyl viologen (BV)	-360
Methyl viologen (MV)	-440

al., 1984) and is large enough to drive the synthesis of ATP and uptake of the acids alanine, glycine, and serine to a concentration of 300 times that of the medium (Hugenholtz and Ljungdahl, 1989, 1990b). Amino acid uptake is independent of sodium. The rate of the amino acid transport is dependent on the generated Δp and thus directly related to the level of carbon monoxide dehydrogenase of the membrane vesicles.

The membrane-associated carbon monoxide dehydrogenase mediated CO-dependent reduction and CO_2-dependent oxidation of the *b*-type cytochromes, menaquinone, flavoprotein, and other components of the electron transport chains of *C. thermoaceticum* and *C. thermoautotrophicum* (Ivey, 1987; Hugenholtz et al., 1987; Das et al., 1989). In freshly prepared right-side-out membrane vesicles the cytochromes are partially reduced. When these vesicles were exposed to CO for 30 s at 50°C, the cytochromes became fully reduced. Subsequent treatment with CO_2 at 50°C oxidized the low-potential Cyt b_{559} but not the high-potential Cyt b_{554}. In addition to the spectral changes of the cytochromes, material absorbing in the 435–500-nm region underwent reduction and oxidation. Flavins and iron-sulfur proteins may be involved in these redox reactions.

Menaquinone has been placed between the Cyt b_{559} and Cyt b_{554} in the electron transport chain postulated for the two clostridia. Support for this are (a) oxidized menaquinone preferentially oxidizes Cyt b_{559} over Cyt b_{554}, (b) Cyt b_{554} is not reduced by CO when the membranous menaquinone has been destroyed by UV treatment; (c) addition of menaquinone to UV-treated membranes restores the reduction of Cyt b_{554} by CO; and (d) the redox potentials of Cyt b_{559} (E'_0-215 mV), menaquinone (E'_0-74 mV), and Cyt b_{554} (E'_0-57 mV) suggest such a sequence. A more complete sequence can be proposed on the basis of these observations and on the redox potentials of the carriers and reactions involved (Table 2.1). Starting with the oxidation of CO to CO_2, the sequence may be as follows: $CO/CO_2 \rightarrow$ ferredoxin $\rightarrow$ flavoprotein $\rightarrow$ Cyt $b_{559} \rightarrow$ methylene-THF/methyl-THF. A branch may involve Cyt $b_{559} \rightarrow$ menaquinone $\rightarrow$ Cyt $b_{554} \rightarrow$ rubredoxin (Fig. 2.3). The reduction of methylene-THF to methyl-THF by Cyt b_{559} is based solely on the redox potential for the methylene-THF/methyl-THF couple which is -120 mV.

The menaquinone of the menaquinone branch may serve as a proton carrier over the membrane, and it was suggested by Ivey (1987) that the menaquinone in *C. thermoaceticum* and *C. thermoautotrophicum* may operate in a cycle similarly to what is proposed for ubiquinone of the bc_1 complex of mitochondria and chloroplasts (Mitchell, 1976; Slater, 1983). In such a cycle, the menaquinone of the interior of the membrane will be reduced concomitant with the uptake of two protons. It will then diffuse to the exterior of the membrane, donate the electrons to the cytochromes, and release the protons. The process allows the translocation of two protons for each electron transferred.

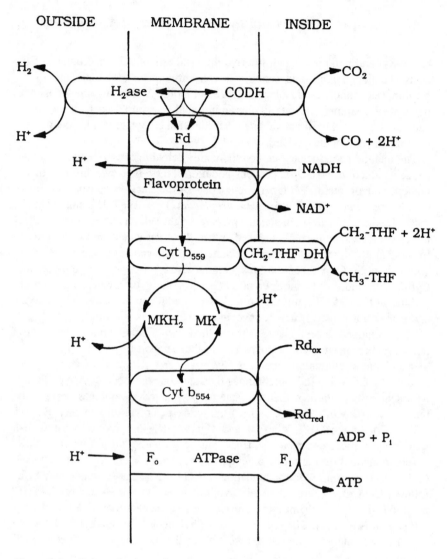

Figure 2.3. Proposed scheme for electron transport and Δp generation in the membrane of *C. thermoaceticum* and *C. thermoautotrophicum*. MK=menaquinone; Fd=ferredoxin; Rd=rubredoxin; CH$_2$-THF DH=methylenetetrahydrofolate reductase.

Growth substrates for *C. thermoaceticum* and *C. thermoautotrophicum* include molecular hydrogen plus CO_2 and methanol plus CO_2. With membrane vesicles prepared with lysozyme, the reduction of the cytochromes occurs when H_2, methanol, formaldehyde, and NADH are provided as electron donors (Hugenholtz and Ljungdahl, 1990a). This indicates the presence of hydrogenase, methanol dehydrogenase, and NADH dehydrogenase in the membranes. The NADH dehydrogenase, as was discussed in the previous section, has been isolated (Ivey, 1987). It was suggested to serve as a transhydrogenase between NADH and $NADP^+$ and also to function as an electron donor for chemiosmosis during growth on sugars.

The presence of hydrogenase in the membranes has been directly demonstrated by using benzyl viologen as electron acceptor, and it is apparently located on the exterior side of the membrane. Shiu-Mei Liu (National Taiwan Ocean University), when at the University of Georgia, purified a hydrogenase of *C. thermoaceticum* grown on glucose under CO_2. She used the supernatant fraction of extracts prepared by the French pressure cell. It appears to be an uptake hydrogenase having an activity of 111 μmol/min^{-1}/mg^{-1}. The H_2 evolution activity was only 2.48 U/mg. Gel filtration indicated that it has a $M_r = 340,000$, and SDS-PAGE demonstrated that it consists of two different subunits with M_r values of 69,000 and 54,000. The best enzyme preparation contained iron and nickel (about 10 and 0.3 g/atoms/mol, respectively) but also significant amounts of tungsten, zinc, and copper. At the present time, it has not been shown if this enzyme is associated with the membrane fraction. It should be noted that Kellum and Drake (1984) have found two hydrogenase in *C. thermoaceticum;* cells grown on glucose with CO_2 as gas phase had only one, whereas two were found in cells grown on glucose with CO in the gas phase.

C. thermoautotrophicum (Wiegel et al., 1981) and *C. thermoaceticum* (Wiegel and Garrison, 1985) grow on methanol in the presence of CO_2. *C. thermoautotrophicum* has a methanol dehydrogenase, which apparently is pyrroloquinoline quinone (PQQ)-dependent (Winters and Ljungdahl, 1989). The purified enzyme contains 2 mol each of PQQ and Zn. The PQQ is tightly but not covalently bound to the enzyme. Addition of PQQ to the apoenzyme is required for activity. It cannot be replaced with NAD, NADP, FAD, or FMN. In addition to oxidizing methanol, the enzyme oxidizes formaldehyde, several primary alcohols, and aldehydes. In fact, the principal product of the oxidation of methanol is formate. The enzyme appears to be loosely connected to the cytoplasmic membrane; however, a substantial amount of activity was found in the cytoplasmic fraction. The metabolic pathway of methanol in acetogens remains unknown. Conceivably, methanol can be oxidized to formaldehyde and formate. The former may enter the acetyl-CoA pathway forming methylene-THF, which, in turn, is reduced to methyl-THF for incorporation as the methyl group of acetyl-CoA. Formate may be further oxidized to CO_2 and then reduced to CO to form the carboxy group of acetyl-CoA. (See Chapter 6 for further information on methanol utilization.)

2.6   Concluding Remarks

This chapter covers some observations related to energy generation and electron transport in *C. thermoaceticum* and *C. thermoautotrophicum*. Some facts are known, but much is still just speculation. For instance, one of the main ideas is that energy generation is coupled to the reduction of methylenetetrahydrofolate to methyltetrahydrofolate. Although this a very likely, ATP production as a result of this reduction has yet to be shown. Several new observations have not been covered. They include growth on methoxylated aromatics, alcohols, oxalate, and aldehydes. Other observations needing work is the replacement of CO_2 with other electron acceptors including dimethylsulfoxide, thiosulfate, sulfide, and nitrate.

The resolution of the acetyl-CoA pathway and the discovery of the metabolic diversity of acetogens are rather recent developments. Advance within the field of acetogenesis has been explosive. It is not possible to cover all advances in a single chapter or even a book. Between 1942 and 1977, 45 publications and 22 abstracts covering work in *C. thermoaceticum* were published. Today, one can count several hundred publications and as many abstracts on *C. thermoaceticum* alone. More significantly, in 1977 two acetogenic bacteria were known, whereas today they number well over 50 (see Table 1.2).

The search started by Harland G. Wood to find the pathway of acetate synthesis from CO_2 was with *C. thermoaceticum*, known initially as a heterotroph with the limited capacity of fermenting glucose, fructose, xylose, and pyruvate to acetate as the only or main product. He has shown that acetate was totally synthesized from CO_2 (Wood, 1952a) and he had the vision that this synthesis could be a significant autotrophic mechanism for CO_2 fixation. He was right, but he could hardly have imagined that the search for the pathway would turn out as exciting as it did. What was discovered was a pathway involving tetrahydrofolates and corrinoids, enzymes dependent on metals such Mo, W, Ni, Fe, Co, and Se, enzymes with several functions, energy conservation, and ATP synthesis. In addition, the metabolic diversity of acetogens and related microorganisms and the apparent ecological significance of these microbes are clearly linked to the Wood acetyl-CoA pathway. Wood started the search out of curiosity and developed an intriguing observation. We now recognize its importance.

The story of acetogenesis, not yet completed, is a celebration of the power and benefits of basic research. Harland stood up for that. He was a true pioneer, and he got the job done.

Acknowledgment

I expressing appreciation to my colleagues, P. S. Beaty, A. Hugenholtz, D. M. Ivey, S.-M. Liu, and S.-S. Yang who have contributed unpublished material and to H.L. Drake for suggestions regarding this manuscript. Work on acetogens at the University of Georgia is supported by Public Health Service Grant No. AM 5R01 DK 27323-15 from the National Institute of Health and by Project 05-93ER20127.000DE-FG from the U.S. Department of Energy. A Georgia Power Professorship is also gratefully appreciated.

References

Adamse, A. D. 1980. New isolation of *Clostridium aceticum* (Wieringa). *Antonie van Leeuwenhoek* **46**:523–531.

Andreesen, J. R., G. Gottschalk, and H. G. Schlegel. 1970. *Clostridium formicoaceticum* nov. spec. isolation, description and distinction from *C. aceticum* and *C. thermoaceticum*. *Arch. Microbiol.* **72**:154–174.

Andreesen, J. R., and L. G. Ljungdahl. 1973. Formate dehydrogenase of *Clostridium thermoaceticum:* Incorporation of selenium-75, and the effects of selenite, molybdate, and tungstate on the enzyme. *J. Bacteriol.* **116**:867–873.

Andreesen, J. R., A. Schaupp, C. Neurauter, A. Brown, and L. G. Ljungdahl. 1973. Fermentation of glucose, fructose and xylose by *Clostridium thermoaceticum*. Effects of metals on growth yield, enzymes, and the synthesis of acetate from CO_2. *J. Bacteriol.* **114**:743–751.

Balch, W. E., S. Schoberth, R. S. Tanner, and R. S. Wolfe. 1977. *Acetobacterium*, a new genus of hydrogen-oxidizing, carbon dioxide-reducing anaerobic bacteria. *J. Syst. Bacteriol.* **27**:355–361.

Barker, H. A., and M. D. Kamen. 1945. Carbon dioxide utilization in the synthesis of acetic acid by *Clostridium thermoaceticum*. *Proc. Natl. Acad. Sci. USA* **31**:219–225.

Baronofsky, J. J., W. J. A. Schreurs, and E. R. Kashket. 1984. Uncoupling by acetic acid limits growth of an acetogenesis by *Clostridium thermoaceticum*. *Appl. Environ. Microbiol.* **48**:1134–1139.

Bateson, M. M., J. Wiegel, and D. M. Ward. 1989. Comparative analysis of 16S RNA sequences of thermophilic fermentative bacteria isolated from hot spring cyanobacterial mats. *Syst. Appl. Microbiol.* **12**:1–7.

Beaty, P. S., and L. G. Ljungdahl. 1991. Growth of *Clostridium thermoaceticum* on methanol, ethanol, propanol, and butanol in medium containing either thiosulfate or dimethylsulfoxide, Abstr. K-131, p. 236. *Ann. Meet. Am. Soc. Microbiol.* 1991.

Braun, M., F. Mayer, and G. Gottschalk. 1981. *Clostridium aceticum* (Wieringa), a microorganism producing acetic acid from molecular hydrogen and carbon dioxide. *Arch. Microbiol.* **128**:288–293.

Bruschi, M. 1979. Amino acid sequence of *Desulfovibrio gigas* ferredoxin: Revisions. *Biochem. Biophys. Res. Commun.* **91**:623–628.

Clark, J. E., and L. G. Ljungdahl. 1984. Purification and properties of 5, 10-methylenetet-rahydrofolate reductase, an iron-sulfur flavoprotein from *Clostridium formicoaceticum*. *J. Biol. Chem.* **259**:10845–10849.

Daniel, S. L., T. Hsu, S. I. Dean, and H. L. Drake. 1990. Characterization of the H_2- and CO-dependent chemolithotrophic potentials of the acetogens *Clostridium thermoaceticum* and *Acetogenium kivui. J. Bacteriol.* **172**:4464–4471.

Daniel, S. L., E. S. Keith, H. Yang, Y-S. Lin, and H. L. Drake. 1991. Utilization of methoxylated aromatic compounds by the acetogen *Clostridium thermoaceticum:* Expression and specificity of the CO-dependent O-demethylating activity. *Biochem. Biophy. Res. Commun.* **180**:416–422.

Das, A., J. Hugenholtz, H. Van Halbeek, and L. G. Ljungdahl. 1989. Structure and function of a menaquinone involved in electron transport in membranes of *Clostridium thermoautotrophicum* and *Clostridium thermoaceticum. J. Bacteriol.* **171**:5823–5829.

Das, A., and L. G. Ljungdahl. 1993. F_0 and F_1 parts of ATP synthases from *Clostridium thermoautotrophicum* and *Escherichia coli* are not functionally compatible. *FEBS Lett.* **317**:17–21.

Diekert, G. B., and R. K. Thauer. 1978. Carbon monoxide oxidation by *Clostridium thermoaceticum* and *Clostridium formicoaceticum. J. Bacteriol.* **136**:597–606.

Diekert, G. B., and M. Ritter. 1983. Purification of the nickel protein carbon monoxide dehydrogenase of *Clostridium thermoaceticum. FEBS Lett.* **151**:41–44.

Diekert, G. B., E. Schrader, and W. Harder. 1986. Energetics of CO formation and CO oxidation in cell suspensions of *Acetobacterium woodii. Arch. Microbiol.* **144**:386–392.

Drake, H. L. 1982. Demonstration of hydrogenase in extracts of the homoacetate-fermenting bacterium *Clostridium thermoaceticum. J. Bacteriol.* **150**:702–709.

Drake, H. L. 1992. Acetogenesis and Acetogenic Bacteria. In: Encyclopedia of Microbiology, Lederberg, J. (ed.), p. 1–15, Academic Press, Inc., San Diego.

El Ghazzawi, E. 1967. Neuisolierung von *Clostrium aceticum* Wieringa und stoffwechselphysiologische Untersuchungen. *Arch. Mikrobiol.* **57**:1–19.

Elliott, J. I., and L. G. Ljungdahl. 1982. Isolation and characterization of an Fe_8-S_8 ferredoxin (ferredoxin II) from *Clostridium thermoaceticum. J. Bacteriol.* **151**:328–333.

Elliott, J. I., S.-S. Yang, L. G. Ljungdahl, J. Travis, and C. F. Reilly. 1982. Complete amino acid sequence of the 4Fe-4S thermostable ferredoxin from *Clostridium thermoaceticum. Biochemistry* **21**:3294–3298.

Fan, C., C. M. Gorst, S. W. Ragsdale, and B. M. Hoffman. 1991. Characterization of the Ni-Fe-C complex formed by reaction of carbon monoxide with the carbon monoxide dehydrogenase from *Clostridium thermoaceticum* by Q-band ENDOR. *Biochemistry* **30**:431–435.

Fischer, F., R. Lieske, and K. Winzer. 1932. Biologische Gasreaktionen. II. Mitteilung: Über die Bildung von Essigsäure bei der biologischen Umsetzung vom Kohlenoxyd und Kohlensäure mit Wasserstoff zu Methan. *Biochem. Z.* **245**:2–12.

Fontaine, F. E., W. H. Peterson, E. McCoy, M. J. Johnson, and G. J. Ritter. 1942. A new type of glucose fermentation by *Clostridium thermoaceticum* n. sp. *J. Bacteriol.* **43**:701–715.

Fuchs, G. 1986. CO_2 fixation in acetogenic bacteria: variations on a theme. *FEMS Microbiol. Rev.* **39**:181–213.

Geerligs, G., P. Schönheit, and G. Diekert. 1989. Sodium dependent acetate formation from CO_2 in *Peptostreptococcus productus* (strain Marburg). *FEMS Microbiol. Lett.* **57**:253–258.

Gottwald, M., J. R. Andreesen, J. LeGall, and L. G. Ljungdahl. 1975. Presence of cytochrome and menaquinone in *Clostridium formicoaceticum* and *Clostridium thermoaceticum*. *J. Bacteriol.* **122**:325–328.

Hase, T., N. Ohmiya, H. Matsbara, R. N. Mullinger, K. K. Rao, and D. O. Hall. 1976. Amino acid sequence of a four-iron-four-sulphur ferredoxin isolated from *Bacillus stearothermophilus*. *Biochem. J.* **159**:55–63.

Heise, R., V. Müller, and G. Gottschalk. 1989. Sodium dependence of acetate formation by the acetogenic bacterium *Acetobacterium woodii*. *J. Bacteriol.* **171**:5473–5478.

Hermolin, J., and R. H. Fillingame. 1989. H^+-ATPase activity of *Escherichia coli* F_1F_0 is blocked after reaction of dicyclohexylcarbodiimide with a single proteolipid (subunit c) of the F_0 complex. *J. Biol. Chem.* **264**:3896–3903.

Hu, S.-I., H. L. Drake, and H. G. Wood. 1982. Synthesis of acetyl coenzyme A from carbon monoxide, methyltetrahydrofolate, and coenzyme A by enzymes from *Clostridium thermoaceticum*. *J. Bacteriol.* **149**:440–448.

Hugenholtz, J., D. M. Ivey, and L. G. Ljungdahl. 1987. Carbon monoxide driven electron transport in *Clostridium thermoautotrophicum* membranes. *J. Bacteriol.* **169**:5845–5847.

Hugenholtz, J., and L. G. Ljungdahl. 1989. Electron transport and electrochemical proton gradient in membrane vesicles of *Clostridium thermoautotrophicum*. *J. Bacteriol.* **171**:2873–2875.

Hugenholtz, J., and L. G. Ljungdahl. 1990a. Metabolism and energy generation in homoacetogenic clostridia. *FEMS Microbiol. Rev.* **87**:383–390.

Hugenholtz, J., and L. G. Ljungdahl. 1990b. Amino acid transport in membrane vesicles of *Clostridium thermoautotrophicum*. *FEMS Microbiol. Lett.* **69**:117–122.

Hugenholtz, J., T. V. Morgan, and L. G. Ljungdahl. 1989. EPR studies of electron transport in membranes of *Clostridium thermoautotrophicum*, Abstr. K-147. *Ann. Meeting Am. Soc. Microbiol.* 1989.

Irion, E., and L. Ljungdahl. 1965. Isolation of Factor III_m coenzyme and cobyric acid coenzyme plus other B_{12} factors from *Clostridium thermoaceticum*. *Biochemistry* **4**:2780–2790.

Ivey, D. M. 1987. Generation of energy during CO_2 fixation in acetogenic bacteria, dissertation, Dept. of Biochemistry, University of Georgia, Athens, GA.

Ivey, D. M. and L. G. Ljungdahl. 1986. Purification and characterization of the F_1-ATPase from *Clostridium thermoaceticum*. *J. Bacteriol.* **165**:252–257.

Karlsson, J. L., B. E. Volcani, and H. A. Barker. 1948. The nutritional requirements of *Clostridium aceticum*. *J. Bacteriol.* **56**:781–782.

Kellum, R., and H. L. Drake. 1984. Effects of cultivation gas phase on hydrogenase of the acetogen. *Clostridium thermoaceticum*. *J. Bacteriol.* **160**:466–469.

Kerby, R., and J. G. Zeikus. 1983. Growth of *Clostridium thermoaceticum* on H_2/CO_2 or CO as energy source. *Curr. Microbiol.* **8**:27–30.

Lentz, K., and H. G. Wood. 1955. Synthesis of acetate from formate and carbon dioxide by *Clostridium thermoaceticum*. *J. Biol. Chem.* **215**:645–654.

Li, L.-F., L. Ljungdahl, and H. G. Wood. 1966. Properties of nicotinamide adenine dinucleotide phosphate-dependent formate dehydrogenase from *Clostridium thermoaceticum*. *J. Bacteriol.* **92**:405–412.

Ljungdahl, L. G. 1986. The autotrophic pathway of acetate synthesis in acetogenic bacteria. *Annu. Rev. Microbiol.* **40**:415–450.

Ljungdahl, L., and H. G. Wood. 1963. Two thermophilic exchange reactions between pyruvate-CO_2 and formate-CO_2 in *Clostridium thermoaceticum*. *Bacteriol. Proc.* 109.

Ljungdahl, L., and H. G. Wood. 1965. Incorporation of C^{14} from carbon dioxide into sugar phosphates, carboxylic acids, and amino acids by *Clostridium thermoaceticum*. *J. Bacteriol.* **80**:1055–1064.

Ljungdahl, L., E. Irion, and H. G. Wood. 1965. Total synthesis of acetate from CO_2. I. Co-methylcobyric acid and Co-(methyl)-5-methoxy-benzimidazolyl-cobamide as intermediates with *Clostridium thermoaceticum*. *Biochemistry* **4**:2771–2779.

Ljungdahl, L. G., and E. Irion. 1966. Photolytic and reductive cleavage of Co-carboxy-methylcobalamin. *Biochemistry* **5**:1846–1850.

Ljungdahl, L., E. Irion, and H. G. Wood. 1966. Role of corrinoids in the total synthesis from CO_2 by *Clostridium thermoaceticum*. *Fed. Proc.* **25**:1642–1648.

Ljungdahl, L. G., W. E. O'Brien, M. R. Moore, and M.-T. Liu. 1980. Methylene-tetrahydrofolate dehydrogenase from *Clostridium formicoaceticum* and methylenetet-rahydrofolate dehydrogenase, methenyltetrahydrofolate cyclohydrolase (combined) from *Clostridium thermoaceticum*. *Methods Enzymol.* **66**:599–609.

Ljungdahl, L. G., and H. G. Wood. 1982. Acetate biosynthesis. In: D. Dolphin (ed.), B_{12}, Vol. 2. *Biochemistry and Medicine*. pp. 165–202. Wiley, New York.

Lundie, L. L. Jr., and H. L. Drake. 1984. Development of a minimally defined medium for the acetogen *Clostridium thermoaceticum*. *J. Bacteriol.* **149**:700–703.

Mayer, F. 1993. Principles of functional and structural organization in the bacterial cell: "Compartments" and their enzymes. *FEMS Microbiol. Rev.* **104**:327–346.

Mayer, F., D. M. Ivey, and L. G. Ljungdahl. 1986. Macromolecular organization of F_1-ATPase isolated from *Clostridium thermoaceticum* as revealed by electron microscopy. *J. Bacteriol.* **166**:1128–1130.

Mitchell, P. 1976. Possible molecular mechanisms of the proton motive function of the cytochrome systems. *J. Theoret. Biol.* **62**:327–367.

Morton, T. A., J. A. Runquist, S. W. Ragsdale, T. Shanmugasundaram, H. G. Wood, and L. G. Ljungdahl. 1991. The primary structure of the subunits of carbon monoxide dehydrogenase/acetyl-CoA synthase from *Clostridium thermoaceticum*. *J. Biol. Chem.* **266**:23824–23828.

O'Brien, W. E., and L. G. Ljungdahl. 1972. Fermentation of fructose and synthesis of acetate from carbon dioxide by *Clostridium formicoaceticum*. *J. Bacteriol.* **109**:626–632.

Park, E. Y., J. E. Clark, D. V. DerVartanian, and L. G. Ljungdahl. 1990. 5,10-Methylenetetrahydrofolate reductases: Iron-sulfur-zinc flavoproteins of two acetogenic clostridia. In: F. Muller (ed.), *Chemistry and Biochemistry of Flavoenzymes*, Vol. I, pp. 389–400. CRC Press, Boca Raton, FL.

Pezacka, E., and H. G. Wood. 1984. Role of carbon monoxide dehydrogenase in the autotrophic pathway used by acetogenic bacteria. *Proc. Natl. Acad. Sci. USA* **81**:6261–6265.

Poston, J. M., K. Kuratomi, and E. R. Stadtman. 1966. The conversion of carbon dioxide to acetate. I. The use of cobalt-methylcobalamin as a source of methyl groups for the synthesis of acetate by cell-free extracts of *Clostridium thermoaceticum*. *J. Biol. Chem.* **241**:4209–4216.

Poston, J. M., and E. R. Stadtman. 1967. The conversion of carbon dioxide to acetate. III. Demonstration of ferredoxin in the system converting Co-^{14}CH$_3$-cobalamin to acetate. *Biochem. Biophys. Res. Commun.* **26**:550–555.

Poston, J. M., K. Kuratomi, and E. R. Stadtman. 1964. Methyl-vitamin B$_{12}$ as a source of methyl groups for the synthesis of acetate by cell-free extracts of *Clostridium thermoaceticum*. *Ann. N.Y. Acad. Sci.* **112**:804–806.

Rabinowitz, J. 1972. Preparation and properties of clostridial ferredoxins. *Methods Enzymol.* **24**:431–446.

Ragsdale, S. W. 1991. Enzymology of the acetyl-CoA pathway of CO$_2$ fixation. *Crit. Rev. Biochem. Mol. Biol.* **26**:261–300.

Ragsdale, S. W., J. E. Clark, L. G. Ljungdahl, L. L. Lundie, and H. L. Drake. 1983. Properties of purified carbon monoxide dehydrogenase from *Clostridium thermoaceticum*, a nickel, iron-sulfur protein. *J. Biol. Chem.* **258**:2364–2369.

Ragsdale, S. W. and L. G. Ljungdahl. 1984. Hydrogenase from *Acetobacterium woodii*. *Arch. Microbiol.* **139**:361–365.

Ragsdale, S. W., and H. G. Wood. 1985. Acetate biosynthesis by acetogenic bacteria. Evidence that carbon monoxide dehydrogenase is the condensing enzyme that catalyzes the final step of the synthesis. *J. Biol. Chem.* **260**:3970–3977.

Savage, M. D., Z. Wu, S. L. Daniel, L. L. Lundie, Jr., and H. L. Drake. 1987. Carbon monoxide-dependent chemolithotrophic growth of *Clostridium thermoautotrophicum*. *Appl. Environ. Microbiol.* **53**:1902–1906.

Schulman, M., D. Parker, L. G. Ljungdahl, and H. G. Wood. 1972. Total synthesis of acetate from CO_2. V. Determination by mass analysis of the different types of acetate formed from $^{13}CO_2$ by heterotrophic bacteria. *J. Bacteriol.* **109**:633–644.

Schulman, M., R. K. Ghambeer, L. G. Ljungdahl, and H. G. Wood. 1973. Total synthesis of acetate form CO_2. VII. Evidence with *Clostridium thermoaceticum* that the carboxyl of acetate is derived from the carboxyl of pyruvate by transcarboxylation and not by fixation of CO_2. *J. Biol. Chem.* **248**:6255–6261.

Seifritz, C., S. L. Daniel, and H. L. Drake. 1993. Nitrate as a preferred electron sink for the acetogen *Clostridium thermoaceticum*. J. Bacteriol. **175**:8008–8013.

Shanmugasundaram, T., and H. G. Wood. 1992. Interaction of ferredoxin with carbon monoxide dehydrogenase from *Clostridium thermoaceticum*. *J. Biol. Chem.* **267**:897–900.

Slater, E. C. 1983. The Q-cycle, an ubiquitous mechanism of electron transfer. *Trends Biochem. Sci.* **8**:239–242.

Taylor, R. T. 1982. B_{12}-dependent methionine biosynthesis. In: D. Dolphin (ed.), B_{12} *Vol. 2. Biochemistry and Medicine*, pp. 307–355. Wiley, New York.

Thauer, R. K. 1972. CO_2-reduction to formate by NADPH, the initial step in the total synthesis of acetate from CO_2 in *Clostridium thermoaceticum*. *FEBS Lett.* **27**:111–115.

Thauer, R. K., K. Jungermann, and K. Decker. 1977. Energy conservation in chemotropic anaerobic bacteria. *Bacteriol. Rev.* **41**:100–180.

Utter, M. F., and H. G. Wood. 1951. Mechanisms of fixation of carbon dioxide by hetrotrophs and autotrophs. *Adv. Enzymol.* **12**:41–151.

Wiegel, J., M. Braun, and G. Gottschalk. 1981. *Clostridium thermoautotrophicum*. Species novum, a thermophile producing acetate from molecular hydrogen and carbon dioxide. *Curr. Microbiol.* **5**:255–260.

Wiegel, J., and R. Garrison. 1985. Utilization of methanol by *Clostridium thermoaceticum*, Abstr. I 115. *Ann. Meet. Am. Soc. Microbiol.* 1985.

Wieringa, K. T. 1936. Over het verdwijnen van waterstof en koolzuur onder anaerobe voorwerden. *Antonie van Leeuwenhoek* **3**:263–273.

Wieringa, K. T. 1940. The formation of acetic acid from carbon dioxide and hydrogen by anaerobic spore-forming bacteria. *Antonie van Leeuwenhoek* **6**:251–262.

Winters, D. K., and L. G. Ljungdahl. 1989. PQQ-dependent methanol dehydrogenase from *Clostridium thermoautotrophicum*. In: J. A. Jongejan, and J. A. Duine (eds.), *PQQ and Quinoproteins*. pp. 35–39. Kluwer Academic Publishers, Boston.

Wood, H. G. 1952a. A study of carbon dioxide fixation by mass determination of the types of C^{13}-acetate. *J. Biol. Chem.* **199**:579–583.

Wood, H. G. 1952b. Fermentation of 3,4-C^{14}- and 1-C^{14}-labeled glucose by *Clostridium thermoaceticum*. *J. Biol. Chem.* **199**:579–583.

Wood, H. G., and C. H. Werkman. 1935. The utilization of CO_2 by the propionic acid bacteria in the dissimilation of glycerol. *J. Bacteriol.* **30**:332.

Wood, H. G., C. H. Werkman, A. Hemingway, and A. O. Nier. 1941. Heavy carbon as a tracer in heterotrophic carbon dioxide assimilation. *J. Biol. Chem.* **139**:365–376.

Wood, H. G., and R. L. Stjernholm. 1962. Assimilation of carbon dioxide by heterotrophic organisms. *In:* I. C. Gunsalus and R. Y. Stainer (eds.), *The Bacteria, Volume III.* pp. 41–117. Academic Press, New York.

Wood, H. G., and L. G. Ljungdahl. 1991. Autotrophic character of the acetogenic bacteria. In: J. M. Shively, and L. L. Barton (eds.), *Variations in Autotrophic Life.* pp. 201–250. Academic Press, New York.

Yamamoto, I., T. Saiki, S.-M. Liu, and L. G. Ljungdahl. 1983. Purification and properties of NADP-dependent formate dehydrogenase from *Clostridium thermoaceticum,* a tungsten-selenium-iron protein. *J. Biol. Chem.* **258**:1826–1832.

Yang, S.-S. 1977. A study of electron-transfer proteins from *Clostridium thermoaceticum,* dissertation, Dept. of Biochemistry, University of Georgia. Athens, GA.

Yang, S.-S., and L. G. Ljungdahl. 1977. Properties of two rubredoxins and a NAD(P)H-rubredoxin reductase from *Clostridium thermoaceticum,* Abstr. K-135. *Ann. Meet. Am. Microbiol. Soc.* 1977.

Yang, S.-S., L. G. Ljungdahl, and J. LeGall. 1977. A four-iron, four-sulfide ferredoxin with high thermostability from *Clostridium thermoaceticum.* *J. Bacteriol.* **130**:1084–1090.

Yang, S.-S., L. G. Ljungdahl, D. V. DerVartanian, and G. D. Watt. 1980. Isolation and characterization of two rubredoxins from *Clostridium thermoaceticum.* *Biochim. Biophys. Acta* **590**:24–33.

Yang, H., and H. L. Drake. 1990. Differential effects of sodium on hydrogen- and glucose-dependent growth of the acetogenic bacterium *Acetogenium kivui.* *Appl. Environ. Microbiol.* **56**:81–86.

3

CO Dehydrogenase and the Central Role of This Enzyme in the Fixation of Carbon Dioxide by Anaerobic Bacteria

Stephen W. Ragsdale

3.1 Introduction

Our planet requires a continual source of fixed carbon because heterotrophic organisms produce energy by oxidizing the organic carbon to CO_2, thereby depleting the organic carbon. Reconversion of CO_2 to organic carbon occurs by a reductive process called CO_2 fixation. Many strict anaerobes perform CO_2 fixation by a pathway which is called the *reductive acetyl-CoA or the Wood/ Ljungdahl pathway* (see Chapter 1). Formation of acetyl-CoA via the Wood/ Ljungdahl pathway occurs in many anaerobic environments including marine and freshwater sediments, in the soil, and in landfills and waste-treatment sites. Autotrophic anaerobes are also common in the rumen of cows, horses, and sheep, and in the hindgut of termites and the large intestine of humans (Breznak and Kane, 1990; Breznak and Switzer, 1986; Lajoie et al., 1988) (see Part IV of this book).

The reductive acetyl-CoA pathway involves the condensation of two one-carbon units to form the two-carbon compound, acetate. Although this process is conceptually very simple, the mechanism of this series of reactions has intrigued chemists, biochemists, and microbiologists since 1945 when some of the first biochemical experiments using radioactive isotope tracer methods were per-formed (Barker and Kamen, 1945). Elucidation of a pathway requires identifica-tion of the intermediates. For the acetyl-CoA pathway, identification of most of these has been accomplished only recently through the use of a number of different biochemical techniques including enzyme kinetics, electrochemistry, and spectroscopy. A general review which includes a historical treatment and mechanistic description of all the enzymes involved in acetyl-CoA synthesis has

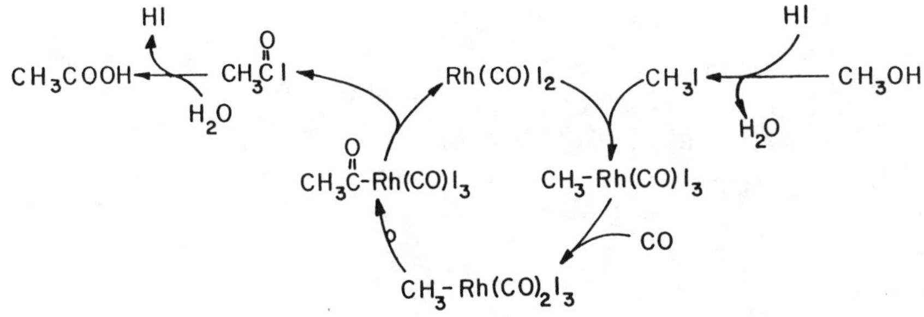

Figure 3.1. *The Monsanto process for the industrial synthesis of acetic acid* [modified from (Forster, 1979)].

recently appeared (Ragsdale, 1991). Probably the most interesting and unique chemistry occurs on an enzyme named carbon monoxide dehydrogenase (CODH), which is the focus of this chapter.

Before discussing the mechanism of CODH, it is interesting to consider a mechanism used by industrial chemists to synthesize acetate from methanol and CO using a rhodium catalyst, the "Monsanto process" (Forster, 1976; Forster, 1979) (Fig. 3.1). After conversion of methanol to methyl iodide, the rhodium center is methylated and then carbonylated. Then a methyl migration is proposed to occur, forming an acetyl-rhodium intermediate which is cleaved by iodide to form acetyliodide, regenerating the active catalyst. Finally, hydrolysis of acetyliodide yields acetate and hydrogen iodide.

In a mechanism which is remarkably similar to the Monsanto process, anaerobic bacteria appear to synthesize acetyl-CoA via methyl-metal, metal-carbonyl, and acetyl-metal intermediates at the active site of CODH. Industrial chemists choose iodide as the acyl group acceptor and final leaving group; anaerobic bacteria use coenzyme A.

3.2 Summary of the Acetyl-CoA Pathway

The acetyl-CoA pathway has been most thoroughly studied in acetogenic bacteria and is outlined in Figure 3.2. Two moles of either CO or CO_2 plus CoA are converted to acetyl-CoA which serves as a precursor of the cell's macromolecules. In addition, enzymatic cleavage of the high-energy thioester bond to form acetate is coupled to ATP synthesis via the phosphotransacetylase and acetate kinase reactions. This pathway then is a mechanism for converting CO_2 or the toxic gas, carbon monoxide, into acetate and into cell material. Overall, this is an irreversible noncyclic pathway which can be divided into three

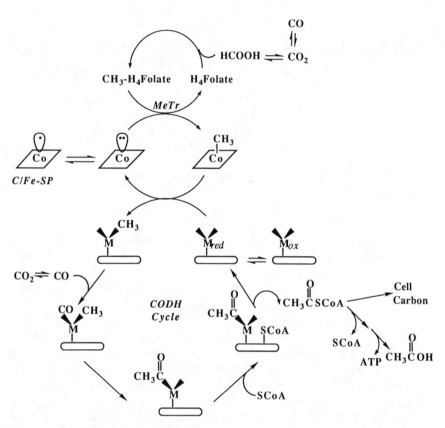

Figure 3.2.   *Scheme of the pathway of acetyl-CoA synthesis.* The square with Co in the center represents the corrinoid ring of the C/Fe-SP; the elongated oval represents CODH; M designates a redox active metal center involved in the assembly of the acetyl group of acetyl-CoA.

cyclic steps. In the first cycle, 1 mol of CO_2 or CO is reduced to the oxidation state of a methyl group. This set of reactions involves H_4folate-dependent enzymes, and the product is CH_3-H_4folate (reaction 3.1). Mechanistic information on the enzymes involved in formation of CH_3-H_4folate has been summarized (Ragsdale, 1991) and will not be discussed here. The rest of the pathway involves the synthesis of acetyl-CoA from CH_3-H_4folate, CO_2 or CO, and CoA (reaction 3.2 or 3.3).

$$H^+ + CO_2 + 3H_2 + H_4\text{folate} + ATP \rightarrow \qquad (3.1)$$
$$CH_3\text{-}H_4\text{folate} + 2H_2O + ADP + PO_4^{3-}$$

$$CH_3\text{-}H_4\text{folate} + CO_2 + H_2 + CoA \rightarrow \qquad (3.2)$$
$$H_2O + CH_3\text{-}CO\text{-}SCoA + H_4\text{folate} + H^+$$
$$CH_3\text{-}H_4\text{folate} + CO + CoA \rightarrow CH_3\text{-}CO\text{-}SCoA + H_4\text{folate} \qquad (3.3)$$

These reactions consist of one-carbon and two-carbon transformations which occur via enzyme-bound intermediates. In the second cyclic step of the pathway, the methyl group of CH_3-H_4folate is transferred to the cobalt active site of a corrinoid/iron-sulfur protein (C/Fe-SP). Mechanistic studies of the C/Fe-SP also have been summarized recently (Ragsdale, 1991). In the third and final minicycle, acetyl-CoA is assembled from the precursor methyl, carbonyl, and CoA moieties, a remarkable series of reactions which occur on CODH. The mechanism of CODH is the subject of this chapter.

3.3 Properties of CODH

Discovery of an enzyme which oxidized CO to CO_2 was first reported by Yagi (1959). Diekert and Thauer (1978) first identified such an enzyme in acetogenic bacteria. It was almost a decade before CODH was purified to homogeneity, due mainly to its extreme oxygen sensitivity. Purification and manipulation of CODH requires a strictly anaerobic environment, and I was a graduate student in Lars Ljungdahl's laboratory when I volunteered to purify the CODH from *C. thermoaceticum*. In Ljungdahl's lab, one of the most oxygen-sensitive proteins known, formate dehydrogenase, was routinely purified (Yamamoto et al., 1983). Treading into the CODH area was begun because of a postulate that formation of formyl-H_4folate may not require formate dehydrogenase but could involve a direct transfer of a formate oxidation level intermediate from CODH to formyl-H_4folate synthetase, thereby saving the ATP required for the synthetase reaction and bypassing formate dehydrogenase (Hu et al., 1982). After working for decades on formate dehydrogenase, such a suggestion that the enzyme may not be essential for acetate synthesis was strongly felt to require a prompt and definitive response. From my standpoint, my enthusiasm as a young graduate student opened a door on one of the most interesting enzymes in nature. Utilizing methodology which is routine in Ljungdahl's lab of preparing anaerobic buffers and using an anaerobic chamber, I was able to purify the enzyme to homogeneity (Ragsdale et al., 1983). We also showed that the above-mentioned postulate was incorrect and that formate dehydrogenase was, indeed, required (Ragsdale et al., 1983).

CODH is one of the four nickel-containing proteins so far discovered in nature. It was first suggested that CODH contained nickel when addition of nickel to the growth medium was found to stimulate CODH activity (Diekert and Thauer,

1980; Diekert et al., 1979). Stronger evidence was provided when CODH activity and radioactivity were shown to comigrate in polyacrylamide gels of cell extracts of *C. thermoaceticum* grown in the presence of ^{63}Ni (Drake et al., 1980). Then, the *C. thermoaceticum* CODH was purified to homogeneity and shown to contain 2 mol of nickel per mol of $\alpha\beta$ dimeric enzyme (Ragsdale et al., 1983). CODH was shown to have an $(\alpha\beta)_3$ structure with subunits of 77 kDa and 71 kDa and to contain, in addition to 2 nickel, $\sim$12 iron, $\sim$1 zinc, and $\sim$14 acid-labile inorganic sulfide per $\alpha\beta$ dimeric unit (Ragsdale et al., 1983). CODH from *C. thermoaceticum* also was purified to near homogeneity by Diekert and Ritter and shown to contain nickel (Diekert and Ritter, 1983). Currently, in my laboratory, bacteria are grown and harvested anaerobically, and every step in the purification and manipulation of the enzymes is performed in an anaerobic chamber manufactured by Vacuum Atmospheres which maintains the oxygen level at $\sim$ 0.2 ppm.

The amino acid sequence of the two subunits of the *C. thermoaceticum* enzyme have been deduced from the sequences of the cloned genes (Morton et al., 1991). In all of the properties so far studied (including metal content, subunit structure, activity, magnetic properties), the enzyme from *Acetobacterium woodii*, a mesophilic acetogen, is very similar to the one from *C. thermoaceticum*. Genes encoding the CODHs from acetogens, methanogens, photosynthetic bacteria, and carboxydobacteria have been cloned. Recently, Kerby et al. compared the sequences of the nickel-containing CODHs and found little sequence homology but 67% sequence similarity to the small subunit of the *C. thermoaceticum* CODH and 47% similarity to the large subunit of the *M. soehngenii* CODH (Kerby et al., 1992). Strictly conserved regions include a CXXCXXGPC region and a GXXAHXXHGXH region which would be able to provide S- and N-rich environments, respectively, for metal binding.

3.3.1   Activities of CODH

(a) CO OXIDATION

CODH can catalyze several reactions. As indicated by its name, it can oxidize CO to CO_2 in the presence of a suitable electron acceptor. Reduction of CO_2 to CO requires a powerful electron donor as the formal equilibrium reduction potential (E'_0) for the CO_2/CO couple at pH 6.3 is -517 mV [calculated from (Weast (1985)]. Thus, CO can be regarded as an extremely strong reductant. The CO oxidation activity allows acetogenic bacteria to grow on carbon monoxide as a carbon and electron source and to generate the carbonyl group of acetyl-CoA during growth on CO_2 or organic carbon sources. Oxidation of CO to CO_2 also is a property of the nickel-containing CODHs from photosynthetic bacteria such as *Rhodospirillum rubrum* and from the molybdopterin-containing enzyme which lacks nickel from carboxydotrophic bacteria. CO is used as a carbon and electron source for a number of bacteria (Colby et al., 1979; Meyer and Schlegel, 1983; Uffen, 1983; Zavarzin and Nozhevnikova, 1977) including acetogenic bacteria such as *C. thermoaceticum* (Daniel et al., 1990; Kerby and Zeikus, 1983). Ecologically, this reaction is important because it aids in maintaining the atmospheric content of CO below toxic levels.

The mechanistic enzymology of CO oxidation/CO_2 reduction has not received as much attention as one would expect given the simplicity of the reaction and its key role in generation of CO_2 from CO for formation of the methyl group and of CO from CO_2 and other carbon sources for generation of the carbonyl group of acetyl-CoA. Because both the aerobic and anaerobic CODHs are dehydrogenases and not oxidases, the second molecule of oxygen in CO_2 is derived from water. Thus, the mechanism may be envisioned as an oxotransfer, and could be similar to those of nitrite and sulfite reductases and the molybdenum-containing hydroxylases.

The oxidation of CO by CODH is inhibited by cyanide and is reversed with CO; however, inhibition by methyl iodide is irreversible and is stimulated by carbon monoxide in the presence of light (Ragsdale et al., 1983). Inhibition of CODH by cyanide has been most thoroughly studied with the nickel-containing enzyme from *R. rubrum* (Ensign et al., 1989). CN forms a complex with the enzyme which can be isolated by gel filtration but which is not formed with the nickel-deficient enzyme purified from cells grown in the absence of nickel. The pattern of inhibition suggested that cyanide acts as a slow binding active-site directed reversible inhibitor.

Ferredoxin is likely to be the natural electron acceptor of the *C. thermoaceticum* enzyme and has been shown to accept electrons from CO (Ragsdale et al., 1983a; Ragsdale et al., 1983b). Ferredoxin also significantly stimulates the rate of the CO/acetyl-CoA exchange reaction (Ragsdale and Wood, 1985) and the synthesis of acetyl-CoA from methyl-H_4folate, CO, and CoA (Roberts et al., 1992). The mode of activation involves formation of an electrostatically stabilized complex with CODH and a complex between ferredoxin and residues 229–239 of the large subunit was isolated by cross-linking the two proteins with a carbodiimide (Shanmugasundaram and Wood, 1992). In *R. rubrum,* a 22-kDa iron-sulfur protein was isolated which coupled the oxidation of CO by CODH to evolution of H_2 via a CO-induced membrane-bound hydrogenase (Ensign and Ludden, 1991). The gene encoding this Fe-S protein, *cooF,* has been sequenced and is part of the gene cluster containing CODH and the hydrogenase (Kerby et al., 1992). Recent kinetic (Kumar et al., 1993) and cyanide inhibition (Anderson et al., 1993) studies show that CO oxidation is catalyzed by an iron-containing center described below as the $g_{av} = 1.82$ species (see Table 3.1) and that this center is distinct from the Ni/Fe-S cluster involved in the assembly of acetyl-CoA.

Since 1870, chemists have been attempting to find efficient catalysts which can reduce CO_2 to organic compounds; however, so far, even the most efficient methods of electrochemical reduction on metal cathodes requires a large overpotential [see Ulman et al. (1987) for general review]. For example, the most efficient catalyst of which this author is aware was a carbon electrode modified with cobalt phthalocyanine, and reduction of CO_2 to CO required the generation of a redox potential ~ 300 mV more negative than the thermodynamic CO_2/CO redox potential (Lieber and Lewis, 1984). The requirement for a large overpotential makes this process inefficient in consumption of energy. In addition, chemists are excited when the nonenzymatic rates can be expressed in units of h^{-1}. In the presence of the CODH from *C. thermoaceticum,* electrochemical reduction of CO_2 to CO occurs at potentials near the thermodynamic reduction for the CO_2/CO couple (Lindahl et al. 1990a) and with turnover numbers as high as ~ 1800 s^{-1} (Ragsdale et al., 1983b). It is hoped that our goal of better characterizing the chemistry and structure of the site of CO oxidation/CO_2 reduction may aid in the development of a highly efficient metal complex catalyst for activation of CO_2 and CO.

(b) ACETYL-CoA SYNTHASE AND CO/ACETYL-CoA EXCHANGE

Many enzymes with the name CODH or CO oxidase appear to only catalyze the oxidation of carbon monoxide or its reverse reaction. However, the most important role for CODH in acetogenic bacteria is to catalyze the synthesis of acetyl-CoA from a methyl group, CO, and coenzyme A. Various roles for CODH in acetate synthesis were suggested before 1985, including reduction of an electron carrier or enzyme prosthetic group involved in CO_2 reduction to acetate (Diekert and Thauer, 1978) and formation of an enzyme-bound [HCOOH] group from pyruvate (Drake et al., 1981a; Pezacka and Wood, 1984a, 1984b) or CO (Hu et al., 1982). However, the chemistry of formation of the C–C bond of acetyl moiety of acetate had been assumed to occur at the cobalt center of a corrinoid or a corrinoid protein. Reasons for this have been summarized in a recent review (Ragsdale, 1991) and the major controversy was whether the synthesis occurred via an acetylcobalt, acetoxycobalt, or a carboxymethylcobalt intermediate.

CODH catalyzes a remarkable reaction (reaction 3.4)

$$CO + CH_3\text{-}^{14}CO\text{-SCoA} \rightarrow CO + CH_3\text{-}CO\text{-SCoA} \qquad (3.4)$$

which was first discovered by Harold Drake when he was puzzled to find that radioactive acetyl-CoA became nonradioactive when it is incubated with CO and a partially purified enzyme fraction containing CODH (Drake et al., 1981b; Hu et al., 1982). When I joined Harland Wood's laboratory as a postdoctor, I wished to investigate the mechanism of the isotope exchange reaction between CO and the carbonyl group of acetyl-CoA. At this time, most workers in the field and in Wood's laboratory were involved in studying the synthesis of acetyl-CoA from methyl-H$_4$folate, CO or pyruvate, and CoA, a reaction sequence which involved several (the exact number was unknown) enzymes. I recognized that study of the isotope exchange reaction would offer a simplified system for understanding the synthesis of acetyl-CoA because, during this exchange, acetyl-CoA must be disassembled by breakage of the methyl-carbonyl and carbonyl-SCoA bonds and then reassembled. I expected that the corrinoid protein, recently isolated in Wood's lab (Hu et al., 1984), would be involved, because it had been assumed to be the catalyst for formation of the methyl-CO bond. My objective was to determine which additional components besides the corrinoid protein would be required, as I had found that the purified corrinoid protein alone did not catalyze the exchange reaction. In a series of experiments, I became convinced that the corrinoid protein, unexpectedly, played no role in the exchange reaction and that, instead, purified CODH was the only catalyst required (Ragsdale and Wood, 1985). These results necessitated a complete revision of our concepts about the mechanism of acetyl-CoA synthesis because this exchange reaction requires the disassembly of acetyl-CoA followed by the resynthesis. As there are no acceptors of the methyl, CO, or CoA groups in the reaction mixture other than CODH, the results indicated that CODH must bind the methyl, carbonyl, and SCoA groups of acetyl-CoA, equilibrate the carbonyl with CO in solution, and then condense these three groups in the resynthesis of acetyl-CoA. Thus, Wood and I proposed that the synthesis and assembly of acetyl-CoA occur on CODH, that the role of the C/Fe-SP was to donate the methyl group to CODH, and that a more appropriate name for CODH is *acetyl-CoA synthase* (Ragsdale and Wood, 1985).

The discovery that CODH was the acetyl-CoA synthase in the pathway (Ragsdale and Wood, 1985) was a fundamental breakthrough; however, it was still subject to criticism [for example, in (Fuchs, 1986)] because the proposed methyl-CODH, CODH-CO, and CODH-SCoA intermediates had not been detected. A major focus of my laboratory in the ensuing years has been to detect, isolate, characterize, and evaluate the catalytic competence of such intermediates. In summary, it now appears that the assembly of acetyl-CoA does indeed occur on CODH and that much of the assembly process occurs on metal centers and involves organometallic bonds as shown in Figure 3.2.

(c) N$_2$O AND COS REDUCTASE

We have found that, besides its ability to assemble acetyl-CoA and to oxidize CO/reduce CO$_2$, CODH can perform oxidation reduction reactions with CO and CO$_2$ analogues. In the presence of a suitable reductant, it can convert nitrous oxide to N$_2$ at a rate approaching those of the copper-containing nitrous oxide reductases from denitrifying bacteria (Lu and Ragsdale, 1991). The E'_0 of the N_2O/N_2 redox couple is $+ 1175$ mV (Jones, 1982). Interesting, the nitrous oxide reductase from *Pseudomonas stutzerii* also oxidizes CO to CO_2, and CO binding was proposed to occur at the copper site (Riester et al., 1989). Carbonyl sulfide (COS) is a structural analogue of CO_2, in which a sulfur atom replaces one of the oxygens. COS can act both as an inhibitor as well as an alternate substrate of CODH. COS inhibits the oxidation of CO to CO_2 competitively ($K_i = 0.4$ mM) with respect to CO (Gorst and Ragsdale, unpublished). In the presence of reduced methyl viologen, CODH acts as a COS reductase (Gorst and Ragsdale, unpublished). Furthermore, under reducing conditions, COS was found to bind directly to the Ni-Fe-containing center, presumably forming the same complex that is formed when CO_2 or CO are reacted with CODH (Gorst and Ragsdale, unpublished). Thus, COS apparently acts both as a structural and kinetic analogue of CO_2 for the CODH from *C. thermoaceticum*. CODH from *Rhodospirillum rubrum* also is competitively inhibited by COS with respect to CO with a binding constant of 2.2 mM (Hyman et al., 1989).

3.3.2   Properties of the Metal Centers in CODH

It appears that CO oxidation and the assembly of acetyl-CoA are catalyzed by metal centers on CODH. Therefore, efforts have been focused on characterizing the metal sites in the protein. CODH contains 1.7 ± 0.2 nickels, 11 ± 2 irons, 1 ± 1 zinc, and 14 acid-labile inorganic sulfides per $\alpha\beta$ dimer (Ragsdale et al., 1983a). Because of the novelty of being a nickel protein and because of the importance of nickel in catalysis by CODH, the role of nickel in CODH has been the subject of review (Ragsdale et al., 1988). Of the nickel-containing proteins, the role of nickel has only been firmly established for CODH. That nickel is a component of the metal center which binds CO and that the resulting metal-CO complex is the precurser of the carbonyl group of acetyl-CoA has been demonstrated by a combination of spectroscopic and kinetic measurements (see below for details).

Determination of the coordination sphere of nickel has been partially successful using EXAFS spectroscopy. Nickel EXAFS analyses of the *C. thermoaceticum* CODH indicates that the Ni is sulfur rich (Bastian et al., 1988; Cramer et al., 1987). EXAFS studies of the *R. rubrum* CODH indicates a coordination sphere including 2 S (2.23 Å) and 2–3 N/O at 1.87 Å (Tan et al., 1992).

By using electrochemistry and spectroscopy, the redox sites in CODH have been classified into three groups: four one-electron Group I redox sites which have low redox potentials and thus appear to be important in catalysis, one or two one-electron Group II redox sites with redox potentials more positive than Group I, and Group III sites resulting from oxidative inactivation of the protein (Shin et al., 1992). Electron paramagnetic resonance (EPR) spectroscopy has

Table 3.1 Properties of the EPR-active metal centers of the CODH from *Clostridium thermoaceticum*

Species	g Values (Linewidths[a])	Spins per Dimer[b]	E_m (mV)
$g_{av} = 1.86$	1.97 (13), 1.87 (15), 1.75 (19)	0.3 ± 0.1	−530
$g_{av} = 1.94$	2.04 (11), 1.94 (13), 1.90 (15)	0.6 ± 0.1	−440
$g_{av} = 1.82$	2.01 (13), 1.81 (28), 1.65 (30)	0.2 ± 0.1	−220
Ni-Fe-C	2.08 (18), 2.07 (12), 2.03 (9)	0.3 ± 0.2	~−520

[a] Linewidths were determined at half-width at half-height and are in Gauss.
[b] These values were determined under conditions in which the signal was present in maximal amounts.

been essential in characterizing the Ni- and Fe-containing centers in CODH. We have observed at least seven EPR signals from CODH, all of which exhibit hyperfine broadening when ^{57}Fe replaces natural abundance iron, indicating that there are seven distinguishable iron-containing metal complexes in the protein (Lindahl et al., 1990a, 1990b). However, some of these signals certainly result from different conformations of the protein. Tables 3.1 and 3.2 summarize the EPR and electrochemical properties of the CODH from *C. thermoaceticum* and compares them with those of other CODHs.

(a) A HIGH-POTENTIAL FE-S CENTER

A high-potential anisotropic EPR signal, Signal I, has been observed in all CODHs so far studied. The function of this center is unknown although its redox and magnetic properties (and thus, presumably, its structure) has been conserved across quite huge evolutionary distances. In the *R. rubrum* CODH, this center is proposed to be a Ni-Fe complex based on hyperfine broadening of the EPR spectrum of the ^{57}Fe- and ^{61}Ni-containing enzyme and on the absence of this EPR signal in samples of the enzyme lacking nickel (Stephens et al., 1989). In addition, cyanide, which is a competitive inhibitor with respect to CO (Ensign et al., 1989), alters the EPR signals described above (Stephens et al., 1989). It has not been established for acetogenic or methanogenic bacterial enzymes whether or not the center responsible for this EPR signal contains nickel. No broadening of any of the resonances of the $g_{av} = 1.87$ EPR signal was observed for the ^{61}Ni-containing enzyme (Jetten et al., 1991) from *M. soenhgenii*, a result indicating that, in methanogens, the metal center responsible for this signal does not contain nickel.

There is evidence that this EPR signal is from a metal center involved in catalysis during methanogenesis. When cells of *Methanosarcina barkeri* are incubated with CO or when they are performing methane synthesis from acetate,

Table 3.2 Comparison of the iron-sulfur clusters in the CODH from *C. thermoaceticum* with those from methanogenic and photosynthetic bacteria

	Clostridium thermoaceticum			*Rhodospirillum rubrum*		*Methanosarcina thermophila*		*Methanosarcina barkeri*		*Methanothrix soehngenii*	
	g_{av}	g Values	E_m (mV)	g Values	E_m (mV)	g Values	E_m (mV)	g Values	E_m (mV)	g Values	E_m (mV)
I	1.82	2.01, 1.81, 1.65	−220	2.03, 1.89, 1.71	NS[a]	2.02, 1.87, 1.72	−154	2.01, 1.91, 1.76	−35	2.01, 1.89, 1.73	−230
II	1.94	2.04, 1.94, 1.90	−440	2.04, 1.94, 1.89	NS	2.04, 1.93, 1.89	−444	2.05, 1.94, 1.90	−390	2.05, 1.93, 1.86	−410
		2.04, 1.94, 1.90	−500			2.05, 1.95, 1.90	−540				
III	1.86	1.97, 1.87, 1.75	−530	NO[b]		?, ?, 1.79	−540	NO		NO	
NiFeC	2.06	2.08, 2.07, 2.03	−540	NO		2.08, 2.07, 2.03	N	NO		NO	
		2.06, 2.05, 2.03				2.06, 2.05, 2.03	S				

[a] NS = not studied.
[b] NO = not observed.

the $g = 1.76$ resonance shifts to $g = 1.73$ (Krzycki et al., 1989). In the *M. thermophila* enzyme, the $g = 1.72$ feature shifts to 1.79 at low redox potentials in the presence of ethylene glycol. So far, it has not been possible to assign a structure to the species giving rise to the $g_{av} = 1.87$ EPR signal of the Ni/Fe-S component of *M. thermophila* or from the CODHs from other microorganisms. Besides the examples described above, similar anisotropic EPR signals have been observed from binuclear iron-oxo complexes (Emptage et al., 1983; Muhoberac et al., 1980) and the substrate-bound form of the [4Fe-4S] cluster of aconitase. Recent studies indicate that this site catalyzes the oxidation of CO to CO_2. The rate of CO binding to this Fe-containing center is $10^8 M^{-1}s^{-1}$ and electrons are transferred from this center to the $[4Fe-4S]^{2+/1+}$ center (described in the next paragraph) at 2000 s^{-1}, which is the k_{cat} for CO oxidation (Kumar et al., 1993).

(b) A $[4Fe-4S]^{2+/1+}$ Center

A $g_{av} = 1.94$ EPR signal (Signal II) also has been observed in all CODHs so far studied. Based on Mössbauer spectroscopic results of the *C. thermoaceticum* enzyme, it is clear that this signal arises from a $[4Fe-4S]^{2+/1+}$ cluster (Lindahl et al., 1990a). It is presumed that it performs an electron transfer function. Apparently, this cluster can interconvert with another form with similar g values but a lower redox potential. Similar shifts have been seen with the *M. thermophila* enzyme (Lu et al., 1994). The inability to observe the low-potential form of this cluster in CODHs other than those from *C. thermoaceticum* and *M. thermophila* probably underlines the importance of using an electrochemical system which is able to maintain stable redox potentials below -500 mV. Such conditions which were not attained in the studies with the *M. barkeri* and *M. soehngenii* enzymes.

(c) Low-Field EPR Signals

Besides the EPR signals resulting from $S = 1/2$ systems, there are EPR signals which account for significant spin intensity in the region between $g = 4$ and 6 (Lindahl et al., 1990a). These are most likely due to high spin-state forms of one or more of the metal centers just described.

(d) The CO Adduct Which Is the Precurser of the Carbonyl Group of Acetyl-CoA

The most important and thoroughly studied metal center is the nickel-containing site that is responsible for binding CO. After binding CO, a CO adduct of CODH is detectible by EPR spectroscopy that is due to a spin-coupled complex consisting of Ni, Fe, and CO and has been referred to as the NiFeC species. The Ni-Fe-C species gives rise to two interconvertible signals with g values at 2.08, 2.07, and 2.03, and at 2.06, 2.05, and 2.03 (Fig. 3.3) (Ragsdale et al., 1985). These EPR signals can be observed at the relatively high temperatures of liquid N_2 ($\sim$ 77 K) where most Fe-S centers relax too quickly to be detected. The 2.08/2.03 signal is the predominant form, and varying amounts of the 2.05/2.03 signal are observed; thus, most studies have been focused on the 2.08/2.07/2.03 EPR signal. Two important questions have been addressed: (1) Is this species of catalytic relevance? (2) What is the structure of the complex giving rise to this EPR signal?

Kinetic competence of the Ni-Fe-C species in acetyl-CoA synthesis. Several experiments had implied the catalytic relevance of the NiFeC species. The NiFeC EPR signal is markedly influenced when a tryptophan residue involved in CoA binding is modified, suggesting that the NiFeC species is near both a tryptophan residue and the site of CoA binding (Shanmugasundaram et al., 1988). In addition, CoA had been found to have a direct effect on the EPR spectrum of the metal-carbonyl intermediate (Ragsdale et al., 1985).

To establish the kinetic competence of the EPR-detectable NiFeC species, one must demonstrate that its rate of formation is at least as rapid as any reaction in which it is proposed to be involved. Because the exchange reaction between CO and acetyl-CoA involves acetyl-CoA synthesis from bound precursors, it is a useful and convenient assay which can be considered to include some key partial reactions in the overall synthesis. In addition, the rate of this reaction is similar to that of the total synthesis of acetyl-CoA from CH_3-H_4folate, CO, and acetyl-CoA (Roberts et al., 1992). Therefore, demonstration that a species is an intermediate in the CO/acetyl-CoA exchange reaction also establishes its relevance in the overall synthesis of acetyl-CoA.

In a series of experiments, we clearly demonstrated that the NiFe-C species acts as a precurser of the carbonyl group of acetyl-CoA (Gorst and Ragsdale, 1991). The rate of formation of the NiFeC species was found to be tenfold faster than the rate of the CO/acetyl-CoA exchange reaction when measured under identical conditions. Therefore, the NiFeC species clearly forms fast enough to support its role as a catalytically competent precursor of the carbonyl group of acetyl-CoA. In addition, we were successful in measuring the exchange of label from [1-^{13}C] acetyl-CoA into the NiFeC EPR signal. The reaction is reflected in broadening of the NiFeC EPR signal when [1-^{13}C] acetyl-CoA is added to the solution containing CO and CODH. As shown in Figure 3.4, the amount of hyperfine broadening expected for complete exchange was obtained, resulting from replacement of the NiFe^{12}C signal by a NiFe^{13}C EPR signal (reaction 3.5). That this isotope replacement occurred at a rate at least as fast as complete exchange of the C-1 radioactive label from the carbonyl group of acetyl-CoA by CO, provided further kinetic and spectroscopic evidence for the intermediacy of the NiFeC species in the formation of the carbonyl group of acetyl-CoA (Gorst and Ragsdale, 1991).

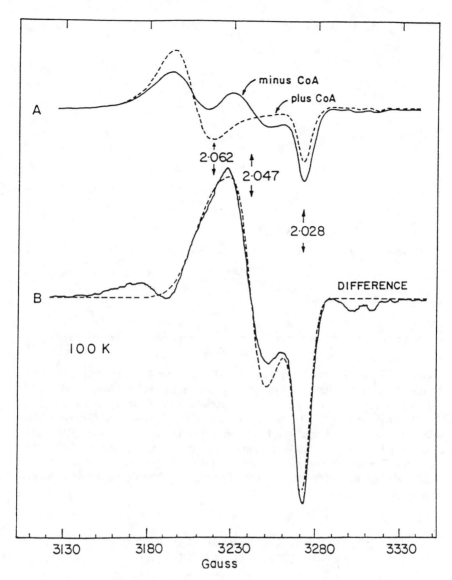

Figure 3.3. *EPR spectrum of the Ni-Fe-CO center showing the effect of CoA binding on the spectrum.* EPR conditions were temperature = 100 K, field set = 3190 G, gain = 3.2 × 10³, power = 10 mW, scan rate = 100 G/min, and frequency = 9.286 GHz [from (Ragsdale et al., 1985)].

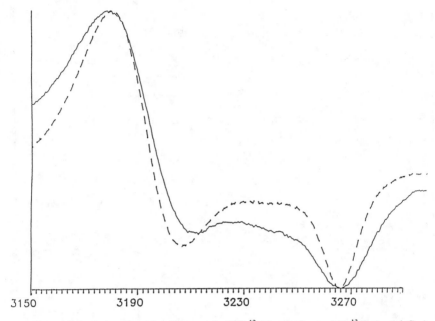

Figure 3.4. *Exchange Reaction between the NiFe*^{12}*CO complex and* [1-^{13}C] *Acetyl-CoA* monitored by EPR. CODH, (200 μl, 67 nmol) was reacted at 25°C with CO (101 kPa) for 2 min, frozen in liquid N_2, and the EPR spectrum recorded (dashed line) after replacing the CO gas phase with N_2. The sample was thawed, [1-^{13}C] acetyl-CoA (540 nmol) was added, the solution was incubated for 2 min, frozen in liquid N_2, and the EPR signal again recorded (solid line). EPR conditions are as in Figure 3.3 except the field center was 3200 G. For the CO-treated sample, the gain was 4×10^3 and the frequency was 9.278 GHz. For the sample after addition of acetyl-CoA, the gain was 1.6×10^4 and the frequency was 9.09 GHz.

$$\begin{aligned} \text{CODH-(NiFe)-CO} + \text{CH}_3\text{-}^{13}\text{CO-SCoA} &\rightarrow \\ \text{CODH-(NiFe)-}^{13}\text{CO} + \text{CH}_3\text{-CO-SCoA} \end{aligned} \qquad (3.5)$$

Further evidence for the intermediacy of the NiFeC species in acetyl-CoA synthesis was obtained by observing the NiFeC signal by reacting acetyl-CoA with CODH at low redox potentials (Gorst and Ragsdale, 1991). If the final steps of acetyl-CoA synthesis are reversible, one would expect to observe the same intermediate from incubation of the product, acetyl-CoA, with CODH as seen when the enzyme is reacted with the physiologically relevant substrates, CO, and CO_2. The EPR signal formed under these conditions has the same *g* values and line shape as the CO-generated signal, implying that the structures of the two complexes are very similar. Line broadening in the 2.03 region of this signal from reaction of CODH with [1-^{13}C] acetyl-CoA clearly demonstrated that the carbonyl from acetyl-CoA is a component of the complex responsible for the EPR signal. This experiment also allowed us to obtain the midpoint potential of the NiFeC intermediate. Reduction of the EPR-silent NiFeC species to its EPR-active form was found to be a one-electron process with a midpoint reduction potential of -541 mV, implying that the NiFeCO species has a midpoint potential of ~ -541 mV. Thus, we predict that -541 mV may be near the true E'_0 of the NiFeC$^{ox/red}$ couple. One might remain cautious even about this estimate, as substrate binding has been shown to alter the E'_0 of redox centers in several enzymes (Adams, 1987; Banerjee et al., 1990; Lenn et al., 1990).

Therefore, combined kinetic, spectroscopic, and electrochemical data strongly support the intermediacy of the NiFe-CO species in the synthesis of acetyl-CoA from CO or CO_2. It has been proposed that formation of methyl-CODH and acetyl-CODH intermediates occur at the same site as formation of the CO adduct.

Structure of the Ni-Fe-C species. The second important question is: What is the structure of the Ni-Fe-CO species? This Ni-containing center is an interesting metal complex which is the focus of synthetic efforts by several inorganic chemical laboratories. The classical way to infer which metals are involved in a spin system is to perform substitution by an isotope which has a nuclear spin (I). The nuclear magnet attracts and alters the properties of the unpaired electron spin (S). This change affects EPR spectrum of the system. The strength of the interaction is described by a tensor A, the magnetic hyperfine coupling tensor, which relates the nuclear and electron spin, $S·A·I$. For example, if a nucleus with a nuclear spin of 1/2 (e.g., ^{1}H) is a component of a system which exhibits an EPR signal, two lines ($2I + 1$) will be observed at each resonance position. Often, instead of distinct lines, only a broadening of the spectrum will be observed. In studying CODH, natural abundance nickel and iron (both predominantly $I = 0$) can be substituted with isotopes containing a nuclear spin. In addition, CO is available as ^{13}CO.

We performed a series of isotope substitutions (Fig. 3.5) (Fan et al., 1991; Ragsdale et al., 1983a). When CODH is reacted with ^{13}CO in place of ^{12}CO, the resulting EPR spectrum exhibits a doublet centered at the g value of the high field resonance of the spectrum of the ^{12}CO-reacted enzyme. This experiment demonstrates that CO is part of the spin system, most likely as a ligand to a metal(s). In order to perform an isotopic substitution with nickel, it was necessary to culture *C. thermoaceticum* in a medium containing ^{61}Ni ($I = 3/2$) in place of natural abundance nickel. After isolation of CODH from these cells and reacting with CO, a significant broadening of the spectrum is observed relative to that of the enzyme-containing natural abundance nickel. This experiment demonstrates that Ni also is part of the spin system. When CODH was isolated from cells grown in the presence of ^{57}Fe ($I = 1/2$) in the place of natural abundance

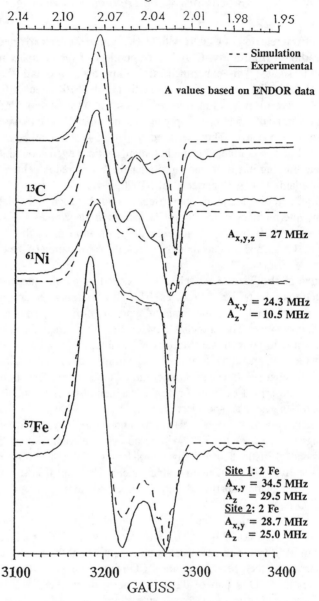

g

2.14   2.10   2.07   2.04   2.01   1.98   1.95

- - - Simulation
—— Experimental

A values based on ENDOR data

^{13}C

$A_{x,y,z} = 27$ MHz

^{61}Ni

$A_{x,y} = 24.3$ MHz
$A_z = 10.5$ MHz

^{57}Fe

<u>Site 1</u>: 2 Fe
$A_{x,y} = 34.5$ MHz
$A_z = 29.5$ MHz
<u>Site 2</u>: 2 Fe
$A_{x,y} = 28.7$ MHz
$A_z = 25.0$ MHz

3100   3200   3300   3400
GAUSS

3.5. *Isotope studies of isotopically enriched CODH after reaction with CO*. Experimental spectra (solid line) and simulated spectra (dashed line) using *A* values derived from the ENDOR experiments of CODH enriched with ^{13}CO, ^{61}Ni, and ^{57}Fe. For simulation of the ^{57}Fe spectra, four irons in two different types of iron were assumed with the *A* values shown [modified from (Gorst and Ragsdale, 1991)].

iron and reacted with CO, the EPR spectrum was significantly broadened relative to that of the [56]Fe-containing enzyme (Fan et al., 1991; Ragsdale et al., 1985). Therefore, CO binds to a site on CODH at a novel metal center containing both nickel and iron, which was called the NiFeC species.

Using electron-nuclear double-resonance (ENDOR) spectroscopy (Fan et al., 1991), we confirmed the EPR evidence for the NiFeC species. ENDOR is a method which can directly reveal the strength of interaction, *A*, between the electronic and nuclear spins. In addition, each magnetic nucleus located in a different environment gives a distinct ENDOR signal. Two ENDOR signals were observed in the [57]Fe-containing enzyme, suggesting that at least two irons were part of the NiFeC complex. A reasonable simulation (dashed lines in Figure 3.5) of the EPR spectrum with the ENDOR-derived coupling constants required the inclusion of at least three iron atoms. [61]Ni and ^{13}C ENDOR signals also were observed from the [61]Ni-containing and ^{13}C-treated enzymes, and the *A* values obtained were used to simulate the EPR spectra. A single Ni and one ^{13}C were sufficient to accurately simulate the EPR spectra. This was the first reported Ni ENDOR study of an enzyme (Fan et al., 1991). Based on the combined ENDOR and EPR results, we proposed that the CODH-CO complex contains CO, Ni, and ~ 3-4 iron atoms (Fan et al., 1991). The coupling constants observed for the [57]Fe-containing enzyme are similar to those obtained for Fe-S clusters, and those for [13]CO are similar to those obtained for the [13]CO-reacted hydrogenase from *C. pasteurianum* in which a unique iron-sulfur cluster, implicated as the site of H_2 activation, binds CO (Tesler et al., 1987) to inhibit the enzyme.

The environment of the iron sites in CODH has been investigated by Mössbauer spectroscopy (Lindahl et al., 1990b). This method requires the substitution of [57]Fe and has been important in the characterization of the metal environment in a number of iron-containing proteins. The hyperfine coupling values used in the simulation of the Mössbauer spectrum of the Ni-Fe-CO complex of CODH are nearly identical to those observed in the ENDOR experiment. The hyperfine coupling constants, isomer shift, and quadrupole splitting parameters used to fit the Mössbauer spectra are very similar to those of [4Fe-4S] centers (Lindahl et al., 1990b).

Based on spectroscopic evidence, several working models for the "M-CO" precursor of the carbonyl of acetyl-CoA can be forwarded (Fig. 3.6). These models have in common at least three irons, one nickel, and a single CO. The M-CO could be a [4Fe-4S] cluster linked to Ni through a ligand bridge, and CO could bind either to Ni or Fe. One possibility, which has been recently ruled out, is that CO could serve as the bridging ligand. Incorporation of Ni into the corner of a cubane Fe_3-S_4 cluster also is possible, in which case CO could bind either to the Ni or Fe component of the complex. Of these possible structures, so far only the Ni-Fe_3-S_4 complex has a purely synthetic precedent (Ciurli et al., 1990; Ciurli et al., 1992). In addition, a Ni-Fe_3-S_4 center has been assembled

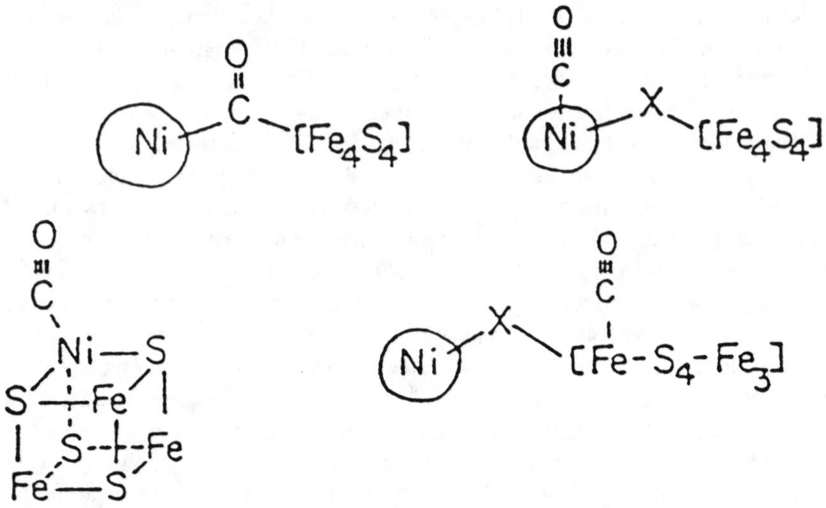

Figure 3.6. *Minimal working models of the Ni-Fe-CO center.*

by incorporation of Ni into the unoccupied edge of a [3Fe-4S] center in a ferredoxin (Conover et al., 1990).

Recent Fourier transform infrared spectroscopic studies have enabled us to firmly establish the mode of binding of CO to CODH in the CODH-CO adduct (Kumar and Ragsdale, 1992). The CO stretching vibration of the CODH-CO complex was observed at 1995.1 cm^{-1}, and when ^{13}CO replaced ^{12}CO, this peak predictably shifted to 1950.8 cm^{-1}. By comparing the position of the peak with those of other infrared (IR)-detectable metal carbonyl complexes, the CODH-CO species could be unambiguously assigned to a metal carbonyl complex in which CO is terminally bound, i.e., metal-C≡O. When we reacted the enzyme with ^{13}CO in the presence of [^{12}C]-acetyl-CoA, the bound CO group which elicits the IR peak was observed to exchange with the carbonyl group of acetyl-CoA. This latter experiment clearly demonstrated the catalytic competence of the enzyme-bound IR-detectable metal carbonyl species. As described earlier, the EPR-detectable Ni-Fe-C species also was found to be catalytically competent in synthesis of acetyl-CoA. Therefore, the IR-detectable and EPR-detectable CO adducts appear to be the same species. Because CO is terminally coordinated to one metal in this complex, it cannot be the bridging ligand, and there must be an endogenous bridge between Ni and Fe. Based on the combined magnetic resonance and IR results, we proposed a working model in which the structure of the CODH-CO complex is described as [Ni-X-Fe$_4$S$_4$]-C≡O, where X is an unknown bridge between the Ni site and a [4Fe-4S] cluster (Kumar and Ragsdale, 1992).

We have recently demonstrated that CO binds to an iron site in the Ni/Fe-S cluster (Qiu et al., 1994). This was unexpected since nickel was assumed to be the site of acetyl-CoA assembly. In order to include an essential role of Ni in catalysis, a bimetallic mechanism has been proposed (Qiu et al., 1994). In this model, CO binds to an iron site in the cluster and the methyl group of the methylated C/Fe-SP is transferred to Ni, forming a methylnickel intermediate. Subsequently, the acetyl-Ni or acetyl-Fe intermediate would be formed by carbonyl insertion or methyl migration and the acetyl group finally transferred to CoA to form acetyl-CoA.

Ni-Fe complexes analogous to the Ni-Fe-C center of acetogens. *Methanosarcina thermophila* contains a multienzyme complex called the CODH complex which has been resolved into a nickel/iron-sulfur and a corrinoid/iron-sulfur component (Abbanat and Ferry, 1990). This complex plays a central role in acetoclastic methanogenesis. The Ni/Fe-S component catalyzes CO oxidation and has been proposed to be involved in cleavage of acetyl-CoA into its methyl, carbonyl, and CoA moieties. As in the acetogenic CODHs, the *M. thermophila* enzyme complex can perform exchange reactions between CO and acetyl-CoA (Raybuck et al., 1991), CoA and acetyl-CoA (Raybuck et al., 1991), and acetyl-CoA synthesis from methyl iodide, CO, and CoA (Abbanat and Ferry, 1990). The protein complex containing CODH from *M. thermophila* was shown to contain a metal center which, when reacted with CO, exhibits two EPR signals very similar to those of the Ni-Fe-CO center in the CODH from *C. thermoaceticum* (Terlesky et al., 1987). These signals can be observed at liquid nitrogen temperatures and have g values of 2.089, 2.078, 2.03 and 2.057, 2.049, 2.027 (Terlesky et al., 1987). Interconversion between these EPR signals is influenced by the presence of acetyl-CoA (Terlesky et al., 1987). Based on isotopic substitution with ^{61}Ni, ^{57}Fe, and ^{13}CO, these EPR signals were shown to result from a Ni-Fe-CO complex (Terlesky et al., 1987). We recently showed that incubation of the Ni/Fe-S component with CO elicits the Ni-Fe-C EPR signal which was observable at temperatures as high as 150 K and had a half-power of saturation of 0.2 mW at 16 K (Lu et al., 1994). This signal had previously been observed only in the CODH complex of *M. thermophila* and the acetyl-CoA synthase from acetate-producing bacteria. Incubation of the CO-reduced Ni/Fe-S component with acetyl-CoA resulted in an increase in intensity of the Ni-Fe-C signal which supports the role of the component in the cleavage of acetyl-CoA. A metal center containing nickel and iron also has been described in the CODH from *Rhodospirillum rubrum* (Stephens et al., 1989) that is similar to the EPR signal observed in methanogenic and acetogenic bacteria. Therefore, it appears that the NiFeC EPR signal is present in CODHs that cleave or synthesize acetyl-CoA, but it is not present in the CODHs that only oxidize CO to CO_2 and lack the synthase activity.

3.4   Enzymology of CODH/Acetyl-CoA Synthase

3.4.1   Methylation of CODH

As shown in Figure 3.2, the methyl group of acetyl-CoA is derived from methyl-CODH which is formed from the methylated C/Fe-SP. Formation of protein-bound methylcobamide from methyl-H_4folate is an extremely interesting reaction which has been recently reviewed (Ragsdale, 1991). The protein-to-protein one-carbon transfer of the methyl group from methylated C/Fe-SP to CODH is the first of a series of reactions in the assembly of acetyl-CoA by CODH.

Before considering the methylation of CODH, it is germane to discuss the stereochemistry of acetyl-CoA synthesis from CH_3-H_4folate, CO, and CoA. Lebertz et al. (1987) found that the conversion of the chiral methyl group of CHTD-H_4folate to the methyl of acetate occurs with overall retention of configuration. The important question is, What is the mechanism of transfer of the methyl group from the methylated C/Fe-SP to CODH? Intermolecular methyl transfer reactions result in inversion of configuration; thus, it is expected that transfer of the methyl group from CH_3-H_4folate to the C/Fe-SP, yielding methylcobamide, results in a single inversion of configuration. Therefore, to gain a net retention of configuration of the methyl group in acetate, an odd number of inversions (most likely, one) must occur in the conversion of methylcobamide to the acetyl-CODH intermediate because migration of a methyl group to a metal-carbonyl (also described as a carbonyl insertion) to form an acetylmetal species and formation of acetyl-CoA and then acetate from an acetylated intermediate would not change the stereochemical configuration of the methyl group. Consequently, the transfer of the methyl group to CODH should involve a reaction with a single inversion of configuration as would be expected if the metal center of CODH displaced the methyl group of the methylated C/Fe-SP.

That CODH per se catalyzes an exchange reaction between CO and acetyl-CoA (Ragsdale and Wood, 1985) first indicated that there is a methyl binding site on CODH. For simplicity, Figure 3.2 depicts CODH, first binding the methyl group of the methylated C/Fe-SP and then CO in an ordered mechanism. In fact, it appears that CODH can bind the methyl and CO groups in a random order. This is based on the observation that (i) methylated CODH can react with CO (Lu et al., 1990), and carbonylated CODH can react with the methylated C/Fe-SP (Ragsdale, Lindahl, and Munck, unpublished) to form acetate, (ii) methylation of CODH occurs in the absence of CO (Lu et al., 1990), and (iii) carbonylation of CODH occurs in the absence of the methylated C/Fe-SP (Ragsdale et al., 1983a; Ragsdale et al., 1985).

What is the mechanism of transfer of the methyl group from the methylated C/Fe-SP to CODH? Studies of the C/Fe-SP by spectroscopy have provided some hints. The protein-bound methylcobamide was found to be the "base-off" conformation (Ragsdale et al., 1987), unusual for model cobalamins at pH values above 3. Analysis of the reactivity of model compounds indicates that the base-off form is more susceptible to nucleophilic attack (Hogenkamp et al., 1987; Kräutler, 1987) and less susceptible to homolytic fission (Chen and Chance, 1990; Grate and Schrauzer, 1979; Pratt, 1982; Schrauzer and Grate, 1981). We recently discovered by EXAFS spectroscopy that the C/Fe-SP maintains the coordination state of the corrinoid in the unusual four-coordinate and five-coordinate forms in the Co^{2+} and methyl-Co^{3+} states, respectively. In addition, pre-steady-state kinetic analysis of the transfer of the methyl group from methylated C/Fe-SP to CODH have established that Co^{1+} is the product of the methyl transfer (Kumar and Ragsdale, unpublished). Therefore, as we postulated earlier (Harder et al., 1989; Lu et al., 1990), it is likely that transfer of the methyl group to CODH occurs by a nucleophilic attack on methylcobamide by CODH. This postulate is consistent with the stereochemical results. One can envision the methyl group of five-coordinate "base-off" methylcobamide, protected against radical attack and activated for nucleophilic displacement, to be poised for the methylation of CODH.

In the first experiments on the methylation of CODH, the enzyme was methylated with $^{14}CH_3I$ or $^{14}CH_3$-C/Fe-SP and a stable $^{14}CH_3$-CODH complex was isolated, hydrolyzed, and identified as S-methyl-cysteine (Pezacka and Wood 1988). In addition, the $^{14}CH_3$-CODH complex was stable enough to locate the radioactivity in the smaller of the two (71 kDa) subunits by SDS gel electrophoresis (Pezacka and Wood, 1988). When $^{14}CH_3$-CODH was treated with CO and CoA, the methyl-cysteine peak was absent from the profile of the radioactive amino acids after hydrolysis (Pezacka and Wood, 1988). These combined results indicated that the mechanism of methyl transfer involves attack of a cysteinyl residue of CODH on the methyl group of the methylated C/Fe-SP, forming an intermediate methyl-cysteine product, and it was further postulated that S-acetylcysteine was a likely intermediate in the final steps of the synthesis (Pezacka and Wood, 1988). Formation of acetyl-cysteine was invoked to satisfy the known stereochemical constraints on acetyl-CoA synthesis from CH_3-H_4folate (Pezacka and Wood, 1988) because migration of the methyl from cysteine to a metal center would result in an additional inversion of configuration.

We have obtained strong evidence that the site on CODH for binding the methyl group is a low-potential metal center rather than a cysteine thiol and that formation of methyl-CODH requires a prior reduction of this metal center (Lu et al., 1990). Much of this evidence relies on investigation of the rate of methyl transfer as a function of reduction potential by a method called controlled potential enzymology which has been discussed in recent articles (Harder et al., 1989; Lu et al., 1990) and in a recent review (Ragsdale 1991). Summarizing several experiments, *first,* methylation of CODH occurs over a wide pH range from 5.5

to 7.5 (Pezacka and Wood, 1988). This pH profile is not consistent with the normal profile for cysteine ionization and could reflect changes in coordination state around a metal center, although there is precedent for cysteine residues in proteins with low pKa values. *Second,* the dependence on redox potential of the methylation of CODH (Lu et al., 1990) is unlike that expected for reduction of a dithiol to an active thiolate because most disulfide/dithiol redox reactions occur in the range of -200 to -300 mV, and the redox center which is methylated on CODH has a redox potential of < -400 mV. *Third,* methylation of CODH at low potentials is not inhibited by thiol reagents (Lu et al., 1990). *Fourth,* based on model chemistry, formation of an acetyl intermediate would be expected to involve a methyl migration (CO insertion), a reaction which has been well documented for organometallic chemistry (Kuhlmann and Alexander, 1980; Moloy and Marks, 1984). Formation of acetyl-cysteine from methyl-cysteine would involve unprecedented chemistry in which there would be a carbonyl insertion into a very stable methyl-thioether bond.

Resolution of the discrepancy between a metal center and cysteine as the site of methylation has not been achieved yet. One possible explanation for the identification of methyl-cysteine from the amino acid hydrolysate (Pezacka and Wood, 1988) is that the methyl group may migrate from a metal center to a cysteinyl residue as an artifact. It is important to spectroscopically identify the methyl-CODH species and to determine the catalytic competence of any spectroscopically observed methylated intermediate.

In summary, as indicated in Figure 3.2, methylation of CODH is proposed to occur by a reductively activated methyl displacement reaction. This reaction would then most likely involve cleavage of the methylmetal bond on the C/Fe-SP to form a methyl-metal bond on CODH. A likely candidate for the methyl acceptor site is the Ni-Fe-C center (discussed earlier) because it appears to be the binding site for the CO which will become the carbonyl of acetyl-CoA and it has a low midpoint reduction potential.

3.4.2   Carbonylation of CODH

When CODH binds CO, an intermediate is formed which is the precursor of the carbonyl group of acetyl-CoA. This intermediate has been probed by spectroscopic and electrochemical methods which indicate that the site of CO binding is the same M shown in Figure 3.2 to be involved in binding the methyl and acetyl groups by CODH. The CO adduct of CODH is paramagnetic and EPR active; the magnetic and spectroscopic properties have been fully described above in Section 3.3.2.4(b). Also discussed earlier are IR experiments which demonstrate that CO binds as a terminal carbonyl. As described earlier, our working model of the CO adduct at the active site metal center, M is [Ni-X-

$Fe_4S_4]-C≡O$, where X is an unknown bridge between the Ni site and a [4Fe-4S] cluster.

3.4.3   Methyl Migration to Form Acetyl-CODH

After carbonylation, there appears to be a methyl migration to form acetyl-CODH. Evidence for this is several-fold. First, acetate is the exclusive product of the reaction of CH_3-CODH with CO in the absence of CoA (Lu and Ragsdale, 1991). Second, acetate has been identified as a product of the reaction of CH_3-C/Fe-SP with CO and either purified CODH [Ragsdale, unpublished results described in (Pezacka and Wood, 1986)] or a protein fraction containing CODH (Hu et al., 1984). Third, acetate has been detected as a minor product of the synthesis of acetyl-CoA from CH_3-H_4folate, CO, and CoA (Hu et al., 1982). Finally, the rate of the CoA/acetyl-CoA exchange reaction is ~200-fold faster than that of the CO/acetyl-CoA exchange reaction (discussed later), indicating that cleavage of the C-S bond of acetyl-CoA, forming the acetyl intermediate, occurs faster than formation/breakage of the C-C (acetyl) bond of acetyl-CoA (Lu and Ragsdale, 1991; Ramer et al., 1989). So far, the acetyl-metal intermediate has not been spectroscopically detected.

3.4.4   Binding of CoA to CODH

Quantitation of the strength of interaction between CoA and CODH has yielded quite disparate values. Kinetic evidence for binding of CoA to CODH was provided by the demonstration that CODH per se catalyzes an exchange reaction between CO and acetyl-CoA and that CoA is a potent inhibitor of this reaction (Ragsdale and Wood, 1985) with a K_i of 7 μM (Raybuck et al., 1988). The K_m for acetyl-CoA in the CO/acetyl-CoA exchange reaction is 600 μM (Raybuck et al., 1988). In the exchange reaction between labeled CoA and acetyl-CoA, the K_m for CoA is 50 μM (Ramer et al., 1989). Equilibrium dialysis experiments indicate that there are two binding sites for CoA with approximately equal occupancy at a low affinity (K_d = 50 μM) and a high affinity (K_d = 2.6 mM) site (Lu and Ragsdale, 1991). In the synthesis of acetyl-CoA from CoA, CO, and either methyl iodide (Lu et al., 1990) or CH_3-H_4folate (Pezacka and Wood, 1984a, 1984b), K_m values for CoA of ~4.7 mM have been measured.

Based on EPR spectroscopy, CoA appears to bind to CODH near the Ni-Fe-CO complex as described earlier. An arginine residue was proposed to be involved in CoA binding possibly via a charge–charge interaction with the pyrophosphate bridge of CoA because the exchange reaction between CO and acetyl-CoA is inhibited by glyoxals (Ragsdale and Wood, 1985). Quenching of the fluorescence of a tryptophan residue on CODH on addition of CoA indicates that a tryptophan residue is near the CoA binding site (Shanmugasundaram et al., 1988a, 1988b).

Further evidence for a CoA-tryptophan interaction has been obtained by a chemical modification approach. When exposed tryptophan residues are oxidized by N-bromosuccinimide (NBS), the CO/acetyl-CoA exchange reaction is significantly inhibited, CoA protects against this inhibition, and concurrent with the inhibition, the NiFeC EPR signal is modified (Shanmugasundaram et al., 1988a). In order to determine which tryptophan is at the CoA binding site, CODH was labeled with 2,4-dinitrophenylsulfenyl chloride (DNPS-Cl), a reagent specific for tryptophan residues, treated with trypsin, and the labeled tryptic peptides were sequenced (Morton et al., 1991). One peptide containing a *trp*-his-thr-gly-gln-*arg* sequence was protected by CoA against DNPS-Cl modification. This peptide was located in the 78-kDa subunit of CODH based on location of the peptide sequence within the deduced amino acid sequence (Morton et al., 1991). The presence of an arginine residue close to the protected tryptophan suggests that it is the residue which is modified by glyoxal (Ragsdale and Wood, 1985).

Further experiments are required to explain the apparent existence of two CoA binding sites per $\alpha\beta$ form of CODH and to resolve the discrepancies between the binding constants described earlier. One possibility is that there are two forms of CODH, a high-activity form that binds CoA tightly and a form that binds CoA weakly. Alternately, a single form of CODH could have a high-affinity catalytic CoA binding site and an additional site that could play a structural or regulatory role. EPR spectroscopic studies (discussed later) also provide evidence for different conformations of CODH (Lindahl et al. 1990a; Ragsdale et al., 1985).

3.4.5   Condensation of the Bound Acetyl and Thiol Groups to Form Acetyl-CoA

If one considers the final steps in the synthesis of acetyl-CoA as reversible reactions, it becomes apparent that binding of CoA and synthesis of the thioester bond can be probed by following an isotope exchange reaction between CoA and the CoA moiety of acetyl-CoA (Lu and Ragsdale, 1991; Pezacka and Wood, 1986; Ramer et al., 1989). Pezacka and Wood demonstrated that CODH catalyzes a slow exchange reaction between [^{3}H]CoA and acetyl-CoA in the presence of a reducing system consisting of either reduced ferredoxin or NADH and disulfide reductase (Pezacka and Wood, 1986). A considerably higher rate of CoA/acetyl-CoA exchange in the presence of CO was reported (Ramer et al., 1989) with rates $\sim$ sevenfold faster than the CO/acetyl-CoA exchange. An even faster rate was measured when the reaction was performed at low redox potentials (Lu and Ragsdale, 1991). The specific activity of the exchange between CoA and the CoA moiety of acetyl-CoA at -520 mV and pH 7.0 at 55°C is 200 μmol/min^{-1}/mg^{-1} (500 s^{-1}), $\sim$ 14,000-fold higher than the values measured earlier (Pezacka and Wood, 1986), presumably due to the stimulatory effect of the low redox potential. The rate at -575 mV is 2000-fold faster than that at -80 mV (Lu and Ragsdale, 1991). Treatment of the kinetic data by a Nernst analysis revealed that a group on CODH with a midpoint reduction potential of ≤ -486 mV must be activated by a one-electron reduction in order to catalyze this exchange reaction (Lu and Ragsdale, 1991). The CoA/acetyl-CoA exchange reaction is inhibited by CO, N_2O, CO_2, CN^-, and CoA analogues (Lu and Ragsdale, 1991).

As shown in Figure 3.7, the redox sensitive CoA/acetyl-CoA exchange reaction is proposed to involve (i) interaction of acetyl-CoA with CODH, (ii) cleavage of the C–S bond forming an acetyl-CODH-CoA intermediate, (iii) release and rebinding of CoA, and (iv) resynthesis of acetyl-CoA (Lu and Ragsdale, 1991). The redox step apparently involves a one-electron reductive activation of a metal center on CODH with a midpoint potential ≤ -486 mV. Interaction of acetyl-CoA with CODH occurs through arginine and tryptophan residues (described earlier). We showed that binding of CoA to CODH is unaffected by the redox potential and presume that acetyl-CoA binding also is redox insensitive. Therefore, if steps (i) and (iii) are not affected by the redox potential, either cleavage

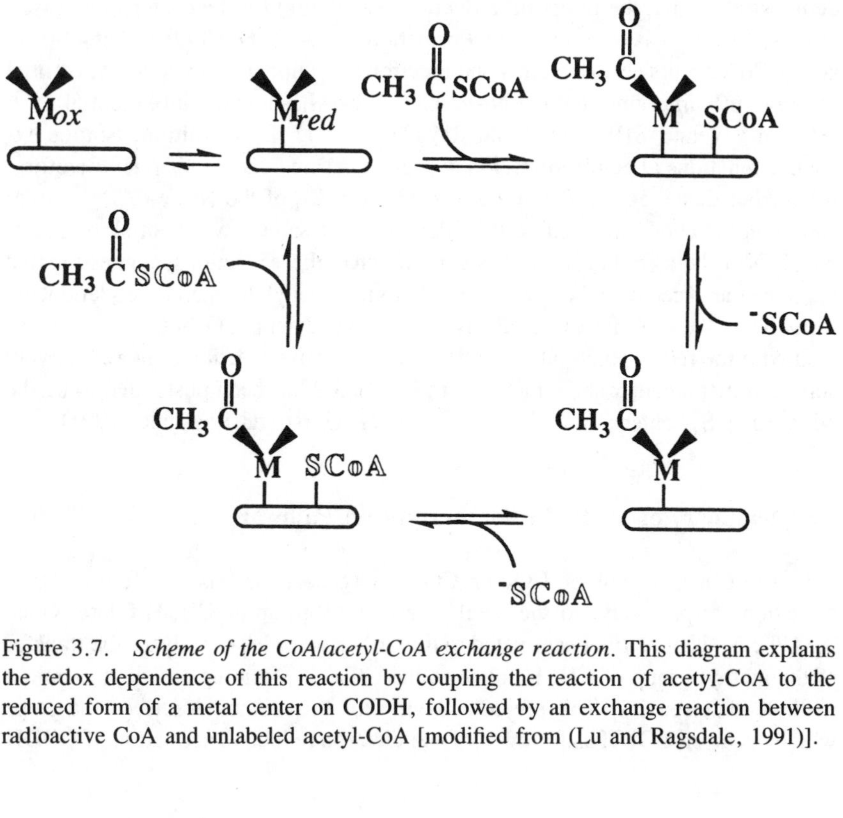

Figure 3.7. *Scheme of the CoA/acetyl-CoA exchange reaction.* This diagram explains the redox dependence of this reaction by coupling the reaction of acetyl-CoA to the reduced form of a metal center on CODH, followed by an exchange reaction between radioactive CoA and unlabeled acetyl-CoA [modified from (Lu and Ragsdale, 1991)].

of the C–S bond of acetyl-CoA, formation of the acetyl-CODH intermediate, or resynthesis of acetyl-CoA is assumed to be the redox sensitive step. Our data would be consistent with reaction of the reduced metal center with acetyl-CoA to cleave the C–S bond and form an acetyl-metal intermediate (Lu and Ragsdale, 1991).

3.4.6 Evidence for the Assembly of Acetyl-CoA at a Single Metal Center on CODH

Based on combined results obtained from electrochemical, spectroscopic, and kinetic studies, a single metal center on CODH is postulated to require reductive activation to form "M_{red}" which is the site of assembly of the methyl, carbonyl, and acetyl moieties of acetyl-CoA. Based on the above considerations, a metal center on CODH with a reduction potential of ≤ -486 mV is reductively activated by one electron before it can form acetyl-CODH. Methylation of CODH required reductive activation of a center with an $E'_0 \leq -450$ mV. The rate of reduction of CO_2 to CO and formation of the EPR signal of the Ni-Fe-C complex from CO_2 exhibit apparent midpoint potentials of ~ -430 mV (Lindahl et al., 1990b), demonstrating that the midpoint reduction potential(s) for the center(s) involved in these reactions is ≤ -430 mV. Formation of the Ni-Fe-CO EPR signal from acetyl-CoA occurs according to a one-electron reduction with a midpoint potential of -541 mV implying that the Ni-Fe-CO species has a midpoint potential of ≤ -541 mV versus SHE (Gorst and Ragsdale, 1991). By definition, isotope exchange reactions at equilibrium do not yield a net free energy change; therefore, we predict that -541 mV may be near the true E'_0 of the Ni-Fe-C$^{ox/red}$ couple. Based on the combined results, it is likely that the same low-potential metal site on CODH is being reductively activated in each of these reactions, suggesting that the same metal center is the site of methylation, carbonylation, and acetylation. As formation of the Ni-Fe-CO EPR signal associated with CO binding correlates well with the redox potential dependence of the above reactions, the reductively activated metal center shown as "M" in Figure 3.2 has been postulated to be the Ni-X-Fe$_{3-4}$-S$_4$ center (Lu and Ragsdale, 1991; Gorst and Ragsdale, 1991).

3.5   Summary of the Reductive Acetyl-CoA Pathway

In summary, 1 mol of CO_2 or CO undergoes a net six- or four-electron reduction, respectively, to the level of a methyl group as CH_3-H_4folate. Once the C/Fe-SP is reductively activated to the Co^{1+} state, MeTr catalyzes the transfer of the methyl group of CH_3-H_4folate to the C/Fe-SP, forming an enzyme-bound methylcobalt species. The prejudiced view of this author describes the steps which occur on CODH as follows. A one-electron reductive activation of CODH precedes a series of reactions involved in the assembly of acetyl-CoA. Formation of a methyl-metal species and a carbonylation step precede the methyl migration to form an acetyl-metal intermediate. After CoA binds, a thiolysis of the acetyl intermediate by CoA yields acetyl-CoA. Insight into the chemistry involved in acetyl-CoA formation would be benefited by the study of biomimetic models including defined nickel- and iron-containing complexes which are analogous to the proposed intermediates. Stable and structurally defined methyl-Ni^{2+}, Ni^{1+}-CO, and acetyl-Ni complexes have been synthesized which can undergo reaction with a thiol to form an acetylthioester (Stavropoulos et al., 1990). However, as described above, we now know that iron plays more than a passive role in catalysis since it is the site of CO binding during acetyl-CoA synthesis. Further study of these complexes is expected to enhance our understanding of the way that nature makes acetyl-CoA.

3.6 Involvement of CODH in the Heterotrophic Growth of Acetogens

So far this discussion has focused on growth of acetogens on CO or H_2/CO_2. However, CODH and the acetyl-CoA pathway also is involved in growth on glucose. Five decades ago, when *C. thermoaceticum* was first isolated, it was found to ferment glucose to approximately 3 mol of acetate (Fontaine et al., 1942). Acetate synthesis occurs by oxidation of glucose to pyruvate by the Embden–Meyerhof pathway, and three acetates are formed from the two pyruvates. Thus, it was recognized that, instead of liberation of the carboxyl groups of the two pyruvates to 2 mol of CO_2, one of the carboxyl groups must be converted to the methyl and one to the carboxyl group of acetate. The accepted mechanism is that one pyruvate reacts with pyruvate ferredoxin oxidoreductase (PFOR, denoted E-TPP), ferredoxin, and CoA to yield CO_2, acetyl-CoA, and reduced ferredoxin (reaction 3.6). CO_2 is reduced to CH_3-H_4folate via the folate enzymes described earlier (reaction 3.7), and the methyl group is transferred to the C/Fe-SP to form the methylated C/Fe-SP (reaction 3.8). Reaction of "M" on CODH with the second mole of pyruvate, the methylated C/Fe-SP, and CoA generates two additional moles of acetyl-CoA (reaction 3.9).

$$CH_3\text{-}CO\text{-}COOH + Fd_{ox} + HSCoA \rightarrow \tag{3.6}$$
$$CH_3\text{-}CO\text{-}SCoA + Fd_{red} + CO_2 + 2H^+$$

$$7H^+ + CO_2 + H_4\text{folate} + 6e^- \rightarrow CH_3\text{-}H_4\text{folate} + 2H_2O \tag{3.7}$$

$$CH_3\text{-}H_4\text{folate} + C/Fe\text{-}SP \rightarrow H_4\text{folate} + CH_3\text{-}C/Fe\text{-}SP \tag{3.8}$$

$$CH_3\text{-}CO\text{-}COO^- + CH_3\text{-}C/Fe\text{-}SP + 2\,HSCoA \rightarrow \qquad (3.9)$$
$$2CH_3\text{-}CO\text{-}SCoA + C/Fe\text{-}SP + H_2O$$

Five protein fractions were isolated which were sufficient to catalyze the synthesis of acetate from $CH_3\text{-}H_4$folate, pyruvate, and CoA (reactions 3.8 and 3.9) and MeTr, PFOR, and Fd were purified to homogeneity (Drake et al., 1981a). Other components have been later shown to include CODH and the C/Fe-SP. Earlier, it had been shown that the carboxyl of pyruvate does not equilibrate with free CO_2 during its conversion to the carbonyl group of acetyl-CoA (Schulman et al., 1973). Thus, it is likely that an enzyme-bound one-carbon intermediate on CODH is formed via direct transfer from PFOR. Pezacka and Wood (1984a) demonstrated the formation of a $^{14}C_1$ complex on CODH on reaction of CODH with PFOR and [1-^{14}C]pyruvate, and, though it was formed in very low yield, this CODH-C_1 formed the C-1 of acetate. The existence of this CODH-C_1 complex apparently obviates the need to postulate a bound C_1 on PFOR. Thus, synthesis of acetyl-CoA from $CH_3\text{-}H_4$folate, pyruvate, and CoA involves the combined actions of CODH, MeTr, and C/Fe-SP, Fd, and PFOR (E-TPP). How PFOR is able to direct the carboxyls of the two pyruvates in different directions is not understood. It is likely that CODH is involved in this partitioning.

3.7 CODH in Other Anaerobic Autotrophs and Organisms That Utilize Acetate

Though the acetyl-CoA pathway has been most thoroughly studied in the acetogenic bacteria, other anaerobes form cell carbon from CO_2 by this pathway. CO_2 is converted to $CH_3\text{-}H_4$MPT via a series of H_4MPT-dependent enzymes analogous to the H_4folate-dependent enzymes of acetogens (reaction 3.10) (Rouviére and Wolfe, 1988). Then $CH_3\text{-}H_4$MPT, CO, and CoA are thought to be converted to acetyl-CoA via a methyltransferase, a corrinoid protein, and CODH (reaction 3.11).

$$CO_2 + 7H^+ + 6e^- + H_4MPT \rightarrow CH_3\text{-}H_4MPT + 2H_2O \qquad (3.10)$$

$$CH_3\text{-}H_4MPT + CO + CoAS^- \rightarrow H_4MPT + CH_3\text{-}CO\text{-}SCoA \qquad (3.11)$$

The most convincing evidence that the acetyl-CoA pathway is involved in autotrophic growth of methanogens was provided by a genetic approach. Loss of the ability of *Methanococcus maripaludis* to grow autotrophically was associated with mutation of CODH, and restoration of autotrophic capacity was provided by recovery of CODH activity (Lapado and Whitman, 1990).

Enzymes of the acetyl-CoA pathway apparently mediate acetyl-CoA oxidation by some sulfate-reducing and methanogenic bacteria [reviewed in (Thauer et al., 1989)]. In the catabolism of acetate by methanogens, the acetate is first converted to acetyl-CoA, and CODH is thought to be involved in the disassembly of acetyl-CoA to form an acetyl-CODH-SCoA intermediate, followed by cleavage of the C–C bond of the acetyl intermediate to generate CH_3-CODH-CO [see (Ferry, 1992) for a recent review]. CO of the CH_3-CODH-CO complex is thought to be oxidized and CO_2 released, and the methyl group is reduced to methane via an enzyme-bound methylcobamide, CH_3-H_4MPT, and methyl-SCoM intermediates. The final step in methane formation is the two-electron reduction of CH_3-SCoM, catalyzed by CH_3-SCoM reductase, a protein containing a nickel porphyrin [see (Bobik et al., 1987; Ellermann et al., 1988) and references therein]. Recently, two corrinoid proteins have been identified which become methylated with the methyl group of acetate, one of which is methylated during methanogenesis and demethylated when methane formation ceases (Xianjun and Krzycki, 1991). In addition, a complex containing CODH and a corrinoid protein has been isolated from *M. thermophila* and the two components resolved (Jones et al., 1987). The complex catalyzes the synthesis of acetyl-CoA from methyl iodide, CO, and CoA (Abbanat and Ferry, 1990), implying the formation of a methylcobamide intermediate, transfer of the methyl group to CODH, followed by assembly of acetyl-CoA by CODH as in the acetogenic system. In a recent study, we were surprised to find a high degree of similarity in EPR spectral morphology and electrochemical properties between the metal sites of the C/Fe-S component and the Ni/Fe-S component of the CODH enzyme complex from *M thermophila* and the C/Fe-SP and CODH from acetogenic bacteria (Jablonski et al., in press; Lu et al., submitted a) given the unusual nature of the centers and the evolutionary distance between *Archae* and *Bacteria*. The similarity between the metal centers and the reactions catalyzed by both methanogenic and acetogenic proteins is of evolutionary significance and provides biochemical evidence that the acetyl-CoA pathway of autotrophic growth as well as the pathway of acetoclastic methanogenesis may have evolved via a divergent pathway from a progenitor of *Archae* and *Bacteria*. Could the acetyl-CoA pathway have been present in an organism which predates the evolutionary branching of *Archae* and *Bacteria*?

3.8 Perspectives and Future Studies

This review has focused on the enzymology of the acetyl-CoA pathway in acetogenic bacteria. A number of questions remain. What reaction(s) limit and regulate the rate of acetyl-CoA formation? This question can be answered by the use of steady-state and pre-steady-state kinetics. One of the most challenging series of experiments remaining is to characterize those intermediates in the

CODH cycle which have been only indirectly identified, for example the methylmetal species. It is important to measure the rates of formation and decay of these intermediates and to establish if the rates are fast enough to account for the overall synthesis of acetyl-CoA. Since Co binds to an iron site, it is important to determine if the methyl group binds the nickel or iron components of the Ni-Fe-S complex. Solution of this problem awaits the successful direct observation of the metal-carbon bond. Recent success in the observation of the carbonyl stretch of the metalcarbonyl of CODH offers hope for use of vibrational spectroscopy to detect the metal-carbon bond. Until then, one can only obtain indirect evidence.

It is important to continue to elucidate the structure of the metal center(s) involved in assembly of the acetyl group of acetyl-CoA. A working model for a $Ni-Fe_{3-4}-S_4$ center has been proposed (discussed earlier). Comparison of spectroscopic data on CODH with similar analyses of model complexes (Ciurli et al., 1990; Conover et al., 1990) are anticipated to help rule out incorrect structures and better understand the role of this unique metal center in the synthesis of acetyl-CoA. Generation of biomimetic models of the active-site centers of CODH (Stavropoulos et al., 1990) and description of their mechanism of performing one-carbon transformations are important in defining the enzymatic mechanism.

How is the Ni-Fe-S center in CODH assembled? So far, this is an uninvestigated area. Assembly of the metal centers could involve a complex series of reactions. For example, formation of the active-site $Mo-Fe_{6-8}S_{6-9}$ center of nitrogenase requires the products of at least six genes in *Klebsiella pneumoniae* besides the structural genes for the two subunits of the protein [see (Downs et al., 1990) and references therein]. Assembly of the nickel site in urease requires at least four genes in addition to the structural genes, all of which are part of a single gene cluster (Lee et al., 1992). In hydrogenase, generation of the nickel active site also appears to require additional gene(s) because a mutation of a nonstructural gene can be complemented by high levels of nickel (Waugh and Boxer, 1986). Recently, the groups of Ludden and Robers have showed that generation of the nickel active site of the CODH from *R. rubrum* also requires genes in addition to CODH structural genes (Kerby et al., 1992).

Development of a genetic system in the acetogens is an important research objective. Cloning of genes into *E. coli* will probably be of limited use due to the complexity of the active site of CODH; however, recent success in reconstitution of the C/Fe-SP from overexpressed genes in *E. coli* (Lu et al., 1993) makes one wonder if a similar approach could be used with CODH. Recently, conjugative transposons have been introduced into *A. woodii* (Stratz et al., 1990). This could prove to be an important method for introducing and deleting genes. Development of a genetic system would allow one to investigate the effects of alteration of specific genes in a systematic way and allow one to perform site-directed mutagenesis.

Acknowledgments

I thank the postdoctors Wei-Ping Lu, Manoj Kumar, Jennifer Runquist, and Scott Harder, students Carol Gorst, Shaying Zhao, Iunia Schiau, David Roberts, and Jacqueline Roberts, and technician Lifei Liu who have studied the pathway of acetogenesis in my laboratory. I am grateful for the opportunity to work with many collaborators, Chao-Lin Fan, Joshua Telser, Brian Hoffman, Paul Lindahl, Eckard Münck, Vincent Huynh, Mike Wirz, Mark Chance, Di Qiu, Tom Spiro, and Michael Johnson, who have performed spectroscopic studies of the metal centers of CODH and the C/Fe-SP. I am thankful to the Department of Energy and the National Institutes of Health for their continued support of our research on the mechanistic enzymology of acetogenesis.

References

Abbanat, D. R., and J. G. Ferry. 1990. Synthesis of acetyl-coenzyme A by carbon monoxide dehydrogenase complex from acetate-grown *Methanosarcina thermophila*. *J. Bacteriol.* **172**:7145–7150.

Adams, M. W. W. 1987. The mechanisms of H_2 activation and CO binding by hydrogenase I and hydrogenase II of *Clostridium pasteurianum*. *J. Biol. Chem.* **262**:15054–15061.

Anderson, M. E., V. J. DeRose, B. M. Hoffmann, and P. A. Lindahl 1993. Identification of a cyanide binding site in CO dehydrogenase from *Clostridium thermoaceticum* using EPR and ENDOR spectroscopies. *J. Am. Chem. Soc.* **115**:12204–12205.

Banerjee, R. V., S. R. Harder, S. W. Ragsdale, and R. G. Matthews. 1990. Mechanism of reductive activation of cobalamin-dependent methionine synthase: an electron paramagnetic resonance spectroelectrochemical study. *Biochemistry* **29**:1129–1135.

Barker, H. A., and M. D. Kamen. 1945. Carbon dioxide utilization in the synthesis of acetic acid by *Clostridium thermoaceticum*. *Proc. Natl. Acad. Sci. USA* **31**:219–225.

Bastian, N. R., G. Diekert, E. G. Niederhoffer, B.-K. Teo, C. P. Walsh, and W. H. Orme-Johnson. 1988. Nickel and iron EXAFS of carbon monoxide dehydrogenase from *Clostridium thermoaceticum*. *J. Am. Chem. Soc.* **110**:5581–5582.

Bobik, T. A., K. D. Olson, K. M. Noll, and R. S. Wolfe. 1987. Evidence that the heterodisulfide of coenzyme M and 7-mercaptoheptanoylthreonine phosphate is a product of the methylreductase reaction in Methanobacterium. *Biochem. Biopohys. Res. Commun.* **149**:455–460.

Breznak, J. A., and M. D. Kane. 1990. Microbial H_2/CO_2 acetogenesis in animal guts: nature and nutritional significance. *FEMS Microbiol. Rev.* **87**:309–314.

Breznak, J. A., and J. M. Switzer. 1986. Acetate synthesis from H_2 plus CO_2 by termite gut microbes. *Appl. Environ. Microbiol.* **52**:623–630.

Chen, E., and M. R. Chance. 1990. Nanosecond transient absorption spectroscopy of coenzyme B_{12}. Quantum yields and spectral dynamics. *J. Biol. Chem.* **265**:12987–12990.

Ciurli, S., S.-B. Yu, R. H. Holm, K. K. P. Srivastava, and E. Münck. 1990. Synthetic

NiFe$_3$Q$_4$ cubane-type clusters (S = 3/2) by reductive rearrangement of linear [Fe$_3$Q$_4$ (SEt)$_4$]$^{3-}$ (Q = S, Se). *J. Am. Chem. Soc.* **112**:8169.

Ciurli, S., P. K. Ross, M. J. Scott, S.-B. Yu, and R. H. Holm. 1992. Synthetic nickel-containing heterometal cubane-type clusters with NiFe$_3$Q$_4$ cores (Q = S, Se). *J. Am. Chem. Soc.* **114**:5415–5423.

Colby, J., H. Dalton, and R. Whittenbury. 1979. Biological and biochemical aspects of microbial growth in C1 compounds. *Annu. Rev. Microbiol.* **33**:481–517.

Conover, R. C., J.-B. Park, M. W. Adams, and M. K. Johnson. 1990. The formation and properties of a NiFe3S4 cluster in *Pyrococcus furiosus* ferredoxin. *J. Am. Chem. Soc.* **112**:4562–4564.

Cramer, S. P., M. K. Eidsness, W.-H. Pan, T. A. Morton, S. W. Ragsdale, D. V. DerVartanian, and L.G. Ljungdahl. 1987. X-ray absorption spectroscopic evidence for a unique nickel site in *Clostridium thermoaceticum* carbon monoxide dehydrogenase. *Inorg. Chem.* **26**:2477–2479.

Daniel, S. L., T. Hsu, S. I. Dean, and H. L. Drake. 1990. Characterization of the H$_2$- and CO-dependent chemolithotrophic potentials of the acetogens *Clostridium thermoaceticum* and *Acetogenium kivui*. *J. Bacteriol.* **172**:4464–4471.

Diekert, G. B. and R. K. Thauer. 1978. Carbon monoxide oxidation by *Clostridium thermoaceticum* and *Clostridium formicoaceticum*. *J. Bacteriol.* **136**:597–606.

Diekert, G. B., E. G. Graf, and R. K. Thauer. 1979. Nickel requirement for carbon monoxide dehydrogenase formation in *Clostridium thermoaceticum*. *Arch. Microbiol.* **122**:117–120.

Diekert, G., and R. K. Thauer. 1980. The effect of nickel on carbon monoxide dehydrogenase formation in *Clostridium thermoaceticum* and *Clostridium formicoaceticum*. *FEMS Microbiol. Lett.* **7**:187–189.

Diekert, G., and M. Ritter. 1983. Purification of the nickel protein carbon monoxide dehydrogenase of *Clostridium thermoaceticum*. *FEBS Lett.* **151**:41–44.

Downs, D. M., P. W. Ludden, and V. K. Shah. 1990. Synthesis of the iron-molybdenum cofactor of nitrogenase is inhibited by a low-molecular-weight metabolite of *Klebsiella pneumoniae*. *J. Bacteriol.* **172**:6084–6089.

Drake, H. L., S.-I. Hu, and H. G. Wood. 1980. Purification of carbon monoxide dehydrogenase, a nickel enzyme from *Clostridium thermoaceticum*. *J. Biol. Chem.* **255**:7174–7180.

Drake, H. L., S.-I. Hu, and H. G. Wood. 1981a. Purification of five components from *Clostridium thermoacticum* which catalyze synthesis of acetate from pyruvate and methyltetrahydrofolate. Properties of phosphotransacetylase. *J. Biol. Chem.* **256**:11137–11144.

Drake, H. L., S.-I. Hu, and H. G. Wood. 1981b. The synthesis of acetate from carbon monoxide plus methyltetrahydrofolate and the involvement of the nickel enzyme, CO dehydrogenase. Abstr. K42, p. 144. *Abstr. Ann. Meet. Am. Soc. Microbiol.* 1981.

Ellermann, J., R. Hedderich, R. Böcher, and R. K. Thauer. 1988. The final step in methane formation. Investigations with highly purified methyl-CoM reductase (component C) from *Methanobacterium thermoautotrophicum* (strain Marburg). *Eur. J. Biochem.* **172**:669–677.

Emptage, M. H., J.-L. Dreyer, M. C. Kennedy, and H. Beinert. 1983. Optical and EPR characterization of different species of active and inactive aconitase *J. Biol. Chem.* **258**: 11106–11111.

Ensign, S. A., M. R. Hyman, and P. W. Ludden. 1989. Nickel-specific, slow binding inhibition of carbon monoxide dehydrogenase from *Rhodospirillum rubrum* by cyanide. *Biochemistry* **28**:4973–4979.

Ensign, S. A., and P. W. Ludden. 1991. Characterization of the CO oxidation/H_2 evolution system of *Rhodospirillum rubrum*. Role of a 22 kDa iron-sulfur protein in mediating electron transfer between carbon monoxide dehydrogenase and hydrogenase. *J. Biol. Chem.* **266**:18395–18403.

Fan, C., C. M. Gorst, S. W. Ragsdale, and B. M. Hoffman. 1991. Characterization of the Ni-Fe-C complex formed by reaction of carbon monoxide with the carbon monoxide dehydrogenase from *Clostridium thermoaceticum* by Q-band ENDOR. *Biochemistry* **30**:431–435.

Ferry, J. G. 1992. Methane from acetate. *J. Bacteriol.* **174**:5489–5495.

Fontaine, F. E., W. H. Peterson, E. McCoy, M. J. Johnson, and G. T. Ritter. 1942. A new type of glucose fermentation by *Clostridium thermoaceticum*. *J. Bacteriol.* **43**:701–715.

Forster, D. J. 1976). On the mechanism of a rhodium-complex-catalyzed carbonylation of methanol to acetic acid. *J. Am. Chem. Soc.* **98**: 846–848.

Forster, D. J. 1979. Mechanistic pathways in the catalytic carbonylation of methanol by rhodium and iridium complexes. *Adv. Organomet. Chem.* **17**:255–266.

Fuchs, G. 1986. CO_2 fixation in acetogenic bacteria: variations on a theme. *FEMS Microbiol. Rev.* **39**:181–213.

Gorst, C. M., and S. W. Ragsdale. 1991. Characterization of a Ni-Fe-CO complex of CO dehydrogenase as a catalytically competent intermediate in the pathway of acetyl-CoA synthesis. *J. Biol. Chem.* **266**:20687–20693.

Grate, J. H., and G. N. Schrauzer. 1979. Sterically induced, spontaneously dealkylation of secondary alkylcobalamins due to axial base coordination and conformational changes of the corrin ligand. *J. Am. Chem. Soc.* **101**:4601–4611.

Harder, S. A., W.-P. Lu, B. F. Feinberg, and S. W. Ragsdale. 1989. Spectroelectrochemical studies of the corrinoid/iron-sulfur protein from *Clostridium thermoaceticum*. *Biochemistry* **28**:9080–9087.

Hogenkamp, H. P. C., G. T. Bratt, and A. T. Kotchevar. 1987. Reaction of alkylcobalamins with thiols. *Biochemistry* **26**:4723–4727.

Hu, S.-I., H. L. Drake, and H. G. Wood. 1982. Synthesis of acetyl coenzyme A from carbon monoxide, methyltetrahydrofolate, and coenzyme A by enzymes from *Clostridium thermoaceticum*. *J. Bacteriol.* **149**:440–448.

Hu, S.-I., E. Pezacka, and H. G. Wood. 1984. Acetate synthesis from carbon monoxide by *Clostridium thermoaceticum*. Purification of the corrinoid protein. *J. Biol. Chem.* **259**:8892–8897.

Hyman, M. R., S. A. Ensign, D. J. Arp, and P. W. Ludden. 1989. Carbonyl sulfide inhibition of CO dehydrogenase from *Rhodospirillum rubrum*. *Biochemistry* **28**:6821–6826.

Jablonski, P. E., W.-P. Lu, S. W. Ragsdale, and J. G. Ferry. 1993. Characterization of the metal centers of the corrinoid/iron-sulfur component of the CO dehydrogenase enzyme complex from *Methanosarcina thermophila* by EPR spectroscopy and spectroelectrochemistry. *J. Biol. Chem.* **268**:325–329.

Jetten, M. S. M., A. J. Pierik, and W. R. Hagen. 1991. EPR characterization of a high-spin system in carbon monoxide dehydrogenase from *Methanothrix soehngenii*. *Eur. J. Biochem.* **202**:1291–1297.

Jones, C. W. 1982. Bacterial respiration and photosynthesis. In: Aspects of Microbiology, vol. 5, J. A. Cole and C. J. Knowles (eds.), p. 41. Thomas Nelson and Sons, Ltd., Hong Kong.

Jones, W. J., D. P. Nagle, Jr., and W. B. Whitman. 1987. Methanogens and the diversity of archaebacteria. *Microbiol. Rev.* **51**:135.

Kerby, R., and J. G. Zeikus. 1983. Growth of *Clostridium thermoaceticum* on H_2/CO_2 or CO as energy source. *Curr. Microbiol.* **8**:27–30.

Kerby, R. L., S. S. Hong, S. A. Ensign, L. J. Coppoc, P. W. Ludden, and G. P. Roberts. 1992. Genetic and physiological characterization of the *Rhodospirillum rubrum* carbon monoxide dehydrogenase system. *J. Bacteriol.* **174**:5284–5294.

Kräutler, B. 1987. Thermodynamic trans-effects of the nucleotide base in the B_{12} coenzymes. *Helv. Chim. Acta* **70**:1268–1278.

Krzycki, J. A., L. E. Mortenson, and R. C. Prince. 1989. Paramagnetic centers of carbon monoxide dehydrogenase from acetoclastic *Methanosarcina barkeri*. *J. Biol. Chem.* **264**:7217–7221.

Kuhlmann, E. J., and J. J. Alexander. 1980. Carbon monoxide insertion into transition metal-carbon sigma-bonds. *Coord. Chem. Rev.* **33**:195–225.

Kumar, M., S. W. Ragsdale, 1992. Characterization of the CO binding site of carbon monoxide dehydrogenase from *Clostridium thermoaceticum* by infrared spectroscopy. *J. Am. Chem. Soc.* **114**:8713–8715.

Kumar, M., W.-P. Lu, L. Liu, and S. W. Ragsdale. 1993. Kinetic evidence that CO dehydrogenase catalyzes the oxidation of CO and the synthesis of acetyl-CoA at separate metal centers. *J. Am. Chem. Soc.* **115**:11646–11647.

Lajoie, S. F., S. Bank, T. L. Miller, and M. J. Wolin. 1988. Acetate production from hydrogen and [^{13}C]carbon dioxide by the microflora of human feces. *Appl. Environ. Microbiol.* **54**:2723–2727.

Lapado, J., and W. B. Whitman. 1990. Method for isolation of auxotrophs in the methanogenic archaebacteria: role of the acetyl-CoA pathway of autotrophic CO_2 fixation in *Methanococcus maripauludis*. *Proc. Natl. Acad. Sci. USA* **87**:5598–5602.

Lebertz, H., H. Simon, L. F. Courtney, S. J. Benkovic, L. D. Zydowsky, K. Lee and H. G. Floss. 1987. Stereochemistry of acetic acid formation from 5-methyl-tetrahydrofolate by *Clostridium thermoaceticum*. *J. Am. Chem. Soc.* **109**:3173–3174.

Lee, M. H., S. B. Mulrooney, M. J. Renner, Y. Markowicz, and R. P. Hausinger. 1992. *Klebsiella aerogenes* urease gene cluster: sequences of ureD and demonstration that four accessory genes (*ureD, ureE, ureF,* and *ureG*) are involved in nickel metallocenter biosynthesis. *J. Bacteriol.* **174**:4324–4330.

Lenn, N. D., M. T. Stankovich, H.-W. Liu. 1990. Regulation of the redox potential of general acyl-CoA dehydrogenase by substrate binding. *Biochemistry* **29**:3709.

Lieber, C. M., and N. S. Lewis. 1984. Catalytic reduction of CO_2 at carbon electrodes modified with cobalt phthalocyanine. *J. Am. Chem. Soc.* **106**:5033–5034.

Lindahl, P. A., E. Münck, and S. W. Ragsdale. 1990a. CO dehydrogenase from *Clostridium thermoaceticum:* EPR and electrochemical studies in CO_2 and argon atmospheres. *J. Biol. Chem.* **265**:3873–3879.

Lindahl, P. A., S. W. Ragsdale, and E. Münck. 1990b. Mössbauer studies of CO dehydrogenase from *Clostridium thermoaceticum*. *J. Biol. Chem.* **265**:3880–3888.

Lu, W.-P., S. R. Harder, and S. W. Ragsdale. 1990. Controlled potential enzymology of methyl transfer reactions involved in acetyl-CoA synthesis by CO dehydrogenase and the corrinoid/iron-sulfur protein. *J. Biol. Chem.* **265**:3124–3133.

Lu, W.-P., P. E. Jablonski, J. G. Ferry, and S. W. Ragsdale. 1994. Characterization of the metal centers of the Ni/Fe-S component of the CO dehydrogenase complex from *Methanosarcina thermophila* by EPR spectroscopy and spectroelectrochemistry. *J. Biol. Chem.* (in press).

Lu, W.-P., I. Schiau, J. R. Cunningham, and S. W. Ragsdale. 1993. Sequence and expression of the gene encoding the corrinoid/iron-sulfur protein from *Clostridium thermoaceticum* and reconstitution of the recombinant protein to full activity. *J. Biol. Chem.* **268**:5605–5614.

Lu, W. P., and S. W. Ragsdale. 1991. Reductive activation of the coenzyme A/acetyl-CoA isotopic exchange reaction catalyzed by carbon monoxide dehydrogenase from *Clostridium thermoaceticum* and its inhibition by nitrous oxide and carbon monoxide. *J. Biol. Chem.* **266**: 3554–3564.

Meyer, O., and H. G. Schlegel. 1983. Biology of aerobic carbon monoxide-oxidizing bacteria. *Ann. Rev. Microbiol.* **37**:277–310.

Moloy, K. G., and T. J. Marks. 1984. Carbon monoxide activation by organoactinides. A comparative synthetic, thermodynamic, kinetic, and mechanistic investigation of migratory CO insertion into actinide-carbon and actinide-hydrogen bonds to yield η2-acyls and η2-formyls. *J. Am. Chem. Soc.* **106**:7051–7064.

Morton, T., J. A. Runquist, S. W. Ragsdale, T. Shanmugasundaram, H. G. Wood, and L. G. Ljungdahl. 1991. The primary structure of the subunits of CO dehydrogenase/acetyl-CoA synthase from *Clostridium thermoaceticum*. *J. Biol. Chem.* **266**:23824–23838.

Muhoberac, B. B., D. C. Wharton, L. M. Babcock, P. C. Harrington, and R. G. Wilkins.

1980. EPR spectroscopy of semi-methemerythin. *Biochim. Biophys Acta* **626**:337–347.

Pezacka, E., and H. G. Wood. 1984a. Role of carbon monoxide dehydrogenase in the autotrophic pathway used by acetogenic bacteria. *Proc. Natl. Acad. Sci. USA* **81**:6261–6265.

Pezacka, E., and H. G. Wood. 1984b. The synthesis of acetyl-CoA by *Clostridium thermoaceticum* from carbon dioxide, hydrogen, coenzyme A and methyltetrahydrofolate. *Arch. Microbiol.* **137**:63–69.

Pezacka, E., and H. G. Wood. 1986. The autotrophic pathway of acetogenic bacteria. Role of CO dehydrogenase disulfide reductase. *J. Biol. Chem.* **261**:1609–1615.

Pezacka, E., and H. G. Wood. 1988. Acetyl-CoA pathway of autotrophic growth. Identification of the methyl-binding site of the CO dehydrogenase. *J. Biol. Chem.* **263**:16000–16006.

Pratt, J. M. 1982. Coordination chemistry of the B_{12} dependent isomerase reactions. In: *Vitamin B_{12}*, D. Dolphin (ed.), pp. 325–392. Wiley, New York.

Qiu, D., M. Kumar, S. W. Ragsdale, and T. Spiro. 1994. Nature's carbonylation catalyst: Raman spectroscopic evidence that CO binds to iron, not nickel, in carbon monoxide dehydrogenase. *Science* (in press).

Ragsdale, S. W. 1991. Enzymology of the acetyl-CoA pathway of autotrophic CO_2 fixation. *CRC Crit. Rev. Biochem. Mol. Biol.* **26**:261–300.

Ragsdale, S. W., J. E. Clark, L. G. Ljungdahl, L. L. Lundie, and H. L. Drake. 1983a. Properties of purified carbon monoxide dehydrogenase from *Clostridium thermoaceticum*, a nickel, iron-sulfur protein. *J. Biol. Chem.* **258**:2364–2369.

Ragsdale, S. W., L. G. Ljungdahl, and D. V. DerVartanian. 1983b. [13]C and [61]Ni isotope substitution confirm the presence of a nickel(III)-carbon species in acetogenic CO dehydrogenases. *Biochem. Biophys. Res. Commun.* **115**:658–665.

Ragsdale, S. W., L. G. Ljungdahl, and D. V. DerVartanian. 1983c. Isolation of the carbon monoxide dehydrogenase from *Acetobacterium woodii* and comparison of its properties with those of the *Clostridium thermoaceticum* enzyme. *J. Bacteriol.* **155**:1224.

Ragsdale, S. W., and H. G. Wood. 1985. Acetate biosynthesis by acetogenic bacteria: evidence that carbon monoxide dehydrogenase is the condensing enzyme that catalyzes the final steps of the synthesis. *J. Biol. Chem.* **260**:3970–3977.

Ragsdale, S. W., H. G. Wood, and W. E. Antholine. 1985. Evidence that an iron-nickel-carbon complex is formed by reaction of CO with the CO dehydrogenase from *Clostridium thermoaceticum*. *Proc. Natl. Acad. Sci. USA* **82**:6811–6814.

Ragsdale, S. W., P. A. Lindahl, and E. Münck. 1987. Mössbauer, EPR, and optical studies of the corrinoid/Fe-S protein involved in the synthesis of acetyl-CoA by *Clostridium thermoaceticum*. *J. Biol. Chem.* **262**:14289–14297.

Ragsdale, S. W., H. G. Wood, L. G. Ljungdahl, T. Morton, and D. V. DerVartanian. 1988. Nickel in CO dehydrogenase. In: *Bioinorganic Chemistry of Nickel*, J. R. Lancaster (ed.), pp. 311–332. VCH Publishers, New York.

Ramer, S. E., S. A. Raybuck, W. H. Orme-Johnson, and C. T. Walsh. 1989. Kinetic characterization of the [3'-^{32}P]coenzyme A/acetyl coenzyme A exchange catalyzed by a three-subunit form of the carbon monoxide dehydrogenase/acetyl-CoA synthase from *Clostridium thermoaceticum*. *Biochemistry* **28**:4675–4680.

Raybuck, S. A., N. R. Bastian, W. H. Orme-Johnson, and C. T. Walsh. 1988. Kinetic characterization of the carbon monoxide-acetyl-CoA (carbonyl group) exchange activity of the acetyl-CoA synthesizing CO dehydrogenase from *Clostridium thermoaceticum*. *Biochemistry* **27**:7698–7702.

Raybuck, S. A., S. E. Ramer, D. R. Abbanat, J. W. Peters, W. H. Orme-Johnson, J. G. Ferry, and C. T. Walsh. 1991. Demonstration of carbon–carbon bond cleavage of acetyl-CoA by using isotopic exchange catalyzed by the CO dehydrogenase complex from acetate-grown *Methanosarcina thermophila*. *J. Bacteriol.* **173**:929–932.

Riester, J., W. G. Zumft, and P. M. H. Kronek. 1989. Nitrous oxide reductase from *Pseudomonas stutzeri*. Redox properties and spectroscopic characterization of different forms of the multicopper enzyme. *Eur. J. Biochem.* **178**:751–762.

Roberts, J. R., W.-P. Lu, and S. W. Ragsdale. 1992. Acetyl-CoA synthesis from methyl-tetrahydrofolate, CO and CoA by enzymes purified from *Clostridium thermoaceticum:* Attainment of *in vivo* rates and identification of rate limiting steps. *J. Bacteriol.* **174**:4667–4676.

Rouviére, P. E., and R. S. Wolfe. 1988. Novel biochemistry of methanogenesis. *J. Biol. Chem.* **263**:7913–7916.

Schrauzer, G. N., and J. H. Grate. 1981. Sterically induced, spontaneous Co–C bond homolysis and β-elimination reaction of primary and secondary organocobalamins. *J. Am. Chem. Soc.* **103**:541–546.

Schulman, M., R. K. Ghambeer, L. G. Ljungdahl, and H. G. Wood. 1973. Total synthesis of acetate from CO_2. VII. Evidence with *Clostridium thermoaceticum* that the carboxyl of acetate is derived from the carboxyl of pyruvate by transcarboxylation and not by fixation of CO_2. *J. Biol. Chem.* **248**:6255–6261.

Shanmugasundaram, T., G. K. Kumar, and H. G. Wood. 1988a. Involvement of trypto-phan residues at the coenzyme A binding site of carbon monoxide dehydrogenase from *Clostridium thermoaceticum*. *Biochemistry* **27**:6499–6503.

Shanmugasundaram, T., S. W. Ragsdale, and H. G. Wood. 1988b. Role of carbon monoxide dehydrogenase in acetate synthesis by the acetogenic bacterium, *Acetobacterium woodii*. *BioFactors* **1**:147–152.

Shanmugasundaram, T., H. G. Wood. 1992. Interaction of ferredoxin with carbon monoxide dehydrogenase from *Clostridium thermoaceticum*. *J. Biol. Chem.* **267**:897–900.

Shin, W., P. R. Stafford, and P. A. Lindahl. 1992. Redox titrations of carbon monoxide dehydrogenase from *Clostridium thermoaceticum*. *Biochemistry* **31**:6003–6011.

Stavropoulos, P., M. Carrié, M. C. Muetterties, and R. H. Holm. 1990. Reaction sequence related to that of carbon monoxide dehydrogenase (acetyl-CoA synthase): thioester formation mediated at structurally defined nickel centers. *J. Am. Chem. Soc.* **112**:5385–5387.

Stephens, P. J., M.-C. McKenna, S. A. Ensign, D. Bonam, and P. W. Ludden. 1989. Identification of a Ni- and Fe-containing cluster in *Rhodospirillum rubrum* carbon monoxide dehydrogenase. *J. Biol. Chem.* **264**:16347–16350.

Stratz, M., G. Gottschalk, and P. Durre. 1990. Transfer and expression of the tetracycline resistance transposon TN 925 in *Acetobacterium woodii*. *FEMS Microbiol. Lett.* **68**:171–176.

Tan, G. O., S. A. Ensign, S. Ciurli, M. J. Scott, B. Hedman, R. H. Holm, and P. W. Luden. 1992. On the structure of the nickel/iron/sulfur center of the carbon monoxide dehydrogenase from *Rhodospirillum rubrum:* An x-ray absorption spectroscopy study. *Proc. Natl. Acad. Sci. USA* **89**:4427–4431.

Terlesky, K. C., M. J. Barber, D. J. Aceti, and J. G. Ferry. 1987. EPR properties of the Ni-Fe-C center in an enzyme complex with carbon monoxide dehydrogenase activity from acetate-grown *Methanosarcina thermophila*. Evidence that acetyl-CoA is a physiological substrate. *J. Biol. Chem.* **262**:15392–15395.

Tesler, J., M. J. Benecky, M. W. W. Adams, L. E. Mortenson, and B. M. Hoffman. 1987. EPR and electron nuclear double resonance investigation of oxidized hydrogenase II (uptake) from *Clostridium pasteurianum* W5. Effects of carbon monoxide binding. *J. Biol. Chem.* **263**:6589–6594.

Thauer, R. K., D. Möller-Zinkhan, and A. M. Spormann. 1989. Biochemistry of acetate catabolism in anaerobic chemotrophic bacteria. *Annu. Rev. Microbiol.* **43**:43–67.

Uffen, R. L. 1983. Metabolism of carbon monoxide by *Rhodopseudomonas gelatinosa:* cell growth and properties of the oxidative system. *J. Bacteriol.* **155**:956–965.

Ulman, M., B. Aurian-Blajeni, and M. Halmann. 1987. Fuel from CO_2: An electrochemical study. *Chemtech.* April: 235–239.

Waugh, R., and D. H. Boxer. 1986. Pleitropic hydrogenase mutants of *Escherichia coli* K-12: growth in the presence of nickel can restore hydrogenase activity. *Biochimie* **68**:157–166.

Weast, R. C. 1985. *Handbook of Chemistry and Physics,* 67th ed., p. D-58. CRC Press, Cleveland, OH.

Xianjun, C., and J. A. Krzycki. 1991. Acetate-dependent methylation of two corrinoid proteins by acetate in extracts of *Methanosarcina barkeri*. *J. Bacteriol.* **173**:5439–5448.

Yagi, T. 1959. Enzymic oxidation of carbon monoxide. *Biochim. Biophys. Acta* **30**:194–195.

Yamamoto, I., T. Saiki, S.-M. Liu, and L. G. Ljungdahl. 1983. Purification and properties of NADP-dependent formate dehydrogenase from *Clostridium thermoaceticum,* a tungsten-selenium-iron protein. *J. Biol. Chem.* **258**:1826–1832.

Zavarzin, G. A., and A. N. Nozhevnikova. 1977. Aerobic carboxydobacteria. *Microbiol. Ecol.* **3**:305–326.

4

The Sodium Ion Cycle in Acetogenic and Methanogenic Bacteria: Generation and Utilization of a Primary Electrochemical Sodium Ion Gradient

Volker Müller and Gerhard Gottschalk

4.1 Introduction

Acetogenic bacteria are strictly anaerobic bacteria which use a wide variety of organic substrates for growth and acetate formation. Glucose, for example, is metabolized via the Embden–Meyerhoff pathway to pyruvate which is then split in the phosphoroclastic reaction to CO_2 and acetyl CoA; the latter is converted to acetate via acetyl phosphate. The term "acetogen" is used for a variety of different organisms which produce acetate as a major fermentation end product. Homoacetogenic bacteria (which are referred to in the following as "acetogens") differ from other organisms in that the CO_2 formed is not an end product; the reducing equivalents obtained during glycolysis are used to reduce the 2 mol of CO_2 produced in the phosphoroclastic reaction to acetate via the acetyl-CoA pathway. Therefore, homoacetogenic bacteria convert 1 mol of glucose to 3 mol of acetate via glycolysis and acetyl-CoA pathway. As the pathway is well established, one can calculate a net formation of 4 mol of ATP per mole of glucose fermented by the mechanism of substrate-level phosphorylation (SLP) (Fuchs, 1986; Ljungdahl, 1986; Wood et al., 1986), which is sufficient to ensure the energy supply of the cells during heterotrophic acetogenesis.

However, the energy balance is different when acetogens rely solely on the acetyl-CoA pathway for energy conservation, i.e., during autotrophic growth. There is no net ATP formation by substrate-level phosphorylation during this reaction sequence: one ATP is produced in the acetate kinase reaction and one ATP is consumed in the formyl-tetrahydrofolate (H_4F) synthetase reaction. Because the pathway is well established and the nature of the intermediates involved is known, an additional, hypothetical site for substrate-level phosphorylation can be excluded.

But how is the acetyl-CoA pathway coupled to net ATP formation? From theoretical considerations it is concluded that ATP can only be formed by a chemiosmotic mechanism. This mechanism is based on an electrogenic transport of an ion across the membrane, resulting in the generation of an ion motive force which, in turn, drives ATP synthesis via an ATP synthase (Mitchell, 1961). To establish such a mechanism, the reaction(s) leading to the generation of the ion gradient as well as the presence of an ATP synthase has to be demonstrated.

Besides the determination of all components necessary for such a mechanism the nature of the ion translocated has to be determined. In the past few years, evidence has accumulated that the "classical chemiosmotic ion," the proton, can be substituted by sodium ions in various bioenergetic reactions in different aerobic and anaerobic bacteria (Skulachev, 1989). In recent experiments with methanogenic bacteria, which also use part of the acetyl-CoA pathway in their energy metabolism (Rouvière and Wolfe, 1988; Thauer, 1990), and with the acetogenic bacterium *Acetobacterium woodii* the important role of Na^+ in the bioenergetics of strictly anaerobic organisms using the acetyl-CoA pathway was documented (Müller et al., 1990). *Methanosarcina* strains as well as *A. woodii* couple the acetyl-CoA pathway with the formation of primary electrogenic sodium ion gradients. Furthermore, ATP synthesis in the course of acetogenesis, as carried out by *A. woodii,* is driven by an electrochemical sodium ion gradient. These studies gave an answer to the question of energy coupling during the operation of the acetyl-CoA pathway.

In this chapter we summarize the experiments which led to the discovery of primary sodium ion gradients and describe the reactions coupled to the generation and utilization of $\Delta\mu Na^+$ during the course of acetogenesis or methanogenesis. For a discussion of the proton energetics in acetogens the reader is referred to Chapter 2 and to a recent review by Hugenholtz and Ljungdahl (1990).

4.2 The acetyl-CoA Pathway: A Comparison of Methanogenesis and Acetogenesis

Methanogenic, acetogenic, and some sulfidogenic bacteria derive their energy, exclusively or in part, by the operation of the acetyl-CoA pathway [for recent reviews on the biochemistry of the acetyl-CoA pathway in methanogens and acetogens see (Keltjens et al., 1990; Ragsdale, 1992)]. The pathway is reversible; the reductive pathway is used by acetogenic and methanogenic bacteria to couple energy conservation to the formation of acetate and methane, respectively. Some methanogenic and sulfidogenic bacteria employ this pathway in the oxidative direction and conserve energy during the formation of CO_2 + CH_4 or CO_2 + H_2S.

The first reactions of acetogenesis and methanogenesis from H_2 + CO_2 which lead from CO_2 to the formation of a methylated intermediate are very similar and involve the binding of the C_1 moiety to a carrier and the subsequent reduction of the carrier-bound intermediate (Fig. 4.1). Whereas the C_1-carrier in acetogens is H_4F (Fig. 4.2), methanogens use tetrahydromethanopterin (H_4MPT) (Fig. 4.3), a compound structurally and functionally analogous to H_4F. Another difference

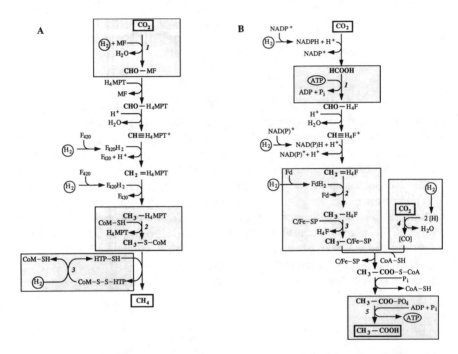

Figure 4.1. Pathway of methanogenesis (A) and acetogenesis (B) from H_2 + CO_2. MF, methanofuran; H_4MPT, tetrahydromethanopterin; F_{420}, oxidized form of coenzyme F_{420}; $F_{420}H_2$, reduced form of coenzyme F_{420}; HTP-SH, 7-mercaptoheptanoylthreonine phosphate; CoM-SH, coenzyme M (2-mercaptoethanesulfonate); CoM-S-S-HTP, heterodisulfide of coenzyme M and 7-mercaptoheptanoylthreonine phosphate; H_4F, tetrahydrofolate; Fd, oxidized form of ferredoxin; FdH₂, reduced form of ferredoxin; C/Fe-SP, corrinoid-iron-sulfur protein; [CO], enzyme-bound CO. Enzymes in (A): *1*, formyl methanofuran dehydrogenase; *2*, methyl-H_4MPT : CoM-SH methyltransferase; *3*, H_2 : heterodisulfide oxidoreductase. Enzymes in (B): *1*, formyl tetrahydrofolate synthetase; *2*, methylene tetrahydrofolate reductase; *3*, methyl-H_4F : corrinoid-iron-sulfur protein methyltransferase; *4*, CO dehydrogenase complex; *5*, acetate kinase. Bioenergetic relevant reactions are boxed in gray.

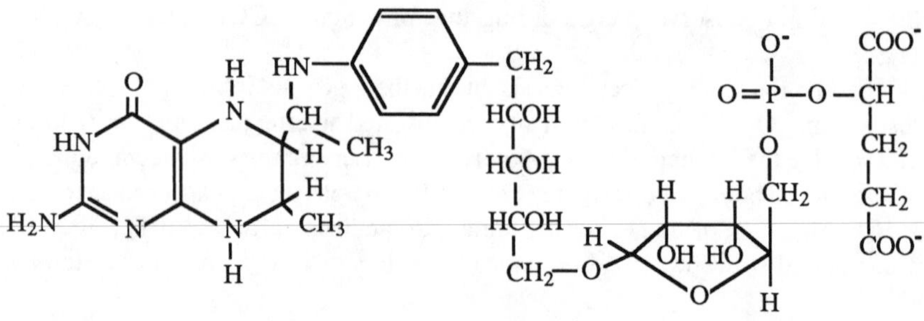

5, 6, 7, 8-tetrahydromethanopterin

Figure 4.2.   Structure of tetrahydromethanopterin (H_4MPT).

is the activation of CO_2: in acetogens, CO_2 is first reduced to formate by the action of formate dehydrogenase, and in the next step, formyl-H_4F is formed from H_4F and formate by formyl-H_4F synthetase; this endergonic reaction ($\Delta G'_0$ = +21.5 kJ/mol with NADH + H^+ as reductant) is driven by ATP hydrolysis. In the case of methanogenesis, CO_2 is bound to the coenzyme methanofuran (MF) and subsequently reduced to give rise to formyl-MF; free formate is not an intermediate. This endergonic reaction ($\Delta G'_0$ = +16 kJ/mol) is most probably driven by an electrochemical ion gradient across the membrane (see Section 4.6.1). The reaction is catalyzed by the formyl-MF dehydrogenase and is not yet fully understood; the enzyme contains a pterin cofactor as well as iron-sulfur centers. The formyl group is then transferred from formyl-MF to H_4MPT giving rise to formyl-H_4MPT. Following the formation of a methenyl intermediate by

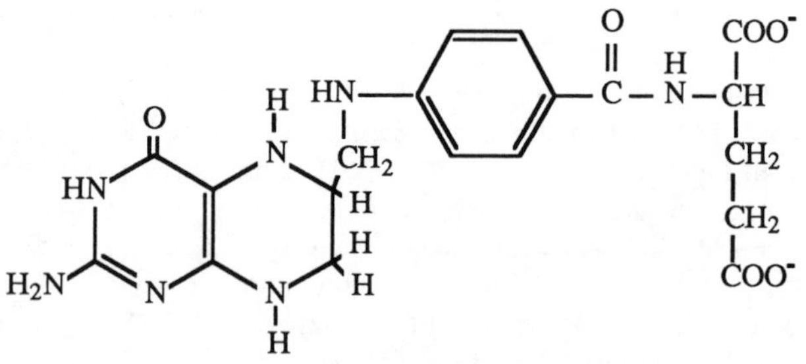

5, 6, 7, 8-tetrahydrofolate

Figure 4.3.   Structure of tetrahydrofolate (H_4F).

the cyclohydrolase, the methenyl group is reduced via a methylene intermediate to the methyl level by the corresponding dehydrogenase and reductase.

The methyl intermediate is the branching point of the different routes leading to methane and acetate. In the case of acetogenesis, the methyl group of methyl-H_4F is first transferred via a corrinoid/iron-sulfur protein (C/Fe-SP) to the enzyme CO dehydrogenase, on which it is condensed with a bound CO, derived from a second molecule of CO_2, to give rise to acetyl-CoA; the latter is further converted to acetate via acetyl phosphate.

As in acetogenic bacteria, the acetyl-CoA formed by this pathway is the starting material for further biosynthetic reactions also in methanogenic bacteria. However, during methanogenesis, the methyl group of methyl-H_4MPT does not undergo a condensation reaction but is transferred to coenzyme M; the so-formed methyl-CoM reacts with 7-mercaptoheptanoylthreonine phosphate (HTP-SH) to yield methane and a disulfide of CoM-SH and HTP-SH (heterodisulfide; CoM-S-S-HTP) which is subsequently reduced by H_2 or $F_{420}H_2$; this reduction is coupled to the generation of a primary proton gradient across the membrane which, in turn, drives ATP synthesis (Deppenmeier et al., 1990, 1991).

The common sequence of reactions of the acetyl-CoA pathway, which is operative during both acetogenesis and methanogenesis, is the conversion of a carrier-bound formyl group to a methyl intermediate and, as we shall see later, this central sequence is coupled to energy transduction. For a long time there was no experimental evidence for the involvement of any of these enzymes in energy transduction. However, thermodynamic considerations bring into focus the methylene-H_4F ($-H_4MPT$) reductase reaction [Eqs. (4.1) and (4.3)] and the methyltransferase reaction [Eqs. (4.2) and (4.4)]: whereas the reactions leading to the methylene intermediate are near equilibrium, the standard free-energy change associated with the reductase and transferase reaction are sufficiently large to transduce energy [$\Delta G'_0$ values used throughout the text are from Keltjens and van der Drift (1986) and Fuchs (1986) unless stated otherwise]:

$$\text{methylene-}H_4F + 2\ [H] \rightarrow \text{methyl-}H_4F \tag{4.1}$$
$$\Delta G'_0 = -22 \text{ kJ/mol with NADH} + H^+ \text{ as reductant}$$
(Wohlfarth and Diekert, 1991)

$$\text{methyl-}H_4F + CO + \text{CoA-SH} \rightarrow CH_3COO\text{-S-CoA} + H_4F \tag{4.2}$$
$$\Delta G'_0 = -21.8 \text{ kJ/mol}$$

$$\text{methylene-}H_4MPT + 2\ [H] \rightarrow \text{methyl-}H_4MPT \tag{4.3}$$
$$\Delta G'_0 = -5.2 \text{ kJ/mol with } F_{420}H_2 \text{ as reductant (Ma and Thauer, 1990)}$$

$$\text{methyl-}H_4MPT + \text{CoM-SM} \rightarrow \text{methyl-S-CoM} + H_4MPT \tag{4.4}$$
$$\Delta G'_0 = -29.7 \text{ kJ/mol}$$

[Note that the $\Delta G'_0$ value for Eq. (4.3) results from the experimentally determined use of $F_{420}H_2$ as a reductant; the value -19.8 kJ/mol reported earlier by several authors was based on H_2 as reductant. The reactions according to Eqs. (4.1) and (4.2) are carried out by acetogens, whereas methanogens catalyze the reactions according to Eqs. (4.3) and (4.4).]

In view of the thermodynamic data, the reductase and/or the transferase reaction were assumed to be involved in the generation of a $\Delta\bar{\mu}_{ion}$. Historically, the first experimental evidence for the involvement of this reaction sequence of the acetyl-CoA pathway in energy transduction came from a study with methanogens using *Methanosarcina barkeri* and *Methanosarcina strain* Göl as model organisms (Müller et al., 1990). As we will see, this reaction sequence is coupled to the generation of a primary electrochemical Na^+ gradient across the membrane.

4.3   Identification of a Primary Sodium Ion Pump Coupled to Methanogenesis from Formaldehyde + H_2

Growth as well as methane formation by resting cells of methanogenic bacteria from all substrates known is strictly dependent on sodium ions and, in analogy to other systems, the involvement of Na^+ in energy transduction (Perski et al., 1981, 1982) was assumed. The elucidation of the sodium ion-dependent step in methanogenesis was very much facilitated by the use of different substrate combinations which feed the C_1 moiety into the central pathway at different redox levels. This strategy was later also used for the localization of sodium ion-dependent reactions in acetogenic bacteria.

Methanogenic bacteria are able to convert $CO_2 + H_2$, $HCHO + H_2$ or $CH_3OH + H_2$ to methane according to

$$CO_2 + 4H_2 \rightarrow CH_4 + 2H_2O, \quad \Delta G'_0 = -130.4 \text{ kJ/mol} \quad (4.5)$$
$$HCHO + 2H_2 \rightarrow CH_4 + H_2O, \Delta G'_0 = -157.3 \text{ kJ/mol} \quad (4.6)$$
$$CH_3OH + H_2 \rightarrow CH_4 + H_2O, \Delta G'_0 = -112.5 \text{ kJ/mol} \quad (4.7)$$

Formaldehyde reacts spontaneously in a nonenzymatic reaction with H_4MPT to yield methylene-H_4MPT (Escalante-Semerena et al., 1984), and methanol enters the central pathway at the level of methyl-CoM (van der Meijden et al., 1983). Therefore, by comparing the substrate combinations $CO_2 + H_2$, $HCHO + H_2$ and $CH_3OH + H_2$ one can investigate partial reaction sequences.

The study of the hypothetical involvement of Na^+ in ATP synthesis was hindered for a long time by the unknown mechanism of ATP synthesis and the components involved; the mechanism of ATP synthesis was discovered in 1984 using resting cells of *Ms. barkeri* and $CH_3OH + H_2$ as a substrate which is converted to methane according to

$$CH_3OH + CoM-SH \rightarrow CH_3-S-CoM + H_2O, \Delta G'_0 = -27.5 \text{ kJ/mol} \quad (4.8)$$

$$CH_3\text{-}S\text{-}CoM + 2\,[H] \rightarrow CH_4 + CoM\text{-}SH,\ \Delta G_0 = -85\ \text{kJ/mol} \qquad (4.9)$$

By determining the effect of different inhibitors on $\Delta\tilde{\mu}_{H^+}$ and on the intracellular ATP content as well as on methane formation, it was demonstrated that ATP is synthesized by a chemiosmotic mechanism (Blaut and Gottschalk, 1984). This was the first evidence that the last step of methanogenesis, the reduction of methyl-CoM to methane [Eq. (4.9)], is an energy-conserving reaction in the pathway of methanogenesis. Because this reaction is common to all methanogenic substrates, this mechanism is suggested to be common in all methanogens. The reduction step was later on shown to consist of two reactions:

$$CH_3\text{-}S\text{-}CoM + HTP\text{-}SH \rightarrow CoM\text{-}S\text{-}S\text{-}HTP + CH_4 \qquad (4.10)$$
$$\Delta G'_0 = -45\ \text{kJ/mol with } H_2 \text{ as reductant (Thauer, 1990)}$$

$$CoM\text{-}S\text{-}S\text{-}HTP + 2[H] \rightarrow CoM\text{-}SH + HTP\text{-}SH \qquad (4.11)$$
$$\Delta G'_0 = -40\ \text{kJ/mol with } H_2 \text{ as reductant (Thauer, 1990)}$$
$$\Delta G'_0 = -27\ \text{kJ/mol with } F_{420}H_2 \text{ as reductant (Thauer, 1990)}$$

In a first step, methyl-CoM condenses with HTP-SH as catalyzed by the methylreductase to yield methane and a mixed disulfide of CoM-SH and HTP-SH (Bobik et al., 1987; Ellermann et al., 1988) which is then reduced by the heterodisulfide reductase to the corresponding thiols (Hedderich and Thauer, 1988). The reduction of the disulfide with either H_2 or $F_{420}H_2$ is the actual proton-motive system (Deppenmeier et al., 1990, 1991).

Using this system, the possible involvement of Na^+ in ATP synthesis could be investigated. Methanogenesis from $CH_3OH + H_2$ as well as ATP synthesis coupled to this reaction does not depend on sodium ions but on protons as coupling ions (Blaut et al., 1985), excluding an involvement of Na^+ in energy conservation as coupled to the reduction of the heterodisulfide.

On the other hand, methanogenesis from $HCHO + H_2$ was strictly dependent on the sodium ion concentration, indicating that the reductive conversion of methylene-H_4MPT to the formal redox level of methanol, i.e., methyl-CoM, is the sodium ion-dependent step during this central pathway (Müller et al., 1989). The nature of the Na^+ dependence was further investigated using $^{22}Na^+$. Upon addition of HCHO to resting cells of *Ms. barkeri* incubated under an atmosphere of H_2, $^{22}Na^+$ is extruded from the cells, resulting in the generation of a ΔpNa of approximately -70 mV (Müller et al., 1989). Sodium ion extrusion is also observed in the presence of protonophores and inhibitors of the Na^+/H^+ antiporter, indicating a primary mechanism. That the primary transport is electrogenic can be concluded from the observed generation of a protonophore-resistant $\Delta\Psi$ and the formation of an inversed ΔpH (inside acidic) during Na^+ transport in the presence of a protonophore. As these effects were not observed with methanol $+ H_2$ as a substrate, the Na^+ pump has to be connected to the conversion of methylene-H_4MPT to methyl-CoM. Experiments performed with *Methanosarcina Sarkeri strain Fusaro* gave the same results, and a Na^+/HCHO stoichiometry of 3–4 mol Na^+/CH_4 was determined (Kaesler and Schönheit, 1989b).

4.3.1   The Methyl-H_4MPT : CoM-SH Methyltransferase of Methanogenic Bacteria Is a Primary Sodium Ion Pump

By comparing the substrate combinations HCHO + H_2 and CH_3OH + H_2, the primary sodium ion pump was shown to be connected to the conversion of formaldehyde, i.e., methylene-H_4MPT, to methyl-CoM. This reaction sequence is catalyzed by two enzymes: the methylene-H_4MPT reductase and the methyl-H_4MPT : CoM-SH methyltransferase. Which of these is involved in Na^+ translocation cannot be determined by the approach mentioned earlier.

The Na^+-translocating reaction was identified very recently using $^{22}Na^+$ and an everted vesicle system of *Methanosarcina* strain Göl. Washed everted vesicles of *Methanosarcina* strain Göl catalyzed the formation of methyl-CoM from HCHO + H_2 + CoM-SH; this reaction is coupled to a primary, electrogenic sodium ion transport into the lumen of the vesicles (Becher et al., 1992a). The methylene-H_4MPT reductase was ruled out as the site for Na^+ translocation: In the absence of CoM-SH and the presence of H_4MPT, the methylene-H_4MPT reductase was active, but a sodium ion transport could not be observed; addition of CoM-SH led to the formation of methyl-CoM and to a restoration of $^{22}Na^+$ transport. Interestingly, 86% of the methyl-H_4MPT : CoM-SH methyltransferase activity was found in the membrane fraction, indicating that the methyltransferase is membrane bound in strain Göl. Furthermore, the methyltransferase reaction was strictly dependent on Na^+ and, most important, was coupled to a primary and electrogenic $^{22}Na^+$ transport which is evidence for the function of the methyltransferase as a sodium ion pump. Methyl-CoM formation from methyl-H_4F as well as the simultaneous sodium ion transport was inhibited by propyliodide (a known inhibitor of corrinoid enzymes), but both activities were restored after illumination of the sample (which is known to abolish the inhibition of corrinoid enzymes caused by propyliodide). These experiments clearly demonstrate that the methyl-H_4MPT : CoM-SH methyltransferase is a membrane-bound, corrinoid-containing enzyme which functions as a primary sodium ion pump in *Methanosarcina* strain Göl. This is the first demonstration of a primary sodium ion pump coupled to a methyltransferase reaction. Because a sodium ion transport coupled to methanogenesis from HCHO + H_2 was also observed with *Ms. barkeri* (Kaesler and Schönheit, 1989b), a common mechanism for Na^+ translocation can be assumed to be present in methanogens.

Little is known about the biochemistry of the methyl-H_4MPT : CoM-SH methyltransferase. The enzyme activity was determined in cell-free extracts indirectly by analyzing the reaction sequence leading from formaldehyde to methyl-

CoM (Poirot et al., 1987) but recently by a direct method using methyl-H₄MPT as a substrate (Becher et al., 1992a; Fischer et al., 1992). The involvement of a corrinoid in the reaction was demonstrated by inhibitor studies as well as by the direct demonstration of methyl-5-hydroxybenzimidazolyl cobamide as an intermediate (Poirot et al., 1987); 5-hydroxybenzimidazolyl cobamide (B_{12}HBI) is the predominant corrinoid found in methanogens (Pol et al., 1982). The methyltransferase as measured in a cell-free extract of *Mb. thermoautotrophicum* as well as *Ms. barkeri* has to be reductively activated by H_2 + ATP and/or Ti(III) (Kengen et al., 1988, 1990; van de Wijngaard et al., 1991b). The same is true for membrane preparations that are able to transfer the methyl group from methyl-H₄MPT to CoM-SH (Becher et al., 1992b; Fischer et al., 1992). In the absence of the methyl-group acceptor CoM-SH, the membranes became methylated and subsequently demethylated on addition of CoM-SH (Fischer et al., 1992).

Recently, a membrane-bound methyl-H₄MPT : CoM-SH methyltransferase has been purified from *Mb. thermoautotrophicum* strain ΔH (Kengen et al., 1992). The enzyme consists of three subunits in a $\alpha\beta\gamma$ configuration with apparent molecular masses of 35, 33, and 31 kDa. The enzyme contains 2 nmol B_{12}HBI/ mg of protein and B_{12}HBI co-purifies with the enzyme. The 33-kDa subunit might be identical to the 33-kDa corrinoid protein of unknown function isolated previously from membranes of *Mb. thermoautotrophicum* (Schulz and Fuchs, 1986; Schulz et al., 1988). Membrane-bound corrinoid proteins were detected in a number of methanogenic bacteria (Dangel et al., 1987). Although a function could not be assigned to these proteins experimentally, it was suggested that they are involved in energy transduction (Schulz et al., 1988). Interestingly, antibodies against a membrane-bound corrinoid isolated from *Mb. thermoautotrophicum* cross-react with the methyl-H₄MPT : B_{12}HBI methyltransferase (Stupperich et al., 1990). Therefore, and under consideration of the results described earlier, it is quite likely that the membrane-bound corrinoid proteins are identical with the sodium ion-motive methyl-H₄MPT : CoM-SH methyltransferase.

This methyltransferase system represents a second site of energy conservation during methanogenesis from H_2 + CO_2 and a novel coupling site in the acetyl-CoA pathway. From the data available, a working model for this reaction has been developed (Fig. 4.4). A nucleophilic attack of the methyltransferase-bound Co(I) species on the methyl carbon at N^5 of the methyl-H₄MPT is assumed, which results in the formation of H₄MPT and a membrane-bound methyl-Co(III) species. In a next step, the methyl-Co(III) is demethylated to Co(I) by the attack of the nucleophilic sulfonium anion of CoM-SH whereby methyl-CoM is produced and the Co(I) species is regenerated. This model is analogous to the reaction mechanism of the methionine synthase which catalyzes the transfer of a methyl group from methyl-H₄F to homocysteine (Banerjee and Matthews, 1990). The need for a reductive activation is also observed for the methionine

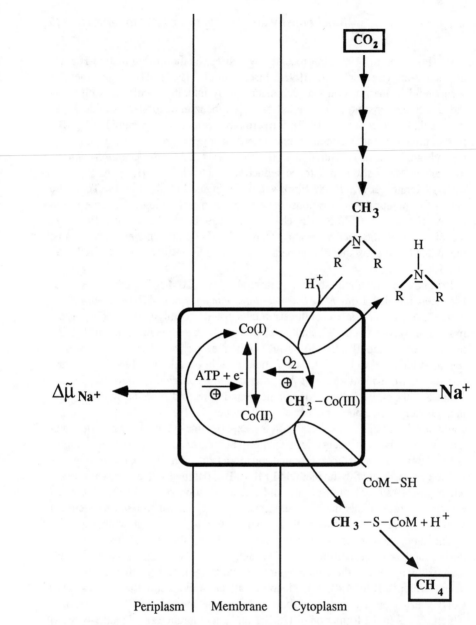

Figure 4.4. Hypothetical scheme showing the sodium ion-motive methyl-H₄MPT : CoM-SH methyltransferase involved in methanogenesis from H_2 + CO_2. In the structural formula of tetrahydromethanopterin, only the N^5 is indicated; R denotes the rest of the molecule. Co(I), Co(II), and methyl-Co(III) indicate the demethylated, partly oxidized, and methylated form of the methyl-H₄MPT : CoM-SH methyltransferase respectively. CoM-SH, 2-mercaptoethanesulfonate.

synthase and reflects the fact that an inactive Co(II) species, which is formed occasionally by autoxidation, has to be reduced to the catalytically active Co(I) species. This is done by reduction and subsequent trapping of Co(I) by methylation with S-adenosylmethionine (SAM) as a methyl-group donor (Fujii and Huennekens, 1974). The free-energy change associated with the methylation by SAM has been shown to shift the redox potential of the methionine synthase-bound Co(I)/Co(II) couple from -526 mV to -82 mV (Banerjee et al., 1990). Inactivation of the methyl-H_4MPT : CoM-SH methyltransferase can occur by traces of oxygen and/or disulfides present in the anaerobic assay mixture. An activation can then be achieved by H_2 + ATP or Ti(III). However, such a reductive activation is only catalytic and not involved in the reaction cycle. The most important question of how the methyltransfer reaction is coupled to a sodium ion extrusion remains open and has to await more data.

The discovery of a sodium ion-motive methyl-H_4MPT : CoM-SH methyltransferase has important implications also for methane formation from acetate. Acetate is converted by *Methanosarcina* and *Methanothrix* strains according to

$$CH_3COOH \rightarrow CH_4 + CO_2, \Delta G'_0 = -32.3 \text{ kJ/mol} \tag{4.12}$$

This reaction is sodium ion dependent (Perski et al., 1982) and accompanied by the generation of $\Delta\bar{\mu}_{Na^+}$ (Peinemann et al., 1988). Acetate is converted to acetyl-CoA (Fig. 4.5) (Fischer and Thauer, 1988; Grahame and Stadtman, 1987) which is then split by CO dehydrogenase to a bound CO and a methyl-group (Eikmanns and Thauer, 1984; Abbanat and Ferry, 1991; Raybuck et al., 1991). The bound CO is oxidized to CO_2; this reaction is accompanied by the generation of $\Delta\bar{\mu}_{H^+}$ (Bott et al., 1986). The reducing equivalents are transported via a ferredoxin and membrane-bound electron carriers to the heterodisulfide reductase (Fischer and Thauer, 1990; Terlesky and Ferry, 1988). The methyl group is transferred from the corrinoid-CO dehydrogenase complex to H_4MPT to give rise to methyl-H_4MPT (Terlesky et al., 1986; Fischer and Thauer, 1989; Grahame, 1991). The latter is converted via the central pathway to methane with the reducing equivalents obtained from oxidation of the bound CO. This sequence of reactions clearly demonstrates that methyl-H_4MPT is an intermediate in methanogenesis from acetate and, therefore, readily explains the sodium ion-dependence of this fermentation; as with H_2 + CO_2 as a substrate, the methyl-H_4MPT : CoM-SH methyltransferase is expected to be sodium ion dependent and to generate a primary electrochemical sodium ion gradient. This is in accordance with the observed failure of protonophores to completely dissipate the electrical potential during methanogenesis from acetate which is only observed on the combined action of protonophores and Na^+ ionophores (Peinemann et al., 1988). The $\Delta\bar{\mu}_{Na^+}$ produced is a second site of energy transduction during this fermentation.

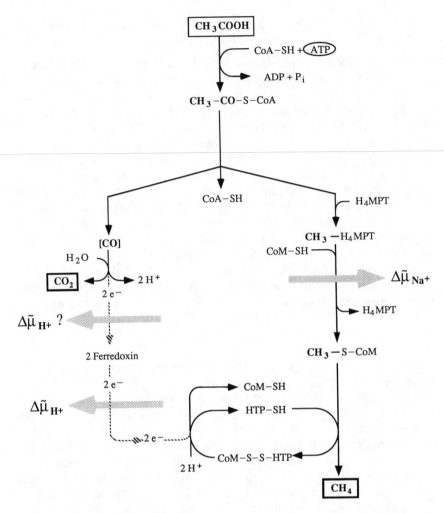

Figure 4.5. Pathway of methanogenesis from acetate. H_4MPT, tetrahydromethanopterin; HTP-SH, 7-mercaptoheptanoylthreonine phosphate; CoM-SH, coenzyme M (2-mercaptoethanesulfonate); CoM-S-S-HTP, heterodisulfide of coenzyme M and 7-mercaptoheptanoylthreonine phosphate. Carbon flow is indicated by solid lines, electron flow is indicated by dashed lines, and reactions thought to be involved in the generation of transmembrane H^+ and Na^+ gradients are indicated by horizontal arrows.

4.4 Sodium Ion Dependence of Acetate Formation in Acetogenic Bacteria

The finding that in methanogenic bacteria a reaction of the acetyl-CoA pathway, i.e., the methyltransferase reaction, is coupled to a primary sodium ion extrusion led to the idea that the corresponding reaction in acetogenic bacteria might also be involved in the generation of a transmembrane Na^+ gradient (Gottschalk, 1989). Such a system would readily explain the mystery of energy coupling during the operation of the acetyl-CoA pathway in acetogens.

Therefore, several laboratories started to investigate the effect of Na^+ on acetate formation as carried out by several acetogenic bacteria. In *A. woodii*, growth and acetate formation from fructose was not dependent on sodium ions in the medium, but growth was stimulated by the addition of Na^+. Increasing the Na^+ concentration from 0.2 to 20 mM led to a decrease in the doubling time from 7.1 to 4.2 h; correspondingly, the cell yield increased. A decrease in the Na^+ concentration was paralleled by a decrease in the acetate : fructose ratio from 2.7 to 2.1, and, simultaneously, the H_2 production was stimulated (Heise et al., 1989). These experiments indicated that the acetyl-CoA pathway is a sodium ion-dependent reaction sequence during heterotrophic acetogenesis. Indeed, growth and acetate formation from $H_2 + CO_2$ was strictly dependent on Na^+ in the medium. Similar effects were observed with other organisms.

In resting cells of *Peptostreptococcus productus* (Geerligs et al., 1989) precultivated on CO, the addition of NaCl to a final concentration of 10 mM stimulated acetate formation from $CO + CO_2$ and $H_2 + CO_2$ by a factor of 2 and 2.7, respectively. The K_m for Na^+ during the conversion of $CO + CO_2$ to acetate was determined to be approximately 2 mM. In the absence of Na^+, formate was produced indicating that the Na^+-dependent step is in between the level of formate and acetate. Similar observations were made with *Acetogenium kivui*: Whereas acetogenesis from glucose was sodium ion independent, growth and acetate formation from $H_2 + CO_2$ was dependent on Na^+, and formate was an end product in Na^+-deficient media (Yang and Drake, 1990).

To localize the sodium ion-dependent step, experiments with *A. woodii* were performed using resting cells and different substrate combinations. *A. woodii* can use $CO_2 + H_2$, $HCHO + H_2 + CO$, or $CH_3OH + CO$ (under an atmosphere of H_2 to prevent oxidation of the methanol) as substrates according to the following equations:

$$2CO_2 + 4H_2 \rightarrow CH_3COO^- + 2H_2O + H^+, \Delta G'_0 = -94.9 \, \text{kJ/mol} \quad (4.13)$$
$$HCHO + H_2 + CO \rightarrow CH_3COO^- + H^+, \quad \Delta G'_0 = -141.6 \, \text{kJ/mol} \quad (4.14)$$
$$CH_3OH + CO \rightarrow CH_3COO^- + H^+, \qquad \Delta G'_0 = -96.4 \, \text{kJ/mol} \quad 4.15)$$

[$\Delta G'_0$ values were calculated from Thauer et al. (1977)].

The organic carbon is fed at different redox levels into the central path: CO_2

enters at the most oxidized, HCHO at the methylene level, and CH_3OH at the methyl level. Because CO is directly incorporated into the carboxyl group of acetate, one can then investigate individual reactions of the central pathway by comparing acetogenesis from these substrates.

Acetogenesis from $CO_2 + H_2$ as carried out by resting cells of *A. woodii* was strictly dependent on Na^+ in the buffer. This is also true for acetogenesis from $HCHO + H_2 + CO$ but not for acetate formation from $CH_3OH + CO$ (under an H_2 atmosphere). Because formaldehyde enters the pathway at the level of methylene-H_4F, these experiments demonstrate that one of the steps leading from methylene-H_4F to the methyl level is a Na^+-dependent step of the acetyl-CoA pathway.

In the absence of H_2, methanol is disproportionated to CO_2 and acetate according to

$$4CH_3OH + 2CO_2 \rightarrow 3CH_3COOH + 2H_2O, \ \Delta G'_0 = -211.9 \ \text{kJ/mol} \ (4.16)$$

This reaction is strictly dependent on Na^+ (Heise et al., 1989). Taking also into account the sodium ion-dependent conversion of formaldehyde to the level of methanol as well as the analogy to methanogenic bacteria, a reversible sodium ion-motive system involved in both methanol oxidation and formaldehyde reduction can be envisaged in *A. woodii*.

In the absence of CO, the carboxyl group of acetate is generated from CO_2 via enzyme-bound CO through the action of CO dehydrogenase (Ragsdale and Wood, 1985). It is known for some time that the conversion of CO_2 to CO is an energy-requiring reaction in both methanogens and acetogens (Bott and Thauer, 1987; Diekert et al., 1986). The reverse reaction, i.e., the oxidation of CO, which has a physiological significance only during growth on CO, is accompanied by the generation of an ion-motive force across the membrane (Bott et al., 1986; Diekert et al., 1986; Hugenholtz et al., 1987). In methanogens, this reaction is not stimulated by Na^+ and was shown to translocate protons across the membrane, thereby generating a $\Delta\bar{\mu}_{H^+}$ (Bott and Thauer, 1989; Peinemann et al., 1988). In view of these experiments, it was of interest to study the effect of Na^+ on CO oxidation in *A. woodii*. In the presence of propyliodide (to prevent acetate formation) CO is converted according to

$$CO + H_2O \rightarrow CO_2 + H_2, \ \Delta G'_0 = -20.1 \ \text{kJ/mol} \quad (4.17)$$

Upon addition of CO to resting cells of *A. woodii*, H_2 was formed irrespective of the presence or absence of Na^+, indicating that the CO dehydrogenase reaction is sodium ion independent (Heise et al., 1989). This was supported by the following observation: Acetate formation from $H_2 + CO_2$ is Na^+ dependent. Upon addition of methanol, acetate formation from $CH_3OH + CO_2 + H_2$ becomes sodium ion independent (Heise et al., 1989). This again indicates that the reduction of CO_2 to CO does not depend on Na^+ but does not prove it because the Na^+ concentration in the buffer might have been well above the K_m value of the system for Na^+.

By the experiments described earlier, the conversion of methylene-H_4MPT to the methyl level was identified as a sodium ion-dependent step of the acetyl-CoA pathway.

4.5 Generation of a Primary Sodium Ion Gradient during Acetogenesis

After the discovery of a sodium ion-dependent acetyl-CoA pathway in acetogens, it was of interest to determine the presence and to study the generation of transmembrane sodium ion gradients in *A. woodii*. Resting cells of *A. woodii* incubated in buffer containing $^{22}Na^+$ maintained a transmembrane Na^+ gradient of -90 mV during acetogenesis from H_2 + CO_2. In the absence of a substrate, $^{22}Na^+$ equilibrated across the membrane but was actively extruded on addition of H_2 + CO_2 until a ΔpNa of -90 mV was restored (Heise et al., 1989). These experiments demonstrated the presence of a transmembrane Na^+ gradient in *A. woodii*.

Sodium ion extrusion can be achieved by primary or secondary mechanisms. A secondary sodium ion extrusion is catalyzed by Na^+/H^+ antiporters (Krulwich, 1983); such cation exchangers are widespread among bacteria and are also present in methanogens (Schönheit and Beimborn, 1985; Müller et al., 1987). A Na^+/H^+ antiporter was also reported to be present in the acetogen *Clostridium thermoaceticum* (Terraciano et al., 1987). Preliminary experiments performed in our laboratory also argue for the presence of a Na^+/H^+ antiporter in *A. woodii* (Heise, Müller, and Gottschalk, unpublished). Primary sodium ion extrusion can be coupled to decarboxylases in some anaerobic bacteria (Dimroth, 1987), NADH-ubiquinon reductases in marine and alkaliphilic bacteria (Tokuda and Unemoto, 1985) and to the methyltransferase in methanogenic bacteria (see Section 4.3.1). In order to further characterize the sodium ion dependence in *A. woodii*, it was necessary to study the mechanism of sodium ion extrusion.

Resting cells of *A. woodii* were incubated in buffer containing $^{22}Na^+$ and the protonophore SF6847, which was shown to be effective in this organism. Upon addition of H_2 + CO_2 to such cell suspensions, acetate was formed; simultaneously, a ΔpNa was generated and ATP was synthesized (Heise, Reidlinger, Müller, and Gottschalk, unpublished). These experiments revealed two features of the acetyl-CoA pathway in *A. woodii*: first, the generation of the $\Delta\bar{\mu}_{Na^+}$ is catalyzed by a primary mechanism and, second, a sodium ion-translocating ATPase should be present in this organism to couple ATP synthesis to the $\Delta\bar{\mu}_{Na^+}$. Sodium ion extrusion was further analyzed in a vesicular system of *A. woodii*.

Upon addition of methylene-H_4F + CO to an inverted vesicle system, $^{22}Na^+$ was transported into the lumen of the vesicles even in the presence of protonophores (Heise, Müller, and Gottschalk, unpublished), again indicating a primary mechanism. Although in *A. woodii* it is not known yet whether the methylene-H_4F reductase reaction or the methyl-H_4F : C/Fe-SP methyltransferase reaction is involved in Na^+ transport, it is conceivable that, in analogy to methanogens, the primary sodium ion pump is connected to the methyltransferase reaction.

4.6. Functions of the Electrochemical Sodium Ion Gradient in Methanogenic and Acetogenic Bacteria

The energy stored in electrochemical ion gradients across membranes is used for the various energy-dependent reactions of a living cell. Solute transport, motility, and ATP synthesis were believed for a long time to be driven exclusively by $\Delta\bar{\mu}_{H^+}$, but in recent years, a number of examples were identified in which the classical driving force $\Delta\bar{\mu}_{H^+}$ is substituted by $\Delta\bar{\mu}_{Na^+}$ (Dimroth, 1991). This is, at least for the reactions investigated so far, also the case for methanogenic and acetogenic bacteria (Blaut et al., 1989).

4.6.1 $\Delta\bar{\mu}_{Na^+}$-Driven Reactions in Methanogenic Bacteria

In addition to the aforementioned reactions known to be driven by $\Delta\bar{\mu}_{Na^+}$, a novel function was demonstrated in methanogenic bacteria; there, the $\Delta\bar{\mu}_{Na^+}$ is used as driving force for a reaction involved in methyl transfer which leads from methanol to methylene-H_4MPT. From the recent finding that the methyl-H_4MPT : CoM-SH methyltransferase is a primary sodium ion pump it is now clear that the reverse reaction, the transfer of the methyl group from methyl-CoM to H_4MPT, is responsible for the observed dependence of the overall reaction sequence on $\Delta\Psi$ and Na^+. This endergonic reaction ($\Delta G'_0 = +29.7$ kJ/mol) is part of the pathway of methanogenesis from methanol and is driven by $\Delta\bar{\mu}_{Na^+}$ (Müller et al., 1988). During methane formation from H_2 + CO_2, the $\Delta\bar{\mu}_{Na^+}$ is believed to be the driving force for the endergonic formation of formyl-MF (Kaesler and Schönheit, 1989b). Interestingly, the analogous reaction in acetogenic bacteria is driven by ATP hydrolysis.

Furthermore, sodium ions are involved in pH regulation under both alkaline (Müller et al., 1987) and acidic conditions (Schönheit and Beimborn, 1985). Whereas the latter function is dependent on the presence of a $\Delta\bar{\mu}_{Na^+}$ as the driving force for proton extrusion, the former uses $\Delta\bar{\mu}_{H^+}$ and results in the generation of a secondary ΔpNa. As observed with other organisms, $\Delta\bar{\mu}_{Na^+}$ is used as the driving force for amino acid uptake (Ekiel et al., 1985; Jarrell et al., 1984).

The most crucial question, if $\Delta\bar{\mu}_{Na^+}$ is used as driving force for ATP synthesis, cannot be answered unequivocally. For the marine organism *Mc. voltae*, the presence of a vanadate-sensitive, Na^+-translocating ATPase has been reported (Carper and Lancaster, 1986; Dharmavaram and Konisky, 1987). This enzyme consists of one subunit of 74 kDa; on the basis of inhibitor studies and biochemical data, it is characterized as an E_1E_2-type ATPase. The corresponding gene has been cloned and sequenced but the deduced amino sequence does not show significant homology to known ATPases (Dharmavaram et al., 1991; Dharmavaram and Konisky, 1989); this enzyme is supposed to be involved in Na^+ extrusion and not ATP synthesis (Crider et al., 1985). Whereas indications for the presence of a Na^+-translocating ATP synthase were not obtained in experiments with *Ms. barkeri* (Müller et al., 1989) and *Mb. thermoautotrophicum* (Kaesler and Schönheit, 1989a), other authors presented indications for the presence of a sodium ion-inducible, Na^+-translocating ATPase in *Mb. thermoautotrophicum* (Al-Mahrouq et al., 1986; Smigan et al., 1988). A final conclusion to this question has to await ion translocation studies with purified systems.

4.6.2 The Acetyl-CoA Pathway in A. woodii *Is Coupled to ATP Synthesis via a Chemiosmotic Mechanism*

Although direct experimental proof for ion gradient-driven ATP synthesis during the operation of the acetyl-CoA pathway in acetogens is not available, there is accumulating evidence in favor of such a mechanism. The high molar growth yields determined during heterotrophic growth of some acetogens already suggested an additional mechanism of ATP synthesis (Andreesen et al., 1973; Godley et al., 1990). The presence of soluble and membrane-bound electron carriers, the demonstration of $\Delta\Psi$ formation during oxidation of CO or H_2, as well as $\Delta\Psi$-driven uptake of amino acids and the presence of a membrane-bound ATPase [for a review see Chapter 2 and (Hugenholtz and Ljungdahl, 1990)] are indications for a chemiosmotic mechanism but do not prove it.

Further indications came from studies with *A. woodii*, where it was shown that the oxidation of CO is coupled to the formation of ATP; based on inhibitor studies, a chemiosmotic mechanism was postulated (Diekert et al., 1986). This indicated that the organism is, in principle, able to synthesize ATP by such a mechanism. However, it has to be pointed out that the energy-generating CO oxidation is of importance only in organisms growing on CO. During autotrophic acetogenesis from $H_2 + CO_2$, this reaction is $\Delta\bar{\mu}_{H^+}$ consuming, and, therefore, the energy requirement during the operation of the acetyl-CoA pathway is even more pronounced.

Hansen et al. (1988) demonstrated ATP synthesis coupled to the reduction of caffeate in *A. woodii*. ATP synthesis was sensitive to valinomycin + nigericin, but the rate of caffeate reduction was increased by these ionophores. Monensin reduced the extent of ATP formed by a factor of 1.8; unfortunately, protonophores like FCCP, TTFB, or TCS did not affect ATP synthesis or the generation of a $\Delta\Psi$. This was explained by a possible inhibition of the electrophoretic action of the protonophores by the special membrane composition of this bacterium. Nevertheless, these experiments provided substantial evidence that ATP is produced by electron transport phosphorylation although they did not reveal the nature of the coupling ion. As caffeate reduction is not involved in the acetyl-CoA pathway, this mechanism is not necessarily representative for acetogenesis from $H_2 + CO_2$ but at least shows that ATP can be synthesized by electron transport phosphorylation.

Indications for ion gradient-driven ATP synthesis coupled to the acetyl-CoA pathway came recently from studies with *A. woodii*. Resting cells of this organism coupled the conversion of $HCHO + H_2 + CO$ to acetate with the generation of a primary $\Delta\bar{\mu}_{Na^+}$. Upon addition of monensin, the ΔpNa is decomposed; simultaneously, acetogenesis is stimulated (Heise et al., 1989), a phenomenon resembling respiratory control. Furthermore, acetogenesis from $H_2 + CO_2$ is coupled to ATP synthesis; generation of $\Delta\bar{\mu}_{Na^+}$ and the synthesis of ATP is not affected by protonophores but by sodium ionophores (Heise, Müller, and Gottschalk, unpublished). These experiments can be taken as an indication of a chemiosmotic mechanism of ATP synthesis in *A. woodii* with Na^+ as the coupling ion; this is further substantiated by the presence of a membrane-bound Na^+-translocating ATPase in this organism (see Section 4.6.3). A scheme of energy conservation coupled to the acetyl CoA pathway in *A. woodii* is shown in Figure 4.6.

4.6.3   Presence of a $\Delta\bar{\mu}_{Na^+}$-Utilizing ATP Synthase in A. woodii

Due to the presence of a primary electrogenic sodium ion gradient in *A. woodii* and the inability of protonophores to prevent ATP synthesis as coupled to the acetyl-CoA pathway, the idea arose that the $\Delta\bar{\mu}_{Na^+}$ may serve as a direct energy source for ATP synthesis. To verify this hypothesis, the presence of a sodium ion-translocating ATP synthase present in *A. woodii* had to be demonstrated.

Resting cells of *A. woodii* incubated in buffer responded to an artificial ΔpH by addition of HCl with the synthesis of ATP (Heise et al., 1991). ATP synthesis was transient and dependent on the magnitude of the artificial ΔpH. Interestingly, ATP synthesis could also be achieved by applying artificial sodium ion gradients to resting cells. ΔpNa-driven ATP formation was transient; it was not inhibited by the protonophore SF6847 but by the Na^+ ionophores ETH 2120 or monensin. The extent of ATP formation was small as compared to ΔpH as the driving force but was significantly enhanced by the membrane-permeable anion tetraphenylborate, indicating the formation of a membrane potential (inside positive) on the movement of Na^+, which results in a thermodynamic back pressure and an inhibition of further ion movements.

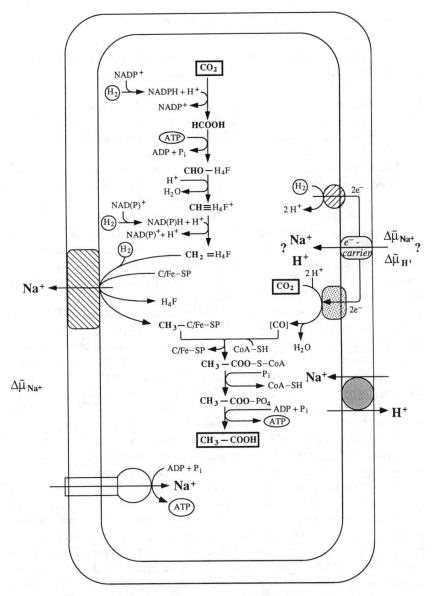

Figure 4.6. Hypothetical scheme of energy conservation and ion flow coupled to aceto-genesis from $H_2 + CO_2$ as carried out by *A. woodii*. H_4F, tetrahydrofolate; Fd, oxidized form of ferredoxin; FdH_2, reduced form of ferredoxin; C/Fe-SP, corrinoid-iron-sulfur protein; [CO], enzyme-bound CO. As it is not known whether the methylene tetrahydrofo-late reductase reaction or the methyl tetrahydrofolate : corrinoid-iron-sulfur protein methyl-transferase reaction is the actual sodium ion-translocating reaction, the overall reaction sequence leading from methylene tetrahydromethanopterin to methyl-corrinoid-iron-sulfur protein is indicated as the site for Na^+ translocation. The electron flow from H_2 to CO_2 giving rise to CO dehydrogenase-bound CO is shown to proceed via membrane-bound electron carriers and is thought to be driven by an electrochemical ion gradient ($\Delta\tilde{\mu}_{Na^+}$ or $\Delta\tilde{\mu}_{H^+}$). Also indicated is a Na^+-translocating ATP synthase and a Na^+/H^+ antiporter.

A proton-diffusion potential could also serve as driving force for ATP synthesis in *A. woodii* (Heise et al., 1991). ATP synthesis was rapid as compared to ΔpH as a driving force, transient and dependent on the magnitude of the potential applied. Furthermore, the extent and the rate of ATP synthesis was strictly dependent on the presence of Na^+ in the buffer system. Taken together, these results are in good agreement with the presence of a sodium ion-translocating ATP synthase in *A. woodii*. That this ATP synthase is of the F_1F_0-type can be concluded from the inhibition pattern: ATP synthesis was inhibited by N,N'-dicyclohexylcarbodiimide (DCCD), venturicidin, and tributyltin (Heise et al., 1991), known potent inhibitors of the F_1F_0-type ATPase but not by vanadate, an inhibitor of E_1E_2-type enzymes (Pedersen and Carafoli, 1987).

The final proof for the presence of a sodium-ion-translocating ATPase came from a study using everted vesicles in which the ATP-dependent transport of $^{22}Na^+$ across the membrane was analyzed. Upon addition of ATP to everted vesicles of *A. woodii*, $^{22}Na^+$ was transported into the lumen of the vesicles; in the presence of 0.2 mM $^{22}Na^+$, a ΔpNa of -84 mV was obtained (Heise et al., 1992). Sodium ion transport was inhibited by the sodium ionophore ETH 2120 but not by the protonophores SF6847 or TCS, indicating a primary event. Because the $^{22}Na^+$ transport was stimulated by valinomycin $+ K^+$, i.e., by a dissipation of the membrane potential, it can be concluded that this transport is electrogenic. Taken together, these experiments clearly demonstrate a primary, electrogenic ATP-dependent Na^+ transport. That this transport is catalyzed by an ATP synthase of the F_1F_0-type can again be concluded from the inhibitor studies: $^{22}Na^+$ translocation is inhibited by DCCD and tributyltin but not by vanadate.

The Na^+-translocating ATPase of *A. woodii* shares some properties with other Na^+-translocating systems. ATP-dependent sodium ion transport in *A. woodii* is inhibited by benzamil, a compound known to inhibit sodium ion channels in eukaryotes (Kleyman and Cragoe, 1988) and the sodium ion-driven flagellar motor in some alkaliphilic bacteria (Sugiyama et al., 1988). Furthermore, the ATPase is inhibited by DCCD, a compound known to modify critical carboxyl groups in ion-translocating proteins including the eukaryotic, renal Na^+/H^+ antiporter. Interestingly, the renal Na^+/H^+ antiporter is protected from DCCD inhibition by addition of Na^+ and it is assumed that the binding of the substrate Na^+ restricts the access of DCCD to the critical carboxyl group of the transporter (Igarashi and Aronson, 1987). Na^+ protection from the DCCD inhibition is also observed with the Na^+-ATPase of *A. woodii* (Heise et al., 1992). On the other hand, Na^+ does not show any protection against the F_1-directed inhibitor azide. In analogy to the renal Na^+/H^+ antiporter, this result can be interpreted by the same or physically closely related sites for the interaction of DCCD and Na^+ with the F_0 part of the enzyme. This assumption is substantiated by the fact that both DCCD and Na^+ are known to interact with the F_0 part of the sodium ion-translocating F_1F_0-ATPase of *Propionigenium modestum* (Laubinger et al., 1990).

As stated earlier, ΔpH-driven ATP synthesis is also observed in *A. woodii*. However, it is not clear at the moment whether this organism contains two enzymes with different ion specificities (H^+ or Na^+) or just one enzyme which can transport both ions alternatively. A further discussion of this interesting enzyme has to await its purification.

4.7 Distribution of the Sodium Ion-Motive Methyltransferase in Organisms Using the Acetyl-CoA Pathway

The acetyl-CoA pathway is widespread among anaerobic bacteria. Acetogenic bacteria such as *A. woodii*, *P. productus*, *Clostridium formicoaceticum*, *Sporomusa ovata*, and *Clostridium thermoaceticum* use this pathway for acetate formation and energy metabolism. All known methanogenic bacteria use at least the central reaction sequence, including the methyltransferase, for methane and energy production. On the other hand, the reverse pathway, i.e., acetate conversion to 2 CO_2 is involved in some sulfate reducers such as *Desulfotomaculum acetoxidans*, *Desulfobacterium thermoautotrophicum*, and *Desulfococcus niacini*. Again, some methanogenic bacteria, for example *Methanosarcina* and *Methanothrix* species, use the central part of this pathway for the conversion of acetate to CO_2 and CH_4 [see Fuchs (1986)]. Considering the wide use of this pathway, the question arises whether a sodium ion-motive methyltransferase is obligatory for the acetyl-CoA pathway. Because of the lack of relevant data for a number of organisms or whole groups of organisms such as the sulfate reducers, a definite answer to this question cannot be given at the moment, but, nevertheless, from the data available for methanogens and acetogens we can develop some rules.

In methanogenic bacteria, members of the Methanosarcinaceae (*Ms. barkeri*, *Methanosarcina* strain Göl) (Müller et al., 1990) use a Na^+-translocating methyltransferase. Furthermore, they have considerable amounts of membrane-bound corrinoids (Dangel et al., 1987) and methane formation in all species tested so far is Na^+-dependent. The presence of the Na^+-translocating methyltransferase has not been determined experimentally for all methanogenic families, but one can certainly assume it to generally function in methanogenic bacteria.

In acetogens, the situation is different. In *A. woodii* (Heise et al., 1989), *P. productus* (Geerligs et al., 1989), and *A. kivuii* (Yang and Drake, 1990), sodium ions play a major role in their bioenergetics and, using *A. woodii* as a model organism, it was demonstrated that ATP is synthesized by $\Delta\bar{\mu}_{Na^+}$ produced during the operation of the acetyl-CoA pathway (Heise et al., 1992). An exception to the rule of a sodium ion-motive acetyl-CoA pathway in acetogens is *C. thermoaceticum*. Acetogenesis from $H_2 + CO_2$ is sodium ion independent (Yang and Drake, 1990). The bioenergetics of this organism has been studied intensively and the involvement of $\Delta\bar{\mu}_{H^+}$ in electron transport phosphorylation has been proposed [see Chapter 2 and (Hugenholtz and Ljungdahl, 1990)]. On theoretical grounds, the methylene-H_4F reductase was proposed to be involved in energy conservation. The enzyme has been shown to be present in the membrane fraction of *C. thermoaceticum* (Hugenholtz et al., 1987), and the standard free-energy change was estimated to be -22 kJ/mol (Wohlfarth and Diekert, 1991), which is sufficiently large to be involved in energy conservation.

Is there a correlation between the presence of membrane-bound corrinoids and the presence of a sodium ion-motive methyltransferase? From the data available, one would expect that organisms which contain membrane-bound corrinoids have a sodium ion-motive methyltransferase, and organisms without corrinoids in their membranes do not have such a system. An exception to this rule is *Methanosphaera stadtmanae*. This organism grows only on $H_2 + CH_3OH$ (Miller and Wolin, 1983). This pathway does not include the methyl-H_4MPT : CoM-SH methyltransferase; consequently, the enzyme could not be measured in a cell-free extract (van de Wijngaard et al., 1991a) and methane formation from this substrate combination was shown to be sodium ion independent (Sparling and Gottschalk, unpublished). On the other hand, the organism contains high amounts of membrane-bound corrinoids (Dangel et al., 1987) whose function remains speculative, but due to the absence of cytochromes, one might consider a function of the corrinoids analogous to the porphyrins in cytochromes.

Interestingly, the sodium ion-independent organism *C. thermoaceticum* does not contain membrane-bound corrinoids, but it contains cytochromes and menaquinones (Gottwald et al., 1975). The reverse, absence of cytochromes but presence of membrane-bound corrinoids, is observed in the Na^+-dependent organisms *A. woodii* and *P. productus*. Therefore, it is tempting to divide acetogenic bacteria into two groups with respect to their bioenergetics. In both groups, the acetyl-CoA pathway is coupled to ATP synthesis by a chemiosmotic mechanism. The first group, with *C. thermoaceticum* as a model organism, generates a $\Delta\bar{\mu}_{H^+}$, most probably in the course of the methylene-H_4F reductase reaction; cytochromes are involved in the proton-motive electron transport chain which leads from the electron donor to methylene-H_4F. $\Delta\bar{\mu}_{H^+}$ is then, in turn, used to drive ATP synthesis via a membrane-bound, H^+-translocating ATP synthase. In the second group, with *A. woodii* as a model organism, a $\Delta\bar{\mu}_{Na^+}$ is produced, most probably in the course of the methyltransferase reaction; corrinoids are involved in the methyl transfer reaction and the sodium ion translocation. $\Delta\bar{\mu}_{Na^+}$ is then, in turn, used to drive ATP synthesis via a membrane-bound, Na^+ translocating ATP synthase.

4.8   Concluding Remarks

The acetyl-CoA pathway in methanogenic and some acetogenic bacteria such as *A. woodii* is coupled to the generation of a primary electrochemical sodium ion gradient. In methanogenic bacteria, the methyl-H_4MPT : CoM-SH methyl-transferase was identified as the sodium ion-motive enzyme, and it is quite likely that the corresponding enzyme in *A. woodii,* the methyl-H_4F : C/Fe-SP methyltransferase, is responsible for the observed sodium ion translocation in this organism. The methyltransferase of the acetyl-CoA pathway is unique with respect to its function as a sodium ion pump, and the study of its structure and function will hopefully contribute to the exciting question of ion transport through membrane proteins.

Whereas an involvement of $\Delta\bar{\mu}_{Na^+}$ in ATP synthesis in methanogens is contro-versial and deserves further investigations, a novel function of $\Delta\bar{\mu}_{Na^+}$ as the driving force for an endergonic reaction other than ATP synthesis was demon-strated. On the other hand, *A. woodii* synthesizes ATP by means of $\Delta\bar{\mu}_{Na^+}$ and is, therefore, besides *P. modestum* (Dimroth, 1991), the second organism known to rely on a $\Delta\bar{\mu}_{Na^+}$ for ATP synthesis. It is not quite clear why these organisms have evolved a sodium ion cycle to conserve energy, but it is conceivable that a $\Delta\bar{\mu}_{Na^+}$ is better suited in anaerobic habitats rich in organic acids and of low pH, conditions which decrease the $\Delta\bar{\mu}_{H^+}$ and, therefore, favor a mechanism of energy conservation which is not based on protons as the coupling ion.

Acknowledgments

The authors are indebted to their co-workers for their excellent contributions. We are grateful to B. Averhoff, J. Reidlinger, and D. Westenberg for a critical review of the manuscript. The work of the authors' laboratory was supported by a grant from the Deutsche Forschungsgemeinschaft.

References

Abbanat, D. R., and J. G. Ferry. 1991. Resolution of component proteins in an enzyme complex from *Methanosarcina thermophila* catalyzing the synthesis or cleavage of acetyl-CoA. *Proc. Natl. Acad. Sci. USA* **88:**3272–3276.

Al-Mahrouq, H. A., S. W. Carper, and J. R. Lancaster, Jr. 1986. Discrimination between transmembrane ion gradient-driven and electron transfer-driven ATP synthesis in the methanogenic bacteria. *FEBS Lett.* **207:**262–265.

Andreesen, J. R., A. Schaupp, C. Neurauter, A. Brown, and L. G. Ljungdahl. 1973. Fermentation of glucose, fructose and xylose by *Clostridium thermoaceticum:* effect of metals on growth yield, enzymes and the synthesis of acetate from CO_2. *J. Bacteriol.* **114:**743–751.

Banerjee, R. V., S. C. Harder, S. W. Ragsdale, and R. G. Matthews. 1990. Mechanism of reductive activation of cobalamin-dependent methionine synthase: an electron para-magnetic resonance spectroelectrochemical study. *Biochemistry* **29**:1129–1135.

Banerjee, R. V., and R. G. Matthews. 1990. Cobalamin-dependent methionine synthase. *FASEB J.* **4**:1450–1459.

Becher, B., V. Müller, and G. Gottschalk. 1992a. The methyl-tetrahydromethanopterin : coenzyme M methyltransferase of *Methanosarcina* strain Göl is a primary sodium pump. *FEMS Microbiol. Lett.* **91**:239–244.

Becher, B., V. Müller, and G. Gottschalk, 1992b. N^5-methyl-tetrahydromethanopterin : coenzyme M methyltransferase of *Methanosarcina* strain Göl is an Na^+ translocating membrane protein. *J. Bacteriol.* **174**:7656–7660.

Blaut, M., and G. Gottschalk. 1984. Coupling of ATP synthesis and methane formation from methanol and molecular hydrogen in *Methanosarcina barkeri*. *Eur. J. Biochem.* **141**:217–222.

Blaut, M., V. Müller, K. Fiebig, and G. Gottschalk. 1985. Sodium ions and an energized membrane required by *Methanosarcina barkeri* for the oxidation of methanol to the level of formaldehyde. *J. Bacteriol.* **164**:95–101.

Blaut, M., V. Müller, and G. Gottschalk. 1989. Energetics of methanogens. In: *Bacterial Energetics,* T. A. Krulwich (ed.), pp. 505–537. Academic Press, New York.

Bobik, T. A., K. D. Olson, K. M. Noll, and R. S. Wolfe. 1987. Evidence that the heterodisulfide of coenzyme M and 7-mercaptoheptanoylthreonine phosphate is a prod-uct of the methylreductase reaction in *Methanobacterium*. *Biochem. Biophys. Res. Commun.* **149**:455–460.

Bott, M., B. Eikmanns, and R. K. Thauer. 1986. Coupling of carbon monoxide oxidation to CO_2 and H_2 with the phosphorylation of ADP in acetate-grown *Methanosarcina barkeri*. *Eur. J. Biochem.* **159**:393–398.

Bott, M., and R. K. Thauer. 1987. Proton-motive-force-driven formation of CO from CO_2 and H_2 in methanogenic bacteria. *Eur. J. Biochem.* **168**:407–412.

Bott, M., and R. K. Thauer. 1989. Proton translocation coupled to the oxidation of carbon monoxide to CO_2 and H_2 in *Methanosarcina barkeri*. *Eur. J. Biochem.* **179**:469–472.

Carper, S. W., J. R. Lancaster, Jr. 1986. An electrogenic sodium-translocating ATPase in *Methanococcus voltae*. *FEBS Lett.* **200**:177–180.

Crider, B. P., S. W. Carper, and J. R. Lancaster. 1985. Electron transfer-driven ATP synthesis in *Methanococcus voltae* is not dependent on a proton electrochemical gradi-ent. *Proc. Natl. Acad. Sci. USA* **82**:6793–6796.

Dangel, W., H. Schultz, G. Diekert, H. Hönig, and G. Fuchs. 1987. Occurrence of corrinoid-containing membrane proteins in anaerobic bacteria. *Arch. Microbiol.* **148**:52–56.

Deppenmeier, U., M. Blaut, A. Mahlmann, and G. Gottschalk. 1990. Membrane-bound $F_{420}H_2$-dependent heterodisulfide reductase in methanogenic bacterium strain Göl and *Methanolobus tindarius*. *FEBS Lett.* **1**:199–203.

Deppenmeier, U., M. Blaut, and G. Gottschalk. 1991. H_2: heterodisulfide oxidoreductase, a second energy-conserving system in the methanogenic strain Göl. *Arch. Microbiol.* **155**:272–277.

Dharmavaram, R., and J. Konisky. 1987. Identification of a vanadate-sensitive, membrane-bound ATPase in the archaebacterium *Methanococcus voltae*. *J. Bacteriol.* **169**:3921–3925.

Dharmavaram, R., and J. Konisky. 1989. Characterization of a P-type ATPase of the archaebacterium *Methanococcus voltae*. *J. Biol. Chem.* **264**:14085–14089.

Dharmavaram, R., P. Gillevet, and J. Konisky. 1991. Nucleotide sequence of the gene encoding the vanadate-sensitive membrane-associated ATPase of *Methanococcus voltae*. *J. Bacteriol.* **173**:2131–2133.

Diekert, G., E. Schrader, and W. Harder. 1986. Energetics of CO formation and CO oxidation in cell suspensions of *Acetobacterium woodii*. *Arch. Microbiol.* **144**:386–392.

Dimroth, P. 1987. Sodium ion transport decarboxylases and other aspects of sodium ion cycling in bacteria. *Microbiol. Rev.* **51**:320–340.

Dimroth, P. 1991. Na^+-coupled alternative to H^+-coupled primary transport systems in bacteria. *Bioessays* **13**:463–468.

Eikmanns, B., and R. K. Thauer. 1984. Catalysis of an isotopic exchange between CO_2 and the carboxyl group of acetate by *Methanosarcina barkeri* grown on acetate. *Arch. Microbiol.* **138**:365–370.

Ekiel, I., K. F. Jarrell, and G. D. Sprott. 1985. Amino acid biosynthesis and sodium-dependent transport in *Methanococcus voltae*, as revealed by ^{13}C NMR. *Eur. J. Biochem.* **149**:437–444.

Ellermann, J., P. Hedderich, R. Böcher, and R. K. Thauer. 1988. The final step in methane formation—investigations with highly purified methyl-CoM reductase (component C) from *Methanobacterium thermoautotrophicum* (strain Marburg). *Eur. J. Biochem.* **171**:669–677.

Escalante-Semerena, J. C., J. A. Leigh, K. L. Rinehart, and R. S. Wolfe. 1984. Formaldehyde activation factor, tetrahydromethanopterin, a coenzyme of methanogenesis. *Proc. Natl. Acad. Sci. USA* **81**:1976–1980.

Fischer, R., and R. K. Thauer. 1988. Methane formation from acetyl phosphate in cell extracts of *Methanosarcina barkeri*. Dependence of the reaction on coenzyme A. *FEBS Lett.* **228**:249–253.

Fischer, R., and R. K. Thauer. 1989. Methyltetrahydromethanopterin as an intermediate in methanogenesis from acetate in *Methanosarcina barkeri*. *Arch. Microbiol.* **151**:459–465.

Fischer, R., and R. K. Thauer. 1990. Ferredoxin-dependent methane formation from acetate in cell extracts of *Methanosarcina barkeri* (strain MS). *FEBS Lett.* **269**:368–372.

Fischer, R., P. Gärtner, A. Yeliseev, and R. K. Thauer. 1992. N^5-methyltetrahydromethanopterin : coenzyme M methyltransferase in methanogenic archaebacteria is a membrane protein. *Arch. Microbiol.* **158**:208–217.

Fuchs, G. 1986. CO₂ fixation in acetogenic bacteria: variations on a theme. *FEMS Microbiol. Rev.* **39**:181–213.

Fujii, K., and F. M. Huennekens. 1974. Activation of methionine synthetase by a reduced triphosphopyridine nucleotide-dependent flavoprotein system. *J. Biol. Chem.* **249**:6745–6753.

Geerligs, G., P. Schönheit, and G. Diekert. 1989. Sodium dependent acetate formation from CO₂ in *Peptostreptococcus productus* (strain Marburg). *FEMS Microbiol. Lett.* **57**:253–258.

Godley, A. W., P. E. Linnett, and J. P. Robinson. 1990. The effect of carbon dioxide on the growth kinetics of fructose-limited chemostat cultures of *Acetobacterium woodii* DSM 1030. *Arch. Microbiol.* **154**:5–11.

Gottschalk, G. 1989. Bioenergetics of methanogenic and acetogenic bacteria. In: *Autotrophic Bacteria,* H. G. Schlegel and B. Bowien (eds.), pp. 383–396. Science Tech Publishers, Madison, WI.

Gottwald, M., J. R. Andreesen, J. LeGall, and L. G. Ljungdahl. 1975. Presence of cytochrome and menaquinone in *Clostridium formicoaceticum* and *Clostridium thermoaceticum. J. Bacteriol.* **122**:325–328.

Grahame, D. A. 1991. Catalysis of Acetyl-CoA cleavage and tetrahydrosarcinapterin methylation by a carbon monoxide dehydrogenase-corrinoid enzyme complex. *J. Biol. Chem.* **266**:22227–22233.

Grahame, D. A., and T. C. Stadtman. 1987. *In vitro* methane and methyl-coenzyme M formation from acetate: evidence that acetyl-CoA is the required intermediate activated form of acetate. *Biochem. Biophys. Res. Commun.* **147**:254–258.

Hansen, B., M. Bokranz, P. Schönheit, and A. Kröger. 1988. ATP formation coupled to caffeate reduction by H₂ in *Acetobacterium woodii* NZva16. *Arch. Microbiol.* **150**:447–451.

Hedderich, R., and R. K. Thauer. 1988. *Methanobacterium thermoautotrophicum* contains a soluble enzyme system that specifically catalyzes the reduction of the heterodisulfide of coenzyme M and 7-mercaptoheptanoylthreonine phosphate with H₂. *FEBS Lett.* **234**:223–227.

Heise, R., V. Müller, and G. Gottschalk. 1989. Sodium dependence of acetate formation by the acetogenic bacterium *Acetobacterium woodii. J. Bacteriol.* **171**:5473–5478.

Heise, R., J. Reidlinger, V. Müller, and G. Gottschalk. 1991. A sodium-stimulated ATP synthase in the acetogenic bacterium *Acetobacterium woodii. FEBS Lett.* **295**:119–122.

Heise, R., V. Müller, and G. Gottschalk. 1992. Presence of a sodium-translocating ATPase in membrane vesicles of the homoacetogenic bacterium *Acetobacterium woodii. Eur. J. Biochem.* **206**:553–557.

Hugenholtz, J., D. M. Ivey, and L. G. Ljungdahl. 1987. Carbon monoxide-driven electron transport in *Clostridium thermoautotrophicum* membranes. *J. Bacteriol.* **169**:5845–5847.

Hugenholtz, J., and L. G. Ljungdahl. 1990. Metabolism and energy generation in homoacetogenic clostridia. *FEMS Microbiol. Rev.* **87**:383–390.

Igarashi, P., and Aronson. 1987. Covalent modification of the renal Na^+/H^+ exchanger by N,N'-dicyclohexylcarbodiimide. *J. Biol. Chem.* **62**:860–868.

Jarrell, K. F., S. E. Bird, and G. D. Sprott. 1984. Sodium-dependent isoleucine transport in the methanogenic archaebacterium *Methanococcus voltae*. *FEBS Lett.* **166**:357–361.

Kaesler, B., and P. Schönheit. 1989a. The role of sodium ions in methanogenesis. Formaldehyde oxidation to CO_2 and $2H_2$ in methanogenic bacteria is coupled with primary electrogenic Na^+ translocation at a stoichiometry of 2–3 Na^+/CO_2. *Eur. J. Biochem.* **184**:223–232.

Kaesler, B., and P. Schönheit. 1989b. The sodium cycle in methanogenesis. CO_2 reduction to the formaldehyde level in methanogenic bacteria is driven a primary electrochemical potential of Na^+ generated by formaldehyde reduction to CH_4. *Eur. J. Biochem.* **186**:309–316.

Keltjens, J. T., and C. van der Drift. 1986. Electron transfer reactions in methanogens. *FEMS Microbiol. Rev.* **39**:259–303.

Keltjens, J. T., B. W. te Brommelstroet, S. W. M. Kengen, C. van der Drift, and G. D. Vogels. 1990. 5,6,7,8-Tetrahydromethanopterin-dependent enzymes involved in methanogenesis. *FEMS Microbiol. Rev.* **87**:327–332.

Kengen, S. W. M., J. J. Mosterd, R. L. H. Nelissen, J. T. Keltjens, C. van der Drift, and G. D. Vogels. 1988. Reductive activation of the methyl-tetrahydromethanopterin : coenzyme M methyltransferase from *Methanobacterium thermoautotrophicum* strain ΔH. *Arch. Microbiol.* **150**:405–412.

Kengen, S. W. M., P. J. H. Daas, J. T. Keltjens, C. van der Drift, and G. D. Vogels. 1990. Stimulation of the methyltetrahydromethanopterin : coenzyme M methyltransferase reaction in cell free extracts of *Methanobacterium thermoautotrophicum* by the heterodisulfide of coenzyme M and 7-mercaptoheptanoylthreonine phosphate. *Arch. Microbiol.* **154**:156–161.

Kengen, S. W. M., P. J. H. Daas, E. F. G. Duits, J. T. Keltjens, C. van der Drift, and G. D. Vogels. 1992. Isolation of a 5-hydroxybenzimidazolyl cobamide-containing enzyme involved in the methyltetrahydromethanopterin : coenzyme M methyltransferase reaction in *Methanobacterium thermoautotrophicum*. *Biochim. Biophys. Acta* **1118**:249–260.

Kleyman, T. R., and E. J. Cragoe, Jr. 1988. Amiloride and its analogs as tools for the study of ion transport. *J. Membrane Biol.* **105**:1–21.

Krulwich, T. A. 1983. Na^+/H^+ antiporters. *Biochim. Biophys. Acta* **726**:245–264.

Laubinger, W., G. Deckers-Hebestreit, K. Altendorf, and P. Dimroth. 1990. A hybrid adenosinetriphosphatase composed of F_1 of *Escherichia coli* and F_0 of *Propionigenium modestum* is a functional sodium ion pump. *Biochemistry* **29**:5458–5463.

Ljungdahl, L. G. 1986. The autotrophic pathway of acetate synthesis in acetogenic bacteria. *Annu. Rev. Microbiol.* **40**:415–450.

Ma, K., and R. K. Thauer. 1990. Purification and properties of N^5, N^{10}-methylenetetrahydromethanopterin reductase from *Methanobacterium thermoautotrophicum* (strain Marburg). *Eur. J. Biochem.* **191**:187–193.

Miller, T. L., and M. J. Wolin. 1983. Oxidation of hydrogen and reduction of methanol to methane is the sole energy source for a methanogen isolated from human feces. *J. Bacteriol.* **153**:1051–1055.

Mitchell, P. 1961. Coupling of phosphorylation to electron and hydrogen transfer by a chemiosmotic mechanism. *Nature* **191**:144–148.

Müller, V., M. Blaut, and G. Gottschalk. 1987. Generation of a transmembrane gradient of Na^+ in *Methanosarcina barkeri*. *Eur. J. Biochem.* **162**:461–466.

Müller, V., M. Blaut, and G. Gottschalk. 1988. The transmembrane electrochemical gradient of Na^+ as driving force for methanol oxidation in *Methanosarcina barkeri*. *Eur. J. Biochem.* **172**:601–606.

Müller, V., C. Winner, and G. Gottschalk. 1989. Electron transport-driven sodium extrusion during methanogenesis from formaldehyde + H_2 by *Methanosarcina barkeri*. *Eur. J. Biochem.* **178**:519–525.

Müller, V., M. Blaut, R. Heise, C. Winner, and G. Gottschalk. 1990. Sodium bioenergetics in methanogens and acetogens. *FEMS Microbiol. Rev.* **87**:373–377.

Pedersen, P. L., and E. Carafoli. 1987. Ion motive ATPases. I. Ubiquity, properties, and significance to cell function. *Trends Biochem. Sci.* **12**:146–150.

Peinemann, S., V. Müller, M. Blaut, and G. Gottschalk. 1988. Bioenergetics of methanogenesis from acetate by *Methanosarcina barkeri*. *J. Bacteriol.* **170**:1369–1372.

Perski, H. J., J. Moll, and R. K. Thauer. 1981. Sodium dependence of growth and methane formation in *Methanobacterium thermoautotrophicum*. *Arch. Microbiol.* **130**:319–321.

Perski, H. J., P. Schönheit, and R. K. Thauer. 1982. Sodium dependence of methane formation in methanogenic bacteria. *FEBS Lett.* **143**:323–326.

Poirot, C. M., S. W. M. Kengen, E. Valk, J. T. Keltjens, C. van der Drift, and G. D. Vogels. 1987. Formation of methylcoenzyme M from formaldehyde by cell free extracts of *Methanobacterium thermoautotrophicum*. Evidence for the involvement of a corrinoid-containing methyltransferase. *FEMS Microbiol. Lett.* **40**:7–13.

Pol, A., C. van der Drift, and G. D. Vogels. 1982. Corrinoids from *Methanosarcina barkeri*: Structure of the α-ligand. *Biochem. Biophys. Res. Commun.* **108**:731–737.

Ragsdale, S. W. 1992. Enymology of the acetyl-CoA pathway of autotrophic CO_2 fixation. *Crit. Rev. Biochem. Mol. Biol.* **26**:261–300.

Ragsdale, S. W. and H. G. Wood. 1985. Acetate biosynthesis by acetogenic bacteria. Evidence that carbon monoxide dehydrogenase is the condensing enzyme that catalyzes the final steps in the synthesis. *J. Biol. Chem.* **260**:3970–3977.

Raybuck, S. A., S. E. Ramer, D. R. Abbanat, J. W. Peters, W. H. Orme-johnson, J. G. Ferry, and C. T. Walsh. 1991. Demonstration of carbon–carbon bond cleavage of acetyl coenzyme A by using isotopic exchange catalyzed by the CO dehydrogenase complex from acetate-grown *Methanosarcina thermophila*. *J. Bacteriol*. **173**:929–932.

Rouvière, P., and R. S. Wolfe. 1988. Novel biochemistry of methanogenesis. *J. Biol. Chem*. **263**:7913–7916.

Schönheit, P., and D. B. Beimborn. 1985. Presence of a Na^+/H^+ antiporter in *Methanobacterium thermoautotrophicum* and its role in Na^+-dependent methanogenesis. *Arch. Microbiol*. **142**:354–361.

Schulz, H., and G. Fuchs. 1986. Cobamide-containing membrane protein complex in *Methanobacterium*. *FEBS Lett*. **198**:279–282.

Schulz, H., S. P. J. Albracht, J. M. C. Coremans, and G. Fuchs. 1988. Purification and some properties of the corrinoid-containing membrane protein from *Methanobacterium thermoautotrophicum*. *Eur. J. Biochem*. **171**:589–597.

Skulachev, V. P. 1989. Bacterial Na^+ energetics. *FEBS Lett*. **250**:106–114.

Smigan, P., L. Horovska, and M. Greksak. 1988. Na^+-driven ATP synthesis in *Methanobacterium thermoautotrophicum* can be modulated with sodium ion concentrations in the growth medium. *FEBS Lett*. **242**:85–88.

Stupperich, E., A. Juza, C. Eckerskorn, and L. Edelmann. 1990. An immunological study of corrinoid proteins from bacteria revealed homologous antigenic determinants of a soluble corrinoid-dependent methyltransferase and corrinoid-containing membrane proteins from *Methanobacterium* species. *Arch. Microbiol*. **155**:28–34.

Sugiyama, S., E. J. Cragoe, Jr., and Y. Imae. 1988. Amiloride, a specific inhibitor for the Na^+-driven flagellar motors of alkalophilic *Bacillus*. *J. Biol. Chem*. **263**:8215–8219.

Terlesky, K. C., M. J. K. Nelson, and J. G. Ferry. 1986. Isolation of an enzyme complex with carbon monoxide dehydrogenase activity containing corrinoid and nickel from acetate-grown *Methanosarcina thermophila*. *J. Bacteriol*. **168**:1053–1058.

Terlesky, K. C., and J. G. Ferry. 1988. Ferredoxin requirement for electron transport from the carbon monoxide dehydrogenase complex to a membrane-bound hydrogenase in acetate-grown *Methanosarcina thermophila*. *J. Biol. Chem*. **263**:4075–4079.

Terraciano, J. S., W. J. A. Schreurs, and E. R. Kashket. 1987. Membrane H^+ conductance of *Clostridium thermoaceticum* and *Clostridium acetobutylicum:* evidence for electrogenic Na^+/H^+ antiport in *Clostridium thermoaceticum*. *Appl. Environ. Microbiol*. **53**:782–786.

Thauer, R. K. 1990. Energy metabolism of methanogenic bacteria. *Biochim. Biophys. Acta* **1018**:256–259.

Thauer, R. K., K. Jungermann, and K. Decker. 1977. Energy conservation in chemotrophic anaerobic bacteria. *Bacteria Rev*. **41**:100–180.

Tokuda, H., and T. Unemoto. 1985. The Na^+-motive respiratory chain of marine bacteria. *Microbiol. Sci*. **2**:65–71.

van de Wijngaard, W. M. H., J. Creemers, G. D. Vogels, and C. van der Drift. 1991a.

Methanogenic pathways in *Methanosphaera stadtmanae*. *FEMS Microbiol. Lett.* **80**:207–212.

van de Wijngaard, W. M. H., R. L. Lugtigheid, and C. van der Drift. 1991b. Reductive activation of the corrinoid-containing enzyme involved in methyl group transfer between methyl-tetrahydromethanopterin and coenzyme M in *Methanosarcina barkeri*. *Antonie van Leeuwenhoek* **60**:1–6.

van der Meijden, P., H. J. Heythuysen, F. P. Pouwels, F. P. Houwen, C. van der Drift, and G. D. Vogels. 1983. Methyltransferase involved in methanol conversion by *Methanosarcina barkeri*. *Arch. Microbiol.* **134**:238–242.

Wohlfarth, G., and G. Diekert. 1991. Thermodynamics of methylenetetrahydrofolate reduction to methyltetrahydrofolate and its implications for the energy metabolism of homoacetogenic bacteria. *Arch. Microbiol.* **155**:378–381.

Wood, H. G., S. W. Ragsdale, and E. Pezacka. 1986. The acetyl-CoA pathway of autotrophic growth. *FEMS Microbiol. Rev.* **39**:345–362.

Yang, H., and H. L. Drake. 1990. Differential effects of sodium on hydrogen- and glucose-dependent growth of the acetogenic bacterium *Acetogenium kivui*. *Appl. Environ. Microbiol.* **56**:81–86.

5

Energetics of Acetogenesis from C_1 Units

Gabriele Diekert and Gert Wohlfarth

5.1 Introduction to Homoacetogenic Bacteria

Homoacetogenic bacteria (also called "acetogenic bacteria") are strictly anaerobic eubacteria, which catalyze the synthesis of acetate from C_1 units in their catabolism. The homoacetogens have proved to be a very heterogenous group of microorganisms: they are gram positive as well as gram negative, spore- or non-spore forming, thermophilic and mesophilic, motile and nonmotile, and so on. Usually they are able to grow on a variety of substrates, some of which are quite unusual for anaerobic microorganisms. Because most of the homoacetogens are able to grow at the expense of H_2 plus CO_2 as the sole energy source, the reduction of 2 CO_2 to acetate must be coupled to the synthesis of ATP. This chapter will deal mainly with the utilization of C_1 substrates and the energy conservation coupled to the synthesis of acetate from these substrates. For general overviews on homoacetogenic bacteria, the reader is referred to Chapters 1 and 7, as well as other recent reviews (Diekert, 1992; Schink and Bomar, 1992; Drake, 1992).

5.1.1 Physiology of Homoacetogens Relative to C_1 Substrates

Homoacetogenic bacteria utilize complex substrates, e.g., sugars and sugar alcohols, as well as different C_1 substrates. Upon growth with hexoses, they form stoichiometrically 3 acetate from one hexose consumed (Fontaine et al., 1942). Hexoses are converted to 2 acetate plus 2 CO_2 via glycolysis. The third acetate is formed in the acetyl-CoA pathway (or carbon monoxide dehydrogenase pathway) from the 2 CO_2 plus the reduction equivalents generated upon the oxidation of the sugars (Figure 1.4). Evidence was provided that the carboxyl group of the third acetate, i.e., the acetate formed from two C_1 units, can be supplied by C-1 of pyruvate without previous decarboxylation to free CO_2 (Schulman et al., 1973). During the last decade it was shown that the carboxyl group of acetate can also be derived from carbon monoxide (Lynd et al., 1982; Lorowitz and Bryant, 1984; Ma et al., 1987), which appears to be an intermediate in the formation of C-1 of acetate (in a bound form) (Diekert et al., 1984). Precursors for the synthesis of the methyl group of acetate can be channeled into the homoacetogenic pathway at different redox levels: as a bound methanol (or methyl$^+$) (CH_3OH, methoxylated aromatic compounds, CH_3Cl) (Genthner et al., 1981; Lynd and Zeikus, 1983; Bache and Pfennig, 1981; Traunecker et al., 1991), as other methoxylated compounds like methoxyethanol and methoxy-acetate (Tanaka and Pfennig, 1988; Schuppert and Schink, 1990), as free formate, which is an intermediate in the formation of the methyl group, as CO_2 in either a free form or in a bound form (carboxyl group of benzoate derivatives or pyruvate) (Hsu et al., 1990; Schulman et al., 1973), or as carbon monoxide via CO oxidation to CO_2. The fate of C_1 substrates for acetate synthesis is shown in a tentative scheme in Figure 5.1. It should be noted here that the acetate synthesis from CO_2 serves two purposes in the homoacetogens: On the one hand, it is coupled to the generation of ATP and the conservation of energy; on the other hand, it provides C_1 and C_2 units as precursors for cell carbon (chemolithoau-totrophic growth) (Fuchs, 1986; Wood and Ljungdahl, 1991).

5.1.2   The Acetyl-CoA Pathway: Flow of Carbon from CO_2

The total synthesis of acetate from CO_2 involves the synthesis of a methyl group and of a carboxyl group. The methyl group is derived from a bound methanol (i.e., a methyl$^+$, CH_3-X in Fig. 5.1); the carboxyl group is derived from a bound carbon monoxide (carbonyl group $\equiv$ [CO]) (Fig. 5.1). Therefore, the synthesis of the methyl group from CO_2 requires 6 [H] (reduction equivalents), the carboxyl group synthesis 2[H]. The methyl group of acetate is formed via formate and tetrahydrofolate-bound C_1 intermediates. Most probably, the methyl group is then transferred to a corrinoid enzyme prior to its incorporation into the methyl group of acetyl-CoA. The bound carbonyl is supplied by the enzyme carbon monoxide dehydrogenase, which mediates the reduction of CO_2 to CO. The carbon monoxide dehydrogenase is a key enzyme in the homoacetogenic pathway and probably, in addition, catalyzes the synthesis of acetyl-CoA from the bound carbonyl, the methyl group, and coenzyme A. The pathway of acetate formation from CO_2 is outlined in Figure 5.2. It should be mentioned here that the elucidation of this pathway was mainly performed by the groups of Wood and Ljungdahl. For reviews on the biochemistry of homoacetogenic fermentation,

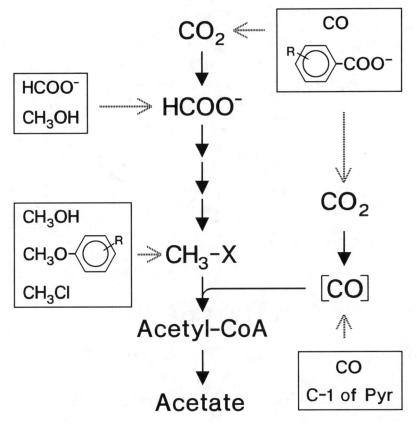

Figure 5.1. Fate of C_1 substrates in the homoacetogenic pathway. [CO] = enzyme-bound carbonyl.

the reader is referred to Chapters 2 and 3 [see also (Wood, 1991; Wood and Ljungdahl, 1991; Ragsdale, 1991; Ljungdahl, 1986; Fuchs, 1986)].

The transfer of reducing equivalents (from e.g., H_2) to CO_2 for the acetate formation involves different electron carriers, including pyridine nucleotides and ferredoxin. However, evidence was accumulated that other compounds like cytochromes, quinones, and iron-sulfur proteins different from ferredoxin might play an important role in the electron transfer chain of homoacetogenesis (Hugenholtz and Ljungdahl, 1990). Because at least some of the homoacetogens apparently lack cytochromes and/or quinones, it is feasible that the mechanism of energy conservation via homoacetogenesis from CO_2 varies within the different groups of homoacetogenic bacteria. Recently, evidence was presented with *Spor-*

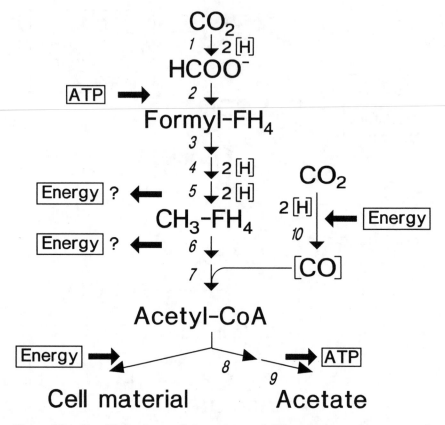

Figure 5.2.   Simplified scheme of the pathway of CO_2 reduction to acetate or cell material. The enzymes involved are the following: *1*: formate dehydrogenase; *2*: formyl tetrahydrofolate synthetase; *3*: methenyl tetrahydrofolate cyclohydrolase; *4*: methylene tetrahydrofolate dehydrogenase; *5*: methylene tetrahydrofolate reductase; *6*: methyltransferase; *7*: acetyl-CoA synthase (= CO dehydrogenase); *8*: phosphotransacetylase; *9*: acetate kinase; *10*: carbon monoxide dehydrogenase. FH_4 = tetrahydrofolate. For details, see text.

omusa sphaeroides that cytochromes might be involved in the oxidation of the methyl group to CO_2 (Kamlage and Blaut, 1993).

5.2   Energetics of Acetate Formation from CO_2

The homoacetogens are extremely versatile with regard to their substrate spectrum. Growth on hydrogen plus CO_2, however, is limited by the hydrogen concentration in the natural environment. In freshwater sediments, homoaceto-gens compete for the hydrogen with the methanogenic bacteria. The methane formation from H_2 plus CO_2 is themodynamically more favorable than the acetate formation from the same substrates:

$$4H_2 + CO_2 \rightarrow CH_4 + 2H_2O, \Delta G'_0 = -130 \text{ kJ/mol} \qquad (5.1)$$
$$4H_2 + 2CO_2 \rightarrow CH_3COO^- + H^+ + 2H_2O, \Delta G'_0 = -95 \text{ kJ/mol} \qquad (5.2)$$

The hydrogen concentration in freshwater sediments is usually about 10^{-4}–10^{-5} atm. Under these conditions, and with regard to the CO_2, CH_4, and acetate concentrations in these environments, the free energy of acetate formation from H_2 plus CO_2 is hardly sufficient to provide the energy for growth of homoacetogens on these substrates. The dependence of the free energy of acetate formation from H_2 plus CO_2 on the H_2 concentration under the environmental conditions is shown in Figure 5.3. The minimal free-energy change required for ATP synthesis by a chemiosmotic mechanism is near -20 kJ/mol (corresponding to one electrogenic proton translocated across the cytoplasmic membrane). It can be calculated that the hydrogen concentration in the freshwater sediment must be about 10^{-4} atm to support growth of homoacetogens. There is evidence for an involvement of homoacetogens in the hydrogen consumption in freshwater sediments at low temperatures (Conrad et al., 1989). In marine sediments, the concentration of H_2 is even lower. Therefore, it can be expected that homoacetogens in these environments are not able to utilize H_2 plus CO_2 as the energy sources; sulfate-reducing bacteria rather than homoacetogens keep the hydrogen partial pressure at the constantly low levels observed for the marine sediments.

5.2.1   Formation of the Methyl Group of Acetate

The methyl group of acetate is formed from CO_2 via formate and tetrahydrofol-ate-bound C_1 intermediates (Ljungdahl, 1986) (see Figure 1.6). The first step in methyl-group synthesis is catalyzed by *formate dehydrogenase,* which contains either tungsten or molybdenum:

$$CO_2 + 2 [H] \rightarrow HCOO^- + H^+ \qquad (5.3)$$

The electron donor for this reaction can be pyridine nucleotides ($E'_0 \approx -320$ mV) or possibly ferredoxin ($E'_0 \approx -420$ mV), depending on the bacterium. In some homoacetogens, the electron donor is still unknown. As the standard redox potential of the pair $CO_2/HCOO^-$ is near -430 mV, the reduction with ferredoxin under standard conditions imposes no thermodynamical problems. Assuming a high concentration of CO_2 and of reduced pyridine nucleotides, and low concen-trations of free formate and $NAD(P)^+$, the reduction of CO_2 with pyridine nucleo-

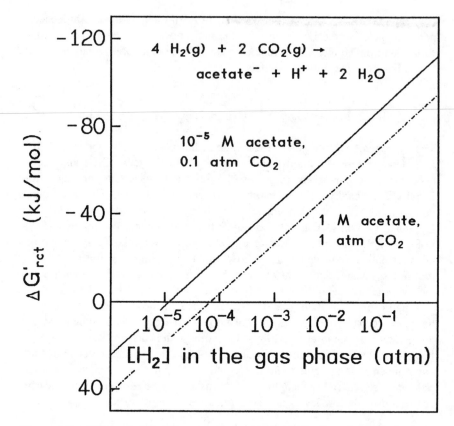

Figure 5.3.   Influence of the H_2 concentration on the $\Delta G'$ value of acetate formation from H_2 plus CO_2. Solid line: conditions similar to those in sediments; dashed line: standard conditions.

tides is thermodynamically feasible. There is, so far, no evidence for an energy-driven CO_2 reduction to formate.

A prerequisite for the formation of the methyl group, which may be considered as a bound methanol, is the reduction of formate to the redox level of formaldehyde. This reaction, however, is highly endergonic (e.g., with pyridine nucleotides as the electron donor):

$$NAD(P)H + HCOO^- + 2H^+ \rightarrow NAD(P)^+ + HCHO + H_2O, \qquad (5.4)$$
$$\Delta G'_0 = +42.7 \text{ kJ/mol}$$

There are two possible ways of circumventing such an energetical problem: either the electrons are "activated" via a chemiosmotic mechanism, i.e., the reaction is driven by an electrochemical kation gradient (H^+ or Na^+) in a reverse electron flow, or the substrate is activated and bound to a carrier, which enhances its redox potential. In methanogenic bacteria the formation of methane from CO_2 involves similar reactions to the synthesis of the methyl group of acetate in homoacetogens (Thauer, 1990). Methanogens use reverse electron flow in the reduction of CO_2 to formaldehyde (in a bound form) (Schönheit, 1993). Homoacetogens activate the substrate by binding formate to tetrahydrofolate (FH_4) prior to reduction of the C_1 unit. This reaction is catalyzed by the enzyme *formyl tetrahydrofolate synthetase:*

$$\text{formate} + FH_4 + ATP \rightarrow \qquad (5.5)$$
$$N^{10}\text{-formyl-}FH_4 + ADP + P_i, \Delta G'_0 = -8.4 \text{ kJ/mol}$$

In the following reaction, the formyl-FH_4 is dehydrated by the enzyme N^5,N^{10}-*methenyl tetrahydrofolate cyclohydrolase:*

$$N^{10}\text{-formyl-}FH_4 + H^+ \rightarrow \qquad (5.6)$$
$$N^5,N^{10}\text{-methenyl-}FH_4^+ + H_2O, \Delta G'_0 = -4.0 \text{ kJ/mol}$$

In methenyl-FH_4, the C_1 unit is still on the redox level of formate. The C_1 unit is now further reduced to a bound formaldehyde by the N^5,N^{10}-*methylene tetrahydrofolate dehydrogenase:*

$$N^5,N^{10}\text{-methenyl-}FH_4^+ + NAD(P)H \rightarrow N^5,N^{10}\text{-methylene-}FH_4 + \quad (5.7)$$
$$NAD(P)^+, \Delta G'_0 = -4.9 \text{ kJ/mol}$$

It is evident that this reaction is thermodynamically much more favorable than the reduction of formate to formaldehyde according to reaction 5.4.

The following step is catalyzed by the N^5,N^{10}-*methylene tetrahydrofolate reductase:*

$$N^5,N^{10}\text{-methylene-}FH_4 + 2\,[H] \rightarrow N^5\text{-methyl-}FH_4 \qquad (5.8)$$

The electron donor can be NADH or a still unknown component, possibly ferredoxin. The thermodynamics of this reaction was controversially discussed [see (Fuchs, 1986; Keltjens and van der Drift, 1986)]. Recently the standard redox potential of the pair methylene/methyl tetrahydrofolate has been determined to be near -200 mV (Wohlfarth and Diekert, 1991), indicating that the reaction is sufficiently exergonic to be coupled to ATP generation by a chemiosmotic mechanism ($\Delta G'_0$ of the reaction (5.8) for $2\,[H] = $ NADH is approximately -22 kJ/mol; for H_2 it is -42 kJ/mol).

Although the C_1 unit in methyl tetrahydrofolate is on the redox level of methanol, it is not directly incorporated into the methyl group of acetyl-CoA. Prior to the incorporation, it is transferred to a corrinoid/iron-sulfur protein (E-[Co] for the enzyme-bound cobalamin) by the enzyme *methyltransferase:*

$$N^5\text{-methyl- } FH_4 + E\text{-[Co]} \rightarrow FH_4 + E\text{-[Co]-CH}_3 \qquad (5.9)$$

Relatively little is known about the thermodynamics of this reaction. This will be discussed, however, in Section 5.2.3. The corrinoid-bound methyl group serves as the precursor for C-2 of acetate.

5.2.2   Formation of the Carboxyl Group of Acetate

The origin of the carboxyl group of acetate was enigmatic for a long period of time. Schulman et al. (1973) presented evidence that the C-1 of pyruvate rather than free CO_2 was the precursor for the carboxyl group of acetate in heterotrophically grown *Clostridium thermoaceticum*. In the meantime, it has been demonstrated that the carboxyl group of acetate can also be derived from CO_2, especially upon autotrophic growth of the homoacetogens. An essential clue to understanding the mechanism for the synthesis of the C-1 of acetate came from the discovery of a highly active *carbon monoxide dehydrogenase* in homoacetogens (Diekert and Thauer, 1978), which *in vitro* catalyzes the oxidation of CO to CO_2 with methyl viologen (MV^{2+}) as the artificial electron acceptor:

$$CO + 2 MV^{2+} + H_2O \rightarrow CO_2 + 2 MV^+ + 2 H^+ \qquad (5.10)$$

The enzyme, which was demonstrated to contain nickel (Diekert et al., 1979; Diekert and Thauer, 1980; Diekert and Ritter, 1983b; Drake et al., 1980; Ragsdale et al., 1983a), *in vivo* probably mediates the reverse reaction, i.e., the reduction of CO_2 to CO (in a bound form) with a still unknown physiological electron donor (Diekert et al., 1984):

$$CO_2 + 2 [H] \rightarrow CO + H_2O \qquad (5.11)$$

Evidence was presented that carbon monoxide is an intermediate in the formation of the carboxyl group of acetate (Diekert et al., 1984).

The standard redox potential (E'_0) of the pair CO_2/CO is -524 mV. When the homoacetogens grow with H_2 plus CO_2 as the energy sources, the reducing equivalents for the CO_2 reduction must be derived from the oxidation of H_2 ($E'_0 = -414$ mV). Therefore, the reaction described by reaction 5.11 is endergonic with a $\Delta G'_0$ value of approximately $+21$ kJ/mol. Evidence is available for *Acetobacterium woodii* that the reduction of CO_2 to (free) carbon monoxide with

H_2 is probably driven by an electrochemical proton gradient (Diekert et al., 1986). From the $\Delta G'_0$ value of reaction 5.11 it can be estimated that at least one electrogenic proton has to be translocated across the cytoplasmic membrane to drive this reaction. The requirement of a proton gradient for CO_2 reduction to CO is substantiated by the finding that the CO dehydrogenase of *Clostridium thermoautotrophicum* is probably membrane associated (Hugenholtz et al., 1987).

The energetics of the CO dehydrogenase reaction depends on the physiological electron donor for CO_2 reduction to CO. From experiments with purified CO dehydrogenase, it was deduced that either ferredoxin or rubredoxin (Ragsdale et al., 1983b) might be electron acceptors for CO oxidation. From thermodynamics, it is unlikely that rubredoxin [$E'_0 = -57$ mV (Eaton and Lovenberg, 1973)] functions in the reverse reaction, i.e., the reduction of CO_2 to CO. Since a prerequisite for energy coupling during the formation of CO from CO_2 would be the involvement of a membrane-associated or -integral electron carrier, it is also doubtful whether ferredoxin provides the electrons for this reaction *in vivo*.

5.2.3 Formation of Acetate

Acetate is formed from the two C_1 units, carbon monoxide (in a bound form) and the corrinoid enzyme-bound methanol via acetyl-CoA and acetyl phosphate as the intermediates. The key reaction in acetate synthesis from CO_2 is the formation of acetyl-CoA from [CO] and E-[Co]-CH_3 according to

$$[CO] + E\text{-}[Co]\text{-}CH_3 + CoA \rightarrow acetyl\text{-}CoA + E\text{-}[Co] \qquad (5.12)$$

For *C. thermoaceticum* evidence is available that the *carbon monoxide dehydrogenase* (in this respect also called *acetyl-CoA synthase*) mediates this reaction in addition to reaction 5.11 (Hu et al., 1982; Roberts et al., 1992). Therefore, the CO dehydrogenase may be considered as a bifunctional enzyme. It was shown that the reaction requires a low redox potential (Lu and Ragsdale, 1991) (Chapter 3).

The thermodynamics of the methyltransferase reaction and that of the acetyl-CoA synthase is still unclear. The overall $\Delta G'_0$ of the acetyl-CoA formation from CO_2 plus H_2 is -91 kJ/mol. The synthesis of methyl tetrahydrofolate from CO_2 and H_2 has a $\Delta G'_0$ of -74 kJ/mol; the CO_2 reduction to CO with H_2 requires almost $+21$ kJ/mol. From these data it can be calculated that under standard conditions at pH 7 the formation of acetyl-CoA from methyl tetrahydrofolate plus carbon monoxide as a sum of reactions 5.9 and 5.12 is exergonic with -38 kJ/mol. As acetyl-CoA as the product of the acetyl-CoA synthase reaction is an energy-rich intermediate, it can be assumed that possibly the major part of the free energy is released in the methyl transferase reaction. This hypothesis, however, is speculative and remains to be proven.

The acetyl-CoA is further converted to acetate in the catabolism by the enzymes *phosphotransacetylase* (reaction 5.13) and *acetate kinase* (reaction 5.14):

$$\text{acetyl-CoA} + P_i \rightarrow \text{acetyl phosphate} + \text{CoA}, \Delta G'_0 = +9 \text{ kJ/mol} \quad (5.13)$$
$$\text{acetyl phosphate} + \text{ADP} \rightarrow \text{acetate} + \text{ATP}, \Delta G'_0 = -13 \text{ kJ/mol} \quad (5.14)$$

The ATP, which was required for the activation of formate (reaction 5.5) is gained back in reaction 5.14.

5.2.4 ATP Synthesis Coupled to Acetate Formation from H_2 plus CO_2

From the fact that most homoacetogens are able to grow at the expense of H_2 plus CO_2 as the sole energy sources, it can be deduced that CO_2 reduction to acetate has to be coupled to the net synthesis of ATP. It is still enigmatic which reaction(s) is (are) involved in energy conservation. From the scheme shown in Figure 5.2 it is evident that one ATP is consumed for the activation of formate. One ATP is generated via substrate-level phosphorylation in the acetate kinase reaction. There is no further possibility for ATP generation via substrate-level phosphorylation. Moreover, the CO_2 reduction to CO requires the input of metabolic energy (approximately one-third ATP for the reverse electron transport under standard conditions), and a minor part of the acetyl-CoA is converted to cell carbon rather than to acetate with concomitant formation of ATP.

For an estimation of the net amount of ATP formed upon acetate formation from H_2 plus CO_2, it might be useful to discuss the growth yields obtained with these substrates. From growth yields Y_S and the cell yield per mol ATP (Y_{ATP}) reported by Tschech and Pfennig (1984) for batch cultures of *A. woodii*, it can be calculated that approximately 1 to 1.5 ATP are generated per mole of acetate formed by the bacteria. From the $\Delta G'_0$ value of -95 kJ/mol acetate formed from H_2 plus CO_2 (see reaction 5.2) and assuming a $\Delta G'$ required for ATP synthesis in the cell of -60 to -80 kJ/mol (Thauer et al., 1977), an ATP yield of 1 to 1.5 per mole of acetate formed can be expected under standard conditions. Although the growth yields were determined in batch cultures under nonstandard conditions, the values obtained can be interpreted to indicate that the free energy available from the fermentation has to be almost completely utilized for ATP synthesis (the anabolism can be ignored here for the calculation due to its minor quantitative importance). Other publications even report higher growth yields with H_2 plus CO_2 (Braun, 1979; Lynd and Zeikus, 1983). From this it can be concluded that all sufficiently exergonic reactions in energy metabolism are possibly involved in energy conservation.

Considering the pathway of acetate synthesis (Fig. 5.2) and the free energies

Table 5.1   Free energy changes under standard conditions at pH 7 of reactions involved in acetate synthesis from CO_2

Reaction	$\Delta G'_0$ for 2 [H] $\equiv$ H$_2$ (kJ/mol)	$\Delta G'_0$ for 2 [H] $\equiv$ NAD(P)H (kJ/mol)
CO_2 + 2 [H] $\rightarrow$ HCOO$^-$ + H$^+$	+3.4	+21.5
Formate + FH$_4$ + ATP $\rightarrow$ formyl-FH$_4$ + ADP + P$_i$	-8.4	-8.4
Formyl-FH$_4$ + H$^+$ $\rightarrow$ methenyl-FH$_4$$^+$ + H$_2$O	-4.0	-4.0
Methenyl-FH$_4$$^+$ + 2 [H] $\rightarrow$ methylene-FH$_4$ + H$^+$	-23.0	-4.9
Methylene-FH$_4$ + 2 [H] $\rightarrow$ methyl-FH$_4$	-42.0	-22.0
Methyl-FH$_4$ + CO + CoA $\rightarrow$ acetyl-CoA + FH$_4$	-38.0	-38.0
Acetyl-CoA + P$_i$ $\rightarrow$ acetyl phosphate + CoA	+9.0	+9.0
Acetyl phosphate + ADP $\rightarrow$ acetate + ATP	-13.0	-13.0
CO_2 + 2 [H] $\rightarrow$ CO + H$_2$O	+21.0	+41.0

Note: The data were mainly taken from (Fuchs, 1986); for the energetics of methylene tetrahydrofolate reductase, see (Wohlfarth and Diekert, 1991).

of the reactions involved (Table 5.1), it is obvious that the methylene tetrahydrofolate reductase (reaction 5.8) and the formation of acetyl-CoA from CH_3-FH$_4$ plus carbon monoxide (reaction 5.9 plus reaction 5.12) are sufficiently exergonic to provide metabolic energy via a chemiosmotic mechanism.

That energy conservation is coupled to methyl group synthesis was confirmed by investigations on acetate formation from CO_2 or CO plus formaldehyde with cell suspensions of *A. woodii* (Diekert et al., 1986). Evidence was presented with *Peptostreptococcus productus* (Wohlfarth et al., 1990) and with *C. thermoautotrophicum* (Hugenholtz et al., 1987), that the methylene tetrahydrofolate reductase might be associated with the cytoplasmic membrane, thus supporting the hypothesis of methylene reduction being involved in energy conservation. The free energy of methyl tetrahydrofolate formation from methylene tetrahydrofolate and H$_2$ (E'_0 = -414 mV) is almost -42 kJ/mol, accounting for the translocation of possibly two monovalent cations across the membrane when coupled to this reaction. This is also true for ferredoxin (Fd) as the electron donor [E'_0 for the pair Fd$_{ox}$/Fd$_{red}$ $\approx$ -420 mV; see, for example (Reubelt et al., 1991)]. It should be noted that in the carbon monoxide utilizing *P. productus* the reducing equivalents for this reaction are provided by NADH (E'_0 $\approx$ -320 mV) rather than by ferredoxin. Therefore, due to the more positive $\Delta G'_0$ of methylene reduction with NADH, the reaction could be coupled to the electrogenic transport of just one monovalent cation. As the hydrogenase probably uses ferredoxin as the electron acceptor [see for example (Ragsdale and Ljungdahl, 1984)], the reducing equivalents in *P. productus* have to be transferred from

Table 5.2   Standard redox potentials of compounds and intermediates involved in acetate synthesis from CO_2

Redox Couple	$\Delta E'_0$ (mV)
CO_2/HCOO⁻	−430
Methenyl-FH_4/methylene-FH_4	−295
Methylene-FH_4/methyl-FH_4	−200
CO_2/CO	−524
2 H^+/H_2	−414
Ferredoxin$_{ox}$/ferredoxin$_{red}$	−420
NAD(P)$^+$/NAD(P)H	−320

Note: The data were mainly taken from (Fuchs, 1986); the value given for the methylene-FH_4/methyl-FH_4 couple are from (Wohlfarth and Diekert, 1991).

ferredoxin to NAD^+ to provide the reducing power for methylene reduction. The standard redox potential difference between ferredoxin and NAD^+ is approximately 100 mV, which could also account for an ATP synthesis ($\sim$ one-third ATP corresponding to one electrogenic proton translocated per 2 [H] transferred under standard conditions) via a chemiosmotic mechanism. Until now, there is no evidence for an energy coupling of the pyridine nucleotide : ferredoxin oxidoreductase reaction, although this cannot be excluded. The standard redox potentials of the metabolites and some electron carriers involved in acetate synthesis are summarized in Table 5.2.

It was found with *P. productus* that acetate formation from CO_2 or CO_2 plus carbon monoxide is dependent on sodium (Geerligs et al., 1989). The data indicated that methyl-group synthesis from formate required sodium. This was confirmed in growth studies for *A. woodii* (Heise et al., 1989) and for *Acetogenium kivui* (Yang and Drake, 1990). With *A. woodii*, evidence was presented that sodium probably plays a role in the formation of the (bound) methanol from methylene tetrahydrofolate (Heise et al., 1989). From these findings it can be concluded that the net ATP formed upon acetate formation is generated by a sodium gradient via a chemiosmotic mechanism.

Which reaction in methyl-group formation is involved in sodium translocation is subject to speculation. Either the methylene tetrahydrofolate reductase or the methyltransferase or even both enzymes might couple the enzymatic reaction with a sodium extrusion (see also Section 5.3). If the methyl-group synthesis is considered as analogous to the methane formation from CO_2 as catalyzed by the methanogenic bacteria, it might be useful to compare the energetics of both pathways with respect to the effect of sodium. Recently, evidence was presented with a strain of *Methanosarcina*, that the methyltransferase reaction in this organism generated a sodium gradient across the cytoplasmic membrane (Becher et al., 1992). This could indicate that also in acetogens the methyltransferase reaction might be the sodium-dependent step. However, it cannot be excluded that, in addition, the methylene tetrahydrofolate reductase reaction drives the extrusion of sodium (or of protons). This was supported by experiments performed with the methyl chloride utilizing strain MC (see Section 5.3).

There are two possibilities for the ATP generation by a sodium gradient (see Fig. 5.4): Either the sodium gradient is converted to a proton gradient, which, in turn, drives the ATP formation by a proton-translocating ATP synthase, or the sodium gradient is directly utilized by a sodium-dependent ATP synthase. The former possibility would involve a Na^+/H^+ antiporter for the interconversion of the gradients. Evidence for the presence of such an antiporter is available for *C. thermoaceticum* (Terracciano et al., 1987). This antiporter, however, could also be required to supply a proton gradient, which drives the reduction of CO_2 to CO, i.e., the formation of the carboxyl group of acetate. Therefore, the involvement of a sodium-dependent ATP synthase is feasible. The presence of a sodium-translocating ATP synthase was recently reported for *A. woodii* (Heise

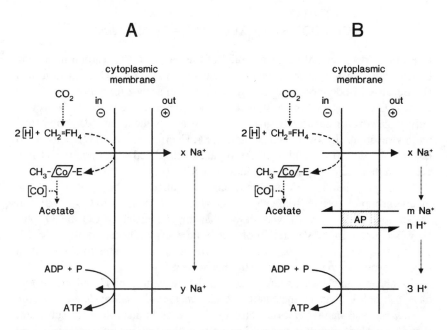

Figure 5.4. Possible chemiosmotic mechanisms of energy conservation via a sodium gradient generated in the methylene tetrahydrofolate reductase/methyltransferase reaction(s). (A) Involvement of a sodium-translocating ATP synthase; (B) involvement of a proton-translocating ATP synthase and a sodium-proton antiporter (AP). $CH_2=FH_4$ = methylene tetrahydrofolate; CH_3-[Co]-E = corrinoid enzyme with a bound methyl group.

et al., 1992). The role of sodium in homoacetogenesis is discussed in more detail in Chapter 4. A tentative scheme of the energy coupling to acetate synthesis from CO_2 is given in Figure 5.4.

5.3   Utilization of Reduced C_1 Substrates

The finding that *carbon monoxide* can be a direct precursor for the synthesis of the carboxyl group of acetate *in vitro* (Hu et al., 1982; Ma et al., 1991; Roberts et al., 1992) and *in vivo* (Diekert and Ritter, 1983a; Ma et al., 1987) provided essential information on the role of CO as an intermediate in acetate synthesis from CO_2 (Diekert et al., 1984). Therefore, it is not surprising that some homoacetogens are able to utilize carbon monoxide as the sole energy source [for example, see (Genthner and Bryant, 1982; Lynd et al., 1982; Lorowitz and Bryant, 1984; Geerligs et al., 1987; Savage et al., 1987; Traunecker et al., 1991)]. Carbon monoxide is converted to acetate according to the following equation:

$$4CO + 2H_2O \rightarrow \qquad\qquad\qquad (5.15)$$
$$1\ CH_3COO^- + H^+ + 2\ CO_2, \Delta G'_0 = -165.6\ \text{kJ/mol}$$

Due to the low standard redox potential of the pair CO_2/CO, carbon monoxide is thermodynamically a more favorable electron donor for CO_2 reduction than H_2. Because carbon monoxide is an intermediate in acetate formation, one could expect CO to be a good growth substrate for all homoacetogens. So far, however, only *P. productus* apparently grows very fast and without any adaptation time on this substrate (Lorowitz and Bryant, 1984; Geerligs et al., 1987). This could possibly be due to the ability of the organism to directly incorporate CO into acetate rather than via previous oxidation to CO_2 (Ma et al., 1987).

To provide the six reducing equivalents required for the synthesis of the methyl group of acetate from CO_2, 3 CO have to be oxidized to CO_2. With cell suspensions of *A. woodii* it was shown that the conversion of CO to CO_2 plus H_2 was coupled to the generation of metabolic energy (Diekert et al., 1986). It can be assumed that CO oxidation in *P. productus* contributes to some extent to ATP synthesis via a chemiosmotic mechanism.

Homoacetogenic bacteria are able to utilize a variety of C_1 substrates that have the redox level of methanol. Besides *methanol, methoxylated aromatic compounds* and *methyl chloride* also can serve as homoacetogenic substrates (Fig. 5.1). Until now it is not yet known how these compounds are channeled into the methyl-group synthesis. C_1 units at the level of methanol (i.e., a methyl$^+$) could be converted either to methyl tetrahydrofolate or to the enzyme-bound methyl corrinoid, or they could be oxidized to formaldehyde or formate prior to being fed into the pool of tetrahydrofolate-bound C_1 intermediates (see Chapters

6 and 17). For the energetics of acetate synthesis from these compounds, it is important to know to which intermediate in methyl-group formation the substrates are converted. The conversion of methyl substrates can be expressed by the following reaction:

$$4 CH_3\text{-}X + 2 CO_2 + 2 H_2O \rightarrow 3 \text{ acetate}^- + 7H^+ + 4X^- \qquad (5.16)$$

One methyl group has to be oxidized to CO_2 to provide the 6 [H] required for the reduction of 3 CO_2 to 3 carbon monoxide (in a bound form). The carbonyl groups are then combined with three methyl groups to yield acetate (Fig. 5.5). As the methyltransferase reaction is considered as a major site in energy conservation, an incorporation of the methyl group into the methyl tetrahydrofolate or the methyl corrinoid pool would imply different energetics of the methyl group conversion. If the methyl group is bound directly to tetrahydrofolate, the methyltransferase would not be involved in the oxidation of the methyl group to CO_2 (Fig. 5.5). Instead, the methyltransferase would mediate the methyl transfer in the exergonic direction, thus being involved in energy conservation. If the target

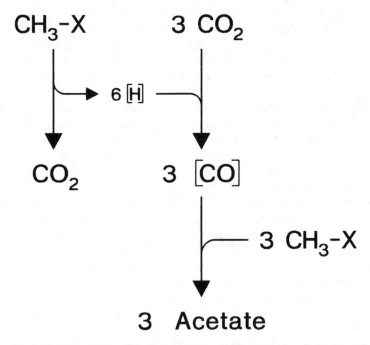

Figure 5.5. Acetate formation from C_1 substrates at the redox level of methanol. Left side: oxidative part; right side: reductive part of the pathway.

of the methyl incorporation is the enzyme-bound corrinoid, one methyl group has to be oxidized (oxidative part of the pathway; Fig. 5.5) via the methyl-transferase reaction in the oxidative, i.e., the endergonic direction, and three methyl groups have to be directly incorporated into C-2 of acetyl-CoA without involvement of the methyltransferase. This implies that the methyl group conversion to methyl tetrahydrofolate would be energetically more favorable than the conversion to an enzyme-bound methyl corrinoid.

With all homoacetogens tested so far, the growth yields with *methanol* ($Y_S \approx$ 5–6 g dry cells/mol) were significantly lower than with *methoxylated benzoate derivatives* ($Y_S \approx$ 8–10 g/mol) (Tschech and Pfennig, 1984; Schuppert and Schink, 1990; Daniel et al., 1991). This could be interpreted to indicate that methanol is directly fed into the methyl corrinoid pool, whereas the methoxyl group of substituted benzoates is converted to methyl tetrahydrofolate. Evidence is available that the incorporation of the methoxyl group into the methyl group of acetate requires tetrahydrofolate (Berman and Frazer, 1992; Doré and Bryant, 1990; Meßmer et al., 1993; Chapter 17) and that the methyl acceptor for methanol conversion is an enzyme-bound cobalamin (van der Meijden et al., 1984; Stupperich, personal communication; Chapter 6). Surprisingly, the methyl group conversion from both substrates appears to require ATP (Berman and Frazer, 1992; van der Meijden et al., 1984), possibly in catalytic amounts (Stupperich, personal communication). Because nothing is known on the energetics of the methyl transfer to the primary acceptor, an ATP requirement of this first step cannot be excluded. It is also feasible that the reaction in the cell-free systems investigated so far exhibits unphysiological properties, and that, *in vivo,* the reaction is not ATP dependent. In addition, it is possible that methanol and/or the methyl group of methoxylated aromatics is oxidized to formaldehyde or formate prior to being channeled into the homoacetogenic pathway. Winters and Ljungdahl (1989) demonstrated the presence of an enzyme in *C. thermoautotrophicum* which catalyzes the oxidation of methanol and formaldehyde to formate with paraquinoline quinone as the prosthetic group. An incorporation at the level of formate would require an additional ATP for the activation of formate.

Growth yields with methanol and other substrates on the redox level of methanol (methoxylated compounds and methyl chloride) are, by far, too great to be explained by all of the possible pathways of metabolism and the Y_{ATP} reported for growth with these substrates (Tschech and Pfennig, 1984). This could either indicate a Y_{ATP} higher than the published value (the value was calculated from the ATP yield of the pathway via substrate-level phosphorylation without consideration of the energy conservation via a chemiosmotic mechanism) or an energy conservation possibly coupled to the transfer of the methyl group to the primary acceptor.

Homoacetogenic bacteria were recently demonstrated to catalyze the transfer of the methyl group of methoxylated compounds to sulfide upon formation of methanethiol and dimethylsulfide in their energy metabolism (Bak et al., 1992). It is still unclear whether the methanethiol formation is a co-metabolic or a metabolic reaction. Studies on the metabolism of these organisms might help to obtain more detailed information on the methyl transfer reaction and its energy coupling.

Recently, it was shown that *methyl chloride* can serve as a homoacetogenic substrate (Traunecker et al., 1991). Growth yields of the bacterium with methyl chloride ($Y_S \approx 7.9$ g/mol) were higher than with methanol ($Y_S \approx 4.9$ g/mol) and in the same order as the growth yield with methoxylated aromatics ($Y_S \approx 7.7$ g/mol) (Traunecker et al., 1991). From the thermodynamics, the methyl chloride conversion to acetate according to

$$4\,CH_3Cl + 2\,CO_2 + 2\,H_2O \rightarrow 3\,CH_3COO^- + 7\,H^+ + 4\,Cl^-, \qquad (5.17)$$
$$\Delta G'_0 = -140 \text{ kJ/mol acetate}$$

is by far more favorable than methanol conversion:

$$4\,CH_3OH + 2\,CO_2 \rightarrow 3\,CH_3COO^- + 2\,H_2O + 3\,H^+, \qquad (5.18)$$
$$\Delta G'_0 = -71 \text{ kJ/mol acetate}$$

Relatively little information is available on methyl chloride metabolism of bacteria. Growth yields may be interpreted to indicate a methyl chloride metabolism similar to that of methoxylated compounds. Cell-free extracts of the methyl chloride utilizing strain MC were demonstrated to catalyze the conversion of methyl chloride plus tetrahydrofolate to methyl tetrahydrofolate, suggesting that FH_4 is the primary acceptor in the CH_3Cl dehalogenation.

However, evidence is available that vitamin B_{12} can catalyze an abiotic dehalogenation of chlorinated methanes including methyl chloride (Krone et al., 1989). It can be speculated that the higher growth yields with CH_3Cl could also be due to an energy conservation coupled to the dehalogenation of methyl chloride. (The free energy of the dehalogenation can be calculated from the difference of the $\Delta G'_0$ values given in reactions 5.17 and 5.18) An energy-yielding dehalogenation was reported for the dechlorination of 3-chlorobenzoate by Dolfing (1990). If the corrinoid protein, which serves as the methyl donor for acetate formation, would be the acceptor in methyl chloride dechlorination, the methyl transferase should be involved in methyl tetrahydrofolate formation from CH_3Cl. This appears unlikely, because the methyl transfer from the corrinoid to FH_4 is an endergonic reaction, which probably would require the input of metabolic energy via a chemiosmotic mechanism and which should not be mediated by cell-free extracts. Therefore, we believe that FH_4 rather than the corrinoid protein is the acceptor of the methyl group in the dehalogenase reaction. Because the growth

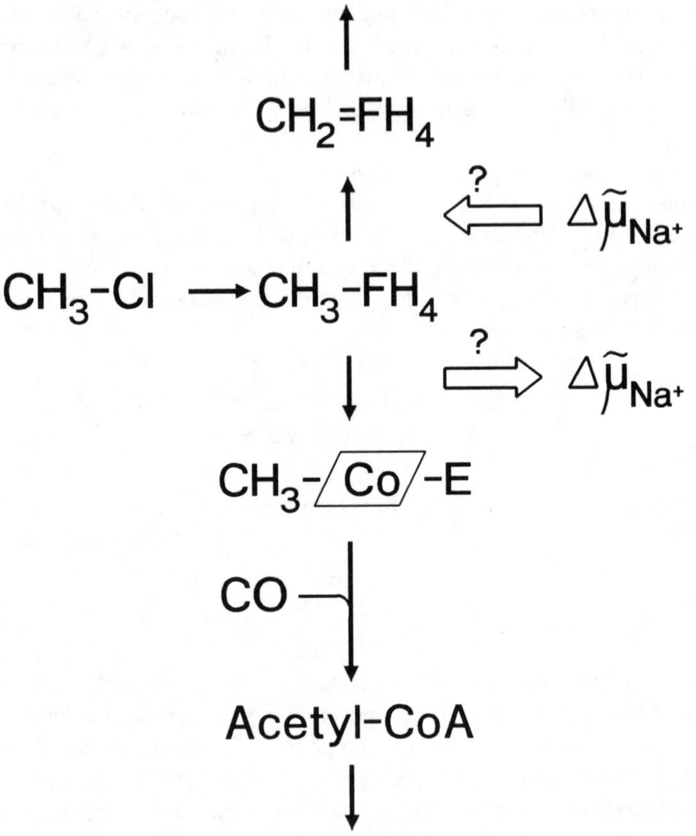

$$CH_2=FH_4$$

$$CH_3-Cl \longrightarrow CH_3-FH_4$$

$$\Leftarrow \ ? \quad \Delta\tilde{\mu}_{Na^+}$$

$$\Rightarrow \ ? \quad \Delta\tilde{\mu}_{Na^+}$$

$$CH_3-\boxed{Co}-E$$

$$CO \dashv$$

$$Acetyl-CoA$$

Figure 5.6. Tentative scheme of methyl chloride conversion in the homoacetogenic pathway. $\Delta\tilde{\mu}_{Na^+}$ = electrochemical sodium gradient. The thick open arrows indicate the input or generation of the sodium gradient. The question marks indicate reactions where sodium may be required.

yields with CH_3Cl are similar to those with methoxylated aromatic compounds, it must be assumed that either both the dehalogenation of CH_3Cl and the O-demethylation or none of both reactions are coupled to energy conservation.

Studies on the effect of sodium and of monensin (an artificial Na^+/H^+ antiporter) on acetate formation from $CH_3Cl \pm CO$ indicated that two sodium-dependent steps could be involved in CH_3Cl metabolism, one of which is in the oxidative part of the pathway and the other in acetyl-CoA synthesis from CH_3Cl plus carbon monoxide (unpublished results; see also (Meßmer et al., 1993). If the methyl group of CH_3Cl is directly transferred to tetrahydrofolate, it is likely that the methylene tetrahydrofolate reductase and the methyltransferase both are sodium-dependent reactions. A tentative scheme of the possible role of sodium in the methyl chloride conversion to acetate is shown in Figure 5.6. Further studies on the metabolism of methoxylated compounds and of methyl chloride might help elucidate the mechanism of energy conservation in homoacetogenic bacteria.

References

Bache, R., and N. Pfennig. 1981. Selective isolation of *Acetobacterium woodii* on methoxylated aromatic acids and determination of growth yields. *Arch. Microbiol.* **130**:255–261.

Bak, F., K. Finster, and F. Rothfuß. 1992. Formation of dimethylsulfide and methanethiol from methoxylated aromatic compounds and inorganic sulfide by newly isolated anaerobic bacteria. *Arch. Microbiol.* **157**:529–534.

Becher, B., V. Müller, and G. Gottschalk. 1992. The methyl-tetrahydromethanopterin : coenzyme M methyltransferase of *Methanosarcina* strain Göl is a primary sodium pump. *FEMS Microbiol. Lett.* **91**:239–244.

Berman, M. H., and A. C. Frazer. 1992. Importance of tetrahydrofolate and ATP in the anaerobic O-demethylation reaction for phenylmethylethers. *Appl. Environ. Microbiol.* **58**:925–931.

Braun, K. 1979. Untersuchungen zum autotrophen, heterotrophen und mixotrophen Wachstum von *Acetobacterium woodii* und *Clostridium aceticum*. Ph.D. thesis, Universität Göttingen, Germany.

Conrad, R., F. Bak, H. J. Seitz, B. Thebrath, H. P. Mayer, and H. Schütz. 1989. Hydrogen turnover by psychrotrophic homoacetogenic and mesophilic methanogenic bacteria in anoxic paddy soil and lake sediment. *FEMS Microbiol. Ecol.* **62**:285–294.

Daniel, S. L., E. S. Keith, H. Yang, Y. S. Lin, and H. L. Drake. 1991. Utilization of methoxylated aromatic compounds by the acetogen *Clostridium thermoaceticum:* expression and specificity of the CO-dependent O-demethylating activity. *Biochem. Biophys. Res. Commun.* **180**:416–422.

Diekert, G. 1992. The acetogenic bacteria. In: *The Prokaryotes,* Vol. I, A. Balows, H. G. Trüper, H. Dworkin, W. Harder, and K. H. Schleifer (eds.), pp. 517–533. Springer-Verlag, New York.

Diekert, G., and R. K. Thauer. 1978. Carbon monoxide oxidation by *Clostridium thermoaceticum* and *Clostridium formicoaceticum*. *J. Bacteriol.* **136**:597–606.

Diekert, G. B., E. G. Graf, and R. K. Thauer. 1979. Nickel requirement for carbon monoxide dehydrogenase formation in *Clostridium pasteurianum*. *Arch. Microbiol.* **122**:117–120.

Diekert, G., and R. K. Thauer. 1980. The effect of nickel on carbon monoxide dehydrogenase formation in *Clostridium thermoaceticum* and *Clostridium formicoaceticum. FEMS Microbiol. Lett.* **7**:187–189.

Diekert, G., and M. Ritter. 1983a. Carbon monoxide fixation into the carboxyl group of acetate during growth of *Acetobacterium woodii* on H_2 and CO_2. *FEMS Microbiol. Lett.* **17**: 299–302.

Diekert, G., and M. Ritter. 1983b. Purification of the nickel protein carbon monoxide dehydrogenase of *Clostridium thermoaceticum. FEBS Lett.* **151**:41–44.

Diekert, G., M. Hansch, and R. Conrad. 1984. Acetate synthesis from 2 CO_2 in acetogenic bacteria: is carbon monoxide an intermediate? *Arch. Microbiol.* **138**:224–228.

Diekert, G., E. Schrader, and W. Harder. 1986. Energetics of CO formation and CO oxidation in cell suspensions of *Acetobacterium woodii. Arch. Microbiol.* **144**:386–392.

Dolfing, J. 1990. Reductive dechlorination of 3-chlorobenzoate is coupled to ATP production and growth in an anaerobic bacterium, strain DCB-1. *Arch. Microbiol.* **153**:264–266.

Doré, J., and P. Bryant. 1990. Metabolism of one-carbon compounds by the ruminal acetogen *Syntrophococcus sucromutans. Appl. Environ. Microbiol.* **56**:984–989.

Drake, H. L., S. I. Hu, and H. G. Wood. 1980. Purification of carbon monoxide dehydrogenase, a nickel enzyme from *Clostridium thermoaceticum. J. Biol. Chem.* **255**:7174–7180.

Drake, H. L. 1992. Acetogenesis and acetogenic bacteria. In: *Encyclopedia of Microbiology, Vol. 1.* J. Lederberg (ed.), pp. 1–15. Academic Press, San Diego, CA.

Eaton, W. A., and W. Lovenberg. 1973. The iron-sulfur complex in rubredoxin. In: *Iron-Sulfur Proteins, Vol. 2*, W. Lovenberg (ed.). pp. 131–162. Academic Press, New York.

Fontaine, F. E., W. H. Peterson, E. McCoy, and M. J. Johnson. 1942. A new type of glucose fermentation by *Clostridium thermoaceticum* n. sp. *J. Bacteriol.* **43**:701–715.

Fuchs, G. 1986. CO_2 fixation in acetogenic bacteria: variations on a theme. *FEMS Microbiol. Rev.* **39**:181–213.

Geerligs, G., H. C. Aldrich, W. Harder, and G. Diekert. 1987. Isolation and characterization of a carbon monoxide utilizing strain of the acetogen *Peptostreptococcus productus. Arch. Microbiol.* **148**:305–313.

Geerligs, G., P. Schönheit, and G. Diekert. 1989. Sodium dependent acetate formation from CO_2 in *Peptostreptococcus productus* (strain Marburg). *FEMS Microbiol. Lett.* **57**:253–258.

Genthner, B. R. S., C. L. Davis, and M. P. Bryant. 1981. Features of rumen and sewage sludge strains of *Eubacterium limosum*, a methanol- and H_2-CO_2-utilizing species. *Appl. Environ. Microbiol.* **42**:12–19.

Genthner, B. R. S., M. P. Bryant. 1982. Growth of *Eubacterium limosum* with carbon monoxide as the energy source. *Appl. Environ. Microbiol.* **43**:70–74.

Heise, R., V. Müller, and G. Gottschalk. 1989. Sodium dependence of acetate formation by the acetogenic bacterium *Acetobacterium woodii*. *J. Bacteriol.* **171**:5473–5478.

Heise, R., V. Müller, and G. Gottschalk. 1992. Presence of a sodium-translocating ATPase in membrane vesicles of the homoacetogenic bacterium *Acetobacterium woodii*. *Eur. J. Biochem.* **206**:553–557.

Hsu, T., S. L. Daniel, M. F. Lux, and H. L. Drake. 1990. Biotransformation of carboxylated aromatic compounds by the acetogen *Clostridium thermoaceticum*: generation of growth-supportive CO_2 equivalents under CO_2-limited conditions. *J. Bacteriol.* **172**:212–217.

Hu, S. I., H. L. Drake, H. G. Wood. 1982. Synthesis of acetyl coenzyme A from carbon monoxide, methyltetrahydrofolate, and coenzyme A by enzymes from *Clostridium thermoaceticum*. *J. Bacteriol.* **149**:440–448.

Hugenholtz, J., D. M. Ivey, and L. G. Ljungdahl. 1987. Carbon monoxide-driven electron transport in *Clostridium thermoautotrophicum* membranes. *J. Bacteriol.* **169**:5845–5847.

Hugenholtz, J., and L. G. Ljungdahl. 1990. Metabolism and energy generation in homoacetogenic clostridia. *FEMS Microbiol. Rev.* **87**:383–390.

Kamlage, B., and M. Blaut. 1993. Isolation of a cytochrome-deficient mutant strain of *Sporomusa sphaeroides* not capable of oxidizing methyl groups. *J. Bacteriol.* **175**:3043–3050.

Keltjens, J. T., and C. van der Drift. 1986. Electron transfer reactions in methanogens. *FEMS Microbiol. Rev.* **39**:259–303.

Krone, U. E., R. K. Thauer, and H. P. C. Hogenkamp. 1989. Reductive dehalogenation of chlorinated C_1-hydrocarbons mediated by corrinoids. *Biochemistry* **28**:4908–4914.

Ljungdahl, L. G. 1986. The autotrophic pathway of acetate synthesis in acetogenic bacteria. *Annu. Rev. Microbiol.* **40**:415–450.

Lorowitz, W. H., and M. P. Bryant. 1984. *Peptostreptococcus productus* strain that grows rapidly with CO as the energy source. *Appl. Environ. Microbiol.* **47**:961–964.

Lu, W. P., and S. W. Ragsdale. 1991. Reductive activation of coenzyme A/acetyl CoA isotopic exchange reaction catalyzed by carbon monoxide dehydrogenase from *Clostridium thermoaceticum* and its inhibition by nitrous oxide and carbon monoxide. *J. Biol. Chem.* **266**:3554–3564.

Lynd, L., R. Kerby, and J. G. Zeikus. 1982. Carbon monoxide metabolism of the methylotrophic acidogen *Butyribacterium methylotrophicum*. *J. Bacteriol.* **149**:255–263.

Lynd, L. H., and J. G. Zeikus. 1983. Metabolism of H_2-CO_2, methanol, and glucose by *Butyribacterium methylotrophicum*. *J. Bacteriol.* **153**:1415–1423.

Ma, K., S. Siemon, and G. Diekert. 1987. Carbon monoxide metabolism in cell suspensions of *Peptostreptococcus productus* strain Marburg. *FEMS Microbiol. Lett.* **43**:367–371.

Ma, K., G. Wohlfarth, and G. Diekert. 1991. Acetate formation from CO and CO_2 by cell extracts of *Peptostreptococcus productus* (strain Marburg). *Arch. Microbiol.* **156**:75–80.

Meßmer, M., G. Wohlfarth, and G. Diekert. 1993. Methyl chloride metabolism of the strictly anaerobic, methyl chloride-utilizing homoacetogen strain MC. *Arch. Microbiol.* **160**:383–387.

Ragsdale, S. W. 1991. Enzymology of the acetyl-CoA pathway of CO_2 fixation. *Crit. Rev. Biochem. Mol. Biol.* **26**:261–300.

Ragsdale, S. W., J. E. Clark, L. G. Ljungdahl, L. L. Lundie, and H. L. Drake. 1983a. Properties of purified carbon monoxide dehydrogenase from *Clostridium thermoaceticum*, a nickel, iron-sulfur protein. *J. Biol. Chem.* **258**:2364–2369.

Ragsdale, S. W., L. G. Ljungdahl, and D. V. Vartanian. 1983b. Isolation of carbon monoxide dehydrogenase from *Acetobacterium woodii* and comparison of its properties with those of the *Clostridium thermoaceticum* enzyme. *J. Bacteriol.* **155**:1224–1237.

Ragsdale, S. W., and L. G. Ljungdahl. 1984. Hydrogenase from *Acetobacterium woodii*. *Arch. Microbiol.* **139**:361–365.

Reubelt, U., G. Wohlfarth, R. Schmid, and G. Diekert. 1991. Purification and characterization of ferredoxin from *Peptostreptococcus productus* (strain Marburg). *Arch. Microbiol.* **156**:422–426.

Roberts, J. R., W. P. Lu, and S. W. Ragsdale. 1992. Acetyl-coenzyme-A synthesis from methyltetrahydrofolate, CO, and coenzyme-A by enzymes purified from *Clostridium thermoaceticum*—attainment of in vivo rates and identification of rate-limiting steps. *J. Bacteriol.* **174**:4667–4676.

Savage, M. D., Z. Wu, S. L. Daniel, L. L. Lundie, and H. L. Drake. 1987. Carbon monoxide-dependent chemolithotrophic growth of *Clostridium thermoautotrophicum*. *Appl. Environ. Microbiol.* **53**:1902–1906.

Schink, B., and M. Bomar. 1992. The genera *Acetobacterium, Acetogenium, Acetoanaerobium* and *Acetitomaculum*. In: The *Prokaryotes*. A Handbook on the Biology of Bacteria: ecophysiology, isolation, identification, applications, 2nd ed., A. Balows, H. G. Trüper, M. Dworkin, W. Harder, and K. H. Schleifer (eds.), Vol. 1, pp. 1925–1936. Springer-Verlag, New York.

Schönheit, P. 1993. The role of Na^+ in the first step of CO_2 reduction to methane in methanogenic bacteria. In: Alkali Kation Transport Systems in Procaryotes, E. Bakker (ed.). pp. 179–202. CRC Press, Boca Raton, FL.

Schulman, M., R. K. Ghambeer, L. G. Ljungdahl, and H. G. Wood. 1973. Total synthesis of acetate from CO_2. VII. Evidence with *Clostridium thermoaceticum* that the carboxyl group of acetate is derived from the carboxyl of pyruvate by transcarboxylation and not by fixation of CO_2. *J. Biol. Chem.* **248**:6255–6261.

Schuppert, B., and B. Schink. 1990. Fermentation of methoxyacetate to glycolate and acetate by newly isolated strains of *Acetobacterium* sp. *Arch. Microbiol.* **153**:200–204.

Tanaka, K., and N. Pfennig. 1988. Fermentation of 2-methoxyethanol by *Acetobacterium malicum* sp. nov. and *Pelobacter venetianus*. *Arch. Microbiol.* **149**:181–187.

Terracciano, J. S., W. J. A. Schreurs, and E. R. Kashket. 1987. Membrane H$^+$ conductance of *Clostridium thermoaceticum* and *Clostridium acetobutylicum:* evidence for electrogenic Na$^+$/H$^+$ antiport in *Clostridium thermoaceticum*. *Appl. Environ. Microbiol.* **53**:782–786.

Thauer, R. K. 1990. Energy metabolism of methanogenic bacteria. *Biochim. Biophys. Acta* **1018**:256–259.

Thauer, R. K., K. Jungermann, and K. Decker. 1977. Energy conservation in chemotrophic anaerobic bacteria. *Bacteriol. Rev.* **41**:100–180.

Traunecker, J., A. Preuß, and G. Diekert. 1991. Isolation and characterization of a methyl chloride utilizing, strictly anaerobic bacterium. *Arch. Microbiol.* **156**:416–421.

Tschech, A., and N. Pfennig. 1984. Growth yield increase linked to caffeate reduction in *Acetobacterium woodii*. *Arch. Microbiol.* **137**:163–167.

van der Meijden, P., C. van der Drift, and G. D. Vogels. 1984. Methanol conversion in *Eubacterium limosum*. *Arch. Microbiol.* **138**:360–364.

Winters, D. K., and L. G. Ljungdahl. 1989. PQQ-dependent methanol dehydrogenase from *Clostridium thermoautotrophicum*. In: *PQQ and Quinoproteins*, J. A. Jongejan and J. A. Duine (eds.), pp. 35–39. Kluwer Academic Publishers, Dordrecht.

Wohlfarth, G., G. Geerligs, and G. Diekert. 1990. Purification and properties of a NADH dependent 5,10-methylenetetrahydrofolate reductase from *Peptostreptococcus productus*. *Eur. J. Biochem.* **192**:411–417.

Wohlfarth, G., and G. Diekert. 1991. Thermodynamics of methylenetetrahydrofolate reduction to methyltetrahydrofolate and its implications for the energy metabolism of homoacetogenic bacteria. *Arch. Microbiol.* **155**:378–381.

Wood, H. G. 1991. Life with CO or CO$_2$ and H$_2$ as a source of carbon and energy. *FASEB J.* **5**:156–163.

Wood, H. G., and L. G. Ljungdahl. 1991. Autotrophic character of the acetogenic bacteria. In: *Variations in Autotrophic Life*, J. M. Shively and L. L. Barton (eds.), pp. 201–250. Academic Press, New York.

Yang, H., and H. L. Drake. 1990. Differential effects of sodium on hydrogen- and glucose-dependent growth of the acetogenic bacterium *Acetogenium kivui*. *Appl. Environ. Microbiol.* **56**:81–86.

6

Corrinoid-Dependent Mechanism of Acetogenesis from Methanol

Erhard Stupperich

6.1 Introduction

More than 13 reactions are catalyzed by corrinoid-containing enzymes in prokaryotes and eukaryotes. These reactions were reviewed previously (Stadtman, 1971; Halpern, 1985) but some additional corrinoid-dependent reactions have been discovered since then. For example, the *methanol* conversion into methane proceeds via a corrinoid-dependent *methyltransferase* (van der Meijden et al., 1984a). The enzyme from *Methanosarcina barkeri* provides the methyl group from methanol to the specific methanogenic cofactor 2-mercaptoethanesulfonic acid (HS-CoM) after the protein is reductively activated by H_2 and ATP. That enzyme revealed an $\alpha_2\beta$ structure and it contained 3-4 mol loosely bound corrinoid per mole of protein. *In vitro* studies with mixed cell-free extracts of *Methanosarcina barkeri* and *Eubacterium limosum* indicated that the acetogenic bacterium also possesses a methylated corrinoid enzyme. This enzyme was demethylated in the presence of methylcobalamin : coenzyme M methyltransferase from the methanogenic bacterium with concomitant formation of methyl-coenzyme M (van der Meijden et al., 1984b).

The mechanism of many corrinoid-dependent reactions is a matter of debate. In particular, the interactions of the corrinoid cofactors with their proteins during the catalytic cycle are not yet understood, despite numerous investigations during the last three decades. This problem is illustrated by two examples. The activation of the C–N bond of N^5-methyltetrahydrofolate by the corrinoid-dependent methionine synthase [E.C. 2.1.1.13] reaction is still a puzzle (Banerjee and Matthews 1990). Three potential activation reactions are postulated: (i) a two-electron oxidation to give rise to a quaternized N^5, (ii) a one-electron oxidation to generate the amine radical, or (iii) a quaternization via protonation at N^5. A similar situation is found with adenosylcorrinoid-dependent enzymes. The formation of protein, cofactor, and substrate radicals during the carbon rearrangement of the methylmalonyl-coenzyme A mutase [E.C. 5.4.99.2] reaction remains to be resolved (Zhao et al., 1992).

Knowledge of the corrinoid-dependent reactions could substantially enhance the development of inhibitory substrate analogues and of modified substrate range of the enzymes. This is of particular interest, because corrinoids form reactive species which are involved in several chemical syntheses (Scheffold et al., 1987) or in dehalogenations (Marks et al., 1989; Krone et al., 1989; Krone and Thauer, 1992). The protein-bound corrinoid may provide a rate enhancement of $10^{12\pm1}$ over the protein-free reaction (Finke and Martin, 1990).

More recently, evidence was presented for a corrinoid-dependent methanol conversion into acetic acid by *Sporomusa ovata* (Stupperich et al., 1992). The reaction mechanism and the physiological methyl group acceptor are currently investigated because this acetogenic system is considered a good choice for tackling corrinoid/protein interactions.

Two main findings prompted the studies of the anaerobic methanol metabolism by *Sporomusa ovata*. (i) The homoacetogenic bacterium produces considerable amounts of a corrinoid-containing protein. The expression of the homomeric peptide is induced by methanol and the enzyme is easily purified to apparent homogeneity. One mole of firmly bound corrinoid cofactor is present per mole of protein. These facts facilitate experiments on the genetic regulation of a corrinoid-containing protein as well as physicochemical studies of the peptide. (ii) The corrinoid cofactor of the *Sporomusa* enzyme is unique. Its chemical structure and perhaps some of its biological features differ significantly from all other corrinoids isolated from organisms so far. We will briefly discuss if this corrinoid structure is a prerequisite of the reaction mechanism.

Sporomusa ovata H1 (DSM2662) was isolated from sugar beet leaf silage using N,N-dimethylglycine as carbon and energy source (Möller et al., 1984). In addition, betaine, H_2/CO_2, fructose, bicarbonate plus either methanol or O-methyl groups of phenylmethylethers are employed in acetic acid formation. The methoxylated aromatics serve exclusively as methyl sources, because the remaining phenols accumulate in the medium.

Interestingly, both methanol and methoxylated aromatics require the activation of a C–O bond as the introducing step of the methyl metabolism. In fact, the *in vitro* conditions for acetyl-coenzyme A formation from both substrates are comparable in *Sporomusa*. Unfavorable reaction conditions are the methanolic C–O bond strength of 375 kJ mol^{-1} (Levis et al., 1989) and its low polarity, which hampers the displacement of either the hydroxyl or the methyl group. The stronger polarity of the C–S bond in CH_3-S^+-adenosylmethionine, for example, facilitates the release of the methyl group and establishes it as a preferred methyl donor in metabolism.

A corrinoid-dependent C–O bond cleavage reaction is catalyzed by 1,2-propanediol dehydratase [E.C. 4.2.1.28]. The enzyme converts the diol, a fermentation product of L-rhamnose in *Salmonella typhimurium* (Obrados et al., 1988), into propionaldehyde. Finally, propionaldehyde dismutates into *n*-propanol and propionate (Toraya and Fukui, 1982):

$$
\begin{array}{ccc}
\text{H} & & \text{O} \\
| & -\,\text{H}_2\text{O} & \| \\
\text{CH}_3\text{-CH-C-OH} & \rightleftharpoons & \text{CH}_3\text{-CH}_2\text{-C—H} \\
| \quad | & & \\
\text{OH} \ \text{H} & &
\end{array}
\qquad (6.1)
$$

The 1,2-*propandiol dehydratase* contains an adenosylcorrinoid cofactor. A *methylcorrinoid*, however, is present in the *Sporomusa* enzymes, which indicates a different C–O cleavage mechanism for the acetogenic methanol utilization.

The following sections describe some properties of corrinoids including special features of the *Sporomusa* cofactor. Thereafter, a proposal of a corrinoid-dependent methanol activation mechanism will be presented, succeeded by findings that demonstrate the involvement of corrinoids in the acetogenic methanol metabolism.

6.2   Properties of Corrinoids

A complete corrinoid molecule is termed a cobamide. Cobamides typically consist of three distinct molecule parts: the planar corrin ring, an upper cobalt ligand, and a lower cobalt ligand (Fig. 6.1). Thus, the oxidized cobalt reaction center carries six ligands. Four ligands derive from the pyrrole nitrogens of the corrin, and a fifth ligand is contributed by a nitrogen of the lower Coα-ligand base, which is typically a benzimidazole or an adenine derivative. In addition to this "base-on" configuration, a "base-off" form has to be distinguished in which the lower Coα ligand is not coordinated to the cobalt. Convincing evidences have been presented that the base-off position supports the heterolytic cleavage of methylcorrinoids with concomitant formation of a reduced Co(I) corrinoid (Lexa and Saveant, 1976; Ragsdale et al., 1987; Wirt et al., 1992).

The sixth ligand position of the cobalt is occupied either by a methyl, adenosyl, hydroxyl, or a cyanoyl function. This upper Coβ ligand indicates the biological role of the corrinoid. A methyl group is present in methyltransferases and the

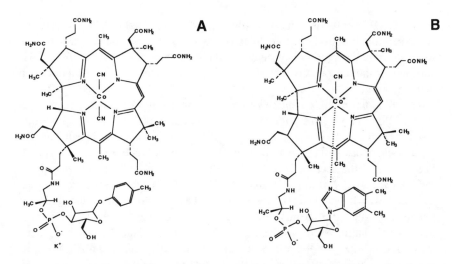

Figure 6.1.   (A) Structural formula of *para*-cresolyl cobamide isolated from *Sporomusa ovata*. (B) Vitamin B_{12} represents those cobamide forms which are frequently detected in prokaryotic and eukaryotic organisms. The cyano ligands of both corrinoids are *in vivo* replaced by a methyl and an histidine residue in *p*-cresolyl cobamide or a methyl group in vitamin B_{12}, respectively.

adenosyl group is found in corrinoid-dependent C–C rearrangement mutases or elimination reactions. The hydroxyl form is frequently found in biological samples. It is also obtained as a degradation product of the light-sensitive methyl and adenosylcorrinoids because those Co–C bonds are readily cleaved by daylight. Hence, protection from light has to be applied in handling the methyl and adenosylcorrinoids.

No biological function can be assigned to CN corrinoids. They are obtained by quantitative corrinoid extractions in the presence of potassium cyanide, or by cyanolysis of adenosylcorrinoids in darkness; methylcorrinoids resist cyanolysis under similar conditions. CN-corrinoids are the preferred commercial preparations due to their stability.

A Co(III) corrinoid can be reduced to Co(II) and Co(I) either by a chemical reductant or enzymatically. The Co(II) corrinoid is paramagnetic, whereas the Co(III) and Co(I) corrinoids are diamagnetic. Hence, the Co(II) corrinoids are accessible to EPR spectroscopy. They possess five ligands and occur after homolytic cleavage of an adenosylcorrinoid. Crystallographic analyses of a "base-on" cob(II)alamin indicates that the cobalt-coordinated 5,6-dimethylbenzimidazole base induces a shortening of the Co–N bond and an upward folding of the corrin with strikingly similarity to the crystallographic conformation of adenosylcobamide (Kräutler et al., 1989).

The Co(I) corrinoids consist of four ligands. Remarkably, these corrinoids are exceptionally strong nucleophiles (Brown, 1982), and, hence, are suitable catalysts in nucleophilic displacements. Another important function of a Co(I) corrinoid is that of a methyl acceptor, which regenerates a Co(I) corrinoid upon demethylation.

The midpoint redox potentials of the aquocobalamin Co(III)/Co(II) couple is $E'_0 = +202$ mV versus NHE and that of the couple Co(II)/Co(I) is $E'_0 = -609$ mV (Lexa et al., 1977). The redox potential is affected by the cobalt ligands, the base-on or base-off configuration, and by the protein's active side. A corrinoid/FeS protein of the acetyl-coenzyme A synthesis by *Clostridium thermoaceticum* revealed $E'_0 = +300$ to $+350$ mV and $E'_0 = -504$ mV, respectively. If the corrinoid was bound to that protein in the base-off form, then the midpoint potentials change to $+207$ mV and -523 mV, respectively. The midpoint potentials of cobalamin bound to the methionine synthase from *E. coli* have been measured to be $E'_0 = -526$ mV for the couple Co(II)/Co(I) and $E'_0 = +273$ mV for the couple Co(III)/Co(II). (Banerjee et al., 1990).

6.3   Peculiarities of the *Sporomusa* Corrinoid

The *Sporomusa* corrinoid was identified as *para*-cresolyl cobamide (Fig. 6.1) by NMR and FAB-MS analyses and by biosynthetic experiments (Stupperich et al., 1988; Stupperich and Eisinger, 1989). The corrinoid exhibits two unusual features in comparison to *vitamin B_{12}* (cobalamin) which is frequently found in prokaryotes and eukaryotes. (i) A heterocyclic base is absent in *p-cresolyl cobamide* and (ii) its nucleotide is not coordinated to the cobalt.

The α-N-glycosidically bound heterocyclic base in vitamin B_{12} and all other corrinoids is substituted by an α-O-glycosidically linked *p*-cresol in *p*-cresolyl cobamide. The aromatic function is not able to coordinate to the cobalt, and, hence, the corrinoid is continuously "base-off," independent on the redox state of the corrinoid. After the addition of potassium cyanide to whole cells, the *p*-cresolyl cobamide is isolated as the dicyano form. This corrinoid is unstable. It decays into two isomers of a monocyano-monoaquo corrinoid, which have been separated by HPLC. Both forms are readily converted into each other. In addition, a dimethyl-*p*-cresolyl cobamide appears unlikely due to the trans-effects of the methyl ligands. Thus, either of the two faces could be methylated *in vivo* and a sixth cobalt ligand should be present on the opposite face.

Interestingly, the biological function of *p*-cresolyl cobamide could not be substituted by vitamin B_{12} in growing cells. Moreover, the unusual corrinoid structure led to the discovery of the Co-N coordinated nucleotide as an important structural element in the recognition mechanism by human intrinsic factor. Intrinsic factors are central proteins of the corrinoid uptake mechanism in eukaryotes

(Guéant and Gräsbeck, 1990). This corrinoid uptake system refuses *p*-cresolyl cobamide (Stupperich and Nexø, 1991).

6.4 Tentative Reaction Mechanism of the Methanol Activation

A *nucleophilic displacement* appears reasonable in the methanol cleavage reaction. An advantageous candidate for a nucleophile is a Co(I) corrinoid, because superreduced corrinoids are considered the strongest nucleophiles in physiology (Brown, 1982). Such a species may attack the methyl carbon of methanol, releasing a hydroxyl ion and a methylcorrinoid in an S_{N2} reaction. Subsequent demethylation of the CH_3 corrinoid regenerates a Co(I) to start a new reaction cycle.

A nucleophilic attack on the methyl group by a Co(I) will readily occur if the carbon is attached to an adjacent *oxonium* ion. A positively charged oxygen could be obtained by *protonation* of the methanol. Thus, one of the mechanisms involving a protonation of the substrate is more likely than just a plain S_{N2} reaction.

Protonation of the alcohol could proceed via two distinct ways in which the proton is supplied. A proton may be shuttled directly on to the methanol via the protein or via a *hydridocorrinoid*. The protein may assist, in either case, the proton transfer by a "charge-relay" mechanism as was suggested for the peptide cleavage by chymotrypsin (Blow, 1971).

The putative hydridocorrinoid reflects a protonated Co(I) species (reactions 6.2–6.4), which might be generated on demethylation of a CH_3 corrinoid. Spectroscopic evidence exists for such a reactive H-Co(III) corrinoid *in vitro* (Chemaly and Pratt, 1984). The H corrinoid could donate a proton to the methanol in a hydrophobic pocket of the enzyme. The overall reaction yields an oxonium ion and a Co(I) corrinoid.

$$H\text{-}Co(III) \rightarrow H^- + Co(III)^+ \tag{6.2}$$
$$H^- \rightarrow H^+ + 2e \tag{6.3}$$
$$Co(III)^+ + 2e \rightarrow Co(I)^- \tag{6.4}$$
$$CH_3OH + H^+ \rightarrow CH_3O^+HH \tag{6.5}$$

The Co(I) nucleophile attacks the protonated methanol (Fig. 6.2A, B) in case the oxonium does not decompose into a *carbenium* and water. Water certainly functions as a weak base, which fits the appropriate nature of a leaving group. The carbenium ion, however, is unstable. Hence, the formation of this methyl cation is not very likely, except the nascent CH_3^+ is trapped immediately after its formation by a Co(I) in generating a CH_3 corrinoid (Fig. 6.2C). Consecutive demethylation of this CH_3 corrinoid by a nucleophilic attack on the CH_3 corrinoid

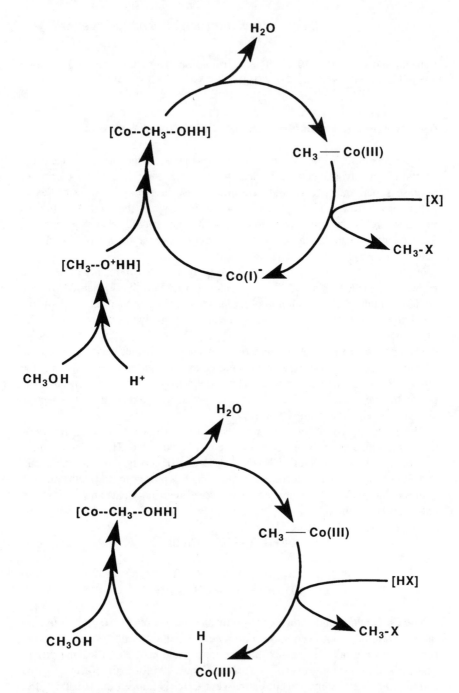

Figure 6.2.   Proposed methanol activation by protonation mechanisms. (A) Protonation mediated by the protein; (B) and (C) proptonation via a hydridocorrinoid. H-Co: hydridocorrinoid; CH$_3$-Co: methylcobamide; X and HX: a not yet identified nucleophile.

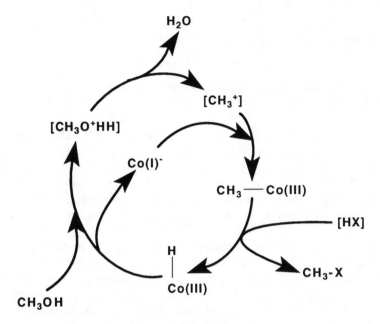

Figure 6.2. *Continued*

(Wood, 1982) recycles the H corrinoid (Fig. 6.2C) or a Co(I) corrinoid (Fig. 6.2A). Thus, the protonation and the trapping of a CH_3^+ should be arranged within one active side of an enzyme.

These types of reactions are enhanced by "base-off" corrinoids like *p*-cresolyl cobamide (Hogenkamp et al., 1987). The postulated Co(I) nucleophile and an unstable carbenium ion is consistent with the S_{N^2} mechanism.

Additional support for the postulated protonation mechanisms during the biological methanol cleavage arises from catalytic chemistry. Recently, the cleavage of the methanolic C–O bond by a Pd {111} catalyst was demonstrated. Interestingly, $Pd(CH_3OH)H^+$, CH_3^+, CHO^+HH, and H_2O have been assigned by X-ray photoelectron spectroscopy and secondary ion mass spectroscopy (Levis et al., 1989). At least some of these species are also rational intermediates of the bacterial methanol activation mechanism.

This outline evidences that the C–O bond cleavage of methanol is certainly distinct to the adenosylcorrinoid-dependent 1,2-propanediol dehydratase reaction. The methanol activation presumably demands a protonation mechanism and a Co(I) species which serves as a methyl acceptor and/or a methyl carrier. Yet, an adenosylcorrinoid is a latent radical. The homolytic cleavage of the cofactor reveals a Co(II) corrinoid, which would not conceive a methyl group. Moreover, two adjacent carbon atoms are necessary for 1,2 migrating substrate radicals in the diol dehydratase reaction (Stubbe, 1989) and, thus, excludes a similar reaction with a one-carbon compound.

6.5 Experimental Evidence for a Corrinoid-Dependent Methanol Cleavage

The data from ongoing experiments are consistent with the above-mentioned methanol activation mechanism. Anaerobic cell-free extracts of *Sporomusa ovata* H1 (DSM 2662) catalyze the *in vitro* formation of [^{14}C] acetyl-coenzyme A from [^{14}C] methanol. The reaction proceeds via enzymes from the soluble cell fraction after reductive activation by carbon monoxide. It was further stimulated by catalytic amounts of ATP, but it was inhibited to some extent by the addition of HO-vitamin B_{12}.

Significant amounts of [^{14}C] radioactivity were detected in three fractions if the reaction was terminated by the injection of propyl iodide. A 40-kDa and a 23-kDa protein contained an extractable [^{14}C]-labeled cofactor, which co-chromatographed with chemically synthesized CH_3-*p*-cresolyl cobamide. The other radioactive fraction consisted of a low molecular compound, which behaved like authentic CH_3-vitamin B_{12}. As the formation of the CH_3-B_{12} progressed, the concentration of HO-B_{12} decreased. Investigations on the radioactive cofactors from the proteins demonstrated that only one out of two possible methyl isomers was formed during the [^{14}C] methanol incubation. The isomer representing the upper Coβ-methyl ligand was identified; the lower Coα-methyl isomer was absent in corrinoid extracts.

Only [^{14}C-methyl] CH_3-vitamin B_{12} was found, but no CH_3-*p*-cresolyl cobamide was detected, if the reaction was ceased by aeration or by rapidly chilling on an ice bath. In addition, corrinoid extractions from whole cells as harvested revealed only traces of CH_3-*p*-cresolyl cobamide. Virtually all of the corrinoids were present as HO-*p*-cresolyl cobamide under extraction conditions that would also retrieve an adenosylcorrinoid. Hence, adenosyl-*p*-cresolyl cobamide is not a predominant corrinoid in *Sporomusa*.

These findings demonstrate that the acetogenic methanol utilization by *Sporomusa ovata* proceeds via two methylated corrinoid proteins. The equilibrium of both methyltransfers favors the demethylation of the corrinoid-containing enzymes, as has been confirmed by the results of the different assay termination experiments.

Cell extracts from 3,4-dimethoxybenzoate or betaine-grown cells were unable to form detectable amounts of [^{14}C] acetyl-coenzyme A from [^{14}C] methanol, indicating that methanol-induced proteins are involved in the methanol conversion into acetyl-coenzyme A. Recently, we have shown by immunological methods that the 40-kDa corrinoid-containing protein is actually induced by methanol

(Stupperich et al., 1992). This corrinoid enzyme, the 23-kDa corrinoid enzyme, or a not yet identified methyltransferase exchanges the methyl group with free cob(I)alamin. The cob(I)alamin is reductively formed from HO-vitamin B_{12} under the assay conditions and hence compete with the physiological CH_3-acceptor for the methyl. This consideration explains the partial inhibition of the acetyl-coenzyme A formation by HO-vitamin B_{12}.

Washed and oxidized extracts also generated [^{14}C]-labeled cofactors at the 40-kDa and 23-kDa proteins as well as CH_3-B_{12} from [^{14}C] methanol. The reaction only proceeds after reduction of the assay by titanium citrate. In addition, tetrahydrofolate was absolutely required besides catalytic amounts of ATP. ATP and tetrahydrofolate have also recently been shown to be necessary for optimal O-demethylation of aromatic compounds by extracts of the acetogen *Acetobacterium woodii* (Berman and Frazer, 1992; see also Chapter 17.) A radioactive low-molecular-weight compound was found under these experimental conditions, which is presumably [^{14}C-methyl]N^5-methyltetrahydrofolate. These results implicate tetrahydrofolate as a physiological CH_3-acceptor during methanol metabolism in *Sporomusa*. (Stupperich and Konle, 1993)

Evidence for a Co(I) corrinoid in whole cells was also presented by EPR spectroscopy (Stupperich et al., 1990). Anaerobic cells as harvested lacked an EPR cobalt signal, but a Co(II) signal appeared after careful oxidation of the sample. This finding demonstrates that a Co(I) was oxidized to the Co(II) species. Yet, the signal vanished after prolonged aeration because the Co(II) was converted into the EPR-silent Co(III) form.

HO-*p*-cresolyl cobamid is the predominant form in corrinoid extractions from whole cells. This HO corrinoid might not occur in its fully oxidized Co(III) state in the cells, as their redox potential should be similar to the redox potential of the medium. Strict anaerobes like acetogens and methanogens need growth media with fairly low midpoint potentials of $E'_0 \leq -350$ mV (Veldkamp, 1970). The HO corrinoid is considered to be reduced to the Co(II) state under these conditions if *p*-cresolyl cobamide-containing proteins have comparable redox properties of the base-off vitamin B_{12}-containing enzymes. An EPR Co(II) signal, however, was lacking in whole cells as harvested. Thus, the emerging Co(II) signal indicates a Co(I) corrinoid, which was stabilized by its protein, or which was temporarily generated by the decomposition of an unstable Co(III) corrinoid.

The Co(I) species are potent methyl group acceptors. Yet, a CH_3-*p*-cresolyl cobamide should possess a sixth cobalt ligand. Previous EPR experiments with [1,3-^{15}N]-*histidine*-labeled cells demonstrated that the *p*-cresolyl cobamide was linked to the protein via a histidine residue (Stupperich et al., 1990). The cobalt–histidine bond explains the stable corrinoid attachment to its protein and the finding that *p*-cresolyl cobamide was not exchangeable by a 100-fold excess of vitamin B_{12} (Stupperich et al., 1992). Also, growing cells did not functionally substitute *p*-cresolyl cobamide with vitamin B_{12}. This suggests that the *p*-cresolyl

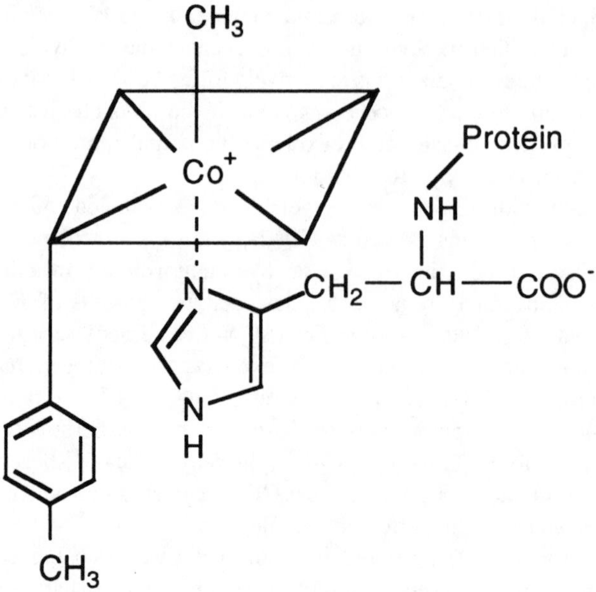

Figure 6.3. *In vivo* state of CH_3-*p*-cresolyl cobamide during the methyltransfer in methanol utilization. The lower cobalt ligand is a histidine residue of the corrinoid-containing enzyme.

cobamide structure is necessary for the physiological function and that a histidine residue of the protein plays a significant role in the corrinoid-reaction mechanism. One possible cofactor state during the corrinoid reaction cycle is depicted in Figure 6.3.

The above-mentioned findings provide credibility to the postulated role of corrinoids in the methanol metabolism of *Sporomusa ovata*. Two pivotal intermediates of the methanol utilization were identified, namely, a Co(I) corrinoid and a protein-bound CH_3 corrinoid. A serious candidate for an additional methyl intermediate in acetyl-coenzyme A formation from methanol is N^5-methyltetrahydrofolate. Thus, the carbon flow from methanol to the carbon monoxide dehydrogenase is consistent with the reaction sequence shown in Figure 6.4.

6.6 Conclusion

Corrinoids are clearly involved in the methanol metabolism of the acetogenic bacterium *Sporomusa ovata*. An adenosylcorrinoid-dependent mechanism like the diol reaction is excluded for the methanolic C–O bond cleavage. Instead, a

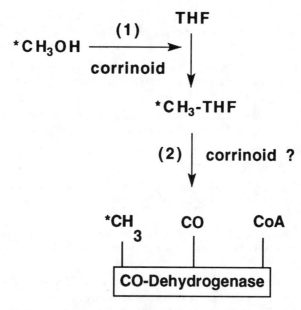

Figure 6.4. Pathway of the methyl carbon from methanol to the carbon monoxide dehydrogenase in *Sporomusa ovata*. Reaction 1 and 2 could include more than one enzymatic step. THF: tetrahydrofolate; CH₃-THF: N⁵-methyltetrahydrofolate.

protonation mechanism is proposed, which polarizes the C–O bond and which involves corrinoid cofactors as methyl carriers. Three prerequisites from an S_{N^2} displacement by corrinoids are met by this mechanism: (i) the equilibrium of the overall reaction favors the corrinoid demethylation, (ii) Co(I) corrinoids supply the very strong nucleophilicity of an entering nucleophile, and (iii) the leaving group water is a weak base. Some postulated intermediates of the methanol metabolism have been identified. Our findings are also consistent with similar results obtained by an analogous reaction in catalytic chemistry. Additional experiments are necessary to verify the role of the unique corrinoid cofactor from *Sporomusa ovata* and the histidine from the protein in this particular reaction mechanism. In general, homologous reactions in diverse bacterial species could use different corrinoid forms.

References

Banerjee, R. V., and G. Matthews. 1990. Cobalamin-dependent methionine synthase. *FASEB J.* **4**:1450–1459.

Banerjee, R. V., S. R. Harder, S. W. Ragsdale, and R. G. Matthews. 1990. Mechanism of reductive activation of cobalamin-dependent methionine synthase: An electron paramagnetic resonance spectroelectrochemical study. *Biochemistry* **29**:1129–1135.

Berman, M. H., and A. C. Frazer. 1992. Importance of tetrahydrofolate and ATP in the anaerobic O-demethylation reaction for phenylmethylethers. *Appl. Environ. Microbiol.* **58**:925–931.

Blow, D. M. 1971. The structure of chymotrypsin. In: *The Enzymes*, 3rd ed., P. D. Boyer (ed.), Vol. 3, pp. 185–212. Academic Press, New York.

Brown, K. 1982. Synthesis of organocobalt complexes. In: *B₁₂*, D. Dolphin, (ed.), Vol. 1, pp. 245–294. Wiley, New York.

Chemaly, S. M., and J. M. Pratt. 1984. The chemistry of vitamin B_{12}. Part 24. Evidence for hydride complexes of the cobalt (III) corrinoids. *J. Chem. Soc. Dalton Trans.* **1984**:595–599.

Finke, R. G., and B. D. Martin. 1990. Coenzyme $AdoB_{12}$ vs $AdoB_{12}$-homolytic Co-C cleavage following electron transfer: A rate enhancement $\geqslant 10^{12}$. *J. Inorg. Biochem.* **40**:19–22.

Guéant, J. L., and R. Gräsbeck. 1990. Assimilation of cobalamins. In: *Cobalamin and Related Binding Proteins in Clinical Nutrition*, J. L. Guéant and J. P. Nicolas (eds.), pp. 33–53. Elsevier, Paris.

Halpern, J. 1985. Mechanisms of coenzyme B_{12}-dependent rearrangements. *Science* **227**:869–875.

Hogenkamp, H. C. P., G. Bratt, and A. T. Kotchevar. 1987. Reaction of alkylcobalamins with thiols. *Biochemistry* **26**:4723–4727.

Kräutler, B., W. Keller, and C. Kratky. 1989. Coenzyme B_{12} chemistry: The crystal and molecular structure of cob(II)alamin. *J. Am. Chem. Soc.* **111**:8936–8938.

Krone, U. E., R. K. Thauer, and H. P. C. Hogenkamp. 1989. Reductive dehalogenation of chlorinated C_1-hydrocarbons mediated by corrinoids. *Biochemistry* **28**:4908–4914.

Krone, U. E., and R. K. Thauer. 1992. Dehalogenation of trichlorofluoromethane (CFC-11) by *Methanosarcina barkeri*. *FEMS Microbiol. Lett.* **9**:201–204.

Lexa, D., and J. M. Saveant. 1976. Electrochemistry of vitamin B_{12}. I. Role of the base-on/base-off reaction in the oxidoreduction mechanism of the B_{12r}-B_{12s} system. *J. Am. Chem. Soc.* **98**:2652–2658.

Lexa, D., J. M. Saveant, and J. Zickler. 1977. Electrochemistry of vitamin B_{12} II. Redox and acid-base equilibria in the B_{12a}/B_{12r} system. *J. Am. Chem. Soc.* **99**:2786–2790.

Levis, R. J., J. Zhicheng and N. Winograd. 1989. Thermal decomposition of CH_3OH adsorbed on Pd{111}: A new reaction pathway involving CH_3 formation. *J. Am. Chem. Soc.* **111**:4605–4612.

Marks, T. S., J. D. Allpress, and A. Maule. 1989. Dehalogenation of lindane by a variety of porphyrins and corrins. *Appl. Environ. Microbiol.* **55**:1258–1261.

Möller, B., R. Ossmer, B. H. Howard, G. Gottschalk, and H. Hippe. 1984. *Sporomusa*, a new genus of Gram-negative anaerobic bacteria including *Sporomusa sphaeroides* spec. nov. and *Sporomusa ovata* spec. nov. *Arch. Microbiol.* **139**:388–396.

Obrados, N., J. Dadía, L. Baldomá, and J. Aguilar. 1988. Anaerobic metabolism of the L-rhamnose fermentation product, 1,2-propanediol in *Salmonella typhimurium*. *J. Bacteriol.* **170**:2159–2162.

Ragsdale, S. W., Lindahl, P. A. and Münck, E. (1987) Mössbauer, EPR and optical studies of the corrinoid/iron-sulfur protein involved in the synthesis of acetyl enzyme A by *Clostridium thermoaceticum*. *J. Biol. Chem.* 262, 14289–14297.

Scheffold, R., S. Abrecht, R. Orlinski, H. R. Ruf, P. Stamouli, O. Tinembart, L. Walder, and C. Weymuth. 1987. Vitamin B_{12}-mediated electrochemical reactions in the synthesis of natural products. *Pure Appl. Chem.* **59**:363–372.

Stadtman, T. C. 1971. Vitamin B_{12}. *Science* **171**:859–861.

Stubbe, J. A. 1989. Protein radical involvement in biological catalysis? *Annu. Rev. Biochem.* **58**:257–285.

Stupperich, E., H. J. Eisinger, and B. Kräutler. 1988. Diversity of corrinoids in acetogenic bacteria. *Eur. J. Biochem.* **172**:459–464.

Stupperich, E., and H. J. Eisinger. 1989. Biosynthesis of *para*-cresolyl cobamide in *Sporomusa ovata*. *Arch. Microbiol.* **151**:372–377.

Stupperich, E., H. J. Eisinger, and S. P. J. Albracht. 1990. Evidence for a super-reduced cobamide as the major corrinoid fraction in vivo and a histidine residue as a cobalt ligand of the *p*-cresolyl cobamide in the actetogenic bacterium *Sporomusa ovata*. *Eur. J. Biochem.* **193**:105–109.

Stupperich, E., and E. Nexø. 1991. Effect of the cobalt-N coordination on the cobamide recognition by the human vitamin B_{12} binding proteins intrinsic factor, transcobalamin and haptocorrin. *Eur. J. Biochem.* **199**:299–303.

Stupperich, E., P. Aulkemeyer, and C. Eckerskorn. 1992. Purification and characterization of a methanol-induced cobamide-containing protein from *Sporomusa ovata*. *Arch. Microbiol.* **158**:370–373.

Stupperich, E., Konle, R. 1993. Corrinoid-dependent methyl transfer reactions are involved in methanol and 3,4-dimethoxybenzoate metabolism by *Sporomusa ovata*. *Appl. Environm. Microbiol.* **59**:3110–3116.

Toraya, T., and S. Fukui. 1982. Diol dehydratase. In: B_{12}, D. Dolphin (ed.), Vol. 2, pp. 233–262. Wiley, New York.

van der Meijden, P, C. van der Drift, and G. D. Vogels. 1984a. Methanol conversion in *Eubacterium limosum*. *Arch. Microboil.* **138**:360–364.

van der Meijden, P., B. W. te Brommelstroet, C. M. Poirot, C. van der Drift, and Vogels. 1984b. Purification and properties of methanol : 5-hydroxybenzimidazolyl-cobamide methyltransferase from *Methanosarcina barkeri*. *J. Bacteriol.* **160**:629–635.

Veldkamp, H. 1970. Enrichment cultures of prokaryotic organisms. In: *Methods in Microbiology*, J. R. Norris and D. W. Ribbons (eds.), Vol. 3A, pp. 305–361. Academic Press, London.

Wirt, M. D., Kumar, M., Ragsdale, S. W. and Chance, M. R. 1993. X-ray absorption spectroscopy of the corrinoid/iron-sulfur protein involved in acetyl coenzyme A synthesis by *Clostridium thermoautotrophicum*. J. Am. Chem. Soc. 115, 2146–2150.

Wood, J. M. 1982. Mechanisms for B_{12}-dependent methyl transfer. In: B_{12}, D. Dolphin (ed.), Vol. 2, pp. 151–164. Wiley, New York.

Zhao, Y., P. Such, and J. Retey. 1992. Radical intermediates in the coenzyme B_{12} dependent methylmalonyl-CoA mutase reaction shown by ESR spectroscopy. *Angew Chem. Int. Educ.* **31**:215–216.

III

ISOLATION, IDENTIFICATION, AND PHYLOGENY OF ACETOGENIC BACTERIA

7

Diversity, Ecology, and Isolation of Acetogenic Bacteria

Bernhard Schink

7.1 Introduction

The characteristic property of homoacetogenic bacteria is their ability to use carbon dioxide as a widespread and easily available electron sink and to reduce it via the carbon monoxide dehydrogenase system to acetate as their typical fermentation product. First note of this activity goes back to Fischer, Lieske, and Winzer (1932) who observed that a sewage sludge sample under an oxygen-free hydrogen atmosphere in the presence of bicarbonate gave rise to the formation of acetic acid. After this, our knowledge on the metabolic versatility of isolated strains of homoacetogenic bacteria has increased considerably, and there is no doubt that homoacetogens are the most versatile physiological group among the anaerobic bacteria we know.

This versatility renders these bacteria interesting, but it also makes an assessment of their ecological role unusually difficult. There is hardly any transformation process in an anoxic environment in which homoacetogens do not participate or with which they do not compete. Because methods for direct specific detection of such bacteria in a natural or seminatural environmental sample through probes are just only being developed (see Chapter 8), our knowledge of their activity in nondefined environments is rather superficial. This chapter intends to survey our understanding of the diversity and ecology of homoacetogens on the basis of our present knowledge of their metabolic abilities. There is no doubt that this approach has its shortcomings and that some of the conclusions drawn today may have to be revised later when more detailed information is available on the basis of direct assays of structure and function of intact anaerobic microbial communities.

7.2   Diversity of Homoacetogenic Bacteria

7.2.1   Taxonomic Groups

Homoacetogenic bacteria do not form a homogenous taxonomic unit. All described strains, however, are eubacteria; bacteria with a homoacetogenic energy metabolism in the Kingdom Archaea have not been reported yet. Nonetheless, the enzymes necessary for this metabolism exist in this Kingdom: Strictly autotrophic methanogenic bacteria build up acetyl-CoA for cell material synthesis through the CO-dehydrogenase (acetyl-CoA synthase) pathway, and acetyl-CoA synthase is being used in the opposite way by acetate-degrading methanogens for acetate cleavage (see Chapters 20–22). However, utilization of this enzyme system for dissimilatory acetate production and energy conservation appears to be, so far, a specific accomplishment within the Kingdom of Eubacteria.

The physiological group of homoacetogenic bacteria comprises Gram-positive and Gram-negative eubacteria, spore formers, and non-spore-formers. They are assigned to 12 different genera. These genera differ considerably with respect to their cytological and physiological properties (Table 7.1). Small cells of less

Table 7.1   Cytological and physiological properties of described genera of homoacetogenic bacteria

	Cell Shape	Spores	Gram Type	% G+C	T_{opt} (°C)	Reduced Products
Acetobacterium	Rods	−	+	38–44	27–30	Acetate
Acetitomaculum	Curved rods	−	+	32–36	38	Acetate
Acetoanaerobium	Rods	−	+[a]	36.8	37	Acetate
Acetohalobium	Rods	−	−	33.6	38–40	Acetate
Acetonema	Long rods	+	−	51.5	30	Acetate, butyrate
Acetogenium	Rods	−	+[b]	38	66	Acetate
Clostridium	Rods	+	+	22–54	30 55–60[c]	Acetate
Eubacterium (Butyribacterium)	Rods	−	+	47–49	30–37	Acetate, butyrate
Peptostreptococcus	Cocci	−	+	44–46	37	Acetate
Sporomusa	Rods	+	−	42–48	30	Acetate
Syntrophococcus	Cocci	−	+	52	35–42	Acetate

[a] Staining reaction Gram (−); in electron microscopic studies characterized as Gram (+).
[b] Originally described as Gram (−); in electron microscopic studies characterized as Gram (+).
[c] Two species are thermophilic (55–60°C).

than 1 μm diameter exist as well as real giants of a size comparable to that of protozoa (Fig. 7.1). Most isolates are mesophilic, but the number of thermophilic isolates with growth optima higher than 55–60°C is increasing. An overview of the described species (see Table 1.2) and their general metabolic capacities (Table 7.2) will be illustrated in the following paragraphs.

The genus *Acetobacterium* was established in 1977 (Balch et al., 1977) to house Gram-positive homoacetogenic bacteria which did not form spores and, therefore, could not be grouped with the already existing forms of homoacetogenic clostridia. The first strain of the type species, *A. woodii*, was enriched and isolated in the summer course on microbial ecology in Woods Hole, Massachusetts from sediment of a small creek estuary. In a setup to enrich for hydrogen-utilizing methanogens at pH 6.7, acetate was formed in considerable amounts, and a new type of rod-shaped bacteria grew that was easy to recognize by its pointed ends which gave it a typical boat-like shape (Fig. 7.1a). Similar bacteria were later obtained in enrichments from freshwater sediment samples. *A. woodii* is, next to *Clostridium thermoaceticum*, the best studied representative among the homoacetogens, and it is still subject to intense research on the biochemistry of energy metabolism (see Chapters 4 and 5). Its popularity for such studies is based on its metabolic versatility and comparably easy cultivation in defined mineral media. A detailed overview of substrate utilization patterns for the genera *Acetobacterium*, *Acetitomaculum*, *Acetoanaerobium*, and *Acetogenium* has been provided recently (Schink and Bomar, 1991).

An isolate obtained from sewage sludge was metabolically nearly identical

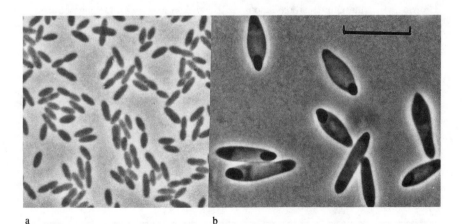

a   b

Figure 7.1. Phase contrast photomicrographs of cells of (a) *Acetobacterium woodii* grown with H_2/CO_2 and (b) *Clostridium magnum* grown with glucose. Bar equals 10 μm for both panels.

Table 7.2   Metabolic capacities of homoacetogenic bacteria. Reference is given only to the first or to a further very essential documentation of the respective activities.

	References
Acetate formation from hydrogen and CO_2	Fischer et al. (1932); Wieringa (1940)
Syntrophic acetate cleavage to hydrogen and CO_2	Zinder and Koch (1984)
Hexose fermentation to three acetate	Fontaine et al. (1942)
Utilization of methanol	Hamlett and Blaylock (1969); Zeikus et al. (1980)
Utilization of carbon monoxide	Diekert and Thauer (1978); Sharak-Genthner and Bryant (1982); Lynd et al. (1982)
Demethylation of phenyl methyl ethers	Bache and Pfennig (1981)
Demethylation of N-methyl compounds	Müller et al. (1981); Möller et al. (1984); Eichler and Schink (1985)
Demethylation of alkyl methyl ethers	Schuppert and Schink (1990)
Methylation of sulfide and methylsulfide	Bak et al. (1992)
Utilization of formaldehyde and hexamethylene tetramine	Schink (unpublished)
Incomplete oxidation of primary aliphatic alcohols	Wieringa (1940); Eichler and Schink (1985)
Incomplete oxidation of mandelate	Dörner and Schink (1991)
Reduction and dismutation of fumarate	Dorn et al. (1978); Matthies et al. (1993)
Reduction of phenyl acrylic acid derivatives	Bache and Pfennig (1981)
Growth energy through decarboxylation of succinate or malonate	Breznak et al. (1988); Dehning et al. (1989)
Utilization of oxalate and glyoxylate	Daniel and Drake (1993)
Decarboxylation of hydroxybenzoates	Hsu et al. (1990)
Energy-conserving reduction of nitrate	Seifritz et al. (1993)
Utilization of aromatic aldehydes	Gößner et al. (1994)
Degradation of gallic acid derivatives	Bak et al. (1992)
Dehalogenation of chloromethanes	Traunecker et al. (1991); Stromeyer et al. (1991, 1992)
Cleavage of ester linkages (triacetin)	Emde and Schink (1987)
Cleavage of ether linkages (ethylene glycol derivatives)	Wagener and Schink (1988); Tanaka and Pfennig (1988); Schramm and Schink (1991)
Fixation of molecular nitrogen	Bogdahn et al. (1983); Schink and Bomar (1991)

with *A. woodii* and was attributed to a new species, *A. wieringae* (Braun and Gottschalk, 1982). Studies on the degradation of primary alcohols and diols in freshwater sediments led to the isolation of *A. carbinolicum* (Eichler and Schink, 1984). This bacterium catalyzes an incomplete oxidation of primary alcohols to their corresponding fatty acids and grows quite well with diols and glycerol. Finally, a malate-utilizing and ether-cleaving strain was isolated from an enrichment culture with methoxyethanol as substrate and placed into a new species, *A. malicum* (Tanaka and Pfennig, 1988).

After establishment of the genus *Acetobacterium*, other genera of non-spore-forming mesophilic homoacetogenic bacteria were defined which are physiologically quite similar to the genus *Acetobacterium* and may even be phylogenetically closely linked to it.

Acetitomaculum cells are slightly curved and the DNA has a guanine-plus-cytosine content which is slightly but not significantly lower than that of *Acetobacterium* DNA (see Table 7.1). Only one species of this genus, *A. ruminis*, has been described so far which was isolated from the bovine rumen (Greening and Leedle, 1989).

Acetoanaerobium differs from *Acetobacterium* only by its reported negative behavior in Gram staining. However, the cell-wall architecture resembles that of a Gram-positive bacterium (Sleat et al., 1985). The only described species, *A. noterae*, requires yeast extract for growth and ferments only a few sugars and H_2/CO_2 as substrates.

Both genera, *Acetitomaculum* and *Acetoanaerobium*, have temperature optima typical of intestinal bacteria. However, this difference to the genus *Acetobacterium* is only of minor taxonomic importance, and a possible reorganization of these genera with the latter genus should be considered.

The genus *Acetonema* was established only recently to accommodate a morphologically quite distinct isolate from the termite gut. Cells of the isolated strain that was assigned to the species *A. longum* are spore-forming motile rods of unusual size [$(0.3–0.4) \times 6 - 60$ μm; Kane and Breznak (1991)]. It grows well with hydrogen and CO_2, as well as with sugars and organic acids. From the latter, it forms acetate and butyrate as reduced fermentation products and resembles with this the genus *Eubacterium*. On the basis of 16 S rRNA comparison, *Acetonema* appears to be related to the genus *Sporomusa*.

Acetogenium kivui is a thermophilic, Gram-negative, nonmotile, non-spore-forming homoacetogen (Leigh et al., 1981) which was isolated from the geothermally heated sediments of Lake Kivu, East Africa. It is a fast-growing ($t_d = 2$ h on H_2/CO_2) thermophile with a growth optimum at 66°C and is characterized by an unusual arrangement of protein units in its cell wall. Metabolically, it resembles the genus *Acetobacterium* and uses several sugars and organic acids for homoacetogenic fermentation.

Acetohalobium arabaticum was described as an extremely halophilic (10–25%

NaCl is necessary) type of homoacetogens that was isolated from salt lagoons of Arabat (East Crimea) (Zhilina and Zavarzin, 1990a, 1990b). The cells are bent rods, (0.7–1.0) μm × (1-5)μm in size, and Gram negative. They can grow either chemolithtrophically with H_2/CO_2 or with CO, demethylate trimethylamine, or ferment simple organic compound such as betaine, lactate, pyruvate, or histidine. Carbohydrates are not utilized. The guanine plus cytosine (G+C) content of the DNA is 33.6 mol%.

The genus *Clostridium* is a rather inhomogenous taxonomic unit characterized only by its ability to form spores and to grow obligately under strictly anoxic conditions. Its phylogenetic heterogeneity has been noted for several years and is documented by the fact that it is composed of species with a wide range of low versus high G+C content of the DNA (Cato et al., 1986). By their cell-wall architecture, they all have to be counted as Gram positive, but some strains tend to behave negative on Gram staining. Metabolically, this genus is extremely heterogenous, comprising saccharolytic, peptolytic, ureolytic, and other species with sometimes broad and sometimes very narrow substrate specificities. In addition, the ability of homoacetogenic CO_2 reduction is encountered within this rather artificial "genus" and is found among mesophilic and thermophilic as well as "low G+C" and "high G+C" representatives.

The first isolate of homoacetogenic clostridia, *Clostridium aceticum* (Wieringa, 1936, 1940), was obtained at a time right after acetate formation from hydrogen and CO_2 had been discovered (Fischer et al., 1932). This species has a G+C content of 33 mol%, forms terminal spherical spores, and ferments several sugars, organic acids including fumarate, malate, and glutamate, as well as alcohols and diols. Unfortunately, the original isolate was lost a few years after its first description and was not reisolated until 1979 (Adamse, 1980). At the same time, an old stock culture preserved on dry sterile soil was discovered by G. Gottschalk in R. Hungate's laboratory and was redescribed soon after (Braun et al., 1981). The failure to reisolate this bacterium earlier is probably due to the use of phosphate-buffered media with little CO_2 content, as they were typically used in those days (see Section 7.4).

C. thermoaceticum (Fontaine et al., 1942) was the first bacterium found to ferment one molecule of hexose to three molecules of acetate. It has a temperature optimum of 55°C and has been subject to extensive studies on the biochemistry of homoacetogenic fermentation; most of our present knowledge on this matter has been obtained with this fast-growing and easy-to-cultivate strain (see Chapters 1–3).

C. formicoaceticum (Andreesen et al., 1970) was obtained by enrichment with hydrogen and CO_2 from pasteurized sediment samples. Different from *C. aceticum,* it cannot grow in pure culture with hydrogen as sole reductant but depends on formate (which must have been supplied in the enrichment culture by other anaerobes, e.g., through formate hydrogen lyase activity). The reported substrate range of this isolate is a little bigger than that of *C. aceticum* and includes lactate, methanol, and glycerol. This species was found recently to be able to grow chemolithotrophically with carbon monoxide (Lux and Drake, 1992).

C. thermoautotrophicum (Wiegel et al., 1981) has a temperature optimum at 55–60°C, with a maximum at 70°C, and a G+C content of 53–55%. Like the other species, it forms terminal, spherical spores. It is motile by peritrichously arranged flagella and required originally at least 0.2% yeast extract for growth. It was shown later that this bacterium can also grow in mineral medium (Savage and Drake, 1986). The substrate range is narrow, comprising H_2/CO_2, glucose, fructose, glycerate, and methanol.

C. magnum was isolated from pasteurized freshwater creek sediment with 2,3-butanediol as substrate (Schink, 1984). It differs from all other clostridia by its unusual size, especially after growth with sugars (Fig. 7.1b). Acetoin, a few sugars, as well as malate and citrate are used for growth. The G+C content of the DNA is 29%.

C. pfennigii is a highly specialized, mesophilic species which uses only the methyl residues of methoxylated aromatics or pyruvate and ferments them with CO_2 to acetate (Krumholz and Bryant, 1985). Carbon monoxide is fermented to a mixture of acetate and butyrate.

C. ljungdahlei is a Gram-positive, motile, spore-forming anaerobe that grows with carbon monoxide, H_2/CO_2, ethanol, pyruvate, pentoses, and hexoses (Tanner et al., 1993). The G+C value is unusually low (22–23 mol%). Acetate is the main and nearly only fermentation product; an originally reported production of ethanol from carbon monoxide (Barik et al., 1988)—which would be rather exciting from a biochemical point of view—could not be reproduced (Tanner et al., 1993). Thus, this bacterium is not unusual with respect to its metabolism.

C. mayombei was isolated from the gut of an African soil-feeding termite and has a low G+C content of 25.6 mol% (Kane et al., 1991). It is metabolically rather versatile and uses H_2/CO_2, several sugars and sugar alcohols, organic acids, and amino acids for growth.

C. fervidus (sic! a correct name would be "fervidum") was described as a thermophilic (optimum temperature 68°C) "acetogen" that grows preferentially with peptides and amino acids and appears not to be able to use H_2/CO_2 or formate (Patel et al., 1987). As neither utilization of one-carbon compounds nor key enzymes of homoacetogenic metabolism have been documented, this bacterium should not be counted among the homoacetogens.

Eubacterium limosum (Prévot, 1938) as well as *Butyribacterium rettgeri* (Barker and Haas, 1944) [which both are taxonomically identical entities (Moore and Cato, 1965; Tanner et al., 1981)] can be differentiated from most other homoacetogens (except for *Acetonema longum* and *Clostridium pfennigii*) by production of butyrate, together with acetate. In addition, *Butyribacterium meth-*

ylotrophicum including the "Marburg strain" (Zeikus et al., 1980) belongs to this genus. The G+C content of the DNA is 47–49 mol%. Besides sugars, these species can grow well with methanol, carbon monoxide, or mixtures of both (Sharak-Genthner et al., 1981; Sharak-Genthner and Bryant, 1982; Lynd and Zeikus, 1983). The ratio of acetate to butyrate formation depends to a large extent on the oxidation/reduction state of the substrates provided, including the relative availability of CO_2 as electron acceptor (Kerby et al., 1983). *E. limosum* also uses betain and ferments one of its methyl groups with CO_2 to butyrate, acetate, and dimethylglycine as products (Müller et al., 1981).

   Peptostreptococcus productus cells are coccoid, Gram positive, and non-spore-forming. The G+C content of the DNA is 45 mol%. It uses H_2/CO_2 and many different sugars and sugar alcohols, as well as glycerol and pyruvate. Its most outstanding property is its fast growth, (t_d = 3 h) with carbon monoxide as sole substrate which it tolerates up to 1 bar partial pressure in the gas phase (Lorowitz and Bryant, 1984; Geerligs et al., 1987). Because this species differs metabolically entirely from other *Peptostreptococcus* species, it should be grouped either with another or as a new taxonomic entity.

   The genus *Sporomusa* was created to comprise homoacetogenic spore-forming bacteria which are definitively Gram negative (Möller et al., 1984). *S. sphaeroides* and *S. ovata* differ mainly by morphological properties including spore shape and use N-methyl compounds, primary alcohols, 1,2-diols, 2,3-butanediol, H_2/CO_2, and the amino acids alanine, serine, and glycine as substrates (Möller et al., 1984). *S. acidovorans* (Ollivier et al., 1985) has a G+C content of 42 mol%. It is more versatile and uses several sugars as well as fumarate and succinate for growth. *S. termitida* was isolated from termite guts (Breznak et al., 1988). Its substrate range resembles that of *S. sphaeroides* and *S. ovata*. *S. malonica* (Dehning et al., 1989) is characterized, among others, by its ability to obtain all its energy for growth from decarboxylation of simple dicarboxylic acids, e.g., succinate or malonate.

   Syntrophococcus sucromutans was isolated from rumen contents as the predominant utilizer of methoxylated aromatics (Krumholz and Bryant, 1986). It carries out only an incomplete homoacetogenic fermentation and depends either on formate as a secondary electron acceptor or on a syntrophic hydrogen-utilizing partner organism for oxidation of sugars. Thus, this bacterium appears to lack an efficient formate dehydrogenase enzyme and depends on formate instead of CO_2 as electron acceptor.

   Strains AOR and TMBS 4 are treated here, although they have not been attributed yet to valid genera or species. However, they both represent quite interesting new metabolic capabilities which have not been observed yet with other homoacetogenic bacteria.

   Strain AOR is a thermophilic homoacetogenic bacterium with a temperature optimum of 60°C (see Chapter 14). It is Gram positive and has a G+C content of 47 mol%. It was isolated from a thermophilic coculture converting acetate to methane and CO_2 (Zinder and Koch, 1984). Growth in pure culture is possible with ethylene glycol, H_2/CO_2, formate, pyruvate, glycine, or betain (Lee and Zinder, 1988a). In the coculture on acetate, strain AOR oxidizes acetate to CO_2 through the CO dehydrogenase pathway (Lee and Zinder, 1988b), thus reversing the "normal" direction of homoacetogenic acetate formation. For this reason, this bacterium has been nicknamed *Reversibacter* (Thauer et al., 1989), but a valid name of this bacterium is still missing. Reversal of acetate formation becomes possible only at very low hydrogen partial pressures which are maintained in the syntrophic coculture by a hydrogen-oxidizing methanogen. The fact that this bacterium can grow either through hydrogen oxidation and acetate formation or vice versa demonstrates how close to the thermodynamic equilibrium an anaerobic energy metabolism can operate.

Strain TMBS 4 was isolated recently from freshwater sediments as the predominant strain growing with trimethoxybenzoate as substrate (Bak et al., 1992). Different from all other homoacetogenic bacteria, this strain not only demethylates trimethoxybenzoate but also ferments the aromatic residue gallic acid completely to acetate and CO_2. It also transfers methyl groups to sulfide and forms methyl sulfide and dimethylsulfide. Because it does not show any relationship to existing homoacetogenic bacteria on the basis of 16S rRNA sequence data, it will be assigned to a new genus and species ("*Holophaga foetida*"; Bak et al. (unpublished).

Comparison of 16S rRNA homologies has revealed that *A. woodii* exhibits some phylogenetic relatedness to *Eubacterium limosum* and some *Clostridium* species (Tanner et al., 1981). *Acetogenium kivui* is also related to some *Clostridium* species (Leigh et al., 1981), and the Gram-negative spore former *Sporomusa* was found to be related on the basis of 16S rRNA similarities to clostridia (Stackebrandt et al., 1985) as well as to *Acetonema* (Kane and Breznak, 1991). These findings indicate that the obvious physiological similarities among these main groups of homoacetogens may have, to some extent, a phylogenetic basis.

7.2.2 Metabolic Versatility

Homoacetogenic bacteria are experts in the utilization and transformation of one-carbon compounds. Moreover, they can carry out incomplete oxidations of reduced fermentation products released by other fermenting bacteria, and they can grow with monomeric sugars that are fermented completely to acetate with a comparably high ATP yield. An overview of the metabolic capacity of homoacetogenic bacteria is presented in Table 7.2. This table lists only some main metabolic capabilities without intending to specify those for every single species. Rather, Table 7.2 illustrates the history of our broadening knowledge of the metabolic versatility of homoacetogens.

(a) ACETATE SYNTHESIS AND DEGRADATION

Growth with hydrogen and CO_2 according to the reaction

$$4\,H_2 + 2\,CO_2 \rightarrow CH_3COO^- + H^+ + 2\,H_2O,\ \Delta G'_0 = -95\ \text{kJ/mol} \quad (7.1)$$

has been reported for nearly all homoacetogens, except for *Clostridium formicoaceticum, Syntrophococcus sucromutans,* strain TMBS4, and *Clostridium magnum.* The latter was found to grow by this reaction if some yeast extract was provided in the growth medium (Bomar et al., 1991). In addition, formate is used by all described homoacetogens tested.

The reversal of the acetate formation reaction to form hydrogen and CO_2 has been documented so far only for the thermophilic strain AOR (Lee and Zinder, 1988a), and some reports exist of mesophilic cocultures with similar metabolism (Schnürer and Svensson, personal communication; Galouchko, Rozanova, and Zavarzin, personal communication). Reversal of reaction 7.1 becomes thermodynamically feasible at 25°C only at a hydrogen partial pressure of 10^{-4} bar (10 Pa) and can yield a minimum amount of energy for ATP formation [about 20 kJ/mol reaction; Schink (1990)] only at 10^{-5} bar (1 Pa). This is also the minimum hydrogen partial pressure which can be maintained by a methanogenic partner bacterium under these conditions. In acetate-utilizing cocultures growing at 60°C, hydrogen partial pressures in the range of 20–50 Pa were measured (Lee and Zinder, 1988c). These values which are significantly higher than those predicted above can be explained by the elevated cultivation temperature: According to the van't Hoff equation, the total reaction yields more energy at increasing temperatures. Therefore, a detailed study of the reported mesophilic cocultures would be of great interest for an understanding of anaerobic growth energetics in general.

(b) UTILIZATION OF SUGARS

Sugars are usually converted to acetate as sole product,

$$C_6H_{12}O_6 \rightarrow 3\,CH_3COO^- + 3\,H^+ \quad (7.2)$$

except for *Eubacterium* and *Acetonema* sp., which produces mixtures of acetate and butyrate, depending on the availability of CO_2 and on the pH. Pentoses are converted to 2.5 mol acetate; the pathways engaged for utilization of pentoses have not been studied in detail yet.

(c) Metabolism of Methyl Residues and Other C_1 Compounds

Methanol is converted to acetate, either with CO_2 or with CO or formate as cosubstrate, according to the reactions

$$4\,CH_3OH + 2\,CO_2 \rightarrow 3\,CH_3COO^- + 3\,H^+ + 2\,H_2O \qquad (7.3)$$

or

$$CH_3OH + CO \rightarrow CH_3COO^- + H^+ \qquad (7.4)$$

or

$$CH_3OH + HCOO^- \rightarrow CH_3COO^- + H_2O \qquad (7.5)$$

In reactions according to reaction 7.4, the methanol carbon is transformed preferentially into the methyl carbon of the acetate molecule formed (Kerby et al., 1983).

Tests for utilization of methanol have mostly been carried out with too high substrate concentrations, which yielded negative results. *A. woodii* was first reported to be unable to grow with this substrate (Balch et al., 1977); reevaluation of this finding revealed good growth with 5–10 mM methanol (Bache and Pfennig, 1981).

Carbon monoxide, as sole substrate, can be converted to acetate, according to

$$4\,CO + 2\,H_2O \rightarrow CH_3COO^- + H^+ + 2\,CO_2 \qquad (7.6)$$

Experts in this reaction are *Peptostreptococcus productus* and *Eubacterium* sp.

Methoxylated aromatic compounds are demethylated to the corresponding phenols; the methyl residue is fermented to acetate, analogous to methanol (Bache and Pfennig, 1981). This reaction was observed later with many other homoacetogenic bacteria. Some *A. woodii* strains reduce the acrylic acid side chain of caffeate derivates to the corresponding dihydrocaffeates; this process appears to be associated with electron transport phosphorylation (Tschech and Pfennig, 1984; Hansen et al., 1988). The ability to reduce acrylic acid side chains exists also among other homoacetogenic bacteria [see, e.g., (Parekh et al., 1992)]. Most homoacetogenic bacteria demethylate only methoxylated aromatic compounds; cleavage of the methyl alkyl ether methoxyacetate was demonstrated so far only with a new *Acetobacterium* isolate (Schuppert and Schink, 1990).

Betaine is demethylated to dimethylglycine (Müller et al., 1981; Eichler and

Schink, 1984). Dimethylglycine is not demethylated any further as far as this was studied quantitatively. It has still to be clarified whether the demethylation reaction is carried out by the same enzyme system which also demethylates the phenylmethyl ethers.

Strain TMBS 4 transfers methyl groups from trimethoxybenzoate to sulfide to form methanethiol and dimethylsulfide (Bak et al., 1992). This reaction does not yield energy to the bacterium (Kreft and Schink, 1993) and has to be considered as a gratuitous side reaction under conditions of excess sulfide supply.

A further one-carbon compound, formaldehyde, can also be used as a growth substrate by homoacetogenic bacteria, but extreme care has to be taken to avoid killing by this rather toxic compound. In enrichment cultures with 1 mM formaldehyde or the formaldehyde derivative hexamethylene tetramine, fluorescing methanogenic bacteria were selected together with *Acetobacterium*-like cells, and acetate and methane were formed as products. Unfortunately, these cultures could never be maintained or purified (Schink, unpublished).

(d) Oxidation of Primary Alcohols

Primary aliphatic alcohols such as ethanol, propanol, or butanol select for enrichment of, e.g., *A. carbinolicum*. The alcohols are converted to the corresponding fatty acids with concomitant formation of acetate from carbon dioxide (Eichler and Schink, 1984):

$$2\,CH_3\,CH_2\,CH_2\,OH + 2\,CO_2 \rightarrow 2\,CH_3\,CH_2\,COO^- + CH_3\,COO^- + 3\,H^+,$$
$$\Delta G_0' = -76\,kJ/mol \tag{7.7}$$

Only a few homoacetogenic bacteria grow with these alcohols. It was found later that *A. woodii* and *A. wieringae* can use these alcohols if the medium contains at least 100 mM bicarbonate buffer (Buschhorn et al., 1989). With limiting phosphate concentrations, *A. woodii, A. wieringae,* and *A. carbinolicum* fermented excess glucose (50 mM) to acetate mainly as well as ethanol and alanine as side products, which could later be reoxidized (Buschhorn et al., 1989).

Ethylene glycol or 1,2-propanediol are fermented to acetate or acetate plus propionate; 2,3-butanediol and acetoin go to acetate as well (Schink, 1984). Fermentation of glycerol can lead to acetate as sole product (Eichler and Schink, 1984) or to acetate together with 1,3-propanediol that cannot be utilized any further (Emde and Schink, 1987). 1,3-Propanediol is formed via a diol dehydratase which perhaps also attacks the other 1,2-diols mentioned.

Enrichment cultures with mercaptoethanol (2-hydroxyethanethiol) yielded *Acetobacterium*-like homoacetogens which oxidized this substrate incompletely to a mixture of thioglycolate and its oxidized dimer. Obviously, the bacteria did not only oxidize the alcohol function to the corresponding acid but also the thiol group to form (probably) a C-S-S-C linkage (Friedrich and Schink, unpublished).

(e) INCOMPLETE OXIDATION OF ORGANIC ACIDS

Mandelate (phenylglycolate) is oxidized by *Acetobacterium* strains via benzoyl-CoA to benzoate, with acetate as reduced by-product (Dörner and Schink, 1991).

Fumarate can be fermented by *Clostridium formicoaceticum* to succinate plus acetate, whereas the isoelectronic substrate malate is converted through the homoacetogenic pathway to acetate only (Dorn et al., 1978). Fumarate can also be reduced with other substrates as electron donors (Matthies et al., 1993). In *Acetobacterium malicum*, malate is fermented through an NAD-dependent malic enzyme and, therefore, gives rise to higher cell yields than the isoelectronic substrate lactate, which is oxidized through a membrane-bound (quinone-linked?) lactic dehydrogenase (Strohhäcker and Schink, 1991).

Growth by decarboxylation of short-chain dicarboxylic acids has been observed among the homoacetogens only with *Sporomusa termitida, S. acidovorans*, and *S. malonica*. The mechanisms of energy conservation in these reactions have not been studied yet with these bacteria. Decarboxylation of hydroxybenzoates by, e.g., *Clostridium thermoaceticum* (Hsu et al., 1990) appears not to be coupled to energy conservation.

(f) DEHALOGENATIONS AND ETHER CLEAVAGE REACTIONS

Chloromethanes (CH_2Cl_2, CH_3Cl, CCl_4) can be dehalogenated and further metabolized by new strains of homoacetogenic bacteria (Traunecker et al., 1991) and by *Acetobacterium woodii* (Egli et al., 1988; Stromeyer et al., 1991, 1992). Whereas the former two substrates are fermented to acetate in an energy-yielding reaction, CCl_4 is either reduced to CH_2Cl_2 or hydrolyzed cometabolically to CO_2.

Acetobacterium malicum can grow with methoxyethanol, an important industrial solvent. The substrate spectrum and the products formed indicate that this substrate is degraded not via demethylation to ethylene glycol but through a diol dehydratase-analogous reaction releasing methanol and acetaldehyde; both are further oxidized to acetate (Tanaka and Pfennig, 1988). Other homoacetogenic bacteria were found to ferment polyethylene glycols through a similar reaction sequence to acetate (Wagener and Schink, 1988; Schramm and Schink, 1991).

(g) NITROGEN FIXATION

Nitrogen fixation by a homoacetogenic bacterium was first observed with *Clostridium formicoaceticum* (Bogdahn et al., 1983). Later it was found that all

Acetobacterium species and many other homoacetogenic bacteria are able to fix molecular nitrogen (Schink and Bomar, 1991), but, as yet, this finding has not been documented in detail. The ecological importance of nitrogen fixation in the usually ammonia-rich environments typically inhabited by these bacteria is still a matter of speculation.

(h) "Missing Capabilities"

Looking at the enormous metabolic versatility of homoacetogenic bacteria, one should ask the question, "Which reactions are actually *not* being catalyzed by homoacetogens?" Obviously, homoacetogens are superior to other fermenting bacteria by their ability to use CO_2 as an external electron acceptor which gives them a special advantage in utilization of electron-rich compounds such as primary fermentation products. However, these specific chances are not always used as the subsequent examples should demonstrate.

Obviously, isolated strains of homoacetogenic bacteria do not ferment fatty acids, e.g., according to the reaction

$$2\,CH_3CH_2CH_2COO^- + 2\,CO_2 + 2\,H_2O \rightarrow 5\,CH_3COO^- + 3\,H^+ \quad (7.8)$$
$$,\Delta G_0{}' = +1.7\;kJ/mol$$

Contrary to reaction 7.7, this acetogenic conversion of a primary fermentation product cannot yield energy to allow energy conservation and growth under conditions comparable to standard conditions. Nonetheless, indications of such a conversion have been found in slightly acidic sediments, but homoacetogenic bacteria catalyzing this reaction have not been isolated yet.

Aromatic compounds (except for gallic acid fermentation by strain TMBS 4) are not degraded by homoacetogens, although such conversions are sufficiently exergonic to allow bacterial growth, as shown here for benzoate and phenol:

$$4\,C_6H_5COO^- + 2\,CO_2 + 18\,H_2O \rightarrow 15\,CH_3COO^- + 11\,H^+ \quad (7.9)$$
$$\Delta G_0{}' = -87.1\;kJ/mol\;\text{or}\;-22\;kJ/mol\;benzoate$$
$$2\,C_6H_5OH + 2\,CO_2 + 8\,H_2O \rightarrow 7\,CH_3COO^- + 7\,H^+ \quad (7.10)$$
$$\Delta G_0{}' = -74.4\;kJ/mol\;\text{or}\;-37.2\;kJ/mol\;phenol$$

In the absence of nitrate or sulfate as alternative electron acceptors of anaerobic oxidation, these substrates are degraded only by syntrophic associations of fermenting bacteria and hydrogen-oxidizing methanogens.

Last but not least, it is surprising to realize that only monomeric or dimeric, low-molecular-weight substrates appear to be used by homoacetogens, only few nitrogen compounds, and so far only one amino acid (histidine utilization by *Acetohalobium*). No substrates that require extracellular hydrolysis, such as polysaccharides, proteins, nucleic acids, or lipids [except for triacetin; Emde and Schink (1987)] are attacked by these bacteria, and they depend on extracellular hydrolases produced by other fermenting anaerobes for utilization of these important substrates. Lack of own hydrolytic enzymes forces them to cooperate closely with partner organisms in the complex anaerobic food-chain system.

7.3 Ecology

7.3.1 Habitats

Most known homoacetogenic bacteria were isolated from strictly anoxic environments, typically black sediments of estuaries, marine sources, freshwater ponds, and anoxic sewage sludge. In the latter, they were found to make up about 1% of the total hydrogen-oxidizing community (Braun et al., 1979). Homoacetogenic activities were also detected in the gastrointestinal tracts of, e.g., termites (Breznak and Switzer, 1986; Chapter 11) and of higher animals and man (Prins and Lankhorst, 1977; Lajoie et al., 1988; Chapters 12 and 13). A first search for homoacetogens in the rumen turned out negative (Braun et al., 1979), but, later, considerable numbers (10^7–10^8 cells/ml) were detected in the rumen of steers (Greening and Leedle, 1989). Homoacetogenic activities were also reported from flooded soils and rice paddies (Conrad et al., 1989; Krumboeck and Conrad, 1991), but no studies on population sizes in these habitats were reported. Recent studies have also identified acetogenic populations in low-temperature and haline habitats (see Chapters 15 and 16).

Of special interest is the relationship of homoacetogens to salt. The first strain of *Acetobacterium woodii* was isolated from an estuarine sediment. All *Acetobacterium*-like strains isolated in our laboratory were rather independent of the prevailing salt concentration and grew equally well in freshwater, brackish water, or saltwater medium. The recently discovered involvement of sodium ions in the energy metabolism and the presence of sodium-proton antiporters in the cytoplasmic membrane (see Chapters 4 and 5) may allow these bacteria to convert proton motive force into sodium motive force and vice versa and, with this, render them excellently adapted to environments of periodically changing salinity, as typical of an estuary. The same feature may also allow them to cope with comparably broad variations in the pH regime; this is especially important in soil in which microenvironments may change the proton activity on small scales depending on the availability of fermentable sugars, or in sewage digestors and other areas of technical transformation of substrates at high concentrations.

The importance of homoacetogens in the digestive tracts of animals depends to some extent on the respective animal studied. Although reports on their occurrence in the cow rumen are contradictory (Braun et al. 1979; Greening and

Leedle, 1989), homoacetogens appear to be the predominant hydrogen utilizers in the hindgut of certain termites and cockroaches (Breznak and Switzer, 1986). It is not yet understood why homoacetogens can outcompete the energetically more favored methanogens in the gastrointestinal system of these insects. Perhaps the insect itself moderates the establishment of the anaerobic population in its guts by excretion of, e.g., certain digestive enzymes or other kinds of substances, tensides, etc., which may be inhibitory to methanogens.

On the other hand, recent studies have disclosed that the hindgut of studied termites is not entirely a strictly anoxic environment. Due to its small size (about 1 μl volume), the resulting high surface-to-volume ratio, and lack of efficient insulation against oxygen penetration, anoxic conditions in this microenvironment can be maintained only by intense oxygen consumption through aerobic respiration. The hindgut flora contains significant numbers of aerobic and facultatively aerobic bacteria (Brune and Breznak, personal communication; König, personal communication) that may be responsible for maintenance of anoxic conditions in the hindgut's interior. Traces of oxygen penetrating into the hindgut's outer layers may be sufficient to inhibit methanogenic bacteria that, in general, appear to be more oxygen-sensitive than homoacetogens. The same may be the reason for the preference of homoacetogenic over methanogenic activities in periodically flooded soils and rice paddies (Krumboeck and Conrad, 1991).

7.3.2   *Metabolic Interactions*

(a) Degradation of Sugars and Other More-Carbon Compounds

In all the environments mentioned above, homoacetogens have to compete with other anaerobes for electron sources. As utilization of carbon dioxide as electron sink allows complete conversion of sugars to acetate, homoacetogens should have a considerable advantage over classical fermenters in carbohydrate fermentation. However, homoacetogens grow slower on sugars than, e.g., a *Clostridium butyricum* or *Escherichia coli* does, and no homoacetogen has so far been described which degrades polysaccharides, the main sources for sugars in biomass. In the complex methanogenic food chain depicted in Figure 7.2, homoacetogenic bacteria compete with primary fermenting bacteria for monomeric sugars, and with secondary fermenters (and, e.g., in marine sediments, with sulfate-reducing bacteria) for primary fermentation products such as lactate and ethanol.

In oxidation of hydrogen and formate, homoacetogens compete with methanogens and sulfate reducers which, due to higher substrate affinity and energy yields, usually win this competition (Schink, 1987; Zehnder and Stumm, 1988; Cord-Ruwisch et al., 1988). Homoacetogens fill a special function in the anaerobic food web (Fig. 7.2) by balancing between the pool of hydrogen and one-

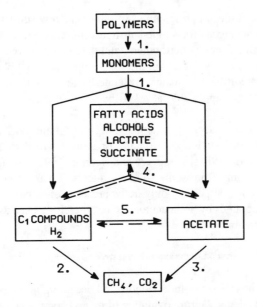

Figure 7.2. Carbon and electron flow in methanogenic degradation of complex organic matter in a neutral freshwater lake sediment or sewage sludge [redrawn after Zehnder et al. (1982)]. Solid lines: main flows; dashed lines: backward reactions occurring only under special conditions. Bacterial groups involved (1) classical "primary" fermenting bacteria; (2) hydrogen- and formate-oxidizing methanogenic bacteria; (3) acetotrophic methanogenic bacteria; (4) syntrophic "secondary" fermenting bacteria; (5) homoacetogenic bacteria

carbon compound on the one side and of acetate on the other, both being the key substrates of the two main physiological groups of methanogenic bacteria.

In utilization of methanol and methylamines, homoacetogens and methanogens are similarly successful (Winfrey and Ward, 1983; King et al., 1983; Lovley and Klug, 1983). Oxidation of primary alcohols and diols is efficiently carried out by sulfate reducers or by fermenting bacteria cooperating syntrophically with methanogens (Eichler and Schink, 1985).

In general, homoacetogens appear to be, in every single case, inferior to the respective specialists. Their success in a natural anoxic environment appears to be based mainly on their metabolic versatility, i.e., their ability to change between various substrates or to use them simultaneously. Recent studies with pure cultures of *Escherichia coli* in chemostats under energy limitation have shown that simultaneous utilization of more than one substrate enhances the affinity for every single substrate in play (Lendenmann et al., 1992). A metabolically versatile homoacetogen can use many more substrates simultaneously than can a highly specialized competitor which is bound to only very few substrates. The actual affinity of homoacetogens for single substrates may, therefore, be much higher in a complex multisubstrate environment, and their competitiveness has probably been underestimated considerably in the past: Some microbial physiologists tend to look at substrate utilization mainly by a one-bacterium–one-substrate concept, a concept that is not applicable at all to bacterial living conditions in natural environments.

The only metabolic activity in which homoacetogens are specialists themselves is, so far, the demethylation of aromatic or aliphatic methylether compounds (Bache and Pfennig, 1981; Schuppert and Schink, 1990). So far, competition in this field came only from some sulfate reducers (Tasaki et al., 1992) which, however, are strong competitors only in the presence of sulfate. Efficient utilization of these compounds by homoacetogens has, therefore, been used successfully for selective enrichment.

(b) INTERSPECIES TRANSFER OF REDUCING EQUIVALENTS

Homoacetogens can act both as donors and acceptors in interspecies transfer of reducing equivalents. During oxidation of methyl compounds or other organic substrates, homoacetogenic bacteria produce hydrogen at low levels which is either used up again by the homoacetogens themselves (Seitz et al., 1988; Bomar et al., 1991) or by other anaerobic hydrogen oxidizers. Interspecies hydrogen transfer to methanogenic or sulfate-reducing bacteria unable to use methanol was demonstrated in cocultures with *Sporomusa acidovorans* growing on methanol, or with other homoacetogens (Cord-Ruwisch and Ollivier, 1986; Heijthuisen and Hansen, 1986).

In other systems of interspecies hydrogen transfer, *Acetobacterium* sp. acts as hydrogen scavenger rather than hydrogen producer: Figure 7.3a shows bacterial colonies grown upon dilution in agar shake cultures of an enrichment culture from marine sediment with polyethylene glycol as substrate. A big central colony of *Pelobacter venetianus* degrading the polymer is surrounded by and mixed with colonies of *Acetobacterium* sp. which utilizes hydrogen released by the bacteria in the central colony. The colonies of the homoacetogen decrease in size with increasing distance from the central colony and outnumber those of the polymer degrader by far, indicating that they take the lion's share of energy available from polyethylene glycol degradation in the enrichment culture (Schink and Stieb, 1983). In other enrichment cultures from marine sediments with ethanol as substrate (Fig. 7.3b), *Acetobacterium*-like homoacetogens cooperated with syntrophically ethanol-oxidizing anaerobes as hydrogen scavengers (Eichler and Schink, 1985). It is worth emphasizing that in both these marine enrichment cultures, homoacetogens outcompeted methanogens for hydrogen or other re-

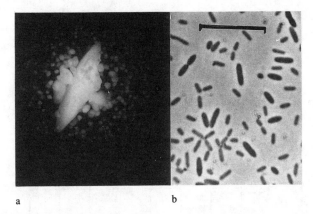

Figure 7.3.   (a) Mixed colony of *Pelobacter venetianus* (central) and *Acetobacterium* sp. (satellites) growing with polyethylene glycol and CO_2 [from Schink and Stieb (1983), with permission]. Diameter of the central colony is about 0.5 mm. (b) Phase-contrast photomicrograph of cells of a mixed culture of *Pelobacter* sp. (faint) and *Acetobacterium* sp. (more refractile) growing syntrophically with ethanol and CO_2. Bar equals 10 μm.

duced electron carriers. Enhanced salt concentrations may have a strong selective effect in these cultures. On the other hand, enrichments from freshwater sources with glycolate as the electron donor led first to the isolation of a homoacetogenic binary coculture (Friedrich et al., 1991) before a methanogenic co-culture could be established.

   The relative advantage of sulfate reducers and methanogens over homoaceto-gens in hydrogen utilization has been explained usually on the basis of higher energy yields and higher substrate affinity (Kristjansson and Schönheit, 1983; Robinson and Tiedje, 1984). This advantage becomes even more important with low hydrogen partial pressures as demonstrated in Figure 7.4: Even at low acetate concentrations (10 μM) typical of sediments or sewage sludge, homoacetogenic hydrogen oxidation approaches thermodynamic equilibrium at 10^{-5} bar H_2 and releases the required minimum energy equivalent for synthesis of one-third of an ATP [about 20 kJ/mol reaction; Schink (1990)] with 10^{-4} bar H_2 at minimum. Methanogenesis at conditions typical of a freshwater sediment or a sewage sludge digestor reaches thermodynamic equilibrium only at $10^{-5.5}$ bar H_2, and sulfate reduction is in all cases better off, regardless of whether sulfate concentrations typical of freshwater or of saltwater sediments are applied (for details, see Fig. 7.4). The minimum thresholds of hydrogen uptake have been found to be higher with homoacetogens (with CO_2 as acceptor) than with any other hydrogenotrophic anaerobes; they were considerably lower, even with homoacetogens themselves,

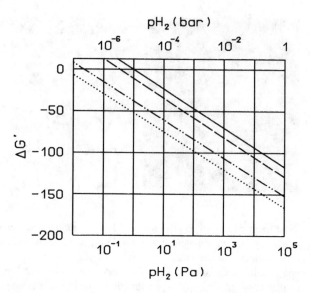

Figure 7.4. Dependence of free-energy change of hydrogen oxidation by homoacetogenic (solid line), methanogenic (dashed line), and sulfate-reducing bacteria in a freshwater (dash-dot line), and a marine (dotted line) sediment. Standard Gibbs free-energy changes (-95 kJ, -131 kJ, and -150 kJ, per 4 mol H_2, respectively) were corrected for the following conditions: CO_2: 0.3 bar; CH_4:0.7 bar; acetate: 10 μM; sulfide: 100 μM; sulfate (freshwater): 100 μM; sulfate (marine): 30 mM.

if another electron acceptor such as caffeate was reduced, indicating that energetics and kinetics of hydrogen oxidation may be linked (Cord-Ruwisch et al., 1988).

Nonetheless, lower substrate affinity and higher threshold values are only part of the whole story. It is obvious that also the spatial arrangement of the partner bacteria, their relative numbers to each other, and, with this, the diffusion distance for the transferred hydrogen tolerance to temporal oxygen exposure, etc., are further selecting factors that will direct the preference of certain partners over others.

Experiments have been reported that indicate that in certain cases not hydrogen but formate is the electron carrier transferred between syntrophic partner bacteria (Thiele and Zeikus, 1988a, 1988b). Energetically, formate formation from CO_2 and hydrogen formation from protons pose the same problems as both have about the same standard redox potentials ($E_0' = -430$ and -414 mV, respectively). At a CO_2 concentration typical of sediment or sludge, a hydrogen partial pressure of 10^{-4} bar is in equilibrium with a formate concentration of 10 μM, and both are at about the same redox potential as the NAD^+/NADH couple (Fig. 7.5). All oxidation reactions in, e.g., sugar fermentation to acetate can easily donate electrons to these carriers if the concentrations are maintained as postulated earlier. Because measurement of formate at low concentrations has become possible only recently with ion chromatography, reliable data on pools of formate versus hydrogen in anoxic habitats are scarce, and no data at all exist on comparison of fluxes through these pools (Thiele and Zeikus, 1988b). Moreover, nearly all partner bacteria involved in interspecies electron transfer have active enzymes both for hydrogen and for formate activation, and usually couple both redox reactions with the same low-potential electron carriers in the cell. Therefore, it is extremely difficult to distinguish between both carriers experimentally and both may be employed simultaneously to varying extent from case to case (Schink, 1991a).

Evidence has been provided that homoacetogenic bacteria can outcompete methanogens for hydrogen and hydrogen equivalents under certain conditions, e.g., in a mildly acidic lake sediment [pH 6.1 in Knaack Lake sediment; Phelps and Zeikus (1984)]. Tracer studies have revealed that nearly the entire electron

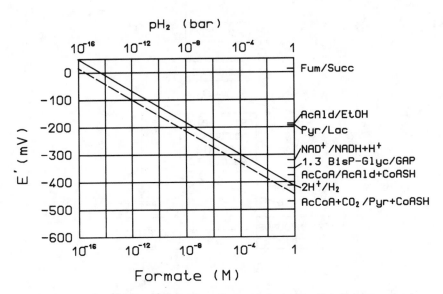

Figure 7.5. Redox potentials in equilibrium with hydrogen and formate at various concentrations, compared to standard potentials of important redox reactions in fermentation of biomass. Values were calculated for pH 7.0 and 0.3 bar CO_2. Fum: fumarate; Succ: succinate; AcAld: acetaldehyde; EtOH: ethanol; Pyr: pyruvate; Lac: lactate; 1.3-BisP-Glyc: 1,3-*bis*-phosphoglycerate; GAP: glyceraldehyde-3-phosphate; AcCoA; acetyl-CoA; CoASH: free coenzyme A. Solid line, hydrogen; broken line, formate.

flux from biomass to methane goes through the acetate pool, due to specific inhibition of hydrogen-oxidizing methanogens by the enhanced proton activity. Under these conditions, the scheme of overall electron flow changes from that of Figure 7.2 to that depicted in Figure 7.6, indicating that homoacetogenic bacteria would take over the function of hydrogen-oxidizing methanogens in the utilization of hydrogen and one-carbon compounds. This implies that fatty acids derived from lipid degradation also would be subject to syntrophic oxidation with homoacetogens as hydrogen scavengers, a reaction that is slightly endergonic at standard conditions (reaction 7.8). However, this reaction becomes exergonic at lower acetate and, e.g., butyrate concentrations, and yields -65 kJ per reaction at 100 μM concentration of both acids (and 0.7 bar CO_2). This would leave about -22 kJ for every partial reaction in this syntrophic cooperation, the energy minimum for synthesis of ATP fractions (Schink, 1990, 1991a, 1991b). The measured acetate concentration in Knaack lake sediment was in the range of 100 μM (Phelps and Zeikus, 1984), indicating that the partner bacteria actually operated with this minimum amount of energy.

At low temperatures ($< 20°C$), homoacetogens can take over a significant part

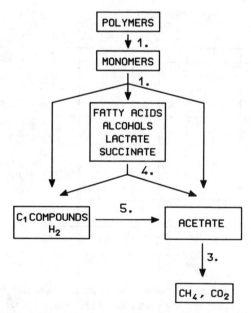

Figure 7.6.   Carbon and electron flow in methanogenic degradation of complex organic matter in a slightly acidic (pH 6.1) limnic sediment (Knaack Lake, Wisconsin). Bacterial groups are the same as in Figure 7.2. Group 2 (hydrogen-oxidizing methanogens) is missing.

in hydrogen oxidation (Conrad et al., 1989; Conrad and Wetter, 1990). The known species of methanogenic bacteria are not significantly active at low temperatures (Zeikus and Winfrey, 1976); obviously homoacetogens are less restricted in this respect. This may explain why they can win the game in several cases, especially if, as in soils, periodic oxygenation and acidification selects in a similar direction. Unfortunately, systematic studies on population sizes of homoacetogens versus methanogens in such environments are lacking so far (see also Chapters 10 and 15).

(c) OTHER TYPES OF METABOLIC COOPERATION

Experiments with binary mixed cultures of homoacetogens and acetate-utilizing methanogens have shown that complete methanogenic degradation of hexoses is possible in such defined cultures (Winter and Wolfe, 1979, 1980). Cooperation of both partners in these cultures is only metabiotic; the methanogen depends on the activity of the acetogen but does not significantly influence its metabolism and vice versa.

Slightly more complex is the cooperation between *Clostridium thermocellum* and *Acetogenium kivui* in thermophilic conversion of cellulose to acetate (Le Ruyet et al., 1984): In this case, the homoacetogenic partner modifies the metabolism of the cellulolytic *Clostridium* by maintaining a low hydrogen partial pressure. No ethanol is formed under these conditions, therefore, and the fermentation leads to nearly stoichiometric formation of 2.7 acetate units per hexose monomer.

Another type of metabolic coupling was established in a reaction chain of *Acetobacterium woodii*, *Pelobacter acidigallici*, and *Desulfobacter postgatei* for complete sulfate-dependent oxidation of trimethoxybenzoate to carbon dioxide (Fig. 7.7). In this case, the homoacetogen starts the reaction sequence by demethylating the original substrate to gallic acid, which is subject to acetogenic fermentation by *P. acidigallici*. Both bacteria form acetate as the sole fermentation product which is finally oxidized by the sulfate reducer. This experiment shows that homoacetogens can also act as key organisms, making a substrate available to other bacteria in the anaerobic food chain.

(d) METABOLISM OF O-, N-, AND S-METHYL COMPOUNDS

As pointed out earlier, homoacetogens are specialists in one-carbon metabolism in which they have to compete with methanogens or sulfate reducers, as is true with formate, methanol, or methylamines. Only methyl groups of methoxylated carbon compounds appear to be a food source which they really have to their own. Cleavage of the ether linkage between an aromatic ring carbon and the methyl carbon is chemically difficult and employs, in oxic environments, participation of an oxygenase reaction (Bernhardt et al., 1970). Anaerobic cleavage of

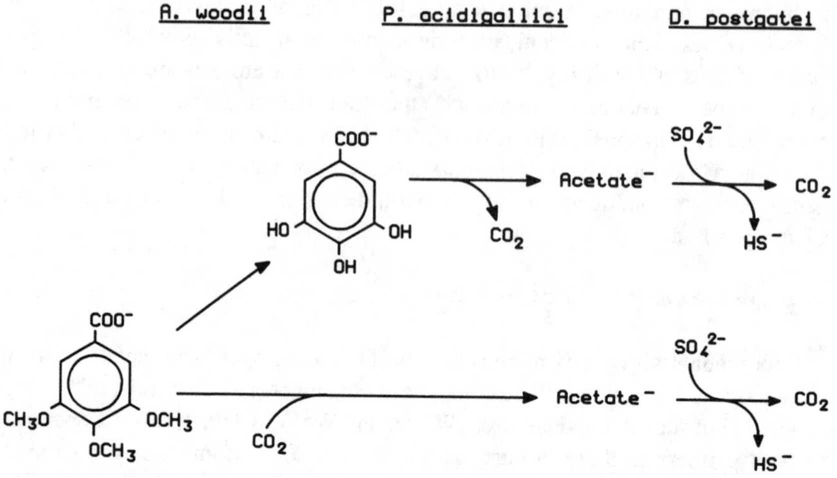

A. woodii   P. acidigallici   D. postgatei

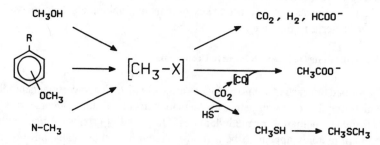

Figure 7.7.   Complete anaerobic oxidation of trimethoxybenzoate to CO_2 by a triculture of *Acetobacterium woodii, Pelobacter acidigallici,* and *Desulfobacter postgatei* [after Kreikenbohm and Pfennig (1985)].

this linkage has been known now for more than 12 years (Bache and Pfennig, 1981), but the chemistry of this reaction has not been resolved yet. Recent work indicates that the methyl group is transferred to tetrahydrofolate or a cobamide as first carrier (Berman and Frazer, 1992; Stupperich et al., 1992; Kreft and Schink, 1993). Whether methyl groups from N-methyl compounds or from methanol are transferred to the same carrier, and whether the same or different enzyme proteins are involved in these transfer reactions has still to be elucidated. The bound methyl group can be metabolized in various ways as Figure 7.8 illustrates:

Figure 7.8.   Transfer and conversion of methyl residues by homoacetogenic bacteria. X: a methyl carrier, either a cobalamine or tetrahydrofolate.

In pure cultures, it is condensed with a bound carbonyl residue to form acetyl-CoA and, finally, acetate as fermentation product. In mixed culture with hydrogenotrophic methanogens or sulfate reducers, partial or complete oxidation to CO_2 and/or formate can occur as demonstrated with *Sporomusa acidovorans* (Cord-Ruwisch and Ollivier, 1986) and other species of homoacetogenic bacteria (Heijthuisen and Hansen, 1986).

A new type of methyl metabolism was discovered recently with the homoacetogenic bacterium strain TMBS 4: This bacterium methylates sulfide to methanethiol and further to dimethyl sulfide [Fig. 7.8; Bak et al. (1992); Finster et al. (1990)]. This reaction obviously does not allow energy conservation (Kreft and Schink, 1993) and has to be considered as a gratuitous side reaction occurring in the presence of sulfide at high concentrations. Nonetheless, this reaction makes the gallic residue of trimethoxybenzoate accessible to degradation by the same bacterium, thus combining the activities of *A. woodii* and *P. acidigallici* (Fig. 7.7) in one bacterium.

Methanethiol and dilmethyl sulfide are the most important forms in which reduced sulfur is released from marine sediments into the atmosphere (Andreae, 1986; Kelly and Smith, 1990). Although dimethyl sulfide is formed also by reduction of dimethyl sulfoxide or fermentative degradation of dimethyl sulfoniopropionate (Kiene and Capone, 1988), formation of methanethiol and dimethyl sulfide from methoxylated aromatics by homoacetogenic bacteria may be of global importance (Finster et al., 1990). Population-size estimations by dilution of sediment material in agar shake cultures have shown that homoacetogens of the metabolic type of strain TMBS 4 outnumber *Acetobacterium*-like homoacetogens by several orders of magnitude (Bak. unpublished). Their significance in limnic or marine sediments appears to be high, therefore. One should bear in mind that other methylated compounds such as lignin derivatives could act as well as methyl donors for methanethiol formation by homoacetogens.

7.4 Enrichment and Isolation

7.4.1 Preparation of Growth Media

Homoacetogenic bacteria depend on the availability of sufficient amounts of carbon dioxide/bicarbonate in their growth medium. Lack of bicarbonate in boiled phosphate-buffered media has prevented successful enrichment and isolation of homoacetogens for many years. For enrichment, isolation, and cultivation of mesophilic homoacetogenic bacteria, the following carbonate-buffered mineral medium is recommended [after Widdel and Pfennig (1981); Schink and Pfennig (1982)]:

Dissolve in 1 liter of distilled water:

	Freshwater	Brackish water	Saltwater
KH_2PO_4	0.2 g	0.2 g	0.2 g
NH_4Cl	0.5 g	0.5 g	0.5 g
NaCl	1.0 g	8.0 g	20.0 g
$MgCl_2 \times 6\ H_2O$	0.4 g	1.2 g	3.0 g
KCl	0.5 g	0.5 g	0.5 g
$CaCl_2 \times 2\ H_2O$	0.15 g	0.15 g	0.15 g

Autoclave the complete mineral medium in a vessel equipped with

- a filter inlet to allow flushing of the headspace with sterile oxygen-free gas
- screw cap inlets for addition of thermally unstable additives after autoclaving
- a silicon tubing connection from the ground of the vessel out to a dispensing tap (if possible with a protecting bell) for sterile dispensing of the medium (do not use latex tubing; it releases in the autoclave compounds which are highly toxic to many anaerobes)
- a stirring bar

After autoclaving, connect the vessel with the still-hot medium to a line of oxygen-free nitrogen/carbon dioxide mixture (90% N_2/10% CO_2) at low pressure (< 100 mbar), flush the headspace, and cool it under this atmosphere to room temperature, perhaps with the help of a cooling water bath.

The mineral medium is amended with the following additions from stock solutions which have been sterilized separately (amounts per liter medium):

- 30 ml of 1 M $NaHCO_3$ solution (autoclaved in a *tightly closed* screw-cap bottle with about 30% headspace; the bottle should be autoclaved in a further protecting vessel, e.g., a polypropylene beaker, to avoid spills of carbonates if the bottle breaks at high internal pressure in the autoclave)
- 2 ml of 0.5 M $Na_2S \times 9\ H_2O$ solution (autoclaved separately under oxygen-free gas atmosphere as above)
- 1 ml of trace element solution, e.g., SL 10 (Widdel et al., 1983) or SL 9 (Tschech and Pfennig, 1984)
- 0.5 ml of tenfold concentrated, filter-sterilized vitamin solution (Pfennig, 1978)
- adequate amounts of sterile 1 M HCl or 1 M Na_2CO_3 to adjust the pH to 7.1–7.3.

The complete medium is dispensed into either screw-cap bottles or screw-cap tubes which are filled completely to the top, leaving a lentil-sized air bubble for pressure equilibration. Enrichment cultures usually produce gas in the first enrichment stages and are better cultivated in half-filled serum bottles (50–100 ml volume) under a headspace of a nitrogen/carbon dioxide mixture (90%/10%). Growth with hydrogen/carbon dioxide can be tested either in half-filled tubes sealed with butyl rubber stoppers (e.g., Hungate tubes) which are incubated lying, or in serum bottles which are incubated horizontally on a slow shaker to assure sufficient gas supply in the culture fluid.

This mineral medium is amended with the respective organic substrates for enrichment and cultivation of pure cultures. The vitamin mixture is not really needed by all strains. For example, *Acetobacterium carbinolicum* does not require any vitamins; others depend on only few vitamins. *A. wieringae* and many other homoacetogens have been reported to depend on yeast extract additions to the medium; with the medium described above, yeast extract is usually not required, but growth yields are higher in its presence. Special care should be taken with the trace-element solutions used. We observed that copper and borate concentrations in conventional trace-element solutions can easily be toxic to *Acetobacterium* species; delays in growth initiation in freshwater medium could be overcome by enhancing the sodium chloride content to about 50 mM. *Acetoanaerobium noterae* uses yeast extract also as an energy source.

For enrichment and cultivation of thermophilic homoacetogens, the following medium is recommended which differs from the above medium mainly in a higher phosphate content to assure better buffering [values in gram per liter medium; after Leigh et al. (1981)]:

K_2HPO_4	0.80	g
KH_2PO_4	0.80	g
NH_4Cl	0.31	g
$(NH_4)_2SO_4$	0.22	g
NaCl	1.00	g
$MgSO_4 \times 7\ H_2O$	0.09	g
$CaCl_2 \times 2\ H_2O$	0.006	g
$NaHCO_3$	4.5	g
L-cysteine $\times$ HCl $\times$ H_2O	0.5	g
$Na_2S \times 9\ H_2O$	0.5	g
SL 10	1.0	ml

It is advisable to autoclave cysteine, sodium sulfide, and the trace-element solution SL 10 separately as concentrated stock solutions in tightly closed vials. Cultures can be grown with 2–4 bar of hydrogen/carbon dioxide mixture (80%/20%) or 1 bar of nitrogen/carbon dioxide mixture (70%/30%) at the respective optimum temperatures of the desired strains. Enhanced CO_2 partial pressures are required to maintain the pH in the neutral range because solubility of CO_2 decreases rapidly with rising temperature.

7.4.2 Enrichment Cultures

For enrichment of mesophilic homoacetogens, a H_2/CO_2 mixture (80%/20%) in the culture headspace was usually a sufficiently selective enrichment substrate. Yeast extract or other complex medium additions should not be used in the enrichment stage to prevent growth of (often pathogenic!) peptolytic bacteria. To avoid growth of hydrogen-utilizing sulfate reducers, sulfate should be omitted from the medium. Growth of competing methanogens can be suppressed by lowering the pH to 6.0–6.5 or by addition of substances which are inhibitory for methanogens (e.g., 5–20 mM 2-bromoethanesulfonate in the medium, 0.5% acetylene, or 2% ethylene in the headspace). Small amounts of dithionite have been used for the same purpose (Balch et al., 1977) but did not select satisfactorily in our hands. The reducing agent cysteine can be used also as the energy source by several fermenting bacteria and should, therefore, be avoided in early stages of enrichments.

Ethylene glycol (10 mM) has proven as a rather selective substrate for *Acetobacterium* sp., although it is a good substrate for many *Pelobacter* species as well (Schink, 1991b). *A. woodii* can be selectively enriched with methoxylated aromatic compounds such as vanillate, ferulate, trimethoxycinnamate, syringate, etc. (5–10 mM) (Bache and Pfennig, 1981). Although other *Acetobacterium* sp. can use these substrates as well, *A. woodii* appears to grow the fastest with them. *A. carbinolicum* can be easily enriched and isolated from freshwater and brackish water sediments with 10–20 mM ethanol, *n*-propanol, or *n*-butanol as sole substrates; methoxyethanol or malate select for *A. malicum*. A selective enrichment procedure for *A. wieringae* is not known; the type strain was isolated with H_2/CO_2 from anoxic sewage sludge.

Acetoanaerobium noterae appears to be more fastidious than *Acetobacterium* sp. in its dependence on higher amounts of yeast extract (0.2% w/v). It was originally isolated from sediment of an oil-drilling site with pH 8.0, with H_2/CO_2 as substrate; perhaps the sampling site itself selected sufficiently for this organism which resembles *Acetobacterium* in nearly all properties.

The same may apply to *Acetitomaculum ruminis:* it is, besides *Eubacterium limosum,* so far the only non-spore-forming homoacetogen isolated from intestinal material of higher animals and appears to depend on at least 1% rumen fluid in the medium.

It should be mentioned at this point that enrichments with many other substrates have yielded pure cultures of homoacetogenic bacteria which resemble *Acetobacterium* strains both in their morphology and physiology. Among these substrates are triacetin (Emde and Schink, 1987), the formaldehyde-yielding compound hexamethylene tetramine (Schink, 1987), short-chain polyethylene glycols and their surface-active derivatives (Wagener and Schink, 1988), methoxyacetate (Schuppert and Schink, 1990), DL-mandelate (Dörner and Schink, 1991), and furfuryl alcohol (unpublished results from our lab).

For enrichment and isolation of specific types of homoacetogens, the original enrichment substrates should be used as specified in Chapter 2. In general, fairly reduced, electron-rich substrates such as alcohols, polyols, 2,3-butanediol, etc., select rather specifically for homoacetogens that can use these substrates due to their ability to reduce CO_2 as electron acceptor.

Spore-forming homoacetogens (genera *Acetonema, Clostridium,* and *Sporomusa*) are successfully enriched in the same way as described earlier for other homoacetogens but with autoclaved inoculum material (heat treatment for 10 min *in a sterilized tube* at 80°C in a water bath. The tube neck should be flamed before the inoculum is withdrawn). The pasteurization process can be repeated before isolation to make sure that only spore formers will grow. It should be assured microscopically that sufficient numbers of mature spores are present in the inoculum material! In addition, dried inocula such as soil samples or samples from thermal environments will preferentially give rise to enrichment of spore-forming homoacetogens.

7.4.3 Isolation Procedure

After one to two transfers in the above-mentioned enrichment media, assay of the typical fermentation product acetate, hydrogen consumption without significant methane formation, and appearance of rod-shaped, nonfluorescent cells of, e.g., *Acetobacterium*-like shape (see Fig. 1a) should indicate that homoacetogenic bacteria are present and, perhaps, are predominant. Purification can be carried out with organic substrates in agar shake dilution series (Pfennig, 1978) or, especially with hydrogen as substrate, in roll tubes (Balch et al., 1979) either diluted in the liquid agar or streaked on the agar surface. A handy variation of roll tubes are flat bottle plates in which a thin agar film is kept under a defined gas mixture (Braun et al., 1979). With bottle plates and roll tubes, the sulfide content of the media should be enhanced to 2–3 mM to balance for losses of hydrogen sulfide into the gas phase. At this stage, yeast extract or other undefined complex nutrients can be added to obtain a spectrum of homoacetogens as broad as possible. The use of calcium carbonate in the medium ("chalk agar") as an indicator of acid formation can be helpful for differentiation of acetic acid formers from other anaerobes (Braun et al., 1979), but the slightly elevated pH of such media may inhibit certain types of homoacetogens.

Colonies should be picked with either sterile Pasteur pipettes that have been pulled in the flame to thin capillaries of about 100 μm inner diameter, or with a platinum inoculation loop, and again purified. Only after at least two subsequent dilution/streaking steps, the obtained cultures, if morphologically homogenous, can be regarded as pure. Purity should be checked after growth in complex medium, e.g., "AC-medium" (Difco, Ann Arbor, Michigan). In our experience, AC-medium prepared by the recommended recipe is so concentrated that it inhibits many sediment bacteria. It should be applied not only at its original concentration but also at a 1 : 5 or 1 : 10 dilution, provided that the agar content is corrected with additional agar. Instead of AC-medium, a self-made complex medium with 0.05% (w/v) of each glucose, fructose, malt extract, yeast extract, and peptone can be applied also.

7.5   Conclusions

In this chapter, I have tried to give an overview of the morphological, taxonomic, and metabolic diversity of homoacetogenic bacteria, of their metabolic function in natural or seminatural (e.g., sewage sludge digestors) environments, and of techniques on how to cultivate these bugs in the laboratory. Due to their enormous metabolic versatility, homoacetogens are involved in many processes of substrate turnover, either as primary catalyzing agents or as partner bacteria influencing the metabolic activities of other anaerobes. These bacteria are favored over other fermenting bacteria growing on complex organic substrates by their ability to use CO_2 as an external electron acceptor, which is usually available in excess. They depend on such bacteria for utilization of polymeric substrates that make up the main part of biomass in nature. On the other hand, these bacteria are hampered on energetical grounds in their competition for electron sources against methanogenic and sulfate-reducing bacteria. They are experts in metabolism of one-carbon compounds and, as with methoxylated organics, even specialists in a limited field of their own.

Our knowledge of the metabolic capacities of these bacteria has increased steadily through the recent 15 years, and I am confident that more unusual metabolic features of homoacetogens will be discovered in the near future, especially if hot, acidic, alkaline, or saline environments, or digestive tracts of lower animals are checked in more detail for homoacetogenic inhabitants. For a better understanding of their actual metabolic activity in a natural environment, we would need much more information about population sizes and structures, in combination with information on substrate turnover in situ on a microscale.

Application of molecular probes and of microelectrodes in combination with data on tracer transformations, and new techniques for cultivation even of so far noncultivated bacteria will probably provide means that could at least pave the way for a better understanding of the metabolic activities in nature of this fascinating physiological group of bacteria.

Acknowledgments

This chapter is based in part on a review on the genus *Acetobacterium* and related species, which was published recently in *The Prokaryotes,* 2nd ed. (Schink and Bomar, 1991). The author gratefully acknowledges numerous fruitful discussions with Nobert Pfennig, Rudolf K. Thauer, Fritz Widdel, Gabi Diekert, Ralf Conrad, and Friedhelm Bak on the physiology and ecology of homoacetogenic bacteria. Further thanks are due to Andreas Brune and Karin Denger for help with literature surveys and preparation of the figures.

This chapter is dedicated to Friedhelm Bak who died recently and much too early of cancer. The microbiological scientific community of our country has lost one of its most promising young representatives.

References

Adamse, A. D. 1980. New isolation of *Clostridium aceticum* (Wieringa). *Antonie van Leeuwenhoek* **46:**523–531.

Andreae, M. O. 1986. The ocean as a source for atmospheric sulfur compounds. In: *Biogenic Sulfur in the Environment,* W. S. Saltzman and W. J. Cooper (eds.), pp. 2–14, American Chemical Society, Washington, D.C.

Andreesen, J. R., G. Gottschalk, and H. G. Schlegel. 1970. *Clostridium formicoaceticum* nov. spec. Isolation, description and distinction from *C. aceticum* and *C. thermoaceticum. Arch. Mikrobiol.* **72:**154–174.

Bache, R., and N. Pfennig. 1981. Selective isolation of *Acetobacterium woodii* on methoxylated aromatic acids and determination of growth yields. *Arch. Microbiol.* **130:**255–261.

Bak, F., K. Finster, and F. Rothfuß. 1992. Formation of dimethylsulfide and methanethiol from methoxylated aromatic compounds and inorganic sulfide by newly isolated anaerobic bacteria. *Arch. Microbiol.* **157:**529–534.

Balch, W. E., S. Schoberth, R. S. Tanner, and R. S. Wolfe. 1977. *Acetobacterium,* a new genus of hydrogen-oxidizing, carbon dioxide-reducing, anaerobic bacteria. *Int. J. Syst. Bacteriol.* **27:**355–361.

Balch, W. E., G. E. Fox, L. J. Magrum, C. R. Woese, and R. S. Wolfe. 1979. Methanogens: reevaluation of a unique biological group. *Microbiol. Rev.* **43:**260–296.

Barik, S., S. Prieto, S. B. Harrison, E. C. Clausen, and J. L. Gaddy. 1988. Biological production of alcohols from coal through indirect liquefaction. *Appl. Biochem. Biotechnol.* **18**:363–378.

Barker, H. A., and V. Haas. 1944. *Butyribacterium*, a new genus of Gram-positive, nonsporulating anaerobic bacteria of intestinal origin. *J. Bacteriol.* **47**:301–305.

Berman, M. H., and A. C. Frazer. 1992. Importance of tetrahydrofolate and ATP in the anaerobic O-demethylation reaction for phenylmethylethers. *Appl. Environ. Microbiol.* **58**:925–931.

Bernhardt, F.-H., H. Staudinger, and V. Ulrich. 1970. Eigenschaften einer *p*-Anisat-O-Demethylase im zellfreien Extrakt von *Pseudomonas* sp. Hoppe-Seyler's. *Z. Physiol. Chem.* **351**:467–478.

Bogdahn, M., J. R. Andreesen, and D. Kleiner. 1983. Pathways and regulation of N_2, ammonium and glutamate assimilation by *Clostridium formicoaceticum*. *Arch. Microbiol.* **134**:167–169.

Bomar, M., H. Hippe, and B. Schink. 1991. Lithotrophic growth and hydrogen metabolism by *Clostridium magnum*. *FEMS Microbiol. Lett.* **83**:347–350.

Braun, M., S. Schoberth, and G. Gottschalk. 1979. Enumeration of bacteria forming acetate from H_2 and CO_2 in anaerobic habitats. *Arch. Microbiol.* **120**:201–204.

Braun, M., F. Mayer and G. Gottschalk. 1981. *Clostridium aceticum* (Wieringa), a microorganism producing acetic acid from molecular hydrogen and carbon dioxide. *Arch. Microbiol.* **128**:288–293.

Braun, M., and G. Gottschalk. 1982. *Acetobacterium wieringae* sp. nov., a new species producing acetic acid from molecular hydrogen and carbon dioxide. *Zbl. Bakt. Hyg., I. Abt. Orig.* **C3**:368–376.

Breznak, J. A., and J. M. Switzer. 1986. Acetate synthesis from H_2 plus CO_2 by termite gut microbes. *Appl. Environ. Microbiol.* **52**:623–630.

Breznak, J. A., J. M. Switzer, and H.-J. Seitz. 1988. *Sporomusa termitida* sp. nov., an H_2/CO_2-utilizing acetogen isolated from termites. *Appl. Environ. Microbiol.* **150**:282–288.

Buschhorn, H., P. Dürre, and G. Gottschalk. 1989. Production and utilization of ethanol by the homoacetogen *Acetobacterium woodii*. *Appl. Environ. Microbiol.* **55**:1835–1840.

Cato, E. P., W. L. George, and S. M. Finegold. 1986. Genus *Clostridium*. In: *Bergey's Manual of Systematic Bacteriology*, P. H. A. Sneath, N. S. Mair, M. E. Sharpe, and J. G. Holt (eds.), Vol. 2, pp. 1141–1207. Williams and Wilkins, Baltimore, MD.

Conrad, R., F. Bak, H. J. Seitz, B. Thebrath, H. P. Mayer, and H. Schütz. 1989. Hydrogen turnover by psychotrophic homoacetogenic and mesophilic methanogenic bacteria in anoxic paddy soil and lake sediment. *FEMS Microbiol. Ecol.* **62**:285–294.

Conrad, R., and B. Wetter. 1990. Influence of temperature on energetics of hydrogen metabolism in homoacetogenic, methanogenic, and other anaerobic bacteria. *Arch. Microbiol.* **155**:94–98.

Cord-Ruwisch, R., and B. Ollivier. 1986. Interspecies hydrogen transfer during methanol degradation by *Sporomusa acidovorans* and hydrogenophilic anaerobes. *Arch. Microbiol.* **144**:163–165.

Cord-Ruwisch, R., H.-J. Seitz, and R. Conrad. 1988. The capacity of hydrogenotrophic anaerobic bacteria to compete for traces of hydrogen depends on the redox potential of the terminal electron acceptor. *Arch. Microbiol.* **149**:350–357.

Daniel, S. L., and H. L. Drake. 1993. Oxalate- and glyoxylate-dependent growth and acetogenesis by *Clostridium thermoaceticum*. *Appl. Environ. Microbiol.* **59**:3062–3069.

Dehning, I., M. Stieb, and B. Schink. 1989. *Sporomusa malonica* sp. nov., a homoacetogenic bacterium growing by decarboxylation of malonate or succinate. *Arch. Microbiol.* **151**:421–426.

Diekert, G. B., and R. K. Thauer. 1978. Carbon monoxide oxidation by *Clostridium thermoaceticum* and *Clostridium formicoaceticum*. *J. Bacteriol.* **136**:597–606.

Dörner, Ch., and B. Schink. 1991. Fermentation of mandelate to benzoate and acetate by a homoacetogenic bacterium. *Arch. Microbiol.* **156**:302–306.

Dorn, M., J. R. Andreesen, and G. Gottschalk. 1978. Fermentation of fumarate and L-malate by *Clostridium formicoaceticum*. *J. Bacteriol.* **133**:26–32.

Egli, C., T. Tschan, R. Scholtz, A. M. Cook, and Th. Leisinger. 1988. Transformation of tetrachloromethane to dichloromethane and carbon dioxide by *Acetobacterium woodii*. *Appl. Environ. Microbiol.* **54**:2819–2824.

Eichler, B., and B. Schink. 1984. Oxidation of primary aliphatic alcohols by *Acetobacterium carbinolicum* sp. nov., a homoacetogenic anaerobe. *Arch. Microbiol.* **140**:147–152.

Eichler, B., and B. Schink. 1985. Fermentation of primary alcohols and diols and pure culture of syntrophically alcohol-oxidizing anaerobes. *Arch. Microbiol.* **143**:60–66.

Emde, R., and B. Schink. 1987. Fermentation of triacetin and glycerol by *Acetobacterium* sp. No energy is conserved by acetate excretion. *Arch. Microbiol.* **149**:142–148.

Finster, K., G. M. King, and F. Bak. 1990. Formation of methylmercaptan and dimethylsulfide from methoxylated aromatic compounds in anoxic marine and freshwater sediments. *FEMS Microbiol. Ecol.* **74**:295–302.

Fischer, F., R. Lieske, and K. Winzer. 1932. Biologische Gasreaktionen. II. Mitteilung: Über die Bildung von Essigsäure bei der biologischen Umsetzung von Kohlenoxyd und Kohlensäure mit Wasserstoff zu Methan. *Biochem. Z.* **245**:2–12.

Fontaine, F. E., W. H. Peterson, E. McCoy, M. J. Johnson, and G. J. Ritter. 1942. A new type of glucose fermentation by *Clostridium thermoaceticum* n. sp. *J. Bacteriol.* **43**:701–715.

Friedrich, M., U. Laderer, and B. Schink. 1991. Fermentative degradation of glycolic acid by defined syntrophic cocultures. *Arch. Microbiol.* **156**:398–404.

Geerligs, G., H. C. Aldrich, W. Harder, and G. Diekert. 1987. Isolation and characterization of a carbon monoxide utilizing strain of the acetogen *Peptostreptococcus productus*. *Arch. Microbiol.* **148**:305–313.

Gößner, A., S. L. Daniel, and H. L. Drake. 1994. Acetogenesis coupled to the oxidation of aromatic aldehyde groups. *Arch. Microbiol.* (in press).

Greening, R. C., and J. A. Z. Leedle. 1989. Enrichment and isolation of *Acetitomaculum ruminis* gen. nov., sp. nov.: acetogenic bacteria from the bovine rumen. *Arch. Microbiol.* **151**:399–406.

Hamlett, N. U., and B. A. Blaylock. 1969. Synthesis of acetate from methanol. *Bacteriol. Proc.* 207.

Hansen, B., M. Bokranz, P. Schönheit, and A. Kröger. 1988. ATP formation coupled to caffeate reduction by H_2 in *Acetobacterium woodii* NZval6. *Arch. Microbiol.* **150**:447–451.

Heijthuijsen, J. H. F. G., and T. A. Hansen. 1986. Interspecies hydrogen transfer in co-cultures of methanol-utilizing acidogens and sulfate-reducing or methanogenic bacteria. *FEMS Microbiol. Ecol.* **38**:57–64.

Hsu, T., M. F. Lux, and H. L. Drake. 1990. Expression of an aromatic-dependent decarboxylase which provides growth-essential CO_2 equivalents for the acetogenic (Wood) pathway of *Clostridium thermoaceticum*. *J. Bacteriol.* **172**:5901–5907.

Kane, M. D., and J. A. Breznak. 1991. *Acetonema longum* gen. nov. sp. nov., an H_2/CO_2 acetogenic bacterium from the termite, *Pterotermes occidentis*. *Arch. Microbiol.* **156**:91–98.

Kane, M. D., A. Brauman, and J. A. Breznak. 1991. *Clostridium mayombei* sp. nov., an H_2/CO_2 acetogenic bacterium from the gut of the African soil-feeding termite, *Cubitermes speciosus*. *Arch. Microbiol.* **156**:99–104.

Kelly, D. P., and N. A. Smith. 1990. Organic sulfur compounds in the environment. In: *Advances in Microbial Ecology*, K. C. Marshall (ed.), Vol. 11, pp. 345–385. Plenum Press, New York.

Kerby, R., W. Niemczura, and J. G. Zeikus. 1983. Single-carbon catabolism in acetogens: analysis of carbon flow in *Acetobacterium woodii* and *Butyribacterium methylotrophicum* by fermentation and ^{13}C nuclear magnetic resonance measurement. *J. Bacteriol.* **155**:1208–1218.

Kiene, R. P., and D. G. Capone. 1988. Microbiol transformations of methylated sulfur compounds in anoxic salt marsh sediments. *Microb. Ecol.* **15**:275–291.

King, G. M., M. J. Klug, and D. R. Lovley. 1983. Metabolism of acetate, methanol, and methylated amines in intertidal sediments of Lowes Cove, Maine. *Appl. Environ. Microbiol.* **45**:1848–1853.

Kreft, J.-U., and B. Schink. 1993. Demethylation and degradation of phenylmethylethers by the sulfide-methylating homoacetogenic bacterium strain TMBS4. *Arch. Microbiol.* **159**:308–315.

Kreikenbohm, R., and N. Pfennig. 1985. Anaerobic degradation of 3.4.5-trimethoxybenzoate by a defined mixed culture of *Acetobacterium woodii*, *Pelobacter acidigallici*, and *Desulfobacter postgatei*. *FEMS Microbiol. Ecol.* **31**:29–38.

Kristjansson, J. K., and P. Schönheit. 1983. Why do sulfate-reducing bacteria outcompete methanogenic bacteria for substrates? *Oecologia (Berlin)* **60**:264–266.

Krumboeck, M., and R. Conrad. 1991. Metabolism of position-labelled glucose in anoxic methanogenic paddy soil and lake sediment. *FEMS Microbiol. Ecol.* **85**:247–256.

Krumholz, L. R., and M. P. Bryant. 1985. *Clostridium pfennigii* sp. nov. uses methoxyl groups of mono-benzenoids and produces butyrate. *Int. J. Syst. Bacteriol.* **35**:454–456.

Krumholz, L. R., and M. P. Bryant. 1986. *Syntrophococcus sucromutans* sp. nov. gen. nov. uses carbohydrates as electron donors and formate, monobezenoids or *Methanobrevibacter* as electron acceptor systems. *Arch. Microbiol.* **143**:313–318.

Lajoie, S. F., S. Bank. T. L. Miller, and M. J. Wolin. 1988. Acetate production from hydrogen and [^{13}C] carbon dioxide by the microflora of human feces. *Appl. Environ. Microbiol.* **54**:2723–2727.

Lee, M. J., and S. H. Zinder. 1988a. Isolation and characterization of a thermophilic bacterium which oxidizes acetate in syntrophic association with a methanogen and which grows acetogenically on H_2-CO_2. *Appl. Environ. Microbiol.* **54**:124–129.

Lee, M. J., and S. H. Zinder. 1988b. Carbon monoxide pathway enzyme activities in a thermophilic anaerobic bacterium grown acetogenically and in a syntrophic acetate-oxidizing coculture. *Arch. Microbiol.* **150**:513–518.

Lee, M. J., and S. H. Zinder. 1988c. Hydrogen partial pressures in a thermophilic acetate-oxidizing methanogenic coculture. *Appl. Environ. Microbiol.* **154**:1457–1461.

Leigh, J. A., F. Mayer, and R. S. Wolfe. 1981. *Acetogenium kivui,* a new thermophilic hydrogen-oxydizing, acetogenic bacterium. *Arch. Microbiol.* **129**:275–280.

Lendenmann, U., M. Snozzi, and T. Egli. 1992. Simultaneous utilization of diauxic sugar mixtures by *Escherichia coli,* Abstract P2-04-13. 6th International Symposium on Microbial Ecology, Barcelona, Spain.

Le Ruyet, P., H. C. Dubourgier, and G. Albagnac. 1984. Homoacetogenic fermentation of cellulose by a coculture of *Clostridium thermocellum* and *Acetogenium kivui. Appl. Environ. Microbiol.* **48**:893–894.

Lorowitz, W. H., and M. P. Bryant. 1984. *Peptostreptococcus productus* strain that grows rapidly with CO as the energy source. *Appl. Environ. Microbiol.* **47**:961–964.

Lovley, D. R., and M. J. Klug. 1983. Methanogenesis from methanol and methylamines and acetogenesis from hydrogen and carbon dioxide in the sediments of a eutrophic lake. *Appl. Environ. Microbiol.* **45**:1310–1315.

Lux, M. F., and H. L. Drake. 1992. Reexamination of the metabolic potentials of the acetogens. *Clostridium aceticum* and *Clostridium formicoaceticum:* chemolithoautotrophic and aromatic-dependent growth. *FEMS Microbiol. Lett.* **95**:49–56.

Lynd, L. H., R. Kerby, and J. G. Zeikus. 1982. Carbon monoxide metabolism of the methylotrophic acidogen *Butyribacterium methylotrophicum. J. Bacteriol.* **149**:255–263.

Lynd, L. H., and J. G. Zeikus. 1983. Metabolism of H_2-CO_2, methanol, and glucose by *Butyribacterium methylotrophicum. J. Bacteriol.* **153**:1415–1423.

Matthies, C., A. Freiberger, and H. L. Drake. 1993. Fumarate dissimilation and differential reductant flow by *Clostridium formicoaceticum* and *Clostridium aceticum*. *Arch. Microbiol.* **160**:273–278.

Möller, B., R. Oßmer, B. H. Howard, G. Gottschalk, and H. Hippe. 1984. *Sporomusa,* a new genus of gram-negative anaerobic bacteria including *Sporomusa sphaeroides* spec. nov. and *Sporomusa ovata* spec. nov., *Arch. Microbiol.* **139**:388–396.

Moore, W. E. C., and E. P. Cato. 1965. Synonymy of *Eubacterium limosum* and *Butyribacterium rettgeri: Butyribacterium limosum* comb nov. *Int. Bull. Bacteriol. Nomenclature Taxonomy* **15**:69–80.

Müller, E., K. Fahlbusch, R. Walther, and G. Gottschalk. 1981. Formation of N,N-dimethylglycine, acetic acid, and butyric acid from betaine by *Eubacterium limosum*. *Appl. Environ. Microbiol.* **42**:439–445.

Ollivier, B., R. Cord-Ruwisch, A. Lombardo, and J. L. Garcia. 1985. Isolation and characterization of *Sporomusa acidovorans* sp. nov., a methylotrophic homoacetogenic bacterium. *Arch. Microbiol.* **142**:307–310.

Parekh, M., E. S. Keith, S. L. Daniel, and H. L. Drake. 1992. Comparative evaluation of the metabolic potentials of different strains of *Peptostreptococcus productus:* utilization and transformation of aromatic compounds. *FEMS Microbiol. Lett.* **94**:69–74.

Patel, B. K. C., C. Monk, H. Littleworth, H. W. Morgan, and R. M. Daniel. 1987. *Clostridium fervidus* (sic!) sp. nov., a new chemoorganotrophic acetogenic thermophile. *Int. J. Syst. Bacteriol.* **37**:123–126.

Pfennig, N. 1978. *Rhodocyclus purpureus* gen. nov. and sp. nov., a ring-shaped, vitamin B_{12} requiring member of the family Rhodospirillaceae. *Int. J. Syst. Bacteriol.* **28**:283–288.

Phelps, T. J., and J. G. Zeikus. 1984. Influence of pH on terminal carbon metabolism in anoxic sediments from a mildly acidic lake. *Appl. Environ. Microbiol.* **48**:1088–1095.

Prévot, A. R. 1938. Étude de systématique bactérienne. *Ann. Inst. Pasteur (Paris)* **60**:285–307.

Prins, R. A., and A. Lankhorst. 1977. Synthesis of acetate from CO_2 in the cecum of some rodents. *FEMS Microbiol. Lett.* **1**:255–258.

Robinson, J. A., and J. H. Tiedje. 1984. Competition between sulfate-reducing and methanogenic bacteria for H_2 under resting and growing conditions. *Arch. Microbiol.* **137**:26–32.

Savage, M. D., and H. L. Drake. 1986. Adaptation of the acetogen *Clostridium thermoautotrophicum* to minimal medium. *J. Bacteriol.* **165**:315–318.

Schink, B. 1984. *Clostridium magnum* sp. nov., a non-autotrophic homoacetogenic bacterium. *Arch. Microbiol.* **137**:250–255.

Schink, B. 1987. Ecology of C_1-metabolizing anaerobes. In: *Microbial Growth on C_1-Compounds,* H. W. van Verseveld and J. A. Duine (eds.), pp. 81–85. Martinus Nijhoff Publishers, Dordrecht.

Schink, B. 1990. Conservation of small amounts of energy in fermenting bacteria. In: *Biotechnology in Focus* 2, R. K. Finn and P. Präve (eds.), pp. 63–89. Hanser, Munich.

Schink, B. 1991a. Syntrophism among prokaryotes. In: The Prokaryotes, 2nd ed., A. Balows, H. G. Trüper, M. Dworkin, W. Harder, and K. H. Schleifer (eds.), Chapter 11, pp. 276–299. Springer-Verlag, New York.

Schink, B. 1991b. The genus *Pelobacter.* In: *The Prokaryotes,* 2nd ed., A. Balows, H. G. Trüper, M. Dworkin, W. Harder, and K. H. Schleifer (eds.), Chapter 186, pp. 3393–3399. Springer-Verlag, New York.

Schink, B., and M. Bomar. 1991. The genera *Acetobacterium, Acetogenium, Acetoanaerobium,* and *Acetitomaculum.* In: *The Prokaryotes,* 2nd ed., A. Balows, H. G. Trüper, M. Dworkin, W. Harder, and K. H. Schleifer (eds.), Chapter 87, pp. 1925–1936. Springer-Verlag, New York.

Schink, B., and N. Pfennig. 1982. Fermentation of trihydroxybenzenes by *Pelobacter acidigallici* gen. nov. sp. nov., a new strictly anaerobic, Gram-negative, nonsporeforming bacterium. *Arch. Microbiol.* **133**:195–201.

Schink, B., and M. Stieb. 1983. Fermentative degradation of polyethylene glycol by a strictly anaerobic, nonsporeforming bacterium, *Pelobacter venetianus* sp. nov. *Appl. Environ. Microbiol.* **45**:1905–1913.

Schramm, E., and B. Schink. 1991. Ether-cleaving enzyme and diol dehydratase involved in anaerobic polyethylene glycol degradation by an *Acetobacterium* sp. *Biodegradation* **2**:71–79.

Schuppert, B., and B. Schink. 1990. Fermentation of methoxyacetate to glycolate and acetate by newly isolated strains of *Acetobacterium* sp. *Arch. Microbiol.* **153**:200–204.

Seifritz, C., S. L. Daniel, A. Gößner, H. L. Drake. 1993. Nitrate as a preferred electron sink for the acetogen *Clostridium thermoaceticum.* *J. Bacteriol.* **175**:8008–8013.

Seitz, H.-J., B. Schink, and R. Conrad. 1988. Thermodynamics of hydrogen metabolism in methanogenic cocultures degrading ethanol or lactate. *FEMS Microbiol. Lett.* **55**:119–124.

Sharak-Genthner, B. R., and M. P. Bryant. 1982. Growth of *Eubacterium limosum* with carbon monoxide as the energy source. *Appl. Environ. Microbiol.* **43**:70–74.

Sharak-Genthner, B. R., and M. P. Bryant. 1987. Additional characteristics of one-carbon-compound utilization by *Eubacterium limosum* and *Acetobacterium woodii.* *Appl. Environ. Microbiol.* **53**:471–476.

Sharak-Genthner, B. R., C. L. Davis, and M. P. Bryant. 1981. Features of rumen and sewage sludge strains of *Eubacterium limosum,* a methanol and H_2-CO_2-utilizing species. *Appl. Environ. Microbiol.* **42**:12–19.

Sleat, R., R. A. Mah, and R. Robinson. 1985. *Acetoanaerobium noterae* gen. nov. sp. nov.: an anaerobic bacterium that forms acetate from H_2 and CO_2. *Int. J. Syst. Bacteriol.* **35**:10–15.

Stackebrandt, E., H. Pohla, R. Kroppenstedt, H. Hippe, and C. R. Woese. 1985. 16S rRNA analysis of *Sporomusa, Selenomonas,* and *Megasphaera:* on the phylogenetic origin of Gram-positive eubacteria. *Arch. Microbiol.* **143**:270–276.

Strohhäcker, J., and B. Schink. 1991. Energetic aspects of malate and lactate fermentation by *Acetobacterium malicum. FEMS Microbiol. Lett.* **90**:83–88.

Stromeyer, S. A., W. Winkelbauer, H. Kohler, A. Cook, and Th. Leisinger. 1991. Dichloromethane utilized by an anaerobic mixed culture: acetogenesis and methanogenesis. *Biodegradation* **2**:129–137.

Stromeyer, S. A., K. Stumpf, A. M. Cook, and Th. Leisinger. 1992. Anaerobic degradation of tetrachloromethane by *Acetobacterium woodii. Biodegradation* **3**:113–123.

Stupperich, E., P. Aulkemeyer, and C. Eckerskorn. 1992. Purification and characterization of a methanol-induced cobamide-containing protein from *Sporomusa ovata. Arch. Microbiol.* **158**:370–373.

Tanaka, K., and N. Pfennig. 1988. Fermentation of 2-methoxyethanol by *Acetobacterium malicum* sp. nov. and *Pelobacter venetianus. Arch. Microbiol.* **149**:181–187.

Tanner, R. S., E. Stackebrandt, G. E. Fox, and C. R. Woese. 1981. A phylogenetic analysis of *Acetobacterium woodii, Clostridium barkeri, Clostridium butyricum, Clostridium lituseburense, Eubacterium limosum,* and *Eubacterium tenue. Curr. Microbiol.* **5**:35–38.

Tanner, R. S., L. M. Miller, and D. Yang. 1993. *Clostridium ljungdahlii* sp. nov., an acetogenic species in clostridial rRNA homology group I. *Int. J. Syst. Bacteriol.* **43**:232–236.

Tasaki, M., Y. Kamagota, K. Nakamura, and E. Mikami. 1992. Utilization of methoxylated benzoates and formation of intermediates by *Desulfotomaculum thermobenzoicum* in the presence or absence of sulfate. *Arch. Microbiol.* **157**:209–212.

Thauer, R. K., D. Möller-Zinkhan, and A. M. Spormann. 1989. Biochemistry of acetate catabolism in anaerobic chemotrophic bacteria. *Annu. Rev. Microbiol.* **43**:43–67.

Thiele, J., and J. G. Zeikus. 1988a. Control of interspecies electron flow during anaerobic digestion: The role of formate versus hydrogen transfer during syntrophic methanogenesis in flocs. *Appl. Environ. Microbiol.* **54**:20–29.

Thiele, J., and J. G. Zeikus. 1988b. Interactions between hydrogen- and formate-producing bacteria and methanogens during anaerobic digestion. In: *Handbook on Anaerobic Fermentations,* L. E. Erickson, and D. Y.-C. Fung (eds.), pp. 537–595. Marcel Dekker Inc., New York.

Traunecker, J., A. Preuß, and G. Diekert. 1991. Isolation and characterization of a methyl chloride utilizing, strictly anaerobic bacterium. *Arch. Microbiol.* **156**:416–421.

Tschech, A., and N. Pfennig. 1984. Growth yield increase linked to caffeate reduction in *Acetobacterium woodii. Arch. Microbiol.* **137**:163–167.

Wagener, S., and B. Schink. 1988. Fermentative degradation of nonionic surfactants by enrichment cultures and by pure cultures of homoacetogenic and propionate-forming bacteria. *Appl. Environ. Microbiol.* **54**:561–565.

Widdel, F., and N. Pfennig. 1981. Studies on dissimilatory sulfate-reducing bacteria that decompose fatty acids. I. Isolation of new sulfate-reducing bacteria enriched with acetate from saline environments. Description of *Desulfobacter postgatei* gen. nov. sp. nov. *Arch. Microbiol.* **129**:395–400.

Widdel, F., G. W. Kohring, and F. Mayer. 1983. Studies on dissimilatory sulfate-reducing bacteria that decompose fatty acids. III. Characterization of the filamentous gliding *Desulfonema limicola* gen. nov. sp. nov., and *Desulfonema magnum* sp. nov. *Arch. Microbiol.* **134**:286–294.

Wiegel, J., M. Braun, and G. Gottschalk. 1981. *Clostridium thermoautotrophicum* species novum (sic!), a thermophile producing acetate from molecular hydrogen and carbon dioxide. *Curr. Microbiol.* **5**:255–260.

Wieringa, K. T. 1936. Over net verdwijnen van waterstof en koolzuur onder anaerobe voorwarden. *Antonie van Leeuwenhoek J. Microbiol. Serol.* **3**:263–273.

Wieringa, K. T. 1940. The formation of acetic acid from CO_2 and H_2 by anaerobic spore-forming bacteria. *Antonie van Leeuwenhoek J. Microbiol. Serol.* **6**:251–262.

Winfrey, M. R., and D. M. Ward. 1983. Substrates for sulfate reduction and methane production in intertidal sediments. *Appl. Environ. Microbiol.* **45**:193–199.

Winter, J., and R. S. Wolfe. 1979. Complete degradation of carbohydrate to carbon dioxide and methane by syntrophic cultures of *Acetobacterium woodii* and *Methanosarcina barkeri*. *Arch. Microbiol.* **121**:97–102.

Winter, J., and R. S. Wolfe. 1980. Methane formation from fructose by syntrophic associations of *Acetobacterium woodii* and different strains of methanogens. *Arch. Microbiol.* **124**:73–79.

Zehnder, A. J. B., and W. Stumm. 1988. Geochemistry and biogeochemistry of anaerobic habitats. In: *Biology of Anaerobic Microorganisms*, A. J. B. Zehnder (ed.), pp. 1–38. Wiley, New York.

Zehnder, A. J. B., K. Ingvorsen, and T. Marti. 1982. Microbiology of methane bacteria. In: *Anaerobic Digestion 1981*, D. E. Hughes, D. A. Stafford, B. I. Wheatley, W. Baader, G. Lettinga, E. J. Nyns, and W. Verstraete (eds.), pp. 45–68 Elsevier, Amsterdam.

Zeikus, J. G., L. H. Lynd, T. E. Thompson, J. A. Krzycki, P. J. Weimer, and P. W. Hegge. 1980. Isolation and characterization of a new, methylotrophic acidogenic anaerobe, the Marburg strain. *Curr. Microbiol.* **3**:381–386.

Zeikus, J. G., and M. Winfrey. 1976. Temperature limitation of methanogenesis in aquatic sediments. *Appl. Environ. Microbiol.* **31**:99–107.

Zhilina, T. N., and G. A. Zavarzin. 1990a. A new extremely halophilic homoacetogenic bacterium *Acetohalobium arabaticum*, gen. nov. sp. nov. *Dokl. Akad. Nauk SSSR* **311**:745–747.

Zhilina, T. N., and G. A. Zavarzin. 1990b. Extremely halophilic, methylotrophic, anaerobic bacteria. *FEMS Microbiol. Rev.* **87**:315–322.

Zinder, S. H., and M. Koch. 1984. Non-aceticlastic methanogenesis from acetate: acetate oxidation by a thermophilic syntrophic coculture. *Arch. Microbiol.* **138**:263–272.

8

Development of DNA Probes for the Detection and Identification of Acetogenic Bacteria

Charles R. Lovell

8.1 Introduction

Acetogenic bacteria may play an important role in anaerobic environments by supplying acetate, a major substrate for terminal carbon metabolism (Christensen, 1984; Parkes et al., 1989; Smith and Klug, 1981; Zeikus, 1977). Unfortunately, although acetogens are readily isolated and appear to occur in significant numbers in many anaerobic habitats (Braun et al., 1979; Laanbroek et al., 1983; Phelps and Zeikus, 1984), much less is known about their distribution and activities than about many other types of organisms. This is due, in part, to difficulties encountered in identifying, isolating, and characterizing acetogens.

Natural bacteria have frequently proven refractory to isolation on artificial culture media (Brock, 1987; Staley and Konopka, 1985). This would certainly be expected of the acetogens, a diverse group with a broad range of temperature, pH, carbon source, and micronutrient and vitamin requirements (Ljungdahl, 1986; Drake 1992) (see Chapters 1 and 7). As there is no reason to believe that acetogens can be recovered more efficiently than other groups of bacteria, there are likely to be unidentified species which have thus far resisted cultivation. For the acetogens and many other groups of bacteria, there is a demonstrable need for isolation and characterization methods which are based less on traditional pure culture approaches and more on molecular strategies.

Much of the current research in molecular microbial ecology is aimed at developing and utilizing DNA probes for the study of undisturbed natural populations of bacteria. DNA probes can be used to examine DNA or intact cells recovered from natural samples (Holben and Tiedje, 1988; Pace et al., 1986; Sayler and Layton, 1990). The probes used can be species-specific, allowing detection of a particular species, or functional-group-specific, allowing simultaneous detection of all species belonging to the group studied (Lovell and Hui, 1991). When a functional-group-specific probe is hybridized to restriction-digested DNA samples, information on the number of species or strains present in the sample can be obtained on the basis of restriction fragment length polymorphism (RFLP) at the locus homologous to the probe. Targeted enrichment of functional group species can also be performed, using the probe as an aid in determining appropriate enrichment conditions. Finally, sequences unique to a particular species can be amplified using polymerase chain reaction (PCR), cloned into an appropriate vector, and sequenced. This approach allows identification of the species in question on the basis of the sequence data obtained but will be most useful when many more sequences from known bacterial species are available for comparison.

In this chapter, some recent advances in detection and identification of acetogens using a functional-group-specific DNA probe, targeted enrichment, and PCR amplification of phylogenetically important DNA sequences are described. Because these approaches are new and depend on methods for recovering and analyzing natural DNA which are still in their infancy, few data are currently available. However, the dramatic growth of interest in this area of research supports an appraisal of current progress and prospects for future development. Several recent reviews detail some of the specific techniques employed in molecular microbial ecology and should be consulted for more detailed technical background (Holben and Tiedje, 1988; Pace et al., 1986; Sayler and Layton, 1990; Steffan and Atlas, 1991).

8.2   Functional-Group-Specific Acetogen Probes

8.2.1   Background

The ideal group-specific probe for acetogens would hybridize with a specific DNA sequence found in all acetogens but not in other organisms. The acetogens are phylogenetically quite diverse (Cato and Stackebrandt, 1989; Stackebrandt et al., 1985; Tanner et al., 1982). This makes the development of a group-specific probe based on conserved 16S ribosomal RNA sequences, the sequences most frequently used as probes, difficult or impossible. The development of genus-specific probes would be hindered by the nonacetogenic species common to many genera containing acetogens. This problem is particularly evident for the acetogenic Clostridium species. A mixture of species-specific probes might also be of limited usefulness because new acetogens are isolated quite frequently and no upper limit on the total number of acetogen species can be set at present. It is reasonable to expect that a wide variety of acetogen species remain to be isolated and that these currently unknown species have important ecological roles. A practical solution to these problems would be a DNA probe based on the signature property of the acetogens, the production of acetate from C_1 compounds. This probe would allow the acetogens to be treated as a functional group.

Acetate production from C_1 compounds in acetogens occurs via the acetyl-CoA pathway (Ragsdale, 1991; Wood and Ljungdahl, 1991). Many of the key enzymes of this pathway, including the ATP-dependent formyltetrahydrofolate synthetase (FTHFS) (Figure 8.1), are structurally and catalytically very similar among the known acetogens (Ljungdahl, 1986; Lovell et al., 1990; O'Brien et al., 1976). In addition, although readily detected in all acetogens tested to date, FTHFS is absent from many other eubacteria (Dev and Harvey, 1978; Whitehead et al., 1988). The structural gene-encoding FTHFS has been previously cloned from an acetogen, *Clostridium thermoaceticum,* into *Escherichia coli* (Lovell et al., 1988), and its complete nucleotide sequence determined (Lovell et al., 1990). The FTHFS gene from a purine fermenter, *C. acidiurici,* has also been cloned and sequenced (Whitehead and Rabinowitz, 1986, 1988), but the two genes have a low level of nucleotide sequence homology, with no significant stretches of sequence identity (Lovell et al., 1990). In order to be useful as a functional-group probe, the *C. thermoaceticum* FTHFS gene must be highly conserved among the acetogens. The FTHFS gene should also be very different from genes in nonacetogens to allow DNA-DNA hybridization between them. Lovell and Hui (1991) recently developed a functional-group-specific probe for acetogens and tested it against DNA purified from cultures of acetogenic and nonacetogenic bacteria (Table 8.1). The probe has also been used for detection of acetogens from a few environmental samples.

8.2.2   Conservation of the Functional-Group Probe Sequence among the Acetogens

When blots of restriction-digested DNA from acetogens were hybridized with the acetogen probe, specific hybrids between the probe and DNA fragments in the digests were formed (Lovell and Hui, 1991). A composite blot of DNA from several acetogens hybridized with the acetogen probe is shown in Figure 8.2. The digests used were selected on the basis of producing a single hybridizing fragment (i.e., no internal restriction sites) in the size range of 2–7 kbase pairs. A hybrid was also obtained with *C. aceticum* DNA but was somewhat less stable than those formed with DNA fragments from the other acetogens. As this hybrid was not observed in all experiments, *C. aceticum* was not included in Figure 8.2. The FTHFS structural gene occurs as a single-copy gene, as indicated by single hybrids in these and other digests, in all nine acetogens tested, as well as in *Desulfotomaculum orientis*. Additional faint signals due to incompletely digested DNA were often visible in blots of *C. thermoaceticum* and *C. thermoautotro-*

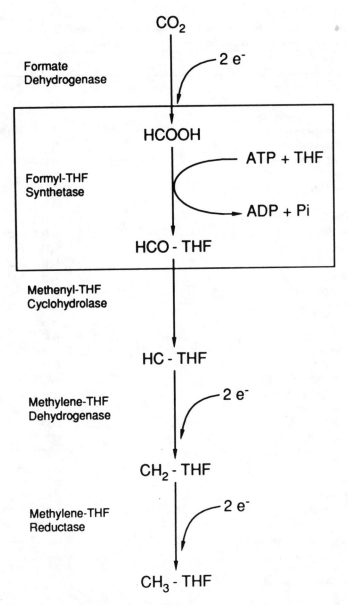

Figure 8.1. Acetyl-CoA pathway reactions responsible for the reduction of formate to methyltetrahydrofolate. The reaction catalyzed by formyltetrahydrofolate synthetase is boxed. THF = tetrahydrofolate.

Table 8.1. Bacterial strains used in testing the acetogen probe. Organism sources, media, and growth conditions are detailed in Lovell and Hui (1991)

Organism	Strain
Acetobacterium woodii[a]	ATCC 29683
Acetogenium kivui[a]	DSM 1428
Acinetobacter calcoaceticus	ATCC 19606
Arthrobacter globiformis	ATCC 8010
Azospirillum lipoferum	Sp59b
Azotobacter chroococcum	ATCC 9043
A. vinlandii	UW
Bacillus brevis	ATCC 8186
B. cereus	ATCC 14579
B. circulans	ATCC 61
B. licheniformis	ATCC 14580
B. subtilis	ATCC 6051
Clostridium aceticum[a]	DSM 1496
C. acidiurici	ATCC 7906
C. cylindrosporum	ATCC 7905
C. formicoaceticum[a]	ATCC 23439
C. magnum[a]	WoBdP1
C. pasteurianum	ATCC 6013
C. perfringens	876
C. thermoaceticum[a]	DSM 521
C. thermoautotrophicum[a]	JW701/5
Desulfotomaculum orientis	ATCC 19365
Desulfovibrio baarsii	2st14
D. bacculatis	DSM 1743
D. desulfuricans	G100A
D. gigas	ATCC 19364
D. vulgaris	ATCC 29579
Escherichia coli	B
Eubacterium limosum[a]	RF
Klebsiella oxytoca	ATCC 50231
Micrococcus luteus	ATCC 4698
Peptostreptococcus productus[a]	U1
Proteus vulgaris	ATCC 13315
Pseudomonas aeruginosa	ATCC 27853
Rhizobium leguminosarum	USDA 2370
Biovar viceae	
R. Meliloti	USDA 1025
Rhodospirillum rubrum	Molisch S1
Sporomusa ovata[a]	H1
S. termitida[a]	JSN-2 (DSM 4440)
Staphylococcus aureus	ATCC 12600
Streptococcus faecalis	ATCC 19433
Xanthomonas maltophila	ATCC 13637

[a] Known acetogens.

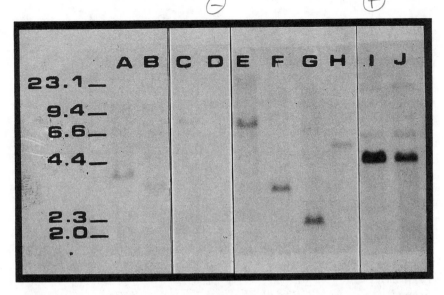

Figure 8.2. Autoradiogram of a composite blot of restriction-digested DNA from several acetogenic bacteria hybridized with the acetogen probe. Two-microgram DNA samples from the acetogens were digested with restriction endonucleases (see Table 8.2) and the fragments fractionated on 0.8% agarose gels. *Kpn I* digested DNA from *Clostridium thermoaceticum* and *Eco RI* digested DNA from *C. magnum* serve as positive and negative controls (see below), respectively. Southern blots (Southern, 1975; Wahl et al., 1979) were probed by hybridization (Maniatis et al., 1982), using 10^6 cpm of ^{35}S-labeled acetogen probe per milliliter of hybridization solution. Hybridization and wash temperatures were empirically determined and were 65°C in the 50% formamide hybridization solution of Maniatis et al. (1982) with 75°C washes for *C. thermoaceticum* and *C. thermoautotrophicum*, 30°C in 50% formamide solution with 40°C washes for *Desulfotomaculum orientis*, *Eubacterium limosum*, *Peptostreptococcus productus*, and *Sporomusa ovata*, and 30°C in 40% formamide solution (Lovell and Hui, 1991) with washes at 40°C for *Acetobacterium woodii*, *Acetogenium kivui*, and *C. formicoaceticum*. The blots were washed twice for a total of 40 min with 2X SSC (0.3 M NaCl and 30 mM sodium citrate), air dried, and exposed to X-OMAT AR film (Eastman Kodak Company, Rochester, NY) at room temperature for 48 h. The negative control (*C. magnum*) lane from the least stringent experimental conditions is shown to demonstrate the low background typical of these hybridizations. Molecular weights are given in kilobase pairs. The order of the digests is (A) *Acetobacterium woodii,* (B) *Acetogenium kivui,* (C) *Clostridium formicoaceticum,* (D) *C. magnum* (negative control), (E) *Desulfotomaculum orientis,* (F) *Eubacterium limosum,* (G) *Peptostreptococcus productus,* (H) *Sporomusa ovata,* (I) *C. thermoaceticum* (positive control), (J) *C. thermoautotrophicum.* The faint bands visible in the *C. thermoaceticum* and *C. thermoautotrophicum* are due to incomplete digestion.

Table 8.2   Restriction digests used, sizes of fragments detected, maximum wash temperatures used, and calculated sequence homologies of hybrids formed between the acetogen probe, derived from the formyltetrahydrofolate synthetase structural gene, and restriction-digested DNA samples from acetogens in hybridization experiments[a]

Organism	Restriction Digest Used	Fragment Size (kb)	Maximum Wash Temperature (°C)	Percent Homology[b]
A. woodii	Pst I	3.9	50	67
A. kivui	Pst I	3.1	60	74
C. aceticum	Cla I	7.0	40	60
C. formicoaceticum	Hind III	8.5	60	74
C. thermoaceticum	Kpn I	5.2	93	100[c]
C. thermoautotrophicum	Kpn I	4.2	93	96
D. orientis	Hind III	8.4	70	80
E. limosum	Eco RI	3.4	70	80
P. productus	Eco RI	2.4	70	80
S. ovata	Hind III	5.2	70	80

[a] Maximum wash temperatures were determined from blots hybridized with the acetogen probe in 40% or 50% formamide hybridization solution, following procedures given in the legend to Figure 8.2.
[b] Percent homology values were calculated according to Meinkoth and Wahl (1984), assuming 1°C above each maximum tested wash temperature to be the melting temperature of each duplex.
[c] Homology between the cloned FTHFS gene and the original genomic form was assumed to be 100%.

phicum DNA, particularly at low stringency. The specific hybrids formed were much more stable than any product of nonspecific hybridization between the probe and DNA from a wide variety of negative control organisms. Background levels under the experimental conditions used were uniformly low. The results of hybridization experiments using DNA from acetogens are summarized in Table 8.2. A rough value for percentage sequence homology between the probe DNA and each hybridizing fragment was calculated (Meinkoth and Wahl, 1984), assuming the duplex melting temperature to be 1°C higher than the maximum wash temperature used in our hybridization experiments. These sequence homology values are given in Table 8.2.

8.2.3   Specificity of the Acetogen Probe for Acetogens

DNA samples from a wide variety of control organisms (Table 8.1) were screened in order to detect any homology between the acetogen probe and nonacetogen sequences (Lovell and Hui, 1991). Only one nonacetogen showed such homology. The nonacetogens tested included several organisms encoding FTHFS or other folate-binding enzymes. Although many of the bacteria used were distantly related to Clostridium group organisms, four non-acetyl-CoA pathway utilizing Clostridium species, including *C. acidiurici, C. cylindrosporum, C. pasteurianum,* and *C. perfringens,* and an acetogenic heterotroph, *C. magnum,* were also examined. *C. pasteurianum* and *C. perfringens* are nonacetogenic. Yeast extract supplemented *C. magnum* cultures were recently shown to grow on H_2 + CO_2, producing acetate (Bomar et al., 1991), but no hybridization between DNA from this organism and the probe was ever observed. FTHFS activity has not been reported in *C. magnum;* if a structural gene-encoding FTHFS is present in this organism, it must be substantially different from those of the other acetogens used in these studies. *C. acidiurici* and *C. cylindrosporum* are known to utilize FTHFS in purine fermentation (Ljungdahl, 1984). Previous studies (Ljungdahl, 1986; O'Brien et al., 1976) have shown strong structural and functional similarities between the acetogen and purine fermenter formyltetrahydrofolate synthetases, but direct comparison of the nucleotide sequences revealed major differences between the FTHFS genes in *C. acidiurici* and *C. thermoaceticum* (Lovell et al., 1990). As all of these organisms are species in the genus *Clostridium* and both *C. acidiurici* and *C. cylindrosporum* are known to contain FTHFS, they served well as negative control organisms in our experiments. Although background in our autoradiograms of blots prepared from restriction-digested DNA samples was uniformly low, background from dot blots (Kafatos et al., 1979) was very high under all but the most stringent conditions.

8.2.4 Detection of Acetogens in Natural Samples

Blots of restriction-digested DNA purified from bacteria physically separated from horse manure were also hybridized with the acetogen probe (Lovell and Hui, 1991) (Fig. 8.3). Blots prepared with *Eco RI*-digested DNA and hybridized with the acetogen probe under conditions of moderate stringency resulted in a profile of six visible hybrids representing DNA fragments of varying sizes and abundance in the digests. These six hybrids represent six "species" (based on RFLP) of acetogens of unknown phylogenetic affiliation present in this sample. It should be noted that *Eco RI* digests of *C. thermoaceticum* and *C. thermoautotrophicum* DNA showed no RFLP at this locus and that our use of moderately stringent conditions would preclude detection of hybrids stable under lower stringency conditions. It is thus likely that six detected "species" is an underestimate and that species diversity among acetogens in the fresh horse manure sample was higher than our results indicated. None of these signals were observed when hybridizations were done under higher stringency conditions. Although *C. thermoaceticum,* the source of the probe, was originally isolated from a horse

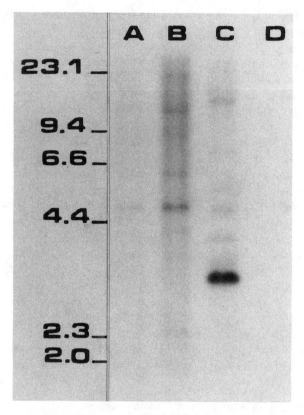

Figure 8.3.   Autoradiogram of a blot of *Eco RI*-digested DNA samples from natural bacteria recovered from horse manure (Lovell and Hui, 1991). The blot was hybridized with the acetogen probe at 30°C in 50% formamide hybridization solution (Maniatis et al., 1982) and washed twice for a total of 40 min at 40°C in 2X SSC. Lane A is 2 μg horse manure DNA; lane B is 5 μg horse manure DNA; lane C is *Eubacterium limosum* DNA (positive control); lane D is *Clostridium magnum* DNA (negative control). Molecular weights are given in kilobase pairs. Faint bands in the positive control lane are due to incomplete digestion of the *E. limosum* DNA.

manure pile (Ljungdahl, 1986), no highly homologous DNA fragment was observed in this sample. Apparently neither *C. thermoaceticum* or any closely related organisms were present at detectable levels in this fresh manure sample.

8.3   The Acetogen Functional-Group Probe. Prospects and Problems

Hybridization studies revealed three levels of homology between DNA fragments from the acetogens tested and the functional-group probe. The highest levels of homology to the probe were obtained from the source organism *C. thermoaceticum* and the closely related *C. thermoautotrophicum* (Bateson et al., 1989; Cato and Stackebrandt, 1989; Weigel et al., 1981). *D. orientis, Eubacterium limosum, Peptostreptococcus productus,* and *Sporomusa ovata* showed approximately 80% sequence homology with the probe. *E. limosum* is clearly related to the thermophilic clostridia, although a fairly deep division between these organisms exists (Cato and Stackebrandt, 1989). The other organisms in this group are more distantly related both to the thermophilic clostridia and to each other (Cato and Stackebrandt, 1989; Tanner et al., 1982). The only significant common characteristics of these organisms are acetogenesis and the moderate mol% G + C contents of their genomes, ranging from 42 for *S. ovata* to 58 for *C. thermoautotrophicum.* The other acetogens tested had lower mol% guanine plus cytosine (G + C) values, ranging from 34 (*C. formicoaceticum*) to 39 (*A. woodii*), and showed lesser degrees of homology with the probe, ranging from approximately 60% to 74%. Of these organisms, *A. kivui* is closely related to the thermophilic clostridia (all are members of subline VI), and *C. formicoaceticum* and *C. aceticum* much more distantly related. Results from *C. aceticum* were inconsistent, although a specific hybrid with the probe was observed on several occasions. The calculated homology level between the *C. aceticum* DNA fragment and the probe was about 60%. That this represents the lower limit of homology yielding an interpretable result under the experimental conditions used is supported by failure of a stable hybrid to form under the same conditions between the probe and *C. acidiurici* DNA. The nucleotide sequence homology between the *C. thermoaceticum* and *C. acidiurici* FTHFS genes is 61% (Lovell et al., 1990). DNA from another acetogenic Sporomusa species, *S. termitida* (Breznak and Switzer, 1986; Breznak et al., 1988; Breznak and Blum, 1991), was also tested in our experiments. Although this species was recently found to have FTHFS activity (Breznak, personal communication), our experiments revealed no substantial homology between the probe and the *S. termitida* FTHFS gene. Thus, the sequence conservation of the FTHFS gene among acetogens, although very high, is not universal.

Sequence homologies among acetogen FTHFS genes are also not completely dependent on the known phylogenetic relationships of the acetogens. Great disparity in G + C contents of the probe and DNA fragments from acetogens would be expected to reduce hybrid stability. However, the absence of probe hybridization with negative control DNA samples ranging from 26.5 mol% G + C (*C. perfringens*) to 72 mol% G + C (*Micrococcus luteus*) shows that sample G + C content does not critically affect discrimination between acetogen and nonacetogen sequences by the probe. Although the sequence homologies found for the acetogen FTHFS genes provide little direct phylogenetic information on the acetogens, these results indicate differential rates of sequence divergence between the FTHFS genes and the 5S and 16S rRNA genes used in determining acetogen phylogeny (Cato and Stackebrandt, 1989; Tanner et al., 1982). This is most graphically demonstrated by the failure of a stable hybrid to form between the probe and the *S. termitida* FTHFS gene, whereas the gene from the closely related *S. ovata* (Breznak, personal communication) showed substantial sequence homology (80%) to the probe. The degree of homology between the probe and the *A. kivui* FTHFS gene was also less than would be predicted on the basis of phylogeny, and the uniformity of the homology values calculated for several phylogenetically diverse organisms provides further evidence for differential rates of sequence divergence between FTHFS genes and rRNA genes. Complete nucleotide sequences of FTHFS genes from more acetogens are needed to clarify the relationship between acetogen phylogeny and sequence homology among acetogen FTHFS genes.

In addition to the acetogens tested, one sulfate-reducing bacterium also formed a stable hybrid with the acetogen probe. *D. orientis* is capable of autotrophic growth, using CO_2 as an electron acceptor in the absence of sulfate (Klemps et al., 1985). No direct evidence for the utilization of the acetyl-CoA pathway by this organism has been published, although some aceticlastic species of sulfate reducers have been found to use a reversed form of the acetyl-CoA pathway for acetate cleavage (Lange et al., 1989; Schauder et al., 1989; Spormann and Thauer, 1989). Approximately 80% homology was observed between the acetogen probe and a DNA fragment from *D. orientis,* implying capacity of the organism to encode FTHFS similar to the acetogen enzyme. The genes encoding FTHFS in other aceticlastic sulfate reducers may also have homology to the probe.

Among the many nonacetogens tested, only DNA from *Proteus vulgaris* formed a stable hybrid with the probe. Although low levels of FTHFS activity have been reported for *P. mirabilis* (Whitehead et al., 1988), enterics typically do not have this enzyme and are so distantly related to the acetogens that sequence conservation between the acetogen FTHFS genes and a *P. vulgaris* DNA fragment was quite surprising. No such hybrids were formed between the probe and DNA samples from other Gram-negative facultative anaerobes, supporting the findings by Whitehead et al. (1988) of low or nonexistent FTHFS activity in these organisms. Low levels of FTHFS activity have been reported for *Bacillus subtilis, C. pasteurianum,* and *S. faecalis* (Whitehead et al., 1988), but no hybridization of the acetogen probe to DNA from these organisms was observed.

Dot blots consistently resulted in much higher background signals from nonacetogens than were observed in Southern blots. Many ecological studies have relied on dot or slot blot hybridization techniques to examine DNA isolated from natural samples. An important problem in using hybridization techniques with DNA extracted from environmental samples is the complex nature of this material (Holben et al., 1988; Saylor and Layton, 1990). The consistently high and potentially misleading background signals observed in our dot blot experiments prompted the use of Southern blots of DNA digests instead.

The acetogen probe allows complex natural samples to be examined for the presence of acetogens. Probing DNA digests allowed characterization of the "species" of acetogens in the sample on the basis of RFLP, providing information on species diversity. This approach also avoided the high background seen in dot blot experiments. Our results show that the FTHFS gene is highly conserved in the acetogenic bacteria and that this sequence can be used as a probe for the study of acetogen ecology.

8.4 PCR Amplification of Acetogen-Specific DNA Sequences

8.4.1 Background

The polymerase chain reaction (PCR) technique has become a widely used method for amplifying specific DNA fragments for cloning, sequencing, and other types of analysis. PCR methods are substantially more sensitive, less expensive, and less time-consuming than most alternative procedures. Various published protocols, including a current review, are widely available and should be consulted for technical details (Oste, 1988; Innis et al., 1990; Steffan and Atlas, 1991). The basic PCR method consists of three steps repeated sequentially through several cycles. The double-stranded DNA sample to be amplified, called the template, is first denatured at high temperature. The temperature is then lowered and a pair of extension primers, oligonucleotides complementary to sites flanking the region to be amplified, are allowed to anneal to opposite template strands with their 3' ends facing each other. These primers should have noncomplimentary sequences and are added in large molar excess over the template DNA to favor primer annealing over template strand reannealing. The third step is DNA polymerase catalyzed synthesis, starting at the primers, of daughter DNA strands complementing the template strands. This step uses one of several commercially available thermostable DNA polymerases. One set of the three steps outlined earlier is referred to as a cycle. The cycle is repeated several times, allowing accumulation of PCR products to great excess over the original template and any other DNA present in the sample.

The usefulness of the PCR method for both phylogenetic studies of pure cultures and for amplification and recovery of 16S ribosomal copy DNA (rcDNA) from uncultured species has been demonstrated repeatedly (Britschgi and Giovannoni, 1991; Schmidt et al., 1991; Ward et al., 1990; Weller and Ward, 1989). The major limitation on this method is the availability of sequence data for the design of suitable primers. Fortunately, there are several domains of the 16S rDNA sequence which are highly conserved among eubacteria and can be targeted for primer annealing.

8.4.2   Selection of Primers

PCR primers useful for amplification of acetogen 16S rDNA sequences can be derived from universally conserved procaryote sequences (Lane et al., 1985; Weisburg et al., 1985; Weisburg et al., 1991). This approach has been successfully used by Bateson et al. (1989) to examine the phylogenetic affiliations of pure cultures of thermophilic acetogens isolated from hot spring cyanobacterial mats. Libraries of cloned PCR amplified 16S rDNA sequences from hot spring communities have been constructed and screened for uncultured organisms (Ward et al., 1990). This approach could be used to recover uncultured acetogens. However, the complexity of such a library is dependent on the species diversity of the community. For communities having high species diversity, primers which are more specific to acetogens would be very advantageous. Unfortunately, because the acetogens are phylogenetically diverse, there is little chance that an acetogen-specific 16S rDNA primer pair could be designed. It is possible that a useful acetogen primer pair could be based on another conserved gene, following the example of the *nifH* gene (encoding dinitrogenase reductase) of the nitrogen-fixing bacteria (Kirshtein et al., 1991). More nucleotide sequences and, particularly, more FTHFS sequences from acetogens will have to be examined to assess the practicality of this approach.

8.5   Prospects for Future Research

The physiologic and phylogenetic diversity of the acetogens make the use of group-specific DNA probes for their study highly desirable. Acetogen detection in natural samples can be readily achieved through functional-group DNA probe hybridization. Insight into acetogen-species diversity in natural samples can be obtained through RFLP analysis and nucleotide sequence homology determinations using the probe. These capabilities make the acetogen functional-group DNA probe a valuable tool in the study of acetogen distribution and ecology.

Another useful application of the group-specific DNA probe would be the targeted enrichment of acetogen species from natural samples. DNA is first purified from a sample, hybridized with the probe, and the profile of hybridizing target DNA fragments, each representing an acetogen "species," is obtained. The intensity of the signal obtained from a hybrid is dependent on the degree of nucleotide sequence homology between the target fragment and the probe and on the abundance of the target fragment in the sample. If the degree of sequence homology is determined from hybridization experiments following the methods described earlier, the abundance of a given DNA fragment can be estimated qualitatively and hybrids representing numerically important "species" in the sample identified. The sample can be used to inoculate a series of enrichment cultures, covering a wide range of substrates and culture conditions. After incubation, DNA recovered from aliquots of these enrichment cultures is screened with the probe. Enrichments showing the hybrid of interest, indicating the presence of the desired "species," are successively subcultured and the "species" tracked through each subculturing by probe hybridization. The stable enrichments are then plated for colony isolation and a pure culture of the "species" obtained. This approach offers a significant improvement over blind enrichment methods by allowing appropriate growth conditions for the "species" of interest to be determined directly. Use of targeted enrichment based on the acetogen probe may promote the isolation of previously uncultivated acetogen species.

Phylogenetic analysis based on the nucleotide sequences of PCR amplified 16S rcDNA can be used to establish the identities of new acetogen isolates, following the approach of Bateson et al. (1989). Such studies can be performed more rapidly than classical physiological characterization and yield either readily determined identities for new isolates, based on similarities to known 16S rDNA sequences, or indicate previously unknown organisms. This type of analysis can also be performed on sequences derived from uncultivated species of acetogens by amplifying DNA recovered from natural samples. The physiological and phylogenetic diversity of the acetogens make 16S rcDNA analysis particularly desirable.

Functional-group DNA probe hybridization for acetogen-species diversity determination and preliminary species characterization, probe-targeted enrichment for new species isolation, and PCR-driven phylogenetic analysis of new isolates offer the ability to isolate and characterize previously unknown acetogen species from any environment. These approaches will certainly extend our knowledge of the ecology and phylogeny of the acetogens and introduce a variety of new acetogens into active study.

Acknowledgments

I thank the many investigators who have very generously provided strains, cells, or DNA samples. Their courtesy is greatly appreciated. I also thank John Breznak for providing information prior to publication and Lars Ljungdahl for helpful comments and advice at many stages of my research on the acetogens. Support for work on the acetogen functional-group probe from the Westinghouse Savannah River Company and the Carolina Venture Fund is also gratefully acknowledged.

References

Bateson, M. M., J. Wiegel, and D. M. Ward. 1989. Comparative analysis of 16S ribosomal RNA sequences of thermophilic fermentative bacteria isolated from hot spring cyanobacterial mats. *Syst. Appl. Microbiol.* **12**:1–7.

Bomar, M., H. Hippe, and B. Schink. 1991. Lithotrophic growth and hydrogen metabolism by *Clostridium magnum*. *FEMS Microbiol. Lett.* **83**:347–350.

Braun, M., S. Schoberth, and G. Gottschalk. 1979. Enumeration of bacteria forming acetate from H_2 and CO_2 in anaerobic habitats. *Arch. Microbiol.* **120**:201–204.

Breznak, J. A., and J. M. Switzer. 1986. Acetate synthesis from H_2 plus CO_2 by termite gut microbes. *Appl. Environ. Microbiol.* **52**:623–630.

Breznak, J. A., J. M. Switzer, and H.-J. Seitz. 1988. *Sporomusa termitida* sp. nov., an H_2/CO_2-utilizing acetogen isolated from termites. *Arch. Microbiol.* **150**:282–288.

Breznak, J. A., and J. S. Blum. 1991. Mixotrophy in the termite gut acetogen, *Sporomusa termitida*. *Arch. Microbiol.* **156**:105–110.

Britschgi, T. B., and S. J. Giovannoni. 1991. Phylogenetic analysis of a natural marine bacterioplankton population by rRNA gene cloning and sequencing. *Appl. Environ. Microbiol.* **57**:1707–1713.

Brock, T. D. 1987. The study of microorganisms *in situ:* progress and problems. In: *Ecology of Microbial Communities,* M. Fletcher, T. R. G. Gray, and J. G. Jones (eds.), pp. 1–20. Cambridge University Press, New York.

Cato, E. P., and E. Stackebrandt. 1989. Taxonomy and phylogeny. In: *Clostridia,* N. P. Minton and D. J. Clarke (eds.), pp. 1–26. Plenum Press, New York.

Christensen, D. 1984. Determination of substrates oxidized by sulfate reduction in intact cores of marine sediments. *Limnol. Oceanogr.* **29**:189–192.

Dev, I. K., and R. J. Harvey. 1978. A complex of N^5N^{10}-methylene-tetrahydrofolate dehydrogenase and N^5N^{10}-methenyltetrahydrofolate cyclohydrolase in *Escherichia coli*. Purification, subunit structure, and allosteric inhibitions by N^{10}-formyltetrahydrofolate. *J. Biol. Chem.* **253**:4245–4253.

Drake, H. L. 1992. Acetogenesis and acetogenic bacteria. In: *Encyclopedia of Microbiology,* J. Lederberg (ed.), Vol. 1, pp. 1–15. Academic Press, San Diego, CA.

Holben, W. E., and J. Tiedje. 1988. Applications of nucleic acid hybridization in microbial ecology. *Ecology* **69**:561–568.

Holben, W. E., J. K. Jansson, B. K. Chelm, and J. M. Tiedje. 1988. DNA probe method for the detection of specific microorganisms in the soil bacterial community. *Appl. Environ. Microbiol.* **54**:703–711.

Innis, M. A., D. H. Gelfand, J. J. Sninsky, and T. J. White. 1990. *PCR Protocols,* pp. 1–482. Academic Press, San Diego, CA.

Kafatos, F. C., C. W. Jones, and A. Efstratiadis. 1979. Determination of nucleic acid sequence homologies and relative concentrations by a dot hybridization procedure. *Nucl. Acids Res.* **7**:1541–1552.

Kirshtein, J. D., H. W. Paerl, and J. Zehr. 1991. Amplification, cloning, and sequencing of a *nifH* segment from aquatic microorganisms and natural communities. *Appl. Environ. Microbiol.* **57**:2645–2650.

Klemps, R., H. Cypionka, F. Widdel, and N. Pfennig. 1985. Growth with hydrogen, and further physiological characteristics of *Desulfotomaculum* species. *Arch. Microbiol.* **143**:203–208.

Laanbroek, H. J., H. J. Geerligs, A. A. C. M. Peijnenburg, and J. Siesling. 1983. Competition for L-lactate between *Desulfovibrio*, *Veillonella*, and *Acetobacterium* species isolated from anaerobic intertidal sediments. *Microb. Ecol.* **9**:341–354.

Lane, D., B. Pace, G. J. Olsen, D. A. Stahl, M. L. Sogin, and N. R. Pace. 1985. Rapid determination of 16S ribosomal RNA sequences for phylogenetic analysis. *Proc. Natl. Acad. Sci. USA* **82**:6955–6959.

Lange, S., R. Scholtz, and G. Fuchs. 1989. Oxidative and reductive acetyl CoA/carbon monoxide dehydrogenase pathway in *Desulfobacterium autotrophicum*. 1. Characterization and metabolic function of the cellular tetrahydropterin. *Arch. Microbiol.* **151**:77–83.

Ljungdahl, L. G. 1984. Other functions of folates. In: *Folates and Pterins*, R. L. Blakley and S. J. Benkovic (eds.), pp. 555–579. Vol. 1, Wiley, New York.

Ljungdahl, L. G. 1986. The autotrophic pathway of acetate synthesis in acetogenic bacteria. *Annu. Rev. Microbiol.* **40**:415–450.

Lovell, C. R., A. Przybyla, and L. G. Ljungdahl. 1988. Cloning and expression in *Escherichia coli* of the *Clostridium thermoaceticum* gene encoding thermostable formyltetrahydrofolate synthetase. *Arch. Microbiol.* **149**:280–285.

Lovell, C. R., A. Przybyla, and L. G. Ljungdahl. 1990. Primary structure of the thermostable formyltetrahydrofolate synthetase from *Clostridium thermoaceticum*. *Biochemistry* **29**:5687–5694.

Lovell, C. R., and Y. Hui. 1991. Design and testing of a functional group-specific DNA probe for the study of natural populations of acetogenic bacteria. *Appl. Environ. Microbiol.* **57**:2602–2609.

Maniatis, T., E. F. Fritsch, and J. Sambrook. 1982. *Molecular Cloning: A Laboratory Manual*, pp. 1–545. Cold Spring Harbor Laboratory, Cold Spring Harbor, NY.

Meinkoth, J., and G. Wahl. 1984. Hybridization of nucleic acids immobilized on solid supports. *Anal. Biochem.* **138**:267–284.

O'Brien, W. E., J. M. Brewer, and L. G. Ljungdahl. 1976. Chemical, physical, and enzymatic comparisons of ormyltetrahydrofolate synthetases from thermo- and mesophilic clostridia. In: *Enzymes and Proteins from Thermophilic Microorganisms, Structure and Functions*, H. Zuber (ed.), pp. 245–262. Birkhauser Verlag, Boston.

Oste, C. 1988. Polymerase chain reaction. *BioTechniques* **6**:162–167.

Pace, N. R., D. A. Stahl, D. J. Lane, and G. T. Olsen. 1986. The analysis of natural microbial populations by ribosomal RNA sequences. *Adv. Microb. Ecol.* **9**:1–55.

Parkes, R. J., G. R. Gibson, I. Mueller-Harvey, W. H. Buckingham, and R. A. Herbert. 1989. Determination of the substrates for sulphate-reducing bacteria within marine and estuarine sediments with different rates of sulphate reduction. *J. Gen. Microbiol.* **135**:175–187.

Phelps, T. J., and J. G. Zeikus. 1984. Influence of pH on terminal carbon metabolism in anoxic sediments from a mildly acidic lake. *Appl. Environ. Microbiol.* **48**:1088–1095.

Ragsdale, S. W. 1991. Enzymology of the acetyl-CoA pathway of CO_2 fixation. *CRC Crit. Rev. Biochem. Mol. Biol.* **26**:261–300.

Sayler, G. S., and A. C. Layton. 1990. Environmental application of nucleic acid hybridization. *Annu. Rev. Microbiol.* **44**:625–648.

Schauder, R., A. Preuss, M. Jetter, and G. Fuchs. 1989. Oxidative and reductive acetyl CoA/carbon monoxide dehydrogenase pathway in *Desulfobacterium autotrophicum.* 2. Demonstration of the enzymes of the pathway and comparison of CO dehydrogenase. *Arch. Microbiol.* **151**:84–89.

Schmidt, T. M., E. F. DeLong, and E. R. Pace. 1991. Analysis of a marine picoplankton community by 16S rRNA gene cloning and sequencing. *J. Bacteriol.* **173**:4371–4378.

Smith, R. L., and M. J. Klug. 1981. Electron donors utilized by sulfate-reducing bacteria in eutrophic lake sediments. *Appl. Environ. Microbiol.* **42**:116–121.

Southern, E. M. 1975. Detection of specific sequences among DNA fragments separated by gel electrophoresis. *J. Mol. Biol.* **98**:503–517.

Spormann, A. M., and R. K. Thauer. 1989. Anaerobic acetate oxidation to CO_2 by *Desulfotomaculum acetoxidans.* Isotopic exchange between CO_2 and the carbonyl group of acetyl-CoA and topology of enzymes involved. *Arch. Microbiol.* **152**:189–195.

Stackebrandt, E., H. Pohla, R. Kroppenstedt, H. Hippe, and C. R. Woese. 1985. 16S rRNA analysis of *Sporomusa, Selenomonas,* and *Megasphaera:* on the phylogenetic origin of Gram-positive eubacteria. *Arch. Microbiol.* **143**:270–276.

Staley, J. T., and A. Konopka. 1985. Measurement of in situ activities of nonphotosynthetic microorganisms in aquatic and terrestrial habitats. *Annu. Rev. Microbiol.* **39**:321–346.

Steffan, R. J., and R. M. Atlas. 1991. Polymerase chain reaction: Application in environmental microbiology. *Annu. Rev. Microbiol.* **45**:137–161.

Tanner, R. S., E. Stackebrandt, G. E. Fox, R. Gupta, L. J. Magrum, and C. R. Woese. 1982. A phylogenetic analysis of anaerobic eubacteria capable of synthesizing acetate from carbon dioxide. *Curr. Microbiol.* **7**:127–132.

Wahl, G. M., M. Stern, and G. R. Stark. 1979. Efficient transfer of large DNA fragments from agarose gels to diazobenzyloxymethal-paper and rapid hybridization by using dextran sulfate. *Proc. Natl. Acad. Sci. USA* **76**:3683–3687.

Ward, D. M., R. Weller, and M. M. Bateson. 1990. 16S rRNA sequences reveal uncultured inhabitants of a well-studied thermal community. *FEMS Microbiol. Rev.* **75**:105–116.

Weisburg, W. G., Y. Oyaizu, H. Oyaizu, and C. R. Woese. 1985. Natural relationships between bacteroides and flavobacteria. *J. Bacteriol.* **164**:230–236.

Weisburg, W. G., S. M. Barns, D. A. Pelletier, and D. J. Lane. 1991. 16S ribosomal DNA amplification for phylogenetic study. *J. Bacteriol.* **173**:697–703.

Weller, R., and D. M. Ward. 1989. Selective recovery of 16S rRNA sequences from natural microbial communities in the form of cDNA. *Appl. Environ. Microbiol.* **55**:1818–1822.

Whitehead, T. R., and J. C. Rabinowitz. 1986. Cloning and expression in *Escherichia coli* of the gene for 10-formyltetrahydrofolate synthetase from *Clostridium acidiurici* ("*Clostridium acidi-urici*"). *J. Bacteriol.* **167**:205–209.

Whitehead, T. R., and J. C. Rabinowitz. 1988. Nucleotide sequence of the *Clostridium acidiurici* ("*Clostridium acidi-urici*") gene for 10-formyltetrahydrofolate synthetase shows extensive amino acid homology with the trifunctional enzyme C_1-tetrahydrofolate synthase from *Saccharomyces cerevisiae*. *J. Bacteriol.* **170**:3255–3261.

Whitehead, T. R., M. Park, and J. C. Rabinowitz. 1988. Distribution of 10-formyltetrahydrofolate synthetase in eubacteria. *J. Bacteriol.* **170**:995–997.

Wiegel, J., M. Braun, G. Gottschalk. 1981. *Clostridium thermoautotrophicum* species novum, a thermophile producing acetate from molecular hydrogen and carbon dioxide. *Curr. Microbiol.* **5**:255–260.

Wood, H. G., and L. G. Ljungdahl. 1991. Autotrophic character of the acetogenic bacteria. In: *Variations in Autotrophic Life*, J. M. Shively and L. L. Barton (eds.), pp. 201–250. Academic Press, San Diego, CA.

Zeikus, J. G. 1977. The biology of methanogenic bacteria. *Bacteriol. Rev.* **41**:514–541.

9

A Phylogenetic Assessment of the Acetogens

Ralph S. Tanner and Carl R. Woese

9.1 Introduction

Our first collaborative foray into molecular systematics was a phylogenetic examination of the acetogens, beginning in 1979. Following the discovery that a physiologically defined group of microorganisms, the methanogens, corresponded to a phylogenetically coherent group (Balch et al., 1977), it was natural to ask the question whether the acetogens, then limited to a handful of species, formed a phylogenetically distinct unit. The initial assessment based on examination of SSU (small subunit; 16S) rRNA catalogs of six acetogens and several other anaerobic microorganisms (Tanner et al., 1981; Tanner et al., 1982) indicated that the acetogens did not form a phylogenetically distinct group per se but that they were all members of a larger group of eubacteria more or less defined by the clostridia (Fox et al., 1980; Tanner et al., 1982). Today there exist at least 27 more or less well-described species of acetogens scattered among at least 12 different genera of eubacteria, based on the criteria currently used in microbial taxonomy. Our understanding of microbial systematics has kept pace with the great increase in described species of anaerobes over the past decade, and a phylogenetic assessment of the acetogens can be performed. The systematics of the acetogens and the clostridia are still basically intertwined.

9.2 Importance of Phylogeny

Biology is in large measure an historical science. Every organism is the product of a unique evolutionary history, which it embodies. In other words, an organism is not comprehensible apart from its history. Every comprehensive explanation of a biological entity must include an understanding of its evolution. It follows that classification, the framework within which biology is structured, must reflect this evolutionary, historical principle to the greatest extent possible.

Perhaps the central problem with, the main weakness of, microbiology has been the lack of a phylogenetic framework within which to operate: a natural taxonomy. This problem has largely gone unnoticed in recent decades as microbiologists made substantial progress in using microorganisms to examine biochemical or molecular systems, in dealing with pathogens, etc. A determinative classification of microorganisms was established which was functional, within its limits. For most of us, intellectually, "microbiology is what microbiology is." We tended to overlook the fact that a taxonomy which does not include phylogenetic relationships: (1) does not suggest that what is known about organism X can be useful for the study of organism Y, its phylogenetic relative, (2) renders comparative study of organisms, whose phylogenetic relationships are unknown, a (for the most part) waste of time and money, and (3) makes difficult, if not impossible, the construction of useful generalizations about groups of organisms.

Today we can determine microbial phylogenies and construct a natural taxonomy, a truly useful microbial classification. We are beginning to see the benefits. of so doing in studies like unraveling the phylogeny of the acetogens, the subject of this chapter.

9.3 Methodology

Our fundamental understanding of the phylogeny of the bacteria is based on analysis of the sequences of SSU (small subunit; 16S) rRNA (Lane et al., 1985; Oyaizu et al., 1987). The SSU and LSU (large subunit; 23S) rRNAs have been key to phylogenetic studies. As integral parts of the translation apparatus, they are found in all cellular forms of life. Both types are highly conserved functionally and structurally. This fact makes them useful in recognizing and defining the higher levels of bacterial taxa. Because of its much smaller size, the 5S rRNA is of only limited utility for phylogenetic studies.

A brief discussion of the methods that were used to examine the phylogeny of the acetogens follows. More extensive descriptions have been published in the works cited above and those of Fox and Stackebrandt (1987), Johnson and Francis (1975), and Stackebrandt (1992).

9.3.1 LSU rRNA Hybridization

The earliest rRNA research relevant to the phylogeny of the acetogens was a hybridization study of LSU chromosomal DNA of clostridia (Johnson and Francis, 1975). Two LSU rRNA homology groups were defined. Group I included species showing homology with *Clostridium butyricum*, whereas group II included species showing homology with *Clostridium lituseburense*. A number of the clostridial species examined in this study did not belong to either LSU rRNA homology group and did not show homology with one another as well. *Clostridium barkeri* was selected as representative of these nongroup clostridia (Johnson and Francis, 1975).

9.3.2   SSU rRNA Oligonucleotide Cataloging

The phylogeny of the known acetogens per se was first seriously examined by the method known as rRNA oligonucleotide cataloging (Tanner et al., 1981; Tanner et al., 1982). P^{32}-labeled SSU rRNA was isolated (Woese et al., 1976) from actively growing cultures of these and other microorganisms. The SSU rRNAs were digested with T_1 RNAase, which cleaves specifically following G residues, and the resulting oligonucleotides separated by two-dimensional paper electrophoresis, with subsequent sequence determination of the resolved oligonucleotides (Uchida et al., 1974; Woese et al., 1976). The oligonucleotide catalogs, which by definition included only hexamers and larger, from all possible pairs of organisms were compared to obtain a set of (so-called) binary association coefficients, a measure of sequence similarity. A table of such coefficients could then be used to construct a dendogram, or phylogenetic tree (Fox et al., 1977). This method utilizes about 40% of the nucleotides in a SSU rRNA. Quantitatively, the cataloging method is more precise than hybridization studies. It will detect more distant phylogenetic relationships than hybridization studies can. An oligonucleotide catalog comprises many independent taxonomic characters, i.e., each oligonucleotide, whereas hybridization studies generate a single number to characterize a relationship. On the other hand, the catalog is derived from a single gene, whereas hybridization measures an average property of the entire genome. The cataloging method produced many important phylogenetic results, including discovery of the Archae [archaebacteria; Balch et al. (1977)].

The acetogens analyzed by the cataloging method were *Acetobacterium woodii, Acetogenium kivui, Clostridium aceticum, Clostridium formicoaceticum, Clostridium thermoaceticum,* and *Eubacterium limosum* (Tanner et al., 1981; Tanner et al., 1982). *A. woodii, C. barkeri* [representative of clostridial species which do not belong to either of the two LSU homology groups defined by Johnson and Francis (1975)] and *E. limosum* formed a phylogenetically distinct cluster. Further evidence that these microorganisms were specifically related to each other was the fact that all had cell walls containing a rare type of murein cross-linkage (Tanner et al., 1981). *C. aceticum* and *C. formicoaceticum* were shown to be members of the same genus. *A. kivui* and *C. thermoaceticum*, the two thermophiles examined in the study, showed some degree of phylogenetic relatedness as well. However, the acetogens as a whole did not form a phylogenetically coherent group. None of these acetogens belonged to either of the two LSU rRNA homology groups. Yet, the acetogens all belonged to the larger taxonomic unit of Gram-positive eubacteria whose DNA showed a relatively low mol% G+C, roughly defined by the clostridial phenotype. Based on the results of SSU rRNA oligonucleotide cataloging (Cato and Stackebrandt, 1989; Fox et al., 1980), it was clear that the taxonomy of the clostridia and their relatives needed a comprehensive reevaluation.

9.3.3 SSU rRNA Sequences

Today, SSU rRNA sequencing has replaced oligonucleotide cataloging for phylogenetic assessment. Most of the sequence information used for the phylogenetic assessment of the acetogens presented here was generated as follows. Total RNA was isolated from the species of interest. Various oligodeoxynucleotides complementary to highly conserved regions of the SSU rRNA sequence were used to prime a reverse transcriptase dideoxynucleotide-terminating sequencing of the SSU rRNA (Lane et al., 1985; Oyaizu et al., 1987; Sanger et al., 1987). A nearly complete (>95%) SSU sequence results from this procedure, which employs seven to eight of these primers.

Recently, many workers have turned to chromosomal SSU rDNA sequencing, which is done by amplification of the rDNA gene by the polymerase chain reaction, usually followed by cloning of the reaction products, often into bacteriophage M13, and sequencing of the individual cloned genes (Sogin, 1990; Stackebrandt, 1992).

9.3.4 LSU rRNA Sequences

A typical SSU rRNA sequence comprises roughly 1500 nucleotides. A LSU rRNA sequence is about 2800 nucleotides. The larger rRNA is clearly under a different set of functional constraints than the smaller, for the nature of evolutionary changes in the two cases is different, at least in a quantitative sense. For example, addition/deletion events are more frequent among LSU rRNAs than SSU rRNAs (Ribosome Database Project, University of Illinois). More importantly, LSU rRNA provides a larger collection of highly variable positions and so proves the more useful of the two in determining phylogenetic relationships among closely related species. Fewer than 20 complete LSU sequences of the clostridial type have been published (Ludwig et al., 1992) compared to about 400 SSU sequences (Olsen et al., 1992). At least for the near future SSU data will be the main source of phylogenetic information on the acetogens and their relatives.

9.3.5   Sequence Analysis

Evolutionary distance analysis of aligned SSU rRNA sequences [distance matrix analysis; De Soete (1983); Jukes and Cantor (1969); Weisburg et al. (1989)] was used to construct the phylogenetic trees presented in Figures 9.1 and 9.2. Evolutionary distance analysis is the simplest of the three methods for inferring phylogenies from sequences now in common use. The other two are known as maximum parsimony and maximum likelihood. Distance analysis is the most degenerative of the information in a sequence alignment (of the three analysis methods), for it utilizes only one number for each pair of sequences: the average number of changes per sequence position. Maximum parsimony and maximum likelihood both take into account the changes at individual positions in the sequences. One might think, therefore, that evolutionary distance analysis would prove the least effective in inferring the correct phylogenetic tree from sequence data. However, this is not the case either in simulation studies or in practice (Swofford and Olsen, 1990). Evolutionary distance has been shown to be approximately as good as maximum parsimony in inferring close relationships and better, under certain circumstances, for inferring distant relationships. Maximum likelihood, which should be the most effective of all methods in the sense that it combines the virtues of both evolutionary distance and maximum parsimony, would probably be the analysis method of choice had it not proven to be computationally intractable (for reasonable numbers of taxa). The recent recasting of the maximum likelihood algorithm by G. J. Olsen (personal communication) removes that obstacle, making the method computationally feasible for relatively large numbers of rRNA sequences. Unfortunately, maximum likelihood has not been used in practice long enough yet for an understanding of its relative weaknesses (if any).

We would caution against the use of cluster analyses (which are sometimes used) to infer trees from sequence data. These methods by their nature treat all lineages as evolving at the same, constant rate. Because this is rarely the actual case, cluster analyses tend to place highly diverged lineages more remotely in a branching order than is actually the case. Helpful reviews of tree inference methods have been published by Swofford and Olsen (1990) and by Li and Graur (1991).

9.4   Phylogeny of the Acetogens

It is difficult to separate a discussion of the phylogeny of the acetogens from that of the clostridia. Indeed, the phylogeny of acetogens is understood properly only in the context of general clostridial phylogeny.

9.4.1 Clostridial Phylogeny

The LSU-rRNA hybridization study by Johnson and Francis (1975) gave the first indication that the clostridia were a large and diverse group of microorganisms. Of the 52 species of *Clostridium* for which hybridization data were tabulated, 41 of them sorted into two homology groups, but the remaining 11 species were not detectably related to either group, nor to one another (Johnson and Francis, 1975).

SSU oligonucleotide cataloging studies of Gram-positive bacteria showed clearly the futility of trying to define phylogenetically valid taxa in terms of phenotypic properties (Cato and Stackebrandt, 1989; Fox and Stackebrandt, 1987; Fox et al., 1980). There is little agreement between the taxa defined phenotypically and their genotypically defined counterparts, taxa defined on the basis of molecular sequence information. The contrast of the two alerts us to the problem of reconciling phenotype with "phylotype" (rRNA genotype). A few examples follow.

The possession of heat-resistant endospores is a good, if very broad, phylogenetic indicator of phylogeny in one sense. Indeed, the known genera of endospore-forming bacteria, *Bacillus, Clostridium, Desulfotomaculum, Heliobacterium, Sporohalobacter, Sporolactobacillus, Sporomusa, Sporosarcina, Syntrophospora,* and *Thermoanaerobium,* all belong to a class of bacteria in the division Firmicutes whose DNA has a relatively low mol% G+C content: the clostridia (Cato and Stackebrandt, 1989; Cook et al., 1991; Madigan, 1992; Stackebrandt et al., 1985; Zhao et al., 1993). However, the group defined, even within the currently defined genus *Clostridium,* is paraphyletic in that many non-spore-forming specific relatives of species of *Clostridium* are excluded from the genus. These excluded species do not form one or a few taxa in their own right; they are very much intermixed with spore formers.

The Gram-positive characteristic, even when refined by definition in terms of cell-wall composition, is similarly defective. A number of groups such as *Sporomusa,* the heliobacteria, and the mycoplasmas are excluded phenotypically from the class of bacteria they are related to phylogenetically, the clostridia. Cell morphology may be the worst offender of all, for it has produced out and out polyphyletic taxa. Few taxa defined primarily by cell shape have withstood the phylogenetic test (examples within the clostridia include *Sarcina, Peptostreptococcus,* and *Peptococcus*).

The two LSU rRNA homology groups defined by Johnson and Francis (1975) are phylogenetically valid. This is not surprising because the basis of their definition lay in actual, although undetermined, molecular sequence. However, the groups as originally defined are paraphyletic because the study was restricted by design only to members of the genus *Clostridium.*

A classification for a clostridial class of the division Firmicutes based on SSU rRNA cataloging was tentatively proposed by Cato and Stackebrandt (1989). Of the five suggested orders, acetogens were found in only two, the Clostridiales and the Acetobacteriales.

A phylogenetic tree of the clostridia based on SSU rRNA sequences is shown in Figure 9.1. Most of the sequences used for this and the phylogenetic tree for the acetogens (Fig. 9.2) have been published (Devereux et al., 1989; Green et al., 1985; Kane et al., 1991; Kane and Breznak, 1991; Mountfort et al. 1993; Paster et al., 1993; Rainey and Stackebrandt, 1993; Tanner et al., 1993; Weisburg

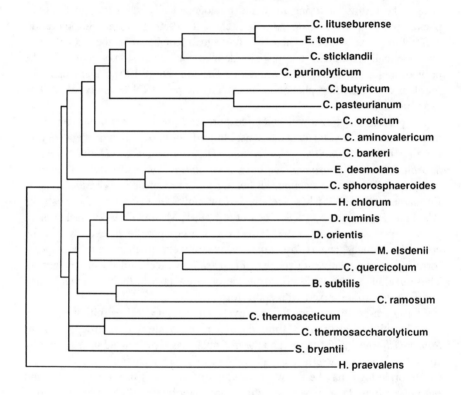

Figure 9.1.   Phylogenetic relationships of the clostridia, showing the major divisions in the proposed class Clostridia, as determined from a SSU rRNA sequence analysis. Genus names are given in the text. The total horizontal distance between two species indicates the difference in their sequences. The bar indicates a distance corresponding to a 10% difference (0.1 evolutionary distance unit).

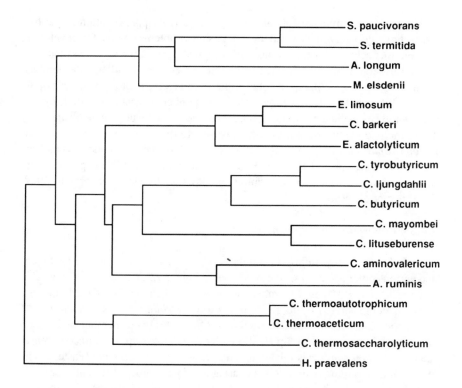

Figure 9.2. Phylogenetic relationships of the acetogens, as determined from a SSU rRNA sequence analysis. Genus names are given in the text. The total horizontal distance between two species indicates the difference in their sequences. The bar indicates a distance corresponding to a 10% difference (0.1 evolutionary distance unit).

et al., 1989; Zhao et al., 1993), and all have been deposited in GenBank or are available through the Ribosomal Database Project (University of Illinois, Urbana). The genera included in this phylogenetic tree are *Clostridium, Eubacterium, Heliobacterium, Desulfotomaculum, Megasphaera, Bacillus, Syntrophospora,* and *Haloanaerobium.* At this time, the clostridia can be divided into at least eight orders, where all members of a given order show no more than 15% difference in sequence. A phylogenetic tree for the clostridia inferred from LSU rRNA sequences agrees with that based on SSU rRNA sequences (Ludwig et al., 1992).

Although this phylogeny should be taken as a first approximation to the true clostridial phylogeny, it clearly demonstrates the problems involved in developing a taxonomic scheme for this group. Phenotypically, the clostridial class includes spore formers, non-spore-formers, obligate anaerobes, facultative anaerobes, facultative aerobes, obligate aerobes, sulfate reducers, phototrophs, syntrophs, thermophiles, halophiles, acetogens, etc. A number of genera not traditionally considered clostridia are, indeed, members of the group by phylogenetic measure. These include the mycoplasmas, which belong to the order represented by *Clostridium ramosum* in Figure 9.1, species of *Bacillus* and their relatives, and the lactic acid bacteria and their relatives, which implies that these groups were derived from anaerobic, spore-forming ancestors (Cato and Stackebrandt, 1989; Weisburg et al., 1989).

9.4.2 Phylogeny of Acetogens

A phylogenetic tree inferred from the SSU rRNA sequences of nine acetogens and various other bacterial species is shown in Figure 9.2. *Clostridium ljungdahlii* (Tanner et al., 1993) is a member of Johnson and Francis's rRNA homology group I, demonstrated by its relationship to *Clostridium tyrobutyricum* and *C. butyricum*. *Clostridium mayombei* (Kane et al., 1991) is a member of Johnson and Francis's homology group II, shown by its relationship to *C. lituseburense*. *Acetitomaculum ruminis* is related to *Clostridium aminovalericum* (Paster et al., 1993); this grouping will likely be assigned to an order different from that including both of Johnson and Francis's rRNA homology groups. The thermophilic acetogens are related to each other and to *Clostridium thermosaccharolyticum*. SSU rRNA oligonucleotide cataloging showed that *A. kivui*, another thermophile, falls within this cluster as well (Tanner et al., 1982). The group of syntrophic bacteria, represented by *Syntrophospora* and *Syntrophomonas* (Zhao et al., 1993), may be members of the same higher-level taxon as are these thermophilic acetogens/clostridia and their relatives.

Acetonema longum, Sporomusa termitida, and *Sporomusa paucivorans* are members of yet another cluster of acetogens, which joins *Megasphaera elsdenii, Clostridium quercicolum,* and others to form yet another clostridial order (that may or may not include the heliobacteria and species of *Desulfotmaculum*). Oligonucleotide cataloging evidence demonstrated that *Sporomusa ovata* and *Sporomusa sphaeroides* are related to *S. paucivorans* (Hermann et al., 1987; Stackebrandt et al., 1985).

Oligonucleotide cataloging also demonstrated that two acetogens, *A. woodii* and *E. limosum,* were relatives of *C. barkeri,* a relationship that was supported phenotypically by the common presence of a rare type of murein cross-linkage in their cell walls (Tanner et al., 1981). A preliminary analysis of their SSU rRNAs indicates that *E. limosum* and *Butyribacterium methylotrophicum* are strains of the same species, consistent with the fact that their phenotypes are essentially identical (Genthner et al., 1981; Zeikus et al., 1980). Because the mureins of *Acetobacterium carbinolicum* and *Acetobacterium wieringae* also contain this unusual B-type cross-linkage (Braun and Gottschalk, 1982; Eichler and Schink, 1984), these species are likely members of this same cluster, which also contains *Eubacterium alactolyticum* (Mountfort et al., 1993).

Acetogens have been identified in five of the eight (tentatively) proposed orders of the clostridia. The acetogenic phenotype does not predict a species' phylogeny other than to place it within the general clostridial group.

9.5   Unanswered Questions

9.5.1   Classification of Acetogens

Without a doubt, over the last decade molecular phylogenetic approaches have drastically altered the classification of bacteria. The trend has been, and rightly so, to classify bacteria more on the basis of their phylogenetic status rather than according to their phenotypic characters, moving microbiology in the direction of botany and zoology, where taxonomy is logically based on phylogeny. This fulfills the vision of Kluyver and van Niel, who stated in 1936: ". . . the only truly scientific foundation of classification is to be found in appreciation of the available facts from a phylogenetic point of view. Only in this way can the natural interrelationships of the various bacteria be properly understood. . . . A true reconstruction of the course of evolution is the ideal of every taxonomist." (Kluyver and van Niel, 1936).

The number of changes in bacterial taxonomic nomenclature, especially at the genus level, sometimes appears overwhelming even to microbiologists whose primary research interests lie in microbial systematics. Yet, this is to be expected, given the vast determinative (and, basically, nonphylogenetic) system that has been built up over the decades, which now must be largely dismantled. Further taxonomic reorderings can be expected in the near future, and it is unlikely that the acetogens will be exempt from this.

In all probability, the designation *Clostridium* will become restricted to species which belong to Johnson and Francis' rRNA homology group I. The status of the genus *Eubacterium*, whose species are scattered throughout the clostridial class (and beyond) even more so than acetogens, has been questioned ever since the beginning of the rRNA studies, if not before. It is unlikely that genera containing only a single species, such as *Acetitomaculum*, *Acetogenium*, and *Butyribacterium*, will withstand phylogenetic reevaluation.

The acetogen longest studied by microbiologists and biochemists, *C. thermoaceticum*, is phylogenetically distant from the type species of *Clostridium*, *C.*

butyricum, and all the other species in the rRNA homology group I, indicating that it ultimately will be assigned to a different new genus. However, just as we have come to appreciate the metabolic capabilities of *C. thermoaceticum* that were unknown at the time of its discovery (the same is true for *A. woodii* and *E. limosum*), we can learn to appreciate this interesting microorganism's phylogeny.

Our experience with classifying acetogens, which has a parallel in numerous other prokaryotic groups, should alert us to the problems and promises in all microbial classification. Classification of the higher eukarya, plants and animals, has from the start (at least from the time of Darwin) involved a balance between phylotype (the evolutionary relationship with other species) and phenotype (an organism's properties). Indeed, this combination of organismal history, as it were, and organismal characteristics is the necessary basis for our understanding of life. Unfortunately, microbial classification has never achieved this balance, despite the acute awareness of the problem by some microbiologists. Microbial classification has been, by default, solely based on phenotype. Phenotype is determined under laboratory conditions; it is probably desirable to assess the phenotype-in-nature, the ecotype, as well.

Microbial classification can achieve a balance of phylotype and phenotype/ ecotype through the use of molecular sequence methods. Microbiologists should go about the task of redefining microbial taxa in these terms without reluctance. A striking example of what is meant here is seen in *Epulopiscium fishelsoni* (Angert et al., 1993). This organism and its relatives inhabit the intestinal tract of surgeonfish; it has not yet been cultured in the laboratory. The organism is large even for a eukaryotic cell, sometimes larger than 600 μm by 80 μm, and there was considerable debate whether it was actually eukaryotic. The debate was settled by determining the organism's phylotype, which turns out to be clostridial (Angert et al., 1993); *E. fishelsoni* belongs to the clostridial order which includes *C. aminovalericum*. This knowledge should have a pronounced effect on how the organism should be studied. On the other hand, the phylotype alone would have never hinted at the fact that *E. fishelsoni* is so large, underscoring the obvious importance of understanding phenotype in assessing uniqueness; phenotype/ecotype remains valuable in classification.

The determination of the phylotype of microorganisms remains problematic in that the expertise required for proper rRNA sequence analysis is confined to a handful of laboratories. Given the general benign neglect that microbial systematics (and microbial physiology, useful for the determination of phenotype/ecotype) has suffered at the hands of funding agencies and university curricula, necessitating the ad hoc approaches to systematics that most microbiologists take, classification and taxonomic criteria will remain in a state of flux for the near term. However, phylogenetic studies of the clostridia based on SSU and LSU rRNA, which have yielded consistent results since their inception, are being used by more and more microbiologists and will continue to increase their importance for the classification of acetogens/clostridia for the foreseeable future.

9.5.2 Evolutionary Questions

Is acetogenesis among the clostridia the result of divergent evolution from an ancestral species or has it appeared independently among different groups of the clostridia? A case has been made that phototrophy and thermophily are ancient bacterial traits (Woese, 1987). It is interesting that acetogenic species seem most prevalent in those clostridial lineages that might (for other reasons) be considered as being more "ancestral" in their properties. Acetogens are prevalent in the presumably ancestrally thermophilic, deeply branching lineage represented by microorganisms such as *C. thermoaceticum*, *C. thermoautotrophicum*, *C. thermosaccharolyticum*, and (probably) *A. kivui*. They are also prevalent on the deeply branching lineage of clostridia that do not possess typical Gram-positive cell walls, including *Sporomusa*, *Acetonema*, *Selenomonas*, and *Megasphaera*. Our analysis here shows that acetogens are found in at least five of the (tentatively) eight proposed orders within the clostridial class. The fact that species found in the deeper branches of the clostridial tree often have cell walls that are not the normal Gram-positive type, yet they still form endospores, such as *Sporohalobacter* [related to *H. praevalens;* Oren et al. (1987)] and *Syntrophospora*, suggests that the latter property arose prior to the full-blown development of the Gram-positive cell wall. Thus, the ancestral form of this large phylogenetic unit was likely to be anaerobic, Gram-negative, and spore forming. The role of acetogenesis in the ancestral clostridia remains to be resolved.

Phylogenetic studies can suggest which phenotypic characters are the more useful for determining a species' phylogenetic status, suggest new ways in which species should be characterized, and suggest ways to further integrate the phenotypic and phylogenetic characterization of microorganisms. The Acetobacteriales, primarily defined by the *A. woodii*, *E. limosum*, and *C. barkeri* cluster, examined to date have in common an atypical cross-linkage in their cell wall. One may predict on the basis of phylogenetic analysis that *E. alactolyticum*, *B. methylotrophicum*, and, presumably, *Acetobacterium malicum* also possess this property. The discovery that *E. limosum* is an acetogen was made at about the same time as its relationship to *A. woodii* was uncovered (Genthner et al., 1981; Tanner et al., 1981). Could *C. barkeri* or *E. alactolyticum* be cultured as acetogens? How similar are the physiologies of *C. barkeri* and *E. limosum*? Similar questions could be asked of the two closely related species in the Clostridiales, *C. ljungdahlii* and *C. tyrobutyricum*.

Other acetogens will be discovered as additional studies of anaerobic microorganisms are conducted. It may be useful to test the known species within the clostridia for the ability to carry out an acetogenic metabolism as another way to discover acetogenic species. In any case, determination of the phylogeny/ taxonomy of the acetogens/clostridia will continue to be an interesting, and sometimes controversial, area of research during the near future.

Acknowledgments

We wish to acknowledge *in toto* our numerous colleagues who have contributed to the work in the phylogeny of the acetogens over the years. Carl R. Woese's contribution was supported by grant BSR-87-05352 from the National Science Foundation.

References

Angert, E. R., K. D. Clements, and N. R. Pace. 1993. The largest bacterium. *Nature* **362**:239–241.

Balch, W. E., L. J. Magrum, G. E. Fox, R. S. Wolfe, and C. R. Woese. 1977. An ancient divergence among the bacteria. *J. Mol. Evol.* **9**:305–311.

Braun, M., and G. Gottschalk. 1982. *Acetobacterium wieringae* sp. nov., a new species producing acetic acid from molecular hydrogen and carbon dioxide. *Zbl. Bakt. Hyg. I. Abt. Orig.* **C3**:368–376.

Cato, E. P., and E. Stackebrandt. 1989. Taxonomy and phylogeny. In: *Clostridia*, N. P. Minton and D. J. Clarke (eds.), pp. 1–26. Plenum Press, New York.

Cook, G. M., P. H. Janssen, and H. W. Morgan. 1991. Endospore formation by *Thermoanaerobium brockii* HTD4. *Syst. Appl. Microbiol.* **14**:240–244.

De Soete, G. 1983. A least squares algorithm for fitting additive trees to proximity data. *Psychometrika* **48**:621–626.

Devereux, R., M. Delaney, F. Widdel, and D. A. Stahl. 1989. Natural relationships among sulfate-reducing eubacteria. *J. Bacteriol.* **171**:6689–6695.

Doolittle, R. F. ed. 1990. Molecular evolution: computer analysis of protein and nucleic acid sequences. In: *Methods in enzymology*. Academic Press, San Diego, CA. Vol. 183.

Eichler, B., and B. Schink. 1984. Oxidation of primary alcohols by *Acetobacterium carbinolicum* sp. nov., a homoacetogenic anaerobe. *Arch. Microbiol.* **140**:147–152.

Fox, G. E., K. J. Pechman, and C. R. Woese. 1977. Comparative cataloging of 16S ribosomal ribonucleic acid: molecular approach to procaryotic systematics. *Int. J. Syst. Bacteriol.* **27**:44–57.

Fox, G. E., E. Stackebrandt, R. B. Hespell, J. Gibson, J. Maniloff, T. A. Dyer, R. S. Wolfe, W. E. Balch, R. S. Tanner, L. Magrum, L. B. Zablen, R. Blakemore, R. Gupta, L. Bonen, B. J. Lewis, D. A. Stahl, K. R. Luehrsen, K. N. Chen, and C. R. Woese. 1980. The phylogeny of the prokaryotes. *Science* **209**:457–463.

Fox, G. E., and E. Stackebrandt. 1987. The application of 16S rRNA cataloguing and 5S rRNA sequencing in bacterial systematics. In: *Methods in Microbiology* vol. 19, R. R. Colwell and R. Grigorova (eds.), pp. 405–458. Academic Press, New York.

Genthner, B. R. S., C. L. Davis, and M. P. Bryant. 1981. Features of rumen and sewage sludge strains of *Eubacterium limosum,* a methanol- and H_2-CO_2-utilizing species. *Appl. Environ. Microbiol.* **42**:12–19.

Green, C. J., G. C. Stewart, M. A. Hollis, B. S. Vold, and K. F. Bott. 1985. Nucleotide sequence of *Bacillus subtilis* ribosomal RNA operon, *rrnB. Gene* **37**: 261–266.

Hermann, M., M.-R. Popoff, and M. Sebald. 1987. *Sporomusa paucivorans* sp. nov., a methylotrophic bacterium that forms acetic acid from hydrogen and carbon dioxide. *Int. J. Syst. Bacteriol.* **37**:93–101.

Johnson, J. L., and B. S. Francis. 1975. Taxonomy of the clostridia: ribosomal ribonucleic acid homologies among the species. *J. Gen. Microbiol.* **88**:229–244.

Jukes, T. H., and C. R. Cantor. 1969. Evolution of protein molecules. In: *Mammalian Protein Metabolism,* H. N. Munro (ed.), pp. 21–132. Academic Press, New York.

Kane, M. D., A. Brauman, and J. A. Breznak. 1991. *Clostridium mayombei* sp. nov., an H_2/CO_2 acetogenic bacterium from the gut of the African soil-feeding termite, *Cubitermes speciosus. Arch. Microbiol.* **156**:99–104.

Kane, M. D., and J. A. Breznak. 1991. *Acetonema longum* gen. nov. sp. nov., an H_2/CO_2 acetogenic bacterium from the termite, *Pterotermes occidentis. Arch. Microbiol.* **156**:91–98.

Kluyver, A. J., and C. B. van Niel. 1936. Prospects for a natural system of classification of bacteria. *Zbl. Bakteriol. Parasit. Infektion. II* **94**:369–403.

Lane, D. J., B. Pace, G. J. Olsen, D. A. Stahl, M. L. Sogin, and N. R. Pace. 1985. Rapid determination of 16S ribosomal RNA sequences for phylogenetic analyses. *Proc. Natl. Acad. Sci. USA* **82**:6955–6959.

Li, W.-H., and D. Graur. 1991. *Fundamentals of Molecular Evolution.* Sinauer Associates, Inc., Sunderland, MA.

Ludwig, W., G. Kirchhof, N. Klugbauer, M. Weizenegger, D. Betzl, M. Ehrmann, C. Hertel, S. Jilg, R. Tatzel, H. Zitzelsberger, S. Liebl, M. Hochberger, J. Shah, D. Lane, P. R. Wallnofer, and K. H. Scheifer. 1992. Complete 23S ribosomal RNA sequences of Gram-positive bacteria with a low DNA G+C content. *Syst. Appl. Microbiol.* **15**:487–501.

Madigan, M. T. 1992. The family Heliobacteriaceae. In: *The Prokaryotes, Vol. 2,* 2nd ed., A. Balows, H. G. Truper, M. Dworkin, W. Harder, and K.-H. Schleifer (eds.), pp. 1981–1992. Springer-Verlag, New York.

Mountfort, D. O., W. D. Grant, H. Morgan, F. A. Rainey, and E. Stackebrandt. 1993. Isolation and characterization of an obligately anaerobic, pectinolytic, member of the genus *Eubacterium* from mullet gut. *Arch. Microbiol.* **159**:289–295.

Olsen, G. J., R. Overbeek, N. Larsen, T. L. Marsh, M. J. McCaughey, M. A. Maciukenas, W. M. Kuan, T. J. Macke, Y. Xing, and C. R. Woese. 1992. The ribosomal database project. *Nucleic Acid Res.* **20**:2199–2200.

Oren, A., H. Pohla, and E. Stackebrandt. 1987. Transfer of *Clostridium lortetii* to a new genus *Sporohalobacter* gen. nov. as *Sporohalobacter lortetii* comb. nov., and description of *Sporohalobacter marismortui* sp. nov. *Syst. Appl. Microbiol.* **9**:239–246.

Oyaizu, H., B. Debrunner-Vossbrinck, L. Mandelco, J. A. Studier, and C. R. Woese. 1987. The green non-sulfur bacteria: a deep branching in the eubacterial line of descent. *Syst. Appl. Microbiol.* **9**:47–53.

Paster, B. J., J. B. Russell, C. M. J. Yang, J. M. Chow, C. R. Woese, and R. S. Tanner. 1993. Phylogeny of the ammonia-producing ruminal bacteria *Peptostreptococcus anaerobius, Clostridium sticklandii,* and *Clostridium aminophilum* sp. nov. *Int. J. Syst. Bacteriol.* **43**:107–110.

Rainey, F. A., and E. Stackebrandt. 1993. Phylogenetic analysis of the bacterial genus *Thermobacteroides* indicates an ancient origin of *Thermobacterroides proteolyticus. Lett. Appl. Microbiol.* **16**:282–286.

Sanger, F., S. Nicklen, and A. R. Coulson. 1977. DNA sequencing with chain-terminating inhibitors. *Proc. Natl. Acad. Sci. USA* **74**:5463–5467.

Sogin, M. L. 1990. Amplification of ribosomal RNA genes for molecular evolution studies. In: *PCR Protocols, a Guide to Methods and Applications.* M. A. Innis, D. H. Gelfand, J. J. Sninsky, and T. J. White (eds.), pp. 307–314. Academic Press, San Diego, CA.

Stackebrandt, E. 1992. Unifying phylogeny and phenotypic diversity. In: *The Prokaryotes, Vol. 1,* 2nd ed. A. Balows, H. G. Truper, M. Dworkin, W. Harder, and K.-H. Schleifer (eds.), pp. 19–47. Springer-Verlag, New York.

Stackebrandt, E., H. Pohla, R. Kroppenstadt, H. Hippe, and C. R. Woese. 1985. 16S rRNA analysis of *Sporomusa, Selenomonas,* and *Megasphaera:* on the phylogenetic origin of Gram-positive eubacteria. *Arch. Microbiol.* **143**:270–276.

Swofford, D. L., and G. J. Olsen. 1990. Phylogenetic reconstruction. In: *Molecular Systematics,* D. Hillis and C. Moritz (eds.), pp. 411–501. Sinauer Associates, Inc., Sunderland, MA.

Tanner, R. S., E. Stackebrandt, G. E. Fox, and C. R. Woese. 1981. A phylogenetic analysis of *Acetobacterium woodii, Clostridium barkeri, Clostridium butyricum, Clostridium lituseburense, Eubacterium limosum,* and *Eubacterium tenue. Curr. Microbiol.* **5**:35–38.

Tanner, R. S., E. Stackebrandt, G. E. Fox, R. Gupta, L. J. Magrum, and C. R. Woese. 1982. A phylogenetic analysis of anaerobic eubacteria capable of synthesizing acetate from carbon dioxide. *Curr. Microbiol.* **7**:127–132.

Tanner, R. S., L. M. Miller, and D. Yang. 1993. *Clostridium ljungdahlii* sp. nov., an acetogenic species in clostridial rRNA homology group I. *Int. J. Syst. Bacteriol.* **43**:232–236.

Uchida, T., L. Bonen, H. W. Schaup, B. J. Lewis, L. Zablen, and C. R. Woese. 1974. The use of ribonuclease U_2 in RNA sequence determinations. Some corrections in the catalog of oligomers produced by ribonuclease T_1 digestion of *Escherichia coli* 16S ribosomal RNA. *J. Mol. Evol.* **3**:63–77.

Weisburg, W. G., J. G. Tully, D. L. Rose, J. P. Petzel, H. Oyaizu, D. Yang, L.

Mandelco, J. Sechrest, T. G. Lawrence, J. Van Etten, J. Maniloff, and C. R. Woese. 1989. A phylogenetic analysis of the mycoplasmas: basis for their classification. *J. Bacteriol.* **171**:6455–6467.

Woese, C. R. 1987. Bacterial evolution. *Microbiol. Rev.* **51**:221–271.

Woese, C. R. 1992. Prokaryotic systematics: the evolution of a science. In: *The Prokaryotes, Vol. 1,* 2nd ed. A. Balows, H. G. Truper, M. Dworkin, W. Harder, and K.-H. Schleifer (eds.), pp. 3–18. Springer-Verlag, New York.

Woese, C. R., M. Sogin, D. Stahl, B. J. Lewis, and L. Bonen. 1976. A comparison of the 16S ribosomal RNAs from mesophilic and thermophilic bacilli: some modifications in the Sanger method for RNA sequencing. *J. Mol. Evol.* **7**:197–213.

Zeikus, J. G., L. H. Lynd, T. E. Thompson, J. A. Krzycki, P. J. Weimer, and P. W. Hegge. 1980. Isolation and characterization of a new, methylotrophic, acidogenic anaerobe, the Marburg strain. *Curr. Microbiol.* **3**:381–386.

Zhao, H., D. Yang, C. R. Woese, and M. P. Bryant. 1993. Assignment of fatty acid-β-oxidizing syntrophic bacteria to *Syntrophomonadaceae* fam. nov. on the basis of 16S rRNA sequence analyses. *Int. J. Syst. Bacteriol.* **43**:278–286.

IV

PHYSIOLOGICAL AND ECOLOGICAL POTENTIALS OF ACETOGENIC BACTERIA

10

Acetogenesis: Reality in the Laboratory, Uncertainty Elsewhere

Harold L. Drake, Steven L. Daniel, Carola Matthies, and Kirsten Küsel

10.1 Introduction

This chapter focuses on recent work in our research group that further extends our awareness of the diverse metabolic potentials of acetogens and, consequently, broadens our uncertainty in making accurate predictions of the role acetogens actually play at the ecosystem level (i.e., "elsewhere" per the title of this chapter). Without debating what ecosystems are, acetogens are difficult to study in their natural habitat. This difficulty stems largely from the fact that the main product we think they make (i.e., acetate) is not easily assessed (a gaseous product minimizes this complication) and likely turns over rapidly *in vivo*. Likewise, many of the substrates they may consume are also problematic to assess. In addition, approaches such as the [^{3}H]thymidine incorporation method to assess the productivity of acetogens may greatly underestimate their magnitude (Winding, 1992; Wellsbury et al., 1993). Thus, although enrichment and physiological studies have been somewhat elegant in recent years relative to defining acetogenic potentials in the laboratory, comparatively little is known about what they really do "elsewhere" (as emphasized in Chapter 7). Clearly, native ecosystems such as forests have little in common with test-tube cultures. In the present chapter and those that follow in Part IV these realities and uncertainties are addressed.

10.2 Diverse Reductant Sinks: Where Electrons go in Laboratory Cultures

The new potentials outlined below were discovered as a consequence of evaluating the effects of CO_2 limitation on the growth of seemingly well-characterized acetogens. This approach has necessitated defining the nutritional and base metabolic requirements of certain model organisms (Lundie and Drake, 1984; Savage and Drake, 1986; Lux and Drake, 1992) and is in contrast to the use of enrichment techniques for the isolation of new acetogens with selective substrates that might reveal previously unrecognized metabolic potentials within this bacteriological group (see Chapter 7). Carbon dioxide is the classic terminal electron acceptor of acetogens, and approximately 10 years ago we began to evaluate the metabolic consequence of CO_2 deprivation. Although the main goal was initially to simply determine if acetogens could cope with this problem, such experiments led to the discovery of new acetogenic potentials relative to the flow of carbon and reductant.

10.2.1   Effect of CO_2 Deprivation on Chemolithoautotrophic and Heterotrophic Growth

As illustrated in Figure 10.1, the methyl and carbonyl sites of acetyl-CoA synthase are major "transitory depots" for the flow of both carbon and reductant from numerous acetogenic substrates. Thus, the acetyl-CoA pathway can be envisioned as a large, seemingly inescapable funnel into which everything is routed toward acetate. In this sense, acetogenic substrates appear similar: They are all oxidized, and both the reductant and CO_2 produced are largely used (conserved) in the synthesis of acetyl-CoA. Consequently, the potential importance of CO_2 is somewhat masked by schemes that illustrate the acetyl-CoA pathway.

However, if one examines *when* a particular process occurs, i.e., what process occurs first, the potential importance of CO_2 to acetogens becomes a little clearer. As shown in Figure 10.2, the oxidation of glucose to the level of pyruvate yields reduced electron carriers prior to the production of CO_2 via decarboxylation of pyruvate. It might, thus, be imagined that exogenous CO_2 would be of some benefit to the cell relative to recycling electron carriers (as depicted). Depending on the substrate and acetogen, this is, in fact, the case. Indeed, even with the classic homoacetogenic substrates glucose or fructose, poor growth and extensive lag phases are often encountered when acetogens are grown with these substrates in the absence of CO_2. This was eloquently shown in early studies with *Clostridium formicoaceticum*, in which both growth and fructose utilization were markedly curtailed by depriving the cells of exogenous CO_2 (Andreesen et al., 1970; O'Brien and Ljungdahl, 1972).

As shown in reaction I, the carbon balance (i.e., stoichiometry) for glucose- or fructose-dependent homoacetogenesis highlights the fact that all glucose-derived carbon and reductant are recovered in acetate:

$$C_6H_{12}O_6 \rightarrow 3\ CH_3COOH \tag{I}$$

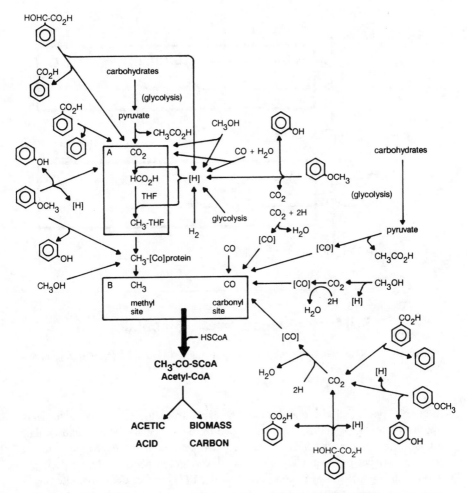

Figure 10.1. Flow of carbon and reductant toward the acetyl-CoA pathway. Box A: tetrahydrofolate pathway. Box B: methyl and carbonyl sites of acetyl-CoA synthase (Drake, 1992).

Although this reaction has made acetogens grant-worthy and kept many of us employed, it is argumentatively misleading. As written, reaction I hides the fact that CO_2 is produced from glucose during growth and that exogenous CO_2 is largely used instead, as suggested in Figure 10.2. Consistent with this concept, when *Clostridium thermoaceticum* is cultivated with growth-limiting concentrations of uniformly labeled [U-^{14}C]-glucose in the presence of unlabeled CO_2 (thus minimizing the extent to which glucose-derived $^{14}CO_2$ can recycle via the

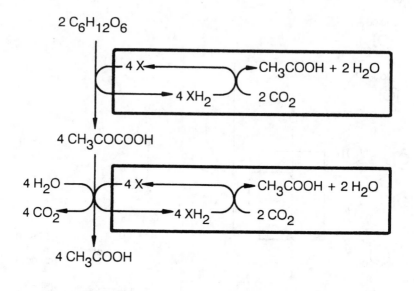

SUM: 2 glucose ➡ 6 acetate

Figure 10.2.   Role of CO_2 in the recycling of reduced electron carriers during aceto-genesis.

free CO_2 pool), almost one-third of the label from the consumed glucose is recovered as $^{14}CO_2$ and the remaining two-thirds as acetate [Martin and Drake, unpublished; see also Martin et al. (1985)]. Although the exchange reaction known to occur between CO_2 and the carboxyl group of pyruvate via pyruvate oxidoreductase of this acetogen (Drake et al., 1981) makes definitive interpretation of such observations difficult, it is noteworthy that (i) this distribution of ^{14}C is, in fact, predicted from the scheme in Figure 10.2 and (ii) early ^{14}C-labeling studies show quite clearly that carbons 3 and 4 of glucose are not routed directly into acetate but instead enter mostly into the CO_2 pool (Wood, 1952; O'Brien and Ljungdahl, 1972). Although potential transcarboxylation reactions may nonetheless prioritize the flow of carboxyl donors toward the carboxyl group of acetate under certain conditions (Schulman et al., 1973; Martin et al., 1985), such a process has not been definitely proven.

The consequence of CO_2 limitation is accentuated (and thus more easily observed) with methanol- and methoxyl-level reductant. *C. thermoaceticum* and *Clostridium formicoaceticum* (and many other acetogens) essentially do not grow with such substrates in the absence of supplemental CO_2 or a CO_2 replacement

(Hsu et al., 1990a; Matthies et al., 1993). Even with carbon monoxide, on paper an excellent acetogenic substrate as it offers the cell a preformed CO-level carbon and also yields excess CO_2 as a product during acetate synthesis according to reaction II, the initial CO-dependent growth response of an acetogen may nonetheless be significantly affected by the availability of exogenous CO_2 (Savage et al., 1987; Fig. 10.3). Likewise, the capacity of acetogens to consume CO (i.e., rate of CO consumption) may also be negatively influenced by the absence of CO_2 (Savage et al., 1987; Ma et al., 1987).

$$4\,CO + 2\,H_2O \rightarrow CH_3COOH + 2\,CO_2 \qquad (II)$$

Collectively, such results suggest that there are both CO_2-dependent and CO_2-independent processes in acetogens. Because CO_2 is considered "the" terminal acceptor for acetogens (they can be thought of as CO_2 respirers), this conclusion may, on the surface, seem somewhat trivial relative to acetogenesis, especially under chemolithoautotrophic conditions. However, regardless of their purpose in doing so, what is not trivial is that acetogens have many ways of dealing with this situation. Indeed, in some cases, CO_2 and the acetyl-CoA pathway is not the preferred terminal electron-accepting process.

10.2.2 Use of Aromatic Carboxyl Groups as CO_2 Equivalents

C. thermoaceticum has the ability to decarboxylate aromatic compounds (Hsu et al., 1990a, 1990b). This potential is inducible and is discussed in further detail in Chapter 17. What is most important about this decarboxylation is that the cell readily uses the carboxyl group as a growth-essential CO_2 equivalent in acetate synthesis under conditions of CO_2 deprivation. This was shown for both methanol- and methoxyl-group-dependent acetogenesis (Hsu et al., 1990a). Although induced resting cells are competent in the decarboxylation reaction (Hsu et al., 1990a), during growth, the aromatic carboxyl group still partitions itself about 30% into the acetate pool, suggesting that it is not totally diluted out by exogenous CO_2 (Hsu et al., 1990b). Although we do not know how tightly coupled decarboxylation is to its subsequent utilization, acetate is totally synthesized from this CO_2 equivalent when CO_2 is limiting (Hsu et al., 1990b). It would be important to determine if acetogens have diverse decarboxylation (not just aromatic compounds) potentials relative to the generation of CO_2 equivalents and acetogenesis.

10.2.3 Dissimilation of the Carbon–Carbon Double Bonds of Fumarate and Aromatic Acrylates

The capacity of acetogens to conserve energy via the reduction of a carbon-carbon double bond was first observed with fumarate-dependent growth of *C.*

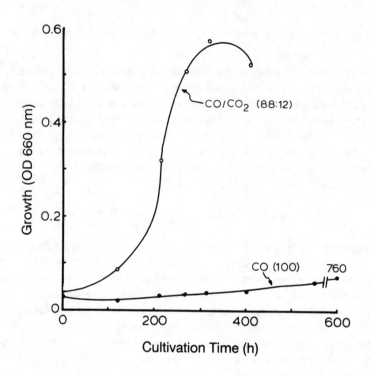

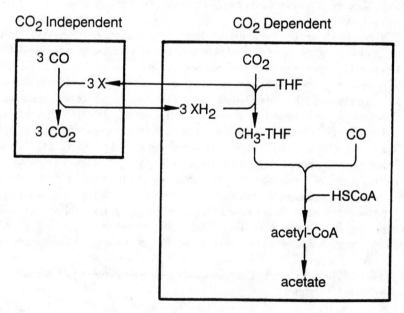

Figure 10.3.   Effect of CO_2 on the CO-dependent growth of *Clostridium thermoautotrophicum* (Savage et al., 1987). Hypothetical scheme illustrates the role of CO_2 in the recycling of electron carriers.

formicoaceticum (Dorn et al. 1978a, 1978b); this potential has also been established for *Clostridium aceticum* (Braun et al. 1981; Matthies et al. 1993). Fumarate supports the growth of these acetogens via dismutation:

Oxidation: $\text{fumarate}^{2-} + 2H_2O + H^+ \rightarrow \text{acetate}^{1-} + 2CO_2 + 4[H]$ (III)

Reduction: $2\ \text{fumarate}^{2-} + 4[H] \rightarrow 2\ \text{succinate}^{2-}$ (IV)

Sum: $3\ \text{fumarate}^{2-} + H^+ + 2H_2O \rightarrow 2\ \text{succinate}^{2-}$
       $+\ \text{acetate}^{1-} + 2CO_2$ (V)

Energy conservation during furmarate dismutation by *C. formicoaceticum* can occur by (i) substrate-level phosphorylation via acetate kinase and (ii) via fumarate reduction to succinate. Fumarate reductase activity is membranous (Dorn et al., 1978b), and a *b*-type cytochrome and menaquinone have been identified in fumarate-grown cells (Gottwald et al., 1975).

By cosubstrate growth experiments in which an acetogenic substrate (e.g., methanol) is used simultaneously with fumarate, we have evaluated this potential in detail to determine (i) if fumarate is, in fact, a preferred terminal reductant sink (relative to the reduction of CO_2) and (ii) if fumarate reduction is definitely coupled to energy conservation (Matthies et al., 1993). We have found that fumarate can, under some conditions, replace CO_2. Although methanol or the O-methyl group of vanillate does not support the growth of *C. formicoaceticum* or *C. aceticum* under CO_2-limited conditions, when fumarate is provided concomitantly, the oxidation of these acetogenic substrates becomes growth supportive under defined conditions and is coupled to the reduction of fumarate to succinate (fumarate dissimilation). Concomitantly, fumarate oxidation (reaction III) and, thus, dismutation does not occur (Fig. 10.4A). Similar results are obtained with H_2 and *C. aceticum* under CO_2-limited conditions; fumarate provides access to growth-supportive, H_2-derived reductant via dissimilation. These results demonstrate that the reduction of fumarate to succinate is, indeed, energy conserving. However, when both CO_2 and fumarate are available under defined conditions, acetogenesis and the reduction of CO_2 is the preferred electron-accepting process (Matthies et al., 1993). Thus, fumarate reduction is not preferred. Significantly, under yeast extract-enriched conditions, neither of these sinks is selectively utilized as an electron sink. This is illustrated in Figure 10.4B in which *C. formicoaceticum* simultaneously oxidizes both fumarate and methanol and engages the concomitant usage of both CO_2 and fumarate as reductant sinks.

In contrast to these processes, an unclassified syntrophic bacterium (MPOB) has recently been shown to grow at the expense of fumarate according to reaction VI (see Chapter 22):

$$7\ \text{Fumarate} + 2H^+ + 4H_2O \rightarrow 6\ \text{succinate} + 4CO_2 \qquad \text{(VI)}$$

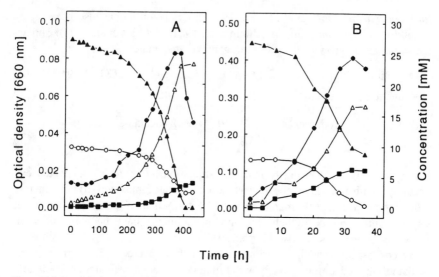

Figure 10.4.   Methanol- and fumarate-derived product and growth profiles of *Clostridium formicoaceticum* cultivated in CO_2-limited defined medium (A) and CO_2-enriched undefined medium (B). Symbols: methanol, ○; fumarate, ▲; succinate, △; acetate, ■; growth, ●.

This bacterium oxidizes fumarate completely to CO_2 and appears to engage the acetyl-CoA pathway in the reverse direction. The nonacetogens *Proteus rettgeri* and *Malonomonas rubra* ferment fumarate by the same stoichiometry, but fumarate oxidation is via the citric acid cycle (Kröger, 1974; Dehning and Schink, 1989a).

The carbon–carbon double bond of the acrylate side chain of certain aromatic compounds is also reduced by some acetogens (Table 10.1). This potential was first demonstrated with *Acetobacterium woodii* and was shown to be energy conserving under CO_2-enriched conditions (Bache and Pfennig, 1981; Tschech and Pfennig, 1984; Hansen et al., 1988). Although the mechanism is unknown, energy conservation is likely chemiosmotic and coupled to electron transport. Thus, like fumarate, aromatic acrylates can be alternative energy-conserving reductant sinks that compete with CO_2 for reductant (Fig. 10.5). However, unlike fumarate, with *A. woodii* under CO_2-enriched conditions, methanol-derived reductant appears to flow preferentially toward the acrylate rather than CO_2 (Tschech and Pfennig, 1984). *Peptostreptococcus productus* also transforms aromatic acrylates (Parekh et al., 1992), and we have examined the consequence of acrylate usage under CO_2 limitation to more fully establish if acrylates support growth in the absence of acetogenic potentials (Misoph, Daniel, and Drake, unpublished results; Daniel et al., 1992). Under such conditions, we have found that ferulate

Table 10.1 Reduction of the acrylate side chain of aromatic acrylates during growth of acetogens

Acetogen	Aromatic Acrylate Utilized	Aromatic Product(s) Formed	Reference
Acetobacterium			
woodii (strain NZva16)	Ferulate	Hydrocaffeate	Bache and Pfennig (1981)
	Sinapate	3,4,5-Trihydroxy-3-phenylpropionate	Bache and Pfennig (1981)
	3,4,5-Trimethoxycinnamate	3,4,5-Trihydroxy-3-phenylpropionate	Bache and Pfennig (1981)
	3-Methoxycinnamate	3-Hydroxyhydrocinnamate	Bache and Pfennig (1981)
	Caffeate[a]	Hydrocaffeate	Tschech and Pfennig (1984)
Clostridium formicoaceticum	Ferulate	Hydrocaffeate	Lux and Drake (1992)
Clostridium aceticum	Ferulate	Hydrocaffeate	Lux and Drake (1992)
Clostridium pfennigii	Ferulate	Caffeate, hydrocaffeate[b]	Krumholz and Bryant (1985)
Peptostreptococcus	Ferulate	Hydrocaffeate	Parekh et al. (1992)
productus (ATCC 35244)	Hydroferulate	Hydrocaffeate	Parekh et al. (1992)
	2-Methoxycinnamate[c]	2-Methoxyhydrocinnamate	Parekh et al. (1992)
	Caffeate[c]	Hydrocaffeate	Parekh et al. (1992)
	2-Hydroxycinnamate[c]	2-Hydroxyhydrocinnamate	Parekh et al. (1992)
	4-Hydroxycinnamate[c]	4-Hydroxyhydrocinnamate	Parekh et al. (1992)
Syntrophococcus	Ferulate	Caffeate, hydrocaffeate[b]	Krumholz and Bryant (1986)
sucromutans	Caffeate[d]	Hydrocaffeate	Krumholz and Bryant (1986)
Strain TH-001	Ferulate	Hydrocaffeate	Frazer and Young (1985)

[a] Aromatic acrylate is utilized during syringate-, formate-, H$_2$-, or methanol-dependent growth.
[b] Caffeate is formed when ferulate is O-demethylated and the acrylate side chain is reduced; hydrocaffeate is formed when ferulate is only O-demethylated.
[c] Aromatic acrylate is utilized during CO-dependent growth.
[d] Aromatic acrylate is utilized during fructose-dependent growth.

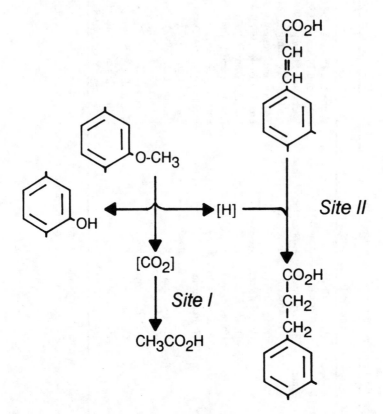

Figure 10.5.   Two potential sites for the conservation of energy during the metabolism of aromatic methoxyl and acrylate groups.

is growth supportive as a consequence of the initial reduction of the acrylate group rather than CO_2. Only when the acrylate group is nearly fully reduced and CO_2 is synthesized and thus available via the oxidation of the methoxyl group is CO_2 reduced and acetate formed.

Aromatic acrylates and fumarate are sometimes considered as very similar compounds relative to bioenergetics (Tschech and Pfennig, 1984). Although these two alternative reductant sinks can be energy conserving under conditions of CO_2 deprivation, acetogens capable of using them seem to utilize them differently. Acrylates can only be dissimilated (reduced), whereas fumarate is primarily reduced under conditions of CO_2 limitation but dismutated when otherwise possible. In general, acrylates are preferentially used over CO_2, whereas fumarate is not. However, this is a large generalization that is more accurately stated as: The selective utilization of fumarate and acrylates as reductant sinks depends on

(i) the origin of the reductant, (ii) the acetogen, and (iii) the environmental conditions.

10.2.4 Dismutation of Pyruvate to Lactate and Acetate

P. productus is considered homoacetogenic (see Chapter 7). Indeed, with CO or H_2 as reductant, acetate is generally the sole reduced end product formed (Lorowitz and Bryant, 1984; Geerligs et al., 1987). However, in the literature, lactate has been indicated to be produced in trace levels from glucose; unfortunately, quantitative information on this potential was not shown (Lorowitz and Bryant, 1984). We reasoned that under conditions of CO_2 deprivation, lactate fermentation might replace acetogenesis with this acetogen. This was indeed the case (Misoph, Daniel, and Drake, unpublished; Drake, 1993). When this acetogen is cultivated with glucose or fructose in the absence of supplemental CO_2, growth is not appreciably affected and lactate is synthesized as the predominant reduced end product. When lactate is formed, it appears that pyruvate replaces CO_2 during the oxidation of glucose (Fig. 10.2). Significantly, decarboxylation of pyruvate nonetheless yields CO_2 that is then used in acetogenesis, i.e., lactate fermentation predominates, but acetogenesis occurs as a secondary but simultaneous process. The integration of these two systems is a previously unreported, growth-supportive process for an acetogen. We have screened several other acetogens and have not seen this potential expressed; so far it is restricted to *P. productus*. Indeed, as indicated above, many acetogens struggle when grown with carbohydrates in the absence of CO_2 or a CO_2 replacement.

10.2.5 Energy-Conserving Dissimilation of Nitrate

The preceding metabolic potentials depend on the reduction of carbon in the form of either CO_2 or a more complex organic structure. Nitrate can also assume the role of reductant sink for *C. thermoaceticum* (Seifritz et al., 1993). This process is strongly preferred over the reduction of CO_2 and is coupled to the conservation of energy. In the presence of nitrate, reductant flow yields only nitrate and ammonium as reduced end products from the oxidation of otherwise acetogenic substrates (such as aromatic methoxyl groups, methanol, or CO). The presence of CO_2 makes no difference in this regard; the acetyl-CoA pathway is simply turned off or not engaged relative to acetate synthesis.

The capacity to use aromatic methoxyl groups may be of particular importance to the ecological roles and trophic links of acetogens. It is, therefore, noteworthy that nitrate might provide a mechanism for generating a greater amount of utiliz-

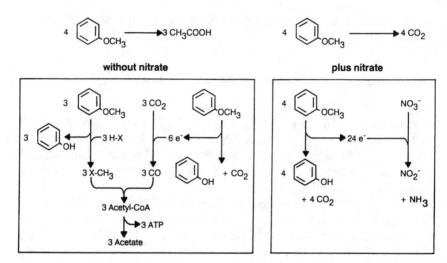

Figure 10.6.   Effect of nitrate on the usage of aromatic methoxyl groups by *Clostridium thermoaceticum* (Seifritz et al., 1993).

able methoxyl-derived reductant than does CO_2 [Fig. 10.6; Seifritz et al. (1993)]. In this regard, methoxyl-derived growth yields are significantly higher when nitrate is used as a reductant sink. Thermodynamically, this stands to reason: the CO_2/acetate half-cell approximates -0.34 V under standard conditions (Fuchs, 1990), whereas that of the NO_3^-/NO_2^- half-cell approximates $+0.42$ (Loach, 1976). This large difference clearly offers the cell greater potential to conserve energy.

Although we do not, at the present time, know why reductant flow is selectively channeled away from CO_2, it is nonetheless surprising that the model acetogen *C. thermoaceticum* actually prefers to *not* synthesize acetate. This fact illustrates the potential importance of the utilization of alternative sinks by acetogens in their native environments.

10.3   Diverse Origins of Reductant: Where Electrons Originate in Laboratory Cultures

From the above studies, we have learned that seemingly well-characterized acetogens harbor many previously undetected metabolic potentials relative to the destination (and importance) of reductant flow. From recent studies, previously unrecognized acetogenic potentials (i.e., use of new substrates for the synthesis of acetate) have also been discovered. In some cases, we have merely "updated" organisms relative to their autotrophic, lithotrophic, and heterotrophic potentials (e.g., Daniel et al., 1990, 1991; Parekh et al., 1992; Lux and Drake, 1992). Sections 10.3.1 and 10.3.2 focus on acetogenic potentials not previously recognized and not well characterized.

10.3.1   Acetogenic Oxidation of Aromatic Aldehydes

The turnover of aromatic materials under anaerobic conditions has been intensely studied during the last 10 years (Evans and Fuchs, 1988; Schink et al., 1992). However, a potential that has not been addressed in much detail is the anaerobic fate of aromatic aldehydes. Such compounds can be derived from the breakdown of lignin (Kirk and Farrell, 1987). Although several species of *Desulfovibrio* have been shown recently to oxidize aromatic aldehydes to the corresponding aromatic carboxylates (sulfate is concomitantly reduced to sulfide) (Zellner et al., 1990), little more has been reported in the literature relative to the beneficial usage of aromatic aldehydes by obligately anaerobic bacteria.

Numerous transformations of aromatic aldehydes (both oxidation and reduction) have nonetheless been reported for acetogens; however, none of these transformations were shown to be growth supportive (Krumholz and Bryant, 1985, 1986; Lux et al., 1990; Sembiring and Winter, 1990; Daniel et al. 1991; Lux and Drake, 1992). Subsequent to these studies, we discovered that oxidation of the aldehyde group of 4-hydroxybenzaldehyde or vanillin can be coupled to the acetyl-CoA pathway, energy conservation, and growth of *C. formicoaceticum* and *C. aceticum* (Gößner et al., 1994). Acetogenesis via the oxidation of 4-hydroxybenzaldehyde and vanillin approximates the stoichiometries of reactions VII and VIII, respectively:

$$4 \text{ 4-Hydroxybenzaldehyde} + 2CO_2 + 2H_2O \rightarrow \qquad \text{(VII)}$$
$$4 \text{ 4-hydroxybenzoate} + \text{acetate}$$

$$\text{Vanillin} \rightarrow \text{protocatechuate} + \text{acetate} \qquad \text{(VIII)}$$

In the case of vanillin, the aldehyde and methoxyl groups are hypothesized to be utilized according to the scheme in Figure 10.7. We do not know how the aldehyde-level reductant is activated. It may occur via a coenzyme A intermediate, although we have no evidence of this. During the turnover of vanillin, significant concentrations of vanillate are detected extracellularly, suggesting that the aldehyde oxidoreductase might be membranous (Gößner et al., 1994). If this is true, the enzyme might be coupled to an energy-conserving electron transport system that is integrated to the acetyl-CoA pathway. As *C. formicoaceticum* apparently lacks hydrogenase, the aldehyde-derived reductant likely does not pass through free H_2 enroute to acetate.

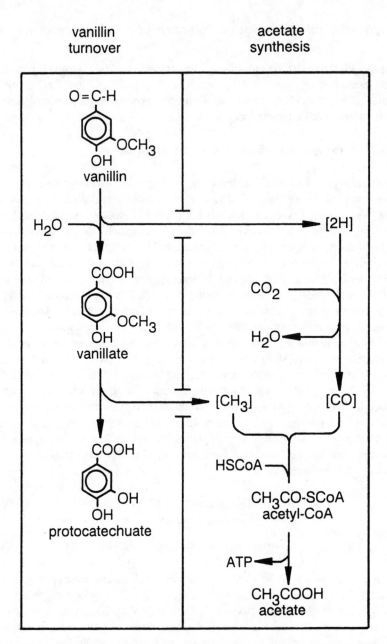

Figure 10.7. Growth-supportive, acetogenic usage of both aldehyde and methoxyl groups of vanillin by *Clostridium formicoacetium* (Gößner et al., 1994).

10.3.2 C_1 versus C_2 Metabolism: Oxalate and Glyoxylate as Acetogenic Substrates

Information concerning the growth-supportive properties of C_2 compounds for acetogenic bacteria and how C_2 compounds are integrated into the C_1-based acetyl-CoA (Wood) pathway is limited. A number of acetogenic bacteria are able to grow at the expense of ethanol or ethylene glycol (Table 10.2). When coupled to acetogenesis, ethanol and ethylene glycol are metabolized according to the following stoichiometries (Buschhorn et al., 1989; Eichler and Schink, 1984; Tanaka and Pfennig, 1988):

Ethanol

Oxidation: $2CH_3CH_2OH + 2H_2O \rightarrow 2CH_3COO^- + 2H^+ + 8[H]$ (IX)

Reduction: $2HCO_3^- + 8[H] + H^+ \rightarrow CH_3COO^- + 4H_2O$ (X)

Sum: $2CH_3CH_2OH + 2HCO_3^- \rightarrow 3CH_3COO^- + H^+ + 2H_2O$ (XI)

Ethylene glycol

Oxidation: $4HOCH_2CH_2OH \rightarrow 4CH_3COO^- + 4H^+ + 8[H]$ (XII)

Reduction: $2HCO_3^- + 8[H] + H^+ \rightarrow CH_3COO^- + 4H_2O$ (XIII)

Sum: $4HOCH_2CH_2OH + 2HCO_3^- \rightarrow$
                $5CH_3COO^- + 3H^+ + 4H_2O$ (XIV)

During the oxidation of ethanol and ethylene glycol, acetaldehyde (CH_3COH) is formed as an intermediate (Buschhorn et al., 1992; Schramm and Schink, 1991). The resulting reducing equivalents generated are subsequently utilized for the reduction of CO_2 to acetate. Although *C. thermoaceticum* is unable to couple the oxidation of ethanol to acetogenesis, it is competent in ethanol-dependent growth when ethanol oxidation is coupled to the reduction of dimethyl-sulfoxide or thiosulfate (Beaty and Ljungdahl, 1991).

Surprisingly, acetate is another two-carbon substrate which has the potential to support the growth of the acetogen strain AOR when the proper conditions are present (Lee and Zinder, 1988a, 1988b). This intriguing syntroph is competent in either acetate-or H_2-dependent growth and is described in detail in Chapter 14.

Recently, the two-carbon substrates, oxalate and glyoxylate, have been shown to support the growth of *C. thermoaceticum* and *Clostridium thermoautotrophicum* (Daniel and Drake, 1993). Other acetogenic bacteria tested but not capable of growth with these substrates include *Acetobacterium woodii*, *Acetogenium kivui*, *C. aceticum*, *C. formicoaceticum*, *Eubacterium limosum*, and *P.*

Table 10.2   Acetogenic bacteria that grow at the expense of the two-carbon compounds ethanol and ethylene glycol

Acetogen	Strain	Growth		Reference
		Ethanol	Ethylene Glycol	
Acetoanaerobium romashkovii	12036	ND[a]	+	Davydova-Charakhch'yan et al. (1993)
Acetobacterium sp.	AmMan1	−	+	Dörner and Schink (1991)
Acetobacterium sp.	HA1	−	+	Schramm and Schink (1991)
Acetobacterium sp.	HP4	ND	+	Conrad et al. (1989)
Acetobacterium sp.	KoB58	−	+	Wagener and Schink (1988)
Acetobacterium sp.	MrTac1	+	+	Emde and Schink (1987)
Acetobacterium carbinolicum	WoProp1	+	+	Eichler and Schink (1984)
Acetobacterium malicum	LiME1[b]	−	+	Tanaka and Pfennig (1988)
Acetobacterium wieringae	DSM 1911[c]	−[d]	+	Braun and Gottschalk (1982); Eichler and Schink (1984)
Acetobacterium woodii	DSM 1030	−[d]	+	Balch et al. (1977); Schink and Bomar (1992)
Acetonema longum	APO-1	−	+	Kane and Breznak (1991)
Clostridium aceticum	DSM 1496	+	+	Braun et al. (1981)
Clostridium formicoaceticum	ATCC 27076	+	+	Braun et al. (1981); Daniel and Drake (1993)
Clostridium ljungdahlii	ATCC 49587	+	ND	Tanner et al. (1993)
Sporomusa malonica	DSM 5090	+	−	Dehning et al. (1989)
Sporomusa ovata	DSM 2662	+	−	Möller et al. (1984)
Sporomusa paucivorans	DSM 3637	+	+	Hermann et al. (1987)
Sporomusa sphaeroides	DSM 2875	+	+	Möller et al. (1984)
Sporomusa termitida	DSM 4440	+	ND	Breznak et al. (1988)
AOR		−	+	Lee and Zinder (1988b)

[a] ND, not determined.

[b] *A. malicum* strain MuME1 does not grow at the expense of ethanol or ethylene glycol (Tanaka and Pfennig, 1988).

[c] *A. wieringae* also grows at the expense of the two-carbon compound ethanolamine (Tanaka and Pfennig, 1988).

[d] Ethanol is a growth-supportive two-carbon substrate for this acetogenic bacterium when the cultivation medium contains greater than 90 mM bicarbonate (Buschhorn et al. 1989).

productus. Oxalate and glyoxylate can serve as the sole energy source for *C. thermoaceticum* and are converted to acetate and CO_2 by the following reactions [kJ/mol calculated from Thauer et al. (1977)]:

$$4^-OOC\text{-}COO^- + 5H_2O \rightarrow CH_3COO^- + 6HCO_3^- + OH^- \qquad (XV)$$
$$(-41.4 \text{ kJ/mol oxalate})$$

$$2HOC\text{-}COO^- + 2H_2O \rightarrow CH_3COO^- + 2HCO_3^- + H^+ \qquad (XVI)$$
$$(-85.7 \text{ kJ/mol glyoxylate})$$

In the case of oxalate, consumption of protons (rather than production of OH^- as indicated in reaction XV) would also account for the increase in pH observed during oxalate-dependent growth (Daniel and Drake, 1993). The use of oxalate and glyoxylate has not been widely demonstrated in obligately anaerobic bacteria (Table 10.3).

Oxalate and glyoxylate support the rapid growth of *C. thermoaceticum* under CO_2^- limited conditions, suggesting that "decarboxylation" might precede or be tightly coupled to oxidative steps in the catabolism of these substrates (Fig. 10.8). The participation of CoA in these two schemes is based on the reported role of oxalyl-CoA decarboxylase and formyl-CoA transferase in oxalate degradation by *Oxalobacter formigenes* (Allison et al., 1985; Baetz and Allison, 1989, 1990, 1992). With *O. formigenes*, formate is an end product of oxalate metabolism (Allison et al., 1985; Dawson et al., 1980), whereas with *C. thermoaceticum*, it is likely oxidized via formate dehydrogenase to the level of CO_2 and utilizable reductant. We are more speculative with glyoxylate oxidation; it may yield carbon monoxide (CO) as an intermediate (Fig. 10.8B). If this is true, the CO-level intermediate is likely bound because CO is not detected in the headspace of either oxalate or glyoxylate cultures (Daniel and Drake, 1993).

The molar cell yields of glyoxylate-cultivated cells of *C. thermoaceticum* are approximately twice those of oxalate-cultivated cells (Table 10.3). This is consistent with the differences observed in the changes in Gibbs free energy for glyoxylate- and oxalate-derived acetogenesis (reactions XV and XVI). That glyoxylate cultures yield higher cell yields than oxalate cultures may also be explained by the potential of glyoxylate-cultivated cells to catalyze a decarbonylation reaction (Fig. 10.8B), thereby generating a preformed (thermodynamically favored) carbonyl unit for the synthesis of acetyl-CoA. Although this appears to make sense on paper, oxalate- and glyoxylate-dependent growth by *C. thermoaceticum* are nearly the same relative to the amount of biomass synthesized per unit of theoretical reductant consumed (Daniel and Drake, 1993).

Nonetheless, the molar cell yields of oxalate- and glyoxylate-cultivated cells of *C. thermoaceticum* are significantly greater than those of other anaerobic

Table 10.3 Anaerobic oxalate- or glyoxylate-utilizing bacteria and their mechanisms for the conservation of oxalate- and glyoxylate-derived energy

Substrate	Organism	Source	Energy-Conserving Process Coupled to Oxalate or Glyoxylate Utilization	Molar Cell Yield (g/mol)	Reference
Oxalate	Clostridium thermoaceticum	Horse manure[a]	Acetogenesis	4.9	Daniel and Drake (1993)
	Clostridium thermoautotrophicum	Mud[b]	Acetogenesis	ND[c]	Daniel and Drake (1993)
	Desulfovibrio vulgaris subsp. oxamicus	Mud	Dissimilatory sulfate reduction	ND	Postgate (1963)
	Oxalobacter formigenes[e]	Sheep rumen[e]	Electrogenic oxalate^{2-} : formate^{-1} antiport system	1.1	Dawson et al. (1980), Allison et al. (1985), Anantharam et al. (1989)
	Oxalobacter vibrioformis[d]	Freshwater sediment	ND	1.6	Dehning and Schink (1989b)
	Clostridium oxalicum[d]	Freshwater sediment	ND	1.8	Dehning and Schink (1989b)
	Strain Ox-8[d]	Freshwater sediment	ND	0.07	Smith et al. (1985)
	Clostridium sp.	Donkey dung	ND	ND	Chandra and Shethna (1975)
Glyoxylate	Clostridium thermoaceticum	Horse manure[a]	Acetogenesis	9.4	Daniel and Drake (1993)
	Clostridium thermoautotrophicum	Mud[b]	Acetogenesis	ND	Daniel and Drake (1993)
	Strain PerGlx1	Marine sediment	Fermentation	1.8	Friedrich and Schink (1991)

[a] Originally isolated by Fontaine et al. (1942).
[b] Originally isolated by Wiegel et al. (1981).
[c] ND, Not determined.
[d] These mesophilic anaerobes convert oxalate to formate and CO_2 and require acetate for the synthesis of cell biomass.
[e] Strains of O. formigenes have also been isolated from human feces (Allison et al., 1985), the cecal contents of swine (Allison et al., 1985), freshwater lake sediments (Smith et al., 1985), and the cecal contents of rats (Daniel et al., 1987).

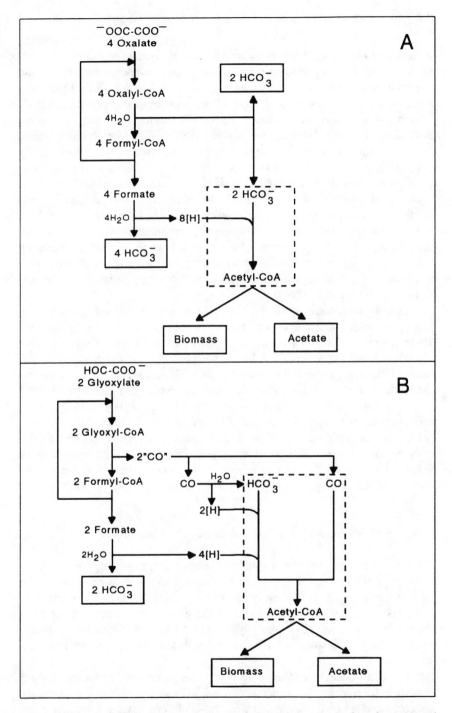

Figure 10.8. Hypothetical pathways for the acetogenic turnover of oxalate (A) and glyoxylate (B) by *Clostridium thermoaceticum* [modified from Daniel and Drake (1993)]. The dashed boxes represent the acetyl-CoA pathway.

oxalate- and glyoxylate-utilizing bacteria (Table 10.3). Lower molar cell yields would be anticipated given the smaller change in free energy for oxalate and glyoxylate metabolism by these nonacetogens (-26.7 and -35.7 kJ per mole of oxalate and glyoxylate, respectively). These anaerobes do not conserve oxalate-derived energy via either substrate-level or electron-transport phosphorylation (Allison et al., 1985; Dehning and Schink, 1989b). In *O. formigenes*, oxalate decarboxylation, a proton-consuming reaction catalyzed in the cytoplasm by oxalyl-CoA decarboxylase (Allison et al., 1985; Baetz and Allison, 1989) appears to be coupled to a membrane-bound, oxalate^{2-}/formate^{1-} antiport protein (Anantharam et al., 1989; Ruan et al., 1992); together, these two enzyme systems function to conserve the energy of oxalate decarboxylation in the form of a transmembrane proton gradient. Alternatively, certain decarboxylation reactions can be coupled to the generation of a Na$^+$ ion gradient via sodium-pumping decarboxylases (Dimroth, 1987). Determining if such energy-conserving processes are engaged during the metabolism of oxalate or glyoxylate by *C. thermoaceticum* will require further study.

Oxalate is widespread in nature, occurring in soils, in aquatic sediments, in many plant, algal, and fungal species, and in the diets of humans and animals (Hodgkinson, 1977). In soils, oxalate is often the most common organic anion present (Fox and Comerford, 1990) and interacts with a variety of soil minerals, thereby influencing mineral availability in soils (Graustein et al., 1977; Johnston and Vestal, 1993). Oxalate is degraded under anaerobic conditions by bacterial populations in aquatic sediments (Smith and Oremland, 1983) and in the gastrointestinal tracks of humans and animals (Allison and Cook, 1981; Allison et al., 1977, 1986; Daniel et al., 1987). Because acetogenic bacteria are present in most anaerobic environments, it is possible that they are involved in the degradation of oxalate in nature and that their metabolic activity may influence the turnover of oxalate and the cycling of minerals in these habitats.

10.4   The Distance between Test Tubes and the Ecosystem

The above findings have been gratifying because they have added new purpose to our physiological endeavors. However, such studies only point to metabolic capacities rather than prove their importance or function at the ecosystem level. In the last year, we have begun to assess the potential existence and activities of acetogens in forest soils of our region around Bayreuth, Germany. Although acetogens have been isolated from soils, what they do in such complex habitats is not known.

In our first studies, we discovered that beech forest soils had a tremendous capacity to form acetate from endogenous materials when incubated under strictly anaerobic conditions (Küsel and Drake, 1994). In one case, soils (collected in mid-August) that had an organic carbon content of approximately 60 g per kg dry wt soil synthesized 15 g and 8 g acetate carbon per kg dry wt soil at 20°C and 5°C, respectively. At 20°C, following an 18-day lag period, rates of 0.07 mmol acetate synthesized per g dry wt soil per day were observed. Although acetate is known to occur in trace concentrations in forest foils (Fox and Comerford, 1990; Huang and Violante, 1986; Stevenson, 1967), we were, not to say the least, surprised at the high levels of acetate observed. These values fluctuated from soil type to soil type and over time. Its formation is likely coupled to complex processes that give rise to acetate precursors, the natures of which remain unknown.

To determine if this acetate-forming capacity was associated with acetogenic bacteria, soils were supplemented with H_2 or CO. Both of these substrates were converted to acetate in stoichiometries that approximate those associated with H_2- or CO-dependent acetogenesis (e.g., 4 CO consumed per acetate formed) (Figure 10.9). This acetate was formed during the periods that closely paralleled the times during which acetate was formed from endogenous materials.

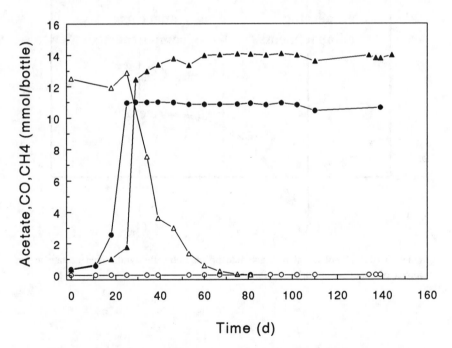

Figure 10.9. Synthesis of acetate by soil suspensions (1:10 dilution) from a Bavarian beech forest soil (Chromic Luvisol, 30 g wet weight) at 20°C. Symbols: acetate formed under N_2/CO_2, ●; acetate formed under CO/CO_2, ▲; CO consumed under CO/CO_2, △; methane formed under either N_2/CO_2 or CO/CO_2, O.

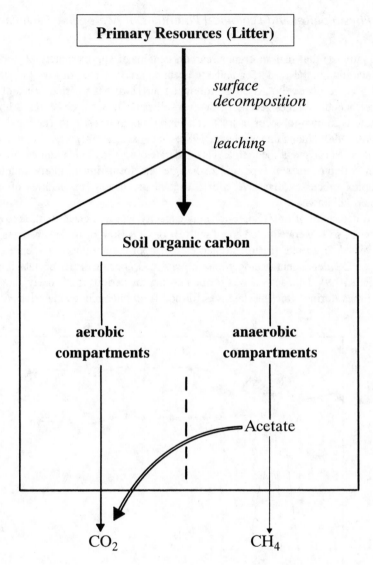

Figure 10.10. Hypothetical role and fate of anaerobically synthesized acetate in the turnover of endogenous carbon in forest soil.

As discussed in Chapters 7, 12, 15, and 16, acetate is an important methanogenic trophic link in many anaerobic habitats. Acetogens appear to contribute to the turnover of H_2 at low temperatures in anoxic sediments and rice paddy soil, acetate then serving as a precursor for methane in these methanogenic habitats (Lovley and Klug, 1983; Jones and Simon, 1985; Conrad et al., 1989; chapter 15). However, in contrast to such systems, the acetate formed by this beech forest soil was not subject to immediate turnover to methane. This was independent of moisture content and acetate concentrations formed and appeared to be due to the absence of significant methanogenic capabilities. Indeed, supplemental H_2 was always turned over to acetate rather than methane. Forest soils are not considered to be very significant in the global production of methane (Boone, 1992; Tyler, 1992).

Because of the apparent stability of acetate under anaerobic conditions and because soils are complex tapestries of fluxuating aerobic and anaerobic zones, we reasoned that acetate might be subject to aerobic rather than anaerobic turnover. To test this hypothesis, oxygen (O_2) was introduced to anaerobically incubated soils that had produced stable concentrations of acetate. Upon addition of oxygen, both acetate and oxygen were subject to immediate usage according to the following stoichiometry:

$$CH_3COOH + 2O_2 \rightarrow 2CO_2 + 2H_2O \qquad (XVII)$$

From this result, we postulate that, under *in vivo* conditions, the acetate that is formed in the anaerobic compartments of these soils is subject to aerobic turnover by virtue of (i) the diffusion of oxygen into formerly anaerobic zones or (ii) the transport of acetate with the soil solution into aerobic zones. These processes are illustrated by the broken line in Figure 10.10. Should this hypothesis be correct, acetate would constitute a trophic link that forms an interface between anaerobic and aerobic communities.

Although we have some distance to go before we know the actual magnitude or consequence of acetogenesis in these forest soils, such experiments illustrate that acetogenesis might be an underlying, previously unappreciated process within such complex poorly described ecosystems. On a practical level, our physiological studies and 2-day growth curves reinforce our belief that acetogens are the most metabolically talented group of obligate anaerobes in test tubes. However, on a more philosophical and less defined level, our ecological studies suggest that accurately assessing these potentials and putting these beliefs to the test at the ecosystem level will not be overnight jobs.

Acknowledgments

The authors express their appreciation to Lars G. Ljungdahl for critical review of the manuscript. Current studies are supported by funds from the Bundesministerium für Forschung und Technologie (0339476A0).

References

Allison, M. J., E. T. Littledike, and L. F. James. 1977. Changes in ruminal oxalate degradation rates associated with adaptation to oxalate ingestion. *J. Anim. Sci.* **45**:1173–1179.

Allison, M. J., and H. M. Cook. 1981. Oxalate degradation by microbes of the large bowel of herbivores: the effect of dietary oxalate. *Science* **212**:675–676.

Allison, M. J., K. A. Dawson, W. R. Mayberry, and J. G. Foss. 1985. *Oxalobacter formigenes* gen. nov., sp. nov.: oxalate-degrading anaerobes that inhabit the gastrointestinal tract. *Arch. Microbiol.* **141**:1–7.

Allison, M. J., H. M. Cook, D. B. Milne, S. Gallagher, and R. V. Clayman. 1986. Oxalate degradation by gastrointestinal bacteria from humans. *J. Nutr.* **116**:455–460.

Anantharam, V., M. J. Allison, and P. C. Maloney. 1989. Oxalate : formate exchange: the basis for energy coupling in *Oxalobacter*. *J. Biol. Chem.* **264**:7244–7250.

Andreesen, J. R., G. Gottschalk, and H. G. Schlegel. 1970. *Clostridium formicoaceticum* nov. spec. isolation, description and distinction from *C. aceticum* and *C. thermoaceticum*. *Arch. Microbiol.* **72**:154–174.

Bache, R., and N. Pfennig. 1981. Selective isolation of *Acetobacterium woodii* on methoxylated aromatic acids and determination of growth yields. *Arch. Microbiol.* **130**:255–261.

Baetz, A. L. and M. J. Allison. 1989. Purification and characterization of oxalyl-coenzyme A decarboxylase from *Oxalobacter formigenes*. *J. Bacteriol.* **171**:2605–2608.

Baetz, A. L., and M. J. Allison. 1990. Purification and characterization of formyl-coenzyme A transferase from *Oxalobacter formigenes*. *J. Bacteriol.* **172**:3537–3540.

Baetz, A. L., and M. J. Allison. 1992. Localization of oxalyl-coenzyme A decarboxylase, and formyl-coenzyme A transferase in *Oxalobacter formigenes* cells. *Sys. Appl. Microbiol.* **15**:167–171.

Balch, W. E., S. Schoberth, R. S. Tanner, and R. S. Wolfe. 1977. *Acetobacterium*, a new genus of hydrogen-oxidizing, carbon dioxide-reducing, anaerobic bacteria. *Int. J. Syst. Bacteriol.* **27**:355–361.

Beaty, P. S., and L. G. Ljungdahl. 1991. Growth of *Clostridium thermoaceticum* on methanol, ethanol, or dimethylsulfoxide. Abstr. K-131, p. 236. 1991. *Ann. Meet. Am. Soc. Microbiol.*

Boone, D. R. 1992. Ecology of methanogenesis. In: *Microbial Production and Consumption of Greenhouse Gases: Methane, Nitrogen Oxides, and Halomethanes*, J. E. Rogers and W. B. Whitman (eds.), pp. 57–70. American Society for Microbiology, Washington, D.C.

Braun, M., F. Mayer, and G. Gottschalk. 1981. *Clostridium aceticum* (Wieringa), a microorganism producing acetic acid from molecular hydrogen and carbon dioxide. *Arch. Microbiol.* **128**:288–293.

Braun, M., and G. Gottschalk. 1982. *Acetobacterium wieringae* sp. nov., a new species producing acetic acid from molecular hydrogen and carbon dioxide. *Zbl. Bakt., 1. Abt. Orig.* **C3**:368–376.

Breznak, J. A., J. M. Switzer, and H.-J. Seitz. 1988. *Sporomusa termitida* sp. nov., an H$_2$/CO$_2$-utilizing acetogen isolated from termites. *Arch. Microbiol.* **150**:282–288.

Buschhorn, H., P. Dürre, and G. Gottschalk, 1989. Production and utilization of ethanol by the homoacetogen *Acetobacterium woodii*. *Appl. Environ. Microbiol.* **55**:1835–1840.

Buschhorn, H., P. Dürre, and G. Gottschalk. 1992. Purification and properties of the coenzyme A-linked acetaldehyde dehydrogenase of *Acetobacterium woodii*. *Arch. Microbiol.* **158**:132–138.

Chandra, T. S., and Y. I. Shethna. 1975. Isolation and characterization of some new oxalate-decomposing bacteria. *Antonie van Leeuwenhoek* **41**:101–111.

Conrad, R., F. Bak, H. J. Seitz, B. Thebrath, H. P. Mayer, and H. Schütz. 1989. Hydrogen turnover by psychrotrophic homoacetogenic and mesophilic methanogenic bacteria in anoxic paddy soil and lake sediment. *FEMS Microbiol. Ecol.* **62**:285–294.

Daniel, S. L., P. A. Hartman, and M. J. Allison. 1987. Microbial degradation of oxalate in the gastrointestinal tracts of rats. *Appl. Environ. Microbiol.* **53**:1793–1797.

Daniel, S. L., T. Hsu, S. I. Dean, and H. L. Drake. 1990. Characterization of the H$_2$- and CO-dependent chemolithotrophic potentials of the acetogens *Clostridium thermoaceticum* and *Acetogenium kivui*. *J. Bacteriol.* **172**:4464–4471.

Daniel, S. L., E. S. Keith, H. Yang, Y. Lin, and H. L. Drake. 1991. Utilization of methoxylated aromatic compounds by the acetogen *Clostridium thermoaceticum:* expression and specificity of the CO-dependent O-demethylating activity. *Biochem. Biophys. Res. Commun.* **180**:416–422.

Daniel, S. L., M. Misoph, A. Gößner, and H. L. Drake. 1992. Growth of acetogenic bacteria in the absence of autotrophic CO$_2$ fixation to acetate. Abstr. C133. *7th Int. Symp. Microbial Growth on C$_1$ Compounds*, Warwick.

Daniel, S. L., and H. L. Drake. 1993. Oxalate- and glyoxylate-dependent growth and acetogenesis by *Clostridium thermoaceticum*. *Appl. Environ. Microbiol.* **59**:3062–3069.

Davydova-Charakhch'yan, I. A., A. N. Mileeva, L. L. Mityushina, and S. S. Belyaev. 1993. Acetogenic bacteria from oil fields of Tataria and western Siberia. Translated from *Mikrobiologiya*, **61**, 306–315, 1992.

Dawson, K. A., M. J. Allison, and P. A. Hartman. 1980. Isolation and some characteristics of anaerobic oxalate-degrading bacteria from the rumen. *Appl. Environ. Microbiol.* **40**:833–839.

Dehning, I., and B. Schink. 1989a. *Malonomonas rubra* gen. nov. sp. nov., a microaerotolerant anaerobic bacterium growing by decarboxylation of malonate. *Arch. Microbiol.* **151**:427–433.

Dehning, I., and B. Schink. 1989b. Two new species of anaerobic oxalate-fermenting bacteria, *Oxalobacter vibrioformis* sp. nov. and *Clostridium oxalicum* sp. nov., from sediment samples. *Arch. Microbiol.* **153**:79–84.

Dehning, I., M. Stieb, and B. Schink. 1989. *Sporomusa malonica* sp. nov., a homoaceto-genic bacterium growing by decarboxylation of malonate or succinate. *Arch. Microbiol.* **151**:421–426.

Dimroth, P. 1987. Sodium ion transport decarboxylases and other aspects of sodium ion cycling in bacteria. *Microbiol. Rev.* **51**:320–340.

Dorn, M., J. R. Andreesen, and G. Gottschalk, 1978a. Fermentation of fumarate and L-malate by *Clostridium formicoaceticum*. *J. Bacteriol.* **133**:26–32.

Dorn, M., J. R. Andreesen, and G. Gottschalk. 1978b. Fumarate reductase of *Clostridium formicoaceticum*. *Arch. Microbiol.* **119**:7–11.

Dörner, C., and B. Schink. 1991. Fermentation of mandelate to benzoate and acetate by a homoacetogenic bacterium. *Arch. Microbiol.* **156**:302–306.

Drake, H. L., S.-I. Hu, and H. G. Wood. 1981. Purification of five components from *Clostridium thermoaceticum* which catalyze synthesis of acetate from pyruvate and methyltetrahydrofolate. *J. Biol. Chem.* **256**:11137–11144.

Drake, H. L. 1992. Acetogenesis and acetogenic bacteria. In: *Encyclopedia of Microbiology, Vol. 1,* J. Lederberg (ed.), pp. 1–15. Academic Press, San Diego, CA.

Drake, H. L. 1993. CO_2, reductant, and the autotrophic acetyl-CoA pathway: alternative origins and destinations. In: *Microbial Growth on C_1 Compounds,* J. C. Murrell and D. P. Kelly (eds.), Intercept Ltd., Andover, U.K.

Eichler, B., and B. Schink. 1984. Oxidation of primary aliphatic alcohols by *Acetobacterium carbinolicum* sp. nov., a homoacetogenic anaerobe. *Arch. Microbiol.* **140**:147–152.

Evans, C. W., and G. Fuchs. 1988. Anaerobic degradation of aromatic compounds. *Annu. Rev. Microbiol.* **42**:289–317.

Emde, R., and B. Schink. 1987. Fermentation of triacetin and glycerol by *Acetobacterium* sp. No energy is conserved by acetate excretion. *Arch. Microbiol.* **149**:142–148.

Fontaine, F. E., W. H. Peterson, E. McCoy, M. J. Johnson, and G. J. Ritter. 1942. A new type of glucose fermentation by *C. thermoaceticum* n. sp. *J. Bacteriol.* **43**:701–715.

Fox, T. R., and N. B. Comerford. 1990. Low-molecular-weight organic acids in selected forest soils of the southeastern USA. *Soil Sci. Soc. Am. J.* **54**:1139–1144.

Frazer, A. C., and L. Y. Young. 1985. A gram-negative anaerobic bacterium that utilizes O-methyl substituents of aromatic acids. *Appl. Environ. Microbiol.* **49**:1345–1347.

Friedrich, M., and B. Schink. 1991. Fermentative degradation of glyoxylate by a new strictly anaerobic bacterium. *Arch. Microbiol.* **156**:392–397.

Fuchs, G. 1990. Alternatives to the Calvin cycle and the Krebs cycle in anaerobic bacteria: pathways with carbonylation chemistry. In: *The Molecular Basis of Bacterial Metabolism,* G. Hauska, and R. Thauer (eds.), pp. 13–20. Springer-Verlag, Berlin.

Geerligs, G., H. C. Aldrich, W. Harder, and G. Diekert. 1987. Isolation and characterization of a carbon monoxide utilizing strain of the acetogen *Peptostreptococcus productus*. *Arch. Microbiol.* **148**:305–313.

Gößner, A., S. L. Daniel, and H. L. Drake. 1994. Acetogenesis coupled to the oxidation of aromatic aldehyde groups. *Arch. Microbiol* **161**:126–131.

Gottwald, M., J. R. Andreesen, J. LeGall, and L. G. Ljungdahl. 1975. Presence of cytochrome and menaquinone in *Clostridium formicoaceticum* and *Clostridium thermoaceticum*. *J. Bacteriol.* **122**:325–328.

Graustein, W. C., K. Cromack, Jr., and P. Sollins. 1977. Calcium oxalate: occurrence in soils and effects on nutrient and geochemical cycles. *Science* **198**:1252–1254.

Hansen, B., M. Bokranz, P. Schönheit, and A. Kröger. 1988. ATP formation coupled to caffeate reduction by H_2 in *Acetobacterium woodii* NZva16. *Arch. Microbiol.* **150**:447–451.

Hermann, M., M.-R. Popoff, and M. Sebald. 1987. *Sporomusa paucivorans* sp. nov., a methylotrophic bacterium that forms acetic acid from hydrogen and carbon dioxide. *Int. J. Syst. Bacteriol.* **37**:93–101.

Hodgkinson, A. 1977. *Oxalic Acid in Biology and Medicine*. Academic Press, New York.

Hsu, T., S. L. Daniel, M. F. Lux, and H. L. Drake. 1990a. Biotransformations of carboxylated aromatic compounds by the acetogen *Clostridium thermoaceticum:* generation of growth-supportive CO_2 equivalents under CO_2-limited conditions. *J. Bacteriol.* **172**:212–217.

Hsu, T., M. F. Lux, and H. L. Drake. 1990b. Expression of an aromatic-dependent decarboxylase which provides growth-essential CO_2 equivalents for the acetogenic (Wood) pathway of *Clostridium thermoaceticum*. *J. Bacteriol.* **172**:5901–5907.

Huang, P. M., and A. Violante. 1986. Influence of organic acids on crystallization and surface properties of precipitation products of aluminum. In: *Interactions of Soil Minerals with Natural Organics and Microbes*, P. M. Huang and M. Schnitzer (eds.), pp. 159–221. Soil Science Society of America, Inc., Madison, WI.

Johnston, C. G., and J. R. Vestal. 1993. Biogeochemistry of oxalate in the antarctic cryptoendolithic lichen-dominated community. *Microb. Ecol.* **25**:305–319.

Jones, J. G., and B. M. Simon. 1985. Interactions of acetogens and methanogens in anaerobic freshwater sediments. *Appl. Environ. Microbiol.* **49**:944–948.

Kane, M. D., and J. A. Breznak. 1991. *Acetonema longum* gen. nov., an H_2/CO_2 acetogenic bacterium from the termite, *Pterotermes occidentis*. *Arch. Microbiol.* **156**:91–98.

Kirk, T. K., and R. L. Farrell. 1987. Enzymatic "combustion": the microbial degradation of lignin. *Annu. Rev. Microbiol.* **41**:465–505.

Krumholz, L. R., and M. P. Bryant. 1985. *Clostridium pfennigii* sp. nov. uses methoxyl groups of monobenzoids and produces butyrate. *Int. J. Syst. Bacteriol.* **35**:454–456.

Krumholz, L. R., and M. P. Bryant. 1986. *Syntrophococcus sucromutans* sp. nov. gen. nov. uses carbohydrates as electron donors and formate, methoxymonobenzenoids or *Methanobrevibacter* as electron acceptor systems. *Arch. Microbiol.* **143**:313–318.

Kröger, A. 1974. Electron-transport phosphorylation coupled to fumarate reduction in anaerobically grown *Proteus rettgeri*. *Biochem. Biophys. Acta* **347**:273–289.

Küsel, K., and H. L. Drake. 1994. Acetate synthesis by soil from a Bavarian beech forest. *Appl. Environ. Microbiol.* **60**:1370–1373.

Lee, M. J., and S. H. Zinder. 1988a. Carbon monoxide pathway enzyme activities in a thermophilic anaerobic bacterium grown acetogenically and in a syntrophic acetate-oxidizing coculture. *Arch. Microbiol.* **150**:513–518.

Lee, M. J., and S. H. Zinder. 1988b. Isolation and characterization of a thermophilic bacterium which oxidizes acetate in syntrophic association with a methanogen and which grows acetogenically on H_2/CO_2. *Appl. Environ. Microbiol.* **54**:124–129.

Loach, P. A. 1976. Oxidation-reduction potentials, absorbance bands and molar absorbance of compounds used in biochemical studies. In: *Handbook of Biochemistry and Molecular Biology, Physical and Chemical Data,* 3rd ed. Fasman, G. D. (ed.), Vol. 1, pp. 122–130. CRC Press, Cleveland, OH.

Lorowitz, W. H., and M. P. Bryant. 1984. *Peptostreptococcus productus* strain that grows rapidly with CO as the energy source. *Appl. Environ. Microbiol.* **47**:961–964.

Lovley, D. R., and M. J. Klug. 1983. Methanogenesis from methanol and methylamines and acetogenesis from hydrogen and carbon dioxide in the sediments of a eutrophic lake. *Appl. Environ. Microbiol.* **45**:1310–1315.

Lundie, L. L., Jr., and H. L. Drake. 1984. Development of a minimal defined medium for the acetogen *Clostridium thermoaceticum*. *J. Bacteriol.* **159**:700–703.

Lux, M. F., E. Keith, T. Hsu, and H. L. Drake. 1990. Biotransformation of aromatic aldehydes by acetogenic bacteria. *FEMS Microbiol. Lett.* **67**:73–78.

Lux, M. F., and H. L. Drake. 1992. Re-examination of the metabolic potentials of the acetogens *Clostridium aceticum* and *Clostridium formicoaceticum*: chemolithoautotrophic and aromatic-dependent growth. *FEMS Microbiol. Lett.* **95**:49–56.

Ma, K., S. Siemon, and G. Diekert. 1987. Carbon monoxide metabolism in cell suspensions of *Peptostreptococcus productus* strain Marburg. *FEMS Microbiol. Lett.* **43**:367–371.

Martin, D. R., A. Misra, and H. L. Drake. 1985. Dissimilation of carbon monoxide to acetic acid by glucose-limited cultures of *Clostridium thermoaceticum*. *Appl. Environ. Microbiol.* **49**:1412–1417.

Matthies, C., A. Freiberger, and H. L. Drake. 1993. Fumarate dissimilation and differential reductant flow by *Clostridium formicoaceticum* and *Clostridium aceticum*. *Arch. Microbiol.* **160**:273–278.

Möller, B., R. Oßmer, B. H. Howard, G. Gottschalk, and H. Hippe. 1984. *Sporomusa,* a new genus of gram-negative anaerobic bacteria including *Sporomusa sphaeroides* spec. nov. and *Sporomusa ovata* spec. nov. *Arch. Microbiol.* **139**:388–396.

O'Brien, W., and L. G. Ljungdahl. 1972. Fermentation of fructose and synthesis of acetate from carbon dioxide by *Clostridium formicoaceticum*. *J. Bacteriol.* **109**:626–632.

Parekh, M., E. S. Keith, S. L. Daniel, and H. L. Drake. 1992. Comparative evaluation of the metabolic potentials of different strains of *Peptostreptococcus productus:* utilization and transformation of aromatic compounds. *FEMS Microbiol. Lett.* **94**:69–74.

Postgate, J. R. 1963. A strain of *Desulfovibrio* able to use oxamate. *Arch. Mikrobiol.* **46**:287–295.

Ruan, Z.-S., V. Anantharam, I. T. Crawford, S. V. Ambudkar, S. Y. Rhee, M. J. Allison, and P. C. Maloney. 1992. Identification, purification, and reconstitution of OxIT, the oxalate : formate antiport protein of *Oxalobacter formigenes*. *J. Biol. Chem.* **267**:10537–10543.

Savage, M. D., and H. L. Drake. 1986. Adaptation of the acetogen *Clostridium thermoautotrophicum* to minimal medium. *J. Bacteriol.* **165**:315–318.

Savage, M. D., Z. Wu, S. L. Daniel, L. L. Lundie, Jr., and H. L. Drake. 1987. Carbon monoxide-dependent chemolithotrophic growth of *Clostridium thermoaceticum*. *Appl. Environ. Microbiol.* **53**:1902–1906.

Schink, B., and M. Bomar. 1992. The genera *Acetobacterium*, *Acetogenium*, *Acetoanaerobium*, and *Acetitomaculum*, In: *The Prokaryotes. A Handbook on the Biology of Bacteria: Ecophysiology, Isolation, Identification, Applications, Vol. II*. A. Balows, H. G. Trüper, M. Dworkin, W. Harder, and K.-H. Schleifer (eds.), pp. 1925–1936. Springer-Verlag, New York.

Schink, B., A. Brune, and S. Schnell. 1992. Anaerobic degradation of aromatic compounds. In: *Microbial Degradation of Natural Products*, G. Winkelmann (ed.), pp. 221–242. VCH, Weinheim.

Schramm, E., and B. Schink. 1991. Ether-cleaving enzyme and diol dehydratase involved in anaerobic polyethylene glycol degradation by a new *Acetobacterium* sp. *Biodegradation* **2**:71–79.

Schulman, M., R. K. Ghambeer, L. G. Ljungdahl, and H. G. Wood. 1973. Total synthesis of acetate from CO_2. VII. Evidence with *Clostridium thermoaceticum* that the carboxyl of acetate is derived from the carboxyl of pyruvate by transcarboxylation and not by fixation of CO_2. *J. Biol. Chem.* **248**:6255–6261.

Seifritz, C., S. L. Daniel, A. Gößner, and H. L. Drake. 1993. Nitrate as a preferred electron sink for the acetogen *Clostridium thermoaceticum*. *J. Bacteriol.* **175**:8008–8013.

Sembiring, T., and J. Winter. 1990. Demethylation of aromatic compounds by strain B10 and complete degradation of 3-methoxybenzoate in co-culture with *Desulfosarcina* strains. *Appl. Microbiol. Biotechnol.* **33**:233–238.

Smith, R. L., and R. S. Oremland. 1983. Anaerobic oxalate degradation: widespread natural occurrence in aquatic sediments. *Appl. Environ. Microbiol.* **46**:106–113.

Smith, R. L., F. E. Strohmaier, and R. S. Oremland. 1985. Isolation of anaerobic oxalate-degrading bacteria from freshwater lake sediments. *Arch. Microbiol.* **141**:8–13.

Stevenson, F. J. 1967. Organic acids in soil. In: *Soil Biochemistry*, A. D. McLaren and G. H. Peterson (eds.), Vol. 1, pp. 119–146. Marcel Dekker, New York.

Tanaka, K., and N. Pfennig. 1988. Fermentation of 2-methoxyethanol by *Acetobacterium malicum* sp. nov. and *Pelobacter venetianus*. *Arch. Microbiol.* **149**:181–187.

Tanner, R. S., L. M. Miller, and D. Yang. 1993. *Clostridium ljungdahlii* sp. nov., an acetogenic species in clostridial rRNA homology group I. *Int. J. Syst. Bacteriol.* **43**:232–236.

Thauer, R. K., K. Jungermann, and K. Decker. 1977. Energy conservation in chemotrophic anaerobic bacteria. *Bacteriol. Rev.* **41**:100–180.

Tschech, A., and N. Pfennig. 1984. Growth yield increase linked to caffeate reduction in *Acetobacterium woodii*. *Arch. Microbiol.* **137**:163–167.

Tyler, S. C. 1992. The global methane budget. In: *Microbial Production and Consumption of Greenhouse Gases: Methane, Nitrogen Oxides, and Halomethanes*, J. E. Rogers and W. B. Whitman (eds.), pp. 57–70. American Society for Microbiology, Washington, D.C.

Wagener, S., and B. Schink. 1988. Fermentative degradation of nonionic surfactants and polyethylene glycol by enrichment cultures and by pure cultures of homoacetogenic and propionate-forming bacteria. *Appl. Environ. Microbiol.* **54**:561–565.

Wellsbury, P., R. A. Herbert, and R. J. Parkes. 1993. Incorporation of [methyl-^{3}H]thymidine by obligate and facultative anaerobic bacteria when grown under defined culture conditions. *FEMS Microbiol. Ecol.* **12**:87–95.

Wiegel, J., M. Braun, and G. Gottschalk. 1981. *Clostridium thermoautotrophicum* species novum, a thermophile producing acetate from molecular hydrogen and carbon monoxide. *Curr. Microbiol.* **5**:255–260.

Winding, A. 1992. [^{3}H]Thymidine incorporation to estimate growth rates of anaerobic bacterial strains. *Appl. Environ. Microbiol.* **58**:2660–2662.

Wood, H. G. 1952. Fermentation of 3,4-C^{14}- and 1-C^{14}-labeled glucose by *Clostridium thermoaceticum*. *J. Biol. Chem.* **199**:579–583.

Zellner, G., H. Kneifel, and J. Winter. 1990. Oxidation of benzaldehydes to benzoic acid derivatives by three *Desulfovibrio* strains. *Appl. Environ. Microbiol.* **56**:2228–2233.

11

Acetogenesis from Carbon Dioxide
in Termite Guts

John A. Breznak

11.1 Introduction

Since the isolation of *Clostridium aceticum* (Wieringa, 1940), the first bacterium ever shown to derive energy for growth by acetate synthesis from H_2 + CO_2, the phenomenon of acetogenesis from C_1 compounds has been of intrinsic interest to microbiologists and biochemists. As seen from other chapters in this volume, work in various laboratories over the years has now led to the isolation of over two dozen different species of such acetogens and to the recognition that these bacteria, united by their unique metabolism, are actually quite diverse phenotypically and phylogenetically. Likewise, detailed studies on the biochemistry of acetogenesis from CO_2, conducted mainly with *Clostridium thermoaceticum* by H. G. Wood and his students, have identified each step in the pathway and resulted in the purification and characterization of the relevant enzymes, and in some cases the genes encoding them. Nevertheless, the ecological significance of acetogenesis from CO_2 has remained obscure. Certainly, the ability of most acetogens to use H_2 as a reductant suggests that they might function as terminal or subterminal "electron sink" organisms in anaerobic microbial food webs, and they are often included in that position in diagrams depicting such webs (e.g., Zinder, 1984). Yet, rarely have habitats been identified in which acetogens outprocess, or are strongly competitive with, other potential H_2 consumers such as methanogens and sulfate-reducing bacteria. Hence, their significance in the flow of carbon and reducing equivalents during anoxic decomposition processes has been debatable. In recent years, however, it has been found that the gastrointestinal tract of vertebrates and invertebrates is one type of habitat in which acetogens often appear to be major H_2 consumers (Breznak and Kane, 1990;

also see Wolin and Miller Chapter 13). During microbial fermentation in the gut of certain termites, in particular, acetogens not only appear to constitute the primary H_2 sink, but their production of acetate from $H_2 + CO_2$ makes a major contribution to termite nutrition.

This chapter reviews the phenomenon of acetogenesis from $H_2 + CO_2$ by termite gut microbes, from the events leading up to its discovery, to the isolation and characterization of some of the acetogenic bacteria. Inasmuch as acetate formation in termite guts can arise by fermentative processes not involving reduction of CO_2, I will hereafter use the prefix "H_2/CO_2" to indicate that reduction of CO_2 (and/or HCO_3^-) to acetate is with H_2 serving as the electron donor. Included in this chapter are speculations on why H_2/CO_2 acetogens are able to outprocess methanogens as H_2 consumers in guts of certain termites. This is still an unresolved issue for which we have only partial answers. First, however, it is appropriate to begin with a brief introduction to the basic biology of termites, especially those aspects relating to their nutritional ecology. More expanded accounts of the intestinal microecology of termites, and the extent to which gut microbes contribute to termite digestion and nutrition, can be found elsewhere (Breznak, 1982, 1984a, 1990; Breznak and Brune, 1994; O'Brien and Slaytor, 1982; Slaytor, 1993).

11.2   Termites and Their Nutritional Ecology

Termites are insects that belong to the order Isoptera, meaning "equal winged" and reflecting the fact that the forewings and hindwings of reproductives are similar in size and veination. They are among the most abundant and important of all soil macroinvertebrates. Roughly two-thirds of the Earth's land surface—that occurring between 45° N and 45° S latitude—is inhabited by one or more termite species (Wood and Sands, 1978). However, their populations are greatest in tropical and subtropical regions, where their numbers can exceed $6000/m^2$ and their biomass densities (up to 50 g/m^2) often surpass that of grazing mammalian herbivores (0.013–17.5 g/m^2; Collins and Wood, 1984; Lee and Wood, 1971).

There exist over 2000 species of termites whose biology, behavior, and nutritional ecology are extremely diverse. Although frequently thought of as feeding on "wood," their diet (depending on the particular species) actually includes a great variety of living or dead plant material (the latter being either sound, or in various stages of decay), as well as dung, and soil rich in organic matter (i.e., humus). Specialized or incidental foods include fungi, algae, lichens, organic-rich portions of termite nests, members (including eggs) of their own colony, and skins or other parts of vertebrate corpses (Wood and Johnson, 1986). Global consumption of plant biomass by termites in all ecological regions has been estimated at 3.36 × 10^{15} g annually, which represents about 4% of the plant material synthesized each year in terrestrial ecosystems (Collins and Wood, 1984; Ljungdahl and Eriksson, 1985).

The gut of termites consists of the foregut (which includes the crop and muscular gizzard), the tubular midgut (which, as in other insects, is a key site for secretion of digestive enzymes and for absorption of soluble nutrients), and the relatively voluminous hindgut, which is also a major site for digestion and absorption of nutrients. Blind-ended Malpighian tubules, which transport urine and urinary metabolites for excretion, empty at or near the junction of the midgut and hindgut. Some termites also possess a so-called "mixed" segment, which is a portion of the gut bounded on one side by microvilli-containing midgut tissue, and on the other side by cuticle-lined hindgut tissue (Noirot and Noirot-Timothée, 1969). Although some microbes colonize the foregut and midgut [some rather distinctive morphotypes being situated between midgut microvilli; Breznak and Pankratz (1977)], the bulk of the intestinal microbiota is found in the hindgut. Owing to the relatively small size of most termites, the physico-chemical conditions existing in the gut have been difficult to define precisely, especially with techniques that impose little or no invasiveness. Nevertheless, the clear tendency is for anoxic conditions to become more apparent as one moves from the foregut to the hindgut, being greatest within microbe-packed portions of the hindgut. Hindguts typically have a relatively low redox potential (ranging from -50 to -270 mV in different species) and a pH around neutrality (6.2–7.6), but some portions of the hindgut of soil-feeding termites may be as high as pH 11.0 (Bignell, 1984; Bignell and Anderson, 1980; Veivers et al., 1980, 1982).

Termites are currently divided into two broad groups (Krishna, 1969, 1970). The so-called "lower" termites comprise five families and harbor in their hindgut a dense population of bacteria and cellulose-digesting, flagellate protozoa. The latter represent unique genera and species of protozoa found virtually nowhere else in nature (Honigberg, 1970). "Higher" termites (family Termitidae), which comprise three-fourths of all termite species, also harbor a dense population of gut bacteria, but they lack cellulolytic protozoa and have a more elaborate external and internal anatomy and social organization than do the lower termites.

The gut microbiota is important for termite vitality, and much of the termites' nutrition derives from microbial metabolism. For example, although it now appears that all termites can synthesize and secrete some of their own cellulase components (Slaytor, 1993), the presence of cellulolytic protozoa is required for efficient digestion of cellulose in the lower termites and is critical to their survival on sound wood or cellulose (Cleveland, 1924, 1925). Acetate, propionate and other organic acids are produced during microbial fermentation of carbohydrate in the hindgut, and these are important oxidizable energy sources for termites (Odelsen and Breznak, 1983) as well as carbon skeletons for biosynthesis (Blomquist et al., 1979, 1982; Prestwich et al., 1981; Mauldin, 1982; Guo et al., 1991). Microbial fermentation in the gut also results in the production of methane (owing to the activity of methanogenic bacteria), and termite emissions contribute up to 4% of global atmospheric methane [20 Tg/yr; Khalil et al. (1990); see also Brauman et al. (1992)]. Gut bacteria are also important to N acquisition of some termites via N_2 fixation (Bentley, 1984; Breznak 1984a) and to N conservation via uric acid-N recycling (Potrikus and Breznak, 1981). Still other termites (higher termites of the subfamily *Macrotermitinae*) have established spectacular symbiotic relationships with fungi (*Termitomyces* sp.), which are cultivated *externally* by termites in elaborate gardens and, when consumed, augment the termite's own cellulase repertoire (Martin, 1987; Veivers et al., 1991; Breznak and Brune, 1994).

The morphological diversity of termite gut microbes is remarkable and has been documented by electron microscopy for both lower and higher termites (Breznak and Pankratz, 1977; To et al., 1978; Bignell et al., 1980, 1983; Czolij et al., 1985). A major component of the bacterial flora are spirochetes, a group of spiral bacteria defined on the basis of distinctive ultrastructural features (Canale-Parola, 1984). Indeed, rarely do spirochetes occur in nature in as great a density and morphological diversity as in the gut of termites (Breznak, 1984b; Margulis and Hinkle, 1992). However, their specific role in the gut is unknown, as none have been isolated and studied in pure culture.

11.3   The Discovery of H_2/CO_2 Acetogenesis in Termite Guts

The discovery of H_2/CO_2 acetogenic activity in termite hindguts was made by Breznak and Switzer (1986). This followed a suggestion that H_2/CO_2 acetogenesis was probably an important feature of microbial fermentation in hindguts, based on studies of volatile fatty acid (VFA) turnover in the hindgut of xylophagous species (Odelson and Breznak, 1983). It is worthwhile to review some of the research leading up to this discovery, as it illustrates the experimental approaches that have been used to elucidate symbiotic interactions between termites and their gut microbiota.

Early work by Hungate (1939, 1943) demonstrated the presence of acetate in hindguts of the lower termites *Zootermopsis* sp. and *Reticulitermes claripennis*. By using mixed and protozoan-enriched microbial suspensions prepared from hindgut contents from *Z. angusticollis* and *Z. nevadensis*, he also showed that fermentative dissimilation of cellulose by the large, cellulolytic protozoa was the primary source of acetate. Protozoan suspensions also formed H_2 and CO_2 during cellulose degradation, and two fermentation patterns were observed: one in which the molar ratio of acetate/CO_2/H_2 was roughly 5 : 1 : 1; the other in which the ratio was 1 : 1 : 2 (Hungate, 1943). The basis for the different fermentation patterns was not clear but was thought to reflect differences in the physiology, or relative numbers, of particular protozoa in the suspensions. [In light of current knowledge, the results might well have reflected greater numbers or activity of H_2/CO_2 acetogenic bacteria in suspensions displaying the 5 : 1 : 1 pattern]. Assuming that the fermentation product ratios exhibited by the protozoa *in vitro* would also be displayed *in situ,* and that any H_2 formed by the protozoa *in situ* would not be consumed by any other member of the gut microbiota or by the termite, Hungate used rates of H_2 emission by live specimens of *Zootermopsis* as an indirect measure of acetate production by the hindgut protozoa *in situ.* He also determined the rate of O_2 consumption by *Zootermopsis.* When this was done, it was found that H_2 emission was 10% that of O_2 consumption for some groups of *Zootermopsis* and 55% for others. If the former were attributed to an *in situ* fermentation pattern of the 5 : 1 : 1 type, and the latter to an *in situ* fermentation pattern of the 1 : 1 : 2 type, it was calculated that the *in situ* rate of acetate production was about one-third to one-half of O_2 consumption by the termites. Inasmuch as 2 mol of O_2 are consumed in the complete oxidation of 1 mol of acetate, i.e.,

$$CH_3COOH + 2\,O_2 \rightarrow 2\,CO_2 + 2\,H_2O \tag{11.1}$$

it, therefore, appeared that acetate formation by the hindgut protozoa could support the bulk, and in some cases all, of the oxygen consumption by the insect. This interpretation was also consistent with the respiratory quotient of *Zootermopsis* (i.e., moles of CO_2 evolved/moles O_2 consumed), which was close to unity, and with the inability of lower termites to survive on sound wood or cellulose if defaunated, i.e., rid of their cellulolytic hindgut protozoa (Cleveland, 1924, 1925; Yamin and Trager, 1979). Hence, Hungate proposed that acetate production by cellulolytic protozoa, and its uptake and oxidation by *Zootermopsis,* was a central feature of the "symbiotic utilization of cellulose" in lower termites (Hungate, 1946).

The studies made by Hungate were elegant and insightful. Although perusal of the original research papers reveals that material balances for protozoan fermentations of cellulose were often incomplete, Hungate's experiments were carefully and painstakingly done, without the aid of sophisticated analytical equipment available today and before the advent of ^{14}C tracer technology which would have greatly enhanced the sensitivity of analyses. Subsequent work with axenic cultures of the protozoa *Trichomitopsis termopsidis* and *Trichonympha sphaerica,* both isolated by Yamin (1978, 1981) from hindgut contents of *Zootermopsis,* has shown that each ferments cellulose in a manner consistent with the following stoichiometry:

$$C_6H_{12}O_6 + 2\,H_2O \rightarrow 2\,CH_3COOH + 2\,CO_2 + 4\,H_2 \tag{11.2}$$

[where $C_6H_{12}O_6$ represents each glucosyl unit of cellulose; Yamin (1980, 1981); Yamin and Trager (1979); Odelson and Breznak (1985)]. H_2 appears to be the sole reduced end product, being formed by *T. termopsidis* from the action of pyruvate : ferredoxin oxidoreductase (and, presumably, 3-phosphoglyceralde-hyde dehydrogenase) coupled to hydrogenase (Yamin, 1980). Formate was not detected as a product. This fermentation pattern, with a product ratio of 1 acetate : 1 CO_2 : 2 H_2, was similar to that observed by Hungate (1943) with some suspensions of mixed protozoa taken directly from hindguts of *Zootermopsis*. Subsequent to Hungate's studies, Kovoor (1967) identified acetate as the principle volatile fatty acid (VFA) in hindguts of the higher termite, *Microcerotermes edentatus*. This suggested that acetate formation in the hindgut may also be important to the nutrition of higher termites, which contain an abundant flora of gut bacteria but which lack the cellulolytic protozoa typically present in lower termites.

Odelson and Breznak (1983) reexamined the issue of VFA production and utilization in wood-feeding termites. They confirmed the identity of acetate in termite guts by using gas chromatography/mass spectroscopy, and they found that acetate dominated the VFA pool in hindguts of the lower termites *Z. angusticollis, Reticulitermes flavipes, Incisitermes schwarzi, Schedorhinotermes lamanianus, Prorhinotermes simplex*, and *Coptotermes formosanus*, and of the higher termite *Nasutitermes corniger*. Smaller amounts of C3–C5 VFAs were also observed. In extracellular hindgut fluid of *R. flavipes* and *Z. angusticollis*, acetate occurred at concentrations of 80.6 and 66.2 mM and constituted 94 and 99 mol%, respectively, of the total VFA pool. By feeding *R. flavipes* wood powder amended with ^{14}C-labeled substrates and determining the specific ^{14}C radioactivity of acetate in hindgut fluid, it was also found that more than three-fourths of the acetate was derived from cellulose carbon; about one-fourth was derived from hemicellulose. Furthermore, direct measurements of acetate in hindguts of *R. flavipes* revealed that its steady-state level (20 nmol per termite) and rate of production (20.2–43.3 nmol/h per termite), were significantly affected by treatments which disrupted the hindgut microbiota. Defaunation, by brief incubation of termites under 100% O_2, markedly decreased the steady-state concentration (1 nmol per termite) and rate of production (4 nmol/h per termite) of acetate in hindguts. Elimination of bacteria (by feeding termites antibacterial drugs) also decreased the steady-state concentrations (8.4 nmol per termite) and production rate of acetate in hindguts (6.8 nmol/h per termite), but to a lesser extent. These results were consistent with the notion that protozoa were the agents primarily responsible for acetate production in hindguts but suggested that bacteria were also important in this activity.

When rates of oxygen uptake by *R. flavipes* were compared to those of acetate formation by their hindgut microbiota *in situ*, it was concluded that 77–100% of the insect's respiratory requirement could be met by uptake and oxidation of microbially produced acetate from the hindgut (Odelson and Breznak, 1983). This conclusion was similar to that made by Hungate years earlier with *Zootermopsis*, and it was consistent with more recent studies showing significant concentrations of acetate in termite hemolymph [9.0–11.6 mM; Odelson and Breznak (1983)], the ability of intact termite hindguts, *in vitro*, to transport acetate to the surrounding medium (Hungate, 1943; Hogan et al., 1985), the presence, in termite tissues, of all the enzymes necessary for complete oxidation of acetate to CO_2 and H_2O (O'Brien and Breznak, 1984), the ability of live termites to evolve $^{14}CO_2$ readily when fed ^{14}C-labeled acetate and other VFAs (Odelson and Breznak, 1983), and a value of 1.00 to 1.05 for the respiratory quotient of *R. flavipes* (Odelson and Breznak, 1983).

The striking dominance of acetate in the total VFA pool of hindguts of lower termites was not surprising, given (i) the relatively large biomass of cellulolytic protozoa in the hindgut (Hungate, 1939; Katzin and Kirby, 1939; Breznak and Pankratz, 1977); (ii) the ability of these cellulolytic flagellates to endocytose, and, hence, dominate the hydrolysis and fermentation of, wood particles entering the gut; and (iii) the fact that either mixed suspensions or axenic cultures of termite hindgut protozoa formed acetate as the sole or major nongaseous product of cellulose fermentation. Recall, however, that such protozoa also form H_2 as the principle reduced product of cellulose fermentation. Thus, if wood polysaccharide fermentation in hindguts of *R. flavipes* were viewed simply as a protozoan fermentation of cellulose followed by a termite oxidation of the resulting acetate, then H_2 emission from *R. flavipes* should equal O_2 uptake, as predicted by reaction 11.3 [= reaction 11.2 + 2 (reaction 11.1)].

$$C_6H_{12}O_6 + 4\,O_2 \rightarrow 6\,CO_2 + 4\,H_2 + 2\,H_2O \qquad (11.3)$$

However, H_2 emission by *R. flavipes* was very low, usually 0.4–0.9 nmol/h per termite (equiv. 0.1–0.2 μmol/g fresh wt/h) and $\leq 0.8\%$ that of O_2 consumption (Odelson and Breznak, 1983). Moreover, the respiratory quotient predicted from reaction 11.3 (i.e., 1.5) would be entirely inconsistent with that measured (1.00–1.05). Either the hindgut protozoa of *R. flavipes* (cellulolytic members of which have not yet been studied *in vitro*) did not form H_2 as a major product, or H_2 produced by them was used by some other member of the gut microbiota or by the termite itself.

One obvious possibility was that reducing equivalents (as H_2) generated during microbial fermentation in the hindgut of *R. flavipes* were taken up by methanogenic bacteria and used for methanogenesis from CO_2 as follows:

$$4\,H_2 + CO_2 \rightarrow CH_4 + 2\,H_2O \qquad (11.4)$$

This overall reaction is the typical, terminal "electron sink" reaction of microbial fermentation in the rumen (Wolin, 1981). If methane emission accounted for the "missing" reducing equivalents in *R. flavipes,* then its magnitude should be 25% that of O_2 consumption as predicted by reaction 11.5 [= reaction 11.3 + reaction 11.4]. Reaction 11.5 also brings the predicted respiratory quotient (1.25) closer to unity:

$$C_6H_{12}O_6 + 4 O_2 \rightarrow 5 CO_2 + CH_4 + 4 H_2O \qquad (11.5)$$

However, methane emission from *R. flavipes* was in the same low range as that of H_2 emission, i.e., 0.4 nmol/h per termite (equiv. 0. 1μmol/g fresh wt/h) and <1% that of termite O_2 consumption. Another possibility, that most of the CH_4 (and/or H_2) produced by the gut microbiota was oxidized aerobically before emanating from the insect, was discounted as being inconsistent with the overall kinetics of O_2 consumption by *R. flavipes.*

The most logical explanation for the data was that the hindgut fermentation in *R. flavipes* was probably essentially homoacetic, consisting of a protozoan fermentation of cellulose (reaction 11.2), accompanied by H_2/CO_2 acetogenesis catalyzed by some other microbes in the hindgut community (presumably bacteria; reaction 11.6) then followed by an aerobic oxidation of the resulting acetate by the termite [3 (reaction 11.1)]

$$C_6H_{12}O_6 + 2 H_2O \rightarrow 2 CH_3COOH + 2 CO_2 + 4 H_2 \qquad (11.2)$$

$$4 H_2 + 2 CO_2 \rightarrow CH_3COOH + 2 H_2O \qquad (11.6)$$

$$3 CH_3COOH + 6 O_2 \rightarrow 6 CO_2 + 6 H_2O \qquad [3(\text{reaction } 11.1)]$$

$$\text{Sum: } C_6H_{12}O_6 + 6 O_2 \rightarrow 6 CO_2 + 6 H_2O \qquad (11.7)$$

This hypothetical scheme was consistent with the dominance of acetate in the hindgut VFA pool, with the relatively low H_2 and CH_4 emission rates from *R. flavipes,* and with the observed respiratory quotient of *R. flavipes,* which, according to reaction 11.7, should be 1.0. It was also consistent with the curious observation that H_2 and CH_4 emission rates of *R. flavipes* increased fourfold to sixfold after the termites were fed antibacterial drugs (Odelson and Breznak, 1983). That would be expected if the dominant H_2-producing organisms in the gut were eukaryotic protozoa, whereas the dominant H_2 consumers were acetogenic Bacteria rather than methanogenic Archaea. However, the above scheme also implied that one out of every three acetates formed in the gut arose from CO_2 reduction, i.e., that 33% of the daily energy needs of *R. flavipes* were being met by oxidation of that acetate formed from gut microbes via H_2/CO_2 acetogenesis. Such a contribution of H_2/CO_2 acetogenic gut bacteria to host nutrition would be remarkable, and unprecedented. H_2CO_2 acetogenesis had been previously demonstrated with the cecal flora of rodents (Prins and Lankhorst, 1977); however, this process was estimated to be able to support only 0.25% of the maintenance energy requirement of rats (Breznak and Kane, 1990). More recently, Lajoie et al. (1988) showed that the fecal flora from "nonmethanogenic" humans carried out H_2/CO_2 acetogenesis, but acetate formed by such microbes in the colon would contribute less than 2% to the basal metabolic requirement of those adults (Breznak and Kane, 1990). Nevertheless, validation of the above scheme for termites came when Breznak and Switzer (1986) confirmed the ability of *R. flavipes* gut microbes to effect a complete synthesis of acetate from $H_2 + CO_2$ at a rate close to one-third that of total hindgut acetate productions, as predicted from reactions 11.2 and 11.6.

11.4 Characteristics of Acetogenesis from CO_2 by the Gut Microbiota of Termites

11.4.1 The Standard Assay System

Acetogenesis from CO_2 by the termite gut microbiota is routinely measured in this author's laboratory by determining the rate of reduction of $^{14}CO_2$ to ^{14}C-acetate. Ideally, radiometric measurements such as these should involve minimal disturbance of the system (Wang et al., 1975); however, this is almost impossible to avoid, owing to the relatively small size of most termites. For example, the total volume of *R. flavipes* hindgut contents is about 0.7 mm^3 (Schultz and Breznak, 1978) with 39% of that as extracellular liquid (Odelson and Breznak, 1983), giving a fluid volume of only 0.3 mm^3 or 0.3 μl. Thus, if one wished to inject $NaH^{14}CO_3$ (as a source of $^{14}CO_2$) into the hindgut region of a gut removed from *R. flavipes*, the $NaH^{14}CO_3$ would have to be contained in an injection volume <0.003 μl in order to change the volume of hindgut fluid <1%. Such injections would have to be perfect each time, with no leakage of isotope from the puncture site, and gut removal and substrate injection, as well as the subsequent incubation (for at least an hour or more), would have to be done under conditions which prevented dessication of the tiny gut. Moreover, corrections would have to be made for the dilution of label by the endogenous CO_2/HCO_3^- pool, and for the loss of $H^{14}CO_3^-$ to the exogenous gas phase, via equilibration of injected $H^{14}CO_3^-$ as $H^{14}CO_3^-/^{14}CO_2$ and from the respiratory generation of $^{14}CO_2$ by termite tissue from the reaction product (^{14}C-acetate). The latter process could be suppressed by using anoxic incubation conditions, which would allow ^{14}C-acetate to accumulate rather than turn over. However, the procedure as described would provide no information about the reductant driving acetogenesis from ^{14}CO$_2$, and it would be impractical when many assays are to be performed, some evaluating the effect of various inhibitors, electron donors, or alternate C$_1$ substrates, and some performed under crude laboratory conditions in the field. Accordingly, a facile assay procedure was developed which measures the H$_2$-dependent reduction of ^{14}CO$_2$ to ^{14}C-acetate by anoxic gut homogenates (Breznak and Switzer, 1986). Some practical aspects of this assay system are described below, including limitations on the interpretation of results.

Generally, termites are introduced into an anoxic chamber, wherein 10–50 guts are removed, pooled, and homogenized in 2.0 ml of a dithiothreitol-reduced anoxic buffered salts solution, after which 0.2-ml portions are distributed in serum vials, made up to 0.4 ml, and sealed under 100% N$_2$ or 100% H$_2$. Reactions are initiated by injection of 0.1 ml of ~ 10 mM NaH^{14}CO$_3$ of suitable specific activity (~ 10^6 dpm/μmol). Reactions are terminated by acidification to convert any unreacted NaH^{14}CO$_3$ to ^{14}CO$_2$, which is swept out of the gas phase. The reaction mixtures are then neutralized, clarified, amended with a mixture of nonradioactive (carrier) organic acids, and the amount of ^{14}C-label in various organic acids is determined after separation of the acids by high-performance liquid chromatography (HPLC). The amount of ^{14}CO$_2$ converted to ^{14}CH$_4$ may also be determined, by sampling the gas phase of reaction mixtures just before acidification and using the procedure of Zehnder et al. (1979).

Under the assay conditions (pH 7.3), the injected NaH^{14}CO$_2$ will, during incubation, equilibrate as a mixture of H^{14}CO$_3^-$ and ^{14}CO$_2$. Thus, although results are usually referred to as "acetogenesis from ^{14}CO$_2$," the actual reactive C$_1$ species (i.e., H^{14}CO$_3^-$ or ^{14}CO$_2$, or both) is not known. The amount of ^{14}C label in a particular product formed in reactions incubated under H$_2$ is corrected for the amount formed under N$_2$, by subtracting the latter from the former. This is done to quantify strictly H$_2$-dependent ^{14}CO$_2$ fixation products. Hence, a stimulation by H$_2$ of ^{14}C-acetate formation from ^{14}CO$_2$ is termed "H$_2$/^{14}CO$_2$ acetogenesis." However, even though the amount of ^{14}CO$_2$ fixed into soluble products under H$_2$ is almost always greater than that fixed under N$_2$ (up to 10- to 20-fold greater, depending on the particular termite species; see Table 11.1), acetate is almost always the major ^{14}CO$_2$ fixation product under N$_2$ as well, its formation under N$_2$ presumably driven by reducing equivalents (e.g., H$_2$) generated endogenously by the microbes in the gut homogenate. The routine use of 100% H$_2$ (initial) as a reservoir of reductant, and NaH^{14}CO$_3$ as a source of ^{14}CO$_2$/H^{14}CO$_3^-$, is done for simplicity and to conserve the otherwise large amount of ^{14}C label that would be needed to achieve sufficient specific activity if reactions were performed under an atmosphere of 80% H$_2$/20% ^{14}CO$_2$ (with pH buffered by inclusion of 52 mM NaH^{14}CO$_3$ in the liquid phase). Nevertheless, control assays with R. flavipes gut homogenates incubated under the latter conditions revealed no significant difference in ^{14}C acetogenesis rates when compared to the standard assay system employing 100% H_2 atmospheres (Brauman et al., 1992).

This standard assay system does not *prove* that H_2 is the relevant reductant used by the gut microbiota *in situ*, it indicates the *potential* of H_2 to serve as the reductant for acetogenesis from CO_2. Likewise, one must be careful about using the numerical data as an estimate of H_2/CO_2 acetogenic rates *in situ*. An *over*estimate may occur if the 100% H_2 atmosphere in reaction vials provides far more reductant than acetogenic microbes have access to *in situ*. On the other hand, an *under*estimate will occur if important cell–cell interactions, which may maximize the efficiency of H_2 transfer between H_2-producing microbes and H_2-consuming acetogens *in situ*, are disrupted by homogenization of guts and are not completely compensated by using 100% H_2 in the gas phase.

It seems clear, then, that assays of $H_2/^{14}CO_2$ acetogenesis by termite gut homogenates have certain limitations and uncertainties, and results from such assays must be interpreted with caution. However, when included as part of a larger body of data derived from other experimental approaches, such assays can indicate the likelihood that H_2/CO_2 acetogenesis is an important feature of the hindgut fermentation, and they can be used to compare the H_2/CO_2 acetogenic activity of hindgut microbiota from different termite taxa.

11.4.2 H_2/CO_2 Acetogenesis by the Gut Microbiota of Xylophagous Termites

By using the assay system described earlier, Breznak and Switzer (1986) showed that gut homogenates of the xylophagous lower termites *R. flavipes, Z. angusticollis,* and *P. simplex,* and the xylophagous higher termites *N. costalis* and *N. nigriceps,* catalyzed H_2/CO_2 acetogenesis at rates ranging from 0.53 μmol acetate formed g termite fresh wt^{-1} h^{-1} (*Z. angusticollis*) to 4.96 μmol g^{-1} h^{-1} (*N. costalis*). The demonstration of H_2/CO_2 acetogenic activity in gut homogenates of *Z. angusticollis* supported the idea that the fermentation product ratio Hungate observed with some suspensions of hindgut protozoa from *Zootermopsis* (i.e., 5 mol acetate : 1 mol CO_2 : 1 mol H_2) may have been due to the presence of H_2/CO_2 acetogens in the protozoan suspensions. Rates of H_2/CO_2 acetogenesis for *R. flavipes* (mean = 1.11 μmol g^{-1} h^{-1}; maximum = 1.76 μmol g^{-1} h^{-1}) were close to one-third that of *total* hindgut acetogenesis (2–4 μmol g^{-1} h^{-1}), supporting the hypothesis by Odelson and Breznak (1983) that H_2/CO_2 acetogenesis was the major electron sink reaction of the hindgut fermentation in this species. Corresponding rates for reduction of $^{14}CO_2$ to $^{14}CH_4$ were up to 10-fold less than those of acetogenesis from $^{14}CO_2$ when reactions were performed under H_2.

Rates of H_2/CO_2 acetogenesis by gut homogenates of the American cockroach

Periplaneta americana (0.02 μmol g^{-1} h^{-1}) were significantly lower than those of the termites tested. Subsequent studies (Kane and Breznak, 1991a) confirmed that rates of H_2/CO_2 acetogenesis by *P. americana* were relatively low (0.002–0.012 μmol g^{-1} h^{-1}) and showed that this was true regardless of whether the cockroaches were fed diets high or low in fiber. By contrast, gut homogenates of *Cryptocercus punctulatus* [a wood-feeding cockroach whose hindgut microbiota is similar to that of lower termites, and which is believed to be closely related to termites phylogenetically (Honigberg, 1970; Krishna, 1969, 1970)] displayed rates of H_2/CO_2 acetogenesis that were generally 10-fold greater than those of *P. americana* [0.14 μmol g^{-1} h^{-1} maximum; Breznak and Switzer (1986)].

Detailed studies with *R. flavipes* (Breznak and Switzer, 1986) revealed that the H_2CO_2 acetogenic activity of gut homogenates was inhibited by oxygen (50% inhibition occurring at 57 Pa; complete inhibition occurring at 1.01 kPa), KCN, $CHCl_3$, and iodopropane, and it declined to undetectable levels after feeding termites cellulose tablets amended with antibacterial drugs [a treatment known to drastically decrease the population of viable bacteria in guts (Potrikus and Breznak, 1981)]. By contrast, only a moderate decrease in H_2/CO_2 acetogenesis was observed after feeding *R. flavipes* on a diet of starch, a treatment which eliminated the cellulolytic protozoa from guts without an obvious effect on the bacterial population. These results implicated bacteria, rather than protozoa, as the agents of H_2/CO_2 acetogenesis, a conclusion supported by the demonstration of H_2/CO_2 acetogenesis in higher termites (i.e., *Nasutitermes* species), which contain only bacteria in their hindguts.

Termite gut homogenates also produced formate from CO_2, and in some cases (e.g., *R. flavipes*) the amount of CO_2 reduced to formate was equal to that converted to acetate when reactions were carried out under the standard assay procedure employing 100% H_2 (initial) in the gas phase. However, the ratio of CO_2 reduction to acetate versus reduction to formate varied with the initial concentration of H_2 (balance N_2) in the gas phase, being 6 : 1 at 1% H_2, and 1 : 1 and 2 : 1 at 100% H_2 and 0% H_2, respectively. Separate experiments showed that nonradioactive formate stimulated the conversion of $^{14}CO_2$ to ^{14}C-acetate by gut homogenates under an atmosphere of N_2. However, ^{14}C-formate, although readily oxidized to $^{14}CO_2$, was a poor C source for acetogenesis, even under an atmosphere of 100% H_2. Recent work with *Sporomusa termitida*, an H_2/CO_2 acetogen isolated from guts of *Nasutitermes costalis*, has shown that cell suspensions produce substantial amounts of formate, in addition to acetate, from H_2 + CO_2 (Breznak and Blum, 1991). However, *S. termitida* exhibits poor growth and acetogenesis with formate (Breznak et al., 1988). Taken together, these results suggest that some of the formate produced from CO_2 by termite gut homogenates may arise from H_2/CO_2 acetogens, but the ability of formate to stimulate acetogenesis from CO_2 under N_2 is most likely a result of its prior oxidation to H_2 by nonacetogens.

11.4.3   Variability of H_2/CO_2 Acetogenesis and Methanogenesis in Termites of Different Feeding Guilds

The initial studies by Breznak and Switzer (1986) were with gut homogenates from diverse lower and higher termites representing three different families. However, all were wood feeders, and it was possible that termites from other feeding guilds had intestinal fermentations in which alternative processes (e.g., methanogenesis) were the principal electron sinks. The first indication that this was so was by Brauman et al. (1990), who found that two soil-feeding termites (*Cubitermes speciosus* and *Thoracotermes macrothorax*) emitted 5- to 10-fold more methane than did various xylophagous species but whose gut microbiota exhibited little or no H_2/CO_2 acetogenic activity. These differences in activity were paralleled by differences in population levels of the relevant H_2-consuming bacteria. For example, numbers of H_2/CO_2 *aceto*genic versus *methano*genic bacteria in the soil feeders were (10^5 cells/ml gut fluid) 0.8 versus 304.0, respectively, for *C. speciosus;* and 0.4 versus 236.0, respectively, for *T. macrothorax.* By contrast, corresponding numbers for two xylophagous species were 1150.0 versus 15.0, respectively, for *Nasutitermes lujae;* and 815.0 versus 156.0, respectively, for *Microcerotermes parvus.*

Recently, Brauman et al. (1992) completed a more comprehensive survey of H_2/CO_2 acetogenic activity and methane emission from 24 species representing 4 different feeding guilds. Results, which are summarized in Table 11.1, confirmed that CO_2-reducing acetogenic bacteria usually outprocessed methanogenic bacteria for reductant (presumably H_2) generated during microbial fermentation in the hindgut of wood-feeding and grass-feeding species. However, acetogenesis from CO_2 was of little significance in soil-feeding and fungus-cultivating termites, which generally emitted more methane than their wood-feeding and grass-feeding counterparts. Interestingly, CO_2-reducing acetogenic activity of one fungus-cultivating (*Macrotermes mülleri*) and the five soil-feeding species was almost nonexistent, even when supplied with exogenous H_2. Conversely, three wood-feeding species (*C. formosanus, C. cavifrons,* and *P. occidentis*) evolved little or no methane.

Regarding acetogenesis and methanogenesis, it is known that microbes catalyzing these processes contain significant amounts of corrinoids, vitamin-B_{12}-like cofactors that appear to be important in methyl-group transfer reactions (Stupperich et al., 1990; Stupperich, 1993; Stupperich and Konle 1993). Hence, it is not surprising that in a survey of 23 insect species, representing 8 phylogenetically diverse orders, termites contained the greatest amount of "vitamin B_{12}," the source of which appeared to be their gut bacteria (Wakayama et al., 1984). The putative vitamin B_{12} was assayed only on the basis of the cyano derivative of termite extracts to compete with authentic [57]Co-cyanocobalamin for binding to intrinsic factor, i.e., the lower α ligand of the derivative was not specifically

Table 11.1 $H_2/^{14}CO_2$ acetogenesis by termite gut homogenates, and CH_4 emission by live termites of different feeding guilds[a]

Feeding Guild[d]	^{14}C-Acetate[b,c] With Exogenously Supplied H_2	From Endogenously Produced H_2	CH_4 Emission[b,c]
Wood-feeding termites			
Coptotermes formosanus	1.66	0.10	0.01
Prorhinotermes simplex	1.18	0.57	0.45[e]
Pterotermes occidentis	2.07	0.48	0.00
Reticulitermes flavipes	0.93 ± 0.43	0.09 ± 0.06	0.10
Zootermopsis angusticollis	0.33 ± 0.25	0.07 ± 0.02	1.30
Amitermes sp.	5.16	1.03	0.13
Gnathamitermes perplexus	1.83	0.13	0.21
Microcerotermes parvus	4.96 ± 1.34	1.16 ± 0.98	0.14
Nasutitermes arborum	2.29	3.00	0.13
Nasutitermes costalis	5.96	0.99	0.49[e]
Nasutitermes lujae	1.91	0.13	0.15
Nasutitermes nigriceps	3.68	0.89	0.24
Tenuirostritermes tenuirostris	0.98	0.05	0.11
Grass-feeding termite			
Trinervitermes rhodesiensis	2.70	2.38	0.18
Grand mean[f]	2.54 ± 0.47	0.79 ± 0.24	0.23 ± 0.10
Fungus-growing termites			
Macrotermes mülleri	0.05	0.01	0.25
Pseudacanthotermes militaris	0.23	0.16	0.67
Pseudacanthotermes spiniger	0.17	0.01	0.36
Grand mean:	0.17 ± 0.03	0.06 ± 0.05	0.43 ± 0.13
Soil-feeding termites			
Crenetermes albotarsalis	0.05	0.02	0.93
Cubitermes fungifaber	0.56	0.21	0.48
Cubitermes speciosus	0.02 ± 0.01	0.01 ± 0.01	0.85
Noditermes sp.	0.03	0.05	0.64
Procubitermes sp.	0.05	0.03	0.39
Thoracotermes macrothorax	0.07	0.01	1.09
Grand mean:	0.13 ± 0.10	0.06 ± 0.03	0.73 ± 0.11

[a] Modified from Brauman et al. (1992).
[b] Given in units of μmol product (g termite)$^{-1}$ h^{-1}.
[c] Results are mean values of duplicate analyses for $n = 1$, except for the following species which are mean values of duplicate analyses for n as indicated: *R. flavipes*, $n = 20$; *Z. angusticollis*, $n = 3$; *M. parvus*, $n = 3$; *N. lujae*, $n = 2$; *C. albotarsalis*, $n = 2$; *C. speciosus*, $n = 3$. For results where $n \geq 3$, results are presented as the mean ± standard deviation.
[d] The first five species listed are lower termites; the remaining species are higher termites.
[e] Mean values of duplicate analyses for $n = 3$ to 5. Values for *P. simplex* and *N. costalis* were determined by measuring $^{14}CO_2$ reduction to $^{14}CH_4$ by gut homogenates in the presence of exogenously supplied H_2.
[f] Grand mean ± standard error were calculated by using all of the individual analyses for each member of the feeding guild.

identified as 5,6-dimethylbenzimidazolyl nucleotide. Thus, it would be interesting to reexamine termite extracts, or to screen pure cultures of termite gut acetogens, by HPLC (Stupperich et al., 1989) to see which of a number of various corrinoids they contain, and whether such analyses might reveal new ones (see chapter 6).

11.5 H$_2$/CO$_2$ Acetogenic Bacteria Isolated from Termite Guts

To date, only three species of H$_2$/CO$_2$ acetogens have been isolated from termite guts and characterized. However, they do not all fall into a single, neatly defined taxon. In fact, each has been found to represent a new species of bacteria, and the phenotypic properties of one has prompted the creation of a new genus to accommodate it (Table 11.2).

The first to be isolated was *Sporomusa termitida,* obtained from the wood-feeding higher termite *Nasutitermes nigriceps* (Breznak et al., 1988). Assignment to *Sporomusa,* a relatively new genus originally described by Möller et al. (1984) [also see Breznak (1992)], was based, in part, on the rather unusual coincidence of two phenotypic properties, i.e., the presence of an outer membrane-containing, Gram-negative type cell wall and the formation by cells of dipicolinic acid-containing, heat-resistant endospores. The other two isolates are *Acetonema longum,* from the dry wood-feeding lower termite *Pterotermes occidentis* (Kane and Breznak, 1991b), and *Clostridium mayombei,* from the African soil-feeding higher termite *Cubitermes speciosus* (Kane et al., 1991). Curiously, *A. longum,* like *S. termitida,* was also a Gram-negative endospore former. However, its morphology was distinctly different from that of sporomusas, being about half as wide and up to 10 times as long (Table 11.2). Moreover, although most sporomusas display obvious, tumbling motility in wet mounts owing to their possession of lateral flagella, motility of *A. longum* was inconspicuous unless cells were allowed to settle on the surface of agar-covered slides, whereupon their peritrichous flagella imparted a rapid, slithering movement to cells. These, as well as other differences with sporomusas in physiology and 16S rRNA nucleotide sequences, supported the creation of the new genus, *Acetonema,* for such acetogens. Similar considerations also supported the creation of the new species, *C. mayombei.* Interestingly, however, analysis of 16S rRNA nucleotide sequences revealed that the closest known relative of *C. mayombei* was *C. lituseburense,* a species also originally isolated from African source material [soil and humus from Ivory Coast; Cato et al., (1986)].

As with all other known H$_2$/CO$_2$ acetogens, each of the termite gut isolates is a strict anaerobe, but none is an obligate chemolithotroph. All are fairly versatile with respect to substrates that can be used to support growth (Table 11.2 includes a partial list). Accordingly, aside from their ability to carry out H$_2$/CO$_2$ acetogen-

Table 11.2   Major properties of H_2/CO_2 acetogens isolated from termite guts

	Acetonema longum	Clostridium mayombei	Sporomusa termitida
Cell shape	Straight rod	Straight rod	Sl. curved rod
Cell size (μm)	(0.3–0.4) × (6–60)	1 × (2–6)	(0.5–0.8) × (2–8)
Cell wall morphology	Gram negative	Gram positive	Gram negative
Endospore shape	Spherical	Oval	Oval–spherical
Endospore position	Terminal	Subterminal	Subterminal–terminal
Endospore-sporangium	Swollen	Sl. swollen	Swollen
Motility (flagellation)	+ (peritrichous)[a]	+ (peritrichous)	+ (lateral)
Catalase	+	−	+
Oxidase	−	−	−
Growth characteristics			
pH optimum (range)	7.8 (6.4–8.6)	7.3 (5.5–9.3)	7.2 (6.2–8.1)
temp. optimum (range)	30–33°C (19–40°C)	33°C (15–45°C)	30°C (19–37°C)
Utilization of[b]			
$H_2 + CO_2$ (dbl. time)	+ (36 h)	+ (5 h)	+ (8 h)
Methanol (+ CO_2)	−	−	+
Formate	−	+	+[c]
Glucose	+	+	−
Fructose	+	+	−
Cellobiose	−	+[c]	−
Maltose	−	+	−
Mannitol	+	−	+
Ribose	+	−	−
Xylose	−	+	−
Lactate	−	−	+
Citrate	+[c]	−	+
Succinate	−	+[d]	+[d]
Betaine	−	−	+
3,4,5,-trimethoxybenzoate	+[c]	−	+
G+C content (mol %)	51.5[e]	25.6[e]	48.6[e]
Reference	Kane and Breznak (1991)	Kane et al. (1991)	Breznak et al. (1988)

[a] Motility unmistakenly observed only for cells placed on agar-covered slides.
[b] For cells grown at 30°C.
[c] Poor growth with this substrate.
[d] Propionate + CO_2 are major fermentation products with this substrate.
[e] Values given for A. longum strain APO-1, C. mayombei strain SFC-5, and S. termitida strain JSN-2.

esis, these bacteria may also contribute to termite nutrition by acetate formation from nongaseous intermediates formed during the degradation of plant polysaccharides in the gut (e.g., glucose and cellobiose from cellulose; xylose from xylan; glucose and maltose from starch; methanol from pectin), and possibly from methoxyl groups of plant aromatics such as lignin (tested by using trimethoxybenzoate as a model methoxylated benzenoid). The formation of endospores (a common property of all termite gut acetogens isolated so far) is undoubtedly of survival value for the bacteria, and possibly for the termites as well. For example, the chitinous lining of the termite hindgut, and with it much of the hindgut contents, is expelled at each molt. Hence, the ability to form endospores should help ensure that viable propagules of these strict anaerobes will be picked up by postmolt termites and recolonize the gut. Reinoculation is further favored by the colonial life-style of termites, which includes social interactions such as grooming and trophallaxis, i.e., transfer of foregut and hindgut contents, and inocula contained therein, between colony mates (La Fage and Nutting, 1978).

Despite the apparent importance of H_2/CO_2 acetogenesis to termite nutrition (particularly the nutrition of wood and grass feeders; Table 11.1) and the current representation in pure culture of termite gut acetogens, little information is available on population levels of *specific* H_2/CO_2 acetogens in termite guts. Moreover, it has not yet been proven conclusively that the acetogens currently in culture are the ones primarily responsible for the H_2/CO_2 acetogenic activity of gut homogenates. For example, consider that cell suspensions of H_2/CO_2-grown *A. longum* exhibit rates of H_2/CO_2 acetogenesis of 85.6 nmol acetate formed (h^{-1}) (mg cell protein)$^{-1}$ (Kane and Breznak, 1991b). Assuming that each cell of *A. longum* has the same dry mass (2.8×10^{-13} g) and protein content (55.0% on a dry weight basis) as a cell of *Escherichia coli* (Neidhardt et al. 1990), the rate of acetogenesis could be expressed as 1.32×10^{-8} nmol acetate formed (h^{-1}) (cell^{-1}). Now consider that *Pterotermes occidentis,* the termite from which *A. longum* was isolated, is a relatively large termite (29 mg fresh wt) with about 5 μl of hindgut fluid and a rate of $H_2/^{14}CO_2$ acetogenesis of 1.59 nmol acetate (h^{-1}) (mg fresh wt^{-1}) (Kane and Breznak, 1991b). From this information, one would predict that a density of 7×10^{11} cells of *A. longum* per milliliter hindgut fluid would be necessary to account for the $H_2/^{14}CO_2$ acetogenic activity of gut homogenates. However, although cells resembling *A. longum* were observed microscopically in hindgut contents, enrichment cultures initiated with serial dilutions of *P. occidentis* gut homogenates suggested that only about 10^6 cells of *A. longum* were present per milliliter of hindgut fluid (Kane and Breznak, 1991b)—a population that is significant but is far less than predicted.

Several factors might account for this discrepancy. First, the H_2/CO_2 acetogenic activity exhibited by *A. longum* cells *in vitro* might be considerably less than that of cells *in situ*. In this regard, it is noteworthy that H_2/CO_2 acetogenic activity of another termite gut acetogen, *S. termitida,* varied from 0.56 to 5.70 (μmol acetate formed) (h^{-1}) (mg protein)$^{-1}$ depending on the growth conditions, being greatest for cells removed from an H_2-limited continuous culture (Breznak et al., 1988; Breznak and Blum, 1991). Similar variation in rates of H_2/CO_2 acetogenesis has also been observed for *A. longum* (M. D. Kane, personal communication). Second, cells of *A. longum,* which measure 0.3–0.4 μm in diameter and 6–60 μm in length, are more voluminous (up to 8 μm^3) than a typical *E. coli* cell (0.4 μm^3). Hence, fewer cells of *A. longum* (perhaps 20-fold fewer than predicted, based on cell size alone) would be required to impart the acetogenic activity observed with hindgut homogenates. That this is probably true is also suggested by the unrealistically large population density (7 $\times$ 10^{11} cells/ml gut fluid) predicted earlier. At such densities, 10-μm/long cells of *A. longum* would occupy about 96% of the gut volume, and that is simply not the case. Finally, dilution/ culture methods, which have long been known to underestimate organism numbers in natural samples (Brock, 1987), may seriously underestimate *in situ* population densities of acetogens as well, especially if such forms are unusually sensitive to currently employed *in vitro* growth conditions. This was probably the case for *S. termitida,* which could only be isolated from *Nasutitermes nigriceps* termites by using H_2 (+ PdCl$_2$ catalyst) as the reducing agent in anoxic media, and even then only from concentrated inocula (1 gut equivalent per culture tube). Only after continued laboratory cultivation could *S. termitida* be "trained" to grow in anoxic media containing more commonly used reducing agents, e.g., cysteine, dithiothreitol, or sodium sulfide (Breznak et al., 1988). Indeed, our ignorance of their specific growth requirements is undoubtedly the reason why H_2/CO_2 acetogens have not yet been isolated from *Reticulitermes flavipes,* the termite used for most of the studies in this author's laboratory [e.g., see Breznak and Switzer (1986)]—and this remains a nagging source of frustration.

Nothing can be said about the significance of *C. mayombei* in the gut of the *Cubitermes speciosus,* as the bacteria were isolated from suspensions of intact guts shipped to this author's laboratory from Africa, unfrozen, over a period of almost 2 weeks. However, *C. speciosus* is a soil-feeding termite whose gut homogenates exhibit little or no H_2/CO_2 acetogenic activity (Table 11.1). Hence, if *C. mayombei* is significant to the nutrition of *C. speciosus,* it is probably so in ways other than H_2/CO_2 acetogenesis.

From the foregoing discussion, it seems clear that more reliable estimates of *in situ* population densities of both total and specific acetogens, within and among termite species, are needed to complement the cultivation/isolation approach and biochemical studies. Quantitative immunofluorescent microscopy, or use of fluorescent-labeled rRNA probes, should be helpful for enumerating individual species, whereas functional-group-specific gene probes would be useful for estimating total numbers of H_2/CO_2 acetogens, which together constitute a phylogenetically diverse group of bacteria. In this regard, Lovell and Hui (1991) developed a relatively specific probe for H_2/CO_2 acetogens by using the gene (cloned from *Clostridium thermoaceticum*) encoding formyltetrahydrofolate synthetase (FTHFS), a key enzyme in the acetyl-CoA pathway. Curiously, however, this probe showed no significant hybridization with DNA from the termite gut acetogen *Sporomusa termitida* (Lovell and Hui, 1991), even though cell extracts of *S. termitida* possess ATP-, formate-, and tetrahydrofolate-dependent FTHFS activity [84.0 ± 40.2 (nmol 5,10 methenyltetrahydrofolate formed) (min^{-1}) (mg protein)$^{-1}$ (n=3); J. A. Breznak, (unpublished)], and even though 80% homology was found to exist between the probe and a *Hin*dIII digest of the DNA of another *Sporomusa, S. ovata.* It appears that the FTHFS of *S. termitida* is encoded by a gene significantly different in nucleotide sequence from that of most other acetogens. Whether this is true for termite gut acetogens in general is not yet known, but it does not diminish the potential usefulness of molecular probe technology for future studies of termite acetogen ecology. More on this issue is presented by Lovell elsewhere in this volume (Chapter 8).

11.6 Competition for H_2 between Termite Gut Microbes

The two physiological groups of bacteria that appear to be competing as terminal "electron sink" organisms in the termite hindgut fermentation are acetogens and methanogens. That H_2 rather than, say, formate, is the reductant for which they are competing *in situ* is suggested by the fact that cellulolytic protozoa, which dominate the hindgut fermentation in lower termites, produce H_2 as their sole or principle reduced end product, but they do not appear to produce formate. Furthermore, exogenous H_2 stimulates methane emission from live termites (Messer and Lee, 1989), as well as acetogenesis from CO_2 by gut homogenates (Table 11.1). Also noteworthy is the fact that *A. longum* and *S. termitida* (Table 11.2), isolated from wood-eating termites whose gut homogenates exhibit high rates of $H_2/^{14}CO_2$ acetogenesis (i.e., *P. occidentis* and *N. nigriceps,* respectively; Table 11.1), use formate poorly or not at all as a substrate for growth and acetogenesis. Yet, in most anoxic habitats low in sulfate (i.e., nonmarine habitats), CO_2-reductive methanogenesis almost always dominates terminal electron flow. This may be related to the standard free-energy change for methanogenesis from H_2 and CO_2 (reaction 11.4; $\Delta G_0' = -135.6$ kJ per reaction), which is more exergonic than that of acetogenesis (reaction 11.6; $\Delta G_0' = -104.6$ kJ per reaction), about three times more so if $\Delta G'$ values are calculated by using concentrations of reactants and products typically found in anoxic habitats (Dolfing, 1988). Exceptions do exist, however, and include sediments of freshwater lakes, such as the mildly acidic Knaack Lake (Phelps and Zeikus, 1984), or Blelham Tarn during periods of limited organic carbon input (Jones and Simon, 1985), as well as certain vertebrate gastrointestinal ecosystems [Breznak and Kane 1990; also see Wolin and Miller (Chapter 13)] and, of course, the gut microbiota of wood- and grass-feeding termites (Table 11.1) and the wood-feeding cockroach, *Cryptocercus punctulatus* (Breznak and Switzer, 1986).

The exceptions listed above are enigmatic, and the basis for them is not clear. Experiments to resolve this issue are still rather limited. On the notion that an intrinsically high affinity for H_2 might account for the competitiveness of termite gut acetogens, the kinetic parameters for H_2 uptake were determined for cells of *Sporomusa termitida* removed from an H_2-limited continuous culture. However, the apparent K_m (6 μM) and V_{max} [380 nmol min^{-1} (mg protein)$^{-1}$] were similar to those reported for various methanogens (Breznak et al., 1988). Furthermore, the threshold for H_2 uptake by *S. termitida* and other acetogens (362–4660 ppm) was found to be 10- to 100-fold higher (not lower) than that of known methanogens (Lovley et al., 1984; Lovley, 1985; Cord-Ruwisch et al., 1988). Thus, in a homogeneous, well-mixed system, acetogens would never be expected to compete effectively with methanogens, unless H_2 concentrations were somehow maintained above their threshold. This may explain why the rumen acetogen, *Acetitomaculum ruminis* [H_2 threshold $\sim$ 4200 ppm; J. A. Robinson (personal communication)], which is present in the rumen at a density of about 10^7–10^8 cells/ml (Greening and Leedle, 1989), fails to outcompete methanogens for H_2 *in situ* where the H_2 concentration is about 1900 ppm (Smolenski and Robinson, 1988).

On the other hand, termite guts (and for that matter most other natural habitats) are not entirely homogeneous nor necessarily well mixed. Indeed, most anoxic ecosystems are spatially organized, with the physico-chemical characteristics of the environment and substrate availability determining the stratification (Zeikus, 1983). Hence, properties other than substrate affinity alone probably contribute to the competitiveness of the acetogens for H_2. One such property might be their physical position in the gut. If acetogens could attach to, or position themselves near, H_2-producing cells, this would expose them to higher H_2 concentrations than exist downfield in the diffusion zone and, hence, favor their success as H_2 consumers. However, we do not yet have information on their pattern of distribution in the gut, or with respect to specific H_2 producers. Here again is an area where microscopy of the termite gut microbiota *in situ*, employing specific antibodies or gene probes, could have a major impact.

The nutritional ecology of the termite itself appears to influence terminal electron flow of its gut microbial fermentation, being directed toward acetogenesis in wood and grass feeders, and toward methanogenesis in the soil feeders and fungus cultivators (Table 11.1). However, it is not yet clear whether this difference is due to the quality of the food eaten or to some anatomical or physiological adaptation of the termites' digestive system which has accompanied its evolution into a particular feeding guild.

Another factor which may contribute to the competitiveness of acetogens as H_2 consumers is their ability to carry out mixotrophy, i.e., to use $H_2 + CO_2$

simultaneously with organic substrates for carbon and energy. Known acetogens are fairly versatile with respect to organic compounds that can be used for growth. Hence, mixotrophy might enable cells to gain more energy per unit time (and/ or per electron transferred) than from $H_2 + CO_2$ alone and perhaps give some acetogens a competitive advantage over methanogens, which have a much narrower spectrum of utilizable substrates (Whitman et al., 1992). Mixotrophy has been demonstrated with the free-living acetogen *Acetobacterium woodii* (Braun and Gottschalk, 1981), and it has also been shown for *S. termitida* when growing on $H_2 + CO_2$ plus lactate or methanol (Breznak and Blum, 1991). Rates of H_2/$^{14}CO_2$ acetogeneseis by cell suspensions of *S. termitida* previously grown (without added H_2) on organic substrates ranged from 5% to 95% that of cells grown on $H_2 + CO_2$, indicating that *S. termitida* was always "primed" to carry out acetogenesis if $H_2 + CO_2$ became available. This was true for cells grown on methanol, mannitol, and lactate, whose fermentation included acetogenesis from C_1 intermediates, as well as for cells grown on glycine, whose fermentation probably does not (McInerney, 1988). In fact, of all organic substrates tested, highest rates of H_2/CO_2 acetogenesis were observed for cells grown on glycine.

11.7 Concluding Remarks

Studies of the termite gut microbiota have revealed a tiny, anoxic, and yet marvelously intriguing microbial ecosystem in which H_2/CO_2 acetogenesis often dominates terminal electron flow during degradation of lignocellulosic plant material, with important nutritional consequences to the host animal. Such studies have also increased our appreciation of the microbial diversity that exists within the acetogens, with the isolation of a number of new species from termites. However, none of the research to date satisfactorily explains why H_2/CO_2 acetogenesis usually outcompetes methanogenesis as the principle electron sink in the hindgut fermentation of wood- and grass-feeding termites, and why the reverse is true in soil-feeding and fungus-cultivating termites. Part of the answer may lie in the nutritional versatility of acetogens and their capacity for mixotrophy, or in their distribution *in situ,* or in their physical association with other microbes (e.g., H_2 producers), with food particles, or with gut tissues. Unfortunately, these are aspects about which we still know very little. However, the current availability of several strains of termite gut acetogens in pure culture, and the anticipated discovery of new strains in the future, provides the material from which specific antibody or gene probes can be prepared to resolve this question. Also promising could be some imaginative experiments, employing mixed cultures of termite gut acetogens and methanogens competing for H_2 in multistage chemostats (Parkes and Senior, 1987), in an effort to define the activity and habitat domains of each under various nutritive and dilution rate regimes. Such studies might prompt us to refine our concept of the principles and limits of microbial biodegradation in anoxic habitats and the role of H_2/CO_2 acetogens in the process.

Acknowledgments

I wish to thank M. D. Kane for helpful comments on this manuscript. Much of the information presented in this chapter was the result of research supported by grants to this author from the National Science Foundation (U.S.A.), for which he is very grateful.

References

Bentley, B. L. 1984. Nitrogen fixation in termites: fate of newly fixed nitrogen. *J. Insect Physiol.* **30**:653–655.

Bignell, D. E. 1984. Direct potentiometric determination of redox potentials of the gut contents in the termites *Zootermopsis nevadensis* and *Cubitermes severus* and in three other arthropods. *J. Insect Physiol.* **30**:169–174.

Bignell, D. E., and J. M. Anderson. 1980. Determination of pH and oxygen status in the guts of lower and higher termites. *J. Insect Physiol.* **26**:183–188.

Bignell, D. E., H. Oskarsson, and J. M. Anderson. 1980. Specialization of the hindgut wall for the attachment of symbiotic micro-organisms in a termite *Procubitermes aburiensis* (Isoptera, Termitidae, Termitinae). *Zoomorphology* **96**:103–112.

Bignell, D. E., H. Oskarsson, J. M. Anderson, and P. Ineson. 1983. Structure, microbial associations and function of the so-called "mixed segment" of the gut in two soil-feeding termites, *Procubitermes aburiensis* and *Cubitermes severus* (Termitidae, Termitinae). *J. Zool. Lond.* **201**:445–480.

Blomquist, G. J., R. W. Howard, and C. A. McDaniel. 1979. Biosynthesis of cuticular hydrocarbons of the termite *Zootermopsis angusticollis* (Hagen). Incorporation of propionate into dimethylalkanes. *Insect Biochem.* **9**:371–374.

Blomquist, G. J., L. A. Dwyer, A. J. Chu, R. O. Ryan, and M. de Renobales. 1982. Biosynthesis of linoleic acid in a termite, cockroach and cricket. *Insect Biochem.* **12**:349–353.

Brauman, A., M. D. Kane, M. Labat, and J. A. Breznak. 1990. Hydrogen metabolism by termite gut microbes. In: *Microbiology and Biochemistry of Strict Anaerobes Involved in Interspecies Hydrogen Transfer,* J.-P. Belaich, M. Bruschi, and J.-L. Garcia. eds., pp. 369–371. Plenum Press, New York.

√ Brauman, A., M. D. Kane, M. Labat, and J. A. Breznak. 1992. Genesis of acetate and methane by gut bacteria of nutritionally diverse termites. *Science* **257**:1384–1387.

√ Braun, K., and G. Gottschalk. 1981. Effect of molecular hydrogen and carbon dioxide on chemo-organotrophic growth of *Acetobacterium woodii* and *Clostridium aceticum*. *Arch. Microbiol.* **128**:294–298.

Breznak, J. A. 1982. Intestinal microbiota of termites and other xylophagous insects. *Annu. Rev. Microbiol.* **36**:323–343.

Breznak, J. A. 1984a. Biochemical aspects of symbiosis between termites and their intestinal microbiota. In: *Invertebrate–Microbial Interactions,* J. M. Anderson, A. D. M. Rayner, and D. W. H. Walton (eds.), pp. 173–203. Cambridge University Press, Cambridge.

Breznak, J. A. 1984b. Hindgut spirochetes of termites and *Cryptocercus punctulatus.* In: *Bergey's Manual of Systematic Bacteriology,* N. R. Krieg and J. G. Holt (eds.), Vol. 1, pp. 67–70. Williams & Wilkins, Baltimore, MD.

Breznak, J. A. 1990. Metabolic activities of the microbial flora of termites. In: *Microbiology in Poecilotherms,* R. Lesel (ed.), pp. 63–68. Elsevier, Amsterdam.

Breznak, J. A. 1992. The genus *Sporomusa.* In: *The Prokaryotes,* Vol. II, 2nd ed., A. Balows, H. G. Trüper, M. Dworkin, W. Harder, and K.-H. Schleifer (eds.), pp. 2014–2021. Springer-Verlag, New York.

Breznak, J. A., and H. S. Pankratz. 1977. In situ morphology of the gut microbiota of wood-eating termites [*Reticulitermes flavipes* (Kollar) and *Coptotermes formosanus* Shiraki]. *Appl. Environ. Microbiol.* **33**:406–426.

✓ Breznak, J. A., and J. M. Switzer. 1986. Acetate synthesis from H_2 plus CO_2 by termite gut microbes. *Appl. Environ. Microbiol.* **52**:623–630.

Breznak, J. A., J. M. Switzer, H.-J. Seitz. 1988. *Sporomusa termitida* sp. nov., an H_2/CO_2-utilizing acetogen isolated from termites. *Arch. Microbiol.* **150**:282–288.

Breznak, J. A., and M. D. Kane. 1990. Microbial H_2/CO_2 acetogenesis in animal guts: nature and nutritional significance. *FEMS Microbiol Rev.* **87**:309–314.

Breznak, J. A., and J. S. Blum. 1991. Mixotrophy in the termite gut acetogen, *Sporomusa termitida. Arch. Microbiol.* **156**:105–110.

Breznak, J. A., and A. Brune. 1994. Role of microorganisms in the digestion of lignocellulose by termites. *Annu. Rev. Entomol.* **39**:453–487.

Brock, T. D. 1987. The study of microorganisms *in situ:* progress and problems. In: *Ecology of Microbial Communities,* M. Fletcher, T. R. G. Gray, and J. G. Jones. (eds.), pp. 1–20. Cambridge University Press, New York.

Canale-Parola, E. 1984. Order I. *Spirochaetales* Buchanan 1917, 163[AL]. In: *Bergey's Manual of Systematic Bacteriology,* N. R. Krieg, and J. G. Holt (eds.), Vol. 1, pp. 38–39. Williams & Wilkins, Baltimore, MD.

Cato, E. P., W. L. George, and S. M. Finegold. 1986. Genus *Clostridium* Prazmowski 1880, 23[AL]. In: Bergey's Manual of Systematic Bacteriology, P. H. A. Sneath, N. S. Mair, M. E. Sharpe, and J. G. Holt (eds.), Vol. 2. pp. 1141–1200. Williams & Wilkins, Baltimore, MD.

Cleveland, L. R. 1924. The physiological and symbiotic relationships between the intestinal protozoa of termites and their host, with special reference to *Reticulitermes flavipes* Kollar. *Biol. Bull.* **46**:178–227.

Cleveland, L. R. 1925. The effects of oxygenation and starvation on the symbiosis between the termite *Termopsis*, and its intestinal flagellates. *Biol. Bull.* **48**:309–326.

Collins, N. M., and T. G. Wood. 1984. Termites and atmospheric gas production. *Science* **224**:84–86.

Cord-Ruwisch, R., H.-J. Seitz, and R. Conrad. 1988. The capacity of hydrogenotrophic anaerobic bacteria to compete for traces of hydrogen depends on the redox potential of the terminal electron acceptor. *Arch. Microbiol.* **149**:350–357.

Czolij, R., M. Slaytor, and R. W. O'Brien. 1985. Bacterial flora of the mixed segment and the hindgut of the higher termite *Nasutitermes exitosus* Hill (Termitidae, Nasutitermitinae). *Appl. Environ. Microbiol.* **49**:1226–1236.

Dolfing, J. 1988. Acetogenesis. In: *Biology of Anaerobic Microorganisms*, A. J. B. Zehnder (ed.), pp. 417–468. Wiley, New York.

Greening, R. C., and J. A. Z. Leedle. 1989. Enrichment and isolation of *Acetitomaculum ruminis*, gen. nov., sp. nov.: acetogenic bacteria from the bovine rumen. *Arch. Microbiol.* **151**:399–406.

Guo, L., D. R. Quilici, J. Chase, and G. J. Blomquist. 1991. Gut tract microorganisms supply the precursors for methyl-branched hydrocarbon biosynthesis in the termite, *Zootermopsis nevadensis*. *Insect Biochem.* **21**:327–333.

Hogan, M. E., M. Slaytor, and R. W. O'Brien. 1985. Transport of volatile fatty acids across the hindgut of the cockroach *Panesthia cribrata* Saussure and the termite *Mastotermes darwiniensis* Froggatt. *J. Insect Physiol.* **31**:587–591.

Honigberg, B. M. 1970. Protozoa associated with termites and their role in digestion. In: *Biology of Termites*, K. Krishna and F. M. Weesner. (eds.), Vol. II, pp. 1–36. Academic Press, New York.

Hungate, R. E. 1939. Experiments on the nutrition of *Zootermopsis*. III. The anaerobic carbohydrate dissimilation by the intestinal protozoa. *Ecology* **20**:230–245.

Hungate, R. E. 1943. Quantitative analyses on the cellulose fermentation by termite protozoa. *Ann. Entomol. Soc. Am.* **36**:730–739.

Hungate, R. E. 1946. The symbiotic utilization of cellulose. *J. Elisha Mitchell Sci Soc.* **62**:9–24.

Jones, J. G., and B. M. Simon. 1985. Interaction of acetogens and methanogens in anaerobic freshwater sediments. *Appl. Environ. Microbiol.* **49**:944–948.

Kane, M. D., and J. A. Breznak. 1991a. Effect of host diet on production of organic acids and methane by cockroach gut bacteria. *Appl. Environ. Microbiol.* **57**:2628–2634.

Kane, M. D., and J. A. Breznak. 1991b. *Acetonema longum* gen. nov. sp. nov., an H_2/CO_2 acetogenic bacterium from the termite, *Pterotermes occidentis*. *Arch. Microbiol.* **156**:91–98.

Kane, M. D., A. Brauman, and J. A. Breznak. 1991. *Clostridium mayombei* sp. nov., an H_2/CO_2 acetogenic bacterium from the gut of the African soil-feeding termite, *Cubitermes speciosus*. *Arch. Microbiol.* **156**:99–104.

Katzin, L. I., and H. Kirby. 1939. The relative weights of termites and their protozoa. *J. Parasitol.* **25**:444–445.

Khalil, M. A. K., R. A. Rasmussen, J. R. J. French, and J. A. Holt. 1990. The influence of termites on atmospheric trace gases: CH_4, CO_2, $CHCl_3$, N_2O, CO, H_2, and light hydrocarbons, *J. Geophys. Res.* **95**:3619–3634.

Kovoor, J. 1967. Presence d'acides gras volatils dans la panse d'un termite superieur (*Microcerotermes edentatus* Was., Amitermitidae). *CR Acad. Sci. (Paris)* **264**:486–488.

Krishna, K. 1969. Introduction. In: *Biology of Termites,* K. Krishna and F. M. Weesner (eds.), Vol. I, pp. 1–17. Academic Press, New York.

Krishna, K. 1970. Taxonomy, phylogeny and distribution of termites. In: *Biology of Termites,* K. Krishna and F. M. Weesner (eds.), Vol. II, pp. 127–152. Academic Press, New York.

La Fage, J. P., and W. L. Nutting. 1978. Nutrient dynamics of termites. In: *Production Ecology of Ants and Termites,* M. V. Brian (ed.), pp. 165–232. Cambridge University Press, New York.

Lajoie, S. F., S. Bank, T. L. Miller, and M. J. Wolin. 1988. Acetate production from hydrogen and [^{13}C]carbon dioxide by the microflora of human feces. *Appl. Environ. Microbiol.* **54**:2723–2727.

Lee, K. E., and T. G. Wood. 1971. *Termites and Soils.* Academic Press, New York.

Ljungdahl, L. G., and K.-E. Eriksson. 1985. Ecology of microbial cellulose degradation. *Adv. Microb. Ecol.* **8**:237–299.

Lovell, C. R., and Y. Hui. 1991. Design and testing of a functional group-specific DNA probe for the study of natural populations of acetogenic bacteria. *Appl. Environ. Microbiol.* **57**:2602–2609.

Lovley, D. R. 1985. Minimum threshold for hydrogen metabolism in methanogenic bacteria *Appl. Environ. Microbiol.* **49**:1530–1531.

Lovley, D. R., R. C. Greening, and J. G. Ferry. 1984. Rapidly growing rumen methanogenic organism that synthesizes coenzyme M and has a high affinity for formate. *Appl. Environ. Microbiol.* **48**:81–87.

Margulis, L., and G. Hinkle. 1992. Large symbiotic spirochetes: *Clevelandina, Cristispira, Diplocalyx, Hollandina,* and *Pillotina.* In: *The Prokaryotes,* 2nd ed., Vol. IV, A. Balows, H. G. Trüper, M. Dworkin, W. Harder, and K.-H. Schleifer (eds.), pp. 3965–3978. Springer-Verlag, New York.

Martin, M. M. 1987. *Invertebrate–Microbial Interactions: Ingested Fungal Enzymes in Arthropod Biology.* Comstock Publishing Assoc., Ithaca, N.Y.

Mauldin, J. K. 1982. Lipid synthesis from [^{14}C]-acetate by two subterranean termites, *Reticulitermes flavipes* and *Coptotermes formosanus. Insect Biochem.* **12**:193–199.

McInerney, M. J. 1988. Anaerobic hydrolysis and fermentation of fats and proteins. In: *Biology of Anaerobic Microorganisms,* A. J. B. Zehnder (ed.), pp. 373–415. Wiley, New York.

Messer, A. C., and M. J. Lee. 1989. Effect of chemical treatments on methane emission by the hindgut microbiota in the termite *Zootermopsis angusticollis*. *Microb. Ecol.* **18**:275–284.

Möller, B., R. Oßmer, B. H. Howard, G. Gottschalk, and H. Hippe. 1984. *Sporomusa*, a new genus of Gram-negative anaerobic bacteria including *Sporomusa sphaeroides* spec. nov. and *Sporomusa ovata* spec. nov. *Arch. Microbiol.* **139**:388–396.

Neidhardt, F. C., J. L. Ingraham, and M. Schaechter. 1990. *Physiology of the Bacterial Cell*. Sinauer Associates, Inc., Sunderland, MA.

Noirot, C., and C. Noirot-Timothée. 1969. The digestive system. In: *Biology of Termites*, K. Krishna and F. M. Weesner (eds.), Vol. I, pp. 49–88. Academic Press, New York.

O'Brien, R. W., and M. Slaytor. 1982. Role of microorganisms in the metabolism of termites. *Aust. J. Biol. Sci.* **35**:239–262.

O'Brien, R. W., and J. A. Breznak. 1984. Enzymes of acetate and glucose metabolism in termites. *Insect. Biochem.* **14**:639–643.

Odelson, D. A., and J. A. Breznak. 1983. Volatile fatty acid production by the hindgut microbiota of xylophagous termites. *Appl. Environ. Microbiol.* **45**:1602–1613.

Odelson, D. A., and J. A. Breznak. 1985. Nutrition and growth characteristics of *Trichomitopsis termopsidis*, a cellulolytic protozoan from termites. *Appl. Environ. Microbiol.* **49**:614–621.

Parkes, R. J., and E. Senior. 1987. Multi-stage chemostats and other models for studying anoxic ecosystems. In: *Handbook of Laboratory Model Systems for Microbial Ecosystem Research*, J. W. T. Wimpenny (ed.), CRC Press, Boca Raton, FL.

Phelps, T. J., and J. G. Zeikus. 1984. Influence of pH on terminal carbon metabolism in anoxic sediments from a mildly acidic lake. *Appl. Environ. Microbiol.* **48**:1088–1095.

Potrikus, C. J., and J. A. Breznak. 1981. Gut bacteria recycle uric acid nitrogen in termites: a strategy for nutrient conservation. *Proc. Natl. Acad. Sci. USA* **78**:4601–4605.

Prestwich, G. D., R. W. Jones, and M. S. Collins. 1981. Terpene biosynthesis by nasute termite soldiers (Isoptera: Nasutitermitinae). *Insect Biochem.* **11**:331–336.

Prins, R. A., and A. Lankhorst. 1977. Synthesis of acetate from CO_2 in the cecum of some rodents. *FEMS Microbiol. Lett.* **1**:255–258.

Schultz, J. E., and J. A. Breznak. 1978. Heterotrophic bacteria present in hindguts of wood-eating termites [*Reticulitermes flavipes* (Kollar)]. *Appl. Environ. Microbiol.* **35**:930–936.

Slaytor, M. 1993. Cellulose digestion in termites and cockroaches: what role do symbionts play? *Comp. Biochem. Physiol.* **103B**:775–784.

Smolenski, W. J., and J. A. Robinson. 1988. In situ rumen hydrogen concentrations in steers fed eight times daily measured using a mercury reduction detector. *FEMS Microbiol. Ecol.* **53**:95–100.

Stupperich, E., H. J. Elsinger, and B. Kräutler. 1989. Identification of phenolyl cobamide from the homoacetogenic bacterium *Sporomusa ovata*. *Euro. J. Biochem.* **186**:657–661.

Stupperich, E., H. J. Elsinger, and S. P. J. Albracht. 1990. Evidence for a super-reduced cobamide as the major corrinoid fraction in vivo and a histidine residue as a cobalt ligand of the *p*-cresolyl cobamide in the acetogenic bacterium *Sporomusa ovata*. *Euro. J. Biochem.* **193**:105–109.

Stupperich, E. 1993. Recent advances in elucidation of biological corrinoid functions. *FEMS Microbiol Rev.* **12**:349–366.

Stupperich, E., and R. Konle. 1993. Corrinoid-dependent methyl transfer reactions are involved in methanol and 3,4-dimethoxybenzoate metabolism by *Sporomusa ovata*. *Appl. Environ. Microbiol.* **59**:3110–3116.

To, L., L. Margulis, and A. T. W. Cheung. 1978. Pillotinas and hollandinas: distribution and behaviour of large spirochaetes symbiotic in termites. *Microbios* **22**:103–133.

Veivers, P. C., R. W. O'Brien, and M. Slaytor. 1980. The redox state of the gut of termites. *J. Insect Physiol.* **26**:75–77.

Veivers, P. C., R. W. O'Brien, and M. Slaytor. 1982. Role of bacteria in maintaining the redox potential in the hindgut of termites and preventing entry of foreign bacteria. *J. Insect Physiol.* **28**:947–951.

Veivers, P. C., R. Mühlemann, M. Slaytor, R. H. Leuthold, and D. E. Bignell. 1991. Digestion, diet and polyethism in two fungus-growing termites: *Macrotermes subhyalinus* Rambur and *M. michaelseni* Sjostedt. *J. Insect Physiol.* **37**:675–682.

Wakayama, E. J., J. W. Dillwith, R. W. Howard, and G. J. Blomquist. 1984. Vitamin B$_{12}$ levels in selected insects. *Insect Biochem.* **14**:175–179.

Wang, C. H., D. L. Willis, and W. D. Loveland. 1975. *Radiotracer Methodology in the Biological, Environmental, and Physical Sciences*. Prentice-Hall, Englewood Cliffs, NJ.

Whitman, W. B., T. L. Bowen, and D. R. Boone. 1992. The methanogenic bacteria. In: *The Prokaryotes,* 2nd ed., Vol. I, A. Balows, H. G. Trüper, M. Dworkin, W. Harder, and K.-H. Schleifer. (eds.), pp. 719–767. Springer-Verlag, New York.

Wieringa, K. T. 1940. The formation of acetic acid from carbon dioxide and hydrogen by anaerobic spore-forming bacteria. *Antonie van Leeuwenhoek J. Microbiol. Serol.* **6**:251–262.

Wolin, M. J. 1981. Fermentation in the rumen and human large intestine. *Science* **213**:1463–1468.

Wood, T. G., and W. A. Sands. 1978. The role of termites in ecosystems. In: *Production Ecology of Ants and Termites,* M. V. Brian (ed.), pp. 245–292. Cambridge University Press, New York.

Wood, T. G., and R. A. Johnson. 1986. The biology, physiology and ecology of termites. In: *Economic Impact and Control of Social Insects,* S. B. Vinson (ed.), pp. 1–68. Praeger, New York.

Yamin, M. A. 1978. Axenic cultivation of the cellulolytic flagellate *Trichomitopsis termopsidis* (Cleveland) from the termite *Zootermopsis*. *J. Protozool.* **25**:535–538.

Yamin, M. A. 1980. Cellulose metabolism by the termite flagellate *Trichomitopsis termopsidis*. *Appl. Environ. Microbiol.* **39**:859–863.

Yamin, M. A. 1981. Cellulose metabolism by the flagellate *Trichonympha* from a termite is independent of endosymbiotic bacteria. *Science* **211**:58–59.

Yamin, M. A., and W. Trager. 1979. Cellulolytic activity of an axenically-cultivated termite flagellate, *Trichomitopsis termopsidis*. *J. Gen. Microbiol.* **113**:417–420.

Zehnder, A. J. B., B. Huser, and T. D. Brock. 1979. Measuring radioactive methane with the liquid scintillation counter. *Appl. Environ. Microbiol.* **37**:897–899.

Zeikus, J. G. 1983. Metabolic communication between biodegradative populations in nature. In: *Microbes in Their Natural Environments*, J. H. Slater, R. Whittenbury, and J. W. T. Wimpenny (eds.), pp. 423–462. Cambridge University Press, Cambridge.

Zinder, S. H. 1984. Microbiology of anaerobic conversion of organic wastes to methane: recent developments. *Am. Soc. Microbiol. News* **50**:294–298.

12

Acetogenesis and the Rumen: Syntrophic Relationships

Roderick I. Mackie and Marvin P. Bryant

12.1 Introduction

Acetate can be formed, as a major or minor product, from the fermentation of various organic substrates. It is also formed by a synthesis from CO_2 and/or other one-carbon precursors. Acetogenic bacteria can be defined as obligate anaerobes that utilize CO_2 as a terminal electron acceptor in energy metabolism, producing acetate and other fatty acids (Fuchs, 1986). This review also describes research concerning proton-reducing acetogens in which protons are used as electron acceptors for the oxidation of certain substrates (alcohols, lactate, lowly substituted monobenzenes, and fatty acids) to acetate with concomitant formation of H_2. The synthesis of acetate by these organisms is thermodynamically favorable only when the partial pressure of H_2 is very low and is normally achieved in the presence of a hydrogen-utilizing methanogen or sulfate reducer. Evidence exists that *Clostridium thermoaceticum* may also reduce protons and, hence, produce H_2 under certain conditions. Thus proton-reducing acetogens may also utilize the acetyl CoA pathway for CO_2 reduction (see Chapter 1).

Acetogens that synthesize acetic acid from CO_2 and H_2 were first detected in 1932 in a bacterial culture obtained from sewage sludge. In 1940, Weiringa described *Clostridium aceticum* which produced acetate from hexoses as well as from CO_2 and H_2. However, not much work was done with the organism and it was considered lost until revived from a spore preparation in the laboratory of H. A. Barker by Braun et al. (1981). Consequently, the most detailed studies of acetate synthesis have been performed on *Clostridium thermoaceticum*, isolated in 1942, and the only acetogen available until 1967 (Fontaine et al. 1942). Over the past 10 years many acetogenic bacteria have been discovered and the reductive acetyl CoA pathway has been well established and widely researched [see reviews by Wood et al. (1986), Fuchs (1986), Ljungdahl (1986), Ragsdale (1991), and Wood and Ljungdahl (1991)]. Furthermore, other anaerobes such as methanogenic and sulfate-reducing bacteria synthesize cell carbon from CO_2 using this pathway. In addition, enzymes of this pathway mediate acetyl-CoA oxidation by sulfate-reducing bacteria, acetate catabolism by methanogenic bacteria, and CO oxidation to CO_2 by many anaerobes [reviewed recently by Ragsdale (1991) and Wood and Ljungdahl (1991); see also chapters 1–3].

Current research on acetogenesis in ruminant animals is based on three related areas of interest and application. Methane production is the primary, or predominant, electron sink product formed during degradation of organic matter in the rumen. The gas formed as a result of ruminal fermentation is subsequently eructated and the loss of energy as CH_4 represents a loss of 5–15% of the feed energy to the host animal. Thus, it would be beneficial to the host animal, increasing energetic efficiency, if this loss of feed energy and carbon could be minimized by competitively fixing H_2 and CO_2 into a fermentation acid instead of CH_4. This has stimulated the search for acetogenic bacteria from the rumen ecosystem. Recent research has demonstrated that the gastrointestinal tract of various invertebrates (termites and cockroaches) and vertebrates (rodents and some humans) are anaerobic habitats where acetogenesis from H_2 and CO_2 often dominates terminal electron flow with important nutritional implications (Brauman et al., 1992). This has focused research not only on the organisms involved but also the mechanism and factors influencing acetogenesis in these habitats and how the hindgut environment differs from that found in the foregut. Currently, there is a great interest in global warming forced by the "greenhouse gases" such as CO_2, CH_4, and NO_x. More specifically, ruminants and other gastrointestinal fermentations account for some 50% of the biogenic CH_4 emissions amounting to 70–100 Tg per year (EPA, 1990). A single cow can produce as much as 200 L of CH_4 per day. It is thought that a 50% reduction in CH_4 emissions from ruminants will contribute about 50–75% of the emissions reductions needed to stabilize atmospheric CH_4 concentrations. This has provided the impetus for an increased research effort, and hopefully some funding, to find solutions for limiting CH_4 emissions from livestock and livestock wastes.

12.2   The Rumen Ecosystem

Feedstuffs consumed by ruminants are all initially exposed to the fermentative activity in the rumen prior to gastric and intestinal digestion. Dietary polysaccharides and protein are generally degraded by the ruminal microorganisms into characteristic end products, which, in turn, provide nutrients for metabolism by the host animal. The extent and type of transformation of feedstuffs thus determines the productive performance of the host. Fermentation of feedstuffs in the rumen yields short-chain VFA (primarily acetic, propionic, and butyric acids), carbon dioxide, methane, ammonia, and, occasionally, lactic acid. Some of the change in free energy ($\Delta G_0'$) is used to drive microbial growth, but heat is evolved also. Ruminants use the organic acids and microbial protein as sources of energy and amino acids, respectively, but methane, heat, and ammonia can cause a loss of energy and N. The quality and quantity of rumen fermentation products is dependent on the types and activities of the microorganisms in the rumen (Mackie and White, 1991). This, in turn, will have an enormous potential impact on nutrient output and performance of ruminant animals. The rumen–microbe relationships are summarized in Figure 12.1.

An early step in the study of microbial ecology is to make a detailed examination of the physical and chemical conditions prevailing in the environment. This knowledge is essential in designing habitat-simulating and specific media for the enumeration, isolation, and characterization of the microorganisms involved in gastrointestinal fermentative digestion. This knowledge also provides important information about the metabolic activities of the microbial population in general. The microbial environment in the rumen has been examined in great detail and, allowing for variation in the nature and amount of food ingested, serves as a good model for other gastrointestinal ecosystems in both herbivores and nonherbivores. A summary of some of the approximate physical, chemical, and microbiological characteristics of the rumen of grazing cattle and sheep is presented in Table 12.1. A more detailed list of characteristics is provided by Clarke (1977). Although the physical and chemical parameters of the rumen environment illustrate the complexities which must be considered in media selection and design, it is significant that some rumen bacteria still remain to be isolated. Indeed, most acetogenic bacteria have been isolated not from habitat-simulating or -selective media for specific nutritional groups but from specialized formulations designed to isolate bacteria from a specialized biochemical niche often in combination with a selective enrichment procedure.

As methanogenic bacteria are considered the dominant utilizers of hydrogen in the rumen (Hungate, 1976; Bryant, 1979) inhibition of the resident methanogens with bromoethanesulfonic acid (BES) was found to be essential in providing selective pressure to cultivate acetogenic bacteria from the rumen (Greening and Leedle, 1989) and other gut ecosystems (Breznak et al., 1988; Kane et al., 1991). BES is an analogue of coenzyme M (2-mercaptoethanesulfonic acid) which is unique to methanogens (Balch and Wolfe, 1979) and is a potent inhibitor of the reduction of methyl coenzyme M in cell extracts and isolated whole cells of several methanogens (Gunsalus et al., 1978; Zehnder et al., 1980; Smith and Mah, 1981). BES was included in the enrichment medium at 50 mM by Greening

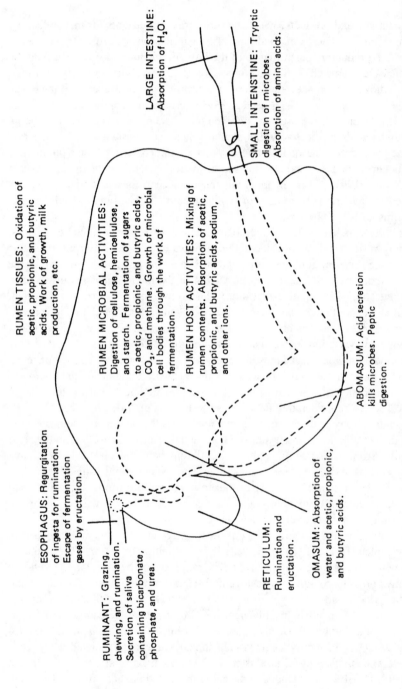

RUMEN TISSUES: Oxidation of acetic, propionic, and butyric acids. Work of growth, milk production, etc.

RUMEN MICROBIAL ACTIVITIES: Digestion of cellulose, hemicellulose, and starch. Fermentation of sugars to acetic, propionic, and butyric acids, CO_2, and methane. Growth of microbial cell bodies through the work of fermentation.

RUMEN HOST ACTIVITIES: Mixing of rumen contents. Absorption of acetic, propionic, and butyric acids, sodium, and other ions.

LARGE INTESTINE: Absorption of H_2O.

SMALL INTENSTINE: Tryptic digestion of microbes. Absorption of amino acids.

ESOPHAGUS: Regurgitation of ingesta for rumination. Escape of fermentation gases by eructation.

RUMINANT: Grazing, chewing, and rumination. Secretion of saliva containing bicarbonate, phosphate, and urea.

RETICULUM: Rumination and eructation.

OMASUM: Absorption of water and acetic, propionic, and butyric acids.

ABOMASUM: Acid secretion kills microbes. Peptic digestion.

Figure 12.1.   Summary diagram showing interrelationships between the rumen, its resident microbial population, and the ruminant animal [from Hungate (1985)].

Table 12.1 Physical, chemical and microbiological characteristics of the rumen

Characteristic	Property
Physical	
pH	5.5–6.9 (mean 6.4)
Redox potential	−350 to −400 mV
Temperature	38–41°C
Osmolality	250–350 mOsmol/kg
Dry matter	10–18%
Chemical	
Gas phase (%)	CO_2, 65; CH_4, 27; N_2, 7; O_2, 0.6; H_2, 0.2
Volatile fatty acids (mM)	Acetic, 68; propionic, 20; butyric, 10; higher, 2
Ammonia	2–12 mM
Amino acids	< 1 mM
Soluble carbohydrates	< 1 mM present 3 h postfeeding
Insoluble polysaccharides	Always present
Lignin	Always present
Minerals	High Na; generally good supply
Trace elements/vitamins	Always present; good source B vitamins
Growth factors	Good supply BCFA, purines, pyrimidines, etc.
Microbiological	
Bacteria	10^{10}–10^{11} g^{-1} (> 200 species)
Ciliate protozoa	10^4–10^6 g^{-1} (predators)
Anaerobic fungi	10^2–10^4 zoospores

and Leedle (1989) because this level was shown to completely inhibit methanogenesis from CO_2 reduction in natural ecosystems (Smith and Mah, 1981; Bouwer and McCarty, 1983; Zinder et al., 1984). Only with methanogens inhibited was it possible to enrich for, and subsequently isolate, other rumen bacteria capable of reducing CO_2 with H_2. The remaining medium components are similar to those used for isolating methanogens with H_2 : CO_2 (80 : 20) gas phase and 5% clarified ruminal fluid. Addition of rumen fluid to growth media for acetogens from the termite gut had little effect but the addition of trypticase (1.0–2.0 g/L) or yeast extract (0.5–1.0g/L) or both were stimulatory (Kane and Breznak, 1991; Kane et al., 1991).

12.3 Acetogenic Bacteria from the Rumen

Large populations of microorganisms inhabit the gastrointestinal tract of all animals. From a microbial ecology viewpoint, the gastrointestinal tract is a rich environment for the development and maintenance of a dense but diverse microbial population. Large numbers of bacteria are present in the rumen (10^{10} viable cells per gram) representing some 200 different bacterial species. Even though it has a high species diversity, only 20–30 species predominate under most feeding conditions. It is interesting to note that the most recent text describing rumen bacteria (Stewart and Bryant, 1988) does not mention acetogens as a property, niche, or group of occasional rumen isolates although the characteristics of two of the organisms are included. This indicates the relatively recent research interest in this group of organisms and the following description of six different acetogenic isolates represent our knowledge to date. Furthermore, it is also worth noting that new species of gut anaerobes are constantly being isolated and described as researchers seek to replicate some of the complex interactions that exist in some of the niches involved.

12.3.1   CO₂-Reducing Acetogens

(a) ACETITOMACULUM RUMINIS

The first report of high numbers [$(2–12) \times 10^8$/g ruminal content] of hydrogen-oxidizing, carbon-dioxide-reducing acetogenic bacteria in the rumen of cattle fed typical low or high forage diets was made by Leedle and Greening (1988). Enrichment of the obligately anaerobic acetogenic bacteria from the rumen of a steer fed a typical high forage diet in media containing BES when incubated under $H_2 : CO_2$ headspace resulted in the successful isolation of five isolates with similar nutritional characteristics. These features included the ability to grow on $H_2 : CO_2$, formate, CO, and glucose, inability to utilize lactate or pyruvate, and similarities in DNA mol% G + C and cell-wall amino acids (Greening and Leedle, 1989). In addition, the type strain (139B; ATCC 43876) grew on cellobiose and fructose and produced acetic acid from esculin. The methoxyl groups of ferulic and syringic acid were also fermented to acetate with formation of hydrocaffeate and gallate, respectively. Rumen fluid was stimulatory and may be required for growth. The cells are Gram-positive, motile, curved rods $(0.8–1.0) \times (2.0–4.0)$ μm.

These ruminal acetogens differed from other hydrogen-oxidizing acetogenic bacteria that have been described. They do not form endospores and, therefore, differ from the genus Clostridium (Braun et al., 1981). They differ from the genera Acetoanaerobium, Sporomusa, and Acetogenium because the isolates stain Gram positive. They differ from Eubacterium in cell morphology, substrates utilized, VFA profiles, and mol% G + C. These isolates most closely resembled the genus Acetobacterium although with a lower mol% G + C (32–36% as compared to 39% for A. woodii and 43% for A. weiringae). The presence of both lysine and diaminopimelic acid as well as amino acid ratios in the cell wall are unusual and also indicate that they differ from other acetogens. Subsequent analysis of the 16S rRNA of the type strain confirmed that these rumen acetogens were generally related to the clostridia but not specifically to any other known acetogens.

(b) *EUBACTERIUM LIMOSUM*

This organism was isolated as the most numerous (mean MPN 6.3 × 10^8 viable cells per ml) methanol-utilizing bacterium (strain RF) in the rumen fluid of sheep fed a diet in which molasses was a major component (Genthner et al., 1981). This was in contrast to the earlier finding by Rowe et al. (1979) who showed that unusually high numbers of *Methanosarcina* (6×10^9/ml) in rumen fluid of sheep fed a molasses-based diet were correlated with acetate degradation. It was also isolated (strain S) from sewage sludge at 9.5 × 10^4 cells/ml. These two strains, strain L34 which was isolated from the rumen of a young calf (Bryant et al., 1958) and the type strain ATCC 8486, grew with H_2-CO_2 as the energy source, although strains RF and S grew much better than the other two strains. Doubling times at 37°C of these two strains (RF and S) in basal medium containing 5% rumen fluid was 7 h with methanol and 14 h on H_2-CO_2 (Genthner et al., 1981).

E. limosum cells are anaerobic, Gram-positive, nonmotile rods 1.2 × 3.3 μm in size. Adonitol, arabitol, erythritol, fructose, glucose, mannitol, and ribose were excellent energy sources, whereas isoleucine, valine, lactate, and methanol supported growth fairly well. Both strains grew well with H_2-CO_2 as the energy and carbon source. After inoculating the H_2-CO_2 medium with a CH_3OH grown culture, a long lag (270 h) was evident; however, subsequent transfers into fresh H_2-CO_2 medium resulted in rapid initiation of growth (Genthner et al., 1981). Acetate (88 mM) and butyrate (2.8 mM) were the products when strain RF was grown to maximum in the chemically defined H_2-CO_2 medium. Strain RF was also able to grow with CO as the sole source of energy and formed acetate and CO_2 as major products. CO_2 was recovered in approximately twice the amount of acetate. The generation time on CO was 7 h and uninhibited growth occurred in cultures containing 50% CO or less in the gas phase (Genthner and Bryant, 1982). *E. limosum* strain RF was also able to grow with formate as the sole energy source and it also replaced the CO_2 requirement for growth on CH_3OH. This strain grew with methoxylated aromatic compounds including vanillate, syringate, and ferulate, and growth was proportional to the number of moles of methoxyl groups present (Genthner and Bryant, 1987).

(c) *SYNTROPHOCOCCUS SUCROMUTANS*

Monobenzenoids are commonly found within the diet of ruminants and nonruminants as derivatives of cinnamic and benzoic acid ester-linked to carbohydrates.

Bacteria within the gastrointestinal tract have been shown to modify phenolic monomers by dehydroxylation, saturation of the propenoate side chain of cinnamate derivatives, decarboxylation of benzoic acid derivatives, and demethoxylation of methyl-ether derivatives. Both *Acetobacterium woodii* and *Eubacterium limosum* have the ability to cleave methyl-ether linkages of monobenzenoids and utilize the methoxyl group as an energy source. These organisms are common in anaerobic digester sludges and fresh water sediments. Because *E. limosum* had been isolated from sheep on an unusual molasses-based diet, an attempt was made to isolate and characterize the predominant rumen bacteria on more normal diets.

Syntrophococcus sucromutans was isolated by Krumholz and Bryant (1986a) from the rumen fluid of an alfalfa hay-fed steer. Most probable number (MPN) estimates indicated that a mean of 4.3×10^7 and 5×10^6 bacteria per milliliter of rumen fluid demethoxylated ferulate and syringate, respectively. Strain S195 (DSM 3224) was isolated after further enrichment from an MPN tube of the highest tube showing demethoxylation of syringate. *S. sucromutans* is an anaerobic, Gram-negative, chemoorganotrophic, nonmotile coccus $1-1.3\mu$m in diameter. Its energy metabolism is unique in that it requires two energy sources for growth; one, an electron donor system can be any one of several sugars or glycosides, and the other, an electron acceptor system can be formate, acrylate side chains of phenolics, an H_2-using methanogen, or any one of several methoxybenzenes which do not directly serve as electron acceptors but, because of stoichiometric considerations, require that some CO_2 or HCO_3^- be reduced in the formation of acetate. It produces acetate, hydroxybenzenoids, and/or propionate side chains from the acrylate side chains from the various electron acceptors, and acetate from the donor (Doré and Bryant, 1990).

The phylogenetic relatedness of *S. sucromutans* to other bacteria was inferred by comparative analysis of 16S rRNA sequencing (Doré, 1989a). It belongs in the Gram-positive phylum in a subdivision with *Roseburia cecicola* (90% sequence similarity), *Butyrivibrio fibrisolvens* strain A38 (88%), and *Lachnospira multipara* strain 40 (89%). This subdivision contains species with either Gram-positive or -negative ultrastructure. It is less closely related to another subdivision of the Gram-positive phylum that includes organisms with Gram-negative ultrastructure, i.e., the *Sporomusa–Megasphaera–Selenomonas* group, and it is very distantly related to the *Syntrophomonas* subdivision of the Gram-positive phylum with similar diversity in species ultrastructure (Zhao et al., 1993).

The requirement for a large amount of rumen fluid and narrow temperature range (30–44°C) suggest that it is a bacterium selected by the rumen environment. The rumen fluid requirement was investigated by Doré and Bryant (1989b). It was found that isomers of octadecenoic acid (C18:1) were highly stimulatory resulting in growth yields similar to those obtained with 30% rumen fluid or crude phosphatidylcoline inclusion in the medium.

(d) CLOSTRIDIUM PFENNIGII

This organism is also capable of fermenting the methoxyl groups of monobenzenoids producing the corresponding hydroxybenzenoid and butyrate. *C. pfennigii* strain V5-2^T (DSM 3222) was isolated after enrichment on vanillate from rumen fluid of a steer fed alfalfa hay and grain in a 70 : 30 ratio (Krumholz and Bryant, 1985). The cells are anaerobic, Gram-positive, motile, slightly curved rods 0.4 μm wide and 1.6–3.5 μm long. Spores are oval, subterminal to terminal. Electron microscopy reveals a single laterally to subterminally attached flagellum.

Strain V5-2^T grows only with monobenzenoids, pyruvate, or CO added to the basal medium. Butyrate and hydroxybenzenoids are the only organic products of methoxybenzenoids. CO is catabolized to acetate and butyrate. Pyruvate is fermented to acetate. Demethoxylation of vanillate to protocatechuate, vanillin to protecatechuic aldehyde and protocatechuate, ferulate to caffeate and hydrocaffeate, and syringate and 3,4,5-trimethoxybenzoate to gallate are the catabolic reactions carried out.

(e) PEPTOSTREPTOCOCCUS PRODUCTUS

P. productus is one of the most numerous Gram-positive bacterial species in the human bowel (Varel et al., 1974) and its presence in sewage sludge (7.5 × 10^6 cells/ml) is not unusual. These organisms are isolated and routinely maintained on media containing sugars as energy sources. However, Lorowitz and Bryant (1984) demonstrated for the first time that a sewage strain (U-1) was capable of growth with CO as energy source and that the production of acetate from CO_2 during growth with CO had a similar stoichiometry to that reported for other acetogenic CO utilizers. Strain U-1 was shown to grow more rapidly ($T_d = 1.5$ h) with much higher levels of CO (90%) than that reported for other species of acetogenic CO utilizers. The ability of two other strains, B43 (isolated from the calf rumen; Bryant et al. 1958) and 2 (isolated from human feces), to grow on 50 or 25% CO suggested that this phenomenon might be widespread in natural ecosystems. However, growth with H_2-CO_2 might be more ecologically significant.

This was the first report of a mesophilic anaerobic acetogenic coccus able to utilize H_2-CO_2 or CO as energy source. This ability suggests that it fits phylogenetically in with the "Clostridium group" of Tanner et al. (1982).

12.3.2 Proton-Reducing Acetogens

(a) SYNTROPHOMONAS WOLFEI

A butyrate-catabolizing bacterium was isolated from sewage digester sludge, aquatic sediments, and rumen digesta in co-culture with an H_2-utilizing bacterium such as a methanogen or sulfate reducer (McInerney et al., 1979; 1981a). The isolation, in co-culture with a single H_2-using species, of a bacterium that catabolizes the normal monocarboxylic, saturated, 4-8C fatty acids with acetate $+H_2$ or acetate, propionate $+ H_2$ as products was the first direct evidence for the existence of a nomethanogenic bacterium that anaerobically degrades any of the fatty acids in the absence of light, sulfate, or nitrate and where protons serve as the electron acceptor. Isoheptanoate was also degraded to acetate, isovalerate $+ H_2$. Recently, *S. wolfei* has been grown in pure culture using crotonate, a monoenoic 4C carboxylic acid, as the carbon source (Beaty and McInerney, 1987). This substrate allows the butyrate dehydrogenation step in the β-oxidation pathway to be bypassed, yielding more free energy for growth.

Syntrophomonas wolfei is a nonsporing, Gram-negative, slightly helical rod 0.5–1.0 by 2.0–7.0 μm in size. Cells possess two to eight flagella that are laterally inserted in a linear fashion on the concave side of the cell about 130 nm apart. Under most conditions, cells usually exhibit only sluggish, twitching motility. Cells have also been shown to accumulate poly-β-hydroxyalkanoates as storage material (McInerney et al., 1981a; Amos and McInerney, 1989). *S. wolfei* was isolated from the rumen fluid of a steer fed corn grain and corn silage in a ratio of 85 : 15 after 3 months of weekly serial transfers in enrichment medium containing 18 mM sodium butyrate (McInerney et al., 1981b). The major H_2-utilizing organism in the enrichment was isolated and shown to be a strain of *Methanosarcina* based on morphology and substrates utilized for CH_4 production ($H_2 + CO_2$, acetate, methanol, methylamine, and trimethylamine). This was unusual because in other methanogenic ecosystems *Methanospirillum hungatei* is usually the main H_2-using methanogen and *Methanothrix (Methanosaeta) soehngenii* is probably the main acetate utilizer. In these enrichments, the main ruminal H_2-utilizing methanogen, *Methanobrevibacter ruminantium*, does not compete with H_2 and acetate-utilizing *Methanosarcina* perhaps because its affinity for H_2 is much less under the very slow growth rates necessary to utilize the H_2 produced from butyrate by *S. wolfei*—this introduces the concept of growth coupling in syntrophs. In addition, it is possible that under the slow growth rates necessary, the maintenance energy requirement of *M. ruminantium* is too high. However, these possibilities require further study.

Phylogenetically, *Syntrophomonas* and similar fatty acid-β-oxidizing syntrophs, regardless of cell-wall ultrastructure (Gram negative or Gram positive) or spore formation (presence or absence), belong in a separate subdivision in the Gram-positive phylum quite distinct from any other eubacteria known (Zhao, et al. 1993).

12.3.3 Other Unusual Acetogens

(a) EUBACTERIUM OXIDOREDUCENS

Trihydroxybenzenoids such as gallate and phloroglucinol are present in free or combined form in a large variety of compounds within the plant kingdom. These include flavonoids, tannins, lignin precursors, and their degradation products and, therefore, are common constituents of the ruminant diet. Ruminal bacteria which are able to degrade phloroglucinol include *Streptococcus bovis* and a *Coprococcus* sp. (Tsai and Jones, 1975) which produce acetate and CO_2 as major products of phloroglucinol catabolism. *Coprococcus* sp. at least partially degrades the flavonols, quercetin, and rhamnetin (Tsai et al., 1976). Although *Eubacterium oxidoreducens* has not been shown to utilize the acetyl-CoA pathway and does not reduce protons, it has a unique acetogenic metabolism which makes its inclusion in this section of the chapter valuable and important.

Eubacterium oxidoreducens was isolated by Krumholz and Bryant (1986b) from the rumen fluid of a hay-fed steer, and strain G41 (DSM 3217) appeared to be the predominant organism (4.3×10^3/ml) in the rumen capable of anaerobic gallate and pyrogallol degradation to nonaromatic products. Based on MPN estimations anaerobic decarboxylation and reductive dehydroxylation of gallate to resorcinol was carried out by other, as yet, unknown organisms. Strain G41 is a Gram-positive, nonmotile, obligately anaerobic, nonsporing, curved rod 0.45 by 1.5–2.2 μm in size.

E. oxidoreducens is a strictly anaerobic chemoorganotroph which requires formate or H_2 as electron donor to catabolize approximately equimolar gallate, pyrogallol, phloroglucinol, or quercetin to acetate, butyrate, and sometimes CO_2; 3,4-dihydroxylphenylacetate is also produced from quercetin. Crotonate is used without formate or H_2 but no other compounds were used with or without formate present (Krumholz and Bryant, 1986b). Because it is unable to use sugars, lactate, amino acids, or peptides for growth, it does not fit any of the described species of *Eubacterium*. However, based on morphology, Gram stain, fermentation end products, and mol% G + C, it is placed in this genus taxonomically.

12.4 Metabolic Activities Involving Acetogenesis

12.4.1 Syntrophic Acetogenesis of Fatty Acids

Under anaerobic conditions in natural and other habitats (aquatic and marine sediments, anoxic soils, marshes, peat bogs, rice paddies, sewage plants, waste digestors), microbial decomposition of degradable organic matter to inorganic products is primarily accomplished by microbial oxidations linked to reduction of protons, sulfur, or carbon atoms as electron sinks. In the absence of light or inorganic hydrogen acceptors, biodegradable organic matter can be completely fermented to CH_4 and CO_2 with only a relatively small cell yield. About 90% of the energy available in the substrate is retained in the gaseous product, CH_4. Substrate conversion is a multistep process involving many different kinds of interacting microbial species, mainly bacteria (Bryant, 1979; McInerney and Bryant, 1981, Dolfing, 1988).

The scheme which best fits our current knowledge on methane fermentation is illustrated in Figure 12.2. In the first stage, fermentative bacteria hydrolyze and ferment carbohydrates, protein, and lipids with the production of acetate, H_2 plus CO_2, propionate and longer-chain fatty acids. Propionate, longer-chain fatty acids, some organic acids, and alcohols are subsequently degraded by a second group of bacteria, called the obligate H_2-producing (proton-reducing) acetogenic bacteria. Finally, methanogens reduce CO_2 to CH_4 using H_2 produced by other bacteria and also cleave acetate to CH_4 and CO_2. Quantitatively, some 70% of the methane produced is derived from the methyl group of acetate. A fourth group of bacteria is able to carry out acetogenic hydrogenation producing acetate from CO_2 plus H_2. As mentioned previously, only a partial methane fermentation occurs in the rumen and gastrointestinal tract due to the short retention time, and, thus, the fatty acids are usually not degraded (Fig. 12.2). In this section of the review we examine and emphasize the contribution of the obligate proton-reducing bacteria to the oxidation of propionate and longer-chain fatty acids in methanogenic anaerobic ecosystems. Under the conditions where sulfate is not limiting, sulfate reducers can exploit these substrates as energy sources utilizing sulfate as electron acceptor and resulting in sulfide production. Several excellent reviews have been published on this subject and should be consulted for further information beyond the scope of the present review (Bryant, 1979; McInerney and Bryant, 1981; Dolfing, 1988; Widdell, 1988; Vogels et al., 1988).

Although only a few species of obligate proton-reducing bacteria have been isolated and studied, this group is composed of many bacterial species, often with different energy-source specificity. As a group, they degrade propionate and long-chain fatty acids, alcohols, aromatic, and organic acids with the production of acetate and H_2, and in the case of odd-numbered C sources, CO_2. Historically, the isolation of the S organism from "*Methanobacillus omelianskii*" was the first documentation of a species of this group (Bryant et al., 1967). This "methanogen" was thought to oxidize ethanol to acetate and to reduce CO_2 to CH_4. However, this fermentation was shown to be carried out by a syntrophic association of two bacterial species. The S-organism catabolized ethanol to acetate and H_2, whereas the methanogen used the H_2 to reduce CO_2. The formation of

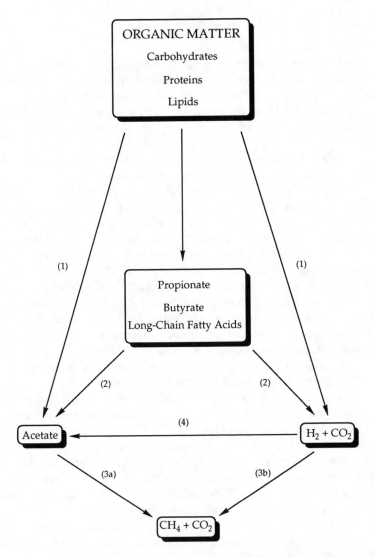

Figure 12.2. Three-stage scheme for the complete anaerobic degradation of organic matter showing the general sequence and major metabolic groups of bacteria: (1) fermentative bacteria; (2) obligate proton-reducing, acetogenic bacteria; (3) acetate decarboxylation (3a) and reductive CH_4 formation (3b) by methanogenic bacteria; and (4) acetogenic hydrogenation by acetogenic bacteria (McInerney and Bryant, 1981). In the rumen, short-chain volatile fatty acids accumulate and are absorbed and utilized by the animal as energy sources. A partial CH_4 fermentation occurs in certain compartments of the gastrointestinal tract such as the rumen, cecum, or large bowel, and CH_4 is produced by CO_2 reduction (3b).

acetate and H_2 from ethanol is energetically unfavorable unless H_2 concentrations are kept low by the H_2 utilizer. The significance of this finding was the probability that alcohols other than methanol and fatty acids other than formate and acetate were not catabolized by methanogens but by an intermediate group of bacteria. This laid the foundation for the discovery of other obligate proton-reducing bacteria.

Strains of *Desulfovibrio desulfuricans* and *D. vulgaris* produce H_2 from lactate and ethanol when grown without sulfate in the presence of H_2-utilizing methanogens (Bryant et al., 1977; McInerney and Bryant, 1981). Lactate is converted to acetate, CO_2, and H_2, and ethanol is degraded to acetate and H_2 if the H_2-utilizing methanogen rapidly reduces CO_2 with the H_2 and allows growth of the acetogen on these substrates. When *D. desulfuricans* is grown in the presence of *M. barkeri*, which utilizes both acetate and H_2 for methanogenesis, lactate is completely degraded to CO_2 and CH_4. Propionate and longer-chained fatty acids are ecologically much more important intermediates in anaerobic degradation than lactate or ethanol, but species that catabolize them have only become documented over the last 14 years. It took until 1979 before McInerney, Bryant, and Pfennig were able to isolate a fatty acid-oxidizing, obligate proton-reducing bacteria in co-culture with a H_2-consuming partner. The bacterium, *Syntrophomonas wolfei*, oxidized butyrate in syntrophic co-culture with either a H_2-utilizing sulfate reducer or methanogen. Boone and Bryant (1980) reported on a syntrophic association of an anaerobic propionate oxidizer in co-culture with a H_2 utilizer. The list of fatty acid-oxidizing, obligate, proton-reducing bacteria has grown slowly and was reviewed by Mackie et al. (1991). There are several reasons for this slow progress in isolating this group of organisms. Growth rates are extremely low ($\mu = 0.1$–0.3 d^{-1}) for these fastidious anaerobes. Chemical methods for removing hydrogen from growing cultures have been ineffective in allowing growth of the obligate H_2 producers in the absence of H_2 consumers. The most successful approach has been to couple the growth of a H_2-utilizing *Desulfovibrio* to the desired H_2 producer, allowing isolation of the obligate syntrophic combination. In some cases, it is possible to replace the *Desulfovibrio* in sulfate-free medium with *Methanospirillum hungatei*.

There is little information available on biochemical pathways in the fatty acid-oxidizers, primarily because they are difficult to grow in pure culture. Enrichment cultures that converted propionate to CH_4 and acetate converted the ^{14}C-label carboxyl group exclusively to CO_2, whereas acetate that accumulated was free of label. Incubation of [2-^{14}C]- and [3-^{14}C]-propionate resulted in the production of labeled acetate with equal radioactivity in the methyl and carboxyl regardless of whether [2-^{14}C] or [3-^{14}C] label was used (Koch et al., 1983). This is consistent with the randomizing pathway as the route for anaerobic propionate degradation. The equation for this reaction is as follows:

$$CH_3CH_2COO^- + 3H_2O \rightarrow CH_3COO^- + HCO_3^- + H^+ + 3H_2, \Delta G_0' + 76.1 \text{ kJ}$$

The anaerobic oxidation of butyrate or even-numbered C fatty acids in *S. wolfei* has been postulated to proceed via β oxidation (McInerney et al., 1981a) as follows:

$$CH_3CH_2CH_2COO^- + 3H_2O \rightarrow 2 CH_3COO^-$$
$$+ H^+ + 2H_2, \quad \Delta G_0' = +48.1 \text{ kJ}$$

or odd-numbered C fatty acids such as valerate:

$$CH_3CH_2CH_2CH_2COO^- + 2H_2O \rightarrow CH_3COO^- + CH_3CH_2COO^- + H^+$$
$$+ 2H_2, \quad \Delta G_0' = +48.1 \text{ kJ}$$

Detailed labeling studies have not been carried out. However, large batches of *S. wolfei* have been grown in the presence of *M. hungatei* and, following selective lysis of *S. wolfei* by lysozyme treatment, analyzed for enzymes of the β-oxidation pathway (Wofford et al., 1986). Cell extracts of *S. wolfei* had high specific activities of acyl-CoA dehydrogenase, enoyl-CoA hydratase, L-3-hydroxyacyl-CoA dehydrogenase, and 3-ketoacyl-CoA thiolase. Fatty acid activation most likely occurs by transfer of CoA from acetyl-CoA because cell extracts had high CoA transferase activity but no detectable acyl-CoA synthetase activity. These workers were able to grow pure cultures of *S. wolfei* on crotonate (Beaty and McInerney, 1987). Crotonate was catabolized by a disproportionation reaction in which crotonate is oxidized to 2 acetate and reduced to butyrate and some caproate. Butyrate catabolism occurred only when *S. wolfei* was reassociated with the H_2 utilizer. Crotonate catabolism would bypass the butyrate dehydrogenation step and increase the amount of free energy available for growth. All of the fatty acid β-oxidizers so far studied can be grown in pure culture on crotonate (Zhao et al. 1993). The degradation of acetate does not normally occur to any significant extent in the rumen (Opperman et al., 1961), unless unusual conditions exist that lead to greatly reduced turnover of the rumen content (Rowe et al., 1979). Under these unusual conditions, *S. wolfei* may become more important in the overall rumen fermentation, and more acetate would also be produced from butyrate and longer-chained fatty acids.

12.4.2 O-Demethylation of Benzenoids

Demethoxylation of phenolic acids occurs in the rumen of sheep after infusion of vanillate and the recovery of predominantly catechol (2-hydroxyphenol) with a small amount of guaiacol (2-methoxyphenol) in the urine (Martin, 1982). When vanillate was incubated anaerobically with a mixed bacterial suspension from the rat cecum, the same products (catechol and guaiacol) together with protocatechuate were found (Scheline, 1966). In similar studies, Scheline (1978) demonstrated that vanillin was catabolized by two pathways: by oxidation to vanillate and by direct reduction to 4-methylcatechol and 4-methylguaiacol.

Syntrophococcus sucromutans appears to be the first known case of an anaerobic bacterium that requires carbohydrate as electron donor and formate, methoxybenzenoids (because of CO_2 requirement), acrylate side chains of benzenoids, or a hydrogenotrophic bacterium as exogenous electron acceptor systems in order to grow effectively (Krumholz and Bryant, 1986a). Electron donors include pyruvate, glucose, fructose, galactose, maltose, cellobiose, lactose, arabinose, ribose, xylose, salicin, and esculin. Electron acceptors (with the major organic products in parentheses) are as follows: formate (acetate), caffeate (hydrocaffeate), ferulate (caffeate, hydrocaffeate, and acetate), syringate and 3,4,5-trimethoxybenzoate (gallate and acetate), vanillin (protocatechuic aldehyde, protocatechuate, and acetate), and vanillate (protocatechuate and acetate). The following reactions summarize the results obtained:

$$\text{Fructose} + \text{formate}^- \leftrightarrows 3.25\,\text{acetate}^- + 0.5\,HCO_3^- + 2.75\,H^+,$$
$$\Delta G_0' = -335.6\,\text{kJ/mol};$$
$$\text{Fructose} + \text{caffeate}^- + H_2O \leftrightarrows 2.75\,\text{acetate}^- + \text{hydrocaffeate}^-$$
$$+ 0.5\,HCO_3^- + 3.25\,H^+, \qquad \Delta G_0' = -361.2\,\text{kJ/mol}$$
$$\text{Fructose} + 1.33\,\text{syringate}^- + 1.33\,HCO_3^- \leftrightarrows 5\,\text{acetate}^- + 1.33\,\text{gallate}^-$$
$$+ 3.67\,H^+, \qquad \Delta G_0' = -484.7\,\text{kJ/mol}$$

Strain 195 grew well in coculture with *Methanobrevibacter smithii* suggesting that it catabolized little fructose unless the methanogen was present to utilize the H_2 as described in the following reactions:

$$\text{Fructose} + 4\,H_2O \leftrightarrows 2\,\text{acetate}^- + 2\,HCO_3^- + 4\,H_2 + 4\,H^+,$$
$$\Delta G_0' = -206.3\,\text{kJ/mol4}\,H_2$$
$$4H_2 + HCO_3^- + H^+ \leftrightarrows CH_4 + 3\,H_2O,$$
$$\Delta G_0' = -135.6\,\text{kJ/mol}$$
$$\text{Fructose} + H_2O \leftrightarrows 2\,\text{acetate}^- + HCO_3^- + 3H^+ + CH_4,$$
$$\Delta G_0' = -341.9\,\text{kJ/mol}$$

S. sucromutans produced only acetate and hydroxylated benzenoids as organic products in its energy catabolism producing up to 2 mol acetate and CO_2 per mol fructose. Formate or methoxyl groups of benzenoids could only be used for growth and acetate production when carbohydrate was present as electron donor.

H_2, CO, and CH_3OH were not utilized under any condition in contrast to most other anaerobic acetogenic bacteria.

The main function of *S. sucromutans* in the rumen is unclear, but methanogens maintain low H_2 concentrations (<1 μm) except shortly after feeding (Robinson et al., 1981; Smolenski and Robinson, 1988). Thus, it is likely to be involved in carbohydrate catabolism with sugars as electron donors and methoxy monobenzenoids as acceptors. It is probably not important in the reduction of acrylate side chains (Chesson et al., 1982) or formate catabolism (Lovly et al., 1984) because of the many other species competing for these substrates.

The metabolism of methoxybenzenoids by *C. pfennigii* producing only hydroxybenzenoids and butyrate are described in the following reactions (Krumholz and Bryant, 1985):

Vanillate$^-$ + 0.2 HCO_3^- $\leftrightarrows$ protocatechuate$^-$ + 0.3 butyrate$^-$ + 0.1 H^+
Vanillin + 0.3 HCO_3^- $\leftrightarrows$ 0.75 protocatechuic aldehyde + 0.25 protocatechuate$^-$
    + 0.325 butyrate$^-$ + 0.275 H^+
Ferulate$^-$ + 0.5 H_2O $\leftrightarrows$ 0.5 caffeate$^-$ + 0.5 hydrocaffeate$^-$ + 0.25 butyrate$^-$
    + 0.25 H^+
Syringate$^-$ + 0.4 HCO_3^- $\leftrightarrows$ gallate$^-$ + 0.6 butyrate$^-$ + 0.2 H^+
Trimethoxybenzoate$^-$ + 0.6 HCO_3^- $\leftrightarrows$ gallate$^-$ + 0.9 butyrate$^-$ + 0.3 H^+

The growth yield of *C. pfennigii* was similar on all five substrates shown above (9.7–10.9 g dry cells per mol substrate) with the exception of ferulate (13.6 g dry cells per mol substrate), suggesting that the reduction of the acrylate side chain provided additional energy for growth. Growth yields were not related to the number of methoxy groups on the substrates. Similar growth yields for *A. woodii* grown on methoxylated benzenoids have been reported (Bache and Pfennig, 1981).

12.4.3 *Acetogenic Metabolism by* Syntrophococcus sucromutans

Experiments were carried out to elucidate the pathways involved in the synthesis of acetate from C_1 compound by *S. sucromutans* (Doré and Bryant, 1990). Cell extracts were shown to possess all the enzymes of the tetrahydrofolate pathway, as well as carbon monoxide dehydrogenase (CODH) at levels similar to those of other acetogens using the acetyl-CoA pathway such as *C. thermoaceticum*, *C. formicoaceticum*, *A. woodii*, and *C. thermoautotrophicum*. However, formate dehydrogenase (FDH) could not be detected in cell extracts, whether formate or a methoxyaromatic was used as electron acceptor for growth of cells on cellobiose. Conversion of ^{14}C-formate to $^{14}CO_2$ was detected, but labeled formate predominantly labeled the methyl group of acetate. $^{14}HCO_3$ readily exchanged with the carboxyl group of pyruvate but not with formate, and both labeled $^{14}HCO_3$ and [1-^{14}C] pyruvate predominantly labeled the carboxyl group of acetate. No $^{14}CO_2$ was formed from O-[methyl-^{14}C] vanillate demethylation and acetate produced was position labeled in the methyl group. From this labeling pattern and specific activities of products formed, complete synthesis of acetate was achieved from the carboxyl group of pyruvate (C_1 of acetate) and the methoxyl group of vanillate (C_2 of acetate).

Several interesting features of C_1 metabolism in *S. sucromutans* are worth noting. CODH activity was good in cell extracts of both cellobiose-formate [2.91 μmol/min^{-1} (mg protein)$^{-1}$] and cellobiose-vanillin (2.31) grown cells. This is further strong evidence for the presence of the Wood pathway and that all of the enzymes, except the methyltransferase system, were similar when the C_1 electron acceptor compound for the methyl group of acetate was supplied by O-demethylation regardless of compound. No attempt was made to demonstrate the involvement of a corrinoid enzyme together with CODH in the transmethylation system. CO could possibly serve as the electron donor in washed cells or cell extracts, or possibly in the absence of carbohydrate.

FDH activity has been detected in many other acetogens (Ljungdahl, 1986) and in control assays with *E. oxidoreducens* but not in *S. sucromutans*. This was confirmed by a lack of rapid exchange between ^{14}C-formate and CO_2 as would occur in an organism having an active FDH. *A. woodii* has an active FDH and can totally synthesize acetate from either formate or H_2-CO_2 (Tanner et al., 1978). Furthermore, only a small amount of $H^{14}CO_3$ was recovered in formate as would be expected if FDH was present, allowing rapid formate–CO_2 exchange. This is clearly an area in which C_1 metabolism by *S. sucromutans* differs from other organisms possessing the Wood pathway.

The catabolism of O-[methyl-^{14}C] vanillate by washed cells of *S. sucromutans* with almost all of the label (97%) recovered in the methyl group of acetate is clearly unique. As it requires an organic electron acceptor system such as vanillate in order to metabolize carbohydrates or pyruvate, it clearly requires a catabolic pathway to reduce its pool of reducing equivalents. It is unable to effectively use H_2 as an electron-sink product of carbohydrate or pyruvate unless grown with a methanogen. Most other acetogens can use the O-methyl groups of vanillate and similar compounds as sole sources of energy to produce acetate and CO_2 (Bache and Pfennig, 1981; Frazer and Young, 1986; Yamamoto et al., 1983). One reason for this is its lack of FDH activity. The O-demethylating system might transfer the methyl group to CH_3-THF and then, via transmethylase, to CODH and, hence, to the methyl group of acetyl-CoA. Lack of FDH would stop any formate formed by reversal of the THF pathway and also from reversal of CODH activity producing CO_2. Another reason for failure to produce CO_2 from the methyl group of vanillate may be methyl transfer and acetate formation requiring intact cell membranes. Further studies are required before a complete understanding of the unusual formate and methoxyl "respiration" and syntrophic growth with a formate-H_2-utilizing methanogen during growth with carbohydrate as the electron donor is obtained.

12.4.4 Flavonoid Degradation

Flavonoids, tannins, lignin precursors (monomers), and their intermediate degradation products are common constituents of the ruminant diet. The fate of these aromatic compounds in anaerobic environments is of increasing interest and importance (Young and Fraser, 1987). Tarvin and Buswell (1934) first reported on the anaerobic degradation of aromatic compounds during methane fermentation and later workers confirmed and extended this work (Clark and Fina, 1952; Fina and Fiskin, 1960; Nottingham and Hungate, 1969). The release of aglycones from many plant secondary compounds can result in the release of compounds, some of which are potent toxins.

The microflora of the bovine rumen has been shown to degrade the flavonoids, rutin, quercetin, naringin, and hesperidin to water-soluble products under anaerobic conditions (Simpson et al., 1969). Subsequently, a *Butyrivibrio* sp. was isolated from rumen contents that cleaved the heterocyclic ring of naringin and rutin producing acetate, butyrate, H_2, CO_2, phloroglucinol and 3,4-dihydroxyphenylacetate (Cheng et al., 1969; Krishnamurty et al., 1970; Cheng et al., 1971). The organism did not metabolize the aglycone, quercetin, or phloroglucinol. Bacteria isolated from the rumen, which slowly degrade phloroglucinol, include *Streptococcus bovis* and *Coprococcus* sp. (Tsai and Jones, 1975). The major products of phloroglucinol catabolism are acetate and CO_2. *Coprococcus* sp. at least partially degrades the flavonoids quercetin and rhamnetin (Tsai et al., 1976).

Trihydroxybenzenoids such as gallate and phloroglucinol are present in free or combined form in a large variety of compounds within the plant kingdom. Krumholz and Bryant (1986b) succeeded in isolating *Eubacterium oxidoreducens* as the most numerous organism capable of rapidly degrading these compounds to water-soluble products. Strain G41 is a strictly anaerobic organotroph requiring formate or H_2 as electron donor to catabolize approximately equimolar gallate, pyrogallol, phloroglucinol, or quercetin to acetate, butyrate, and sometimes CO_2, (3,4-dihydroxyphenylacetate is also produced from quercetin). No exogenous electron donor was required for catabolism (fermentation) of crotonate. These reactions are summarized in the following reactions:

$$\text{Gallate}^- + H_2 + 3\,H_2O \leftrightharpoons 2\,\text{acetate}^- + 0.5\,\text{butyrate}^- + HCO_3^- + 2.5\,H^+,$$
$$\Delta G^0 = -184\,\text{kJ}$$

$$\text{Pyrogallol} + H_2 + 2\,H_2O \leftrightharpoons 2\,\text{acetate}^- + 0.5\,\text{butyrate}^- + 2.5\,H^+,$$
$$\Delta G^0 = 183\,\text{kJ}$$

Phloroglucinol + H_2 + $2 H_2O \leftrightarrows 2$ acetate$^-$ + 0.5 butyrate$^-$ + 2.5 H^+,
$$\Delta G^0 = -183 \text{ kJ}$$
Quercetin + H_2 + $5 H_2O \leftrightarrows 2$ acetate$^-$ + 0.5 butyrate$^-$ + HCO_3^-
$$+ \text{ 3,4-dihydroxyphenylacetate}^- + 4.5 H^+$$
Crotonate$^-$ + $H_2O \leftrightarrows$ acetate$^-$ + 0.5 butyrate$^-$ + 0.5 H^+, $\Delta G^0 = -51$ kJ

A requirement for exogenous electron donors such as H_2 or formate (in approximately a 1 : 1 ratio) for catabolism of these compounds is unusual.

The pathway for the anaerobic catabolism of gallic acid by *E. oxidoreducens* was studied in whole cells and cell-free systems (Krumholz et al., 1987; Krumholz and Bryant, 1988). The proposed pathway for degradation of gallate and phloroglucinol is summarized in Figure 12.3. The first step in the degradation of gallate is decarboxylation followed by conversion of pyrogallol to phloroglucinol. A phloroglucinol reductase (NADP linked) is responsible for reduction of phloroglucinol to dihydrophloroglucinol. Dihydrophloroglucinol hydrolase catalyzes ring cleavage at the keto group yielding 3-hydroxy-5-oxohexanoate. This six-carbon ring cleavage product is hydrolyzed to 3-hydroxybutyryl-CoA, yielding acetate and butyrate as end products via acetoacetyl-CoA and crotonyl-CoA, respectively. Both FDH and hydrogenase were present in cell extracts and both were NADP linked. It is likely that oxidation of formate or hydrogen is coupled indirectly to phloroglucinol reduction. Many features of this pathway are similar in *Pelobacter acidigallici* isolated from sewage and freshwater sediments (Schink and Pfennig, 1982; Brune and Schink, 1990) except that no H_2 or formate are required and butyrate is not produced. In addition, Brune and Schink (1990) showed that the mechanism of the conversion of pyrogallol to phloroglucinol was not an isomerization involving 1,2,3,5-tetrahydroxybenzene but involves intermolecular transfer of the hydroxyl moiety from the cosubstrate (1,2,3,5-tetrahydroxybenzene) to the substrate (pyrogallol), thus forming the product phloroglucinol and regenerating the cosubstrate.

12.5   Factors Affecting Acetogenesis in the Rumen

Our understanding of the ecology of acetogens and of the factors that influence their interaction and competition with other H_2-utilizing bacteria in the rumen is limited. Further information on this aspect in the gastrointestinal tract of termites and other insects and in the hindgut of humans is presented in chapters 11 and 13, respectively (Breznak, 1994; Wolin and Miller, 1994).

12.5.1   *Methanogenesis*

In the rumen, methanogenesis from CO_2 reduction with H_2 dominates terminal electron flow, and little or no acetogenesis from H_2 and CO_2 can be detected in

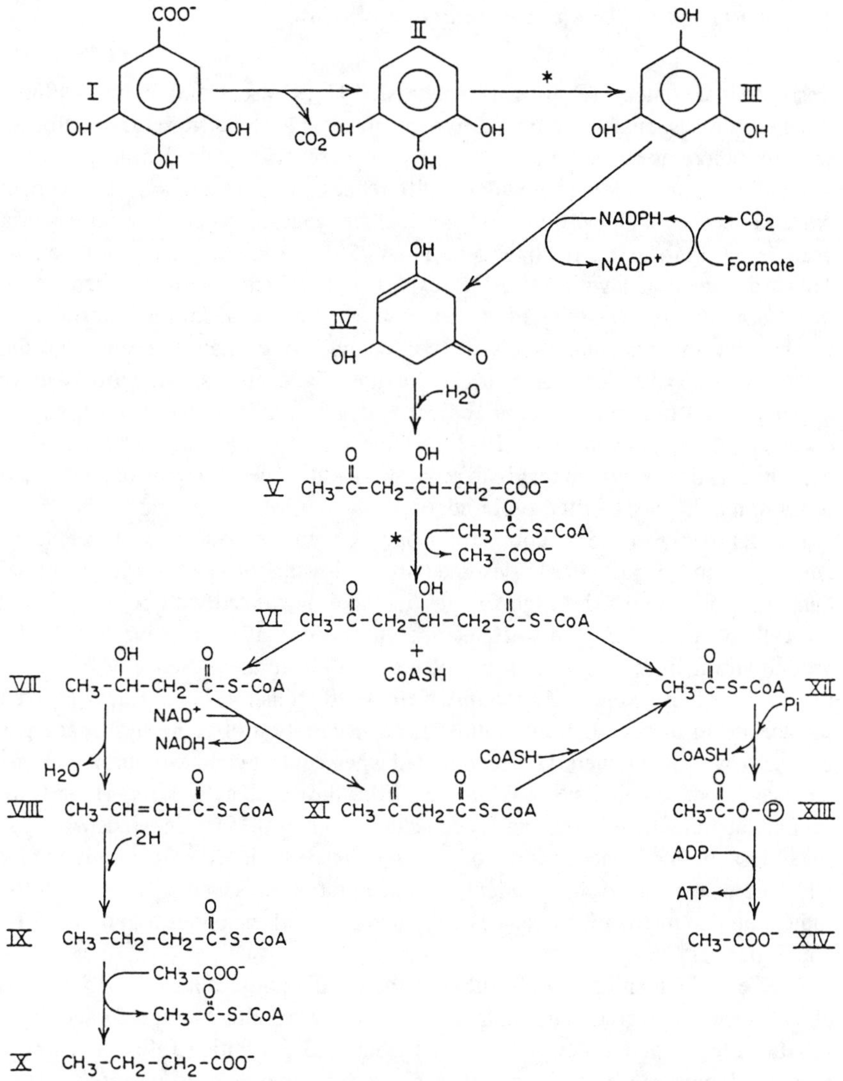

Figure 12.3. Proposed pathway for metabolism of gallate and phloroglucinol by *Eubacterium oxidoreducens* [from Krumholz et al. (1987)]. The compounds represented are gallate (I), pyrogallol (II), phloroglucinol (III), dihydrophoroglucinol (IV), 3-hydroxy-5-oxohexanoic acid (V), 3-hydroxy-5-oxohexanoyl-CoA (VI), 3-hydroxybutyryl-CoA (VII), crotonyl-CoA (VIII), butyryl-CoA (IX), butyrate (X), acetoacetyl-CoA (XI), acetyl-CoA (XII), acetyl phosphate (XIII), and acetate (XIV). Asterisk denotes steps that are still not well understood.

fresh ruminal contents (Prins and Lankhorst, 1977; Breznak and Switzer, 1986). Washed cell suspensions of mixed rumen and cecal bacteria from pigs incubated with cellobiose as substrate in the absence or presence of hydrogen in headspace gas indicated that there were major differences in H_2 utilization. The rumen bacteria used a major portion of H_2 for methanogenesis, whereas cecal bacteria used ~ 33 % for extra methanogenesis and 29% for extra acetate production. This was confirmed with ^{13}C-NMR experiments which clearly showed $^{13}CO_2$ incorporation into acetate and butyrate with cecal bacteria. Propionate was also labeled, but this was most likely the consequence of exchange reactions of the carboxyl group with $^{13}CO_2$. In contrast, the rumen incubation showed low labeling in acetate and butyrate with enrichment values being 10 times lower than pig cecal incubation (Stevani et al., 1991; DeGraeve et al., 1990). Furthermore, these experiments demonstrated that both processes (acetogenesis and methanogenesis) occur simultaneously in the pig hindgut.

In order to investigate the competition between ruminal acetogens and methanogens, Wilfond et al. (1994) designed the following *in vitro* study. Ruminal contents from cattle fed a high-forage diet were inoculated with between 0 and 10^8 cells of *A. ruminis* 190A4 (Greening and Leedle, 1989) and incubated with ground alfalfa under a 100% N_2 gas phase for 43 h. In the presence of BES (2.5 mM), and when levels of *A. ruminis* were $> 10^6$ cells/ml, H_2/CO_2 was utilized for acetate formation and an equilibrium concentration of H_2 in headspace gas of 0.48% was maintained. H_2 accumulated when control vials without *A. ruminis* were assayed. When BES was omitted from the incubation vials, H_2 did not accumulate and the equilibrium H_2 concentration was 0.195% in headspace gas, regardless of the concentration of *A. ruminis* cells added. Accumulation of CH_4 and acetate were independent of concentration of added *A. ruminis* cells, indicating that methanogens successfully outcompeted acetogens for H_2 (Wilfond and Schaeffer, 1990).

Leedle and Greening (1988) followed the population profiles of methanogens and acetogens postprandially in the rumen of cattle fed either a high-forage (75% alfalfa : 25% concentrate) or high-concentrate (25% alfalfa : 75% concentrate) diet. Methanogenic bacterial populations were present at concentrations of (4–8) $\times 10^8$ per g of ruminal content on the high-forage and high-concentrate diets. The acetogenic bacteria were present at concentrations of between 2×10^8 and 12×10^8 per g of ruminal contents with the higher concentration present on the high-concentrate diet. Both the methanogenic and acidogenic populations showed a marked increase 1–2 h after the once daily high-concentrate feed. Of interest was the finding that numbers of acetogens were actually higher than methanogens on the high-concentrate diet, whereas the trend was reversed on the high-forage diet. This suggests that acetogens grow on organic substrates in the rumen but are able to produce acid from H_2 and CO_2 when enumerated using the plating procedure. It is also likely that the procedure used for enumerating acetogens overestimates their numbers. In another study using an MPN technique with a rumen fluid-based medium containing BES (1–2.5 mM), the numbers of H_2/CO_2 utilizing acetogens ranged from 2.6×10^5 to 3.5×10^5 per ml of rumen fluid on a high-forage diet and from 2–75 cells per ml of rumen fluid on a high-grain diet (Wilfond et al. 1991). Furthermore, they constituted, at most, 0.1% of the methanogenic bacterial population on these diets.

It is worthwhile noting that although acetate is a key intermediate in the anaerobic degradation of organic matter by bacteria and that acetogens are common in nature with the discovery of many new acetogenic bacteria, little appears to be known about their ecological significance. Few studies on their overall contribution to acetate production have been published. Mackie and Bryant (1981) published a comprehensive ^{14}C-tracer study which determined the kinetics of fatty acid oxidation and CO_2 reduction and their subsequent quantitative contribution to methanogenesis in stirred fermentors fed semicontinuously with cattle waste. The fermentations were carried out at 40 and 60°C which are considered near optimum for mesophilic and thermophilic digestion, respectively. The synthesis of acetate from $H^{14}CO_3$ was 0.5 and 1.6–2.4 μmol^{-1} min^{-1} in the mesophilic and thermophilic fermentors, respectively. This accounted for 1.4–1.7% and 3.5–5.3% of the acetate synthesized at the two temperatures. For anaerobic sediments, H_2 consumption by acetogens accounted for approximately 5% of all H_2 uptake (Lovley and Klug, 1983). Numbers of homoacetogens of 10^5–10^6 per ml of sewage sludge have been reported (Ohwaki and Hungate, 1977; Braun et al., 1979). However, the ratio of methanogens to acetogens was approximately 100 : 1 in sewage and sediments (Braun et al., 1979).

Breznak and Kane (1990) indicated several factors that influence the competitiveness of acetogens with other hydrogen-consuming bacteria. One factor is that the energy yield for acetogenesis is lower than for methanogenesis from H_2 and CO_2.

$$4\,H_2 + 2\,HCO_3^- + H^+ \rightarrow CH_3COO^- + 4\,H_2O, \qquad \Delta G_0' = -104.6\,kJ$$
$$4\,H_2 + HCO_3^- + H^+ \rightarrow CH_4 + 3\,H_2O, \qquad \Delta G_0' = -135.6\,kJ$$

In fact, at the concentrations of reactants and products typically found in anaerobic environments where H_2 is limiting, methanogenesis is about three times more exergonic than acetogenesis using H_2 and CO_2. This would allow better growth and survival in a dynamic ecosystem like the rumen with constant dilution or washout occurring.

12.5.2 K_m and Threshold for H_2

Another factor advanced for the lack of competition is that the H_2 threshold and K_m for hydrogen of acetogens are much higher than methanogens (Breznak and Kane, 1990). The H_2 threshold of *Sporomusa termitida* (830 ppm or 2.04 μM) isolated from the wood-eating termite *Nasutitermes nigriceps* was about 10-fold higher than that of most methanogens (Cord-Ruwisch et al., 1988; Lovley et al., 1984; Lovley, 1985). The H_2 threshold of approximately 4200 ppm (2.1 μM) reported for *Acetitomaculum ruminis* would help to explain why acetogens are outcompeted by methanogens for H_2 in the rumen, which has an *in situ* concentration of H_2 of about 1900 ppm (0.95 μM).

More recently, four acetogenic bacterial isolates have been obtained from a H_2-limited chemostat inoculated with rumen content from a cow-fed corn silage and corn grain in a 60 : 40 ratio (Bocazzil et al., 1991). All of the isolates stained Gram positive and grew on glucose. Two of the isolates (BA2 and BA10) were coccoids and two (BA4 and BA9) were rods. A methanogenic isolate, similar in morphology to *Methanobrevibacter ruminantium*, was also isolated. The H_2 thresholds for the isolates BA2, BA10, BA4, and BA9 were 2516, 1222, 8061, and 4911 ppm, respectively (mean 4178 ppm or 2.1 μM). The threshold value for the methanogenic isolate was 120 ppm (0.06 μM). These data confirm that these acetogenic isolates would not compete effectively with methanogens for H_2. However, it is important to note that ruminal methanogens may have higher thresholds for H_2 than methanogens isolated from other environments with considerably lower prevailing or *in situ* H_2 concentrations.

Kinetic parameters for H_2 consumption by three methanogenic habitats were determined from progress curve and initial velocity measurements (Robinson and Tiedje, 1982). The K_m values for these habitats were similar, with means of 5.8, 6.0, and 7.1 μM for rumen-fluid, digester sludge and sediment, respectively. V_{max} estimates were as follows: rumen fluid, 14–28 mM h^{-1}; sludge, 0.7–4.3 mM h^{-1}, and sediment, 0.13–0.49 mM h^{-1}. Hungate et al. (1970) reported an average ruminal K_m of 1 μM for H_2 and 30 μM for formate (the value for formate may be lower, Lovley et al., 1984). The K_m of a pure culture of *Methanobrevibacter ruminantium* was 1 μM (Hungate, 1967). These results confirm that metabolism of methanogenic bacteria keeps the rumen concentration of H_2 so low that production of H_2 from NADH is exergonic and H_2 is used to reduce CO_2 and form CH_4 as the primary electron-sink product. These kinetic parameters allow calculations that show that H_2 pool size and turnover can account for all of the CH_4 formed in the rumen (Hungate, 1975).

Although K_m values for ruminal acetogens have not been determined, all evidence indicates that H_2/CO_2 acetogens would not be expected to compete effectively with methanogens unless steady-state H_2 concentrations were maintained above their threshold. However, the gastrointestinal tract cannot be considered as a completely homogeneous system and it is possible that acetogens could inhabit a particular niche or have spatial relationship with other organisms that could affect their competition for substrates such as H_2. An example of this is the episymbiotic association between methanogens and the entodiniomorphid ciliate protozoa in the rumen (Doddema and Vogels, 1978). Between 8 and 55% of the cells of each ciliate species showed this episymbiotic association with methanogens (Vogels et al., 1980). Frequency of association was related to H_2 availability (Stumm et al., 1982). Furthermore, a strong positive correlation with methanogenic activity and the fraction of strained rumen fluid containing the highest number of ciliates has been demonstrated (Krumholz et al., 1983). Further research is required to clarify if acetogens compete with methanogens in this activity as well or if they have some other specialized niche that does not involve competition for H_2/CO_2 with methanogens.

12.5.3 Mixotrophy

The ability to grow by mixotrophy (the simultaneous utilization of H_2/CO_2 and organic substrates) has been advanced as further possibility of enhancing the competitiveness of gastrointestinal H_2/CO_2 acetogens which, unlike methanogens, have a relatively wide substrate utilization range (Breznak and Kane, 1990). Mixotrophy has been demonstrated for the termite gut acetogen *Sporomusa termitida* (Breznak and Blum, 1991). When H_2/CO_2 was included as a growth substrate with either methanol or lactate, both substrates were utilized simultaneously and there was no evidence of diauxie in the growth of cells or in acetate production. Furthermore, the molar growth yield of *S. termitida* was close to that predicted from summation of the yields observed when grown with each substrate alone. Although it is not yet known to what extent this trophic mode is used or is advantageous, *in situ*, such experiments provide data on the possible importance of mixotrophy to the competitive success of gut H_2/CO_2 acetogens. As yet, there is no information on ruminal acetogens, but indirect evidence from the studies of Wilfond et al. (1994) suggest that *A. ruminis* has the ability to grow mixotrophically. In the rumen ecosystem it is possible that the acetogenic bacteria simply use alternative substrates such as carbohydrates and do not compete with methanogens for H_2. The acetogenic bacteria may play an important role in interspecies hydrogen transfer as they can function both as H_2 producers when fermenting organic compounds and as H_2 consumers when growing autotrophically.

12.6 Potential and Implications for Ruminal Manipulation of Acetogenesis

Acetate is a key metabolite in the symbiosis between ruminant animals and their intestinal microbiota. It is the major end product of microbial fermentation in the gut providing a quantitatively major source of oxidizable energy as well as a biosynthetic precursor for the host animal. The molar proportions in which the principal VFA are formed in the rumen are ~ 63 acetic, 21 propionic, 14 butyric and 2 higher acids. In a lactating cow on a hay-concentrate diet, the fermentation can produce 2.4 kg acetic, 0.95 kg propionic, and 0.92 kg butyric acid daily (Hungate, 1966). VFA provide 60–80% of the daily metabolizable energy intake in ruminant animals (Annison and Armstrong, 1970).

Rumen fermentation can be represented as the sum of a set of individual fermentations and the production of CH_4 from H_2/CO_2. The equations that represent the subset activities are based on existing knowledge of pure culture and bacterial fermentations as well as known interactions (Wolin and Miller, 1988). Assuming there is not selective absorption of VFA from the rumen, the relative rates of conversion of hexose into products by each population determine the proportions of VFA formed and the ratio of CH_4 formed per hexose used. Two important principles can be derived from this exercise. First, that combinations of VFA with a lower proportion of acetate have a higher efficiency in the sense that CH_4 produced per mole of hexose fermented is reduced (Table 12.2). Thus, a low-methane–high-propionate pattern in the rumen indicates a fermentation with higher efficiency in terms of gross energy retained in nongaseous end products (Orskov, 1975). Second, relatively small increases of one subset and concomitant decreases of another subset can have a major effect on the relative amounts of the major VFA formed. Because CH_4 production represents a significant loss of energy to the animal, there has been an interest in specific inhibition of the methanogens. Methanogenesis is a process which is easily inhibited by many compounds resulting in simultaneous increases in propionic and sometimes butyric acid production. Examples of these compounds have been listed by Van Nevel and Demeyer (1988). The most practical methane inhibitors/propionate enhancers are the ionophore antibiotics of which monensin is the most widely used (Russell and Strobel, 1989). It may be possible to use a specific inhibitor

Table 12.2   Methane production and VFA proportions in ruminal fermentations as determined from equations and representing stoichiometric relationships for different population subgroups[a]

Molar ratio[b]	Efficiency[c]
70 : 20 : 10	0.64
65 : 20 : 15	0.61
65 : 25 : 10	0.57
60 : 25 : 15	0.54
60 : 30 : 10	0.50
55 : 30 : 15	0.48

[a] From Wolin and Miller (1988).
[b] Molar ratio acetic : propionic : butyric acids.
[c] Moles CH_4 produced per mole hexose fermented.

of methanogenesis such as BES together with an ionophore such as monensin or lasalocid coupled with inoculation of the rumen with acetogenic bacteria. However, Gram-positive acetogens such as *Acetitomaculum ruminis* may also be effectively inhibited by monensin (Drake, personal communication).

However, it is the opinion of the authors that we cannot manage, manipulate, or exploit systems that we do not understand. Understanding of acetogenesis in the rumen and its failure to compete for H_2/CO_2 with methanogenesis will be provided by detailed study of the mechanistic details of acetyl-CoA synthesis and catabolism in pure cultures of methanogens and acetogens which are prominent in the ruminal and other gut ecosystems. We need to study the enzymology of the pathways involved in methanogenesis and acetogenesis to determine which reaction(s) are rate limiting and regulate the rates of methanogenesis and acetogenesis. We also need to have a better understanding of coenzymes, electron carriers, and cellular locations of the different components of these systems.

The ecological importance of the acetyl-CoA and methanogenic pathways can be determined in different organisms under various conditions by screening for gene products encoding enzyme activities such as formate dehydrogenase, carbon monoxide dehydrogenase, corrinoid/iron sulfur protein, and methyltransferase activity. This will provide an understanding of the ecological role and significance of these organisms in their varied metabolic activities and may ultimately lead to successful application in reducing loss of energy as CH_4 from rumen fermentation while at the same time reducing emission of CH_4 from livestock and animals wastes.

References

Amos, D. A., and M. J. McInerney. 1989. Poly-β-hydroxyalkanoate in *Syntrophomonas wolfei*. *Arch. Microbiol.* **152**:172–177.

Annison, E. F., and D. G. Armstrong. 1970. Volatile fatty acid metabolism and energy supply. In: *Physiology of Digestion and Metabolism in the Ruminant*, A. T. Phillipson (ed.), pp. 422–437. Oriel Press, Newcastle.

Bache, R., and N. Pfennig. 1981. Selective isolation of *Acetobacterium woodii* on methoxylated aromatic acids and determination of growth yields. *Arch. Microbiol.* **130**:255–261.

Balch, W. E., and R. S. Wolfe. 1979. Specificity and biological distribution of coenzyme M (2-mercaptoethanesulfonic acid). *J. Bacteriol.* **137**:256–263.

Beaty, P. S., and M. J. McInerney. 1987. Growth of *Syntrophococcus wolfei* in pure culture on crotonate. *Arch. Microbiol.* **147**:389–393.

Boccazil, P., J. A. Patterson, B. J. Wilsey, and D. M. Schaefer. 1991. Hydrogen threshold values for one methanogenic and four acetogenic isolates from the rumen. p. 37. *Abstr. Rumen Function Conf.*, Chicago, IL. 1991.

Boone, D. R., and M. P. Bryant. 1980. Propionate-degrading bacterium, *Syntrophobacter wolinii* sp. nov. gen. nov., from methanogenic ecosystems. *Appl. Environ. Microbiol.* **40**:626–632.

Bouwer, E. J., and P. L. McCarty. 1983. Effects of 2-bromoethanesulfonic acid and 2-chloroethanesulfonic acid on acetate utilization in a continuous-flow methanogenic fixed-film column. *Appl. Environ. Microbiol.* **45**:1408–1410.

Brauman, A., M. D. Kane, M. Labat, and J. A. Breznak. 1992. Genesis of acetate and methane by gut bacteria of nutritionally diverse termites. *Science* **257**:1384–1387.

Braun, M., S. Schoberth, and G. Gottschalk. 1979. Enumeration of bacteria forming acetate from H_2 and CO_2 in anaerobic habitats. *Arch. Microbiol.* **120**:201–204.

Braun, M., F. Mayer, and G. Gottschalk. 1981. *Clostridium aceticum* (Weiringa), a microorganism producing acetic acid from molecular hydrogen and carbon dioxide. *Arch. Microbiol.* **128**:288–293.

Breznak, J. A. 1994. Acetogenesis from carbon dioxide in termite and other insect gastrointestinal systems. In: *Acetogenesis*, H. L. Drake (ed.), pp. 303–330. Chapman and Hall, New York (this volume).

Breznak, J. A., and M. D. Kane. 1990. Microbial H_2/CO_2 acetogenesis in animal guts: nature and nutritional significance. *FEMS Microbiol. Ref.* **87**:309–314.

Breznak, J. A., and J. M. Switzer. 1986. Acetate synthesis from H_2 plus CO_2 by termite gut microbes. *Appl. Environ. Microbiol.* **52**:623–630.

Breznak, J. A., J. M. Switzer, and H.-J. Seitz. 1988. *Sporomusa termitida* sp. nov., an H_2/CO_2-utilizing acetogen isolated from termites. *Arch. Microbiol.* **150**:282–288.

Breznak, J. A., and J. M. Blum. 1991. Mixotrophy in the termite gut acetogen, *Sporomusa termitida*. *Arch. Microbiol.* **156**:105–110.

Brune, A., and B. Schink. 1990. Pyrogallol-to-phloroglucinol conversion and other hydroxyl-transfer reactions catalyzed by cell extracts of *Pelobacter acidigallici*. *J. Bacteriol.* **172**:1070–1076.

Bryant, M. P. 1979. Microbial methane production—theoretical aspects. *J. Anim. Sci.* **48**:193–201.

Bryant, M. P., N. Small, C. Bouma, and I. M. Robinson. 1958. Studies on the composition of the ruminal flora and fauna of young calves. *J. Dairy Sci.* **41**:1747–1767.

Bryant, M. P., E. A. Wolin, M. J. Wolin, and R. S. Wolfe. 1967. *Methanobacillus omelianskii*, a symbiotic association of two species of bacteria. *Arch. Microbiol.* **59**:20–31.

Bryant, M. P., L. L. Campbell, C. A. Reddy, and M. R. Crabill. 1977. Growth of *Desulfovibrio* in lactate or ethanol media low in sulfate in association with H_2-utilizing methanogenic bacteria. *Appl. Environ. Microbiol.* **33**:1162–1169.

Cheng, K. J., G. A. Jones, F. J. Simpson, and M. P. Bryant. 1969. Isolation and identification of rumen bacteria capable of anaerobic rutin degradation. *Can. J. Microbiol.* **15**:1365–1374.

Cheng, K. J., H. G. Krishnamurty, G. A. Jones, and F. J. Simpson. 1971. Identification of products produced by the anaerobic degradation of naringin by *Butyrivibrio* sp. C₃. *Can. J. Microbiol.* **17**:129–131.

Chesson, A., C. S. Stewart, and R. J. Wallace. 1982. Influence of plant phenolics on growth and cellulolytic activity of rumen bacteria. *Appl. Environ. Microbiol.* **44**:597–603.

Clarke, R. T. J. 1977. Methods for studying gut microbes. In: *Microbial Ecology of the Gut*, R. T. J. Clarke and T. Bauchop (eds.), pp. 1–33. Academic Press, New York.

Clark, F. M., and L. R. Fina. 1952. The anaerobic decomposition of benzoic acid during methane fermentation. *Arch. Biochem. Biophys* **36**:26–32.

Cord-Ruwisch, R., H.-J. Seitz, and R. Conrad. 1988. The capacity of hydrogenotrophic anaerobic bacteria to compete for traces of hydrogen depends on the redox potential of the electron acceptor. *Arch. Microbiol.* **149**:350–357.

DeGraeve, K. G., J. P. Grivet, M. Durand, P. Beaumatin, D. Demeyer. 1990. NMR study of ¹³CO₂ incorporation into short-chain fatty acids by pig large-intestinal flora. *Can. J. Microbiol.* **36**:579–582.

Doddema, H. J., and G. D. Vogels. 1978. Improved identification of methanogenic bacteria by fluorescence microscopy. *Appl. Environ. Microbiol.* **36**:752–754.

Dolfing, J. 1988. Acetogenesis. In: Biology of Anaerobic Microorganisms, A. J. B. Zehnder (ed.), Chap. 9, pp. 417–468. Wiley, New York.

Doré, J. M. 1989a. A study of the phylogenetics, nutrition and one-carbon metabolism of *Syntrophococcus sucromutans*, Ph.D. thesis, University of Illinois, Urbana-Champaign.

Doré, J. M., and M. P. Bryant. 1989b. Lipid growth requirement and the influence of lipid supplement on the fatty acid and aldehyde composition of *Syntrophococcus sucromutans*. *Appl. Environ. Microbiol.* **55**:927–933.

Doré, J. M., and M. P. Bryant. 1990. Metabolism of one-carbon compounds by the ruminal acetogen *Syntrophococcus sucromutans*. *Appl. Environ. Microbiol.* **56**:984–989.

EPA. 1990. Methane emissions and opportunities for control. United States Environmental Protection Agency Report No. EPA/400/9-90/007.

Fina, L. R., and A. M. Fiskin. 1960. The anaerobic decomposition of benzoic acid during methane fermentation. II. Fate of carbons one and seven. *Arch. Biochem. Biophys.* **91**:163–165.

Fontaine, F. E., W. H. Peterson, E. McCoy, M. J. Johnson, and G. J. Ritter. 1942. A new type of glucose fermentation by *Clostridium thermoaceticum* n. sp. *J. Bacteriol.* **43**:701–715.

Frazer, A. C., and L. Y. Young. 1986. Anaerobic C₁ metabolism of the O-methyl-¹⁴C-labelled substituent of vanillate. *Appl. Environ. Microbiol.* **51**:84–87.

Fuchs, G. 1986. CO₂ fixation in acetogenic bacteria: variations on a theme. *FEMS Microbiol. Rev.* **39**:181–213.

Genthner, B. R. S., C. L. Davis, and M. P. Bryant. 1981. Features of rumen and sludge strains of *Eubacterium limosum*, a methanol and H_2-CO_2-utilizing species. *Appl. Environ. Microbiol.* **42**:12–19.

Genthner, B. R. S., and M. P. Bryant. 1982. Growth of *Eubacterium limosum* with carbon monoxide as the energy source. *Appl. Environ. Microbiol.* **43**:70–74.

Genthner, B. R. S., and M. P. Bryant. 1987. Additional characteristics of one-carbon compound utilization by *Eubacterium limosum* and *Acetobacterium woodii*. *Appl. Environ. Microbiol.* **53**:471–476.

Greening, R. C., and J. A. Z. Leedle. 1989. Enrichment and isolation of *Acetitomaculum ruminis* gen. nov., sp. nov.: acetogenic bacteria from the bovine rumen. *Arch. Microbiol.* **151**:399–406.

Gunsalus, R. P., J. A. Romesser, and R. S. Wolfe. 1978. Preparation of coenzyme M analogs and their activity in the methyl-coenzyme M reductase in *Methanobacterium thermoautotrophicum*. *Biochemistry* **17**:2374–2377.

Hungate, R. E. 1966. *The Rumen and Its Microbes*. Academic Press, New York.

Hungate, R. E. 1967. Hydrogen as an intermediate in the rumen fermentation. *Arch. Microbiol.* **59**:158–164.

Hungate, R. E. 1975. *The Rumen Microbial Ecosystem*. *Annu. Rev. Ecol. Syst.* **6**:39–60.

Hungate, R. E. 1976. *The Rumen Fermentation*. In: *Microbial Production and Utilization of Gases*, H. G. Schlegel, G. Gottschalk, and N. Pfennig (eds.), pp. 119–124. Goltze, Göttingen.

Hungate, R. E. 1985. Anaerobic biotransformations of organic matter. In: *Bacteria in Nature, Vol. 1*, E. R. Leadbetter and J. S. Poindexter (eds.), pp. 39–95. Plenum, New York.

Hungate, R. E., W. Smith, T. Bauchop, T. Yu, and J. C. Robinowitz. 1970. Formate as an intermediate in the bovine rumen fermentation. *J. Bacteriol.* **102**:389–397.

Kane, M. D., and J. A. Breznak. 1991. *Acetonema longum* gen. nov. sp. nov., an H_2/CO_2 acetogenic bacterium from the termite, *Pterotermes occidentis*. *Arch. Microbiol.* **156**:91–98.

Kane, M. D., A. Brauman, and J. A. Breznak. 1991. *Clostridium mayombei* sp. nov., an H_2/CO_2 acetogenic bacterium from the gut of the African soil-feeding termite, *Cubitermes speciosus*. *Arch. Microbiol.* **156**:99–104.

Koch, M., J. Dolfing, K. Wuhrmann, and A. J. B. Zehnder. 1983. Pathways of propionate degradation by enriched methanogenic cultures. *Appl. Environ. Microbiol.* **45**:1411–1414.

Krishnamurty, H. G., K. J. Cheng, G. A. Jones, F. J. Simpson, and J. E. Watkin. 1970. Identification of products produced by the anaerobic degradation of rutin and related flavenoids by *Butyrivibrio* sp. C_3. *Can. J. Microbiol.* **16**:759–767.

Krumholz, L. R., C. W. Forsberg, and D. M. Veira. 1983. Association of methanogenic bacteria with rumen protozoa. *Can. J. Microbiol.* **29**:676–680.

Krumholz, L. R., M. P. Bryant. 1985. *Clostridium pfennigii* sp. nov. uses methoxyl groups of monobenzenoids and produces butyrate. *Int. J. Syst. Bacteriol.* **35**:454–456.

Krumholz, L. R., and M. P. Bryant. 1986a. *Eubacterium oxidoreducens* sp. nov. requiring H₂ or formate to degrade gallate, pyrogallol, phloroglucinol, and quercetin. *Arch. Microbiol.* **144**:8–14.

Krumholz, L. R., and M. P. Bryant. 1986b. *Syntrophococcus sucromutans* sp. nov. gen. nov. uses carbohydrates as electron donors and formate, methoxybenzenoids or *Methanobrevibacter* as electron acceptor systems. *Arch. Microbiol.* **143**:313–318.

Krumholz, L. R., R. L. Crawford, M. E. Hemling, and M. P. Bryant. 1987. Metabolism of gallate and phloroglucinol in *Eubacterium oxidoreducens* via 3-hydroxy-5-oxohexanoate. *J. Bacteriol.* **169**:1886–1890.

Krumholz, L. R., and M. P. Bryant. 1988. Characterization of the pyrogallol–phloroglucinol isomerase of *Eubacterium oxidoreducens*. *J. Bacteriol.* **170**:2472–2479.

Leedle, J. A. Z., and R. C. Greening. 1988. Methanogenic and acidogenic bacteria in the bovine rumen: postprandial changes after feeding high- or low-forage diets once daily. *Appl. Environ. Microbiol.* **54**:502–506.

Ljungdahl, L. G. 1986. The autotrophic pathway of acetate synthesis in acetogenic bacteria. *Annu. Rev. Microbiol.* **40**:415–450.

Lorowitz, W. H., and M. P. Bryant. 1984. *Peptostreptococcus products* strain that grows rapidly with CO as the energy source. *Appl. Environ. Microbiol.* **47**:961–964.

Lovley, D. R., and M. J. Klug. 1983. Methanogenesis from methanol and methylamines and acetogenesis from hydrogen and carbon dioxide in the sediments of a eutrophic lake. *Appl Environ. Microbiol.* **45**:1310–1315.

Lovley, D. R. 1985. Minimum threshold for hydrogen metabolism in methanogenic bacteria. *Appl. Environ. Microbiol.* **49**:1530–1531.

Lovley, D.R., R. C. Greening, and J. G. Ferry. 1984. Rapidly growing rumen methanogenic organism that synthesizes coenzyme M and has high affinity for formate. *Appl. Environ. Microbiol.* **48**:81–87.

Mackie, R. I., and M. P. Bryant. 1981. Metabolic activity of fatty acid-oxidizing bacteria and the contribution of acetate, propionate, butyrate, and CO₂ to methanogenesis in cattle waste at 40 and 60°C. *Appl. Environ. Microbiol.* **41**:1363–1373.

Mackie, R. I., and B. A. White. 1991. Recent advances in rumen microbial ecology and metabolism: potential impact on nutrient output. *J. Dairy Sci.* **73**:2971–2995.

Mackie, R. I., B. A. White, and M. P. Bryant. 1991. Lipid metabolism in anaerobic ecosystems. *Crit. Rev. Microbiol.* **17**:449–479.

Martin, A. K. 1982. The origin of urinary aromatic compounds excreted by ruminants. 3. The metabolism of phenolic compounds to simple phenols. *Br. J. Nutr.* **48**:497–507.

McInerney, M. J., M. P. Bryant, and N. Pfennig. 1979. Anaerobic bacterium that degrades fatty acids in syntrophic association with methanogens. *Arch. Microbiol.* **122**:129–135.

McInerney, M. J., and M. P. Bryant. 1981. Basic principles of bioconversions in anaerobic digestion and methanogenesis. In: *Biomass Conversion Processes for Energy and Fuels,* S. S. Sofer, and O. R. Zaborsky (eds.), pp. 277–296. Plenum, New York.

McInerney, M. J., M. P. Bryant, R. B. Hespell, and J. W. Costerton. 1981a. *Syntrophomonas wolfei* gen. nov., sp. nov., an anaerobic, syntrophic, fatty acid-oxidizing bacterium. *Appl. Environ. Microbiol.* **41**:1029–1039.

McInerney, M. J., R. I. Mackie, and M. P. Bryant. 1981b. Syntrophic association of a butyrate-degrading bacterium and *Methanosarcina* enriched from bovine rumen fluid. *Appl. Environ. Microbiol.* **41**:826–828.

Nottingham, P. M., and R. E. Hungate. 1969. Methanogenic fermentation of benzoate. *J. Bacteriol.* **98**:1170–1172.

Ohwaki, D., and R. E. Hungate. 1977. Hydrogen utilization by clostridia in sewage sludge. *Appl. Environ. Microbiol.* **33**:1270–1274.

Opperman, R. A., W. O. Nelson, and R. E. Brown. 1961. *In vitro* studies of methanogenesis in the bovine rumen: dissimilation of acetate. *J. Gen. Microbiol.* **25**:103–111.

Orskov, E. R. 1975. Manipulation of rumen fermentation for maximum food utilization. *Wld Rev. Nutr. Diet* **22**:153–182.

Prins, R. A., and A. Lankhorst. 1977. Synthesis of acetate from CO_2 in the cecum of some rodents. *FEMS Microbiol. Lett.* **1**:255–258.

Ragsdale, S. W. 1991. Enzymology of the acetyl-CoA pathway of CO_2 fixation. *Crit. Rev. Biochem. Mol. Biol.* **26**:261–300.

Robinson, J. A., R. F. Strayer, and J. M. Tiedje. 1981. Method for measuring dissolved hydrogen in anaerobic ecosystems: application of the rumen. *Appl. Environ. Microbiol.* **41**:545–548.

Robinson, J. A., and J. M. Tiedje. 1982. Kinetics of hydrogen consumption by rumen fluid, anaerobic digestor sludge and sediment. *Appl. Environ. Microbiol.* **44**:1374–1384.

Rowe, J. B., M. L. Loughman, J. V. Nolan, and R. A. Leng. 1979. Secondary fermentation in the rumen of sheep fed a diet based on molasses. *Br. J. Nutr.* **41**:393–397.

Russell, J. B., and H. J. Strobel. 1989. Effect of ionophores on ruminal fermentation. *Appl. Environ. Microbiol.* **55**:1–6.

Scheline, R. R. 1966. Decarboxylation and demethylation of some phenolic benzoic acid derivatives by rat cecal contents. *J. Pharmacol.* **18**:664–669.

Scheline, R. R. 1978. *Mammalian Metabolism of Plant Xenobiotics.* Academic Press, London.

Schink, B., and N. Pfennig. 1982. Fermentation of trihydroxybenzenes by *Pelobacter acidigallici* gen. nov. sp. nov., a new strictly anaerobic, non-sporeforming bacterium. *Arch. Microbiol.* **133**:195–201.

Simpson, F. J., G. A. Jones, and E. A. Wolin. 1969. Anaerobic degradation of some bioflavonoids by microflora of the rumen. *Can. J. Microbiol.* **15**:972–974.

Smith, M. R., and R. A. Mah. 1981. 2-Bromoethanesulfonate: a selective agent for isolating resistant *Methanosarcina* mutants. *Curr. Microbiol.* **6**:321–326.

Smolenski, W. J., and J. A. Robinson. 1988. *In situ* rumen hydrogen concentrations in steers fed eight times daily, measured using a mercury reduction detector. *FEMS Microbiol. Ecol.* **53**:95–100.

Stevani, J., M. Durand, K. DeGraeve, D. Demeyer, and J. P. Grivet. 1991. Degradative abilities and metabolisms of rumen and hindgut microbial ecosystems. In: *Hindgut '91*, T. Sakata and R. L. Snipes (eds.), pp. 123–135. Senshu University Press, Tokyo.

Stewart, C. S., and M. P. Bryant. 1988. The rumen bacteria. In: *The Rumen Microbial Ecosystem*, P. N. Hobson (ed.), pp. 21–75. Elsevier Applied Science, New York.

Stumm, C. K., H. J. Gijzen, and G. D. Vogels. 1982. Association of methanogenic bacteria with ovine rumen ciliates. *Br. J. Nutr.* **47**:95–99.

Tanner, R. S., R. S. Wolfe, and L. G. Ljungdahl. 1978. Tetrahydrofolate enzyme levels in *Acetobacterium woodii* and their implication in the synthesis of acetate from CO_2. *J. Bacteriol.* **134**:668–670.

Tanner, R. S., E. Stackebrandt, G. E. Fox, R. Gupta, L. J. Magrum, and C. R. Woese. 1982. A phylogenetic analysis of anaerobic eubacteria capable of synthesizing acetate from carbon dioxide. *Curr. Microbiol.* **7**:127–132.

Tarvin, D., and A. M. Buswell. 1934. The methane fermentation of organic acids and carbohydrates. *J. Am. Chem. Soc.* **56**:1751–1755.

Tsai, C. G., and G. A. Jones. 1975. Isolation and identification of rumen bacteria capable of anaerobic phloroglucinol degradation. *Can. J. Microbiol.* **21**:749–801.

Tsai, C. G., D. M. Gates, W. M. Ingledew, and G. A. Jones. 1976. Products of anaerobic phloroglucinol degradation by *Coprococcus* sp. Pe₁5. *Can. J. Microbiol.* **22**:159–164.

Van Nevel, C. J., and D. I. Demeyer. 1988. Manipulation of rumen fermentation. In: *The Rumen Microbiol Ecosystem*, P. N. Hobson (ed.), pp. 387–443. Elsevier Applied Science, New York.

Varel, V. H., M. P. Bryant, L. V. Holdeman, and W. E. C. Moore. 1974. Isolation of ureolytic *Peptostreptococcus productus* from feces using defined medium: failure of common urease tests. *Appl. Microbiol.* **28**:594–599.

Vogels, G. D., W. F. Hoppe, and C. K. Stumm. 1980. Association of methanogenic bacteria with rumen ciliates. *Appl. Environ. Microbiol.* **40**:608–612.

Vogels, G. D., J. T. Keltjens, and C. Van der Drift. 1988. Biochemistry of methane production. In: *Biology of Anaerobic Microorganisms*, A. J. B. Zehnder (ed.), pp. 707–770. Wiley, New York.

Widdell, F. 1988. Microbiology and ecology of sulfate- and sulfur-reducing bacteria. In: *Biology of Anaerobic Microorganisms*, A. J. B. Zehnder (ed.), pp. 469–585. Wiley, New York.

Wilfond, T. L., D. M. Schaefer. 1990. Validation of methods for the enrichment and isolation of H_2-utilizing, CO_2-reducing rumen acetogenic bacteria. *Abstr. Annu. Meet. Am. Soc. Microbiol.*, p. 201.

Wilfond, T. L., J. Ralph, and D. M. Schaefer. 1991. Autotrophic acetogenesis by *Acetitomaculum ruminis* 190A4 reduced methane production during in vitro incubation of rumen contents. *Abstr. Annu. Meet. Am. Soc. Microbiol.*, p. 211.

Wilfond, T. L., J. Ralph, and D. M. Schaefer. 1994. Determination of H_2/CO_2 utilizing acetogenic activity in ruminal contents using ^{13}C-isotope ratio mass spectrometry and NMR. (in preparation).

Wofford, N. Q., P. S. Beaty, and M. J. McInerney. 1986. Preparation of cell-free extracts and the enzymes involved in fatty acid metabolism in *Syntrophomonas wolfei*. *J. Bacteriol.* **167**:179–185.

Wolin, M. J., and T. L. Miller. 1988. Microbe-microbe interactions. In: *The Rumen Microbial Ecosystem*, P. N. Hobson (ed.), pp. 343–359. Elsevier Applied Science, New York.

Wolin, M. J., and T. L. Miller. 1994. Acetogenesis from carbon dioxide in human gastrointestinal systems. In: *Acetogenesis*, H. L. Drake (ed.) pp. 365–385, Chapman and Hall, New York (this volume).

Wood, H. G., S. W. Ragsdale, and E. Pezacka. 1986. The acetyl-CoA pathway of autotrophic growth. *FEMS Microbiol. Rev.* **39**:345–362.

Wood, H. G., and L. G. Ljungdahl. 1991. Autotrophic character of the acetogenic bacteria. In: *Variations in Autotrophic Life*, Barton, L. L. and J. Shively (eds.), pp. 201–250. Academic Press, San Diego.

Yamamoto, I., T. Saiki, S. M. Liu and L. G. Ljungdahl. 1983. Purification and properties of NADP-dependent formate hydrogenase from *Clostridium thermoaceticum*, a tungsten-selenium-iron protein. *J. Bacteriol.* **258**:1826–1832.

Young, L. Y., and A. C. Frazer. 1987. The fate of lignin and lignin-derived compounds in anaerobic environments. *Geomicrobiol J.* **5**:261–293.

Zehnder, A. J. B., B. A. Huser, T. D. Brock, and K. Wuhrmann. 1980. Characterization of an acetate-decarboxylating non-hydrogen oxidizing methane bacterium. *Arch. Microbiol.* **124**:1–11.

Zhao, H., D. Yang, C. R. Woese, and M. P. Bryant. 1993. Assignment of fatty acid-β-oxidizing syntrophic bacteria to Syntyrophomonadeceae fam. nov. on the basis of 16S rRNA sequence analysis. *Int. J. Syst. Bacteriol.* **43**:278–286.

Zinder, S. H., T. Anguish, and S. C. Cardell. 1984. Selective inhibition by 2-bromoethanesulfonate of methanogenesis from acetate in a thermophilic anaerobic digestor. *Appl. Environ. Microbiol.* **47**:1343–1345.

13

Acetogenesis from CO_2 in the Human Colonic Ecosystem

Meyer J. Wolin and Terry L. Miller

13.1 Colonic Fermentation

Figure 13.1 shows the general features of the colonic fermentation. The human diet contains high concentrations of plant polysaccharides (cellulose, hemicellulose, pectin, and starch). Except for starch, they are not digested by host enzymes and pass to the colon where they are fermented by the cooperative metabolism of a large number of different bacterial species (Cummings, 1985; Wolin and Miller, 1983). Significant amounts of starch escape host digestion and are fermented in the colon (Stephen et al., 1983; Thornton et al., 1987). The fermentation of plant polysaccharides produces acetate, propionate, and butyrate and the gases H_2 and CO_2 (Cummings, 1985; McNeil, 1984). CH_4 is a major product of the colonic fermentation of some humans (Bond et al., 1971; Levitt, 1980). They harbor large concentrations of methane-forming Archae that use H_2 to reduce CO_2 to CH_4 (Miller and Wolin, 1982; Weaver et al., 1986).

Acetate is the major acid anion product of the fermentation. On a molar basis, about 3.5–4.5 times more acetate is produced than propionate or butyrate. Approximately 10–30 g of acetate are produced in the colon daily (Royall et al., 1990). A homoacetate bacterium, e.g., *Clostridium thermoaceticum,* has to ferment the same weight of glucose, or a liter of medium containing 1.0–3.0% glucose, to produce the same amount of acetate. Based on the amount of acids per gram dry feces (Weaver et al., 1989) and assuming excretion of 25 g dry feces per day, about 98% of the colonic acid products are absorbed and used by the host (Cummings et al., 1987). The rest of the acids, all of the microorganisms, and 80% of the gases exit in feces or flatus. The remaining gases are absorbed into the blood and are released in breath (Bond et al., 1971).

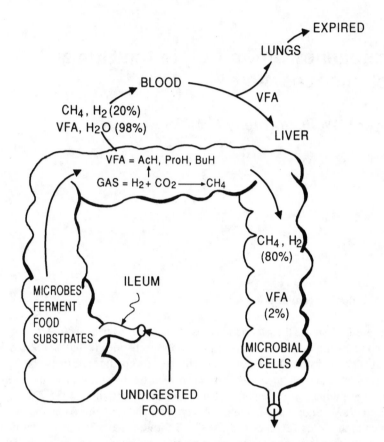

Figure 13.1.   Schematic of the human colonic fermentation. VFA, volatile fatty acids; AcH, acetate; ProH, propionate; BuH, butyrate. Adapted from Figure 1 from Wolin and Miller (1983).

Methane is formed by two species of methanogens. The species responsible for the production of almost all colonic CH_4 is *Methanobrevibacter smithii* (Miller and Wolin, 1982). It uses H_2 to reduce CO_2 to CH_4. A less numerous species, *Methanosphaera stadtmanae,* uses H_2 to reduce methanol to CH_4 and is incapable of producing CH_4 from CO_2 (Miller and Wolin, 1985). Some people have essentially no CH_4 formed in their colon, whereas others have constant daily production of several liters a day (Bond et al., 1971; Miller and Wolin, 1986; Weaver et al., 1986). This large variation is due to large differences in the concentration of colonic methanogens. The concentration of *M. smithii* varies from $<10^0$ to 10^{10} per gram dry matter (gdm) feces (Weaver et al., 1986). Individuals with

10^8–10^{10} *M. smithii* per gdm produce approximately 0.03 to 3 L of methane per day and have detectable ($>$ 1 ppm) breath CH_4 (Miller and Wolin, 1986).

For every liter of CH_4 produced, there are 4 L of H_2 produced and used in the colon. Either much less H_2 is produced when there is no significant methanogenesis or there is an alternative use of the electrons used for reducing CO_2 to CH_4. The use of H_2 to reduce CO_2 to acetate was shown to occur in the ceca of rats (Prins and Lankhorst, 1977) and the hindgut of termites (Breznak et al., 1988). We hypothesized (Wolin and Miller, 1983) and later provided evidence that this alternative path of CO_2 reduction was a major human colonic process in lieu of significant methanogenesis (Lajoie et al., 1988). Reduction of CO_2 to acetate would also account for the large amounts of acetate produced in nonmethanogenic humans.

There is little difference between the ratios of acetate : propionate : butyrate produced in highly methanogenic and nonmethanogenic individuals. We found mean ratios of 56 : 21 : 19 and 55 : 22 : 19 in feces of methanogenic and nonmethanogenic subjects, respectively (Wolin and Miller, 1983). Interspecies H_2 transfer between H_2-forming fermentative bacteria, and *M. smithii* can account for the production of large amounts of acetate in individuals that harbor high concentrations of the methanogen (Wolin, 1982; Wolin and Miller, 1983). However, an alternative mechanism for forming large amounts of acetate, such as CO_2 acetogenesis, has to be present in nonmethanogenic individuals. We use the term "CO_2 acetogenesis" to denote the process of producing acetate by the reduction of CO_2 or other one-carbon compounds to the methyl and carboxyl groups of acetate. We use the term "CO_2 acetogen" to denote an organism that produces acetate by the process. The terms "acetogenesis" and "acetogen" have been applied by others to describe processes and organisms that produce acetate irrespective of the origin of the carbon atoms of acetate.

Table 13.1 presents reactions that describe fermentations of pure cultures of bacteria that form products that are identical to the major products of the colonic

Table 13.1   Component reactions of the colonic fermentation

1. Butyrate-H_2-CO_2
     2 Hexose $\rightarrow$ 2 butyrate + 4 H_2 + 4 CO_2
2. Propionate-acetate-CO_2
     3 Hexose $\rightarrow$ 4 propionate + 2 acetate + CO_2
3. Acetate-H_2
     Hexose $\rightarrow$ 2 acetate + 4 H_2 + 2 CO_2
4. Methane
     8 H_2 + 2 CO_2 $\rightarrow$ 2 CH_4 + 4 H_2O
5. Acetate
     8 H_2 + 4 CO_2 $\rightarrow$ 2 CH_3COOH + 4 H_2O

fermentation. The blending of similar fermentations by individual and interacting species produces the large amounts of acetate and the other acid and gaseous products of the colonic fermentation. The first fermentation is similar to that of a butyrate-forming *Clostridium*. Fermentation of a mole of hexose yields 1 mol of butyrate, 2 mol of H_2, and 2 mol of CO_2. The second describes the formation of propionate, acetate, and CO_2 by a *Propionibacterium*. Although acetate is a product, only 22% of hexose carbon is converted to acetate.

The third fermentation, which yields large amounts of acetate, requires collaboration between H_2-producing and H_2-using organisms. Many anaerobes can ferment a hexose to 2 mol of acetate, 4 mol of H_2 and 2 mol of CO_2 if they are grown with organisms that use H_2. The bacteria that produce acetate and H_2 form a considerable amount of the gas from NADH if the concentration of H_2 is kept below 1 kPa. If a species that uses H_2 is present and uses H_2 as fast as it is produced, the partial pressure is kept at or below 1 kPa. Otherwise, H_2 accumulates and reduces NAD to NADH. NADH must then be reoxidized by dehydrogenases that form products like ethanol or lactate to regenerate NAD necessary for continuing the fermentation. The process that causes the production of acetate and H_2 is called interspecies H_2 transfer (Wolin, 1982). It leads to the formation of 2 mol of acetate per mole of hexose fermented and 67% of hexose carbon is converted to acetate.

Two pathways that may be used in the colon for using H_2 are CH_4 formation from H_2 and CO_2 (reaction 4) and acetate formation from the two gases (reaction 5). When CH_4 is produced, 2 mol of acetate are formed per mol hexose and 67% of hexose carbon yields acetate. When acetate is the product of CO_2 reduction, all of the glucose carbon is converted to carbon atoms of acetate.

There are alternatives to using interspecies H_2 transfer to effect complete conversion of hexose carbon atoms to acetate carbon atoms. Several bacterial species ferment sugars entirely to acetate by themselves (Ljungdahl, 1986). With a hexose, two acetates are formed from the pyruvate produced from the sugar carbon atoms. All of the oxidative reactions of fermentation (NADH and pyruvate oxidation) are coupled to the reduction of CO_2 to acetate. Therefore, the same organism that oxidizes glucose to acetate, CO_2, and electrons reduces CO_2 to acetate. A hexose cannot be converted to acetate and CH_4 by a single organism because methanogens do not use carbohydrates. Nonmethanogenic organisms produce acetate and H_2 that methanogens use to reduce CO_2 to CH_4.

The ratios between the acid end products of the colonic fermentation depend on the relative rates of the indivudal fermentation reactions shown in Table 13.1. For example, if the rates of all reactions were equal, the ratios would be determined by the sum of the reactions as shown in Table 13.1. To achieve ratios of acetate : propionate : butyrate like those formed in the colon, the rate of conversion of hexose to acetate, H_2, and CO_2 must be four to five times greater than the rates of formation of butyrate and propionate.

13.2 Evidence of CO_2 Acetogenesis

We and our colleagues presented evidence that supported the hypothesis that the reduction of CO_2 to acetate is an important feature of the overall fermentation in the colon. We showed that fecal suspensions of humans that harbor very low or moderate concentrations of methanogens use H_2 gas to reduce CO_2 to acetate (Lajoie et al., 1988). We and others showed that reduction of CO_2 to acetate also occurs in the colons of rodents (Lajoie et al., 1988; Prins and Lankhorst, 1977). Additional evidence for the importance of CO_2 acetogenesis comes from our experiments that show production of significant amounts of acetate by CO_2 reduction when vegetable polysaccharides (Table 13.2) are incubated with human fecal suspensions. We incubated $^{14}CO_2$ and a purified amorphous cellulose-containing fraction from cabbage (Weaver et al., 1992) with fecal suspensions from a subject with very low methanogen concentrations. Acetate, propionate, and butyrate were formed and the acetate contained substantial radioactivity (Table 13.2). No lactate, succinate, or formate were detected by HPLC analysis.

The radioactivity of acetate was determined after purification by HPLC. The acetate was then chemically degraded and each carbon was separately trapped as CO_2. The quantitative and radioisotope analyses were consistent with the component pathways shown in Table 13.2. The pathways indicate that most of the labeled acetate originated from the reduction of CO_2 by H_2 or electrons generated from the hexose to acetate fermentation. Some was from H_2 gas formed

Table 13.2 $^{14}CO_2$ reduction to acetate with cellulose as substrate

Incubation
    Human fecal suspension with purified cabbage cellulose for 24 h at 37°C with N_2-CO_2 gas and $^{14}CO_2$

Total fermentation products
    216 μmol acetate, 78 μmol propionate
    35 μmol butyrate, 0 μmol hydrogen
Analysis of ^{14}C
    Acids purified by HPLC, measure specific activity, Schmidt degradation of acetate to determine specific activity of methyl and carboxyl groups
Pathway Analysis:
    1. 60 $C_6H_{12}O_6$ → 80 CH_3CH_2COOH + 40 CH_3COOH + 40 CO_2
    2. 35 $C_6H_{12}O_6$ → 35 $CH_3CH_2CH_2COOH$ + 70 H_2 + 70 CO_2
    3. 54 $C_6H_{12}O_6$ → 108 CH_3COOH + 216 H_2 + 108 CO_2
    4. 286 H_2 + 143 * CO_2 → 71.5 * CH_3 * $COOH$

Sum: 149 $C_6H_{12}O_6$ → 219.5 CH_3COOH (148 CH_3COOH + 71.5 * CH_3 * COOH)
               + 80 CH_3CH_2COOH + 35 $CH_3CH_2CH_2$ COOH + 75 CO_2

in association with the butyrate-H_2 fermentation. Unlabeled acetate arose from the sugar carbon skeleton during the acetate-H_2 and propionate-acetate fermentations. The proportion of acids is 65 acetate : 24 propionate : 11 butyrate, which is similar to proportions found in human feces (Weaver et al. 1989). The unlabeled and labeled acetate of the CO_2 acetogenesis pathways 3 and 4 (Table 13.2) account for 82% of the total acetate formed. The propionate-acetate pathway accounts for the remaining unlabeled acetate formed by fermentation of the purified cabbage cellulose preparation.

13.3   Pure Culture Studies

Although we know that *M. smithii* is the methanogen responsible for reducing CO_2 to CH_4 (Miller and Wolin, 1982; Miller and Wolin, 1986), we cannot specify the bacteria responsible for reducing CO_2 to acetate in the human colon. Figure 13.2 shows the bacteria that are in high concentration in feces of many healthy subjects (Finegold et al., 1983). Many different genera and species are resident in the colon. Almost all are obligate anaerobes. The ability of most of these species to reduce CO_2 to acetate has not been investigated.

CO_2 acetogenesis was demonstrated with strains of *Peptostreptococcus productus* isolated from sewage sludge (Lorowitz and Bryant, 1984). They produced acetate from H_2 and CO_2 or CO. However, strains isolated from human feces do not grow with CO (Lorowitz and Bryant, 1984). Most probable number estimates of bacteria in human feces showed that only 1.1×10^5 bacteria per gram wet feces grew on CO (Lorowitz and Bryant, 1984). The organisms that grew were morphologically similar to *P. productus*. A strain of *P. productus* that grows on CO was also isolated from sewage sludge by Geerligs et al. (1987). It was similar to the strain isolated by Lorowitz and Bryant (1984). Geerligs et al. reported that the type strain of *P. productus* did not grow on CO (Geerligs et al., 1987). Neither the type strain nor the strains from human feces have been tested for growth with H_2 and CO_2. None of the strains were examined for the production of acetate from CO_2 during carbohydrate fermentation.

Another species, *Eubacterium limosum*, formerly called *Butyribacterium rettgeri*, also produces acetate from H_2 and CO_2 and methanol (Genther et al., 1981). Barker et al. (1945) showed that *B. rettgeri* produced acetate from CO_2 during carbohydrate fermentation. *E. limosum* is not found in many subjects and, when present, its concentrations in feces are not exceptionally high (Finegold et al., 1983). The carbohydrate fermentation products of other bacteria shown in Figure 13.2 have been determined (Holderman et al., 1977). Product analyses do not suggest acetate formation from CO_2. Bacterial species isolated from the termite gut (Breznak et al., 1986) or rumen (Greening and Leedle, 1989) that form acetate from CO_2 have not been found in human feces.

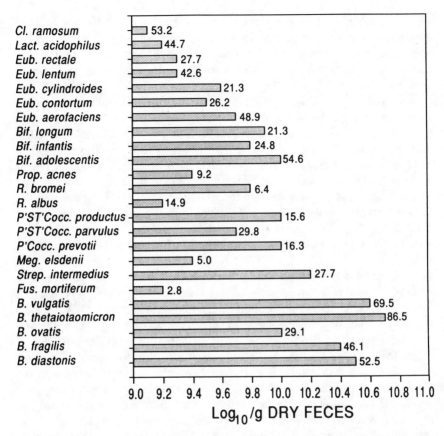

Figure 13.2. Incidence of bacterial species in feces of humans fed a Western diet. The numbers alongside each bar represent the percent of all subjects (62) who harbored the indicated species. The concentrations are the means of the species in those subjects. Cl., *Clostridium;* Lact., *Lactobacillus;* Eub., *Eubacterium;* Bif., *Bifidobacterium;* Prop., *Propionibacterium;* R. *Ruminococcus;* P'ST'Cocc., *Peptostreptococcus;* P'Cocc., *Peptococcus;* Meg., *Megasphaera;* Strep., *Streptococcus;* Fus, *Fusobacterium;* B., *Bacteroides.* Adapted from data in Finegold et al., 1983.

13.3.1   New Isolates from Humans

Because our studies indicate that reduction of CO_2 is an important pathway to acetate formation in the human colon, we initiated experiments designed to selectively isolate and characterize the CO_2 acetogens responsible for the process. We will now describe several CO_2 acetogens we isolated from human feces

(Wolin and Miller, 1993). Unfortunately, we have not had the resources to quantitatively evaluate their significance within and between humans.

We started isolations by using substrates that we thought would be selectively used by CO_2 acetogens. The general strategy for isolating these organisms is shown in Table 13.3. Initially, we enriched relatively low dilutions of fecal suspensions using H_2-CO_2, CO-CO_2, formate, methanol, or vanillate as selective substrates. A control without substrate and a control with 0.5% glucose were inoculated to differentiate substrate-dependent from substrate-independent growth. The basal medium contained salts, vitamins, rumen fluid, yeast extract, and bromethanesulfonate to inhibit growth of methanogens, and small amounts of formate. Formate was added because, as described later, we discovered a CO_2 acetogen that requires formate for growth and for producing acetate. The final concentrations of methanol and sodium formate were 0.25%, and sodium vanillate was 1.0%. The gas phase was 80% N_2–20% CO_2. The gaseous substrate mixtures were 80% H_2–20% CO_2 and 50% CO–50% CO_2. Gases were at 202 kPa.

We inoculated tubes of the medium with 0.5 ml of a 1 : 1000 dilution of a fecal suspension of a subject who harbors less than 1 methanogen per gdm of feces. After incubation with continuous rotation on a rotator at 37°C, growth was measured by determining the optical density at 660 nm. Significant glucose-dependent growth was observed at 24 h. Methanol and vanillate also yielded substrate-dependent growth, but it took about 6–7 days to notice a difference from the control. Although substrate-dependent growth did not occur with H_2, there was slow utilization of the gas. Little H_2 had disappeared at 8 days, but almost all had disappeared after 18 days. No substrate-dependent growth or utilization was observed with CO or formate enrichments.

Methanol-to-methanol and vanillate-to-vanillate transfers were made from the enrichments into sterile enrichment medium. Duplicate transfers of each substrate enrichment were incubated in 80% N_2–20% CO_2 and 80% H_2–20% CO_2. Good substrate-dependent growth was obtained in 1–2 days. Interestingly, H_2 was

---

Table 13.3   General strategy for isolation of CO_2 acetogens

1. Prepare 10-fold dilutions of fecal suspensions.
2. Inoculate selected dilutions into a basal medium containing a substrate for acetogens. The substrates were CO, H_2-CO_2, vanillate, methanol, or formate.
3. Media with glucose and no substrate were included. The growth differential (glucose–no substrate) was a positive control for substrate-dependent growth.
4. When acetogenic substrates yielded substrate-dependent growth, HPLC was used to measure soluble substrates and products and gas changes were measured by GC.
5. If substrate-dependent growth yielded significant acetate, pure cultures were obtained after plating on media with the substrate and agar.

rapidly used by the transfers from the methanol and vanillate enrichments. Large amounts of acetate were formed from methanol and vanillate, and higher amounts were produced when 80% H_2–20% CO_2 was the gas phase. Large amounts of ethanol and small amounts of propionate were also found in all transfers of the methanol and vanillate enrichments.

13.3.2 Vanillate Acetogen

We isolated the organism (CS1Van) responsible for fermentation of vanillate. First we eliminated substantial vanillate-independent growth and fermentation by lowering the yeast extract concentration from 0.2% to 0.04%. An agar medium was prepared using the same medium with 0.04% yeast extract. Dilutions of the enrichment culture were inoculated into vanillate roll tubes and incubated with 101 kPa 80% H_2–20% CO_2. Isolated colonies yielded Gram-positive, short, plump rods with rounded ends that used vanillate as a substrate for growth. The organism grew with methanol but not CO, H_2, or formate added as substrates for growth. They grew with vanillate and methanol without H_2. However, H_2, but not formate, was rapidly used when the isolate was grown with vanillate. A recently isolated acetogen, strain SS1, interestingly requires both H_2 and CO_2 for the acetogenic utilization of aromatic methoxyl groups (Liu and Suflita, 1993). Of a large number of other substrates tested, growth and fermentation were obtained with adonitol, fructose, glucose, lactate, maltose, mannitol, pyruvate, ribose, and xylose. (Table 13.4) and poor growth occurred with pectin and arabinose.

The major product of fermentation of all substrates was acetate. When methanol and CO_2 were substrates, acetate was labeled when either radioactive methanol or CO_2 was supplied (Table 13.5). The specific activity of acetate and its methyl and carboxyl groups indicated that most of the methyl group was formed from methanol and most of the carboxyl group was formed from CO_2 (Table 13.5).

13.3.3 Glucose and Methanol Acetogens

We repeated the dilution-enrichment series with 0.5 ml inocula of 10^6-fold dilutions of a fecal suspension from the same nonmethanogenic subject. The substrates were methanol, vanillate, glucose, and formate at the concentrations used in the first experiment. The same medium was used except the yeast extract was decreased to 0.04% and all enrichments had 80% H_2–20% CO_2 in the gas phase. The no-substrate control was incubated in 80% N_2–20% CO_2 to determine substrate-independent growth. Again, the only substrates that yielded significant growth were glucose, methanol, and vanillate. We isolated a Gram-positive, short, plump rod (CS3MeOH) from the methanol enrichment that grew well with methanol with N_2 and used H_2 rapidly when grown with methanol. It did not

Table 13.4   Substrates fermented by CS1Van[a]

Fermented	Not Fermented	Poorly Fermented
Adonitol	Alanine	Arabinose
Fructose	Avicel	Pectin
Glucose	Carboxymethylcellulose	
Lactate	Dextrin	
Maltose	Dulcitol	
Mannitol	Ferulic acid	
Pyruvate	Fumarate	
Ribose	Galactose	
Xylose	Glycerol	
	Inositol	
	Malate	
	Melezitose	
	Melibiose	
	Raffinose	
	Rhamnose	
	Salicin	
	Serine	
	Starch	
	Succinate	
	Sucrose	
	Trehalose	
	Xylan	

[a]Fermentation produced an optical density (OD) equal to or greater than five times that of a control without added substrate after 24–72 h. Cultures with no fermentation had an OD equal to same control in 192 h. Poor fermentation produced an OD of two to four times the same control in 192 h. The highest OD of the control without substrate was 0.25. OD was measured at 600 nm with a light path of 18 mm. Results were verified by HPLC analysis of products.

Table 13.5   Radioactivity of acetate formed by CS1Van grown with $^{14}CH_3OH$ or $^{14}CO_2$[a]

	Total μmol Acetate	DPM per μmol		
^{14}C Substrate		Acetate	Acetate Methyl	Acetate Carboxyl
Methanol	161	1710	1327	238
Carbon dioxide	422	2603	209	2042

[a] The organism was grown with 202 mM methanol under 80% N_2–20% CO_2 with the indicated radioactive substrate at 37°C with rotation.

Table 13.6   Substrate requirements of CS3Glu

	Optical density (660 nm)	
Addition	N_2-CO_2	H_2-CO_2
0.05% glucose	0.95 (21)[a]	1.90 (45)
None	0.23 (21)	0.50 (45)
0.25% methanol	0.19 (22)	0.49 (48)

[a] Numbers in parentheses show the hours required to reach maximal optical density. All H_2 (1.1 mmol) was used when maximal optical density was measured.

grow with H_2 alone. We have not yet determined if it is the same organism that we isolated from the first vanillate enrichments (CS1Van).

Rapid use of H_2 occurred with the glucose enrichment. Transfers to media with 0.05% glucose grew and used H_2. We proceeded with isolation on the 0.05% glucose agar medium with 80% H_2–20% CO_2 in the gas phase. We isolated a CO_2 acetogen (CS3Glu) that differs from the organisms previously isolated from vanillate or methanol enrichments. It is a Gram-positive coccobacillus that occurs singly, in pairs, and short chains. It grows with 80% H_2–20% CO_2 alone or with glucose with 80% N_2–20% CO_2, but it does not use methanol (Table 13.6). Glucose fermentation yields large amounts of acetate and very small amounts of succinate and lactate (Table 13.7.) Although it does not grow with formate, CS3Glu cometabolizes formate to acetate (Table 13.7).

13.3.4   Second Glucose Acetogen

We tried to isolate a similar glucose- and H_2-using acetogen from higher dilutions of fecal suspensions. We used 0.5 ml inocula of 10^{-6}–10^{-10} dilutions of fecal suspension from the same subject in the same, low yeast extract, medium with 0.25% glucose as substrate. The gas phase was 80% H_2–20% CO_2. Inocula from the 10^{-6}–10^{-9}, but not the 10^{-10}, dilutions produced cultures that used glucose and the formate added to the glucose-containing medium. The cultures formed high concentrations of acetate. The inocula from the 10^{-10} dilutions produced cultures that used glucose and formed considerable amounts of ethanol and produced formate. In contrast to the previous dilution-enrichment series, poor or no H_2 utilization was found in the cultures with inocula from the 10^{-6}–10^{-10} dilutions.

Attempts to isolate pure cultures from the 10^{-9} dilution using the same medium used for the enrichments plus agar with a 80% H_2–20% CO_2 gas phase failed. All colonies picked yielded a Gram-positive coccus that produced acetate, ethanol, and formate. Although spores were not detected by phase-contrast microscopy, we heated transfers of the enrichment culture from the 10^{-9} dilution at

Table 13.7 Fermentation of glucose plus formate by CS3Glu[a]

Compound	Total μmoles[b]	Per 100 mol Glucose	Total Carbon	Oxidation State	Oxidation Level
Substrate					
Glucose	264.3	100.0	600.0	0	0.0
CO_2[c]	29.9	11.3	11.3	-2	-22.6
Formate	395.9	149.8	149.8	-1	-149.8
Sum			761.1		-172.4
Product					
Succinate	59.8	22.6	90.5	1	22.6
Acetate	783.2	296.4	592.7	0	0.0
Lactate	1.4	0.5	1.6	0	0.0
CO_2	198.0	74.9	74.9	2	149.8
Sum			759.7		172.4

[a] Growth was with the total amounts of glucose and formate indicated above with 80% N_2 20% CO_2 at 37°C.
[b] Substrate used or product formed.
[c] CO_2 is calculated from the reactions:

$$3\,C_6H_{12}O_6 + HCO_2H \rightarrow 10\,CH_3COOH + 2\,CO_2 + 2\,H_2O$$
$$3\,C_6H_{12}O_6 + 2\,CO_2 \rightarrow 4\,HOOC\,(CH_2)_2COOH + 2\,CH_3COOH + 2\,H_2O$$

Table 13.8 Fermentation of glucose plus formate by CS7H[a]

Compound	Total μmoles[b]	Per 100 mol Glucose	Total Carbon	Oxidation State	Oxidation Level
Substrate					
Glucose	128.6	100.0	600.0	0	0.0
Formate	607.8	472.5	472.5	-1	-472.5
Sum			1072.5		-472.5
Products					
Acetate	538.2	418.4	836.8	0	0.0
CO_2[c]	303.9	236.2	236.2	2	472.5
Sum			1073.0		472.5

[a] Growth was with the total amounts of glucose and formate indicated above with 80% N_2–20% CO_2 at 37°C.
[b] Substrate used or product formed.
[c] CO_2 is calculated from the reaction.

$$3\,C_6H_{12}O_6 + HCO_2H \rightarrow 10\,CH_3COOH + 2\,CO_2 + 2\,H_2O$$

Table 13.9 Radioactivity of acetate formed by CS7H grown with glucose and $^{14}CO_2$ or ^{14}C-formate

^{14}C-substrate	Total μmoles Acetate	DPM per μmol		
		Acetate	Acetate Methyl	Acetate Carboxyl
Formate[a]	639	1421	1052	68
Carbon dioxide[b]	495	1563	765	809

[a] The organism was grown with 23 mM glucose under 80% N_2–20% CO_2 with 21 mM sodium formate at 37°C with rotation.
[b] The organism was grown in the absence of formate with 23 mM glucose under 80% N_2–20% CO_2 at 37°C with rotation.

80°C for 10 min and then repeated the isolation procedure. A CO_2 acetogen was isolated (CS7H). It is a Gram-positive, short, plump rod that survives 80°C but not 100°C for 10 min. We have not detected spores in the culture. It requires CO_2 for growth. It uses glucose and cometabolizes formate but does not grow with formate as the sole substrate and does not require formate for growth on glucose.

The sole product of fermentation of 0.25% glucose and 0.5% formate is acetate (Table 13.8). Radioisotopic experiments were performed to measure incorporation of CO_2 and formate into acetate. CO_2 was incorporated equally into both carbons of acetate and formate was incorporated primarily into the methyl group (Table 13.9.) No radioactivity from formate was found in CO_2. The isolate does not ferment plant polymers, but it grows rapidly on a large number of sugars including di-, tri-, and tetra-saccharides (Table 13.10).

13.3.5 Formate-Requiring, Cellulose-Using Acetogen

A serendipitous finding concerning reduction of CO_2 to acetate occurred when we isolated a bacterial strain (I52) from the human colon that degrades amorphous but not crystalline cellulose. It is a Gram-negative, nonmotile coccobacillus that forms mainly pairs or short chains. It has an unusual requirement for formate for growth on glucose. It does not grow on formate alone. Four millimolar but not 0.4 mM formate supports growth with glucose. Analyses of glucose fermentations with 4 and 40 mM formate showed all of the formate disappeared during growth. Fermentation products with the two concentrations of formate were distinctly different (Fig. 13.3). The fermentation reactions that account for the product differences are

Table 13.10   Substrates fermented by CS7H

Fermented	Not Fermented
Arabinose	Adonitol
Fructose	Alanine
Galactose	Avicel
Glucose	Carboxymethylcellulose
Lactate	Dextrin
Lactose	Dulcitol
Maltose	Ferulic acid
Mannitol	Fumarate
Melezitose	Glycerol
Melibiose	Inositol
Raffinose	Lactate
Salicin	Malate
Stachyose	Pectin
Sucrose	Pyruvate
Trehalose	Rhamnose
Xylose	Ribose
	Serine
	Starch
	Succinate
	Xylan

Low formate

$$100\ C_6H_{12}O_6 + 21\ HCO_2H + 37\ CO_2 \rightarrow 80\ HO_2C(CH_2)_2CO_2H$$
$$+ 56\ CH_3CHOHCO_2H + 85\ CH_3CO_2H$$

High formate:

$$100\ C_6H_{12}O_6 + 108\ HCO_2H \rightarrow 17\ HO_2C(CH_2)_2CO_2H$$
$$+ 7\ CH_3CHOHCO_2H + 284\ CH_3CO_2H + 51\ CO_2$$

The requirement for low concentrations of formate for growth on glucose indicates that the organism cannot synthesize the formate it needs for biosynthetic reactions. High concentrations of formate shift the fermentation away from succinate and lactate and toward acetate formation. Radioisotopic experiments with high concentrations of formate demonstrated that formate was incorporated into the methyl group of acetate and CO_2 was incorporated into the carboxyl group (Table 13.11). Although similar to CO_2 acetogens, I52 appears to be unable to reduce CO_2 to formate, a prerequisite for producing the methyl group of acetate from CO_2. Both the catabolic formation of acetate from formate and biosynthetic

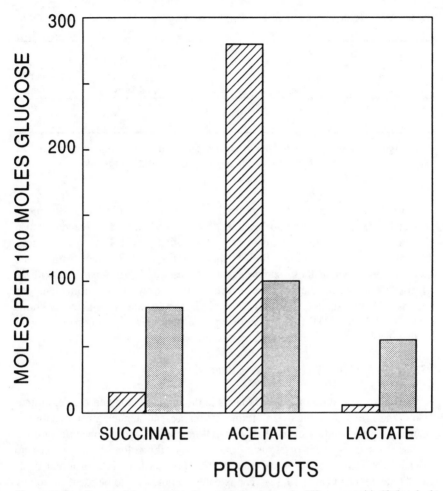

Figure 13.3. Fermentation products formed from glucose by I52 grown with 40 (hatched bar) or 4 mM formate (shaded bar). The initial glucose concentration was 0.5%.

reactions require the formation of formyl tetrahydrofolic acid. The shift to acetate production when high concentrations of formate are available signifies that the organism prefers to use electrons for reduction of formate and CO_2 to acetate rather than for reduction of oxalacetate and fumarate to form succinate or pyruvate to form lactate. The fermentations of I52 suggest the possibility that CO_2 acetogenesis in the colon could involve integration of the reduction of CO_2 to formate by one organism with a fermentation by another organism that produces acetate from formate and CO_2.

Syntrophococcus sucromutans, isolated from the rumen, also requires formate

Table 13.11   Radioactivity of acetate formed by I52 grown with glucose and $^{14}CO_2$ or ^{14}C-formate[a]

^{14}C Substrate	Total µmoles Acetate	DPM per µmol		
		Acetate	Acetate Methyl	Acetate Carboxyl
Formate	1077	762	593	32
Carbon dioxide	1039	858	35	682

[a] The organism was grown with 26 mM glucose and 38 mM sodium formate under 80% N_2–20% CO_2 with the indicated radioactive substrate at 37°C with rotation.

for carbohydrate fermentation (Krumholz and Bryant, 1986). The sole product is acetate. Except for the small amounts of succinate and lactate, reaction 7 also describes its fermentation. Growth of *S. sucromutans* with low concentrations of formate was not examined. Both *S. sucromutans* and I52 reduce formate and CO_2 to acetate when they oxidize sugars to acetate. *S. sucromutans* oxidizes glucose to acetate in the absence of formate if it is grown with a methanogen that uses H_2 to reduce CO_2 to CH_4 (Krumholz and Bryant, 1986) (see reactions 3 and 4 of Table 13.1). I52 does not support growth of a H_2-using methanogen. It also differs morphologically from *S. sucromutans*.

13.3.6   Where Are We with Pure Cultures?

The characterization of CO_2 acetogens in the human colon and the determination of their relation to the overall fermentation presents a much greater challenge than analogous studies of methanogens. Whereas a large number of subjects harbor the same species of H_2-using methanogen, the colon of the person who donated the fecal specimens we examined sustains at least four different types of CO_2 acetogens (Table 13.12). The isolates do not seem to be identical to any of the presently described species of CO_2 acetogens. It is likely that there are additional species of colonic CO_2 acetogens that have not been isolated.

Unlike the situation with H_2-using methanogens, selective isolation and enumeration of CO_2 acetogens is not straightforward. Antibiotics can be used to selectively grow methanogens because many antibiotics that kill the Bacteria do not kill the Archae. In contrast, there is no obvious selective agent that can be used to enumerate and isolate CO_2 acetogens. CO does not work at all, and H_2 is a poor selection substrate. The "second glucose acetogen" does not use H_2. All H_2 users that we isolated grew much better with organic substrates and cometabolized H_2. This is similar to the mixotrophic growth with H_2-CO_2 obtained with CO_2 acetogens isolated from the termite gut (Breznak and Blum, 1991). Further studies are necessary to find out whether methanol and vanillate might

Table 13.12 CO$_2$ acetogens from human feces

Strain	Fecal Dry Matter for Isolation (g)	Morphology	Acetate from	Gram Stain	Labeling in acetate from			Other
					$^{14}CO_2$	H$^{14}CO_2$H	$^{14}CH_3OH$	
I52	1×10^{-11}	Coccobacillus	Glucose, other sugars	−	COOH	CH$_3$	ND[a]	Requires HCOOH. Low HCOOH→ acetate, succinate, lactate. High HCOOH→ acetate. Uses cabbage cellulose.
CS1Van	5×10^{-4}	Short plump rod	Methanol, vanillate, H$_2$, glucose, other sugars	+	COOH & CH$_3$	ND	CH$_3$	Grows slowly with H$_2$ but uses it rapidly when grown with vanillate or methanol.
CS3Glu	5×10^{-7}	Coccobacillus	Glucose, other sugars, H$_2$	+	ND	ND	ND	Grows slowly with H$_2$ but uses it rapidly when grown with glucose.
CS7H	5×10^{-10}	Short plump rod	Glucose, other sugars	+	COOH & CH$_3$	CH$_3$	ND	Survives heating to 80°C for 10 min.

[a] ND = not done.

be useful selective substrates and to examine all isolates for properties that might be used for selection.

The results thus far suggest that reduction of CO_2 to acetate in the colon depends on organisms that use the oxidation of organic substrates such as carbohydrates and methanol to provide the necessary electrons. Some of these organisms cometabolize H_2 or formate and use these substrates to produce acetate from CO_2. The obligatory requirement for formate of I52 and its cometabolism by CS7H indicates that formate produced by other species may be used by CO_2 acetogens as an important precursor of the methyl group of acetate.

13.4   Magnitude of Human CO_2 Acetogenesis

We investigated the quantitative contribution of the formation of acetate by reduction of CO_2 to the overall production of acetate in the human colon. Results of measurements of the specific activities of the methyl and carboxyl carbons of acetate formed from fermentation of $1-^{14}C$-, $3,4-^{14}C$ glucose, and $^{14}CO_2$ by fecal suspensions indicate that reduction of CO_2 accounts for about 35% of the acetate produced in people who have less than 10^8-10^9 methanogens per gdm of feces. As the study of Royall et al. (Royall et al., 1990) suggests that total daily production in the colon is $\sim$ 10–30 g, our results indicate that CO_2 reduction yields $\sim$ 7 g of acetate per day. Our studies of fecal concentrations of methanogens in adults indicate that $\sim$ 90% harbor less than 10^9 methanogens per gdm feces (Weaver et al., 1986). Therefore, in $\sim$ 90% of the world population (5.4×10^9, 35% of colonic acetate is produced by CO_2 reduction. This yields a global daily production of 34,000 metric tons of acetate formed by CO_2 reduction in the human colon. As the oxidative reactions of colonic glycolysis that make acetate directly from the carbon skeletons of carbohydrates are coupled to the reduction of CO_2 to acetate, the homoacetate fermentation of carbohydrates probably accounts for most of the carbohydrate fermentation in 90% of the world's population. These estimates are only approximations. They do not account for the fermentations of unweaned children which differ considerably from adult fermentations or the influences of dietary habits and ethnicity on colonic fermentations.

13.5   Fermentation and Health

The importance of evaluating the relative contribution of the CO_2 acetogenesis fermentation pathway to the overall colonic fermentation is underscored by recent studies of the physiological effects of the fermentation acids. Acetate infused into the colon increases blood cholesterol and infused propionate is gluconeogenic (Wolever et al., 1991). Propionate infusion depresses cholesterol synthesis from acetate. The conversion of propionate to glycogen in the liver may be one of the benefits of the high-fiber diets recommended for diabetes. Colonic infusion of acetate, propionate, and butyrate decreased colonic bleeding of patients with diversion colitis (Royall et al., 1990).

Butyrate is used by colonic epithelial cells as an energy source (Roediger 1982). It has powerful properties as a cell-differentiating agent (Kruh, 1982). Low concentrations of butyrate cause differentiation of mammalian cells including human colon carcinoma cells (Tanaka et al., 1990). Possibly, a colonic cell that loses its intrinsic regulation mechanisms because of changes in its DNA can be extrinsically regulated by butyrate. It then would be phenotypically normal, undergo normal differentiation, and eventually be sloughed off into the lumen and excreted. In the absence of butyrate, the cell would grow as an undifferentiated tumor cell.

Elucidation of the bacterial species responsible for colonic formation of acetate is necessary for understanding how the major colonic fermentation products are formed. The results could aid in the rational design of diets that alter the production of fermentation acids and influence specific features of host physiology and health.

References

Barker, H. A., M. D. Kamen, and B. Haas. 1945. Carbon dioxide utilization in the synthesis of acetic and butyric acids by *Butyribacterium rettgeri*. *Proc. Natl. Acad. Sci. USA.* **31**:355–360.

Bond, J. H., Jr., R. R. Engel, and M. D. Levitt. 1971. Factors influencing pulmonary methane excretion in man. *J. Exp. Med.* **133**:572–578.

Breznak, J. A., J. M. Switzer, and H. Seitz. 1988. *Sporomusa termitida* sp. nov., an H_2/CO_2-utilizing acetogen isolated from termites. *Arch. Microbiol.* **150**:282–288.

Breznak, J. A., and J. S. Blum. 1991. Mixotrophy in the termite gut acetogen, *Sporomusa termitida*. *Arch. Microbiol.* **156**:105–110.

Cummings, J. H. 1985. Fermentation in the human large intestine: evidence and implications for health. *Lancet* **1**:1206–1028.

Cummings, J. H., E. W. Pomare, W. J. Branch, C. P. Naylor, and G. T. MacFarlane. 1987. Short chain fatty acids in human large intestine, portal hepatic and venous blood. *Gut* **28**:1221–1227.

Finegold, S. M., V. L. Sutter, and G. E. Mathisen. 1983. Normal indigenous intestinal flora. In: *Human Intestinal Microflora in Health and Disease*. D. J. Hentges, (ed.). p. 3. Academic Press, New York.

Geerligs, G., H. C. Aldrich, W. Harder, and G. Diekert. 1987. Isolation and characterization of a carbon monoxide utilizing strain of the acetogen *Peptostreptococcus productus*. *Arch. Microbiol.* **48**:305–313.

Genther, B. R. S., C. L. Davis, and M. P. Bryant. 1981. Features of rumen and sewage sludge strains of *Eubacterium limosum,* a methanol- and H_2-CO_2-utilizing species. *Appl. Environ. Microbiol.* **42**:12–19.

Greening, R. C., and J. A. Z. Leedle. 1989. Enrichment and isolation of *Acetitomaculum ruminis,* gen. nov., sp. nov.: acetogenic bacteria from the bovine rumen. *Arch. Microbiol.* **151**:399–406.

Holdeman, L. V., E. P. Cato, and W. E. C. Moore. 1977. *Anaerobic Laboratory Manual.* 4th ed. VPI Anaerobe Laboratory, Virginia Polytechnic Institute and State University, Blacksburg, Virginia.

Kruh, J. 1982. Effects of sodium butyrate, a new pharmacological agent, on cells in culture. *Mol. Cell. Biochem.* **42**:65–82.

Krumholz, L. R., and M. P. Bryant. 1986. *Syntrophococcus sucromutans* sp. nov., uses carbohydrates as electron donors and formate, methoxymonobenzenoids or *Methanobrevibacter* as electron acceptor systems. *Arch. Microbiol.* **143**:313–318.

Lajoie, S. F., S. Bank, T. L. Miller, and M. J. Wolin. 1988. Acetate production from hydrogen and [^{13}C]carbon dioxide by the microflora of human feces. *Appl. Environ. Microbiol.* **54**:2723–2727.

Levitt, M. 1980. Intestinal gas production—recent advances in flatology. *N. Eng. J. Med.* **302**:1474–1475.

Liu, S., and J. M. Suflita. 1993. H_2-CO_2-dependent anaerobic O-demethylation activity in subsurface sediments and by an isolated bacterium. *Appl. Environ. Microbiol.* **59**:1325–1331.

Ljungdahl, L. G. 1986. The autotrophic pathway of acetate synthesis in acetogenic bacteria. *Ann. Rev. Microbiol.* **56**:415–450.

Lorowitz, W. H., and M. P. Bryant. 1984. *Peptostreptococcus productus* strain that grows rapidly with CO as the energy source. *Appl. Environ. Microbiol.* **47**:961–964.

McNeil, N. I. 1984. The contribution of the large intestine to energy supplies in man. *Am. J. Clin. Nutr.* **39**:338–342.

Miller, T. L. and M. J. Wolin. 1982. Enumeration of *Methanobrevibacter smithii* in human feces. *Arch. Microbiol.* **131**:14–18.

Miller, T. L., and M. J. Wolin. 1985. *Methanosphaera stadtmaniae* gen. nov., sp. nov.: a species that forms methane by reducing methanol with hydrogen. *Arch. Microbiol.* **141**:116–122.

Miller, T. L., and M. J. Wolin. 1986. Methanogens in human and animal intestinal tracts. *System. Appl. Microbiol.* **7**:223–229.

Prins, R. A., and A. Lankhorst. 1977. Synthesis of acetate from CO_2 in the cecum of some rodents. *FEMS Microbiol. Lett.* **1**:255–258.

Roediger, W. E. W. 1982. Utilization of nutrients by isolated epithelial cells of the rat colon. *Gastroenterology.* **83**:424–429.

Royall, D., T. M. S. Wolever, K. N. Jeejeebhoy. 1990. Clinical significance of colonic fermentation. *Am. J. Gastroenterol.* **85**:1307–1312.

Stephen, A. M., A. C. Haddad, and S. F. Phillips. 1983. Passage of carbohydrate into the colon. Direct measurements in humans. *Gastroenterology* **85**:589–595.

Tanaka, Y., K. Bush, T. Eguchi, N. Ikekawa, T. Takaguchi, Y. Kobayashi, and P. J. Higgins. 1990. Effects of 1,25-dihydroxyvitamin D_3 and its analogs on butyrate-induced differentiation of HT-29 human colonic carcinoma cells and on the reversal of the differentiated phenotype. *Arch. Biochem. Biophys.* **276**:415–423.

Thornton, J. R., A. Dryden, J. Kelleher, and M. S. Losowsky. 1987. Super-efficient starch absorption. A risk factor for colonic neoplasia. *Dig. Dis. Sci.* 1987. **32**:1088–1091.

Weaver, G. A., J. A. Krause, T. L. Miller, and M. J. Wolin. 1986. Incidence of methanogenic bacteria in a sigmoidoscopy population: An association of methanogenic bacteria and diverticulosis. *Gut* **27**:698–704.

Weaver, G. A., J. A. Krause, T. L. Miller, and M. J. Wolin. 1989. Constancy of glucose and starch fermentation by two different human faecal microbial communities. *Gut* **30**:19–25.

Weaver, G. A., J. A. Krause, T. L. Miller, and M. J. Wolin. 1992. Cornstarch fermentation by the colonic microbial community yields more butyrate than does cabbage fiber fermentation; cornstarch fermentation rates correlate negatively with methanogenesis. *Am. J. Clin. Nutr.* **55**:70–77.

Wolever, T. M. S., P. Spadafora, and H. Eshuis. 1991. Interaction between colonic acetate and propionate in humans. *Am. J. Clin. Nutr.* **53**:681–687.

Wolin, M. J. 1982. Hydrogen transfer in microbial communities. In: *Microbial Interactions and Communities*. A. T. Bull and J. H. Slater (eds.), pp. 323–356. Academic Press, New York.

Wolin, M. J., and T. L. Miller. 1983. Carbohydrate fermentation. In: *Human Intestinal Flora in Health and Disease*, D. A. Hentges (ed.). pp. 147–165. Academic Press. New York.

Wolin, M. J. and T. L. Miller. 1993. Bacterial strains from human feces that reduce CO_2 to acetic acid. *Appl. Environ. Microbiol.* **59**:3551–3556.

14

Syntrophic Acetate Oxidation and "Reversible Acetogenesis"

Stephen H. Zinder

14.1 Introduction

Acetate is an important CH_4 precursor in nature, accounting for two-thirds of the CH_4 produced in many natural habitats and in anaerobic bioreactors. Although microbial methanogenesis from acetate was first described in the early 1900s, the mechanism of methanogenesis from acetate was controversial until 1978, when it was demonstrated that a pure culture of *Methanosarcina barkeri* could grow on acetate (Mah et al., 1978; Smith and Mah, 1978; Weimer and Zeikus, 1978) and convert acetate to CH_4 by a decarboxylation mechanism sometimes called the aceticlastic reaction. With the description of a similar mechanism for *Methanothrix soehngenii* in 1980 (Zehnder et al., 1980), it appeared that acetate decarboxylation was "the" mechanism for methanogenesis from acetate.

However, in 1984, our laboratory published a paper demonstrating that a thermophilic culture converted acetate to CH_4 using a mechanism in which acetate was oxidized to H_2 and CO_2 (Zinder and Koch, 1984). This culture consisted of two organisms syntrophically coupled, an acetate oxidizing rod (AOR), and a thermophilic H_2/CO_2 utilizing methanogen. Thauer et al. (1989) stated that this discovery "surprised the scientific community," although I suspect that only a very small fraction of the scientific community has given the matter any thought. The AOR was eventually isolated in axenic culture and shown to be an acetogen that could convert H_2 and CO_2 to acetate, the opposite reaction from the one it carries out when growing on acetate syntrophically coupled with a methanogen.

This chapter will describe the historical development of concepts concerning the mechanism of methanogenesis from acetate and other early work germane to this area, as well as a brief description of interspecies hydrogen transfer.

Following that will be a somewhat personal description of our studies on the thermophilic acetate oxidizing coculture and the axenic AOR, and the relationship between acetate oxidation by this culture and acetogenesis. Other related examples of acetate oxidation by pure cultures will be discussed as well as the potential role of syntrophic acetate oxidation in natural anaerobic habitats.

14.2 Historical Background

14.2.1 The Rise and Fall of the van Niel/Barker Hypothesis for the Mechanism of Methanogenesis from Acetate

In 1936, H. A. Barker published two important articles on methanogenesis, representing work he did while in Delft, Holland in between his graduate work with C. B. van Niel and his joining the faculty at U.C. Berkeley. One article (Barker, 1936b) described the culture of several methanogens. Many of these cultures were derived from colonies in shake tubes, and although it is likely that the cultures were not axenic, this study represents the first departure from using crude enrichment cultures in the study of microbial methanogenesis.

The other article, entitled "On the biochemistry of the methane fermentation" (Barker, 1936a), presented evidence supporting a proposal attributed to van Niel by Barker that the universal reaction of methanogenesis was

$$4\,H_2A + CO_2 \rightarrow 4\,A + CH_4 + 2\,H_2O \qquad (14.1)$$

This hypothesis, analogous to van Niel's generalized equation for photosynthesis, posited that the universal reaction for methanogenesis was reduction of CO_2 to CH_4 using H_2. Alcohols and fatty acids were considered to be methanogenic substrates at that time (while it was generally agreed that sugars and amino acids were initially acted upon by fermentative bacteria), and it was proposed that none of these was a direct methanogenic substrate, but instead they were all oxidized to provide the H_2 (or electrons) used for CO_2 reduction to CH_4. This theory was attractive because it explained how a wide variety of substrates could be converted into essentially a single reduction product: CH_4.

Thus, Barker proposed that methanogenesis from acetate occurred via a two-step reaction [given according to the current convention that HCO_3^- is the main form of CO_2 at pH = 7; also provided are $\Delta G_0'$ values (Thauer et al., 1977), which were not considered by Barker and van Niel]:

$$CH_3COO^- + 4\,H_2O \rightarrow 2\,HCO_3^- + 4\,H_2 + H^+, \qquad \Delta G_0' =$$
$$+104.6\ \text{kJ/rxn} \qquad (14.2)$$

$$4\,H_2 + HCO_3^- + H^+ \rightarrow CH_4 + 3\,H_2O, \qquad \Delta G_0' =$$
$$-135.6\ \text{kJ/rxn} \qquad (14.3)$$

$$\text{Sum: } CH_3COO^- + H_2O \rightarrow CH_4 + HCO_{3-}, \qquad \Delta G_0' =$$
$$-31.0 \text{ kJ/rxn} \tag{14.4}$$

The net reaction (reaction 14.4) is the equivalent of acetate decarboxylation. In the introductory sections of the article, Barker discussed the evidence that H_2-CO_2 was the universal CH_4 precursor, including Söhngen's demonstration of methanogenesis from H_2 and CO_2. Barker also argued that if acetate was decarboxylated, one might expect ethane from propionate fermentation and propane from butyrate, which was known by that time not to be the case.

Barker showed that ethanol oxidation to acetate was accompanied by disappearance of CO_2 in an amount essentially equal to the CH_4 formed and showed similar CO_2 consumption in the dissimilation butanol to butyrate and acetate. CO_2 consumption could not be examined in cultures converting acetate to CH_4 as there was net CO_2 production. From the results with ethanol, butanol, and butyrate, Barker believed it was logical to extrapolate that other organic compounds could serve as H_2 donors for CO_2 reduction as in reaction 14.1 and stated that the probable mechanism of methanogenesis from acetate involved oxidation (dehydrogenation). Barker concluded his discussion by pointing out the analogy between CO_2 reduction to CH_4 and sulfate and nitrate reduction, reactions which we recognize today as examples of anaerobic respiration.

The mechanism of methanogenesis from acetate would have been readily settled at that time had radioactive carbon isotopes been available. The first carbon radioisotope to become available was ^{11}C, which has a half-life of 20.5 min. Barker and colleagues (1940) were able to use $^{11}CO_2$ because of their proximity to accelerators at Berkeley. Quantitative results demonstrating production of labeled CH_4 from $^{11}CO_2$ by "*M. omelianskii*" were presented. In the case of methanogenesis from acetate by *Methanosarcina*, only a qualitative description of the results was given, and although it was stated that radioactive CH_4 was produced, the amount was only one-tenth the amount incorporated into cells, and the experiment was not considered conclusive. In addition, methanol-grown cells of *Methanosarcina* were used, and it is now known that they would be unlikely to catabolize acetate without an adaptation period (Smith and Mah, 1978; Zinder and Elias, 1985).

It was two sanitary engineers (Buswell and Sollo, 1948) who struck the first blow against the acetate oxidation hypothesis. They incubated material from a digestor-fed acetic acid with unlabeled acetate and with the newly available isotope $^{14}CO_2$ and showed that the initial specific activity of the $^{14}CH_4$ produced was only 1% that of the $^{14}CO_2$ added. However, if the mechanism was that described by reactions 14.2–14.4, one would expect the initial $^{14}CH_4$ specific activity to be equal to that of the $^{14}CO_2$. These results strongly argued that CO_2 reduction was not involved in methanogenesis from acetate by that culture and was in direct contradiction to the mechanism proposed in reactions 14.2–14.4.

The issue was essentially resolved in Barker's laboratory by the classic studies of Stadtman and Barker in 1949 (Stadtman and Barker, 1949). Position-labeled [^{14}C]acetate had become available, and they showed that impure *Methanosarcina* cultures converted $^{14}CH_3COO^-$ primarily to $^{14}CH_4$, whereas $CH_3^{14}COO^-$ was converted primarily to $^{14}CO_2$, giving complete support to a decarboxylation mechanism. They also used $^{14}CO_2$ to verify that CO_2 was the source of CH_4 in the "*M. omelianskii*" culture and, in a subsequent study (Stadtman and Barker, 1951), showed that methanol was a direct CH_4 precursor. To explain how the different substrates could lead to CH_4, they proposed that there was at least one common intermediate in their pathways that served as a CH_4 precursor, a role which today can be attributed to methyl-coenzyme M (DiMarco et al., 1990).

There was still controversy over whether methanogenesis from acetate could support growth because growth of a bona fide pure culture of a methanogen on acetate had not been demonstrated. There were theoretical objections to acetate as a growth substrate because there was not enough energy in the reaction ($\Delta G_0'$ = -31 kJ/rxn) to conserve an entire ATP equivalent (Thauer et al., 1977) and because it was difficult to envision how a decarboxylation could be coupled with energy conservation (Stadtman and Barker, 1951). Growth of pure cultures of *Methanosarcina barkeri* on acetate was demonstrated in 1978 (Smith and Mah, 1978; Mah et al., 1978; Weimer and Zeikus, 1978). These studies, and all subsequent ones, have confirmed decarboxylation as the mechanism for methanogenesis from acetate by pure cultures of methanogens. The pathway has been shown to involve disassembly of acetyl-coenzyme A (CoA) by an enzyme complex with carbon monoxide dehydrogenase activity (Grahame, 1991; Thauer et al., 1989; Ferry, 1992), essentially a reversal of acetate synthesis by CODH in acetogens (Ragsdale et al., 1990).

14.2.2 Interspecies Hydrogen Transfer

Barker and van Niel were correct in their proposal that ethanol oxidation provided reducing power for CO_2 reduction to CH_4 in "*Methanobacillus omelianskii*" cultures. However, they, and others, believed that a single organism was carrying out these coupled reactions. However, in 1967, Marvin Bryant and co-workers (Bryant et al., 1967) resolved this culture into two organisms, called the S (symbiotic) organism and MoH (methanogen oxidizing hydrogen) organism (now *Methanobacterium bryantii*) carrying out the following reactions:

$$2 \text{ Ethanol} + 2 \text{ H}_2\text{O} \rightarrow$$
$$2 \text{ Acetate}^- + 2 \text{ H}^+ + 4 \text{ H}_2, \qquad \Delta G_0' = +19.2 \text{ kJ/rxn} \qquad (14.5)$$

$$4 \text{ H}_2 + \text{HCO}_3^- + \text{H}^+ \rightarrow$$
$$\text{CH}_4 + 3 \text{ H}_2\text{O}, \qquad \Delta G_0' = -135.6 \text{ kJ/rxn} \qquad (14.6)$$

Sum: 2 ethanol + HCO_3^- →
      2 acetate$^-$ + H^+ + CH_4 + H_2O, $G\Delta_0' = -116.3$ kJ/rxn   (14.7)

The reaction carried out by the S organism is thermodynamically unfavorable under standard conditions (solutes at 1 M, gases at 1 atm). However, because the methanogen can consume the H_2 produced, thereby keeping H_2 at a low steady-state partial pressure, it can "pull" the reaction along in this coculture. One can estimate the free energy (in kJ/rxn) available for a reaction in which the products and reactants are at nonstandard conditions using a free-energy form of the Nernst equation; for a hypothetical reaction ($aA + bB \to cC + dD$) at 25°C:

$$\Delta G' = \Delta G_0' + RT \ln \frac{(C)^c (D)^d}{(A)^a (B)^b} = \Delta G_0' + 5.7 \log \frac{(C)^c (D)^d}{(A)^a (B)^b} \qquad (14.8)$$

where (A) is the concentration of A, R is the ideal gas constant, and T is the temperature in °K. Using this relationship, one can calculate that if ethanol and acetate are both 10 mM, a H_2 partial pressure below 10^{-1} atm ($\sim$10 kPa) would be required for ethanol oxidation to be favorable. The *M. omelianskii* culture represented a true mutualistic symbiosis, as the growth of the S organism alone on ethanol was inhibited by its own product, whereas the methanogen was unable to use ethanol in the absence of the S organism. It would be nearly 20 years after the resolution of *M. omelianskii* that it was demonstrated that pure cultures of some methanogens could use ethanol and other alcohols as electron donors for CO_2 reduction (Widdel, 1986), as van Niel and Barker had proposed.

The idea that a methanogen and a nonmethanogen could be metabolically coupled to degrade a single substrate (a form of syntrophism) was a conceptual breakthrough and led, in subsequent years, to demonstration of interspecies hydrogen transfer in the degradation of many substrates, including sugars (Wolin and Miller, 1982), amino acids (Zindel et al., 1988), lactate (Bryant et al., 1977), butyrate (McInerney et al., 1981), and propionate (Boone and Bryant, 1980).

Of particular interest to this discussion is the utilization of fatty acids by methanogenic syntrophic cocultures. A coculture consisting of the butyrate oxidizer *Syntrophomonas wolfei* and a hydrogenotrophic methanogen oxidizes butyrate to acetate by the following two-step process:

2 Butyrate$^-$ + 4 H_2O →
4 acetate$^-$ + H^+ + 4 H_2, $\Delta G_0' = +96.2$ kJ/rxn   (14.9)

$4H_2$ + HCO_{3-} + H^+ →
CH_4 + 3H_2O, $\Delta G_0' = -135.6$ kJ/rxn   (14.10)

Sum: $2 \text{ Butyrate}^- + HCO_3^- + H_2O \rightarrow$
$$4 \text{ acetate}^- + CH_4 + H^+, \quad \Delta_0' = -39.4 \text{ kJ/rxn} \tag{14.11}$$

Note that the energetics of butyrate oxidation are much less favorable than for ethanol oxidation (reaction 14.5). This leads to tighter coupling between the butyrate oxidizer and the methanogen, as butyrate oxidation is only favorable if the H_2 partial pressure is less than $\sim 6 \times 10^{-4}$ atm assuming the concentrations of acetate and butyrate are equal. Because the minimum H_2 threshold for mesophilic methanogens using H_2-CO_2 is $(3-6) \times 10^{-4}$ atm (Cord-Ruwisch et al., 1988), there is a fairly narrow range of H_2 partial pressures at which both organisms can conserve energy during syntrophic metabolism of butyrate; also note that the net energy available, -39.4 kJ/rxn, is barely enough for conservation of a mole of ATP, yet this energy is shared between two organisms and involves the metabolism of several molecules including four pairs of electrons. The energetics of propionate oxidation are even more stringent, and it is not surprising that propionate oxidation is easily perturbed in anaerobic habitats (Boone and Bryant, 1980). As can be seen from reactions 14.2–14.4, the kinetics of syntrophic acetate oxidation are also more stringent than those for butyrate oxidation.

14.3 The Thermophilic–Acetate Oxidizing Coculture

14.3.1 Discovery and Co-isolation

During my stay as a postdoctoral fellow in the Laboratory of Robert A. Mah at University of California, Los Angeles (1977–1980), I isolated *Methanosarcina thermophila* strain TM-1 (Zinder and Mah, 1979; Zinder et al., 1985), the first thermophilic acetotrophic methanogen described. However, while at UCLA, I also developed thermophilic enrichment which was rod shaped and produced CH_4 at 60°C in a growth medium containing calcium acetate as a substrate and ferrous sulfide as a reductant. The rods did not look like *Methanothrix*, but there was some resemblance to a 60°C acetate enrichment culture described previously by Coolhaas (1928). Several attempts at obtaining colonies in roll tubes failed although some CH_4 was produced at low dilutions.

The culture accompanied me to Cornell in 1980. Although I was still unsuccessful in obtaining colonies, I succeeded in obtaining growth in a precipitate-free growth medium containing sodium acetate and a low amount of sodium sulfide (higher amounts were toxic) and put the culture through several 10^{-7} dilutions in liquid culture. There were also some puzzling results: The culture seemed sensitive to antibiotics which inhibit eubacteria and always contained large numbers of an organism which resembled *Methanobacterium thermoautotrophicum*, a methanogen not known to use acetate. I had considered the possibility that the culture was syntrophic and was aware of the van Niel–Barker hypothesis but had thought that it was unlikely that methanogenesis from acetate, which until a few years before had been considered not sufficiently thermodynamically favorable to support a single organism, could actually support two organisms (although, in retrospect, the energetics of the culture were comparable to the propionate and butyrate oxidizing cocultures being described at that time).

The arrival in my laboratory of a gas proportional counter allowed me to readily measure production of labeled $^{14}CH_4$ and $^{14}CO_2$ from [^{14}C]acetate by the culture. Although there were problems concerning quantitative aspects of the results, their pattern was clear: Both methyl- and carboxyl-labeled [^{14}C]acetate were converted mainly to $^{14}CO_2$; there was considerable $^{14}CO_2$ reduction to $^{14}CH_4$, in contrast to Buswell and Sollo's results. The culture was clearly oxidizing acetate and using the electrons to reduce CO_2 to CH_4 and the high numbers of a $M.$ $thermoautotrophicum$-like organism made it likely that the culture was syntrophic. Once the nature of the culture was clarified, we decided to use the ingenious technique from Marvin Bryant's laboratory (McInerney et al., 1979) of "co-isolating" the syntrophic H_2 producer on a lawn of H_2 consumers.

Assisted by technician Tim Anguish, we first isolated a thermophilic H_2-CO_2 utilizing methanogen from high dilutions of the culture, reasoning that it would be prudent to use the H_2 consumer that we knew was already successful in coupling to acetate oxidation. The methanogen seemed to be a typical $Methanobacterium$ $thermoautotrophicum$ until we found that it readily used formate as a methanogenic substrate, an ability which had not been reported for any thermophilic $Methanobacterium$. We, therefore, called it $Methanobacterium$ strain THF (thermophile hydrogen, formate). Soon afterward, several other formate-utilizing thermophilic $Methanobacterium$ cultures were described, and it has been since shown that formate utilization is widespread in thermophilic methanobacteria and does not define a single species (Touzel et al., 1992). It was recently shown that $Methanobacterium$ strain THF contains a plasmid, encoding a restriction-modification system (Nölling and De Vos, 1992).

Using lawns of strain THF, we obtained colonies of the acetate oxidizer at high dilutions, and the culture was found to be "pure", i.e., two membered. As shown in Figure 14.1, the culture contained two morphotypes: a methanogen which was somewhat crooked and thin and showed F_{420} autofluorescence when viewed by epifluorescence microscopy, and a thicker, straighter, rod with pointed ends which was not autofluorescent and was assumed to be the acetate oxidizer. Soon after the isolation, Markus Koch arrived in my laboratory from Switzerland to do a postdoctoral stay, and began studies on the physiology of the acetate-oxidizing coculture.

14.3.2   Acetate Metabolism and Acetate/CO₂ Isotope Exchange

Figure 14.2 shows near stoichiometric conversion of acetate to CH_4 by the culture and that this reaction was coupled to growth measured as cell protein.

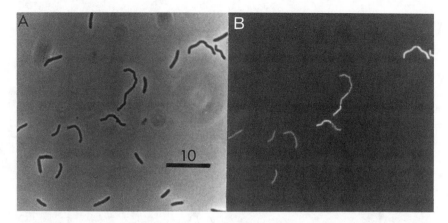

Figure 14.1. (A) Phase-contrast photomicrograph of the acetate oxidizing coculture. Marker represents 10 μm. (B) Epifluorescence micrograph of the same field illuminated to stimulate F_{420} autofluorescence. [From Zinder and Koch (1984).]

The doubling time was estimated as 30–40 h, and a growth yield of 2.7 g dry weight $[\text{mol } CH_4]^{-1}$ was obtained, somewhat higher than those described for the thermophilic acetotrophic methanogens *Methanosarcina thermophila* TM-1 (1.8 g dry wt $[\text{mol } CH_4]^{-1}$ (Zinder and Mah, 1979)) and *Methanothrix* CALS-1 (1.1 g dry wt $[\text{mol } CH_4]^{-1}$) (Zinder et al., 1987). Thus, this syntrophic coculture was essentially as effective as comparable acetotrophic methanogens at converting acetate to CH_4.

Short-term studies on conversion of radio-labeled substrates (Table 14.1) showed that $^{14}CO_2$ was reduced to $^{14}CH_4$ in amounts somewhat lower than predicted, whereas acetate labeled in either position was converted mainly to $^{14}CO_2$ (the small amount of $^{14}CH_4$ formed was due to subsequent reduction of $^{14}CO_2$). However, quantitative aspects of the acetate data were problematic because the amount of label in the products of acetate metabolism was always greater than predicted, especially from the carboxyl position of acetate (Table 14.1). Markus Koch prided himself on his "Swiss precision," as exemplified by his using a barometer in order to account for daily variations of atmospheric pressure when doing gas analyses, and insisted that the results were valid.

One potential explanation for the extra radioactivity we detected was isotopic exchange between acetate and CO_2. Isotopic exchange implies reversibility for reactions. We tested for reversibility by adding $^{14}CO_2$ to a culture actively oxidizing acetate and measuring label incorporation into acetate. We found that there was actually more $^{14}CO_2$ incorporation into acetate than conversion to $^{14}CH_4$ (Zinder and Koch, 1984). Using a *Methanosarcina* culture to split the labeled acetate th CH_4 and CO_2, we estimated that about three-quarters of the label was incorporated into the carboxyl position. This pattern of isotopic exchange would

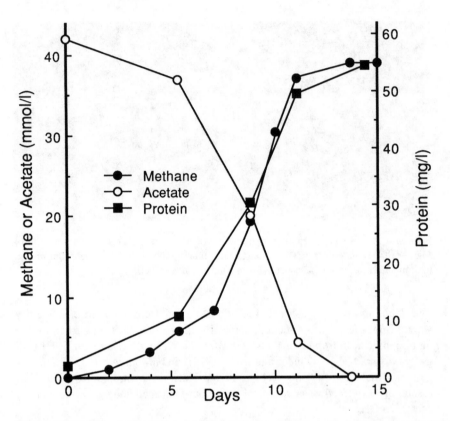

Figure 14.2.   Time course for acetate degradation, methanogenesis, and increase in cell protein by the acetate-oxidizing coculture grown on acetate. Replotted from data in Zinder and Koch (1984).

Table 14.1   Metabolism of radiolabeled substrates by the thermophilic acetate oxidizing coculture.

Label	Hours Incubated	$^{14}CO_2/^{14}CH_4$	^{14}C in Products Actual/Predicted[a]	% Label Recovery
$CH_3{}^{14}COO^-$	26	11.9	2.1	94
$^{14}CH_3COO^-$	26	13.6	1.2	94
$H^{14}CO_{3-}$	11	—	0.8	—

Note: Results are summarized from Zinder and Koch (1984).
[a] Predictions based on specific activities of substrates and on stoichiometries in reactions 14.2–14.4.

not be expected if acetate were oxidized by a pathway related to the citric acid cycle, which at that time was the only known metabolic route for acetate oxidation.

A classic review on acetogenesis (Ljungdahl and Wood, 1969) informed us that acetogens carried out isotopic exchange between CO_2 and acetate, especially at the carboxyl position. This was our first inkling that acetate oxidation by the coculture might be acetogenesis run "backwards." Indeed, acetyl-CoA/CO_2 exchange is now considered to be a hallmark of carbon monoxide dehydrogenase in acetogens (Ragsdale, 1991). On the basis of this idea, we tried to grow the AOR in the coculture acetogenically on H_2-CO_2 by adding 50 mM bromoethane sulfonate to inhibit *Methanobacterium* THF, which otherwise would rapidly use the H_2-CO_2. This attempt failed, as did an attempt to grow the thermophilic acetogen, *Acetogenium kivui*, on acetate coupled to a methanogen (Zinder and Koch, 1984).

14.3.3 H_2 Partial Pressures in the AOR/Methanobacterium THF Coculture

The energetics of the culture were of interest to us, as two organisms were "sharing" only -31 kJ of free energy per reaction. Using reaction 14.8 (with T = 333°K), we estimated (Zinder and Koch, 1984) that, at the midpoint of growth, the H_2 partial pressure in the culture should be greater than 0.9 Pa (1 Pa $\approx 10^{-5}$ atm), if the methanogen was to conserve energy but less than 20 Pa for the AOR to conserve energy. The geometric mean at which the energy available to both organisms would be maximized was near 4 Pa. We pointed out that if both organisms split the -31 kJ/rxn equally, only -16.5 kJ was available to the methanogen, far less than the $\Delta G_0'$ value of -135 kJ/rxn usually quoted for methanogenesis from H_2-CO_2.

Graduate student Monica Lee took over study of the coculture after Markus Koch's return to Switzerland. We measured H_2 partial pressures in the culture headspace and found that they varied between 20 and 40 Pa during growth of the coculture on acetate (Fig. 14.3). At the end of growth, the H_2 partial pressure decreased to 12–14 Pa and remained there, indicating a minimum threshold for H_2 utilization in *Methanobacterium* THF, a phenomenon which had been reported for mesophilic hydrogenotrophic methanogens (Lovley, 1985). The coculture also grew on ethanol, converting it to acetate much like the *M. omelianskii* coculture, and only started using acetate when the ethanol was depleted. The H_2 partial pressure increased to $\sim$ 150 Pa during ethanol consumption, and then decreased sharply to near 40 Pa once the ethanol was consumed followed by acetate disappearance. This suggested that the high H_2 partial pressures from the more energetically favorable ethanol oxidation did not allow acetate oxidation to occur until the H_2 was reduced to $\sim$ 40 Pa.

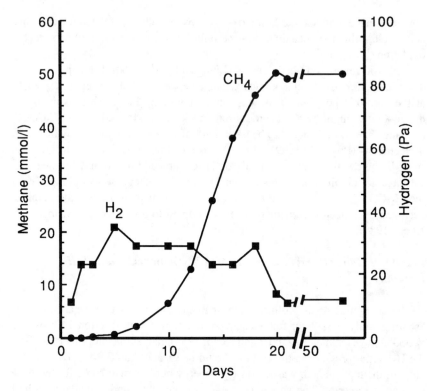

Figure 14.3.   Methanogenesis and H_2 partial pressure in the AOR/*Methanobacterium* THF coculture converting acetate to CH_4. Redrawn from Lee and Zinder (1988b).

The 20–40-Pa H_2 partial pressures detected during growth on acetate were essentially an order of magnitude higher than we had originally predicted from thermodynamic calculations. James Gossett, Department of Civil and Environmental Engineering at Cornell, suggested that we needed to correct the ΔG_f for each of the components in the reaction for entropy change at 60°C using the relationship

$$\Delta G_f = \Delta H_f - T\Delta S \qquad (14.12)$$

in which ΔG_f is the free energy of formation, ΔH_f is the enthalpy of formation, T is the temperature in °K, and ΔS is the entropy change. ΔH_f and ΔS are essentially constant over the temperature range involved (Barrow, 1974). Using this correction, the maximum and minimum predicted values for H_2 were 70 Pa and 2.6 Pa, respectively, so that the measured values of 20–40 Pa H_2 for acetate oxidation were well within this range. Zehnder and Stumm (Zehnder, 1988) calculated a similar temperature correction using the van't Hoff equation, a variant of Eq. (14.12). These calculations indicated a profound effect of temperature on the energetics of syntrophic reactions as shown in Figure 14.4. Predicted H_2 partial pressures in a syntrophic culture at 60°C would be roughly 250-fold higher than at 0°C. Some evidence has been obtained from measurements with other

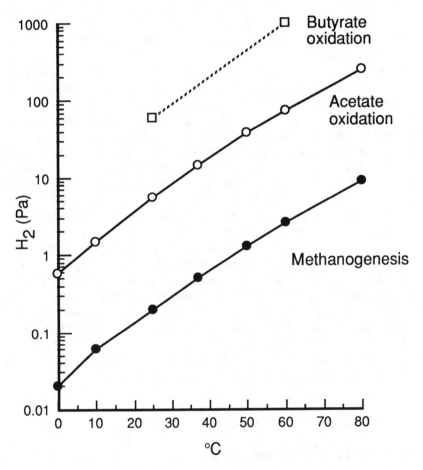

Figure 14.4. Effect of temperature on the H_2 partial pressures at which $\Delta G_0' = 0$ for acetate oxidation (reaction 14.2) or methanogenesis from H_2-CO_2 under conditions approximating the midpoint of growth of a syntrophic acetate-oxidizing methanogenic culture. For both organisms to conserve energy, the H_2 partial pressure must be within the "window" formed by both lines. Also shown is the effect of temperature on butyrate oxidation (reaction 14.9). [From Lee and Zinder, 1988b.]

syntrophic cocultures which corroborated this prediction (Zinder, 1990). Conrad and Wetter (1990) examined the effect of temperature on the minimum threshold value for H_2 consumption by mesophilic and thermophilic methanogens and acetogens and found good correlation with the values predicted using Eq. (14.12).

The minimum threshold for acetogenesis from H_2-CO_2 by *Acetogenium kivui* at 60°C was near 500 Pa. Assuming this value is typical for acetogens at that temperature, anaerobic habitats in which the H_2 partial pressure is greater than 500 Pa should allow acetogenesis from H_2, whereas habitats in which the H_2 partial pressure is less than $\sim$ 40 Pa should allow syntrophic acetate oxidation. Neither process should occur in habitats in which the H_2 partial pressure is between 40 and 500 kPa.

The H_2 partial pressures we detected were in accord with H_2 being an intermediate in the coculture. However, Thiele and Zeikus (1988) had proposed that formate was a more important interspecies electron carrier than H_2 in many syntrophic anaerobic systems because its equilibrium concentrations would be in the micromolar range rather than nanomolar as for H_2, allowing less diffusion limitation of interspecies electron flow (Boone et al., 1989). Because *Methanobacterium* THF could use formate, it was possible that formate was mediating interspecies electron transfer in the acetate oxidizing coculture. The isolation of the AOR from the coculture (section 14.4.1) afforded us the opportunity to examine this possibility because we could now recombine the AOR with a methanogen which could not use formate. We verified that *Methanobacterium thermoautotrophicum* strain ΔH did not use formate, and we combined this culture with the AOR on ethanol/acetate medium (Lee and Zinder, 1988b). The conversion of substrates and partial pressures of H_2 and were indistinguishable from those for cultures reconstituted with *Methanobacterium* THF, indicating that formate was not an obligatory electron carrier in the coculture. The performance of a thermophilic propionate utilizer was also unaffected by whether the coupling methanogen was capable of using formate (Stams et al., 1992). Formate may be important in other syntrophic systems, especially mesophilic ones which would have lower H_2 concentrations and greater diffusion limitation.

14.4   The Isolated Acetate Oxidizing Rod (AOR)

14.4.1   Isolation and Characterization

Our initial attempts to obtain an axenic culture of the AOR failed. We could grow the coculture with ethanol, but the energetics of ethanol oxidation were not sufficiently favorable to allow isolation of the AOR. Pyruvate stimulated growth of the culture but not consistently (in retrospect probably due to toxicity of some pyruvate preparations at high concentrations). Other substrates we tried which would allow axenic growth of the AOR were not utilized (Zinder and Koch, 1984). Bernhard Schink suggested that we try ethylene glycol as a substrate, as he and his coworkers had used it with success in the isolation of other ethanol-oxidizing syntrophs (Eichler and Schink, 1985). Ethylene glycol is essentially a hydrated form of acetaldehyde, a high-energy intermediate, and the its oxidation to acetate is thermodynamically favorable:

$$\text{Ethylene glycol} \rightarrow \text{Acetate}^- + \text{H}^+ + \text{H}_2 \qquad \Delta G'_o = -78.7 \text{ kJ} \quad (14.13)$$

We found that the coculture was able to grow on ethylene glycol, and after one transfer the AOR morphotype represented 80–90% of the bacterial cell numbers in microscopic counts, rather than about half as in the acetate-oxidizing coculture (Lee and Zinder, 1988c). A 10^{-8} dilution of this culture into liquid medium grew on ethylene glycol and did not make CH_4. Although colony formation in ethylene glycol agar was poor, the culture was later found to grow on pyruvate and isolated colonies on pyruvate agar were obtained. The isolated culture could be recombined with *Methanobacterium* THF and the recombined culture was capable of growth on acetate. The lag before acetate utilization commenced by recombined cultures was greatly shortened by adding either 10 mM ethanol or ethylene glycol along with the acetate.

The isolated AOR was 2–3 μm by 0.4–0.6 μm in size, and it stained Gram negative. Thin-section electron micrographs showed a thin peptidoglycan layer surrounded by a densely staining layer which, in some negatively stained preparations, resembled an S layer (Lee and Zinder, 1988c). No evidence for an outer membrane, as is found in most Gram-negative bacteria, was found. A partial sequence of the 16S rRNA (B. White and D. Stahl, unpublished) indicates grouping in the Gram-positive branch of the eubacteria. The AORs DNA had a G+C ratio of 47 mol%.

14.4.2 Reversible Acetogenesis

The isolated AOR used only a very limited range of the substrates we tested. Axenic cultures did not show any evidence for utilization of sugars, amino acids, organic acids, or alcohols. We had suspected that the AOR was an acetate-utilizing sulfate reducer, but no evidence was obtained for utilization of sulfate or any other external electron acceptor. Our initial screen of substrates indicated that the AOR grew on ethylene glycol, pyruvate, or betaine, and that the major product was acetate in amounts suggesting acetogenesis. Thus, it seemed reasonable to test growth on H_2-CO_2. The AOR grew on H_2-CO_2 and produced essentially stoichiometric quantities of acetate (Fig. 14.5), showing unequivocally that it was an acetogen as our early isotopic exchange data had suggested. Acetogenesis from H_2-CO_2 has the reaction

$$2 \, HCO_{3-} + 4 \, H_2 + H^+ \rightarrow CH_3COO^- + 4 \, H_2O, \qquad \Delta G'_0 = \qquad (14.4)$$
$$-104.6 \, kJ/rxn$$

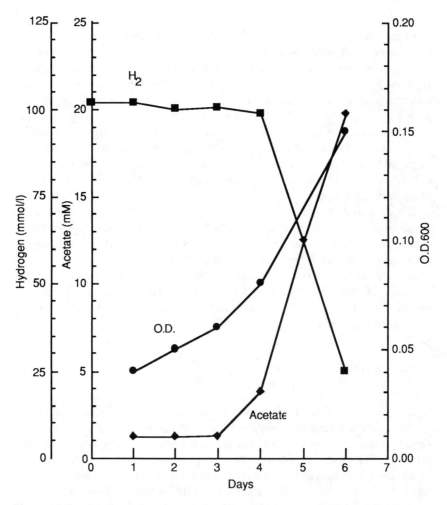

Figure 14.5.   Axenic acetogenic growth of the AOR on H_2-CO_2. O.D.600 = optical density at 600 nm [from Lee and Zinder (1988c)].

This reaction is the reverse of reaction 14.2 and, as such, has the opposite free energy. The implication of this finding is that the AOR can conserve energy carrying out this reaction in either direction depending on the concentration of products and reactants! When the H_2 partial pressure is greater than 500 Pa (0.005 atm, see Section 14.3.3), it will grow acetogenically on H_2-CO_2, whereas H_2 partial pressures below 40 Pa will allow syntrophic growth by oxidizing acetate. Because of this remarkable ability, we have given the AOR the nickname *"Reversibacter,"* a name which has struck the fancy of some reviewers (Thauer et al., 1989). It should also be mentioned that the growth on the AOR on H_2-CO_2 was considerably slower and to lower density than that of *Acetogenium kivui*, which we used as a positive control.

We found that the AOR could also grow acetogenically on formate or CO. When growing on betaine, only one methyl group was fermented to acetate, leaving dimethylglycine (Lee and Zinder, 1988c). Thus, the AOR grew on only a limited number of one- and two-carbon substrates and always produced acetate as its main product.

14.4.3   Biochemistry of the AOR

Showing that the AOR was an acetogen gave more credence to the notion that it was using a pathway resembling a reversal of the Wood acetogenic pathway. In terms of substrate-level phosphorylation, the Wood pathway (Ljungdahl, 1986) involves no net production or consumption of ATP, because the acetate kinase and formyltetrahydrofolate synthetase reactions cancel each other out in the pathway operating either forward or backward. We decided to examine the AOR pure culture and the acetate-oxidizing coculture for evidence of the Wood pathway (Lee and Zinder, 1988a). During the course of these studies, a paper showing that oxidizing sulfate-reducing bacteria could use a reversal of the acetogenic pathway for acetate oxidation was published (Schauder et al., 1986), making the possibility of a reversed Wood pathway in the AOR more likely and less novel.

Carbon monoxide dehydrogenase (CODH) is sensitive to inhibition by cyanide, and because of this, pathways involving CODH are more susceptible to cyanide inhibition than are most other anaerobic processes. For example, 200 μM cyanide inhibited carbon dioxide fixation in *Methanobacterium thermoautotrophicum* Marburg which uses a variant of the Wood pathway for this process, whereas methanogenesis was unaffected (Stupperich and Fuchs, 1984). Similarly, when 200 μM cyanide was added to an active acetate-oxidizing coculture, there was an immediate cessation of $^{14}CH_3COO^-$ oxidation to $^{14}CO_2$ (Fig. 14.6). On the other hand, 200 μM cyanide did not inhibit $^{14}CH_4$ production from $^{14}CO_2$ by the coculture as long as H_2 was also added to compensate for the inhibition of H_2 production via acetate oxidation (Lee and Zinder, 1988a), indicating that

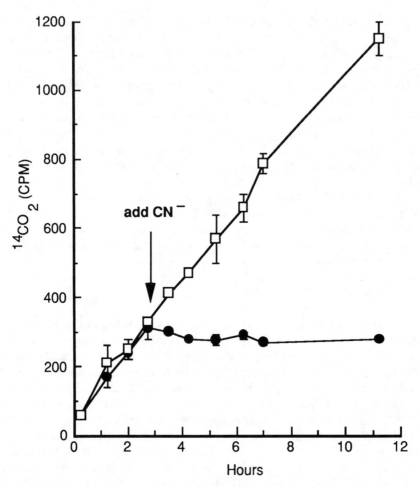

Figure 14.6.   Effect of addition of 200 μM cyanide on $^{14}CO_2$ production from $^{14}CH_3COO^-$ by actively metabolizing AOR/*Methanobacterium* THF cocultures.

methanogenesis from H_2-CO_2 per se was not inhibited by cyanide similar to previous results (Stupperich and Fuchs, 1984). These experiments, therefore, showed that cyanide was inhibitory to acetate oxidation, consistent with the operation of a reversal of the Wood acetogenic pathway.

We examined cell-free extracts from the acetate oxidizing coculture, as well as the isolated AOR and *Methanobacterium* THF grown on H_2-CO_2 for activity of CODH, formate dehydrogenase (FDH), and formyl tetrahydrofolate synthetase (FTHFS), all key enzymes of the Wood pathway. Isocitrate dehydrogenase was used as a marker for the citric acid cycle, because it is not found in *M. thermoauto-trophicum*, which uses an incomplete citric acid cycle for biosynthesis (Zeikus et al., 1977). As shown in Table 14.2, the specific activity of CODH in both the coculture and the axenic AOR were at catabolic levels, whereas CODH activity in strain THF was low, consistent an anabolic role. Since about half of the cell protein in the coculture was from the methanogen, the actual specific activity of CODH due to the AOR was comparable in the monoculture and the axenic culture. Native polyacrylamide gels stained for CODH activity showed bands in lanes from the coculture which were not present in either axenic culture, suggesting either new forms of CODH or associations with different proteins. Formate dehydrogenase was also high in the AOR and the coculture but not in strain THF. Levels of isocitrate dehydrogenase were low in all instances, making acetate catabolism by the citric acid cycle unlikely.

The results of the assay for FTHFS were puzzling because no activity was found in extracts from the AOR coculture or in the acetogenically grown monoculture. High FTHFS activity was found in extracts from *Acetogenium kivui*, indicating that our assay procedures were sound. One possibility was that a pterin derivative other than tetrahydrofolate was utilized by the AOR. We found that extracts showed fluorescence spectra typical for folate derivatives. Bioassays for folate (kindly performed by V. Worrell and D. Nagle) in the AOR monoculture and coculture indicated that folate activities were low, typical of those found in organisms which use folates biosynthetically rather than catabolically. This could be due to low levels of tetrahydrofolate in the cells or to high levels of a pterin which gave slight cross-reactivity with folic acid. Treating the extracts with the enzyme conjugase, which removes extra glutamates from pterin side chains, increased the activity in the bioassay using *Streptococcus fecaelis*, indicating more than one glutamate on the side chain. The identity of the pterin in the AOR

Table 14.2. Activities of enzymes of the Wood pathway and the citric acid cycle in crude extracts of the AOR/*Methanobacterium* THF coculture, the two component organisms, and in *Acetogenium kivui*.

Organism	Growth Substrate	CODH[a]	FDH[a]	FTHF Synthetase[b]	Isocitrate Dehydrogenase[c]
AOR/THF coculture	Acetate	2.0	9.9	<0.01	0.015
AOR monoculture	H_2-CO_2	6.4	29.7	<0.01	0.015
THF	H_2-CO_2	0.055	0.001	<0.01	<0.001
Acetogenium kivui	H_2-CO_2	56.4	ND[d]	7.5	ND

[a] methyl viologen reduced per minute per milligram protein.
[b] Micromoles of formyltetrahydrofolate produced per minute per milligram protein.
[c] NADP reduced per minute per milligram protein.
[d] Not determined.

is unresolved. It is of interest that in *Desulfobacterium autotrophicum,* which can fix CO_2 via the Wood pathway or oxidize acetate via a reversal of this pathway (see Section 14.5), contained a novel tetrahydropterin which had four side-chain glutamates (Länge et al., 1989).

The enzymological evidence was consistent with reversal of the Wood acetogenic pathway in the AOR when oxidizing acetate and suggested that a pterin other than tetrahydrofolate was present in the AOR when grown either acetogenically or acetotrophically. Since the graduation of Monica Lee, our laboratory has done little in terms of further characterization of the AOR.

14.5   Related Examples of Acetate Oxidation via the Wood Acetogenic Pathway in Other Organisms

Although the AOR is presently the only known organism which carries out acetate oxidation coupled to reduction of protons to H_2, it is now known that several organisms use variants of the Wood acetogenic pathway to oxidize acetate. These will be briefly discussed here, and are discussed more completely in Chapter 19.

The best documented cases for the use of a reversal of the Wood acetogenic pathway for acetate oxidation are for the subset of sulfate- and sulfur-reducing bacteria which are capable of complete oxidation of organic substrates to CO_2, meaning that they can oxidize acetyl-CoA and usually exogenously added acetate (Widdel, 1988). These strains are differentiated from other sulfate reducers, such as common *Desulfovibrio* strains, which only partially oxidize their substrates, usually excreting acetate. An enzymological survey of 10 acetyl-CoA oxidizing sulfate reducers (Schauder et al., 1986) showed that eight of them had high levels of CODH and formate dehydrogenase, suggesting the presence of the Wood pathway. *Desulfobacter postgatei,* which uses a variant of the citric acid cycle for acetate oxidation (Thauer et al., 1989), had undetectable levels of these enzymes.

Desulfotomaculum acetoxidans was studied further, and all the enzymes of the Wood pathway were demonstrated in cells grown on acetate and SO_4^{2-} (Spormann and Thauer, 1988). *Desulfobacterium autotrophicum* uses the Wood pathway for CO_2 fixation when growing on H_2 and SO_4^{2-} and was also shown to have similar acetogenic enzyme activities when growing on acetate and SO_4^{2-} (Schauder et al., 1986), thus showing a form of reversible acetogenesis. There was an additional band of CODH activity in native polyacrylamide gels from cells growing on H_2 when compared to those growing on acetate, suggesting a novel enzyme or association (Schauder et al., 1989). Another example of reversible acetogenesis in sulfate reducers was *Desulfotomaculum thermoacetoxidans* (Min and Zinder, 1990), which grew on acetate and SO_4^{2} and apparently oxidized acetate using a reversed Wood pathway as it had high CODH activity. When this culture grew on H_2 and SO_4^{2-} in the presence of CO_2, it produced both acetate and sulfide. Interestingly, it could not grow purely as an acetogen, as it required sulfate for growth on H_2-CO_2.

The extremely thermophilic archaebacterial sulfate reducer, *Archaeoglobus fulgidus*, completely oxidizes pyruvate, but not acetate, to CO_2. Acetyl-CoA produced from pyruvate oxidation is oxidized via a pathway in which the methyl group is carried by tetrahydromethanopterin and methanofuran rather than tetrahydrofolate (Thauer et al., 1989), consistent with a close phylogenetic relationship between *Archaeoglobus* with methanogens. Otherwise, the pathway resembles a reversal of the Wood pathway. Thus, several sulfate reducers have been shown to use a reversal of the Wood pathway for acetate oxidation, and some even show forms of reversible acetogenesis.

Acetate-utilizing methanogens oxidize a small portion of acetate to CO_2 to provide reducing power for biosynthesis, especially reduction of acetyl-CoA to pyruvate (Balch et al., 1979). It was first demonstrated in 1957 (Pine and Vishniac, 1957) that *Methanosarcina* cultures, when growing on methanol in the presence of acetate, oxidized a significant fraction of the methyl group of acetate to CO_2, a finding verified in subsequent studies (Weimer and Zeikus, 1978; Zinder and Mah, 1979). Presumably, the methyl group of acetate is oxidized as tetrahydromethanopterin and methanofuran intermediates in a manner similar to *Archaeoglobus*. Phelps et al. (1985) showed that *Methanosarcina* cultures, when grown on acetate in the presence of SO_4^{2-} and the hydrogenotrophic sulfate-reducing bacterium *Desulfovibrio vulgaris*, produced less CH_4 and showed greater oxidation of $^{14}CH_3COO^-$ to $^{14}CO_2$. The partial pressures of H_2 in the headspace of a pure culture of *Methanosarcina* grown on acetate was near 25 Pa (H_2 production during growth of *Methanosarcina* on acetate is well documented (Ahring et al., 1991; Bhatnagar et al., 1987; Lovley and Ferry, 1985; Zinder and Anguish, 1992)), whereas that of a culture grown in the presence of *D. vulgaris* was near 1.2 Pa, considerably lower. These results indicate that the sulfate reducer pulled the reaction toward acetate oxidation by removing electrons as H_2: a form of interspecies H_2 transfer resembling that in the AOR coculture. There was a slight decrease in the measured biomass of the *Methanosarcina* in these cocultures, suggesting that this interaction was not beneficial to the methanogen.

As will be discussed more completely in Chapter 23, glycine fermentation is a form of reversal of the Wood pathway because glycine decarboxylase transforms glycine to CO_2, ammonia, and methylene tetrahydrofolate. The methylene group is then oxidized to CO_2, and the electrons are used for reductive deamination of glycine via glycine reductase. Thus, glycine serves as an electron acceptor for the oxidation of the tetrahydrofolate derivatives. Attempts to grow the AOR on acetate with glycine or betaine as an electron acceptor failed (Zinder and Koch,

1984). Glycolic acid can be either fermented by pure cultures or oxidized by methanogenic syntrophic cultures (Friedrich et al., 1991). However, glycolate oxidation appears to be by a variant on the citric acid cycle rather than the Wood pathway.

14.6   Ecology of Methanogenic Acetate Oxidation

14.6.1   General Considerations

The key test to determine whether acetate in an anaerobic habitat is decarboxylated to CH_4 and CO_2 or is oxidized to CO_2 is to examine the fate of $^{14}CH_3COO^-$ added to the habitat. A "respiratory index" (RI) can be defined as

$$RI = \frac{^{14}CO_2}{^{14}CO_2 + {}^{14}CH_4} \tag{14.15}$$

Thus, complete acetate conversion to $^{14}CH_4$ would lead to an RI of 0, whereas complete conversion to $^{14}CO_2$ gives an RI of 1.0. In some cases [U-^{14}C]acetate has been used instead of methyl-labeled acetate, so that methanogenic fermentation of acetate yields an RI of 0.5. Clearly, the precision for determination of small amounts of acetate oxidation is greater when using methyl-labeled acetate. An RI value indicating acetate oxidation is often interpreted as indicating the presence of an external electron acceptor, usually sulfate (Sansone and Martens, 1981), and this conclusion may be justified in many cases in which the presence of an electron acceptor is demonstrated. However, the possibility that syntrophic methanogenic acetate oxidation must be considered for some habitats in which an external electron acceptor is absent or in low abundance. Only a few examples exist indicating syntrophic acetate oxidation.

14.6.2   Anaerobic Bioreactors

$^{14}CH_3COO^-$ added to anaerobic bioreactors not receiving sulfate or other potential electron acceptors is usually converted nearly quantitatively to $^{14}CH_4$ (Zinder et al., 1984). One exception involved a study of acetate turnover in the methanogenic reactor in a two-phase mesophilic anaerobic bioreactor system. This reactor received fatty acids from an acidogenic phase treating an artificial complex waste (Weber et al., 1984). It was found that 22–31% of the $^{14}CH_3COO^-$ added to these reactors was detected as $^{14}CO_2$ after incubation with the sludge, whereas 41–51% was converted to $^{14}CH_4$, indicating oxidation of a significant fraction of the acetate added. Complicating these studies was the finding of nearly identical ratio of products from [U-^{14}C]acetate. This finding was rationalized as indicating a high rate of CO_2 conversion to CH_4, but this seems unlikely because turnover of $^{14}CO_2$ produced from $^{14}CH_3COO^-$ was relatively slow. This suggests a flaw in the scheme used for fractionation and analysis of radioactivity.

More recently, Peterson and Ahring (1991) examined methanogenesis from acetate in a thermophilic (60° C) anaerobic bioreactor being fed sewage sludge in which *Methanosarcina* was apparently the dominant acetotrophic methanogen. When digestor contents were incubated with 4.5 mM acetate labeled with $^{14}CH_3COO^-$, the RI was 0.04, whereas incubation with ~ 1.0 mM acetate led to an increase in the RI to 0.14. This was interpreted in light of evidence that *Methanosarcina* shows a minimum threshold for acetate utilization near 1 mM (Min and Zinder, 1989; Ohtsubo et al., 1992; Westermann et al., 1989). *Methanothrix* can use acetate at concentrations as low as to 5–10 μM but was apparently absent from the culture, so it was presumed that an acetate-oxidizing coculture was competing with *Methanosarcina* more successfully at low acetate concentration. A chemostat culture derived from the reactor was eventually dominated by a two-membered culture consisting of a short rod-shaped autofluorescent methanogen, and a rod which was clearly much longer than the AOR. The minimum threshold for acetate utilization by this enrichment culture was near 0.5 mM. Petersen and Ahring (1991) proposed that low acetate concentrations favored the acetate-oxidizing coculture over *Methanosarcina*, although it is not clear why *Methanothrix* was absent from the reactor. An examination by the author of the original data for Figure 14.2 indicates that the AOR/*Methanobacterium* THF coculture was capable of lowering the acetate concentration to less than 0.1 mM, suggesting that it could also outcompete *Methanosarcina* at low acetate concentrations, consistent with the model proposed by Petersen et al.

14.6.3   *Marine and Freshwater Habitats*

Carbon flow in upper layers of marine sediments is clearly driven by sulfate reduction, because of the high amount of sulfate present in seawater (~ 28 mM) and the ability of acetate-oxidizing sulfate reducers to outcompete methanogens for acetate (Schönheit et al., 1982). Therefore, it is not surprising that $^{14}CH_3COO^-$ added to such sediments is usually completely oxidized to $^{14}CO_2$ (King et al., 1983). In fact, only a few methylated compounds appear to be able to serve as CH_4 precursors in sulfate-containing marine sediments. However, in certain marine sediments in which there is sufficient organic matter present to consume all of the sulfate, sulfate may be absent in underlying sediment layers. In these layers, methanogenesis becomes the primary pathway of carbon flow (Oremland, 1988). Studies with [^{14}C]acetate indicate significant acetate oxidation in sulfate-depleted sediments (Sansone and Martens, 1981; Warford et al., 1979). There can also be acetate decarboxylation in some sediments, and marine strains of

Methanosarcina capable of methanogenesis from acetate do exist (Sowers et al., 1984; Sowers and Gunsalus, 1988). Stable isotope fractionation data also support the contention that CO_2 is the major CH_4 precursor in some sulfate-depleted sediments (Claypool and Kaplan, 1974; Witicar et al., 1986). Thus, it is likely that syntrophic methanogenic acetate oxidation occurs in at least some sulfate-depleted marine sediments.

In freshwater sediments, sulfate concentrations are typically ≤ 1 mM in the overlying waters, so that sediments are usually primarily methanogenic. Studies of acetate metabolism are often performed on the upper sediment layers which may contain some sulfate. For example, Winfrey and Zeikus (1977) found that about 20% of the $^{14}CH_3COO^-$ added alone to Lake Mendota sediments was oxidized to $^{14}CO_2$, whereas addition of 10 mM sulfate resulted in complete conversion to $^{14}CO_2$. Thus, there was an active population of acetate-oxidizing sulfate reducers present in the sediments. It is not clear whether the oxidation found in the absence of added sulfate was due to the presence of sulfate or other oxidized forms of sulfur in the sediments or to syntrophic oxidation.

14.7   Summary, Conclusions, and Unanswered Questions

The thermophilic AOR-methanogen coculture clearly converts acetate to CH_4 by the mechanism envisioned by van Niel and Barker. However, a significant difference from their original concept is that instead of a single organism carrying out the two-step process, two syntrophic organisms were involved. Remarkably, acetate oxidation apparently occurs via a reversal of the Wood acetogenic pathway, and, even more remarkably, the acetate oxidizing organism is capable of acetogenic growth on H_2-CO_2.

Presently unknown are whether there are any differences between the pathway operating in one direction versus the other. As the pathway for carbon flow is essentially energy neutral, it is likely that differences between the acetate-oxidizing and acetogenic cells of the AOR involve differences in electron transport components. For example, it would be advantageous for the AOR, when oxidizing acetate, to have hydrogenase operating on the cytoplasmic side of the cell membrane so that protons would be consumed during H_2 production. By the same considerations, acetogenic cells should have their hydrogenases releasing protons outside the cell membranes during H_2 consumption. CODH has been found to be associated with the cell membrane in gently lysed acetogenic bacteria and may play a role in proton motive force generation (Hugenholtz and Ljungdahl, 1990), consistent with the potentially different forms of CODH seen in native polyacrylamide gels of acetate-oxidizing and acetate-producing cells (Lee and Zinder, 1988a; Schauder et al., 1989). Biochemical studies on the AOR/THF coculture are hampered by interference from enzymes of the methanogenic partner, although it is possible that the two cell types might be separated by Percoll gradient centrifugation as has been done for a *Syntrophomonas* coculture (Beaty et al., 1987).

As there was, until recently, only one example of a culture carrying out syntrophic acetate oxidation, it is not clear how widespread a phenomenon it is. With the recent description of a novel syntrophic thermophilic acetate-oxidizing enrichment (Petersen and Ahring, 1991), the number of proton-reducing acetate oxidizers is likely to be doubled in the near future upon isolation of the acetate oxidizer from that culture. It will be interesting to compare the substrate utilization spectrum of this organism with the relatively narrow one of the AOR. The ability to use other substrates may be a competitive advantage for syntrophic acetate oxidizers over the acetotrophic methanogens, which have an even more restricted substrate range.

In terms of mesophilic habitats, there is evidence for syntrophic acetate oxidation in some methanogenic marine sediments, although the microorganisms responsible for the process are presently unknown. There exists the intriguing possibility that acetate-oxidizing sulfate-reducing bacteria are coupling to methanogens upon sulfate depletion in a manner analogous to coupling of certain lactate- or ethanol-utilizing sulfate reducers to methanogens upon sulfate depletion (Bryant et al., 1977). However, attempts by the author and colleagues (Min and Zinder, 1990) and by F. Widdel (personal communication) to obtain syntrophic coupling between acetate-oxidizing sulfate reducers and hydrogenotrophic methanogens have thus far failed.

Finally, little is known about whether other acetogens are capable of "reversible acetogenesis." *Acetogenium kivui* was negative when we tested it (Zinder and Koch, 1984), but in that test, the organism was required to grow in medium with acetate as essentially the only substrate. The AOR showed long lags under those conditions (Lee and Zinder, 1988c), but those lags were much shorter when a readily utilized cosubstrate such as ethanol or pyruvate allowed them to grow and then shift to acetate oxidation. There have been some experiments on coupling between acetogens and methanogens. *Acetobacterium woodii*, when grown on fructose, produced only 2 mol of acetate per mole of hexose when grown in the presence of hydrogenotrophic methanogens (Winter and Wolfe, 1980), because the methanogens could outcompete the acetogens for their own electrons which otherwise would be used to reduce CO_2 to acetate. No evidence for subsequent oxidation of acetate was obtained in these syntrohic cocultures. Likewise, *Sporomusa* cultures, when grown on methanol or fructose, reduced less than CO_2 to acetate when grown together with hydrogenotrophic methanogens or sulfate, nitrate, or fumarate reducers (Cord-Ruwischet al., 1988). No evidence for subsequent acetate oxidation was described in these cultures either. Thus, acetate oxidation may not be common in typical acetogens, but more direct tests need to be made.

The discoveries of a syntrophic acetate-oxidizing methanogenic coculture and that the pathway of acetate oxidation in this culture most likely represented a reversal of the Wood acetogenic pathway were both surprising. However, our studies coincided with findings by others that CODH enzyme complexes could be involved in either acetyl-CoA synthesis in acetogens or acetyl-CoA disassembly in methanogens and sulfate reducers (Grahame, 1991; Ragsdale, 1991; Thauer et al., 1989; Krzycki et al., 1985) which helped put our findings in perspective: Syntrophic acetate oxidation is simply another example of the importance of this fascinating pathway in carbon flow in anaerobic habitats.

Acknowledgments

The author's research on acetate metabolism by anaerobes has been supported by grant DE-FG02-85ER13370 from the U.S. Department of Energy. It is a pleasure to thank my colleagues Markus Koch, Monica Lee, and Tim Anguish for their contributions to this research.

References

Ahring, B. K., P. Westermann, and R. A. Mah. 1991. Hydrogen inhibition of acetate metabolism and kinetics of hydrogen consumption by *Methanosarcina thermophila* TM-1. *Arch. Microbiol.* **157**:38–42.

Balch, W. E., G. E. Fox, M. J. Magrum, C. R. Woese, and R. S. Wolfe. 1979. Methanogens: reevaluation of a unique biological group. *Microbiol. Rev.* **43**:260–296.

Barker, H. A. 1936a. On the biochemistry of the methane fermentation. *Arch. Mikrobiol.* **6**:404–419.

Barker, H. A. 1936b. Studies on the methane-producing bacteria. *Arch. Mikrobiol.* **7**:420–438.

Barker, H. A., S. Ruben, and M. D. Kamen. 1940. The reduction of radioactive carbon dioxide by methane-producing bacteria. *Proc. Natl. Acad. Sci. USA* **26**:426–430.

Barrow, G. M. 1974. Physical Chemistry for the Life Sciences. McGraw Hill Book Co., New York.

Beaty, P. S., N. Q. Wofford, and M. J. McInerney. 1987. Separation of *Syntrophomonas wolfei* from *Methanospirillum hungatei* using Percoll gradients. *Appl. Environ. Microbiol.* **53**:1183–1185.

Bhatnagar, L., J. A. Krzycki, and J. G. Zeikus. 1987. Analysis of hydrogen metabolism in *Methanosarcina barkeri* regulation of hydrogenase and role of CO-dehydrogenase in H_2 production. *FEMS Microbiol. Lett.* **41**:337–343.

Boone, D. R., and M. P. Bryant. 1980. Propionate-degrading bacterium, *Syntrophobacter wolinii* sp. nov. gen. nov., from methanogenic ecosystems. *Appl. Environ. Microbiol.* **40**:626–632.

Boone, D. R., R. L. Johnson, and Y. Liu. 1989. Diffusion of the interspecies electron carries H₂ and formate in methanogenic ecosystems and its implications in the measurement of K_m for H_2 or formate uptake. *Appl. Environ. Microbiol.* **55**:1735–1741.

Bryant, M. P., E. A. Wolin, M. J. Wolin, and R. S. Wolfe. 1967. *Methanobacillus omelianskii*, a symbiotic association of two species of bacteria. *Arch. Microbiol.* **59**:20–31.

Bryant, M. P., L. L. Campbell, C. A. Reddy, and M. R. Crabill. 1977. Growth of *Desulfovibrio* in lactate or ethanol media low in sulfate in association with H_2-utilizing methanogenic bacteria. *Appl. Environ. Microbiol.* **33**:1162–1169.

Bryant, M. P., L. L. Campbell, C. A. Reddy, and M. R. Crabill. 1977. Growth of *Desulfovibrio* in lactate or ethanol media low in sulfate in association with H_2-utilizing methanogenic bacteria. *Appl. Environ. Microbiol.* **33**:1162–1169.

Buswell, A. M., and F. W. Sollo. 1948. The mechanism of the methane fermentation. *J. Am. Chem. Soc.* **70**:1778–1780.

Claypool, G. E., and I. R. Kaplan. 1974. The origin and distribution of methane in marine sediments. In: *Natural Gases in Marine Sediments*, I. R. Kaplan (ed.), pp. 99–140. Plenum Press, New York.

Conrad, R., and B. Wetter. 1990. Influence of temperature on energetics of hydrogen metabolism in homoacetogenic, methanogenic, and other anaerobic bacteria. *Arch. Microbiol.* **155**:94–98.

Coolhaas, V. C. 1928. Zur kenntnis der dissimilation fettsaurer salze und kohlenhydrate durch thermophile bakerien. *Zbl Bakteriol Parasitkenkd Infektionskr Hyg Abt 2* **75**:161–170.

Cord-Ruwisch, R., H.-J. Steitz, and R. Conrad. 1988. The capacity of hydrogenotrophic anaerobic bacteria to compete for traces of hydrogen depends on the redox potential of the terminal electron acceptor. *Arch. Microbiol.* **149**:350–357.

DiMarco, A. A., T. A. Bobik, and R. S. Wolfe. 1990. Unusual coenzymes of methanogenesis. *Ann. Rev. Biochem.* **59**:355–394.

Eichler, B., and B. Schink. 1985. Fermentation of primary alcohols and diols and pure culture of syntrophically alcohol-oxidizing anaerobes. *Arch. Microbiol.* **143**:60–66.

Ferry, J. D. 1992. Methane from acetate. *J. Bacteriol.* **174**:5489–5495.

Friedrich, M., U. Laderer, and B. Schink. 1991. Fermentative degradation of glycolic acid by defined syntrophic cultures. *Arch. Microbiol.* **156**:398–404.

Grahame, D. A. 1991. Catalysis of acetyl-CoA cleavage and tetrahydrosarcinapterin methylation by a carbon monoxide dehydrogenase-corrinoid enzyme complex. *J. Biol. Chem.* **266**:22227–22233.

Hugenholtz, J., and L. G. Ljungdahl. 1990. Metabolism and energy generation in homoacetogenic clostridia. *FEMS Microbiol. Rev.* **87**:383–390.

King, G. M., M. J. Klug, and D. R. Lovley. 1983. Metabolism of acetate, methanol, and methylated amines in intertidal sediments of Lowes Cove, Maine. *Appl. Environ. Microbiol.* **45**:1848–1853.

Krzycki, J. A., L. J. Lehman, and J. G. Zeikus. 1985. Acetate catabolism by *Methanosar-*

cina barkeri: evidence for involvement of carbon monoxide dehydrogenase, methyl coenzyme M and methylreductase. *J. Bacteriol.* **163**:1000–1006.

Länge, S., R. Scholtz, and G. Fuchs. 1989. Oxidative and reductive acetyl CoA/carbon monoxide dehydrogenase pathway in *Desulfobacterium autotrophicum*. 1. Characterization and metabolic function of the cellular tetrahydropterin. *Arch. Microbiol.* **151**:77–83.

Lee, M. J., and S. H. Zinder. 1988a. Carbon monoxide pathway enzyme activities in a thermophilic anaerobic bacterium grown acetogenically and in a syntrophic acetate-oxidizing coculture. *Arch. Microbiol.* **150**:513–518.

Lee, M. J., and S. H. Zinder. 1988b. Hydrogen partial pressures in a thermophilic acetate-oxidizing methanogenic coculture. *Appl. Environ. Microbiol.* **54**:1457–1461.

Lee, M. J., and S. H. Zinder. 1988c. Isolation and characterization of a thermophilic bacterium which oxidizes acetate in syntrophic association with a methanogen and which grows acetogenically on H_2-CO_2. *Appl. Environ. Microbiol.* **54**:124–129.

Ljungdahl, L. G. 1986. The autotrophic pathway of acetate synthesis in acetogenic bacteria. *Annu. Rev. Microbiol.* **40**:415–450.

Ljungdahl, L. G., and H. G. Wood. 1969. Total synthesis of acetate from CO_2 by heterotrophic bacteria. *Annu. Rev. Microbiol.* **23**:515–538.

Lovley, D. R. 1985. Minimum threshold for hydrogen metabolism in methanogenic bacteria. *Appl. Environ. Microbiol.* **49**:1530–1531.

Lovley, D. R., and J. G. Ferry. 1985. Production and consumption of H_2 during growth of *Methanosarcina* spp. on acetate. *Appl. Environ. Microbiol.* **49**:247–249.

Mah, R. A., M. R. Smith, and L. Baresi. 1978. Studies on an acetate fermenting strain of *Methanosarcina*. *Appl. Environ. Microbiol.* **35**:1174–1184.

McInerney, M. J., M. P. Bryant, and N. Pfennig. 1979. Anaerobic bacterium that degrades fatty acids in association with methanogens. *Arch. Microbiol.* **122**:129–135.

McInerney, M. J., M. P. Bryant, R. B. Hespell, and J. W. Costerton. 1981. *Syntrophomonas wolfei* gen. nov. sp. nov., an anaerobic, syntrophic, fatty acid-oxidizing bacterium. *Appl. Environ. Microbiol.* **41**:1029–1039.

Min, H., and S. H. Zinder. 1989. Kinetics of acetate utilization by two thermophilic acetotrophic methanogens: *Methanosarcina* sp. strain CALS-1 and *Methanothrix* sp. strain CALS-1. *Appl. Environ. Microbiol.* **55**:448–491.

Min, H., and S. H. Zinder. 1990. Isolation and characterization of a thermophilic sulfate reducing bacterium. *Desulfotomaculum thermoacetoxidans* sp. nov. *Arch. Microbiol.* **153**:399–404.

Nölling, J., and W. M. De Vos. 1992. Characterization of the archael plasmid-encoded Type II restriction-modification system Mthti from *Methanobacterium thermoformicicum*. THF. Homology to the bacterial Ngopii system from *Neisseria gonorrhoeae*. *J. Bacteriol.* **174**:5719–5726.

Ohtsubo, S., K. Demizu, S. Kohno, I. Miura, T. Ogawa, and H. Fukuda. 1992. Comparison of acetate utilization among strains of an aceticlastic methanogen, *Methanothrix soehngenii*. *Appl. Environ. Microbiol.* **58**:703–705.

Oremland, R. S. 1988. Biogeochemistry of methanogenic bacteria. In: *Biology of Anaerobic Microorganisms*, A. J. B. Zehnder (ed.), pp. 641–706. Wiley Interscience, New York.

Petersen, S. P., and B. K. Ahring. 1991. Acetate oxidation in a thermophilic anaerobic sewage sludge digestor: the importance of non-acetoclastic methanogenesis from acetate. *FEMS Microbiol. Ecol.* **86**:149–158.

Phelps, T. J., R. Conrad, and J. G. Zeikus. 1985. Sulfate dependent interspecies H_2 transfer between *Methanosarcina barkeri* and *Desulfovibrio vulgaris* during coculture metabolism of acetate or methanol. *Appl. Environ. Microbiol.* **50**:589–594.

Pine, M. J., and W. Vishniac. 1957. The methane fermentations of acetate and methanol. *J. Bacteriol.* **73**:736–742.

Ragsdale, S. W. 1991. Enzymology of the acetyl-CoA pathway of CO_2 fixation. *Crit. Rev. Biochem. Mol. Biol.* **26**:261–300.

Ragsdale, S. W., J. R. Baur, C. M. Gorst, S. R. Harder, W.-P. Lu, D. L. Roberts, J. A. Runquist, and I. Schiau. 1990. The acetyl-CoA synthase from *Clostridium thermoaceticum:* from gene cluster to active-site metal clusters. *FEMS Microbiol. Rev.* **87**:397–402.

Sansone, F. J., and C. S. Martens. 1981. Methane production from acetate and associated methane fluxes from anoxic coastal sediments. *Science* **211**:707–709.

Schauder, R., B. Eikmanns, R. K. Thauer, F. Widdel, and G. Fuchs. 1986. Acetate oxidation to CO_2 in anaerobic bacteria via a novel pathway not involving reactions of the citric acid cycle. *Arch. Microbiol.* **145**:162–172.

Schauder, R., A. Preuss, M. Jetten, and G. Fuchs. 1989. Oxidative and reductive acetyl CoA/carbon monoxide dehydrogenase pathway in *Desulfobacterium autotrophicum*. 2. Demonstration of the enzymes of the pathway and comparison of CO dehydrogenase. *Arch. Microbiol.* **151**:84–89.

Schönheit, P., J. K. Kristjansson, and R. K. Thauer. 1982. Kinetic mechanism for the ability of sulfate reducers to out-compete methanogens for acetate. *Arch. Microbiol.* **132**:285–288.

Smith, M. R., and R. A. Mah. 1978. Growth and methanogenesis by *Methanosarcina* strain 227 on acetate and methanol. *Appl. Environ. Microbiol.* **36**:870–879.

Sowers, K. R., S. F. Baron, and J. G. Ferry. 1984. *Methanosarcina acetivorans* sp. nov., an acetotrophic methane-producing bacterium isolated from marine sediments. *Appl. Environ. Microbiol.* **47**:971–978.

Sowers, K. R., and R. P. Gunsalus. 1988. Adaptation for growth at various saline concentrations by the archaebacterium *Methanosarcina thermophila*. *J. Bacteriol.* **170**:998–1002.

Spormann, A. M., and R. K. Thauer. 1988. Anaerobic acetate oxidation to CO_2 by *Desulfotomaculum acetoxidans*. Demonstration of the enzymes required for the opera-

414   Stephen H. Zinder

tion of an oxidative acetyl-CoA/carbon monoxide dehydrogenase pathway. *Arch. Microbiol.* **150**:374–380.

Stadtman, T. C., and H. A. Barker. 1949. Studies on the methane fermentation. VII. Tracer experiments on the mechanism of methane formation. *Arch. Biochem.* **21**:256–264.

Stadtman, T. C., and H. A. Barker. 1951. Studies on the methane fermentation IX. The origin of methane in the acetate and methanol fermentation by *Methanosarcina*. *J. Bacteriol.* **61**:81–86.

Stams, A. J. M., K. C. R. Grolle, C. T. J. J. Fritjers, and J. B. van Lier. 1992. Enrichment of thermophilic propionate-oxidizing bacteria in syntrophy with *Methanobacterium thermoautotrophicum* or *Methanobacterium thermoformicicum*. *Appl Environ. Microbiol.* **58**:346–352.

Stupperich, E., and G. Fuchs. 1984. Autotrophic synthesis of activated acetic acid from two CO_2 in *Methanobacterium thermoautotrophicum*. II. Evidence for different origins of acetate carbon atoms. *Arch. Microbiol.* **139**:14–20.

Thauer, R. K., K. Jungermann, and K. Decker. 1977. Energy conservation in chemotrophic anaerobic bacteria. *Bacteriol. Rev.* **41**:100–180.

Thauer, R. K., D. Möller-Zinkhan, and A. M. Spormann. 1989. Biochemistry of acetate catabolism in anaerobic bacteria. *Annu. Rev. Microbiol.* **43**:43–67.

Thiele, J. H., and J. G. Zeikus. 1988. Control of interspecies electron flow during anaerobic digestion: significance of formate transfer versus hydrogen transfer during syntrophic methanogenesis in flocs. *Appl. Environ. Microbiol.* **54**:20–29.

Trouzel, J. P., E. C. de Macario, J. Nölling, W. M. de Vos, T. Zhilina, A. M. Lysenko. 1992. DNA relatedness among some thermophilic members of the genus *Methanobacterium*. Emendation of the species *Methanobacterium thermoautotrophicum* and rejection of *Methanobacterium thermoformicicum* as a synonym of *Methanobacterium thermoautotrophicum*. *Int. J. Syst. Bacteriol.* **42**:408–411.

Warford, A. L., D. R. Kosiur, and P. R. Doose. 1979. Methane production in Santa Barbara Basin sediments. *Geomicrobiol. J.* **1**:117–137.

Weber, H., K. D. Kulbe, H. Clumiel, and W. Trösch. 1984. Microbial acetate conversion to methane: kinetics, yields and pathways in a two-step digestion process. *Appl. Microbiol. Biotechnol.* **19**:224–228.

Weimer, P. J., and J. G. Zeikus. 1978. Acetate metabolism in *Methanosarcina barkeri*. *Arch. Microbiol.* **119**:175–182.

Westermann, P., B. K. Ahring, and R. A. Mah. 1989. Threshold acetate concentrations for acetate catabolism by acetoclastic methanogenic bacteria. *Appl. Environ. Microbiol.* **55**:514–515.

Widdel, F. 1986. Growth of methanogenic bacteria in pure culture with 2-propanol and other alcohols as hydrogen donors. *Appl. Environ. Microbiol.* **51**:1056–1062.

Widdel, F. 1988. Microbiology and ecology of sulfate- and sulfur-reducing bacteria. In: *Biology of Anaerobic Microorganisms*, A. J. B. Zehnder (ed.), pp. 469–586. Wiley Interscience, New York.

Winfrey, M. R., and J. G. Zeikus. 1977. Effect of sulfate on carbon and electron flow during microbial methanogenesis in freshwater lake sediments. *Appl. Environ. Microbiol.* **33:**275–281.

Winter, J. U., and R. S. Wolfe. 1980. Methane formation from fructose by syntrophic associations of *Acetobacterium woodii* and different strains of methanogens. *Arch. Microbiol.* **124:**73–79.

Witicar, M. J., E. Faber, and M. Schoell. 1986. Biogenic methane formation in marine and freshwater environments. Carbon dioxide reduction vs. acetate fermentation: isotope evidence. *Geochem. Cosmochem. Acta* **50:**693–709.

Wolin, M. J., and T. L. Miller. 1982. Interspecies hydrogen transfer: 15 years later. *ASM News* **48:**561–565.

Zehnder, A. J. B., B. A. Huser, T. D. Brock, and K. Wuhrmann. 1980. Characterization of an acetate-decarboxylating, non-hydrogen-oxidizing methane bacterium. *Arch. Microbiol.* **124:**1–11.

Zehnder, A. J., and W. Stumm. 1988. Geochemistry and biochemistry of anaerobic habitats. In: *Biology of Anaerobic Microorganisms*, A. J. B. Zehnder (ed.), pp. 1–38, John Wiley and Sons, Inc., New York.

Zeikus, J. G., G. Fuchs, W. Kenealy, and R. K. Thauer. 1977. Oxidoreductases involved in cell carbon synthesis of *Methanobacterium thermoautotrophicum*. *J. Bacteriol.* **132:**604–613.

Zinder, S. H., and R. A. Mah. 1979. Isolation and characterization of a thermophilic strain of *Methanosarcina* unable to use H_2-CO_2 for methanogenesis. *Appl. Environ. Microbiol.* **38:**996–1008.

Zinder, S. H., S. C. Cardwell, T. Anguish, M. Lee, and M. Koch. 1984. Methanogenesis in a thermophilic anaerobic digestor: *Methanothrix* sp. as an important aceticlastic methanogen. *Appl. Environ. Microbiol.* **47:**796–807.

Zinder, S. H., and M. Koch. 1984. Non-aceticlastic methanogenesis from acetate: Acetate oxidation by a thermophilic syntrophic coculture. *Arch. Microbiol.* **138:**263–272.

Zinder, S. H., and A. Elias. 1985. Growth substrate effects on acetate and methanol catabolism in *Methanosarcina thermophila* strain TM-1. *J. Bacteriol.* **163:**317–323.

Zinder, S. H., K. R. Sowers, and J. G. Ferry. 1985. *Methanosarcina thermophila* sp. nov., a thermophilic acetotrophic methane producing bacterium. *Int. J. Syst. Bacteriol.* **35:**522–523.

Zinder, S. H., T. Anguish, and T. Lobo. 1987. Isolation and characterization of a thermophilic acetotrophic strain of *Methanothrix*. *Arch. Microbiol.* **146:**315–322.

Zindel, U., W. Freudenberg, M. Rieth, J. R. Andreesen, J. Schnell, and F. Widdel. 1988. *Eubacterium acidaminophilum* sp. nov., a versatile amino acid-degrading anaerobe producing or utilizing H_2 or formate. *Arch. Microbiol.* **150:**254–266.

Zinder, S. H. 1990. Conversion of acetic acid to methane by thermophiles. *FEMS Microbiol. Rev.* **75:**125–138.

Zinder, S. H., and T. Anguish. 1992. Carbon monoxide, hydrogen, and formate metabolism during methanogenesis from acetate by thermophilic cultures of *Methanosarcina* and *Methanothrix*. *Appl. Environ. Microbiol.* **58:**3323–3329.

15

Acetogenesis at Low Temperature

Alla N. Nozhevnikova, Oleg R. Kotsyurbenko, and Marija V. Simankova

15.1 Introduction

Most habitats on earth have a low average annual temperature, and more than 80% of the biosphere is permanently cold. In general, the study of mesophilic processes have dominated the interests of biologists. However, though poorly characterized, microbial processes in cold climatic zones may play important roles in the biological cycles and global ecology.

Acetogenesis is a widespread, microbial metabolic processes that occurs in various anaerobic habitats ranging from soils and sediments to gastrointestinal tracts of many animals (Wood and Ljungdahl, 1991; Schink and Bomar, 1992; Diekert, 1992; Drake, 1992). Acetogens have been studied mostly in pure culture, but in nature, acetogenesis is likely connected with other bacterial groups via complex trophic interactions. These interactions and the competitiveness of certain bacterial groups may be influenced by low-temperature conditions.

Like acetogens, methanogens use CO_2 as a terminal electron acceptor (Whitman, 1985). Under mesophilic conditions, and in the absence of sulfate, methanogenesis competes strongly with acetogenesis for H_2/CO_2. Thermodynamically, H_2/CO_2-methanogenesis is a more energy-yielding process than H_2/CO_2-acetogenesis (Thauer et al., 1977; Dolfing, 1988). In this regard, the activities of methanogenic bacteria can be so high that acetogens are essentially not detectable (Braun et al., 1979).

However, acetogens have diverse metabolic potentials that likely contribute to their competitiveness in nature (Drake, 1993). Indeed, there are several gastrointestinal habitats where acetogenesis has been found to occur at significant levels (Brauman et al., 1992; Breznak and Switzer, 1986; Prins and Lankhorst, 1977;

Lajoie et al., 1988). In addition, acetogenesis is a potentially important process in low-temperature ecosystems such as certain lake sediments (Phelps and Zeikus, 1984; Lovley and Klung, 1983; Jones and Simon, 1985). Conrad et al. (1989) concluded that homoacetogenesis may be a dominant process connected to H_2 turnover at low temperature in anoxic rice paddy soil and sediments of Lake Constance.

Our research has focused on the potential ecological roles of acetogenic bacteria in various low-temperature habitats in northern Russia. Emphasis in the present chapter is placed on the low-temperature anaerobic degradation of organic matter by microbial communities from three environments: (i) a long-standing, wastewater disposal pond, (ii) tundra marsh, and (iii) cattle manure. In these microbial communities, acetogenic organisms are shown to be in competition with methanogens for H_2-supplying substrates and are the likely producers of acetate as a predecessor of acetate-dependent methanogenesis. We show that in the three investigated habitats, the potential role of acetogenesis at the process level is increased fundamentally at low temperatures indicative of our region (Kotsyurbenko et al., 1992a; Nozhevnikova et al., 1992; Parshina et al., 1993).

15.2 Acetogenesis in Pond Sediment

The pond evaluation was man-made and is located in Syktyvkar, Komi Republic (northern portion of the European part of Russia; 62° N 51° E); it was initially used as a polishing step in the Syktyvkar Forest Industrial Complex for wastewater treatment. The pond had been taken out of the purification system for several years and was highly polluted due to the accumulation of organic waste products. The mean depth of the pond is 3 m, and the silt layer thickness is 2 m. The water temperature throughout the year approximates 4–6°C. Despite this low temperature, biogas evolution is indicated by the appearance of gaseous bubbles. The general pattern of microbial activities of such methanogenic communities has been described and is characterized by sequential trophic networks (Zavarzin, 1986).

Samples of silt were collected in the late winter/early spring (1990), put into 5-L bottles, transported to the laboratory, and flushed with N_2 (Kotsyurbenko et al., 1992a). Silt inoculants (i.e., suspensions, 5% v/v) were placed into bottles that contained a mineral salt "Pfennig" solution (final pH was 7.2) and were supplemented with various substrates to study differential community processes. For hydrogen-utilization studies, the initial headspace was adjusted to H_2/CO_2 (4 : 1); for carbon monoxide (CO)-utilization studies, the initial headspace was adjusted to 100% CO. In the studies described below, concentrations of substrate and products, both solutes and gases, were converted to molar concentrations and are expressed as millimoles per liter of the aqueous phase.

15.2.1   Spontaneous Fermentation Patterns in Pond Sediments

Spontaneous anaerobic fermentation of undiluted silt samples resulted in the formation of methane as the predominant product at 28°C and reached concentrations approximating 9 mM after a 100-day incubation period. At this temperature, acetate production was detected but only at negligible levels. A similar though protracted pattern was obtained 15°C. However, when silt was incubated at 6°C, a temperature more indicative of sediment conditions *in vivo*, acetate synthesis increased to a maximum concentration of 1.5 mM and was the sole fatty acid detected. Significantly, methane production fell to levels approximately equal to that of acetate during a 200-day incubation period, following which methane production continued at a low rate and acetate began to be utilized (Kotsyurbenko et al., 1992a; Nozhevnikova et al., 1992). These results suggest that (i) the sediment collection procedure did not radically destroy the native methanogenic capabilities of the sediment and (ii) at 28°C (but not 6°C), premethanogenic processes that converted the native carbon/reductant sources to methanogenic substrates were rate limiting (as they were not detectable) to methane production. These observations further identify acetate as a significant product of endogenous substrates at low temperature. Subsequent studies focused on obtaining more conclusive evidence for the occurrence of acetogenesis is these habitats.

15.2.2   Influence of Temperature on the Formation of Acetate and Methane from Various Organic Substrates

We noted in preliminary work that at low temperatures, supplementing pond sediment with polymers (e.g., cellulose) and saccharides yielded high amounts of volatile fatty acids (mainly acetate and butyrate). Significantly, ethanol was detected as a transient product coincident to acetate formation. Production of fatty acids occurred at a much higher rate than did methanogenesis at 6°C. In addition, hydrogen was also detected as a transient product, and its accumulation and reutilization was associated with acetate formation rather than methanogenesis (Kotsyurbenko et al., 1992a).

Previous studies with pure cultures of acetogens have identified a number of substrates that can be regarded as typical growth-supportive substrates for acetogens, including various methoxylated aromatic compounds, one-carbon compounds, and sugars; homoacetogenesis from certain substrates is evidence of acetogenic bacteria (Drake, 1992; Wood and Ljungdahl, 1991; Diekert, 1992). We set up an extensive series of experiments to assess the inherent capacities of pond sediment to utilize acetogenic substrates and to thus differentiate between bacterial groups and their activities.

At 28°C, the main products of decomposition of gallate and vanillate were acetate, CO_2, and CH_4, and the overall pattern observed at that temperature correlated well with that of Schink and Pfennig (1982). Acetate was formed in 10 days and was parallel to methane accumulation. At 15°C, the overall process was considerably slower, but the ratio of acetate to methane formed began to favor acetate production. At 6°C, acetate was essentially the only product detected from aromatic acids, and acetate turnover was not apparent. For example, with vanillate after 300 days incubation, acetate (at a final stable concentration of 1.5 mM) was in marked excess to methane (at a final concentration of 0.02 mM). Significantly, phenol, catechol, and p-hydroxybenzoate were not degraded, suggesting 3-hydroxy or 3-methoxy substituent groups are essential to ring usage (Kaiser and Hanselmann, 1982; Evans and Fuchs, 1988; Schink et al., 1992; Young and Frazer, 1987).

As noted earlier, ethanol was formed during the decomposition of certain saccharides. Ethanol is a substrate for some acetogenic bacteria (Braun et al., 1981; Lux and Drake, 1992), and the potential turnover of this substrate to acetate was evaluated. At 28°C, ethanol decomposition initially yielded various products, including butyrate, propanol, propionate, caproate, acetate, and H_2. Methane formation during this phase was minimal. However, once this acidophilic phase was complete and ethanol totally consumed, acetoclastic methanogenesis was observed. At 15°C, the formation of butyrate, propanol, propionate, caproate, and H_2 was less apparent, and acetate was synthesized to greater levels; in addition, the turnover of acetate to CH_4 was slow. In contrast, at 6°C, acetate was essentially the only product formed from ethanol during a 150-day incubation period, and methanogenesis was reduced to near background levels (Nozhevnikova et al., 1992).

15.2.3 Acetogenesis from One-Carbon Substrate with Pond Sediments

Anoxic silt sediment incubated under an atmosphere of H_2/CO_2 formed acetate concomitant to the consumption of H_2 at a temperature of 28°C. Following the initial first 20-day period during which H_2 was totally consumed, acetate-dependent methanogenesis was initiated and proceeded from day 20 to day 50. This result was somewhat surprising as H_2/CO_2 methanogenesis was believed to dominate H_2/CO_2 acetogenesis at this temperature. Because acetogenesis rather than methanogenesis was the immediate H_2/CO_2-coupled response observed, we postulated that H_2 turnover was due to acetogenic bacteria, not methanogens. In support of this conclusion, H_2/CO_2-dependent acetogenesis was likewise the sole process observed at 6°C (Kotsyurbenko et al., 1992a; Nozhevnikova et al., 1992).

Under natural conditions, reductant transfer may proceed via formate rather than H_2. When pond sediments were supplemented with formate, acetate was the sole major product formed. Hydrogen was also formed in trace amounts during the initial phase of formate consumption but was subsequently consumed and, apparently, converted to acetate. Methane formation was negligible at 6°C.

Table 15.1   Influence of temperature and substrate on the rate of acetate formation (V_{ac}) by pond sediment[a]

	Temperature					
	28°C		15°C		6°C	
Substrate	lag (d)	V_{ac}	lag (d)	V_{ac}	lag (d)	V_{ac}
Cellulose	0	0.34	9	0.19	37	0.03
Starch	0	1.92	2	0.31	18	0.13
Xylose	0	2.16	0	0.83	0	0.37
Gallate	0	1.00	8	0.19	70	0.11
Ethanol	0	2.33	8	1.35	15	0.41
Formate	0	1.00	0	0.46	19	0.33
H_2+CO_2	0	2.12	0	0.81	33	0.55
CO	6	1.43	33	0.37	80	0.20

[a]Note: V_{ac} measured in millimoles per liter per day.

Rates of acetogenesis from formate were approximately two times lower than from hydrogen (Table 15.1).

Acetogenic bacteria are capable of forming acetate from carbon monoxide (CO); this capacity can be used as a distinguishing feature useful in the identification of acetogens (Diekert, 1992; Drake, 1992). When silt sediments were incubated under an atmosphere of 100% CO at 28°C, CO was rapidly consumed and converted to acetate and H_2; subsequently, H_2 was consumed and apparently converted to methane. At this temperature, acetate was turned over to methane after CO consumption was complete. At 28°C, CO and H_2O appeared to drive two initial processes: (1) the formation of acetate, and (2) the formation of H_2. The CO-dependent formation of H_2 has been observed with both growing cultures and resting cells of the acetogen *Clostridium thermoaceticum* (Martin et al., 1983) and also with the carboxydotroph *Carboxydothermus hydrogenoformans* (Svetlichny et al., 1991). At 6°C, this overall process was considerably slower, and H_2 formation was reduced to very low levels (Fig. 15.1). At 6°C, overall CO consumption yielded the following stoichiometry:

$$115\ CO \rightarrow 32\ \text{acetate} + 61\ CO_2$$

This stoichiometry is in close agreement with the 4 : 1 CO/acetate ratio indicative of CO-dependent acetogenesis and illustrates how dominant chemolithoautotrophic acetogenesis was at low temperature. This potential is in contrast to the hydrolytic transformation of CO and H_2O to H_2 and CO_2 as described by Svetlichny et al. (1991).

Methanol has been shown to be a typical product of pectin decomposition,

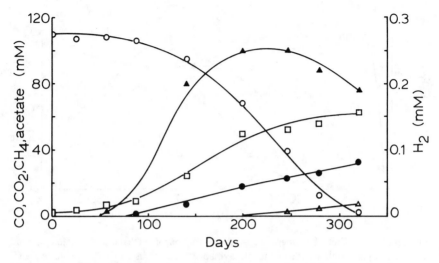

Figure 15.1.   Formation of products from carbon monoxide (CO) by pond sediment at 6°C. Symbols: CO, ○; H_2, ▲; CO_2 □; acetate, ●; methane, △.

and traces of methanol were detected when this polymer was provided as substrate in pond sediment studies (Kotsyurbenko et al., 1992a). At both 28°C and 15°C, methanogenesis proceeded rapidly with methanol, whereas acetate formation was negligible at both temperatures. At 6°C, methanogenesis continued to dominate, although methanol-coupled acetogenesis was more prevalent than at higher temperatures; the acetate formed was subsequently converted to methane during a secondary phase of acetoclastic methanogenesis (Fig. 15.2). Thus, in contrast to other one-carbon metabolic potentials that favored acetogenesis and the formation of acetate at low temperature, methanol consumption by pond sediment favored methanogenesis.

Table 15.1 shows the comparative rates and lag phases for acetate formation from various substrates with pond sediments. In general, the largest temperature-dependent change (i.e., decrease in response to low temperature) in rates was observed with complex polymers rather than more readily utilizable acetogenic substrates. This might be explained, in part, by hydrolytic bacteria being more sensitive to low-temperature conditions than the acetate-forming organisms. Alternatively, the breakdown of polymers kinetically does not favor acetate formation because the acetate-forming organisms are non-rate-limiting.

15.2.4   Acetoclastic Methanogenesis in Pond Sediments

Although acetogens appeared to outcompete methanogens for H_2/CO_2 (as well as formate), acetate was turned over via acetoclastic methanogenesis at all temper-

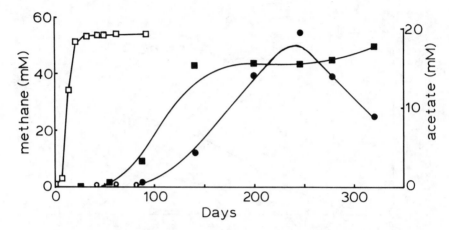

Figure 15.2.   Effect of temperature on the formation of methane and acetate by pond sediment supplemented with methanol. Symbols: methane at 28°C, □; acetate at 28°C, ○; methane at 6°C, ■; acetate at 6°C, ●.

atures evaluated. At higher temperatures, acetoclastic methanogenesis was rapid, and acetate never persisted in such experiments. However, at low temperature (i.e., 6°C), acetate turnover was relatively slow (Kotsyurbenko et al., 1992a). Collectively, these observations suggest that acetate, rather than H_2/CO_2, might be the predominant native substrate of methanogenesis in pond sediment.

15.3   Acetogenesis in Tundra Soil Samples

Soil samples were collected in July 1990 from the tundra wetland of Polar Ural near Vorkuta, Russia (68° N, 65° E). Plant cover was a continuous mat of mosses invaded by lichens, cottongrasses (*Eriophorum* sp.), sedges (*Carex* sp.), and dwarf shrubs. At the time of collection, air temperature ranged between 10°C and 15°C. The soil temperature was 5–6°C at 0.3–0.4 m depth, and the soil solution pH approximated 6.1. The rate of methane emission from the place of sampling was 7.85 mg CH_4 m^{-2} hr^{-1} (Slobodkin et al., 1992).

15.3.1   Spontaneous Fermentation and Evidence of Acetogenesis in Tundra Soil

When tundra soil samples were diluted by mineral solution 1 : 1 (as described in Section 15.2), methane and acetate were formed at all temperatures assessed. At 28°C, 35 mM methane was formed, whereas acetate formation was very low and negligible. In contrast, and similar to the observations with pond sediment described earlier, acetate formation was more apparent at low temperature, with levels reaching 4.1 mM at 6°C; subsequent to its formation, acetate was turned over to methane. Significantly, methane was not formed at low temperature prior to acetate formation.

Tundra soil samples were used to further evaluate potential trophic interactions. Acetate formation was the dominant response at 15°C, 10°C, and 6°C when tundra soils were incubated with cellulose and monosaccarides. Butyrate was also formed in significant quantities. In contrast, methanogenesis was sharply decreased at low temperature. Furthermore, the degradation of most of organic compounds was maximal at 10°–15°C, temperatures at which methanogenesis was greatly retarded (or not optimal).

15.3.2 Acetogenesis from One-Carbon Substrates with Tundra Soil

At 28°C, methane was the primary product detected from H_2/CO_2 (Zarvarzin et al. 1993). However, at lower temperatures (6°C, 10°C, and 15°C), acetate became a more increasingly detectable product (Fig. 15.3). The following stoichiometries were observed for H_2 and acetate:

$$\text{at } 15°C: \quad 130\ H_2 \rightarrow 32\ \text{acetate}$$
$$\text{at } 6°C: \quad 130\ H_2 \rightarrow 24\ \text{acetate}$$

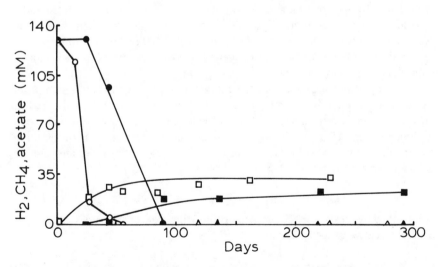

Figure 15.3.   Effect of temperature on the formation of products by tundra soil when supplemented with hydrogen and carbon dioxide (H_2/CO_2). Symbols: H_2, ○ and ●; methane, △ and ▲; acetate, □ and ■. Open symbols, 15°C; closed symbols, 6°C.

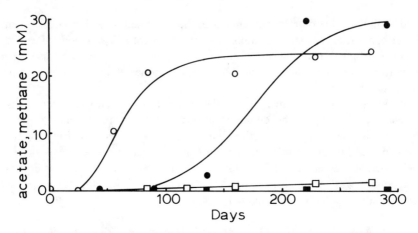

Figure 15.4.   Effect of temperature on the formation of methane and acetate by tundra soil supplemented with methanol. Symbols: methane at 15°C, □; acetate at 15°C, ○; methane at 6°C, ■; acetate at 6°C, ●.

Both of these stoichiometries approximate the theoretical 4 : 1 ratio for H_2 consumption versus acetate production for H_2-dependent acetogenesis (Drake, 1992; Wood and Ljungdahl, 1991). At both of these temperatures, methane was reduced to negligible leevls over the 300-day incubation period. In addition, with either formate or CO, acetate was also the primary product detected under low-temperature conditions. Hydrogen was detected as a transient product at very low concentrations with both formate and CO. At the present time, we are uncertain if CO and formate are converted to hydrogen, which is then rapidly consumed via acetogenic conversion to acetate.

When methanol was provided to tundra soil, methane was not a significant product at either 15°C or 6°C (Fig. 15.4) (see also Zavarzin et al., 1993). This pattern was fundamentally different from the pattern observed with pond sediments (Fig. 15.2), again suggesting community differences between these two habitats. Svensson (1984) also noted a fundamental influence of temperature on methanogenesis from different substrates.

15.4   Acetogenesis during the Anaerobic Digestion of Animal Waste at Low Temperature

Psychrophilic animal waste digestion and methane production has potential economic importance in cold regions (Sutter et al., 1987; Zeeman et al., 1988). However, relative to methane production from animal wastes, one problem identified is the slow rate of methanogenesis at temperatures lower than 15°C (Zeeman et al., 1988). Investigations with cattle and pig manure have shown that acetogenesis (i.e., acetate production rather than methane production) was dominant in fresh manure at low temperature (Kotsyurbenko et al., 1993; Parshina et al., 1993). Acetate was the main fatty acid produced, and the rate of its consumption by methanogens was very low.

Significantly, hydrogen production occurred during the psychrophilic fermentation of pig manure (Parshina et al., 1993). At both 6°C and 10°C, this hydrogen was subsequently utilized until the H_2 concentration approximated 600 ppm. This concentration is near the H_2 threshold for hydrogen consumption by acetogenic bacteria when CO_2 is the terminal electron acceptor (Cord-Ruwisch et al., 1988). This amount of hydrogen (i.e., 600 ppm) was essentially a stable end product for several months, at which time a slow usage of this remaining H_2 was accompanied with the formation of trace levels of methane. Similar results were obtained with cattle manure under psychrophilic conditions (Kotsyurbenko et al., 1993). These observations suggest that acetogenic bacteria may outcompete methanogens for H_2 during the initial phases of manure digestion at low temperature. However, the role of acetogens in the overall turnover of carbon and the synthesis of acetate from animal waste under psychrophilic conditions is not resolved.

15.5 Isolation of Acetogenic Bacteria from Low-Temperature Environments

To verify the existence of acetogens in these habitats, several pure cultures of acetogenic bacteria have been isolated by enrichment under H_2/CO_2 at 6°C (Kotsyurbenko et al., 1992b; Kotsyurbenko et al., in preparation). The four main isolates under study are

(i)   strain ZS, from paper-mill wastewater pond (Section 15.2)

(ii) strain ZT, from tundra soil (Section 15.3)

(iii) strain ZM, from cattle manure digested at low temperature (Section 15.4)

(iv) strain ZB, from sediment of a morass (bog) 100 km north of Moscow

Cells of all strains are oval-shaped, Gram-positive, motile rods. Slime is produced by all strains and spores are not apparent. The pH optima of all strains is between 6–7.

All strains grow on H_2/CO_2 at 6°C and, under these conditions, yield only

Table 15.2   Substrate specificity of new acetogenic isolates at 6°C[a]

Substrate (mM)	Maximal Optical Density at 600 nm			
	Strain ZS	Strain ZB	Strain ZM	Strain ZT[b]
$H_2 + CO_2$ (66/16)	0.12	0.12	0.11	0.12
CO (82)	0.1	nd	0.1	0.05
Formate (74)	0.21	0.1	0.05	0.10
Methanol (123)	0.27	0.32	0	0.06
Maltose (15)	0.02	0.15	0	0.18
Fructose (28)	0.53	0.52	0.03	0.45
Glucose (28)	0.03	0.21	0	0
Xylose (33)	0.04	0.05	0	ND[c]
Cellobiose (15)	0	0.05	0	ND
Lactate (58)	0.13	0.11	0.22	0.32
Dimethylglycine (49)	0	0	0.03	0
Butanediol (56)	0	0	0.04	0
Betaine (43)	0.16	0.23	0.14	0
Vanillate (18)	0.08	0	0.08	ND
Gallate (18)	0.09	0	0	0

[a] All strains were isolated at 6°C with H_2/CO_2 as energy source; see Section 15.5 for more information on strains.
[b] Strain ZT has only recently been obtained in pure culture and its substrate profiles are presently under study.
[c] ND = not determined.

acetate; however, utilization of other substrates is not so uniform (Table 15.2). Strain ZM is a mesophile with a growth-rate temperature optimum of 30°C; however, it displays high growth potentials at low temperature. The growth rate of strain ZS is nearly equivalent between 1°C and 20°C. Strain ZS is, therefore, either psychrotrophic or psychrophilic. Strain ZT displays the best growth potentials at low temperature and may be psychrophilic (Fig. 15.5). Despite these different temperature optima, all four strains have significant growth rates at 1–5°C (most notable in this regard are strains ZT and ZS). Consistent with their different temperature optima, the strains also differ with regard to their lipid complexes.

Strains ZS, ZM and ZB have a guanine plus cytosine (G + C) content of approximately 41 mol% and might be regarded as species in the genus *Acetobacterium* (Schink and Bomar, 1992). The classification of strain ZT remain less certain, although both appear to also be species of the genus *Acetobacterium*. The first psychrotrophic acetogenic bacterium (strain HP4, perhaps a species of *Acetobacterium*) was isolated from an anaerobic lake sediment (Bak, 1988; Conrad et al., 1989) and may be related to one of the isolates (See Addendum).

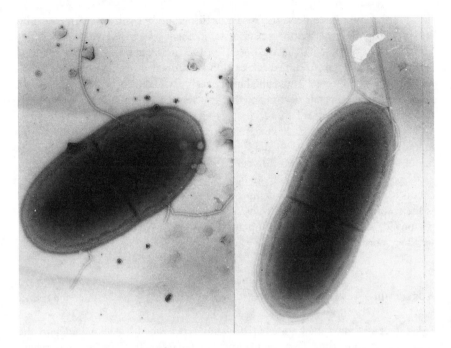

Figure 15.5. Electron micrographs of acetogen strain ZT isolated from tundra soil in Russia.

15.6 Conclusions

Acetate was a predominant anaerobic product of these low-temperature soils and sediments when subjected to temperatures indicative of *in vivo* conditions. Acetate was also a predominant product of animal waste digestion under low-temperature conditions. Assuming that the sampling procedure did not unduly alter the general potentials of the system, it appears likely that native acetogenic bacteria are highly competent and important in carbon and energy transformations at low temperature. In particular, autotrophic acetogenesis appears to be a significant low-temperature potential of these habitats. The observed potential to form butyrate may also be attributable to butyrate-producing acetogens (Zeikus et al., 1985). Acetogens form acetate from diverse substrates but, in addition, catalyze other transformations of carbon substrates (Drake, 1992, 1993). The contribution of acetogens to the total turnover of endogenous carbon in cold climate soils, such as the tundra soils investigated in our studies, remains unknown.

Although we did not characterize the entire microbial community, native methanogens, in general, were only highly active under more mesophilic conditions. This could have been the result of differential sensitivity of mesophilic and psychrophilic/psychrotrophic strains of methanogens; however, this speculation has no experimental basis. Based on the production of either acetate or methane, methanogens seemed to develop subsequent to acetogens under low-temperature conditions. Thus, methanogenesis appeared to be acetoclastic. In addition, because acetate accumulated at low temperature, its formation appears to not limit the system relative to final production of methane.

Numerous investigations have demonstrated the production of methane *in situ* at low temperatures (Svensson and Rosswall, 1984; Harriss et al., 1985; Slobodkin et al., 1992; Panikov and Zelenev, 1992). Despite the apparent high acetogenic potentials observed in our studies, psychrophilic methanogenesis probably occurs in certain balanced, low-temperature ecosystems. A challenge that remains is the resolution of the *in vivo* roles of both acetogens and methanogens in the prolonged, slow mineralization of plant debris indicative of these and other cold weather, northern habitats.

Addendum

We recently submitted some of the strains described in this chapter to the DSM culture collection. They are identified by the following DSM strain numbers and recommended names: strain ZS, DSM 8239 (Z-4391), "Acetobacterium bakii" sp. nov.; strain ZB, DSM 8237 (Z-4092), "Acetobacterium paludosum" sp. nov.; strain ZM, DSM 8238 (Z-4290), "Acetobacterium fimetarium" sp. nov.

Acknowledgments

We thank Professor G. Zavarzin for advice and useful discussions throughout these studies. We also thank the various research groups of the Institute of Microbiology of the Russian Academy of Science for their collegiality and help, and Harold L. Drake for assistance in preparation of the manuscript.

References

Bak, F. 1988. Sulfatreduzierende Bakterien und ihre Aktivität im Litoralsediment der Unteren Gull (Überlinger See) Hartung-Gorre Verlag, pp. 154–158. Konstanz, Germany.

Brauman, A., M. D. Kane, M. Labat, and A. Breznak. 1992. Genesis of acetate and methane by gut bacteria of nutritionally diverse termites. *Science* **257**:1384–1387.

Braun, M., F. Mayer, and G. Gottschalk. 1981. *Clostridium aceticum* (Wieringa), a microorganism producing acetic acid from molecular hydrogen and carbon dioxide. *Arch. Microbiol.* **128**:288–293.

Braun, M., S. Schoberth, and G. Gottschalk. 1979. Enumeration of bacteria forming acetate from hydrogen and carbon dioxide in anaerobic habitats. *Arch. Microbiol.* **120**:201–204.

Breznak, J. A., and J. M. Switzer. 1986. Acetate synthesis from H_2 plus CO_2 by termite gut microbes. *Appl. Environ. Microbiol.* **52**:623–630.

Conrad, R., F. Bak, H. J. Seitz, B. Threbath, H. P. Mayer, and H. Schutz. 1989. Hydrogen turnover by psychrotrophic homoacetogenic and mesophilic methanogenic bacteria in anoxic paddy soil and lake sediment. *FEMS Microbiol. Ecol.* **62**:285–294.

Cord-Ruwisch, R., H.-J. Seitz, and R. Conrad. 1988. The capacity of hydrogenotrophic anaerobic bacteria to compete for traces of hydrogen depends on the redox potential of the terminal electron acceptor. *Arch. Microbiol.* **149**:350–357.

Diekert, G. 1992. The acetogenic bacteria. In: *The Prokaryotes*, 2nd ed., A. Balows, H. G. Trüper, M. Dworkin, W. Harder, K.-H. Schleifer (eds.), Vol. 1, pp. 517–533. Springer-Verlag, New York.

Dolfing, J. 1988. Acetogenesis. In: *Biology of Anaerobic Microorganisms*, A. J. B. Zehnder (ed.), pp. 417–468. Wiley, New York.

Drake, H. L. 1992. Acetogenesis and acetogenic bacteria. In: *Encyclopedia of Microbiology*, Vol. 1, J. Lederberg (ed.), pp. 1–15. Academic Press, San Diego, CA.

Drake, H. L. 1993. CO_2, reductant, and the autotrophic acetyl-CoA pathway: alternative origins and destinations. In: *Microbial Growth on C_1 Compounds*, J. C. Murrell, and D. P. Kelly (eds.), pp. 493–507. Intercept Ltd., Andover, England.

Evans, W. C., and G. Fuchs. 1988. Anaerobic degradation of aromatic compounds. *Annu. Rev. Microbiol.* **42**:289–317.

Harriss, R. C., E. Gorham, D. I. Sebacher, K. B. Bartlett, and P. A. Flebbe. 1985. Methane flux from northern peatlands. *Nature* **315**:652–654.

Jones, J. G., and B. M. Simon. 1985. Interaction of acetogens and methanogens in anaerobic freshwater sediments. *Appl. Environ. Microbiol.* **49**:944–948.

Kaiser, J. P., and K. W. Hanselmann. 1982. Fermentative metabolism of substituted monoaromatic compounds by a bacterial community from anaerobic sediments. *Arch. Microbiol.* **133**: 185–194.

Kotsyurbenko, O. R., A. N. Nozhevnikova, and G. A. Zavarzin. 1992a. Anaerobic degradation of organic matter by psychrophilic microorganisms. *Zh. Obshch. Biol.* **53**:159–175 (in Russian).

Kotsyurbenko, O. R., M. V. Simankova, N. P. Bolotina, T. N. Zhilina, and A. N. Nozhevnikova. 1992. Psychrotrophic homoacetogenic bacteria from several environments, Abstr. C136. *7th Int. Symp. on Microbial Growth on C-1 Compounds*, August 1992, Warwick, UK., 1992.

Kotsyurbenko, O. R., A. N. Nozhevnikova, S. V. Kalyuzhnyy, and G. A. Zavarzin.

1993. Methanogenic fermentation of cattle manure under psychrophilic conditions. *Mikrobiologiya* **62**:761–772 (in Russian).

Kotsyurbenko, O. R., M. V. Simankova, N. P. Bolotina, T. N. Zhilina, A. N. Nozhevnikova, and G. A. Zavarzin. In preparation. New psychrotrophic acetogenic bacterium from several environments.

Lajoie, S. F., S. Bank, T. L. Miller, and M. J. Wolin. 1988. Acetate production from hydrogen and [^{13}C] carbon dioxide by the microflora of human feces. *Appl. Environ. Microbiol.* **54**:2723–2727.

Lovley, D. R., M. J. Klug. 1983. Methanogenesis from methanol and methylamines and acetogenesis from hydrogen and carbon dioxide in the sediments of a eutrophic lake. *Appl. Environ. Microbiol.* **45**:1310–1315.

Lux, M. F., and H. L. Drake. 1992. Re-examination of the metabolic potentials of the acetogens *Clostridium aceticum* and *Clostridium formicoaceticum:* chemolithoautotrophic and aromatic-dependent growth. *FEMS Microbiol. Lett.* **95**:49–56.

Martin, D. R., L. L. Lundie, R. Kellum, and H. L. Drake. 1983. Carbon monoxide-dependent evolution of hydrogen by the homoacetate-fermenting bacterium *Clostridium thermoaceticum. Curr. Microbiol.* **8**:337–340.

Nozhevnikova, A. N., O. R. Kotsyurbenko, and G. A. Zavarzin. 1992. Methanogenesis from organic matter at low temperature. In: *Int. Course on Anaerobic Waste Water Treatment, Case Studies.* IHE Delft, Agricult. Univ. Wageningen, The Netherlands. pp. 72–96.

Panikov, N. S., and V. V. Zelenev. 1992. Emission of CO_2 from northern wetlands to atmosphere: dynamics, controlling factors and tentative mechanisms, pp. 208–216. *1st International Conference on Cryopedology*, Pushchino, 1992 (in Russian).

Parshina, S. N., A. N. Nozhevnikova, and S. V. Kalyuzhnyy. 1993. Degradation of protein substrates by microflora of pig's manure at low temperature. *Mikrobiologiya* **62**:169–180 (in Russian).

Phelps, J., and J. G Zeikus. 1984. Influence of pH on terminal carbon metabolism in anoxic sediments from a mildly acidic lake. *Appl. Environ. Microbiol.* **48**:1088–1095.

Prins, R. A., and A. Lankhorst. 1977. Synthesis of acetate from CO_2 in the cecum of some rodents. *FEMS Microbiol. Lett.* **1**:255–258.

Schink, B., and N. Pfennig. 1982. Fermentation of trihydroxybenzenes by *Pelobacter acidigallici* gen. nov. sp. nov., a new strictly anaerobic non-sporeforming bacterium. *Arch. Microbiol.* **133**:195–201.

Schink, B., and M. Bomar. 1992. The genera *Acetobacterium, Acetogenium, Acetoanaerobium,* and *Acetitomaculum.* In: *The Prokaryotes,* 2nd ed., A. Balows et al. (eds.), Vol. II, pp. 1925–1936. Springer-Verlag, New York.

Schink, B., A. Brune, and S. Schnell. 1992. Anaerobic degradation of aromatic compounds. In: *Microbial Degradation of Natural Products,* G. Winkelmann (ed.), pp. 220–242. VCH Publishers, New York.

Slobodkin, A. I., N. S. Panikov, and G. A. Zavarzin. 1992. Microbiological methane production and consumption in the tundra and middle taiga bogs. *Mikrobiologiya* **61**:683–691 (in Russian).

Sutter, K., K. Egger, and A. Wellinger. 1987. Psychrophilic methane production: a low rate but economically viable technique. In: *Alternative Energy Sources VII*, T. N. Veziroglu (ed.), Vol. 4, pp. 87–98. Hemisphere, Washington, D.C.

Svensson, B. H. 1984. Different temperature optima for methane formation when enrichments from acid peat are supplemented with acetate of hydrogen. *Appl. Environ. Microbiol.* **48**:389–394.

Svensson, B. H., and T. Rosswall. 1984. In situ methane production from acid peat in plant communities with different moisture regimes in a subarctic mire. *Oikos* **43**:341–350.

Svetlichny, V. A., T. G. Sokolova, N. A. Kostrikina, and G. A. Zavarzin. 1991. Anaerobic extremely thermophilic carboxydotrophic bacteria in hydrotherms of Kuril Islands. *Microb. Ecol.* **21**:1–10.

Thauer, R. K., K. Jungermann, and K. Decker. 1977. Energy conservation in chemotrophic anaerobic bacteria. *Bacteriol. Rev.* **41**:100–180.

Young, L. Y., and A. C. Frazer. 1987. The fate of lignin and lignin-derived compounds in anaerobic environments. *Geomicrobiol. J.* **5**:261–293.

Whitman, W. B. 1985. Methanogenic bacteria. In: *The Bacteria*, C. R. Woese and R. S. Wolfe (eds.), Vol. 8, pp. 3–84. Academic Press, San Diego, CA.

Wood, H. G., and L. G. Ljungdahl. 1991. Autotrophic character of the acetogenic bacteria. In: *Variations in Autotrophic Life*, J. M. Shively and L. L. Barton (eds.), pp. 201–250. Academic Press, San Diego, CA.

Zavarzin, G. A. 1986. Trophic relations in methanogenous community. *Izv. AN USSR. Ser.* Biol. **3**:341–360 (in Russian).

Zavarzin, G. A., O. R. Kotsyurbenko, T. I. Soloviova, and A. N. Nozhevnikova. 1993. The temperature threshold in the development of methanogenic versus acetogenic community from the tundra soil. *Dokl AN.* **329**:792–794.

Zeeman, G., T. J. M. Vens, M. E. Koster-Treffers, and G. Lettinga. 1988. Start-up of low temperature digestion of manure. In *Anaerobic Digestion*, E. R. Hall and P. N. Hobson (eds.) 5th Int. Symp. on Anaerobic Digestion, pp. 397–405. Pergamon Press, Elsford, N.Y.

Zeikus, J. G., R. Kerby, and J. A. Krzycki. 1985. Single-carbon chemistry of acetogenic and methanogenic bacteria. *Science* **227**:1167–1173.

16

Halophilic Acetogenic Bacteria

George A. Zavarzin, Tatjana N. Zhilina,
and Margarete A. Pusheva

16.1 Introduction

As outlined in preceding chapters of this book, acetogenic bacteria have a specialized physiological potential for the conservation of energy via the reduction of CO_2 to acetate. They also harbor diverse catabolic processes and are found in unusual habitats. Although their role in nature was initially viewed somewhat restrictively, it is now evident that they might have a large impact on carbon and energy flow in some environments, in particular that of certain gastrointestinal ecosystems (Breznak et al., 1988; Kane and Breznak, 1991a, 1991b). However, specialized (or extreme) terrestrial environments are largely unexplored relative to the involvement of acetogenesis and associated organisms.

Halophilic communities include many procaryobionts, and at extreme salinities these organisms are dominant. Elazari-Volcani (1940, 1944) discovered a number of physiological groups in the Dead Sea. Characterization of the family Halobacteriaceae (aerobic) has received considerable attention and has been reviewed by Larsen (1980, 1982) and Tindall (1992). Anaerobic phototrophs have also been characterized by Trüper and his colleagues (Trüper and Imhoff, 1981). However, in general, anaerobic bacteria that develop in response to hypersaline environments have received relatively little attention despite the fact that such microbial activity in hypersaline water bodies has been known since the early part of this century (Issatchenko, 1927).

The procaryotic diversity in these very productive environments is high. More than 150 microscopically distinguishable bacteria were observed in hypersaline Solar Lake by Hirsch (1980). Because of the low solubility of oxygen in saline habitats and the consumption of oxygen in the upper layers of cyanobacterial mats, anaerobic conditions likely prevail in many niches of such habitats. The isolation and characterization of anaerobic organotrophic bacteria from such habitats has been only recently undertaken after the advent and wider application of the anaerobic techniques of Hungate (1969) (Zeikus et al., 1983; Oren, 1983; Zhilina, 1983, 1986). Haloanaerobic saccharolytic bacteria belong mainly to the family Haloanaerobiaceae (Oren et al., 1984b). It includes the genera *Haloanaerobium* (Zeikus et al., 1983), *Halobacteroides, Sporohalobacter* (Oren et al., 1984a, 1987), *Haloincola* (Zhilina et al., 1992), and celluloytic *Halocella* (Simankova et al., 1993).

A highly specialized halophilic, homoacetogenic bacterium, *Acetohalobium,* was discovered in the course of our studies on the anaerobic degradation of osmoregulatory molecules, compounds which constitute a substantial part of the organic matter in many halophilic bacteria (Zhilina and Zavarzin, 1990a, 1990b, 1991). *Acetohalobium* is a member of the family Haloanaerobiaceae but is functionally similar to *Sporomusa* (Möller et al., 1984). As in the case of *Sporomusa, Acetohalobium* is a Gram-negative, betaine-fermenting homoacetogen. However, as a member of Haloanaerobiaceae, it is quite distant phylogenetically from the freshwater *Sporomusa.* Although their function at the ecosystem level remains uncertain, we postulate that *Acetohalobium* and other unidentified halophilic acetogens may play important roles in the overall turnover of carbon in certain halophilic communities.

16.2   Habitat and Isolation of *Acetohalobium*

Our studies on anaerobic halophiles have focused on the lagoons of the Arabat strait that separates Sivash Lake from the Sea of Azov in eastern Crimea. These lagoons are covered by thick cyanobacterial mats composed primarily of *Microcoleus chthonoplastes* (Bonch-Osmolovskaya et al., 1988; Gerasimenko et al., 1989, 1992). Some were used as solar salterns but were abandoned in 1953 when an irrigating canal was built in northern Crimea. Salinity of the South Sivash gradually decreased from 20% to 5% because of dilution by drainage water. Lagoons are characterized by highly variable regimes that are dependent on weather conditions; in general, after the annual melting of the snow, salinity gradually increases up to saturation with halite by the end of summer. Lagoons on the sandy Arabat strait are quite different from the muddy plains of Crimean shore of Sivash, which is dominated by *Thiocapsa* blooms, sulfate reduction, and decomposition of large masses of *Cladophora sivashensis* with an accompanying fouling odor that is responsible for the trivial name of Sivash, the "Fouling Sea."

For isolation, trimethylamine was used as sole energy substrate at a NaCl concentration of approximately 20% (w/v). A rod-shaped bacterium was observed among flat "broken glass" cells of the halophilic methanogen *Methanohalobium evestigatum*. In the mixed culture, acetate and methane were the sole products detected. An acetogen was isolated from anaerobic roll tubes together with selective inhibition of the methanogen by 10 mM bromoethanesulfonate. The isolate, named *Acetohalobium arabaticum* (type strain is Z-7288 and is available as DSM strain 5502), formed acetate as its main organic product and was the first halophilic acetogen reported (Zhilina and Zavarzin, 1990a).

16.3   Characteristics of *Acetohalobium*

16.3.1   Cellular Properties

A. *arabaticum* Z-7288 is a motile, rod-shaped organism with one to two subterminal flagella (Fig. 16.1). Cells are slightly bent, usually single, but occur sometimes in pairs or short chains. Often palisades are observed because of lengthwise adhesion of cells. Multiplication by binary fission is often nonequal, and cells divide by constriction rather than by septation. The cell-wall structure is typical for Gram-negative bacteria. When initially isolated, spores were sometimes observed, but with continued cultivation, sporulation was no longer observed. After 7 days, colonies (0.5–1 mm diameter) are whitish yellow and round with a slightly hairy border. In broth cultures, rapid lysis occurs at the end of growth; consequently, preservation and maintenance of the culture is tedious.

A. *arabaticum* Z-7288 does not belong to Archaea. Cell lipids contain aliphatic fatty acids and β-hydroxy acids characteristic of the membranes and cell envelopes of Gram-negative eubacteria (Zhilina et al., 1992). About 94% of the fatty acids are straight-chain C_{16} acids (mainly unsaturated D 9 $C_{16:1}$ and D 11 $C_{16:0}$ acids). β-hydroxy fatty acids are represented by $C_{12:1}$ and $C_{12:0}$ in equal amounts, sugars are represented by pentoses and hexoses. Growth is inhibited by streptomycin, canamycin, erythromycin, benzylpenicillin, gentamycin, vankomycin, and tetracycline, and the type strain (DSM 5501) has a guanine plus cytosine (G + C) content of 33.6 ($\pm$ 0.6) mol%.

16.3.2   Physiology

A. *arabaticum* is an obligate halophile. The type strain Z-7288 does not grow at a salinity with NaCl of less than 10% or higher than 25%; the optimum salt concentration approximates 17% NaCl. The optimum pH for growth approximates 7.8, but the pH range is 5.6–8.4. The organism is mesophilic with an optimal growth temperature of 38–40°C; 47°C is maximal.

A. *arabaticum* is an obligate anaerobe and growth is best in the prereduced medium. A. *arabaticum* grows chemolithotrophically with H_2/CO_2, and also with

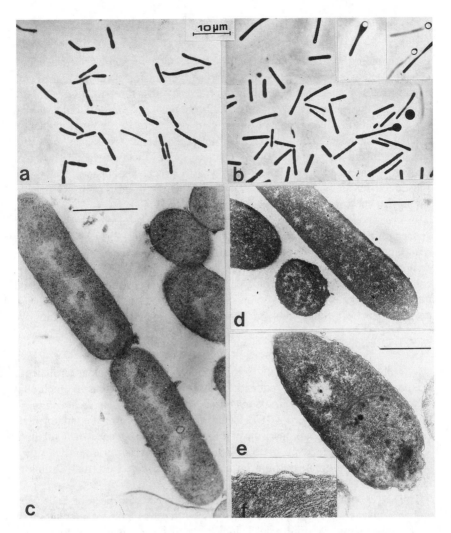

16.1.   Morphology of *Acetohalobium* strains. Panels a and c: strain Z-7288 (type strain) cultivated with formate and 15% NaCl. Panels b, d, e, and f: strain Z-7492 cultivated with trimethylamine and 25% NaCl. Phase contrast is shown in panels a and b, whereas ultrathin sections are shown in panels c–f (bar = 1 μm).

CO. Chemoorganotrophic energy substrates include formate, trimethylamine, betaine, lactate, pyruvate, hystidine, aspartate, glutamate, and asparagine. There is no growth on saccharides (glucose, fructose, galactose, maltose, lactose, saccharose, xylose, arabinose, ramnose, ribose, mannitol, inositol, dulcitol, and sorbitol), monomethylamine and dimethylamine, sarcosine, choline, methanol, ethanol, propanol, butanol, glycerine, fumarate, malate, propionate, butyrate, succinate, meta-metoxybenoic acid, glycine, serine, methionine, or glycogen. Growth is enhanced by the addition of yeast extract; this enhancement is not substituted by amino acids. The organism produces acetate from all growth-supportive substrates and is homoacetogenic with H_2/CO_2.

A. *arabaticum* does not use $S_2O_3^{2-}$, SO_3^{2-}, or SO_4^{2-} as electron acceptors. At concentrations of 1mM sulfite or dithionite inhibit growth, dimethylsulfoxide is reduced to dimethylsulfide and elemental sulfur to H_2S; however, these processes do not appear to be energy yielding (Kevbrin and Zavarzin, 1992). Under certain conditions, the reduction of inorganic sulfur (Beaty and Ljungdahl, 1991) and nitrogen (Seifritz et al., 1993) compounds by the acetogen *Clostridium thermoaceticum* has been found to be energy conserving and coupled to growth.

16.3.3 Phylogenetic Position

The phylogenetic position of *Acetohalobium* within the Haloanaerobiaceae was identified by 5S rRNA sequence. All representative organisms, including *Halobacteroides halobius*, *Halobacteriodes lacunaris*, *Haloincola saccharolytica*, *Halocella cellulolytica* Z-10151, and A. *arabaticum* formed a phylogenetically coherent group. Comparison of 5S rRNA bank sequence data of these strains with representatives of the Archaea and other bacteria suggests they occupy a separate group position on a 5S rRNA phylogenetic tree (Zhilina et al., 1992).

16S rRNA of A. *arabaticum* was aligned with the sequences of eubacteria and archaebacteria. The unrooted phylogenetic tree was built by the method of "maximal topological similarities," indicating that the organism belongs to a separate line of descent on the 16S rRNA tree. The mutation distances between its 16S rRNA and all other rRNAs are comparable to the distances separating the main lineage of eubacteria. The uniqueness of its 16S rRNA suggests that this bacterium diverged at an early stage of evolution with the branching point at the level of *Proteobacteria* (Tourova et al., 1992). This finding is in accordance with results for other Haloanaerobiaceae (Oren et al., 1984b). Recent reviews on the family Haloanaerobiaceae are available (Zhilina et al., 1992; Oren, 1992).

16.3.4 Biochemistry

We have initiated studies on key enzymes and electron transport systems of *Acetohalobium*. Information to date is incomplete, but our goal in these studies is to more completely understand unique processes and their regulation that are important to this new halophilic genus. Our studies have employed anaerobic techniques, including the use of anaerobic chambers, and have focused primarily on CO dehydrogenase, hydrogenase, and electron transport systems.

CO-dehydrogenase in *Acetohalobium* was found both in periplasmic and soluble cell fractions. Salt stimulated activity but was not essential. Maximal activity was observed at around 1–1.5 M NaCl [~ 5.3 μmol CO oxidized (min^{-1}) (mg protein cell extract)$^{-1}$ [using benzyl viologen]. Significantly, salt concentrations approaching saturation (5 M NaCl) were not inhibitory [activities approximating 3.7 μmol CO oxidized (min^{-1}) (mg protein cell extract)$^{-1}$ were observed with both 0 and 5 M NaCl].

When cells were fractionated by lysozyme digestion and differential centrifugation, hydrogenase activity was distributed in many fractions, including the periplasmal and membrane cell fractions. A periplasmic hydrogenase in *Acetobacterium* might indicate adaptation of the bacterium to a low partial pressure of H_2 in hypersaline environments. However, most activity was recovered in the soluble cell fraction. As with CO dehydrogenase, hydrogenase activity was likewise stimulated by salt and reached maximal activities at approximately 1 M LiCl. KCl and NaCl were also greatly stimulatory at 3.0–4.5-M concentrations. However, hydrogenase activity was inhibited at 5 M LiCl. Hydrogenase activities were very high, and the V_{max} observed for the cytoplasmic activity approximated 770 μmol H_2 oxidized (min^{-1}) (mg protein).$^{-1}$ In cell-free extracts of the moderate haloanaerobe *Halobacteroides acetoethylicus*, hydrogenase activity was also stimulated by NaCl until the salt concentration exceeded 2.3 M NaCl, at which point activity decreased (Rengpipat et al., 1988). To date, we have not identified the physiological electron acceptor for hydrogenase in *Acetohalobium*.

Many halophiles need Mg^{2+} or Mn^{2+} and it is possible that divalent cations stabilize proteins under halophilic conditions (Lanyi, 1974; Koesnander et al., 1991). Hydrogenase activity of *Acetohalobium* was stimulated by 20–160 mM Mg^{2+}, 30–50 mM Mn^{2+}, and by 10–70 μM Fe^{2+}.

At the present time we can only speculate on the potential importance of high salt concentrations to the activities of these enzymes *in vivo* (Pusheva et al., 1992a, 1992b). It is well known that enzymes of aerobic halophilic bacteria depend on or are greatly influenced by high intracellular salt concentration (Larsen, 1967). The same appears to be true for saccharolytic haloanaerobes of the family Haloanaerobiaceae (Oren, 1986; Rengpipat et al., 1988; our laboratory, unpublished data).

Neither cytochromes nor chinones were found in the membranes or cell-free extracts of *Acetohalobium* Z-7288. A flavoprotein has been isolated, and various folates and corrinoids (vitamin B_{12} analogue factor III and uroporphirine I and its methylated derivatives) have been detected (Detkova et al., unpublished; Bykhovsky et al., 1994). Such cofactors are found in other acetogens, and complete characterization of the cofactors and electron acceptors of *Acetohalobium* will help resolve its physiological similarities to other acetogens as well as to other members of Haloanaerobiaceae.

16.4 Extreme Halophilic Isolates

After prolonged cultivation on formate, *A. arabaticum* strain Z-7288 lost its ability to utilize trimethylamine. In an attempt to reisolate a trimethylaimine-utilizing organism from the same sample, a new extremely halophilic strain of *Acetohalobium* was isolated; it was designated strain Z-7492 (Fig. 16.1).

Acetohalobium Z-7491 is a spore-forming rod and does not grow at less than 15% salinity; the upper limit of salinity for growth is saturation. At 30% NaCl, growth is very good, and the optimum approximates 25%. Despite their phenotypic differences, *Acetohalobium* Z-7492 has a G + C content (34.6%) similar to that of the type strain Z-7288; there is a high DNA-DNA homology (78%) as well (determined by A. M. Lysenko).

Cells are often swollen at the ends (giving the appearance of budding minicells) (Fig. 16.1). The formation of endospores occurs at stationary phase; cell lyses occurs rapidly in stationary phase. In contrast to *Acetohalobium* Z-7288, dithionite had no inhibitory effect on strain Z-7492.

Cells cultivated in trimethylamine broth have orange to red autofluorescence when oxidized and excited by violet light. Fluorescence strongly depends on the state of the culture and is not observed with lactate-grown cells. This fluorescence is consistent with the presence of flavoproteins with broad emission spectra from 500 to 600 nm and/or tetrapyrrols (corrinoids) which have fluorescence at 640–648 nm.

Strain Z-7492 has a very similar substrate spectrum as that of the type strain *Acetohalobium* Z-7288. The organism is homoacetogenic under certain conditions (e.g., when growing on H_2/CO_2, lactate, or formate). On trimethylamine, traces of methane are detected, and we regard this as a trace by-product under certain conditions. Methane has also been shown to be a trace by-product of *Clostridium thermoautotrophicum* (Savage et al., 1987) and *Acetobacterium woodii* (Buschhorn et al., 1989).

16.5 Position of Haloacetogens in the Halophilic Community

Acetohalobium was discovered during the search for microorganisms responsible for specific reactions leading to methanogenesis in the halophilic community. During these studies, we found that a so-called noncompetitive pathway of methanogenesis in the halophilic community included trimethylamine and betaine as key substrates. This pathway was designated as noncompetitive because sulfate reducers cannot utilize these substrates (King, 1984, 1988). Microorganisms responsible for the metabolic steps in the community were not initially indicated. However, with isolation of halophilic methylotrophic methanogens, a key step for methanogenesis was resolved (Zhilina, 1983, 1986; Zhilina and Zavarzin, 1987). However, these methanogens did not utilize betaine. Significantly, production of trimethylamine from betaine was found to be a metabolic function of *Acetohalobium* (Zhilina and Zavarzin, 1990a, 1990b).

Halophilic communities are unique in that they contain high concentrations of osmoregulatory substances which increase proportionally in response to salinity; such solutes serve to equilibrate osmotic pressure of the medium. Betaine is accumulated as a compatible solute by *Proteobacteria,* which are typical for saline environments (Trüper and Galinski, 1986).

Like many other acetogens, *Acetohalobium* metabolizes a broad range of substrates, including betaine, H_2, formate, lactate, aspartate, and glutamate. However, saccharides do not appear to be utilizable. Betaine decomposed by *Acetohalobium* yields trimethylamine and acetate. Trimethylamine produced by *Acetohalobium* is utilized by methylotrophic methanogens, and interaction between the acetogens and methanogens was confirmed by the production of methane from betaine by combined pure cultures of *Acetohalobium* and *Methanohalobium*. These laboratory models reproduced (confirmed) the noncompetitive pathway of methanogenesis (Zhilina and Zavarzin, 1990b).

Hydrogen may be a key environmental substrate of *Acetohalobium* and is produced, for example, by saccharolytic species of *Halobacteroides* and *Haloincola* that metabolize disaccharides (saccharose and trehalose) which accumulate as compatible solutes by weak halophiles like *Thiocapsa*. These organisms ferment certain heterosides which accumulate as solutes by *Microcoleus,* the main primary producer in the community (Keybrin et al. 1991). Halophilic cellulolytic *Halocella* produces lactate in addition to acetate and hydrogen. Aspartate and glutamate are known to be solutes of weak halophiles, but we have not studied their interaction with *Acetohalobium*.

The fate of acetate produced by halophilic acetogens remains uncertain. Extremely halophilic aceticlastic methanogens are unknown and methane is not readily produced from acetate in hypersaline environments (Oremland et al., 1982; Zhilina, 1986; King, 1988). Acetate-consuming sulfate reducers do not appear to develop well at high salinity (Widdel, 1988); however, one halophilic species recently isolated from Solar Lake had an upper limit of 15% salinity (Gaumette et al., 1991). This would seem to be below the threshold required for a competitive existence under the hypersaline conditions indicative of the habitat of *Acetohalobium*. Extremely halophilic anoxygenic phototrophs typical of the environment might utilize acetate and decompose it anaerobically (Gorlenko et al., 1984). The fate of acetate must be resolved to understand the full meaning of acetogens at the community level. Although we have emphasized the importance of the decomposition of osmoregulatory substances, it must also be realized that there are a number of other metabolic reactions leading to acetate in the halophilic community.

The trophic links outlined in Figure 16.2 highlight the transformations central to the metabolism of the osmoregulatory compounds in halophilic communities. The key position represented by *Acetohalobium*-type acetogens seems obvious. The noncompetitive pathway of methanogenesis is shown to be coupled to the decomposition of osmoregulatory compounds. This overview is certainly incomplete because halophilic acetogens are also integrated to other less-understood community processes. For example, members of Haloanaerobiaceae are capable of producing sulfide and their activity as sulfur-reducers might be partially responsible for sulfidogenesis at high salinities.

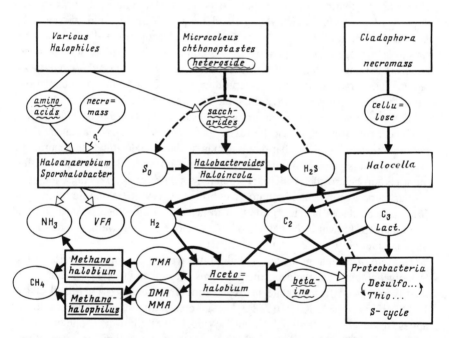

16.2.   Trophic relationships in the halophilic cyanobacterial community, with emphasis on anaerobic decomposition of osmolytes. Heavy arrows indicate transformations observed; organisms catalyzing these transformations are underlined. Osmolytes are indicated by wavy underlines. [From Zhilina and Zavarzin (1991), with permission].

Acknowledgments

The authors express their appreciation to M. V. Simankova and H. L. Drake for communicating unpublished results, to A. M. Lysenko for DNA-DNA homology analysis, and to H. L. Drake for assistance with preparation of the manuscript.

References

Beaty, P. S., and L. G. Ljungdahl. 1991. Growth of *Clostridium thermoaceticum* on methanol, ethanol, propanol, and butanol in medium containing either thiosulfate or dimethylsulfoxide, Abstr. K-131, p. 236. *Ann. Meet. Am. Soc. Microbiol.* 1991.

Bonch-Osmolovskaya, E. A., I. Ya. Vedenina, and G. A. Zavarzin. 1988. Hypersaline lagoons of the Sivash Lake and anaerobic destruction of organic matter in halophilic cyanobacterial mats. *Mikrobiologiya* **57**:442–449 (in Russian).

Breznak, J. A., J. M. Switzer, and H. J. Seitz. 1988. *Sporomusa termitida* sp.nov., an H_2/CO_2-utilizing acetogen isolated from termites. *Arch. Microbiol.* **150**:282–288.

Buschhorn, H., P. Dürre, and G. Gottschalk. 1989. Production and utilization of ethanol by the homoacetogen *Acetobacterium woodii*. *Appl. Environ. Microbiol.* **55**:1835–1840.

Bykhovsky, V. Y. A., M. A. Pusheva, N. I. Zaitseva, T. N. Zhilina, D. B. Pankovskii, and E. N. Detkova. 1994. Biosynthesis of corrinoids and its possible precursers in extremely halophilic homoacetogenic bacterium *Acetohalobium arabaticum* gen nov., sp. nov. Pritladnaya Mikrobiologiya Biochimiya **30**:93–103. (in Russian).

Elazari-Volcani, E. B. 1940. Studies on the microflora of the Dead Sea, Ph.D thesis. Hebrew University of Jerusalem.

Elazari-Volcani, B. 1944. The microorganisms of the Dead Sea. In: *Papers Collected to Commemorate the 70th Anniversary of Dr. Chaim Weizman*, pp. 71–85. Daniel Sieff Research Institute, Rehovoth, Israel.

Gaumette, P., Y. Cohen, and R. Matheron. 1991. Isolation and characterization of *Desulfovibrio halophilus* sp. nov., a halophilic sulfate-reducing bacterium isolated from Solar Lake (Sinai). *Syst. Appl. Microbiol.* **14**:33–38.

Gerasimenko, L. M., V. K. Nekrasova, V. K. Orleansky, S. L. Venetskaya, and G. A. Zavarzin. 1989. The primary production of halophilic cyanobacterium coenoses. *Mikrobiologiya* **58**:507–524 (in Russian).

Gerasimenko, L. M., S. L. Venetskaya, A. V. Dubinin, V. K. Orleansky, and G. A. Zavarzin. 1992. Algo-bacterial communities of hypersaline lagoons of the Sivash (the Crimea) *Algologia* (*Kiev*) **1**:88–94 (in Russian).

Gorlenko, V. M., E. I. Kompantseva, S. A. Korotkov, V. V. Pouchkova, and A. S. Savvichev. 1984. Conditions of development and species of phototrophic bacteria, inhabiting saline shallow reservoirs in Crimea. *Izv. Akad. Nauk SSSR Ser. Biol.* **3**:362–374 (in Russian).

Hirsch, P. 1980. Distribution and pure culture studies of morphologically distinct Solar Lake microorganisms. In: *Hypersaline Brines and Evaporitic Environments*, A. Nissenbaum (ed.), pp. 41–46. Elsevier, Amsterdam.

Hungate, R. E. 1969. A roll tube method for cultivation of strict anaerobes. In: *Methods in Microbiology*, J. R. Norris and D. W. Ribbons (eds.), Vol. 3B, pp. 117–132. Academic Press, New York.

Issatchenko, B. L. 1927. Microbiological investigations on mud lakes. Trudy geol Comiteta new ser., n. 148, p. 154; reprinted in: Issatchko, B. L. 1951. Selected works, vol. 2, A. A. Imshenezky (ed.), USSR Academy of Sciences Press, Moscow-Leningrad (in Russian).

Kane, M. D., and J. A. Breznak. 1991a. *Acetonema longum* gen. nov., sp. nov., an H_2/CO_2 acetogenic bacterium from the termite, *Pterotermus occidentis*. *Arch. Microbiol.* **156**:91–98.

Kane, M. D., and J. A. Breznak. 1991b. *Clostridium mayombei* sp. nov., an H_2/CO_2 acetogenic bacterium from the gut of the African soil-feeding termite, *Cubitermes speciosus*. *Arch. Microbiol.* **156**:99–104.

Kevbrin, V. V., A. V. Dubinin, and G. A. Osipov. 1991. Osmoregulation in the marine cyanobacterium *Microcoleus chtonoplastes*. *Mikrobiologiya* **60**:596–599 (in Russian).

Kevbrin, V. V., and G. A. Zavarzin. 1992. Influence of sulfur compounds on the growth of the halophilic homoacetogenic bacterium. *Acetohalobium arabaticum*. *Mikrobiologiya* **61**:76–81 (in Russian).

King, G. M. 1984. Utilization of hydrogen, acetate and "non-competitive" substrates methanogenic bacteria in marine sediments. *J. Geomicrobiol.* **3**:275–306.

King, G. M. 1988. Methanogenesis from methylated amines in a hypersaline algal mat. *Appl. Environ. Microbiol.* **54**:130–136.

Koesnander, A. S., N. Nishio, and S. H. Nagai. 1991. Effects of the trace metal ions on the growth, homoacetogenesis and corrinoid production by *Clostridium thermoaceticum*. *J. Ferment Bioengineer.* **71**:181–185.

Lanyi, J. K. 1974. Salt-dependent properties of proteins from extremely halophilic bacteria. *Bacteriol. Reviews* **38**:272–290.

Larsen, H. 1967. Biochemical aspects of extreme halophilism. *Adv. Microbiol. Physiol.* **1**:97–132.

Larsen, H. 1980. Ecology of hypersaline environments. In: *Hypersaline Brines and Evaporitic Environments*, A. Nissenbaum (ed.), pp. 23–29. Elsevier, Amsterdam.

Larsen, H. 1982. The Familiy *Halobacteriaceae*. In: *The Prokaryotes*, M. P. Starr, H. Stolp, and H. G. Schlegel (eds.), Vol. 2., pp. 985–994. Springer-Verlag, New York.

Möller, B., R. Ossmer, B. H. Howard, G. Gottschalk, and H. Hippe. 1984. *Sporomusa*, a new genus of Gram-negative anaerobic bacteria including *Sporomusa sphaeroides* sp. nov., and *Sporomusa ovata* sp. nov. *Arch. Microbiol.* **189**:388–396.

Oren, A. 1983. *Clostridium lortetii* sp. nov., a halophilic obligately anaerobic bacterium producing endospore with attached gas vacuoles. *Arch. Microbiol.* **136**: 42–48.

Oren, A. 1986. Intracellular salt concentration of the anaerobic halophilic eubacteria *Haloanaerobium praevalens* and *Halobacteroides halobius*. *Can. J. Microbiol.* **32**:4–9.

Oren, A. 1992. The genera *Haloanaerobium, Halobacteroides,* and *Sporohalobacter*. In: *The Prokaryotes,* A. Ballows, H. G. Truper, M. Dworkin, W. Harder, K.-H. Schleifer (eds.), 2nd ed., pp. 1893–1913. Springer-Verlag, New York.

Oren, A., W. G. Weisburg, M. Kessel, and C. R. Woese. 1984a. *Halobacteroides halobius* gen. nov., sp. nov, a moderately halophilic, anaerobic bacterium from the bottom sediments of the Dead Sea. *System. Appl. Microbiol.* **5**:58–70.

Oren, A., B. J. Paster, and C. R. Woese. 1984b. *Haloanaerobiaceae:* a new family of moderately halophilic, obligately anaerobic bacteria. *System. Appl. Microbiol.* **5**:71–80.

Oren, A., H. Pohla, and E. Stackebrandt. 1987. Transfer of *Clostridium lortetii* to a new genus *Sporohalobacter* gen. nov., as *Sporohalobacter lortetii* comb. nov., and description of *Sporohalobacter marismortui* sp. nov. *System. Appl. Microbiol.* **9**:239–246.

Oremland, R. S., L. Marsh, and D. J. Des Marais. 1982. Methanogenesis in Big Soda Lake, Nevada: an alkaline moderately hypersaline desert lake. *Appl. Environ. Microbiol.* **43**:462–468.

Pusheva, M. A., E. N. Detkova, N. P. Bolotina, and T. N. Zhilina. 1992a. Salt dependent hydrogenase in extremely halophilic methylotrophic homoacetogenic bacterium *Acetohalobium arabaticum*, Abstr. C-135. *7th Int. Symposium on Microbiol Growth on C_1 Compounds*. Warwick, U.K., 1992.

Pusheva, M. A., E. N. Detkova, N. P. Bolotina, and T. N. Zhilina. 1992b. The properties of periplasmic hydrogenase from extremely halophilic homoacetogenic bacterium *Acetohalobium arabaticum*. *Microbiologiya* **61**:933–938 (in Russian).

Rengpipat, S., S. E. Lowe, and J. G. Zeikus. 1988. Effect of extreme salt concentration on the physiology and biochemistry of *Halobacteroides acetoethylicus*. *J. Bacteriol.* **170**:3065–3071.

Savage, M. D., Z. Wu, S. L. Daniel, L. L. Lundie, Jr., and H. L. Drake. 1987. Carbon monoxide-dependent chemolithotrophic growth of *Clostridium thermoautotrophicum*. *Appl. Environ. Microbiol.* **53**:1902–1906.

Seifritz, C., S. L. Daniel, and H. L. Drake. 1993. Nitrate as a preferred electron sink for the acetogen *Clostridium thermoaceticum*. *J. Bacteriol.* **175**:8008–8013.

Simankova, M. V., N. A. Chernych, G. A. Osipov, and G. A. Zavarzin. 1993. *Halocella cellulolytica* gen. nov., sp.nov., a new obligately anaerobic halophilic cellulolytic bacterium. *Syst. Appl. Microbiol.* **16**:385–389.

Tindall, B. J. 1992. The family Halobacteriaceae. In: *The Prokaryotes,* 2nd ed., A. Balows, et al. (eds.), pp. 768–808. Springer-Verlag, New York.

Tourova, T. P., A. Poltoraus, I. Lebedeva, and T. N. Zhilina. 1992. Partial sequence analysis of the 16 S rRNA of *Acetohalobium arabaticum,* a new halophilic acetogenic eubacterium. *International Conference on Taxonomy and Automated Identification of Bacteria.* Prague, 1992.

Trüper, H. G., and E. A. Galinski. 1986. Concentrated brines as habitats for microorganisms. *Experientia* **42**:1182–1187.

Trüper, H. G., and J. F. Imhoff. 1981. The genus *Ectothiorhodospira.* In: M. P. Starr, H. Stolp, H. G. Truper, A. Balows, and H. G. Schlegel (eds.), *The Prokaryotes. A Handbook on Habitats, Isolation, and Identification of Bacteria,* Vol. 1, pp. 274–278. Springer-Verlag, New York.

Widdel, F. 1988. Microbiology and ecology of sulfate and sulfur-reducing bacteria. In: *Biology of Anaerobic Microorganisms,* A. J. B. Zehnder (ed.), pp. 469–587. Wiley, New York.

Zeikus, J. G., P. W. Hegge, T. E. Thomsons, T. J. Phelps, and T. A. Langworthy. 1983. Isolation and description of *Haloanaerobium praevalens* gen. nov., sp. nov., an obligately anaerobic halophile common to Great Salt Lake sediments. *Curr. Microbiol.* **9**:225–234.

Zhilina, T. N. 1983. A new obligately halophilic methane-producing bacteria. *Microbiologiya* **52**:375–382 (in Russian).

Zhilina, T. N. 1986. Methanogenic bacteria from hypersaline environments. *Syst. Appl. Microbiol.* **7**:216–222.

Zhilina, T. N., and G. A. Zavarzin. 1987. *Methanohalobium evastigatum* gen. nov., sp. nov., extremely halophilic methane-producing archaebacteria. *Dolk. Akad. Nauk. SSSR* **293**:464–468 (in Russian).

Zhilina, T. N., and G. A. Zavarzin. 1990a. A new extremely halophilic homoacetogen bacteria *Acetohalobium arabaticum* gen. nov., sp. nov. *Dokl. Akad. Nauk SSSR* **311**: 745–747 (in Russian).

Zhilina, T. N., and G. A. Zavarzin. 1990b. Extremely halophilic, methylotrophic, anaerobic bacteria. *FEMS Microbiol. Rev.* **87**:315–322.

Zhilina, T. N., and G. A. Zavarzin. 1991. Anaerobic bacteria, participating in organic matter destruction in halophilic cyanobacterial community. *J. Obshei Biol.* **52**:302–318 (in Russian).

Zhilina, T. N., G. A. Zavarzin, E. S. Bulygina, V. V. Kevbrin, and G. A. Osipov. 1992. Ecology, physiology and taxonomy studies on a new taxon *Haloanaerobiaceae, Haloincola saccharolytica* gen. nov., sp. nov. *Syst. Appl. Microbiol.* **15**:275–284.

17

O-Demethylation and Other Transformations of Aromatic Compounds by Acetogenic Bacteria

Anne Cornish Frazer

17.1 Introduction

Bache and Pfennig (1981) discovered that *Acetobacterium woodii* strains could be isolated selectively by their capacity to grow methylotrophically on the O-methyl substituents of aromatic compounds. This key article showed that at least 11 methoxylated aromatic compounds supported both growth and acetate formation, without aromatic ring cleavage, by several new isolates and by the type strain of *A. woodii*. These observations provided the impetus for a body of subsequent work aimed at understanding the role that acetogens might play in the anaerobic metabolism of aromatic compounds.

The phenomenon of aromatic compound metabolism by acetogens is of interest because of the great quantity and diversity of these compounds and the importance of understanding the pathways, processes, and microbial catalysts mediating their cycling in anaerobic environments. At present, the types and number of compounds that have been studied with respect to their metabolism by acetogens is not very extensive and is an important subject for future work. The broadest survey of compounds has been done with *Clostridium thermoaceticum* ATCC 39073 which was shown to grow on 20 out of the 42 methoxylated compounds tested (Daniel et al., 1991). Acetogen-mediated biotransformation reactions of aromatic compounds include O-demethylation (which has been studied the most), phenylpropenoate reduction, aldehyde and alcohol oxidative and reductive transformations, decarboxylation of benzoic acid derivatives, decarboxylation of D-mandelic acid, and hydroxylation.

The structures and nomenclature for some of the aromatic compounds which have been the most commonly studied, or which are conceptually important for the purpose of this review, are presented in Figure 17.1. With respect to O-demethylation, aromatic acids related structurally to lignin moieties have been the most widely investigated; these include vanillate (structure 11), ferulate (13), and syringate (18). The corresponding hydroxy derivatives of these compounds are also shown in Figure 17.1: protocatechuate (12), caffeate (14), and gallate (20). The O-demethylation of anisole (methoxylated phenol, 4), guaiacol (2), and veratrol (1) has also been investigated.

With regard to other transformations, some acetogens oxidize an aldehyde group vicinal to the aromatic ring to form the corresponding acid and, thus, can convert vanillin (Fig. 17.1, structure 7) to vanillate. Recently, acetogenesis has been shown to be coupled to the oxidation of aromatic aldehyde groups (e.g., 4-hydroxybenzaldehyde) by *Clostridium formicoaceticum* and *Clostridium aceticum* (Gößner et al., 1993). Certain acetogens saturate the double bond in the side chain of cinnamate derivatives and, thus, convert ferulate or caffeate to hydroferulate or hydrocaffeate (15), respectively. *C. thermoaceticum* ATCC 39073 can decarboxylate aromatic acids like protocatechuate or vanillate to form catechol (3) or its corresponding methoxylated analogue, guaiacol. Recently, it was shown that D-mandelate (16) could be decarboxylated by an acetogen (Dörner and Schink, 1991).

Generally, acetogens do not have the capacity to metabolize the aromatic ring structure and are restricted to transforming various ring substituents. Methoxylated aromatic compounds can be mineralized in defined mixed cultures composed of O-demethylating acetogens acting in concert with other anaerobes (Kreikenbohm and Pfennig, 1985; Sembiring and Winter, 1990). These types of mixed cultures model the role acetogens may serve in anaerobic environments and could be useful for industrial fermentations. In addition, several anaerobes have been isolated which are capable of both acetogenic metabolism of O-methyl substituents and the complete mineralization of certain aromatic compounds. Thus, the

---

Figure 17.1.   Aromatic compounds commonly used in biotransformation studies with acetogens. Trivial and corresponding chemical names are as follows: 1—veratrol (1,2-dimethoxybenzene); 2—guaiacol (2-methoxyphenol); 3—catechol (1,2-benzenediol); 4—anisole (methoxybenzene); 5—*m*-anisic (3-methoxybenzoic) acid; 6—*o*-anisic (2-methoxybenzoic, methoxysalicylic) acid; 7—vanillin (4-hydroxy-3-methoxybenzaldehyde); 8—phenol (hydroxybenzene); 9—veratric (3,4-dimethoxybenzoic) acid; 10—isovanillic (3-hydroxy-4-methoxybenzoic) acid; 11—vanillic (4-hydroxy-3-methoxybenzoic) acid; 12—protocatechuic (3,4-dihydroxybenzoic) acid; 13—ferulic (4-hydroxy-3-methoxycinnamic) acid; 14—caffeic (3,4-dihydroxycinnamic) acid; 15—hydrocaffeic (3,4-dihydroxyphenylpropionic) acid; 16—D-mandelic (α-hydroxybenzeneacetic) acid; 17—3,4,5,-trimethoxybenzoic acid; 18—syringic (4-hydroxy-3,5-dimethoxybenzoic) acid; 19—5-hydroxyvanillic (4,5-dihydroxy-3-methoxybenzoic, methylgallic) acid; 20—gallic (3,4,5-trihydroxybenzoic) acid.

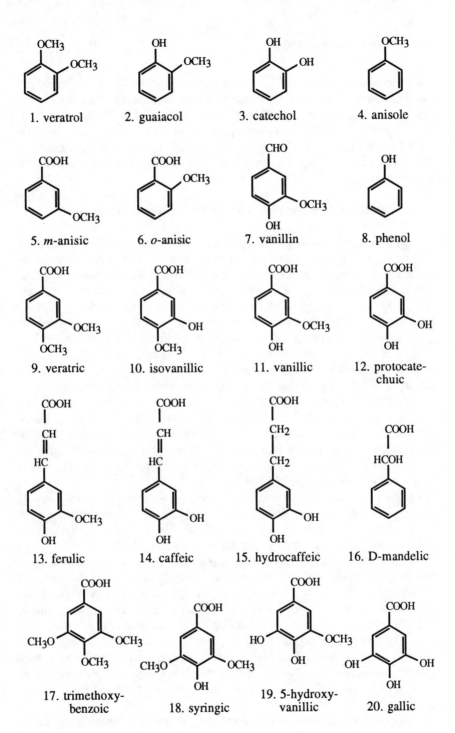

1. veratrol

2. guaiacol

3. catechol

4. anisole

5. *m*-anisic

6. *o*-anisic

7. vanillin

8. phenol

9. veratric

10. isovanillic

11. vanillic

12. protocate-
chuic

13. ferulic

14. caffeic

15. hydrocaffeic

16. D-mandelic

17. trimethoxy-
benzoic

18. syringic

19. 5-hydroxy-
vanillic

20. gallic

sulfate-reducing bacterium, *Desulfotomaculum* sp. strain Groll, can transform vanillyl alcohol to vanillate, O-demethylate and decarboxylate vanillate to catechol, and mineralize the catechol (Kuever et al., 1993). The methanethiol-producing strain, TMBS 4, is also capable of an acetogenic metabolism of O-methyl substituents and has been shown to O-demethylate 3,4,5-trimethoxybenzoate (Fig. 17.1, structure 17) or syringate. The resulting gallate is then mineralized (Bak et al., 1992).

17.1.1   Phenolics from Natural Products and Degradation Processes

Phenolic compounds are present in all types of cells, although plant substances are particularly rich in structurally and functionally diverse phenolic materials. Certain plant phenolics are released as signal molecules in response to mechanical injury, infection, and plant–microbe synergistic or parasitic interactions (Dénarié et al., 1992; Winans, 1992). Many become available to other inhabitants of the biosphere during digestive processes. Hence, plant phenolics are present in rumen or the gastrointestinal tracts of plant-consuming animals and are subject to biotransformation by the resident microflora as well as by the host (Martin, 1982; Scheline, 1973). Plant phenolics are released to the environment as a result of the degradation of lignocellulose by fungi or bacteria (Kirk and Farrell, 1987; Vicuña, 1988; Colberg, 1988). Phenolics may also enter the environment through the release of processing wastes from biomass degradation in fermentors, anaerobic digesters, or pulp and paper production (Kaiser and Hanselmann, 1982b; Hanselmann, 1982; Neilson et al., 1991). Phenolics can partition to soils and sediments and interact chemically with other components in ways that affect their bioavailability (Kaiser and Hanselmann, 1982a).

Among plant phenolics that are widespread and structurally common are (1) the simple phenolics that have a single benzene ring, (2) flavonoids that have two benzene rings connected by a three-carbon chain, and (3) tannins which are polyphenolic materials with a molecular weight range of 500–3000 (Singleton and Kratzer, 1973; Berk, 1976). Simple phenolics and flavonoids may contain O-methyl substituents on the aromatic ring and, thus, are of particular relevance to the subject of this review. Flavonoids exist as glycosides, and upon hydrolysis, component sugars and the aromatic moiety or aglycone are released. Simple phenolics include acids that are either benzenoids, derivatives of benzoate such as vanillate and syringate, or phenylpropanoids, derivatives of cinnamate such as ferulate and caffeate. Cinnamate has the C_6-C_3 skeleton of phenylalanine with a double bond in the side chain. Carbohydrate-phenylpropanoid esters are present in plant cell walls and may be common components in food and feed (Akin et al., 1993).

Lignin, a heterogeneous structural polymer, is one of the major components of wood and, thus, the largest reservoir of aromatic compounds in the global carbon cycle. It is composed of hydroxylated and methoxylated phenylpropanoid units linked through a variety of nonhydrolysable C-C and C-O-C bonds. Degradation of this resistant, water-insoluble material is primarily through the activity of aerobic fungi (Kirk and Farrell, 1987) which effect the release of carbon dioxide and water-soluble oligomers (Reid et al., 1982). Lignin oligomers are further degraded by aerobic and anaerobic bacteria (Vicuña, 1988; Colberg, 1988) to release monomer phenolics that include vanillin, vanillate, syringate, hydroferulate, ferulate, caffeate, cinnamate, benzoate, catechol, and phenylacetate (Colberg and Young, 1985).

17.1.2 The O-Demethylation of Lignin Monomers in Anaerobic Environments

It was shown by Healy and Young (1979, 1980) that methanogenic enrichments, established with inocula from an anaerobic digester-fed municipal sewage, could mineralize 11 ligno-aromatic compounds including syringaldehyde, syringate, vanillin, vanillate, ferulate, catechol, and phenol (Fig. 17.1, structure 8). A more detailed examination of a stable methanogenic consortium converting ferulate to methane showed that acetate was formed as an intermediate (Healy and Young, 1980). A similar stable methanogenic consortium metabolizing syringate was established with an inoculum from anaerobic freshwater lake sediment (Kaiser and Hanselmann, 1982a, 1982b). The consortium could be maintained indefinitely with syringate as the substrate. However, attempts to passage the consortium on vanillate were unsuccessful due to the buildup of catechol, which, in contrast to the consortia of Healy and Young, could not be further metabolized. The isolation of *A. woodii* strains from sediments and sludge, able to grow by O-demethylation of ligno-aromatic compounds, suggested that specialists in O-demethylation might be an important constituent population in consortia-metabolizing ligno-aromatics (Bache and Pfennig, 1981).

It was known from studies of the metabolic capabilities of anaerobic microflora in rodent cecum (Scheline, 1973) and rumen (Martin, 1982; Ivie et al., 1974) that O-demethylation was likely to occur in these environments. Subsequently, O-demethylating acetogens were purified from methanogenic enrichments-metabolizing methoxylated aromatic acids and established with inocula from diverse sources. These included strain TH-001 from the ferulate consortia of Healy and Young (Frazer and Young, 1985), *Clostridium pfennigii* (Krumholz and Bryant, 1985) and *Syntrophococcus sucromutans* (Krumholz and Bryant, 1986) from rumen, and *Eubacterium callandri* (Mountfort and Asher, 1986; Mountfort et al., 1988) from an industrial fermentation plant producing ethanol from wood fiber. Thus, acetogens capable of using O-methyl substituents as a one-carbon source are present in many anaerobic environments. Their activity can convert methoxylated one-carbon substrates to acetate, and the hydroxyl derivative of the aromatic compound is then available to be metabolized by other components in mixed-population ecosystems.

The effectiveness of a degradation chain has been tested with defined mixed cultures of an O-demethylating acetogen grown with other strains capable of metabolizing the aromatic ring of the hydroxyl derivative. Co-cultures of *A. woodii* and *Pelobacter acidigallici* convert syringate completely to acetate (Schink and Pfennig, 1982). If *Desulfobacter postgatei* is included as well in the co-culture, then mineralization of 3,4,5-trimethoxybenzoate is coupled to sulfate reduction (Kreikenbohm and Pfennig, 1985). Co-cultures of strain B10, an O-demethylating acetogen, and *Desulfosarcina* strains were able to couple the degradation of *m*-anisate (Fig. 17.1, structure 5) to the reduction of sulfate (Sembiring and Winter, 1990). Studies on the anaerobic pathways for the complete degradation of aromatic compounds are part of a burgeoning field that has been reviewed elsewhere (Young and Frazer, 1987; Evans and Fuchs, 1988; Schink et al., 1992).

Anaerobic O-demethylation appears to be important in the biotransformation of chlorinated methoxylated catechols. Chloroveratrols and chloroguaiacols are xenobiotics which are released in wastewater effluents from the chemical bleaching of pulp. They were rapidly O-demethylated to the corresponding chlorocatechols prior to reductive dechlorination in an upflow anaerobic sludge blanket reactor (Woods et al., 1989). The inoculum for this reactor was developed from a municipal sludge digester fed a mixture of materials extracted from pulp and paper bleaching effluents plus other carbon sources and nutrients.

Neilson and his colleagues (1991) have intensively studied biotransformations of chloroguaiacols and catechols. They have used anaerobic sediment enrichments, fed a range of environmentally relevant natural aromatic carbon sources, before being challenged with the addition of chloroguaiacols and monitored for subsequent transformations. Among the aromatics used to develop these enrichments are methoxylated compounds such as some flavonoids and 3,4,5-trimethoxybenzoate (Neilson et al., 1991; Allard et al., 1992). A guaiacol cycle has been proposed to conceptualize the transformations observed in sediments, overlying waters, and biota (Remberger et al., 1986). In these studies, anaerobic O-demethylation of chloroveratrols and chlorocatechols preceded anaerobic dechlorination, and it was considered that O-demethylating acetogens might play a role in these biotransformations (Allard et al., 1991). Recently, Häggblom et al. (1993) demonstrated that the acetogens *A. woodii* and *Eubacterium limosum* could indeed O-demethylate a range of chloroguaiacols. No dechlorination of the resulting chlorocatechols was observed.

A considerable effort has been made to isolate novel O-demethylating anaerobes from rumen, gut, digesters, and sediments. Many of these isolates are acetogens that utilize the methyl group as a one-carbon growth substrate. Although the capacity for anaerobic O-demethylation is strongly associated with acetogens, isolates capable of this activity have also been found among other physiological groups of anaerobes including denitrifying, photosynthetic, sulfate reducing, and enteric bacteria (Young and Frazer, 1987). The possible environmental significance of O-demethylation for diverse acetogens was discussed by Ljungdahl (1986). Several other review articles have stressed that most acetogens utilize a broad range of growth substrates, including C-1 sources such as methoxyl substituents (Fuchs, 1986; Heijthuijsen and Hansen, 1990; Diekert, 1992; Drake, 1992). The flexibility of many acetogens with respect to the flow of reductant as well as of carbon has also been stressed (Drake, 1993); see Chapters 1, 7, and 10 for further statements on metabolic diversity.

The capacity of acetogens to grow mixotrophically, forming acetate from H_2/CO_2 while simultaneously utilizing other substrates, has also been suggested as a major factor influencing the competitiveness of these bacteria in different environments (Breznak and Blum, 1991; Breznak and Kane, 1990; see Chapter 11). Strain SS1, isolated from a deep subsurface aquifer sediment, is an unusual acetogen in that obligate mixotrophy seems to be required for growth by O-demethylation of syringate (Liu and Sulfita, 1993). The difficulty of predicting whether acetogens will be prominant in a given anaerobic environment has been discussed (Dolfing, 1988). An example of this with respect to acetogenic O-demethylation was observed even in the initial study by Bache and Pfennig (1981). Acetogenic O-demethylation was only observed in four out of the eight anaerobic freshwater sediment samples from New Zealand that were tested. Thus, the potential for acetogenic O-demethylation in anaerobic environments appears to be common but not universal.

17.2 O-demethylation

17.2.1 Distribution among Acetogens

The capacity of a number of acetogenic strains to grow with aryl O-methyl substituents as a one-carbon source is summarized in Table 17.1. This list of strains cannot be regarded as complete because many acetogens have yet to be examined for their capacity to O-demethylate aromatic compounds, and interesting new isolates continue to be reported. Although a large number of acetogens can utilize O-methyl groups, some strains seem unable to do so, such as *Acetobacterium wieringae* (Eicher and Schink, 1984) and the OMD subline of *C. thermoaceticum*, which is also unusual for its inability to grow autotrophically (Daniel et al., 1991).

The capacity for O-demethylation is tabulated with respect to five substrates that were chosen as the basis of comparison because they are often included in surveys for substrate utilization: 3,4,5-trimethyoxybenzoate, syringate, ferulate,

Table 17.1 Acetogenic strains capable of methylotrophic growth by O-demethylation of aromatic compounds[a]

Strains	TMB	Syr	FA	VA	VN	References
1. *Acetobacterium woodii*						
Strain NZva16	+	+	+	+	+	Bache and Pfennig (1981)
Type strain, ATCC 29683 (DSM 1030)	+	+	+	+	+	Bache and Pfennig (1981)
2. *Acetobacterium carbinolicum*, WoPropl (DSM 2925)	+					Eicher and Schink (1984)
3. *Acetobacterium wieringae*, C (DSM 1911)	−					Eicher and Schink (1984)
4. *Acetobacterium malicum*, MuME1 (DSM 4132)	+	+	+	+		Tanaka and Pfennig (1988)
5. *Acetobacterium* sp., AmMan1	+	+	+	+		Dörner and Schink (1991)
6. Strain B10	+	+	+	+		Sembiring and Winter (1990)
7. *Eubacterium callanderi*, FD (DSM 3662)		+		+	+	Mountfort and Asher (1986)
8. *Eubacterium limosum*						
Strain RF		+	+	+		Genthner and Bryant (1987)
Type strain, ATCC 8486			+	+		DeWeerd et al. (1988), Berman and Frazer (unpublished)
9. Strain B2	+	+	−	+		Cocaign et al. (1991)
Strain TH-001	+	+	+	+		Frazer and Young (1985)
10. *Butribacterium methylotrophicum*	+	+	+	+		Heijthuijsen and Hansen (1990)
11. *Clostridium pfennigii*, V5-2 (DSM 3222)	+	+	+	+	+	Krumholz and Bryant (1985)
12. *Clostridium thermoaceticum*						
ATCC 39073	−	+	+	+	+	Daniel et al. (1988, 1991)
DSM 2955		+		+		Daniel et al. (1988)
13. *Clostridium thermoautotrophicum*						
Type strain, 701/3 (DSM 1974, ATCC 33924)		+		+		Daniel et al. (1988)
Strain 701/5		+		+		Daniel et al. (1988)

14. *Clostridium formicoaceticum*

Organism							Reference
ATCC 27076	+	+	+	+	+	+	Lux et al. (1990), Lux and Drake (1992)
DSM 92	+	+	+	+	+	+	Drake (1992)
15. *Clostridium aceticum*, DSM 1496	+	+	+	+	+	+	Lux and Drake (1992)
16. *Clostridium mayombei*, SFC-5 (DSM 6539)	−	(+)					Kane et al. (1991)
17. *Peptostreptococcus productus* IIb							
Strain U-1 (ATCC 35244)	+	+	+	+			Lux et al. (1990), Drake (1993)
Strain Marburg	+	+					Drake (1993)
Strain MC	+	+	+				Traunecker et al. (1991)
Strain SS1	+	+					Liu and Suflita (1993)
20. *Acetitomaculum ruminis*							
Type strain, 139 B (ATCC 43876)	+	+	+	−			Greening and Leedle (1989)
Strain 190 A4	+	+	+	+			Greening and Leedle (1989)
Strain 40 C	−	−	−	−			Greening and Leedle (1989)
21. *Syntrophococcus sucromutans*, S195 (DSM 3224)	+	+	+	+			Krumholz and Bryant (1986)
22. *Sporomusa termitida*, JSN-2 (DSM 4440)	+	+					Breznak et al. (1988)
23. *Sporomusa ovata* (uses veratric)		−					Konle and Stupperich (1992)
24. *Acetonema longum*, APO-1 (DSM 6540)	(+)	−	+				Kane and Breznak (1991)
25. TMBS 4 (DSM 6591)	+	+	+	−			Bak et al. (1992), Kreft and Schink (1993)
26. *Desulfotomaculum orientis*							
Type strain, Singapore 1 (ATCC 19365, DSM 765)	+	+	+	+			Klemps et al. (1985), Kuever et al. (1993)
27. *Desulfotomaculum thermobenzoicum*, TSB (DSM 6193)	+	+	+				Tasaki et al. (1991, 1992)
28. *Desulfotomaculum* sp., strain Groll	+	+	+				Kuever et al. (1993)
29. *Desulfomonile tiedjei*, DCB-1 (ATCC 49306)	−	−	(+)				DeWeerd et al. (1986, 1990)

[a]Note: Growth supportive: +, tested but not growth supportive: −, reported as poor growth: (+): Abbreviations: DSM (Deutsche Sammlung von Microorganismen, Darmstadt), ATCC (American Type Culture Collection, Rockville, Maryland), TMB (3,4,5-trimethoxybenzoic), Syr (syringic), FA (ferulic), VA (vanillic), and VN (vanillin).

453

vanillate, and vanillin (Fig. 17.1). There is no set of compounds that are invariably used for this purpose, and many alternative compounds could have been included in Table 17.1. For example, veratrate (Fig. 17.1) has been used in many studies. It supports the growth of *Sporomusa ovata* (Konle and Stupperich, 1992) as well as *A. woodii* ATCC 29683 (Berman and Frazer, 1990) and *E. limosum* ATCC 8486 (Frazer and Berman, 1992) but not that of *C. thermoaceticum* ATCC 39073 (Daniel et al., 1991).

Clearly, some strains have a fairly broad substrate range, whereas others are rather restricted. Thus, it is desirable to include several substrates when screening new isolates for the capacity to O-demethylate. For example, *Clostridium mayombei* (Table 17.1, entry 16) uses syringate but not trimethoxybenzoate, whereas just the reverse was observed with *Acetonema longum* (24). Interestingly, *C. thermoaceticum* ATCC 39073 which has a broad substrate range nevertheless fails to use 3,4,5-trimethoxybenzoate (Daniel et al., 1988, 1991).

Substrate concentration is also an important parameter to consider in testing new strains for O-demethylating activity. Some isolates are capable of growth on methoxylated substrates at levels of 5 or 10 mM. However, because of the toxicity of aromatic compounds, it may be advisable to initially use lower levels, in the range of 2 mM for aromatic acids and 0.1–0.5 mM for catechol derivatives. Evidence of O-demethylation should be assessed by high performance liquid chromatography (HPLC) or UV spectroscopy to quantitate substrate loss or product formation. Increase in biomass should be looked for but may be difficult to demonstrate when substrate concentration has to be kept low. Note that if a strain can decarboxylate a compound like vanillate, the potentially more toxic catechol derivative will be formed. This may inhibit growth if the starting substrate concentration is too high.

Strains in the same genus are grouped together in Table 17.1 and listed chronologically by the date of the publication that first described their O-demethylating capacity. Some strains have only been examined in a cursory manner; others were studied fairly intensively. Gram-positive acetogens are included in entries 1 through 20, and Gram-negative acetogens in 21 through 24. Among the isolates not yet taxonomically identified are strain B10 (entry 6) which is probably an *Acetobacterium* sp. (Sembiring and Winter, 1990), strain TH-001 (entry 9) which is related to *Eubacterium* by 16S rRNA sequence analysis [reviewed by Young and Frazer (1987)], strain MC (entry 18) which may be similar to *Peptostreptococcus productus* (Traunecker et al., 1991), and strain SS1 (entry 19), a novel isolate in which O-demethylation is dependent on H_2/CO_2 (Liu and Suflita, 1993). None of these acetogens (1–24) can cleave the aromatic ring, although some are capable of transforming other substituents of aromatic compounds in addition to O-methyl groups.

Strain TMBS 4 (entry 25) is a Gram-negative isolate (Bak et al., 1992) that can metabolize methoxyl carbons acetogenically but, in addition, can mineralize gallate (Kreft and Schink, 1993). Hence, this strain can completely degrade methoxylated aromatic compounds that are precursors of gallate. Entries 26 through 29 include certain sulfate-reducing bacteria with the capacity to metabolize O-methyl substituents acetogenically. The *Desulfotomaculum* species are Gram-positive spore formers. The activity of *Desulfotomaculum orientis* is restricted to O-demethylation without cleavage of the aromatic ring (Klemps et al., 1985). Although *Desulfotomaculum thermobenzoicum* grows on benzoic acid (Tasaki et al., 1991), the aromatic ring of the hydroxylated aromatic acids produced as a consequence of O-demethylation are not further metabolized (Tasaki et al., 1992). By contrast, *Desulfotomaculum sapomandens* which also grows on benzoate is apparently unable to O-demethylate aromatic compounds, for it was unable to grow with either vanillate or syringate (Cord-Ruwisch and Garcia, 1985). *Desulfotomaculum* sp. strain Groll, however, which can metabolize catechol, is able also to O-demethylate vanillate and decarboxylate the resulting protocatechuate to form catechol, thereby completely degrading aromatic substrates that are precursors of catechol or catecholic acids (Kuever et al., 1993). The Gram-negative aryl dechlorinating isolate *Desulfomonile tiedjei* strain DCB-1 is also capable of homoacetogenic growth by O-demethylation (DeWeerd et al., 1986, 1988, 1990; Dolfing and Tiedje, 1991).

Acetogens in Table 17.1 initially isolated by their O-demethylating capability include *A. woodii* strain NZva16 (isolated on 5 mM vanillate), *Eubacterium callanderi* (10 mM ferulate), strain TH-001 (2 mM ferulate), *Clostridium pfennigii* (5 mM vanillate), *Syntrophococcus sucromutans* (5 mM syringate), and strain SS1 (100 μM syringate). In contrast to these are the many acetogens recently isolated by virtue of their capacity to grow on other carbon sources and, subsequently, shown to be capable of growth by O-demethylation. These alternative isolation strategies have included (1) growth methylotrophically on other one-carbon sources (*Butyribacterium methylotrophicum*, *E. limosum* strain RF, and strain MC) or (2) on other organic carbon sources (*Acetobacterium carbinolicum*, *Acetobacterium malicum*, *Acetobacterium* sp. strain AmMan1, and *S. ovata*) or (3) on H_2/CO_2 or CO by acetogenesis (*A. woodii* ATCC 29683, *C. mayombei*, *Clostridium thermoautotrophicum*, *P. productus*, *Acetitomaculum ruminis*, *Sporomusa termitida*, and *A. longum*). Other acetogens listed in Table 17.1 are well-known strains that have been maintained for many years in culture collections and, more recently, have been recognized as capable of growth by O-demethylation of aromatic compounds. These include *E. limosum* ATCC 8486, *C. thermoaceticum*, *Clostridium formicoaceticum*, and *Clostridium aceticum*.

It is interesting that rather different O-demethylating acetogens have been isolated from similar environmental sources, depending on the strategy used for enrichment. For example, when the rumen of steers fed a high-forage diet was examined for O-demethylating activity, *S. sucromutans* was isolated as representative of the most dominant bacterial type (Krumholz and Bryant, 1986). When a similar source was examined for H_2/CO_2 acetogenesis, *A. ruminis* was isolated as representative of the domininat type (Greening and Leedle, 1989). Of course, variation in the endogenous microbial flora of different host animals could also be a factor in the diversity of acetogens isolated.

17.2.2   Types of Compounds

In addition to the data on growth-supportive methoxylated substrates which are summarized in Table 17.1, more detailed studies to assess the diversity of chemical structures acceptable as substrates for O-demethylation by intact cells have been made with six acetogens, *A. woodii* (Bache and Pfennig, 1981; De-Weerd et al., 1988; Häggblom et al., 1993), *E. limosum* (DeWeerd et al., 1988), *C. thermoaceticum* ATCC 39073 (Daniel et al., 1991), *C. formicoaceticum* ATCC 27076 (Lux and Drake, 1992), *C. aceticum* (Lux and Drake, 1992), and strain B10 (Sembiring and Winter, 1990), as well as with sulfate-reducing bacteria that mediate acetogenic O-demethylation, *D. tiedjei* (DeWeerd et al., 1988, 1990), *D. thermobenzoicum* (Tasaki et al., 1992), *D. orientis* and *Desulfotomaculum* sp., strain Groll (Kuenver et al., 1993). The compounds examined have been predominantly either phenolics or phenolic acids, including some chlorinated derivatives. All have been monoaromatic ring compounds with the number of hydroxy or methoxy substituents on the aromatic ring varying from one to three. Flavonoids and dimer or trimer ligno-aromatics have not been examined.

With regard to phenolics (Fig. 17.1), O-demethylation has been examined for a number of substrates including anisole, guaiacol, and veratrol, as well as methoxylated derivatives of resorcinol, 4-hydroxyphenol, and 1,2,3, trihydroxybenzene. *A. woodii* utilizes anisole (Bache and Pfennig, 1981), and both *A. woodii* ATCC 29683 and *E. limosum* ATCC 8486 O-demethylate guaiacol as well as mono-, di-, tri-, or tetrachloroguaiacols (Häggblom et al., 1993). Strain B10 can O-demethylate 3-methoxyphenol to form resorcinol (Sembiring and Winter, 1990). *C. thermoaceticum* ATCC 39073 does not utilize anisole, but many other phenolics are O-demethylated (Daniel et al. (1991): (1) guaiacol and veratrol are converted to catechol, (2) 1,3-dimethoxybenzene and 3-methoxyphenol are converted to resorcinol, (3) 1,4-dimethoxybenzene is converted to 4-hydroxyphenol, and (4) 1,2,3-trimethoxybenzene, 2,6-dimethoxyphenol, 2,3-dimethoxyphenol, and 2-hydroxy-3-methoxyphenol are converted to 1,2,3-trihydroxybenzene. Similarly 1,2,3-trimethoxybenzene is O-demethylated to 1,2,3-trihydroxybenzene by *C. formicoaceticum* ATCC 27076 and *C. aceticum* (Lux and Drake, 1992). *D. orientis* and *Desulfotomaculum* sp. strain Groll are able to O-demethylate anisole, guaiacol, and 2,6-dimethoxyphenol (Kuever et al., 1993).

Methoxylated benzoate or cinnamate derivatives with oxygen substituents at positions 3 and 4 on the aromatic ring have been widely used as model ligno-aromatic compounds: vanillate, syringate, and ferulate (Table 17.1, Fig. 17.1). Other aromatics structurally similar to these have also been studied. *A. woodii* ATCC 29683 and *E. limosum* ATCC 8486 O-demethylate veratrate, isovanillate, isoferulate, (3-hydroxy-4-methoxycinnamate), 3,4-dimethoxycinnamate, synapate (3,5-dimethoxy-4-hydroxycinnamate), and 3,4,5-trimethoxycinnamate (Berman and Frazer, unpublished). *C. thermoaceticum* ATCC 39073 (Daniel et al., 1991), *C. formicoaceticum* ATCC 27076, and *C. aceticum* (Lux and Drake, 1992) also O-demethylate both vanillate and isovanillate. Veratrate is O-demethylated by *C. formicoaceticum* ATCC 27076 and *C. aceticum* but not by *C. thermoaceticum* ATCC 39073. Synapate is O-demethylated by *C. formicoaceticum* ATCC 27076, *C. aceticum*, *C. thermoaceticum* ATCC 39073, and *E. callenderi* (Mountfort and Asher, 1986). These studies support the suggestion that the substrate specificity for acetogenic anaerobic O-demethylation is broad (Bache and Pfennig, 1981).

However, more limited utilization patterns seem to be observed with mono-methoxylated benzoate or cinnamate. Whether this is a consequence of structural specificity in substrate uptake by cells or in O-demethylation, remains to be determined. *A. woodii* NZva-16 utilizes *meta* or *para* anisate, but *E. limosum* ATCC 8486 utilizes only the *meta* isomer. *C. thermoaceticum* ATCC 39073 utilizes *o*-anisate (i.e., methoxysalicylate, Fig. 17.1, structure 6) but not the *meta* or *para* methoxylated analogues (Daniel et al. 1991), whereas strain B10 can utilize all three (Sembiring and Winter, 1990).

Some attention has been focused on the effect of ring substituents *ortho* to methoxy groups. The capacity for growth-supportive O-demethylation with a variety of *o*-anisate analogues was examined in *C. thermoaceticum* ATCC 39073 (Daniel et al., 1991). No growth occurred if the 1-carboxy group of *o*-anisate was replaced by a variety of other substituents including thiol, methyl, nitrile, amide, aldehyde, or amine groups. Additionally, no growth occurred with 2-methoxyacetophenone, 2-methoxyphenylacetate, 2-methoxyphenethyl alcohol, or 2-methoxycinnamate. Strain B10, however, utilizes 3-methoxy- or 3,4,5-trimethoxyphenylacetate (Sembiring and Winter, 1990). With respect to other benzoate derivatives with 2-methoxy substituents, Bache and Pfennig (1981) found that *A. woodii* utilized 2,4-dimethoxybenoate, whereas DeWeerd et al. (1988) found that neither *A. woodii* ATCC 29683 nor *E. limosum* ATCC 8486 utilized 2-hydroxy-4-methoxybenzoate. *C. thermoaceticum* ATCC 39073 converted 2,3-dimethoxybenzoate to 2-hydroxy-3-methoxybenzoate but could not further O-demethylate the product (Daniel et al., 1991).

17.2.3 Carbon Flow to Acetate or Carbon Dioxide

Acetogens use the O-methyl substituent of methoxylated aromatics as a one-carbon growth substrate, and the aromatic ring structure of the hydroxylated product itself is not further degraded. As has been observed in methanol metabolism, the carbon flow for many acetogens probably involves the oxidation of methyl groups to carbon dioxide concomitant with methyl-group conversion to the methyl carbon of acetate. Bache and Pfennig (1981) showed for *A. woodii* that the stoichiometry of O-methyl substituent conversion to acetate, in the presence of carbon dioxide (reaction 17.2), was similar to that of methanol conversion to acetate (reaction 17.1):

$$4 \, CH_3OH + 2 \, CO_2 \rightarrow 3 \, CH_3COOH + 2 \, H_2O \qquad (17.1)$$

$$4 \, \text{Vanillic acid} \, (C_8H_8O_4) + 2 \, CO_2 \rightarrow$$
$$3 \, CH_3COOH + 2 \, H_2O + 4 \, \text{protocatechuic acid} \, (C_7H_6O_4) \qquad (17.2)$$

For methanol metabolism a carbon flow pattern of methyl-group oxidation and concurrent conversion to the methyl carbon of acetate has been established for *B. methylotrophicum* (Zeikus et al., 1980; Kerby et al., 1983) and *Sporomusa acidovorans* (Cord-Ruwisch and Ollivier, 1986). Studies on the mixotrophic growth of *S. termitida* (Breznak and Blum, 1991) showed that some ^{14}C-methanol is oxidized to [14]CO$_2$ during acetogenesis. A hydrogen atmosphere suppressed carbon flow to methyl-group oxidation in favor of acetate formation. Using (methyl ^{14}C) vanillate as the growth substrate for strain TH-001 and an atmosphere of N$_2$/CO$_2$ (70/30), Frazer and Young (1986) observed that, after growth, significant label was present in the gas phase, presumably as carbon dioxide, as well as in the methyl carbon of acetate. Very little label was present in the carboxy carbon of acetate. This is physical evidence for concurrent acetogenesis and carbon dioxide formation in the carbon flow of O-methyl substituents.

As shown in Figure 17.2, the anticipated route for the formation of acetate is through the activity of CODH/acetyl-CoA synthase, with the methyl carbon derived from the methoxy group and the carboxy carbon from carbon dioxide, present under CO$_2$-enriched conditions in the headspace of the cultures. For every 3 mol of acetate formed, 1 mol of methoxy substituents would be oxidized to carbon dioxide, possibly through interconversion of tetrahydrofolate-C-1 carriers. The reducing equivalents generated from methyl-group oxidation would be consumed, in turn, by the reduction of carbon dioxide to the carboxy carbon of acetate.

With *A. woodii*, less acetate is produced per mole of O-methyl substituent metabolized if the propenoate double bond of cinnamate derivatives is present as an alternative electron acceptor (Bache and Pfennig, 1981), as shown in reaction 17.3:

$$2 \, \text{Ferulic acid} \, (C_{10}H_{10}O_4) + 2 \, H_2O \rightarrow CH_3COOH +$$
$$2 \, \text{hydrocaffeic acid} \, (C_9H_{10}O_4) \qquad (17.3)$$

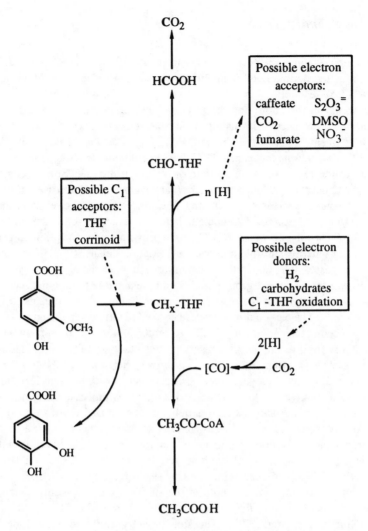

Figure 17.2. Carbon flow for acetogens metabolizing O-methyl substituents. The value of n in the one-carbon oxidation reactions could be 2, 4, or 6, depending on whether a methenyl-, methylene-, or methyl-tetrahydrofolate (THF) product is formed upon O-demethylation. DMSO is dimethylsulfoxide.

Tschech and Pfennig (1984) and Hansen et al. (1988) have shown that the reduction of the double bond is an energy-conserving process and is not just a hydrogen sink reaction.

Depending on the strain and the physiological conditions, O-demethylation can be coupled to acetate formation or to other electron acceptor processes in addition to phenylpropenoate reduction (Fig. 17.2). Beaty and Ljungdahl (1991) showed that *C. thermoaceticum* ATCC 39073 could couple O-demethylation of vanillate to the reduction of thiosulfate or dimethylsulfoxide. Similar coupling may also be possible with other acetogens that have been shown to use thiosulfate as a sulfur source, such as *A. woodii* and *B. methylotrophicum* (Heijthuijsen and Hansen, 1989). Coupling of O-demethylation to fumarate reduction has been shown for *C. formicoaceticum* ATCC 27076 growing under CO_2-limited conditions on the O-methyl substituent of vanillate (Matthies et al., 1993). The stoichiometry suggested that under these conditions, both carbons of acetate were derived from the O-methyl group. This would presumably require the reduction of fumarate coupled with the oxidation of some O-methyl substituents to form carbon dioxide. The carbon dioxide would be used in turn to generate the carboxy group for acetate formation (Fig. 17.2). Recently, nitrate has been shown to be preferentially used as an energy-conserving reductant sink for the oxidation of aromatic methoxyl groups by *C. thermoaceticum* (Seifritz et al., 1993).

Several interesting acetogens have been discovered with unusual features for coupling O-demethylation to reductant and carbon flow processes as summarized in Figure 17.2. *S. sucromutans* uses carbohydrates such as fructose or cellobiose as hydrogen donors and can utilize as electron-acceptor systems either acetogenesis, propenoate reduction of cinnamate derivatives, or co-cultivation with a hydrogenotrophic bacterium (Krumholz and Bryant, 1986; Doré and Bryant, 1989, 1990). When acetogenesis provides the electron-acceptor system, O-methyl substituents or formate are metabolized predominantly to the methyl carbon of acetate, and oxidization of the O-methyl substituent to carbon dioxide does not occur (Doré and Bryant, 1990). The stoichiometry indicated a 1 : 1 ratio of acetate formed per O-methyl substituent utilized (Krumholz and Bryant, 1986). Strain SS1 exhibits an apparently obligatory coupling between O-demethylation and hydrogen/carbon dioxide acetogenesis (Liu and Suflita, 1993). The stoichiometry indicates a 1 : 1 ratio of acetate formed per O-methyl substituent utilized, suggesting that O-methyl substituents are metabolized predominantly to the methyl carbon of acetate, with no carbon flowing to oxidation of the methyl group.

17.2.4   Sources of Radiolabel and Heavy Label Compounds

The O-methyl C^{14} or H^3 labeled derivatives of aromatic compounds can be prepared by several methods. Frazer et al. (1986) used commercially obtained catechol-O-methyltransferase and S-(methyl-^{14}C)adenosyl-L-methionine to prepare C^{14}-labeled vanillate (Fig. 17.1) or ferulate. A similar enzymatic methylation reaction was described by Marigo et al. (1979) using an enzyme extract from poplar plants. Tritium compounds should be as easy to prepare as C^{14} by these methods. Doré and Bryant (1990) used a chemical method to prepare O-methyl-^{14}C-labeled vanillic acid. The O^{18} methylether labeled *m*-anisate used by DeWeerd et al. (1988) was obtained commercially. The (carboxyl-^{14}C)vanillate used in the studies of Hsu et al. (1990a, 1990b) was also commercially obtained.

17.2.5 Sequential O-demethylation

With compounds that have multiple O-methyl substituents, a sequential removal of methyl groups, with the buildup of intermediates in the medium, is likely to be observed. This process has been frequently noted in O-demethylation studies of syringate (Fig. 17.1) in which the transient formation of 5-hydroxyvanillate (Fig. 17.1, structure 19), alternately termed methylgallate, is commonly seen (Cocaign et al., 1991; Kreft and Schink, 1993; Liu and Suflita, 1993; Tasaki et al., 1992; Taylor, 1983). Different strains and physiological conditions show differences in the level of 5-hydroxyvanillate accumulated (Wu et al., 1988).

Sequential O-demethylation of veratrate and 3,4-dimethoxycinnamate has been observed with *A. woodii* ATCC 29683 (Berman and Frazer, 1990) and *E. limosum* ATCC 8486 (Frazer and Berman, 1992). In Figure 17.3, the time course of substrate depletion and product formation is shown for the growth of *A. woodii* in defined medium with veratrate as the sole organic carbon source and an atmosphere of N$_2$/CO$_2$ (70/30). There is transient formation of isovanillate with protocatechuate as the final product. Very little vanillate is formed, suggesting a preference for O-demethylation of veratrate at the *meta* position.

An instance in which sequential O-demethylation apparently occurs, but without the accumulation of intermediates, was studied with regard to the metabolism of 1,2,3-trimethoxybenzene by *C. thermoaceticum* ATCC 39073 (Daniel et al., 1991). Cell suspensions were incubated with mixtures of two components consisting of 1,2,3-trimethoxybenzene plus different di- or monomethoxylated benzenes that were predicted to be intermediates in the sequential O-demethylation of the substrate. The intermediates were utilized preferentially rather than the 1,2,3-trimethoxybenzene. This explains the observation that intermediates do not accumulate during the utilization of this substrate.

17.2.6 Induction

Induction of O-demethylation is of considerable environmental and practical importance but has not yet been investigated in much depth. In all the strains so far examined, O-demethylation is an inducible activity. However, the uninduced

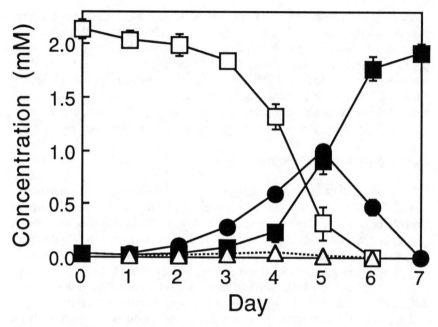

Figure 17.3.   Sequential O-demethylation of veratrate by *Acetobacterium woodii* during growth. Cells were grown in defined medium under an atmosphere of N_2/CO_2 (70/30) with veratrate as the sole carbon source. Levels of veratrate (open squares), protocatechuate (filled squares), isovanillate (filled circles), and vanillate (open triangles) were determined by HPLC.

background level of activity may also be substantial. Induction of anaerobic O-demethylation was reported first with the denitrifying facultative anaerobe, strain PN-1 (Taylor, 1983). It was observed that growth with veratrate (Fig. 17.1) and vanillate, but not isovanillate, induced activity, and it was suggested that protocatechuate might be the inducing compound. Patterns of induction are different in various acetogens and little is known about the structure of the inducers.

The O-demethylating activity in *E. limosum* was shown to be induced by growth on several methoxylated substrates but not induced when cells were grown on methanol or other substrates including glucose, lactate, or isoleucine (DeWeerd et al., 1988). Induction was blocked by the protein synthesis inhibitor tetracycline. Growth with *m*-anisate (Fig. 17.1), vanillate, or isovanillate induced activity, although the rates of O-demethylation of *m*-anisate by cell suspensions grown on vanillate or *m*-anisate were significantly higher [46.9 ± 6.4 or 52.4 ± 0.2 nmoles min^{-1} (mg protein)$^{-1}$, respectively] than for those grown on isovanillate [11.7 ± 0.1 nmoles min⁻¹(mg protein)⁻¹]. The range of substrates that could be O-demethylated by cell suspensions that had been induced by growth on *m*-anisate was seven out of the seven tested for *E. limosum* ATCC 8486, and six out of the seven tested for *A. woodii* strain NZva-16. Differences in the rates of O-demethylation by these two acetogens were seen with *p*-anisate, 3-methoxysalicylate (2-hydroxy-3-methoxybenzoate), and 4-methoxysalicylate (2-hydroxy-4-methoxybenzoate).

Several aspects of the induction of O-demethylation have been investigated in *C. thermoaceticum* ATCC 39073. Cells grown on syringate or vanillate have a prominent protein of approximately 22,000 molecular weight by SDS-PAGE, which is not present if cells are grown on methanol, or glucose (Daniel et al., 1988), or on CO and hydrogen (Wu et al., 1988). Cells grown with syringate had 4.5 to 5.6 times as much capacity to O-demethylate syringate as cells grown on methanol or CO and hydrogen (Wu et al., 1988). Cells grown on glucose had no detectable O-demethylating activity and, thus, were evidently subject to catabolite repression by glucose, and to a lesser extent by methanol. By contrast, the O-demethylating activity of cells grown on a combination of syringate and CO had about the same activity as cells grown on syringate alone. Gallate, the hydroxy derivative of syringate, did not serve as an inducer of O-demethylating activity. A range of substrates including guaiacol and other phenolics, as well as several monomethoxy and dimethoxy benzoate analogues, could be O-demethylated by cells grown with syringate. All of these compounds could also serve as growth substrates.

In our laboratory, the time course of induction for O-demethylation has been studied in several acetogens. For strain TH-001, induction of O-demethylation by ferulate is severely inhibited by glucose (Nessim and Frazer, unpublished). As shown in Figure 17.4, the addition of ferulate to *A. woodii* ATCC 29683 growing in fructose-defined medium induced cells to metabolize ferulate to the product, hydrocaffeate. This required induction of both O-demethylation and reduction of the phenylpropenoate double bond. Adding the RNA polymerase-inhibitor rifampicin (Fig. 17.4), or an inhibitor of protein synthesis like erythromycin, blocked the induction, indicating that de novo synthesis of RNA and protein were required. Similar experiments with *E. limosum* ATCC 8486 growing on lactate showed that the addition of veratrate induced O-demethylating activity and that induction was inhibited by streptomycin (Frazer and Berman, 1992). Thus, O-demethylation in *A. woodii* and *E. limosum* is less affected by catabolite repression than in *C. thermoaceticum* ATCC 39073 or strain TH-001. Strain B10 is another example of an acetogen able to O-demethylate aromatic compounds in the presence of glucose or lactate (Sembiring and Winter, 1990); whereas with *S. sucromutans,* carbohydrate is actually required for O-demethylation (Krumhoz and Bryant, 1986).

The O-demethylating activity of strain TMBS 4 is only present when cells are

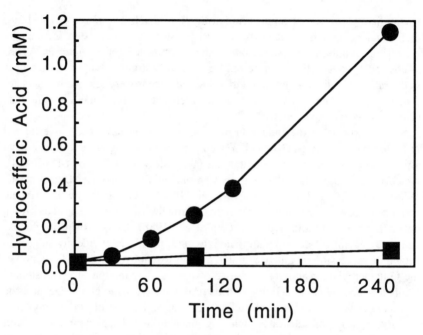

Figure 17.4.   Induction of ferulate metabolism in *Acetobacterium woodii*. Cells were grown to late exponential phase in defined medium under an atmosphere of N_2/CO_2 (70/30) with fructose (0.1%) as the sole carbon source. Induction was initiated by the addition of ferulate (2.4 mM) with (squares) or without (circles) added rifampicin (9 μg ml^{-1}). Formation of the product hydrocaffeate was quantitated by HPLC.

grown on methoxylated substrates (Kreft and Schink, 1993). Cells grown on syringate were also active in O-demethylating trimethoxybenzoate and vice versa. In addition to using sulfide or methylthiol as O-methyl-group acceptors, TMBS 4 could form acetate with CO or carbon dioxide as the carboxy donor. When cells performed an acetogenic O-demethylation, the stoichiometry was 0.75 acetate formed per methoxy group, which is the same as for *A. woodii*.

17.2.7   The Mechanism of O-demethylation and Enzyme Activity in Cell Extracts

An understanding of pathways and enzyme mechanisms whereby acetogens capture methyl groups from aryl-O-methyl compounds and other C-1 growth substrates remains a challenge for the future. The initial steps in O-demethylation and methanol utilization are apparently different (see Chapter 6). At least it is clear that methanol is not an intermediate in the O-demethylation pathway. Many

O-demethylating acetogens do not grow with methanol as a carbon source, and for *A. woodii* strain NZva-16, in which both methanol and O-methyl substituents can support growth, growth yields and rates of growth are higher on methoxylated aromatics than on methanol (Tschech and Pfennig, 1984). It is not known whether there are similarities in the O-demethylation of aromatic substrates and that for the utilization of 2-methoxyethanol in *A. malicum* (Tanaka and Pfennig, 1988).

Anaerobic O-demethylation involves removal of the methyl group (DeWeerd et al., 1988). This was shown for *E. limosum* ATCC 8486, *A. woodii* strain NZva-16, and *D. tiedjei* strain DCB-1, with cells metabolizing 3-(methoxy-^{18}O)methoxybenzoate (i.e., *m*-anisate, Fig. 17.1, structure 5). Analysis of the corresponding hydroxyl product by gas chromatography/mass spectrometry (GC/MS) showed that as the O^{18} label remained intact in 3-hydroxybenzoate, the ether bond was cleaved between the methyl group and the aryl oxygen.

Efforts have been made to elucidate the mechanism of anaerobic O-demethylation in cell extracts of a number of strains. It was shown by Doré and Bryant (1990) that extracts of *S. sucromutans* incubated with O-(methyl ^{14}C)vanillate produced C^{14}-labeled acetate. Although activity was low, a dependence on pyruvate was demonstrated.

Extracts of *C. thermoaceticum* ATCC 39073 prepared by lysozyme digestion of cells (Wu et al., 1988) or by French press disruption (Daniel et al., 1991) had higher levels of O-demethylating activity than intact cells. Activity was dependent on the presence of either CO or pyruvate, either of which could serve as carboxy group donors in the formation of acetyl-CloA by CODH/acetyl-CoA synthase. Thus, there seemed to be a requirement that O-demethylating activity be coupled to acetyl-CoA synthesis in the extract, through the activity of CODH/acetyl-CoA synthase. The specific activity for O-demethylation seemed to be enriched in the membrane fraction following differential centrifugation (Daniel et al., 1991). Cell extracts O-demethylated syringate (Fig. 17.1), *o*-anisate, 1,2,3-trimethoxybenzene, or 2,6-dimethoxyphenol, but not 3,4,5-trimethoxybenzoate. Thus, the inability of *C. thermoaceticum* to O-demethylate this substrate is apparently due to the substrate specificity of the O-demethylating enzyme and not to a deficiency in substrate uptake by cells.

In cell extracts of *A. woodii* ATCC 29683, Berman and Frazer (1992) showed that O-demethylating activity was dependent on tetrahydrofolate (THF) and ATP. As no CO_2, CO, or pyruvate was present to serve as a carboxy-group donor in acetogenesis, the assay for O-demethylating activity was not coupled to acetate formation. This should be an advantage in future studies focusing directly on the mechanism of O-demethylation. Apparent K_m values were estimated to be 0.65 mM for the substrate hydroferulate, 0.17 mM for DL-THF, and 0.27 mM for ATP-Mg^{++}. The V_{max} was 14.2 nmol/min^{-1} (mg protein)$^{-1}$, and the activity was present predominantly in the soluble fraction after centrifugation at 27,000 × *g* for 60 min. The role of THF was thought to be as a cosubstrate, but the putative C-1-THF product has not been identified. Whether ATP is required stoichiometrically or is used catalytically to activate the O-demethylating system is also not known. Because the activity in cell extracts was only 34% of the activity seen with intact cells, it was thought that additional factors remain to be defined in order to optimize the O-demethylation enzyme assay.

A successful O-demethylation enzyme assay has also been devised for cell extracts of strain TMBS 4 (Kreft and Schink, 1993). The reaction mixture uses methanethiol as the methyl-group acceptor and syringate as the methoxylated substrate. Sequential O-demethylation results in the formation of 5-hydroxyvanillate (Fig. 17.1, structure 19) which can be quantitated by a discontinuous HPLC assay, or spectrophotometrically through formation of a yellow complex. This is due to the interaction of vicinal-diols with Ti^{3+}. The V_{max} was about 20 nmoles min^{-1} (mg protein)$^{-1}$, and it was shown that ATP-Mg^{++}, which stimulates the activity, is apparently used catalytically to activate the O-demethylating system and is not required for the O-demethylation reaction directly. Interestingly, the O-demethylating activity was not irreversibly destroyed by a 15-min exposure of cell extracts to the oxygen in air.

The assay of O-demethylating activity has also been reported for crude cell extracts of *S. ovata* grown on veratrate (Konle and Stupperich, 1992). In contrast to the methyl group transfer from methanol to acetic acid, which depends on ATP and a corrinoid-containing enzyme (Stupperich et al., 1992), the O-demethylating activity was ATP independent. Several methoxylated aromatics could be O-demethylated, indicating that substrate specificity for O-demethylation is broad.

Stupperich and Konle (1993) have recently extended their study of methyl transfer reactions in extracts of *S. ovata* induced for methanol or veratrate utilization. The transfer of ^{14}C methyl substituents to corrinoid-containing (i.e., ^{57}Co labeled) protein fractions was demonstrated in enzyme reaction mixtures. DL-tetrahydrofolate was required for activity and 5-(C^{14})methyltetrahydrofolate was formed. The size distributions of proteins labeled during O-demethylation and methanol metabolism were different. Extracts could be reductively reactivated by titanium citrate and catalytic amounts of ATP following oxygen exposure.

The work on cell-free systems from acetogens indicates that these differ from aerobic O-demethylating enzymes in important respects. Aerobic O-demethylating enzymes are hydroxylating monooxygenases requiring NADH plus oxygen, and forming NAD, formaldehyde, and the hydroxylated aromatic product (Bernhardt et al., 1988; Buswell and Ribbons, 1988; Sutherland, 1986). By contrast, anaerobic O-demethylation involves methyl-group transfer and probably requires a C-1 acceptor as a condition of assay (Fig. 17.2). Pyridine nucleotides do not play a role in the reaction (Berman and Frazer, 1992). Studies with purified enzymes are needed to identify unambiguously the reactants and products of the anaerobic O-demethylating enzymes and to determine whether some acetogens may produce distinct isofunctional enzymes. It is very likely that specificity for the methoxylated substrate will be broad among anaerobic O-demethylating enzymes. Questions about the involvement of corrinoid proteins, the role of ATP, and the oxygen sensitivity of enzyme components all need to be reexamined with purified enzyme preparations.

17.3 Reduction of Phenylpropenoate Side Chain

The reduction of the propenoate side chain of cinnamate analogues such as ferulate and caffeate (Fig. 17.1, structures 13 and 14) has been reported for a number of acetogens. These include *A. woodii* (Bache and Pfennig, 1981), strain TH-001 (Frazer and Young, 1985), *C. pfennigii* (Krumholz and Bryant, 1985), *S. sucromutans* (Krumholz and Bryant, 1986), *C. aceticum* and *C. formicoaceticum* ATCC 27076 (Lux and Drake, 1992), and *P. productus* ATCC 35244 (Drake, 1993). Acetogens not capable of mediating this reaction have also been reported, and these include *E. callanderi* (Mountfort and Asher, 1986), *E. limosum* (strain RF, Genthner and Bryant, 1987; ATCC 8486, Berman and Frazer, unpublished), strain B10 (Sembiring and Winter, 1990), and *C. thermoaceticum* ATCC 39073 (Daniel et al., 1991). Further, a number of nonacetogenic rumen bacteria have been found capable of reducing the side chain of ferulate (Chesson et al., 1982). Thus, in contrast to anaerobic O-demethylation, there is no correlation between strains exhibiting the phenylpropenoate reductase capability and the potential for acetogenesis.

Phenylpropenoate reduction, however, can play an important role in the energy metabolism of acetogens able to mediate this reaction. In *S. sucromutans,* caffeate reduction can substitute for acetogenesis as the exogenous electron acceptor system required for growth on carbohydrates (Krumholz and Bryant, 1986). Less acetate or butyrate is formed per O-methyl substituent when ferulate rather than vanillate is metabolized by cells of *A. woodii* (Bache and Pfennig, 1981), *C. pfennigii* (Krumholz and Bryant, 1985), and strain TH-001 (Frazer and Young, 1985). For instance, Bache and Pfennig (1981) showed that the molar ratio of acetate formed per methyl group consumed was 0.75 with methanol, or with the O-methyl substituents of vanillate or syringate. When cells were grown on a combination of syringate and caffeate, the molar ratio decreased to 0.5, the same as that seen for growth with ferulate alone. This result indicates that reduction of the propenoate side chain substitutes for acetate formation as an electron-accepting process. The flow of carbon and reductant can be manipulated physiologically as shown in studies with *P. productus* ATCC 35244 grown on a combination of vanillate and caffeate (in a molar ratio of 1 : 5) in the absence of carbon dioxide (Drake, 1993). Oxidation of O-methyl substituents was coupled to caffeate reduction with very little formation of acetate.

An increase in growth yield per O-methyl substituent has also been reported for cells of *A. woodii* (Bache and Pfennig, 1981) and *C. pfennigii* (Krumholz and Bryant, 1985) grown with ferulate versus vanillate. The growth yield increases for *A. woodii* cells growing with caffeate plus formate or methanol further suggested that caffeate reduction can be coupled to ATP formation (Tschech and Pfennig, 1984). Cells of *A. woodii* can use hydrogen to reduce the double bond of caffeate (Cord-Ruwisch et al., 1988; Hansen et al., 1988). Hydrogen reduction of phenylpropenoates is a more energetically favorable electron-acceptor reaction than hydrogen and carbon dioxide acetogenesis, or the hydrogen reduction of sulfate, thiosulfate, or sulfite (Cord-Ruwisch et al., 1988).

A. woodii apparently can couple the hydrogen reduction of caffeate to the formation of ATP by electron transport phosphorylation (Hansen et al., 1988). Cell suspensions incubated under a hydrogen atmosphere showed a rapid increase in the amount of ATP upon the addition of caffeate. The ATP was rapidly dissipated with the subsequent addition of a mixture of valinomycin and nigericin. Further, uptake of tetraphenylphosphonium cation occurred when cell suspensions were incubated with hydrogen and caffeate, and this also was inhibited in the presence of valinomycin, or a mixture of valinomycin and nigericin. Protonophores had no effect on the uptake. Thus, the nature of the electrochemical gradient presumably driving phosphorylation is not clear.

The specific activity of caffeate reduction by hydrogen was 0.2–1.0 μmol^{-1} min^{-1} (mg protein)$^{-1}$ (Hansen et al., 1988). Cell extracts were fractionated to determine whether a membrane-bound caffeate reductase was present. The caffeate reductase activity in extracts was present in the soluble fraction, and the capacity to couple reduction to hydrogen oxidation was lost during fractionation. These characteristics of the phenylpropenoate reductase activity in *A. woodii* are similar in some respects to those observed for the enoate reductase enzyme in various clostridia (Bühler et al., 1980; Giesel and Simon, 1983). The capacity of cells and crude cell extracts to effect the stereospecific reduction by hydrogen of a broad range of enoates, including crotonate and cinnamate, first drew attention to the phenomenon (Bader et al., 1978). However, purification of the enzyme from *Clostridium* strain La 1 (DSM 1460) showed that the activity was evidently mediated by a combination of two soluble enzymes, a hydrogenase and the enoate reductase, although how this is accomplished is not known (Bader and Simon, 1980). The enoate reductase, highly purified and characterized from several strains, was shown to be an NADH-linked, high-molecular-weight hexomeric iron-sulfur protein that contains FAD (Tischer et al., 1979; Bühler and Simon, 1982).

Work in our laboratory has shown that phenylpropenoate reductase activity in *A. woodii* is inducible by ferulic acid (Fig. 17.4). Cell extracts have a phenylpropenoate reductase activity [230 nmol^{-1} min^{-1} (mg protein)$^{-1}$] that is NADH dependent and found in the soluble fraction (Frazer and Young, unpublished). Both ferulate and caffeate can serve as substrates. Other compounds have not been tested for suitability as substrates; however, Bache and Pfennig (1981) noted that *A. woodii* failed to reduce phenylacrylate, crotonate, acrylate, or fumarate. The enoate reductase from *C. sporogenes* also fails to use crotonate as a substrate, although many of the enzymes studied from clostridia use cinnamate as a substrate (Giesel and Simon, 1983). Whether the NADH-dependent phenylpropenoate reductase acting in concert with the soluble hydrogenase of *A. woodii* (Ragsdale and Ljungdahl, 1984) could play a role in the formation of ATP by electron transport phosphorylation remains to be seen.

Fumarate reductase activity has been studied in *C. formicoaceticum* ATCC 27076 (Dorn et al., 1978a, 1978b). It has recently been shown that O-demethylation in this acetogen can be coupled to fumarate reduction as well as to phenylpropenoate reduction (Matthies et al., 1993). The phenylpropenoate and fumarate reductase activities are attributable to different enzymes, and induction of phenylpropenoate reductase is not repressed by glucose (Drake, 1993).

17.4 Oxidative and Reductive Biotransformations of Aromatic Alcohols, Aldehydes, and Acids

It might be anticipated that benzaldehyde dehydrogenases, mediating the oxidation of aromatic aldehydes to acids, would be present in acetogens as such enzyme activities are widespread among aerobes (Fewson, 1988) and apparently also among anaerobes (Bossert et al., 1989; Häggblom et al., 1990). For example, some *Desulfovibrio* strains can couple sulfate reduction to the oxidation of lignoaromatic aldehydes including vanillin, syringaldehyde, *p*-hydroxybenzaldehyde, 3,4,5-trimethoxybenzaldehyde, and *p*-methoxybenzaldehyde (Zellner et al., 1990). The oxidation of vanillin, *p*-hydroxybenzylaldehyde, and 3,4-dihydroxybenzaldehyde has also been shown with the Groll strain (Table 17.1) of *Desulfotomaculum* (Kuever et al., 1993).

However, oxidative and reductive biotransformations of aromatic alcohols, aldehydes, and acids (Fig. 17.5) have not been widely examined among aceto-

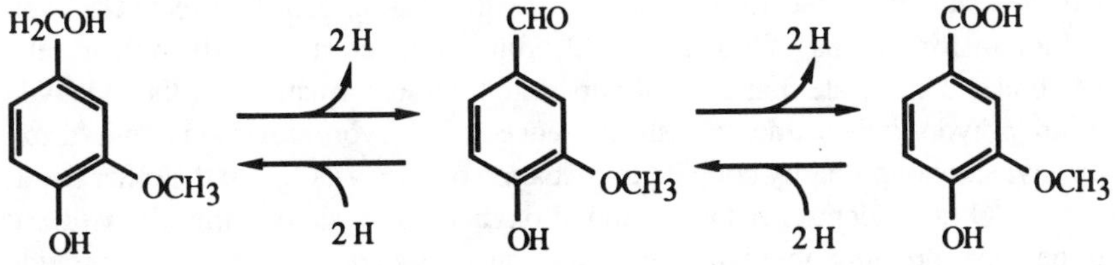

Figure 17.5. Oxidative and reductive biotransformations of vanillyl alcohol, vanillaldehyde, and vanillic acid.

gens. Some acetogens are known to O-demethylate the aromatic aldehyde vanillin (Table 17.1 and Fig. 17.1, structure 7), but whether the aldehyde group itself was reactive has not always been determined (Bache and Pfennig, 1981; Mountfort and Asher, 1986). Oxidation of the aldehyde group of vanillin has been shown for *C. pfennigii* (Krumholz and Bryant, 1985), *S. sucromutans* (Krumholz and Bryant, 1986), *C. formicoaceticum* ATCC 20976 (Lux et al., 1990; Lux and Drake, 1992), and *C. aceticum* (Lux and Drake, 1992). Strain B10 similarly oxidizes the aldehyde group of *o*-vanillin (2-hydroxy-3-methoxybenzaldehyde) or of 2,3-dimethoxybenzaldehyde to the corresponding acid (Sembiring and Winter, 1990). *P. productus* ATCC 35244 and *C. thermoaceticum* ATCC 39073 can either oxidize aromatic aldehydes to the corresponding acids or reduce them to alcohols (Lux et al., 1990). In some cases, the oxidation of aldehyde groups of aromatic compounds does not affect the growth yield of acetogens (Krumholz and Bryant, 1985; 1986). However, in *C. pfennigii* (Krumholz and Bryant, 1985), the stoichiometry of butyrate formation from aryl-O-methyl substituents may be affected by the simultaneous oxidation of the aldehyde group. The only clear evidence that the oxidation of aromatic aldehydes is a growth-supportive transformation by acetogens has been recently obtained with *C. formicoaceticum* and *C. aceticum* (Gößner et al. 1994). These two acetogens obtain energy by oxidizing 4-hydroxybenzaldehyde to 4-hydroxybenzoate, the reducing equivalents concomitantly driving the reductive synthesis of acetate from CO_2 (see chapter 10).

Several studies on the effect of CO, or other cosubstrates, on the biotransformation of the aldehyde group suggest that aldehyde biotransformations are responsive to the flow of reductant. In *C. formicoaceticum* ATCC 27076, CO inhibits the oxidation of vanillin to vanillic acid by 80–85% (Lux and Drake, 1992). In fact, the presence of CO probably favors the reverse reaction, the conversion of vanillic acid to vanillin, since Fraisse and Simon (1988) showed that reduction of carboxylic acids to aldehydes can be coupled to the oxidation of CO. In *P. productus* ATCC 35244, CO stimulates reduction of vanillin or 3,4-dihydroxybenzaldehyde to the corresponding alcohol. However, during CO-dependent growth, *p*-hydroxybenzaldehyde is transformed to both the corresponding acid and the alcohol (Lux et al., 1990). For *C. thermoaceticum* ATCC 39073 metabolizing vanillin, the presence of CO has a very complex metabolic effect because, with this substrate, decarboxylation and hydroxylation can be affected as well as biotransformations of the aldehyde group (Lux et al., 1990). When cells metabolize a substrate that can only undergo biotransformations of the aldehyde group, *p*-hydroxybenzaldehyde, the presence of CO favors reduction to *p*-hydroxybenzyl alcohol probably coupled to oxidation of CO to CO_2. On the other hand, during CO-dependent growth, 3,4-dihydroxybenzaldehyde is primarily oxidized to the corresponding acid which can subsequently be decarboxylated to provide cells with a source of CO_2.

The substrate specificity of aromatic aldehyde oxidation has been examined in cell suspensions of *C. formicoaceticum* ATCC 27076 (Lux and Drake, 1992). Vanillin, isovanillin, 3,4-dihydroxybenzaldehyde, and *o*-, *m*-, or *p*-hydroxybenzaldehyde were all oxidized. However, 2-methoxybenzaldehyde, 2-hydroxy-3-methoxybenzaldehyde (*o*-vanillin), and 2,3,4-trihydroxybenzaldehyde were not oxidized. The background level of aromatic aldehyde dehydrogenase activity was high in uninduced cells grown on CO.

The reversible transformations of aromatic alcohols to aldehydes or of aldehydes to acids (Fig. 17.5) could be mediated by broad specificity alcohol or aldehyde dehydrogenases. Beaty and Ljungdahl (1991) have shown that induced levels of aldehyde dehydrogenase in membrane vesicles of *C. thermoaceticum* ATCC 39073 could be involved in the energy metabolism of cells, as well as functioning in the pathway for the oxidation of short-chain alcohols to fatty acids. Broad specificity aldehyde dehydrogenase/carboxylic acid reductases have been purified from *C. thermoaceticum* DSM 521 (White et al., 1989; Strobl et al., 1992) and *C. formicoaceticum* DSM 92 (Fraisse and Simon 1988; White et al., 1991, 1993). These exhibit activity with some aromatic substrates as well as with two to four carbon-chain-length aliphatic substrates. It would be interesting to know whether the aldehyde dehydrogenase activities that have been assayed in cell extracts of other acetogens, including *C. aceticum* DSM 1496, *C. thermoautotrophicum* DSM 1974, *B. methylotrophicum*, and *E. limosum* DSM 20543 (White and Simon, 1992), are also active with aromatic aldehydes like vanillin.

17.5 Decarboxylation of Benzoic Acid Derivatives

Decarboxylation of benzoate derivatives has only been observed in one acetogen. *C. thermoaceticum* ATCC 39073 actively decarboxylates vanillate, a transformation not observed with *C. formicoaceticum* ATCC 27076 or *C. aceticum* (Hsu et al., 1990a) nor reported for other acetogens capable of O-demethylating vanillate (Table 17.1, Fig. 17.1). However, because decarboxylation converts vanillate to a more toxic catechol derivative, the detection of acetogens with this capability may often be technically difficult. Hence, the distribution of acetogens with the potential to decarboxylate aromatic acids is yet to be determined.

Radiolabeled carbon dioxide is formed when *C. thermoaceticum* is grown with (carboxy-^{14}C)vanillate, and the carbon dioxide requirement for growth on one-carbon substrates such as methanol or methoxy groups can be replaced by a cosubstrate that undergoes decarboxylation (Hsu et al., 1990a). The occurrence of decarboxylation was confirmed by showing that labeled carbon dioxide was the predominant product formed by cell suspensions during short-term incubations with (carboxy-^{14}C)vanillate (Hsu et al., 1990b). That this carbon dioxide was growth supportive through the acetyl-CoA pathway was further demonstrated by growing cells with the carboxy radiolabeled vanillate in a basal-salts medium containing yeast extract but limited for carbon dioxide. Under these conditions, most of the labeled carbon dioxide was converted into acetate in which both carbons were equally labeled (Hsu et al., 1990b).

Decarboxylation is induced by growth with benzoate derivatives having a *para* hydroxy substituent, such as vanillate, protocatechuate, or 4-hydroxybenzoate. No induction results when cells are grown with syringate or gallate, or with 2-hydroxy- or 3-hydroxybenzoate (Hsu et al., 1990a). The specific activity of cells grown in batch culture depends heavily on the growth phase and is highest in the early log phase (Hsu et al., 1990b). Induction of decarboxylation is independent of induction for O-demethylation, and the inducers for each activity must also be suitable substrates for the corresponding activity. For example, syringate (Fig. 17.1) induced only O-demethylation, whereas 4-hydroxybenzoate induced only decarboxylation, and vanillate induced both. Catabolite repression is only observed for O-demethylation.

Cell suspension studies showed that although a *para* hydroxy group is required, specificity at the *meta* position is broad (Hsu et al., 1990b). Relatively good decarboxylation activity occurs with a variety of substituents at the *meta* position including cloro, fluoro, hydroxy, or methoxy derivatives of benzoate. However, strict anaerobiosis was required.

Activity was observed in extracts prepared by French press disruption or lysozyme treatment, followed by centrifugation at $25,000 \times g$ for 30 min (Hsu et al., 1990b). The formation of the decarboxylated aromatic product was quantitated by HPLC after a reaction time of 2 min at 55°C, pH 6, in 50 mM sodium phosphate buffer. The V_{max} was 217 $nmol^{-1}$ min^{-1} (mg protein)$^{-1}$ in the cell-free assay and the apparent K_m for 4-hydroxybenzoate was 3.33 mM, compared to 0.67 mM for cell suspensions.

17.6   Mandelate Decarboxylation

The *Acetobacterium* sp. strain AmMan1 was isolated from an enrichment culture in bicarbonate DL-mandelate medium inoculated with anoxic mud from a creek. The pathway involved in D-mandelate utilization has been studied (Dörner and Schink, 1991). Morphologically similar bacteria were seen in enrichments inoculated with anaerobic sediment from other freshwater and marine sites. Hence, AmMan1-like isolates may be widely distributed, although *A. malicum,* and possibly other previously recognized *Acetobacterium* species, apparently lack the capacity to metabolize mandelate (Fig. 17.1, structure 16). Gas was produced, and the stoichiometry for substrate conversion to products was

$$2 \text{ D-mandelate} \rightarrow 2 \text{ benzoate} + \text{acetate} + H_2 \qquad (17.4)$$

Thus, it seems that the carboxy carbon of mandelate serves as a one-carbon substrate for acetate formation, and hydrogen gas may also be formed.

Enzyme assays were devised, analogous to established assays for lactate utilization. High activities were observed, using crude French press cell extracts, for a D-mandelate NAD-dependent dehydrogenase (4.0 μmol^{-1}/min^{-1} (mg protein)$^{-1}$, and a phenylglyoxylate : acceptor oxidoreductase [4.7 μmol^{-1} min^{-1} (mg protein)$^{-1}$]. These activities would be sufficient to convert D-mandelate via phenylglyoxylate to carbon dioxide and the proposed aromatic intermediate, benzoyl-CoA. Induced enzyme levels were seen with mandelate compared to lactate-grown cells. The proposed pathway for mandelate metabolism in AmMan1 is distinct from the well-characterized aerobic pathway for L-mandelate utilization in *Pseudomonas putida*, in which L-mandelate is converted via phenylglyoxylate (i.e., benzoylformate) to carbon dioxide and benzaldehyde (Fewson, 1988).

17.7   Hydroxylation

In cultures of *C. thermoaceticum* ATCC 39073 grown on vanillin (Fig. 1) plus CO, low levels of gallate were formed, indicating that vanillin can be hydroxylated as well as subject to O-demethylation and aldehyde oxidation (Lux et al., 1990). Hydroxylation of aromatic compounds has not been reported with other acetogens, and the reaction has not been studied. However, *C. barkeri* is well known for its capacity to utilize nicotinic acid, a heterocyclic aromatic compound, and the degradation pathway has been established [see reviews in Gottschalk et al. (1981) and Evans and Fuchs (1988)]. The first step in the fermentation is hydroxylation of the heterocyclic ring to form 6-hydroxynicotinate, mediated by an NADP-dependent nicotinate dehydrogenase which has been purified and characterized (Holcenbeg and Stadtman, 1969). Whether a similar reaction could hydroxylate other aromatic compounds is not known.

17.8   Concluding Remarks

Acetogens can modify several types of aromatic-ring substituents. Anaerobic O-demethylation is the most striking of these reactions because it may be a significant ecological process in certain environments and is a potential determinant affecting microbial community structure. By means of O-demethylation, acetogens can make use of methoxylated ligno-aromatic compounds as one-carbon growth substrates which are widespread in the biosphere. Fully induced O-demethylating activities in acetogens are high and their substrate specificity is broad. Indications are that the fate of the methyl group is the formation of carbon dioxide, or acetate, or both (Fig. 17.2). Conversion to methanethiol or dimethylsulfide should also be considered a possible consequence of anaerobic O-demethylation (Bak et al., 1992). The products formed from O-demethylation will depend on the strain, its potential to use various electron acceptors (e.g., phenylpropenoates or thiosulfate), and the availability of carbon dioxide and hydrogen. The presence of other carbon sources is also a determining factor. For example, with *S. sucromutans,* a carbohydrate cosubstrate is required for O-demethylation, whereas in *C. thermoaceticum* its presence can result in strong catabolite inhibition. It is not known whether pure cultures of acetogens can O-demethylate substrates with flavonoid or dilignol structures, or whether they can metabolize glycosylated phenolics.

Progress has been made in studying the O-demethylating enzyme system in several strains that mediate acetogenic O-demethylation, but success has been slow in coming. Anaerobic O-demethylation is a methyltransfer reaction (De-Weerd et al., 1988) so that appropriate C-1 acceptors must be present in order to demonstrate activity (Berman and Frazer, 1992; Kreft and Schink, 1993). Use of genetic and molecular biology techniques might be of special benefit in better defining the component proteins mediating O-demethylation, although the application of this approach is in its early stages (Berman and Frazer, 1991). Studies on the transport of substrates to the cytosol and the regulation of gene expression are lacking. However, a start has been made on identifying proteins that are specifically induced during O-demethylation (Daniel et al., 1988; Wu et al., 1988).

Activities variably present in acetogens include the capacity to reduce the side chain of phenylpropenoate derivatives, and possibly the ability to hydroxylate aromatic compounds. Oxidative and reductive biotransformations of aromatic alcohols, aldehydes, and acids may be widespread among acetogens as they are in many other aerobes and anaerobes, but this has not yet been widely investigated.

The capacity of some acetogens to decarboxylate mandelate (Dörner and Schink, 1991) or derivatives of benzoate (Hsu et al., 1990a, 1990b) resulting in the generation of growth supportive carbon dioxide is of special interest. Because acetogens are so effective in utilizing other one-carbon sources, it seems unlikely that this source would have been overlooked in evolution. Perhaps these capabilities are just waiting to be discovered among already known acetogenic strains and new isolates.

Acknowledgments

Research in this laboratory has been supported in part by EPA grant 810393-01-0, DOE grant FG02-88ER13924, NIEHS grant 5-P42-ES04895, and NSF grant MCB-9219277.

References

Akin, D. E., W. S. Borneman, L. L. Rigsby, and S. A. Martin. 1993. *p*-Coumaroyl and feruloyl arabinoxylans from plant cell walls as substrates for ruminal bacteria. *Appl. Environ. Microbiol.* **59**:644–647.

Allard, A.-S., P.-A. Hynning, C. Lindgren, M. Remberger, and A. H. Neilson. 1991. Dechlorination of chlorocatechols by stable enrichment cultures of anaerobic bacteria. *Appl. Environ. Microbiol.* **57**:77–84.

Allard, A.-S., P.-A. Hynning, M. Remberger, and A. H. Neilson. 1992. Role of sulfate concentration in dechlorination of 3,4,5-trichlorocatechol by stable enrichment cultures grown with coumarin and flavanone glycones and aglycones. *Appl. Environ. Microbiol.* **58**:961–968.

Bache, R., and N. Pfennig. 1981. Selective isolation of *Acetobacterium woodii* on methoxylated aromatic acids and determination of growth yields. *Arch. Microbiol.* **130**:255–261.

Bader, J., H. Günther, B. Rambeck, and H. Simon. 1978. Properties of two *Clostridia* strains acting as catalysts for the preparative stereospecific hydrogenation of 2-enoic acids and 2-alken-1-ols with hydrogen gas. *Hoppe-Seyler's Z. Physiol. Chem.* **359**:19–27.

Bader, J., and H. Simon. 1980. The activities of hydrogenase and enoate reductase in two *Clostridium* species, their interrelationship and dependence on growth conditions. *Arch. Microbiol.* **127**:279–287.

Bak, F., K. Finster, and F. Rothfuss. 1992. Formation of dimethylsulfide and methanethiol from methoxylated aromatic compounds and inorganic sulfide by newly isolated anaerobic bacteria. *Arch. Microbiol.* **157**:529–534.

Beaty, P. S., and L. G. Ljungdahl. 1991. Growth of *Clostridium thermoaceticum* on methanol, ethanol, propanol and butanol in medium containing either thiosulfate or dimethylsulfoxide. Abstr. K131, *ASM May Meeting*, Dallas, TX 1991.

Berk, K. 1976. Phenolic compounds. In: *Braverman's Introduction to the Biochemistry of Foods*, pp. 235–247. Elsevier, Amsterdam.

Berman, M. H., and A. C. Frazer. 1990. Studies on the O-demethylation of methoxylated aromatics by *Acetobacterium woodii* (ATCC 29683), Abstr. Q165, *ASM May Meeting, Anaheim, CA* 1990.

Berman, M. H., and A. C. Frazer. 1991. Introduction of the tetracycline resistance transposon. Tn916, into *Eubacterium limosum*, Abstr. Q21, *ASM May Meeting*, Dallas, TX 1991.

Berman, M. H., and A. C. Frazer. 1992. Importance of tetrahydrofolate and ATP in the anaerobic O-demethylation reaction for phenylmethylethers. *Appl. Environ. Microbiol.* **58**:925–931.

Bernhardt, F.-H., E. Bill, A. X. Trautwein, and H. Twilfer. 1988. 4-Methoxybenzoate monooxygenase from *Pseudomonas putida*: isolation, biochemical properties, substrate specificity and reaction mechanisms of the enzyme components. *Methods Enzymol.* **161**:281–294.

Bossert, I. D., G. Whited, D. T. Gibson, and L. Y. Young. 1989. Anaerobic oxidation of *p*-cresol mediated by a partially purified methylhydroxylase from a denitrifying bacteria. *J. Bacteriol.* **171**:2956–2962.

Breznak, J. A., J. M. Switzer, and H.-J. Seitz. 1988. *Sporomusa termitida* sp. nov., an hydrogen/carbon dioxide-utilizing acetogen isolated from termites. *Arch. Microbiol.* **150**:282–288.

Breznak, J. A., and M. D. Kane. 1990. Microbial H_2/CO_2 acetogenesis in animal guts: nature and nutritional significance. *FEMS Microbiol. Rev.* **87**:309–314.

Breznak, J. A., and J. S. Blum. 1991. Mixotrophy in the termite gut acetogen, *Sporomusa termitida*. *Arch. Microbiol.* **156**:105–110.

Bühler, M., and H. Simon. 1982. On the kinetics and mechanism of enoate reductase. *Hoppe-Seyler's Z. Physiol. Chem.* **363**:609–625.

Bühler, M., H. Giesel, W. Tischer, and H. Simon. 1980. Occurrence and the possible physiological role of 2-enoate reductases. *FEBS Lett.* **109**:244–246.

Buswell, J. A., and D. W. Ribbons. 1988. Vanillate O-demethylase from *Pseudomonas* species. *Methods Enzymol.* **161**:294–301.

Chesson, A., C. S. Stewart, and R. J. Wallace. 1982. Influence of plant phenolic acids on growth and cellulolytic activity of rumen bacteria. *Appl. Environ. Microbiol.* **44**:597–603.

Cocaign, M., E. Wilberg, and N. D. Lindley. 1991. Sequential demethoxylation reactions during methylotrophic growth of aromatic substrates with *Eubacterium limosum*. *Arch. Microbiol.* **155**:496–499.

Colberg, P. J. 1988. Anaerobic microbial degradation of cellulose, lignin, oligolignols, and monoaromatic lignin derivatives. In: *Biology of Anaerobic Microorganisms,* A. J. B. Zehnder (ed.), pp. 333–372. Wiley, New York.

Colberg, P. J., and L. Y. Young. 1985. Aromatic and volatile acid intermediates observed during anaerobic metabolism of lignin-derived oligomers. *Appl. Environ. Microbiol.* **49**:350–358.

Cord-Ruwisch, R., and J. L. Garcia. 1985. Isolation and characterization of an anaerobic benzoate-degrading spore-forming sulfate-reducing bacterium, *Desulfotomaculum sapomandens* sp. nov. *FEMS Microbiol. Lett.* **29**:325–330.

Cord-Ruwisch, R., and B. Ollivier. 1986. Interspecies hydrogen transfer during methanol degradation by *Sporomusa acidovorans* and hydrogenophilic anaerobes. *Arch. Microbiol.* **144**:163–165.

Cord-Ruwisch, R., H.-J. Seitz, and R. Conrad. 1988. The capacity of hydrogenotrophic anaerobic bacteria to compete for traces of hydrogen depends on the redox potential of the terminal electron acceptor. *Arch. Microbiol.* **149**:350–357.

Daniel, S. L., A. Wu, and H. L. Drake. 1988. Growth of thermophilic acetogenic bacteria on methoxylated aromatic acids. *FEMS Microbiol. Lett.* **52**:25–28.

Daniel, S. L., E. S. Keith, H. Yang, Y.-S. Lin, and H. L. Drake. 1991. Utilization of methoxylated aromatic compounds by the acetogen *Clostridium thermoaceticum*: expression and specificity of the CO-dependent demethylating activity. *Biochem. Biophys. Res. Commun.* **180**:416–422.

Dénarié, J., F. Debellé, and C. Rosenberg. 1992. Signaling and host range variation in nodulation. *Annu. Rev. Microbiol.* **46**:497–531.

DeWeerd, K. A., J. M. Suflita, T. Linkfield, J. M. Tiedje, and P. H. Pritchard. 1986. The relationship between reductive dehalogenation and other aryl substituent removal reactions catalyzed by anaerobes. *FEMS Microbiol. Ecol.* **38**:331–339.

DeWeerd, K. A., A. Saxena, D. P. Nagle, and J. M. Suflita. 1988. Metabolism of the ^{18}O-methoxy substituent of 3-methoxybenzoic acid and other unlabeled methoxybenzoic acids by anaerobic bacteria. *Appl. Environ. Microbiol.* **54**:1237–1242.

DeWeerd, K. A., L. Mandelco, R. S. Tanner, C. R. Woese, and J. M. Suflita. 1990. *Desulfomonile tiedjei* gen. nov. and sp. nov., a novel anaerobic, dehalogenating, sulfate-reducing bacterium. *Arch. Microbiol.* **154**:23–30.

Diekert, G. 1992. The acetogenic bacteria. In: *The Prokaryotes*, 2nd ed., A. Balows, H. G. Truper, M. Dworkin, W. Harder, and K.-L. Schleifer (eds.), Vol. 1, pp. 517–533. Springer-Verlag, New York.

Dolfing, J. 1988. Acetogenesis. In: *Biology of Anaerobic Microorganisms*. A. J. B. Zehnder (ed.), pp. 417–468. Wiley, New York.

Dolfing, J., and J. M. Tiedje. 1991. Acetate as a source of reducing equivalents in the reductive dechlorination of 2,5-dichlorobenzoate. *Arch. Microbiol.* **156**:356–361.

Doré, J., and M. J. Bryant. 1989. Lipid growth requirement and influence of lipid supplement on fatty acid and aldehyde composition of *Syntrophococcus sucromutans*. *Appl. Environ. Microbiol.* **55**:927–933.

Doré, J., and M. P. Bryant. 1990. Metabolism of one-carbon compounds by the ruminal acetogen *Syntrophococcus sucromutans*. *Appl. Environ. Microbiol.* **56**:984–989.

Dorn, M., J. R. Andreesen, and G. Gottschalk, 1978a. Fermentation of fumarate and L-malate by *Clostridium formicoaceticum*. *J. Bacteriol.* **133**:26–32.

Dorn, M., J. R. Andreesen, and G. Gottschalk. 1978b. Fumarate reductase of *Clostridium formicoaceticum*. *Arch. Microbiol.* **119**:7–11.

Dörner, C., and B. Schink. 1991. Fermentation of mandelate to benzoate and acetate by a homoacetogenic bacterium. *Arch. Microbiol.* **156**:302–306.

Drake, H. L. 1992. Acetogenesis and acetogenic bacteria. In: *Encyclopedia of Microbiology*, Vol. 1, J. Lederberg (ed.), pp. 1–15. Academic Press, New York.

Drake, H. L. 1993. CO_2, reductant, and the autotrophic acetyl-CoA pathway: alternative origins and destinations. In: Microbial Growth on C_1 Compounds, J. C. Murrell, and D. P. Kelly (eds.), pp. 493–507. Intercept Ltd., Andover, England.

Eicher, B., and B. Schink. 1984. Oxidation of primary aliphatic alcohols by *Acetobacterium carbinolicum* sp. nov., a homoacetogenic anaerobe. *Arch. Microbiol.* **140**:147–152.

Evans, W. C., and G. Fuchs. 1988. Anaerobic degradation of aromatic compounds. *Annu. Rev. Microbiol.* **42**:289–317.

Fewson, C. A. 1988. Microbial metabolism of mandelate: a microcosm of diversity. *FEMS Microbiol. Rev.* **54**:85–110.

Fraisse, L., and H. Simon. 1988. Observations on the reduction of non-activated carboxylates by *Clostridium formicoaceticum* with carbon monoxide or formate and the influence of various viologens. *Arch. Microbiol.* **150**:381–386.

Frazer, A. C., and L. Y. Young. 1985. A gram-negative anaerobic bacterium that utilizes O-methyl substituents of aromatic acids. *Appl. Environ. Microbiol.* **49**:1345–1347.

Frazer, A. C., and L. Y. Young. 1986. Anaerobic ^{14}C metabolism of the O-methyl-^{14}C-labeled substituent of vanillate. *Appl. Environ. Microbiol.* **51**:84–87.

Frazer, A. C., I. Bossert, and L. Y. Young. 1986. Enzymatic aryl-O-methyl-^{14}C labeling of model lignin monomers. *Appl. Environ. Microbiol.* **51**:80–83.

Frazer, A. C., and M. H. Berman. 1992. Characterization of the veratrate induction of anaerobic O-demethylation in *Eubacterium limosum*, Abstr. Q236, *ASM May Meeting*, New Orleans, LA 1992.

Fuchs, G. 1986. Carbon dioxide fixation in acetogenic bacteria: variations on a theme. *FEMS Microbiol. Rev.* **39**:181–213.

Genthner, B. R. S., and M. P. Bryant. 1987. Additional characteristics of one-carbon-compound utilization by *Eubacterium limosum* and *Acetobacterium woodii*. *Appl. Environ. Microbiol.* **53**:471–476.

Giesel, H., and H. Simon. 1983. On the occurrence of enoate reductase and 2-oxocarboxylate reductase in clostridia and some observations on the amino acid fermentation by *Peptostreptococcus anaerobius*. *Arch. Microbiol.* **135**:51–57.

Greening, R. C., and J. A. Z. Leedle. 1989. Enrichment and isolation of *Acetitomaculum ruminis*, gen. nov., sp. nov.: acetogenic bacteria from the bovine rumen. *Arch. Microbiol.* **151**:399–406.

Gößner, A., S. L. Daniel, and H. L. Drake. 1994. Acetogenesis coupled to the oxidation of aromatic aldehyde groups. *Arch. Microbiol.* (in press).

Gottschalk, G., J. R. Andreesen, and H. Hippe. 1981. The genus *Clostridium* (nonmedical aspects). In: *The Prokaryotes*, M. P. Starr, H. Stolp, H. G. Truper, A. Balows, and H. G. Schlegel (eds.), Vol. 2, pp. 1767–1803. Springer-Verlag, New York.

Häggblom, M. M., M. D. Rivera, I. D. Bossert, J. E. Rogers, and L. Y. Young. 1990. Anaerobic biodegradation of *para*-cresol under three reducing conditions. *Microb. Ecol.* **20**:141–150.

Häggblom, M. M., M. H. Berman, A. C. Frazer, and L. Y. Young. 1993. Anaerobic O-demethylation of chlorinated guaiacols by *Acetobacterium woodii* and *Eubacterium limosum*. *Biodegradation* **4**:107–114.

Hanselmann, K. W. 1982. Lignochemicals. *Experientia* **38**:176–189.

Hansen, B., M. Bokranz, P. Schonheit, and A. Kroger. 1988. ATP formation coupled to caffeate reduction by H₂ in *Acetobacterium woodii* NZva16. *Arch. Microbiol.* **150**:447–451.

Healy, J. B., Jr., and L. Y. Young. 1979. Anaerobic biodegradation of eleven aromatic compounds to methane. *Appl. Environ. Microbiol.* **38**:84–89.

Healy, J. B., Jr., and L. Y. Young. 1980. Methanogenic decomposition of ferulic acid, a model lignin derivative. *Appl. Environ. Microbiol.* **39**:436–444.

Heijthuijsen, J. H. F. G., and T. A. Hansen. 1989. Selection of sulphur sources for the growth of *Butyribacterium methylotrophicum* and *Acetobacterium woodii*. *Appl. Microbiol. Biotechnol.* **32**:186–192.

Heijthuijsen, J. H. F. G., and T. A. Hansen. 1990. One-carbon metabolism in anaerobic non-methanogenic bacteria. In: *Autotrophic Microbiology and One-Carbon Metabolism*, G. A. Codd, L. Dijkhuizen, and F. R. Tabita (eds.), pp. 163–191. Kluwer, Dordrecht.

Holcenberg, J. S., and E. R. Stadtman. 1969. Nicotinic acid metabolism. III. Purification and properties of a nicotinic acid hydroxylase. *J. Biol. Chem.* **244**:1194–1203.

Hsu, T., S. L. Daniel, M. F. Lux, and H. L. Drake. 1990a. Biotransformations of carboxylated aromatic compounds by the acetogen *Clostridium thermoaceticum*: generation of growth-supportive carbon dioxide equivalents under carbon dioxide-limited conditions. *J. Bacteriol.* **172**:212–217.

Hsu, T., M. F. Lux, and H. L. Drake. 1990b. Expression of an aromatic-dependent decarboxylase which provides growth-essential carbon dioxide equivalents for the acetogenic (Wood) pathway of *Clostridium thermoaceticum*. *J. Bacteriol.* **172**:5901–5907.

Ivie, G. W., D. E. Clark, and D. D. Rushing. 1974. Metabolic transformation of disugran by rumen fluid of sheep maintained on dissimilar diets. *J. Agric. Food Chem.* **22**:632–634.

Kaiser, J.-P., and K. W. Hanselmann. 1982a. Fermentative metabolism of substituted monoaromatic compounds by a bacterial community from anaerobic sediments. *Arch. Microbiol.* **133**:185–194.

Kaiser, J.-P., and K. W. Hanselmann. 1982b. Aromatic chemicals through anaerobic microbial conversion of lignin monomers. *Experientia* **38**:167–176.

Kane, M. D., A. Brauman, and J. A. Breznak. 1991. *Clostridium mayombei* sp. nov., an hydrogen/carbon dioxide acetogenic bacterium from the gut of the African soil-feeding termite, *Cubitermes speciosus*. *Arch. Microbiol.* **156**:99–104.

Kane, M. D., and J. A. Breznak. 1991. *Acetonema longum* gen. nov. sp. nov., an hydrogen/carbon dioxide acetogenic bacterium from the termite, *Pterotermes occidentis*. *Arch. Microbiol.* **156**:91–98.

Kerby, R., W. Niemczura, and J. G. Zeikus. 1983. Single-carbon catabolism in acetogens: analysis of carbon flow in *Acetobacterium woodii* and *Butyribacterium methylotrophicum* by fermentation and ¹³C nuclear magnetic resonance measurement. *J. Bacteriol.* **155**:1208–1218.

Kirk, T. K., and R. L. Farrell. 1987. Enzymatic "combustion": the microbial degradation of lignin. *Annu. Rev. Microbiol.* **41**:465–505.

Klemps, R., H. Cypionka, F. Widdel, and N. Pfennig. 1985. Growth with hydrogen, and further physiological characteristics of *Desulfotomaculum* species. *Arch. Microbiol.* **143**:203–208.

Konle, R., and E. Stupperich. 1992. Anaerobic demethoxylation and methanol metabolism in *Sporomusa ovata*. Abstr. P163, *VAAM March Meeting*, Dusseldorf, 1992.

Kreft, J.-U., and B. Schink. 1993. Demethylation and degradation of phenylmethylethers by the sulfide-methylating homoacetogenic bacterium strain TMBS 4. *Arch. Microbiol.* **159**:308–315.

Kreikenbohm, R., and N. Pfennig. 1985. Anaerobic degradation of 3,4,5-trimethoxybenzoate by a defined mixed culture of *Acetobacterium woodii, Pelobacter acidigallici,* and *Desulfobacter postgatei. FEMS Microbiol. Ecol.* **31**:29–38.

Krumholz, L. R., and M. P. Bryant. 1985. *Clostridium pfennigii* sp. nov. uses methoxyl groups of monobenzenoids and produces butyrate. *Int. J. Syst. Bacteriol.* **35**:454–456.

Krumholz, L. R., Bryant, M. P. 1986. *Syntrophococcus sucromutans* sp. nov. gen. nov. uses carbyhydrates as electron donors and formate, methoxymonobenzenoids or *Methanobrevibacter* as electron acceptor systems. *Arch. Microbiol.* **143**:313–318.

Kuever, J., J. Kulmer, S. Jannsen, U. Fischer, and K.-H. Blotevogel. 1993. Isolation and characterization of a new spore-forming sulfate-reducing bacterium growing by complete oxidation of catechol. *Arch. Microbiol.* **159**:282–288.

Liu, S., and J. M. Suflita. 1993. H_2-CO_2-dependent anaerobic O-demethylation activity in subsurface sediments and by an isolated bacterium. *Appl. Environ. Microbiol.* **59**:1325–1331.

Ljungdahl, L. G. 1986. The autotrophic pathway of acetate synthesis in acetogenic bacteria. *Annu. Rev. Microbiol.* **40**:415–450.

Lux, M. F., E. Keith, T. Hsu, and H. L. Drake. 1990. Biotransformations of aromatic aldehydes by acetogenic bacteria. *FEMS Microbiol. Lett.* **67**:73–78.

Lux, M., and H. L. Drake. 1992. Re-examination of the metabolic potential of the acetogens *Clostridium aceticum* and *Clostridium formicoaceticum:* chemolithoautotrophic and aromatic-dependent growth. *FEMS Microbiol. Lett.* **95**:49–56.

Marigo, G., D. Riviere, and A. M. Boudet. 1979. Synthese biologique des acides ferulique et sinapique marques au ^{14}C. *J. Labelled Compounds Radiopharmaceut.* **18**:695–702.

Martin, A. K. 1982. The origin of urinary aromatic compounds excreted by ruminants 2. The metabolism of phenolic cinnamic acids to benzoic acid. *Br. J. Nutr.* **47**:155–164.

Matthies, C., A. Freiberger, and H. L. Drake. 1993. Regulation of electron sink utilization during aromatic methoxyl- and methanol-dependent growth of the acetogen *Clostridium formicoaceticum,* Abstr. K93, *ASM May Meeting*, Atlanta GA 1993.

Mountfort, D. O., and R. A. Asher. 1986. Isolation from a methanogenic ferulate degrading consortium of an anaerobe that converts methoxyl groups of aromatic acids to volatile fatty acids. *Arch. Microbiol.* **144**:55–61.

Mountfort, D. O., W. D. Grant, R. Clarke, and R. A. Asher. 1988. *Eubacterium callanderi* sp. nov. that demethoxylates O-methylated aromatic acids to volatile fatty acids. *Int. J. Syst. Bacteriol.* **38**:254–258.

Neilson, A. H., A.-S. Allard, P.-A. Hynning, and M. Remberger. 1991. Distribution, fate and persistence of organochlorine compounds formed during production of bleached pulp. *Toxicol. Environ. Chem.* **30**:3–41.

Ragsdale, S. W., and L. G. Ljungdahl. 1984. Hydrogenase from *Acetobacterium woodii*. *Arch. Microbiol.* **139**:361–365.

Reid, I. D., G. D. Abrams, and J. M. Pepper. 1982. Water-soluble products from the degradation of aspen lignin by *Phanerochaete chrysosporium*. *Can. J. Bot.* **60**:2357–2364.

Remberger, M., A.-S. Allard, and A. H. Neilson. 1986. Biotransformations of chloroguaiacols, chlorocatechols, and chloroveratroles in sediments. *Appl. Environ. Microbiol.* **51**:552–558.

Scheline, R. S. 1973. Metabolism of foreign compounds by gastrointestinal microorganisms. *Pharmacolog. Rev.* **25**:451–523.

Schink, B., and N. Pfennig. 1982. Fermentation of trihydroxybenzenes by *Pelobacter acidigallici* gen. nov. sp. nov., a new strictly anaerobic, non-sporeforming bacterium. *Arch. Microbiol.* **133**:195–201.

Schink, B., A. Brune, and S. Schnell. 1992. Anaerobic degradation of aromatic compounds. In: *Microbial Degradation of Natural Products*, G. Winkelmann (ed.), pp. 219–242. VCH, New York.

Seifritz, C., S. L. Daniel, A. Gößner, and H. L. Drake. 1993. Nitrate as a preferred electron sink for the acetogen *Clostridium thermoaceticum*. *J. Bacteriol.* **175**:8008–8013.

Sembiring, T., and J. Winter. 1990. Demethylation of aromatic compounds by strain B10 and complete degradation of 3-methoxybenzoate in co-culture with *Desulfosarcina* strains. *Appl. Microbiol. Biotechnol.* **33**:233–238.

Singleton, V. L., and F. H. Kratzer. 1973. Plant phenolics. In: *Toxicants Occurring Naturally in Foods*, 2nd ed. National Academy of Sciences, Washington, D.C.

Strobl, G., R. Feicht, H. White, F. Lottspeich, and H. Simon. 1992. The tungsten-containing aldehyde oxidoreductase from *Clostridium thermoaceticum* and its complex with a viologen-accepting NADPH oxidoreductase. *Biol. Chem. Hoppe-Seyler* **373**:123–132.

Stupperich, E., P. Aulkemeyer, and C. Eckerskorn. 1992. Purification and characterization of a methanol-induced cobamide-containing protein from *Sporomusa ovata*. *Arch. Microbiol.* **158**:370–373.

Stupperich, E., R. Konle. 1993. Corrinoid-dependent methyl transfer reactions are involved in methanol and 3,4-dimethoxybenzoate metabolism by *Sporomusa ovata*. *Appl. Environ. Microbiol.* **59**:3110–3116.

Sutherland, J. B. 1986. Demethylation of veratrole by cytochrome P-450 in *Streptomyces setonii*. *Appl. Environ. Microbiol.* **52**:98–100.

Tanaka, K., and N. Pfennig. 1988. Fermentation of 2-methoxyethanol by *Acetobacterium malicum* sp. nov. and *Pelobacter venetianus*. *Arch. Microbiol.* **149**:181–187.

Taskaki, M., Y. Kamagata, K. Nakamura, and E. Mikami. 1991. Isolation and characterization of a thermophilic benzoate-degrading, sulfate-reducing bacterium, *Desulfotomaculum thermobenzoicum* sp. nov. *Arch. Microbiol.* **155**:348–352.

Tasaki, M., Y. Kamagata, K. Nakamura, and E. Mikami. 1992. Utilization of methoxylated benzoates and formation of intermediates by *Desulfotomaculum thermobenzoicum* in the presence or absence of sulfate. *Arch. Microbiol.* **157**:209–212.

Taylor, B. F. 1983. Aerobic and anaerobic catabolism of vanillic acid and some other methoxy-aromatic compounds by *Pseudomonas* sp. strain PN-1. *Appl. Environ. Microbiol.* **46**:1286–1292.

Tischer, W., J. Bader, and H. Simon. 1979. Purification and some properties of a hitherto-unknown enzyme reducing the carbon-carbon double bond of α,β-unsaturated carboxylate anions. *Eur. J. Biochem.* **97**:103–112.

Traunecker, J., A. Preuss, and G. Diekert. 1991. Isolation and characterization of a methyl chloride utilizing strictly anaerobic bacterium. *Arch. Microbiol.* **156**:416–421.

Tschech, A., and N. Pfennig. 1984. Growth yield increase linked to caffeate reduction in *Acetobacterium woodii*. *Arch. Microbiol.* **137**:163–167.

Vicuña, R. 1988. Bacterial degradation of lignin. *Enzyme Microb. Technol.* **10**:646–655.

White, H., G. Strobel, R. Feicht, and H. Simon. 1989. Carboxylic acid reductase: a new tungsten enzyme catalyses the reduction of non-activated carbosylic acids to aldehydes. *Eur. J. Biochem.* **184**:89–96.

White, H., R. Feicht, C. Huber, F. Lottspeich, and H. Simon. 1991. Purification and some properties of the tungsten-containing carboxylic acid reductase from *Clostridium formicoaceticum*. *Biol. Chem. Hoppe-Seyler* **372**:999–1005.

White, H., and H. Simon. 1992. The role of tungstate and/or molybdate in the formation of aldehyde oxidoreductase in *Clostridium thermoaceticum* and other acetogens; immunological distances of such enzymes. *Arch. Microbiol.* **158**:81–84.

White, H., C. Huber, R. Feicht, and H. Simon. 1993. On a reversible molybdenum-containing aldehyde oxidoreductase from *Clostridium formicoaceticum*. *Arch. Microbiol.* **159**:244–249.

Winans, S. C. 1992. Two-way chemical signaling in *Agrobacterium*–plant interactions. *Microbiol. Rev.* **56**:12–31.

Wu, Z., S. L. Daniel, and H. L. Drake. 1988. Characterization of a CO-dependent O-demethylating enzyme system from the acetogen *Clostridium thermoaceticum*. *J. Bacteriol.* **170**:5747–5750.

Woods, S. L., J. F. Ferguson, and M. M. Benjamin. 1989. Characterization of chlorophenol and chloromethoxybenzene biodegradation during anaerobic treatment. *Environ. Sci. Technol.* **23**:62–68.

Young, L. Y., and A. C. Frazer. 1987. The fate of lignin and lignin-derived compounds in anaerobic environments. *Geomicrobiol. J.* **5**:216–293.

Zeikus, J. G., L. H. Lynd, T. E. Thompson, J. A. Krzycki, P. J. Weimer, and P. W. Hegge. 1980. Isolation and characterization of a new, methylotrophic, acidogenic anaerobe, the Marburg strain. *Curr. Microbiol.* **3**:381–386.

Zellner, G., H. Kneifel, and J. Winter. 1990. Oxidation of benzaldehydes to benzoic acid derivatives by three *Desulfovibrio* strains. *Appl. Environ. Microbiol.* **56**:2228–2233.

18

Acetate and the Potential of Homoacetogenic Bacteria for Industrial Applications

Juergen Wiegel

18.1 Introduction

In this chapter, the practical uses of acetogens will be addressed. The discussion will focus mainly on homoacetogens or homoacetogenic capacities (see Chapter 1, Section 1.3, for definitions and applications of these terms). Thus, not all potentials of acetogenic bacteria, such as solvent production from mixed fermenters like *Eubacterium limosum* or *Butyribacterium methylotrophicum*, will be included in the present statement [see Lowe et al. (1993), Grethlein and Jain (1992), and Chapter 1 for further reviews of this topic]. It is clear from the preceding chapters that acetogenic and homoacetogenic bacteria can no longer be regarded as a small, insignificant group of physiological "odd balls." They are important organisms in various anaerobic environments such as the intestinal tracts of insects, the human gut, sewage sludge, and sediments. This statement is based mainly on their recently discovered wide distribution and abundance in these environments, and also on the fact that 75% of the methane produced in sediments is formed from acetate (Braun et al., 1979; Wiegel et al., 1981; chapter 7; unpublished results). As further environments are evaluated for the presence of homoacetogenic and acetogenic bacteria, new homoacetogenic organisms will certainly be isolated. This liklihood is illustrated with the recent isolation of *Clostridium ljungdahlii* (Tanner et al., 1993) which belongs to a different phylogenetic branch (clostridial rRNA homology group I) than the other homoacetogens (which belong to neither homology group I nor II).

The potential industrial application of acetogens has provided incentives for research on the diversity of these bacteria. In the past, homoacetogens have been regarded as a highly specialized physiological group that is characterized by the use of the autotrophic acetyl-CoA Wood–Ljungdahl pathway. Initial focus on these organisms suggested that their substrate spectrum was relatively narrow. However, recent studies indicate that members of this group are significantly more versatile and diverse in their metabolic capabilities than previously thought (see Chapters 1, 7, and 10). With respect to potential commercial applications, a challenging question with respect to their diversity is whether extreme thermophilic (growth above 80°C) homoacetogenic organisms exist, or whether the involvement of the tetrahydrofolate pathway limits the temperature range of homoacetogens to growth below 80°C. Furthermore, it is puzzling that no psychrophilic (optimum growth temperature below 15°C) homoacetogenic organisms have been isolated from either spoiled and unspoiled refrigerated food or from arctic land and marine sediments. It is, therefore, of related interest that low-temperature acetogens, including perhaps a psychrophilic species, have been recently isolated from tundra wetland soil in northern Russia (Chapter 15). These specialized high- and low-temperature homoacetogens would be of interest for various biomass conversion-based processes and for processes in the food industry. One possibility of detecting homoacetogens in specialized environments would be the use of specific probes for the tetrahydrofolate enzymes (see Chapter 8).

Since the first oil-crisis in 1972–1973, interest in reviving the old fermentative processes and developing new ones has come and gone. The recent new interest in environmentally safe industrial processes makes the conversion of renewable resources, such as biomass into feedstock chemicals, again a target for further research. Among the organic acids formed as major fermentation products by anaerobic bacteria are acetate, lactate, and butyrate (Ljungdahl et al., 1986). These acids have a high potential to be produced as microbially derived feedstock chemicals. Butyrate is of interest as a biologically produced feedstock chemical for the production of banana and similar flavors in the food industry, whereas lactate, besides its importance in food and dairy products, is of interest for lactate-based biodegradable plastics.

Acetate, the fermentation product of main interest for this chapter, is widely used for various applications, including the preservation of grain during storage in large silos to prevent fungal spoilage. Recently, acetate became of interest for the production of a novel, environmentally safer, road de-icer, calcium-magnesium-acetate (Ca-Mg-acetate; CMA). At the industrial scale, however, acetate is presently produced from petrochemicals, except for the acetic acid in the form of vinegar; it is produced mainly from ethanol with aerobic organisms such as *Gluconobacter oxidans* and *Acetobacter* species that exhibit an incomplete oxidation of sugars. According to the U.S. International Trade Commission, 1.3 $\times$ 10^{9}kg of acetate were produced in 1986. It is still expected that the demand for acetate will increase yearly. If Ca-Mg-acetate can be produced economically as a substitute for the rock salt in ecologically sensitive areas and corrosion-proned constructions, acetate could become one of the major microbially produced compounds, equalling or exceeding ethanol and citric acid. In particular, the use of Ca-Mg-acetate produced via microbial processes could become implemented through politically motivated governmental decisions. This would double the demand for acetic acid. One economical reason to microbiologically produce the road de-icer Ca-Mg-acetate from starchy material (e.g., corn starch) is that the United States has a surplus of corn that would provide for the use of large amounts of the de-icer. Consequently, some states have already demanded that the road de-icer should be produced via corn starch-based processes (Wise et al., 1991). Because the homoacetogens can convert sugars to acetate at efficiencies of 90% and above (based on carbon content) and can also utilize several C-1 compounds such as methanol and Co/CO_2 (see Chapter 1), they are obviously ideal candidates for such a process. This potential application is presently the most important one for homoacetogenic bacteria.

Although many organisms form acetate as a major product, most of them produce additional acids or alcohols. The additional costs for separation of these products renders these organisms less desireable relative to fermentation economics. Only homoacetogenic bacteria and a few acetogenic bacteria can produce acetate from certain substrates as the only organic compound; homoacetogens are the most efficient in this regard and can achieve high conversion ratios of nearly 3 acetate per hexose utilized. For example, *Acetothermus paucivorans* and *Acetomicrobium flavidum* form acetate but do not use the acetyl-CoA pathway. Consequently, they produce about 2 mol acetic acid per mol hexose; onethird of the sugar carbons are "lost" as CO_2 as these organisms cannot convert CO_2 plus H_2 (or other reduced equivalents) to the desired product acetate. Although many of the homoacetogens can utilize a wide variety of hexoses, pentoses, and polyols, some homoacetogens such as *Clostridium aceticum*, *Clostridium formicoaceticum*, and *Sporomusa acidovorans*, cannot utilize glucose (although they do use fructose). This somewhat narrow substrate spectrum is assumed to be advantageous in their habitats, i.e., they do not have to compete with the many glucose utilizers; it minimizes, however, their potential application in biomass conversion processes and many industrial applications. But if the other properties favor their industrial application, it should not be to difficult to clone the suitable enzymes for glucose uptake and metabolism into these organisms because the main glycolytic pathway is already in operation.

18.2   Aspects of Using Homoacetogens in Industrial Applications

18.2.1   Thermophilic Versus Mesophilic Homoacetogens

Although questioned by some authors, thermophilic bacteria generally have an advantage over mesophilic bacteria in industrial applications. The mesophilic

and thermophilic homoacetogens have the same basic pathway for acetate formation, likely with similar regulatory processes; thus, with respect to overproduction
of metabolites, the same problems arise with thermophiles as with mesophiles.
However, thermophiles do not require that sterilized medium be cooled down
close to ambient temperatures; cooling of large industrial volumes can become
expensive. Furthermore, the dissipation of heat from mixing and stirring does
not become a problem when working at elevated temperatures, and maintaining
anaerobic conditions becomes easier due to the increasing insolubility of gases
(e.g., oxygen) with increasing temperatures.

The use of thermophiles has additional advantages for the production of the
road de-icer Ca-Mg-acetate: in comparison to contamination of the Ca-Mg-acetate
with spores from mesophilic homoacetogens, contamination of the product with
spores from thermophiles would create a less potential problem for the environment [because soil and sediments would barely reach a temperature (above 40°C)
adequate for the germination and growth of thermophilic spores]. A lower risk that
these homoacetogens would proliferate in the environment would be important if
genetically engineered homoacetogens are used, as it would be less probable
that the engineered traits could be passed on to other organisms. Furthermore,
because pathogenic organisms can be introduced by substrates such as municipal
and agricultural wastes, a thermophilic fermentation will reduce, if not eliminate,
pathogenic organisms including virus particles from the product and the waste
stream of the fermentation. The creation of thermophilic pathogens, which exhibit
a wide range of growth temperatures (i.e., growing between 15 and 65°C) and
are thus able to proliferate in mesobiotic as well as thermobiotic environments,
is a theoretical possibility. This could occur accidentally by transfer of genetic
information (e.g., via a plasmid) from a mesophilic pathogen into the thermophilic
homoacetogen by using pathogen-contaminated waste material as a fermentation
substrate (Wiegel, 1990). However, this theoretical possibility can be regarded
as very slim.

In conclusion, the author believes that, among the presently known homoacetogens, the thermophiles offer greater opportunities for developing economically
feasible processes. However, a thermophilic halophile with the homoacetogenic
properties of *Clostridium thermoautotrophicum* or *Acetogenium kivuii* would
offer even more advantages (in this regard, the reader is directed to Chapter 16
for a description of newly isolated mesophilic, halophilic acetogens).

18.2.2. *Problems of Product Recovery*

A process is only economical if appropriate product separation and recovery
are possible. Although distillation of the free acetic acid is a straightforward and
well-known process, the energy costs required for distillation of dilute fermentation broths are presently regarded as too high to render the production of acetate

of anaerobic homoacetogens economically feasible (Busche, 1985). None of the presently described acetogens is able to produce, at a reasonable rate, acetic acid concentrations above 50 g/L, which is considered the lowest concentration possible for economical distillation. One of the reasons for the low tolerance of acetogens toward acetate/acetic acid (and thus a reason why higher acetate concentrations are difficult to achieve) is that free, nondissociated acetic acid diffuses into the cell, where it dissociates into H^+ and acetate anion due to the higher pH value inside the cell. Consequently, as the total acetate/acetic acid concentration increases, the concentration of undissociated acetic acid in the medium will increase and with that, the influx of undissociated acetic acid into the cell also increases. This leads to an increased proton concentration inside the cell due to the dissociation of the acetic acid and causes an uncoupling of the proton gradient. This uncoupling diminishes the energization of the membrane and, consequently, the energy-dependent metabolism of the cells (Baronofsky et al., 1984, Wang and Wang, 1984; Terracciano et al. 1987; Reed et al. 1987).

Several research groups have tried to adapt homoacetogens to tolerate lower pH values. Strain C5-3 of *C. thermoaceticum* (Reed et al., 1987) exhibited good volumetric activities of 1.3 g undissociated acetic acid/L at pH 5.75 (and a total concentration of 1.45 g acetic acid/L). However, no continuous growth was observed if the acetic acid increased to above 1.5 g undissociated acid/L. The acid tolerant mutant C5-3 maintained growth in continuous culture up to 3.5 and 4 g undissociated acid at pH 5.3–5.8, but below pH 4.7, growth was only maintained at 2.5 g acid/L. Schwarz and Keller (1982a, 1982b) tried to overcome this limitation by mutating *C. thermoaceticum*. Although they were the first to achieve a decent growth with *C. thermoaceticum* on glucose at a pH of 4.5, the cultures grew too slow at lower pH values (mass doubling times at pH 4.5 were 36 h) and the acetate concentrations obtained (4.5 g/L, about 75mM) were too low to be useful for industrial application.

Interestingly, *C. thermoautotrophicum* [all strains tested; Wiegel et al. (1981)] and to a lesser extent *C. thermoaceticum* [strains LJD, Wood, ATCC 39073, and 35608; Wiegel and Garrison (1985)] grow well at pH 4.5 when methanol is used as carbon and energy source. When grown with methanol (or glycerate), both species have a pH optimum for growth and acetate production around 5.8, whereas for glucose-dependent growth, both exhibit an optimal pH around pH 6.8 (Ljungdahl et al., 1985; Wiegel et al., 1981). Apparently, the enzymes of the glycolytic pathway are more sensitive to the acetic-acid-induced effects. However, the use of methanol does not appear to be an economical substrate for acetate production at elevated temperatures. Under such conditions, the vapor pressure of methanol increases, thus requiring increased capital costs for special equipment in order to minimize risk of explosions.

Ljungdahl et al. (1986) and Wiegel et al. (1991) tried a different approach. They developed a rotary fermentor (section 18.2.3c) to both obtain and maintain low-pH-adapted cultures during growth on carbohydrates. Although the combination of the novel fermenter and variation in the fermentation approach yielded faster fermentations at pH values as low as pH 4.3, the acetate concentrations obtained (8.1 g/L at pH 4.8) and production rates (3.9 g/L h^{-1}) were still far too low for an economical process (Wiegel et al., 1991). This rotary fermentation approach, however, appears promising, and further research using combinations of the above approaches, i.e., selection of adapted cultures and modified fermentation methods, may lead to an economically successful process.

At the present time, fermentations have to be carried out under pH control, and the fermentation broth has to be acidified before harvesting. Shimshick (1981) and Yates (1981) suggested using carbon dioxide for acidification and organic solvent extraction. Other suggestions include solvent membrane techniques (Kuo and Gregor 1983) and electrodialysis (Busche, 1983). It appears that in the recent past, the economic requirements on product recovery have prevented the production of acetic acid as a feedstock chemical via fermentation. Busche (1991) concludes that an improved *Acetobacter suboxydans* system as a two-step process could lead to 50g acetic acid/L and, thus, to a marginally economically acceptable process. The assumed process includes the conversion of glucose to ethanol and CO_2 by *Saccharomyces* at a 90% yield, coupled to the conversion of ethanol to acetate with 85–90% yield via recycling of the cells and an organic solvent extraction for product recovery. Busche concludes that the higher acetate concentration obtained with *Acetobacter* outcompetes the advantage of the one-step process with the higher conversion ratio of approximately 3 acetate per glucose of the homoacetogen *C. thermoaceticum* (compared to approximately 2 acetate per glucose for the coupled *Acetobacter* system). However, Busche's calculation was based on the assumption that only 18 g acetate/L was possible with *C. thermoaceticum*. This concentration has been exceeded by several researchers: 108 g/L at pH 5.6–7.2 in a dilution-cycle fermentation mode (Ljungdahl et al. 1986; Wiegel et al., 1991), 56 g/L at pH 6.9 in pH-controlled batch (Wang and Wang, 1984); 46 g/L at pH 6.6 in fed-batch fermentation (Parekh and Cheryan, 1990a, 1990b).

The product recovery of acetic acid for use in Ca-Mg-acetate production is a special case. Using milled dolomite as the neutralizing agent during the fermentation and to also adjust the pH to above 8.5 at the end of fermentation, the harvest of acetic acid can occur as Ca-Mg-acetate salt. The salt is obtained without separation from the components of the fermentation broth through industrial spray-drying. This approach is possible because the other medium components are not diminishing the function of the salt as de-icer, and no environmental side effects were observed during field trials (Horner et al., 1991). However, this approach leads to the requirement that the medium should contain a minimum of phosphate salts. We have shown that the phosphate in yeast extract is sufficient for formal growth of these organisms, but for increased cell density, additional phosphate is needed. The required amount of phosphate should be chosen so that the final concentration at the end of the fermentation is below 1 mM (Wiegel et al., 1991).

18.2.3   Approaches to Reach Higher Acetate Concentrations and Production Rates

(a) Growth-Associated versus Uncoupled Acetate Formation

As apparent from growth curves, the majority of acetate formation is generally coupled to the growth of homoacetogens. However, cell recycling, fed-batch fermentation, and immobilized cell approaches indicate that at elevated product concentrations, *C. thermoaceticum* exhibits to some degree a non-growth-associated acetate formation. The author is, however, not aware of any in-depth investigations on how much non-growth-associated acetate formation can be maintained. However, it is interesting to note that the metabolic inhibitor harmaline prevented the growth of *A. kivui* but did not influence the cell's capacity to synthesize acetate from H_2/CO_2 (Yang and Drake, 1990). Harmaline was shown to increase the acetate-to-biomass ratio approximately 13- and 2-fold for H_2- and glucose-coupled acetogenesis, respectively. The development and implementation of metabolic uncouplers that prevent or retard growth but nonetheless permit acetate synthesis might be of some practical value relative to maintaining low biomass production.

In this regard, an increase in non-growth-associated acetate formation could be important for the industrial production of acetate because (i) it would increase the yield (no substrate utilization for biomass production), (ii) it would decrease the biomass waste/contamination in the final product (especially for Ca-Mg-acetate production), (iii) it might decrease the effect of growth inhibition by acetic acid, and (iv) it would increase the efficiency of immobilized cell approaches. Several authors reported that they maintained continuous (including stepwise continuous) thermophilic cultures from several months up to nearly 3 years (26,000 h) (Yates, 1981). Apparently, one can sustain a thermophilic homoacetate fermentation process for a long time.

(b) Batch and Continuous Cultures

Several approaches have been used to increase the acetic acid/acetate concentration produced in cultures to overcome the above-discussed problems of acetic-acid-induced inhibition and problems in the recovery of acetate. Wang and Wang (1983) used immobilized whole cells of *C. thermoaceticum* on κ-carrageenan. This approach allowed the use of dilution rates (0.4 h^{-1}) that were faster than the maximal growth rate and also achieved a maximal volumetric productivity of 6.9 g^{-1}/h^{-1} at a dilution rate of 0.37 h^{-1} and 19g/L with a steady-state cell concentration of 60 g dry weight per liter gel. The maximal cell concentration obtained was 65 g dry weight/liter gel. Most of the cells grew near or on the surface of the gel particles; within the matrix, the highest concentration was only 35 g dry weight/liter gel. Consequently, the use of a material allowing a higher density within the gel particles or employing particles with a greater surface area should lead to higher productivity. This was to some extent done in the approach by Ljungdahl et al. (1986) and Wiegel et al. (1991) who used the above-mentioned rotary fermentor with felt disks (section 18.2.3c).

Wang and Wang (1984) employed a fed-batch approach with *C. thermoaceticum* DSM 521. Maintaining the glucose concentration between 0.5 and 1.5% (wt/vol), they reached 56 g acetate/L at pH 6.9 after 125 h. A similar approach was taken by Parekh and Cheryan (1990a, 1990b) who found that increasing the calcium and magnesium ion concentrations increased the total amount of acetate produced to 35 g acetic acid/L; however, the final concentration depended on the strain used. The finding of Parekh and Cheryan is in agreement with similar results reported earlier by Ljungdahl et al. (1985). Fed-batch fermentations performed by Parekh and Cheryan (1990b) with an adapted strain of *C. thermoaceticum* ATCC 39829 yielded 46 g acetate/L at pH 6.6 after 192 h. Higher production rates of 14.3 $g/L^{-1}/h^{-1}$ were obtained by Reed and Bogdan (1985) using cell recycling, but they only obtained 7.1 g acetate/L. Using ultraviolet light and ethyl and methanesulfonate (EMS) for mutagenesis (Ljungdahl et al. 1986), efforts were made to increase acetate production by employing repeated mutagenesis and continuous selection in a combined continuous fermentation setup (Reed et al., 1987). N-methyl-N-nitro-N-nitrosoguanidine was ineffective, perhaps due to the strongly reduced conditions. Reed et al. (1987) did not obtain any improvements for *C. thermoautotrophicum* with this mutagenesis approach, perhaps because the wild-type strain already grows at the low pH values of the improved strain *C. thermoaceticum* C5-2. Strain C5-3 (obtained by EMS treatment) could not grow below pH 4.6, but it tolerated 0.35–0.4% free acetic acid at pH 5.3–5.8 and 0.1–0.25% at pH 4.7–5.3.

In our laboratory, using the wild-type strains of *C. thermoaceticum* strain LJD or Wood, *C. thermoautotrophicum*-type strain JW 701/3 or strain 701/3, and *A. kivui*, the cell-recycling approach did not yield the rate and concentrations of Reed and Bogdan (1985); production rates of 4.0, 3.8, and 4.0 $g/L^{-1}/h^{-1}$ were obtained for *C. thermoaceticum* strain LJD, *C. thermoautotrophicum* strain JW701/5, and *A. kivui* strain DSM 171, respectively. The maximal acetate concentrations achieved were around 18g/L (Ljungdahl et al., 1986; Wiegel et al., 1991). The major problem was the tendency of the cells, especially those of *C. thermoaceticum* strain LJD, to lyse; a bleeding rate of about 50% was required to sustain a continued fermentation for more than a month. During continuous cultivation using enzymatic hydrolyzed corn starch as substrate, *A.*

kivui developed into a flocculating culture. A rate of about 4.8 g acetate/L^{-1}/h^{-1} at a concentration of around 9 g acetate/L were obtained (Wiegel et al., 1991). The flocculation was apparently due to the formation of long filamentous cells, which were not observed in cultures fed with glucose obtained from Sigma. In an industrial application, flocculation could be an important advantage because separation techniques developed for flocculating yeast could be adapted for fermentations under cell-recycling and cell-retainment processes for the production of acetate or Ca-Mg-acetate.

However, the author and his co-workers were not able to adapt *A. kivui* to high calcium, magnesium, or acetate concentrations without losing its advantageously fast growth rate. In a continuous culture, longer retention times increased the acetate concentration achieved, but reduced the production rate. An optimal compromise was reached at about 4–6 h for the two clostridia and around 2 h for the faster growing *A. kivui*. The highest acetate concentrations (about 2 M in the presence of dolime, i.e., milled dolomite consisting mainly of Ca-Mg-carbonate, as neutralizing agent) were obtained using mutants *C. thermoaceticum* and *C. thermoautotrophicum*. The mutants were obtained by an EMS treatment of *C. thermoaceticum* and by a 20-min nitrous acid treatment of *C. thermoautotrophicum* grown in the presence of 200 mM pyruvate. The production rates, however, were low (about 12 g/24 h). Unfortunately, when the mutants were grown in a 400-liter fermentor with normal medium, they grew slowly and apparently were not stable under these conditions because they could not produce more than 600 mM acetate (about 36 g/L). Apparently, all the obtained mutants are less efficient than the wild-type strains. Nevertheless, these mutants still could be used for the production of acetate by employing a continuous, partial batch, approach, i.e., harvesting only one-third of the fermentor volume every 3 days and refilling the fermentor with sterilized medium. This simple approach maintained both a relatively high concentration of acetate (above 750 mM) and cell density without requiring equipment for continuous culture).

(c) ROTARY FERMENTOR, A NOVEL APPROACH FOR ACETATE AND CA-MG-ACETATE PRODUCTION

Clyde (1983) developed a horizontal rotary fermentor for a fast release of carbon dioxide during ethanol fermentation by yeast. Ljungdahl et al. (1986) and Wiegel et al. (1991) adapted the principal of the rotary fermentor but with the reverse notion, i.e., for high CO_2 uptake for homoacetogenic fermentation by *C. thermoaceticum* and *C. thermoautotrophicum*. The rotary fermentor combines features of a fixed-film fermentor (attached bacteria in felt pads) and of continuous culture with cell retention. The principal is that fresh media is introduced at the bottom of a rotating stack of pads and exit the tube-like fermentor at the top. The pads were cut out of sheets of felt and Du Pont Reemay 2033. The alternating felt and DuPont Reemay pads (at a ratio of 3:1 with Teflon® washers as spacers) were mounted on the rotating axis (see Wiegel et al., 1991, Fig. 4). The two main ideas behind this approach are that (i) cells attach to the filter pads and then adapt accordingly to the specific conditions within these areas and (ii) new growing cells become permanently modified with time to grow optimally under the combined influence of both product concentration and pH. Cells at the bottom pads adapt to high glucose but low acetate concentrations, optimal pH for growth, and, thus, high productivity. Cells attached to the top pads will adapt to produce further acetic acid at the lower pH (due to the produced acetic acid) and low glucose concentrations. This arrangement results in a pH gradient within the fermentor and a continuous selection for cells more adapted to the special conditions. In the small-scale fermentation employed (i.e., with a relatively short column), the pH gradient was only approximately 0.5 pH unit, but production rates of more than 10 g acetate/L^{-1}/h^{-1} (retention time of 30 min) were achieved; unfortunately, under these fast production rates, the acetate concentration was low. However, higher acetate concentrations were reached when using two rotary fermentors connected in tandem (Wiegel et al., 1991). The author believes that this approach, when further developed, could lead to a thermophilic, homoacetogenic fermentation that would be economically acceptable.

(d) ACETOGENIUM KIVUI

A few approaches have addressed the potential application of this fast growing thermophilic (T_{max} of 75°C) acetogen. Ljungdahl et al. (1985) abandoned work with this organism because they could not improve its resistance to increased acetate or calcium and magnesium concentrations. Based on experiments by Klemps et al. (1987), Eysmondt et al. (1990) used a continuous culture and computer simulation to predict the steady-state concentration of acetate. The maximum acetate concentration obtained (as calculated to permit positive growth) was 34.5 g/L; the conversion of glucose to acetic acid was maximal at 2.35 acetate/glucose (78.7% of the theoretical value). The product concentration was inverted proportional to the growth rate. The highest production obtained was around 3.55 g/L^{-1}/h^{-1} at a dilution rate of around 0.42 h^{-1}. In conclusion, despite its rapid growth rate, this organism has not emerged as the most promising acetate producer.

18.2.4. *Acetate from Noncarbohydrate Substrates*

The characterization of newly isolated homoacetogens and more detailed physiological studies on both new and known strains have shown that their substrate spectrum is much wider than originally assumed from the initial investigations of *C. aceticum*, *C. thermoaceticum*, and *C. formicoaceticum*. The extended range of substrates includes methanol, CO, H_2/CO_2, methoxy groups of substituted aromatic compounds, and decarboxylation of substituted benzoates (see Chapters 1, 7, and 10). *C. thermoaceticum* (DSM 521) also metabolizes formate and lactate in the presence of a second fermentable substrate, and the mutant strain C5-2 grows at pH 6.5–7.2 with lactate as the sole carbon source, converting about 2 lactate to 3 acetate (Bream, 1988). According to Bream (1988), *C. thermoautotrophicum* JW 701/5 and *A. kivui*-type strain were not able to use lactate either in combination with glucose or alone. Wiegel et al. (1981, unpublished results) showed that other strains of *C. thermoautotrophicum* can utilize lactate; strain JW701/5, however, did not. The utilization of these noncarbohydrate substrates by homoacetogens may make acetate production more economical from less competitive substrates, such as syngas, methanol, fermented and nonfermented diary waste streams, and breakdown products of lignin (Bream and Datta, 1985). Lignin-based compounds are produced during biomass-utilizing processes, including the pulping process. The latter application could be used to reduce the waste stream in the pulping process and to increase production of valuable side products.

Bache and Pfenning (1981) were the first to demonstrate that the mesophilic, non-spore-forming *Acetobacterium woodii* is able to O-demethylate various methoxylated aromatic acids, including vanillic, syringic, 3,4,5-trimethoxybenzoic, ferulic, sinapic, and 3,4,5-trimethoxycinnamic acid. Most homoacetogenic bacteria are able to O-demethylate methoxy-substituted aromatic compounds and to use the methoxylsubstitution as methylgroup donor for acetate synthesis. Bache and Pfenning (1981) also demonstrated that the double bond of the acrylic acid side chains of caffeic, ferrulic, sinapic, and 3,4,5-trimethoxycinnamic acid is reduced to the propionic acid side chain. Hsu et al. (1990a, 1990b) have shown that *C. thermoaceticum* (ATCC 39073) not only O-demethylates vanillic acid and similar aromatic compounds but also decarboxylates the acids to the corresponding phenolic compounds. We have also recently observed this activity with *C. thermoaceticum* strains LJD and Wood, and also with *C. thermoautotrophicum* JW701/3 and JW701/5 (Zhang and Wiegel, 1994; unpublished data). The compounds subject to decarboxylation include chlorinated aromatic acids such as 3-chloro-4-hydroxybenzoate (Hsu et al., 1990b). The reader is directed to Chapters 7, 10, and 17 for further discussion of the metabolism of aromatic compounds.

Kuhn et al. (1989) demonstrated that the decarboxylation of 4-hydroxybenzoate and similar compounds such as vanillic and cinamic acid occurs in anaerobic environments. We have also observed this activity and found that the route by which 3-chloro-4-hydroxybenzoate is degraded depended on the cocontamination and prior exposure to other halogenated and nonhalogenated aromatic compounds of the sediment (Zhang and Wiegel, unpublished results). The halogenated hydroxybenzoates were either first dechlorinated and then decarboxylated, or first decarboxylated and then dechlorinated to phenol (or cresol in the case of methylated derivatives). Phenol is then recarboxylated and dehydroxylated to benzoic acid (presumedly via the intermediate 4-hydroxybenzoate) which is then ring-cleaved to acetate and CO_2 plus H_2, all of which are subsequently converted to methane plus CO_2. In nutrient-rich sediments, the compounds are decarboxylated faster than dechlorinated, whereas in nutrient-poor sediments both reactions occur simultaneously. It is of interest that most of our enrichments for transformation of chlorophenols and phenols contained homoacetogenic bacteria, but it is not yet clear whether the homoacetogens were opportunistic, beneficial, or required members of these enriched, phenol-degrading communities. Nevertheless, these observations demonstrate the potential importance of homoacetogenic organisms in the breakdown of compounds in the environment and suggest that homoacetogens may be important members of consortia and microbial communities of use in bioremediation processes.

18.2.5   Conversion of Renewable Materials to Acetate Using Monocultures and Cocultures

The fermentation of carbohydrates to acetic acid by homoacetogens usually requires monomeric sugars, which are more costly substrates than untreated or nonhydrolyzed polymeric material. As none of the known homoacetogenic organisms can utilize cellulosic material directly (i.e., cellulose and hemicellulose) without prior enzymatic or chemical hydrolysis to oligomeric or monomeric sugars, the use of cellulolytic cocultures is a viable alternative. Cellulosic material is still regarded as an economically renewable resource. An example of the fermentation of sulfuric acid pretreated oat spelt xylan and poplar hemicellulose preparations was published by Brownell and Nakas (1991). They used a monoculture of *C. thermoaceticum* ATCC 39073 and obtained about 12 and 14.4 g acetic acid from the oat spelt and poplar xylan preparations, respectively, in 72 h. When the same authors used xylose in a fed-batch fermentation up to 42 g acetic acid/L were obtained within 116 h when the pH and the xylose concentration were maintained at 7.0 and 2% (wt/vol), respectively. If the pH was not controlled, the usual yield was only around 12 g/L. Freier (1981) demonstrated that in thermophilic cocultures of *C. thermocellum* JW20 with *C. thermoaceticum* DSM 521 or *C. thermoautotrophicum* JW701/3, cellulose can be effectively converted to acetate, yielding conversions of up to 1.4 mol acetate/mol glucose-equivalent cellulose. The cellulolytic monoculture produced only 0.3 mol acetate/mol glucose-equivalent cellulose. However, relatively high glucose and cellobiose concentrations were found (0.26 mol glucose and 0.5 mol cellobiose/mol glucose-equivalent cellulose utilized). Furthermore, contrary to the generally accepted assumption, the co-cultures did not exhibit a significantly higher or faster cellulose degradation. More recently, Siman'kowa and Nozhevnikova (1989) demonstrated again that a similar coculture consisting of *C. thermocellum* strain Z-53 and *C. thermoautotrophicum* strain Z-25 effectively converted cellulose to acetate. The employed homoacetogenic strain used glucose and carbon dioxide simultaneously and, thus, exhibited a 2.5-fold higher acetate production compared to growth on either of the substrates alone. The hydrogen concentration in the coculture was negligible. Similar to the results of Freier (1981), the coculture exhibited a cellulose to acetate conversion of around 1.4 mol acetate per glucose equivalent of cellulose utilized. Although the level of reducing sugars in the supernatant of this coculture was lower than in the coculture of Freier, the coculture of strain Z-53 and Z-25 did not exhibit a faster cellulose degradation. Despite the fact that the cellulose is not converted to acetate as efficiently as glucose, the direct transformation of cellulose to acetate (i.e., without first hydrolyzing cellulose to glucose which then is fermented to acetate) seems to be a viable option with respect to product recovery costs, as long as the total amount of ethanol and lactate formed can be kept to minimal levels.

Better conversion ratios were obtained when starch instead of cellulose was used as polymeric substrate. Wiegel et al. (1984) showed that a coculture of a saccharolytic clostridium (*Clostridium "thermosaccharovorum"* RB-9, isolated by Carreira and Ljungdahl (unpublished results), and *C. thermoautotrophicum* converted soluble cornstarch to acetate with a yield of 2.73 acetate/mol glucose equivalent of starch. Acetate concentrations of nearly 300 mM (120 mM sugar equivalent supplied) were obtained without any adaptation or optimization of the system. Unfortunately, this route has not been further investigated. One of the reasons for not further evaluating this coculture was that glucose syrup and glucose from enzyme hydrolyzed cornstarch became available at economically competitive costs (projected costs of $0.10/lb). The same starch hydrolyzing coculture was also used for conversion of steam-exploded hardwood to acetate leading to acetate yields of about 2.4 mol acetate/mol xylose equivalent of utilized hemicellulose. The acetate-to-xylose ratios could be even higher when additional acetate is formed from hemicellulosic material with a high content of acetyl chains and/or methoxylated aromatic compounds. Because xylose is a preferred substrate over glucose, the utilization of hydrolyzed xylans can likely be added to any industrial fermentation substrate to be converted to acetate.

18.3   Ca-Mg-Acetate (CMA) Production
with Thermophilic Homoacetogens

Marynowski et al. (1983, 1985) first suggested the use of *C. thermoaceticum* for the thermophilic production of Ca-Mg-acetate (Chollar 1984). Ljungdahl et al. (1986) and Wiegel et al. (1991) have recently shown that both thermophilic homoacetogens *C. thermoaceticum* and *C. thermoautotrophicum* have a high potential for the technical production of Ca-Mg-acetate from hydrolyzed corn starch in the presence of milled dolime. Another alternative to the cornstarch-based route is the production of acetate from syngas or methanol, both of which are available from industrial processes. Research is underway for both options.

The two thermophilic clostridia, *C. thermoaceticum* strains LJD and Wood, and *C. thermoautotrophicum* JW701/3 and JW701/5, with doubling times of 4–8h, could be adapted, while still maintaining a high ratio of at least 2.5 acetate formed per glucose utilized, to tolerate relatively high concentrations of sodium (up to 850 mM) or potassium acetate (up to 305 mM), dolime (2–3%) and Ca-Mg-acetate (250–300 mM). Thus, the clostridial organisms can be used to produce elevated acetate concentrations in the presence of dolime to neutralize the formed acid. In contrast, the non-spore-forming and fast-growing (doubling time of 2–3h) *A. kivui* could not be adapted to the higher salt concentration and, thus, appears not be a good candidate for an industrial Ca-Mg-acetate fermentation process. On the other hand, the clostridial strains require the presence of yeast extract for a fast and efficient fermentation. Although Lundie and Drake (1984) and Savage and Drake (1986) developed mineral media for both *C. thermoaceticum* and *C. thermoautotrophicum*, respectively, the requirement of costly media supplements such as nicotinic acid and a slower fermentation rate eliminate the advantage of employing a mineral medium for this application. Although the best fermentation was obtained at a concentration of 20–40 mM phosphate, no or very little additional phosphate [besides the phosphate already present in the yeast extract (0.5%)] is required for growth of the clostridia. As discussed in the previous section on product recovery, the most economical way to harvest the Ca-Mg-acetate from fermentation broths is to spray-dry the fermentation broth; high phosphate concentrations have to be avoided to prevent eutrophication of the waterways which potentially receive runoff water from the treated high-ways. High yeast extract (or its technical substitutes) concentrations in the fermentation broths have to be avoided too, because (besides the economical point of view) the extra organic material could act as fertilizer and lead to unwanted plant and microbial growth on the roadside. With this in mind, acceptable fermentations were obtained with a 0.1% yeast extract and 1 mM additional added sodium potassium phosphate. Although the potential for the production of Ca-Mg-acetate via thermophilic homoacetate fermentation of hydrolyzed biomass or from techni-cal-grade methanol is high, more optimized strains and fermentation processes have to be developed before Ca-Mg-acetate can be commercially produced by fermentation at competitive costs (about $0.20–$0.25/lb Ca-Mg-acetate).

At present, no similar data are available for the mesophilic homoacetogens. The use of slightly halophilic mesophiles that form both acetate and butyrate have been suggested (1st Symposium of the Corn Growers Association, 1987). However, a detailed discussion of mixed acid-producing organisms is beyond the scope of this chapter.

18.4   Miscellaneous Uses of Acetogens

18.4.1   Production of Compounds Other Than Acetate

(a) 5-Aminovulinic Acid and B$_{12}$

Several articles were published and two patents were issued to use *C. thermo-aceticum* ATCC 31490 for the production of 5-aminovulinic acid (Koesnander et al., 1989; JP 02,261,389) and for corrinoid production (Koesnander et al., 1991b; JP 02,261,391). The latter application is not surprising, as methyl-B$_{12}$ is part of the acetate- and energy-producing acetyl-CoA Wood-Ljungdahl pathway of homoacetogens. B$_{12}$ concentrations in homoacetogenic are 10-fold and higher compared to other bacteria (Stupperich et al., 1988; Clarke et al., 1982; Ljungdahl, 1986). As a group, acetogens and homoacetogens contain quite a diversity of corrinoids; thus, several corrinoid derivatives can be obtained from the acetogenic bacteria (Stupperich et al., 1988). Koesnandar et al. (1991b) partially optimized the system with respect to the effect of heavy metals on growth and corrinoid production. In addition to using *C. thermoaceticum*, Inoue et al. (1992) suggested the use of a newly isolated marine strain of *Acetobacterium* sp. for the production of cyanocobalamin. During growth on methanol, the wild type and tetrachloromethane-resistant mutant strains produced about 11 mg and up to 23 mg, respectively, cyanocobalamin per g dry weight of cells within 7 days.

(b) Acetate Kinase

The industrial use of immobilized acetate kinase from *C. thermoaceticum* has been patented (Unitika, Ltd, 1982). At 30°C, the enzyme was stable for more than 1 year.

(c) Methane from Hexoses Employing Cocultures

A different application of homoacetogens-containing cocultures is the use of a homoacetogen in combination with acetoclastic methanogens to convert hexoses to methane via acetate. Such combinations probably occur quite frequently in methanogenic sediments. Ibba and Fynn (1991) employed a two-stage fermentation using *A. kivuii* and a newly isolated acetoclastic methanogen. They concluded that such an approach could be used for cleanup and fermentation of acidic carbohydrate containing waste water. However, the author would suggest that for such an application an even more suitable combination, that is, a well-adapted coculture or mixed culture, could be isolated from corresponding waste-stream fermentation broths using a proper inoculum and enrichment scheme.

18.4.2 Transformations

(a) DEHALOGENATION OF CHLORINATED ALIPHATIC COMPOUNDS

Traunecker et al. (1991) reported that a newly isolated homoacetogenic strain dechlorinated methylchloride effectively. However, so far, there is only this one report. But based on this first report and the increasing discoveries of the diversity of homoacetogens, it is tempting to speculate that homoacetogens have a high potential for bioremediation applications. Homoacetogens can grow heterotrophically on breakdown products from lignin containing carboxyl and methoxylgroups as well as chemolithoautotrophically with CO_2 plus H_2 and CO; thus, homoacetogens would have a competitive advantage in environments with low concentrations of utilizable carbohydrate substrates. Further studies are clearly needed, including screening for new and more suitable strains.

(b) CONVERSION OF CYSTINE TO CYSTEINE

L-Cysteine is widely used in the production of pharmaceutical and other products, and as a substitute for thioglycolic acid, an antioxidant in the food industry (Soda et al., 1983). Koesnandar et al. (1991a) tested the use of cell-free extracts of *C. thermoaceticum* (ATCC 31490) in the presence of hydrogen and methylviologen for the reduction of cystine at 60°C and pH 9.0. Under these conditions, high rates were obtained and cysteine was relatively soluble. The hydrolysis of cysteine by cell-free extracts was very low, as cysteine desulfhydrase exhibited only negligible activity. Koesnandar et al. (1991a) concluded that the use of a membrane-retaining enzyme reactor or immobilized enzyme fermentor would be a useful approach because the conversion is complete and cysteine is the only product, thus requiring no costly separation.

(c) SEQUESTERING OF HEAVY METAL IONS

The use of homoacetogens for the precipitation of heavy metals such as cadmium is one of the most recently described potential applications of *C. thermoaceticum* ATCC 39073 (Cunningham and Lundie, 1993). While growing on complex medium in the presence of cysteine, 1-mM Cd concentrations were reduced to undetectable levels. Within 72 h, cadmium precipitated onto the surface of nonstarved cells as well as in the medium surrounding the cells. Cadmium was apparently precipitated to white cadmium carbonate, which was slowly converted to yellow cadmium sulfide. Interestingly, in the presence of cadmium, twice as much protein was formed than in cultures without supplementation of cadmium. At this time, it is not clear whether the additional protein is responsible for the external precipitation. Furthermore, cultures containing cadmium produced about fourfold higher amounts of sulfides than the control cultures, which may have been due to an earlier onset of cysteine sulfhydrase expression in the cadmium-containing cultures. Unfortunately, for a possible industrial application, the presence of 1 mM nickel, which alone was not inhibitory, was toxic in the presence of 1 mM cadmium. In addition, the requirement of relatively high cysteine concentrations (0.1%) and a low tolerance of not much more than 4 mM cadmium in 2 mM glycerol and phosphate-buffered medium are limiting factors relative to the usefulness of this organism for the detoxification of Cd and other heavy metals in oxygen-deficient waste streams. However, the organism might be useful for specific industrial waste streams low in sulfide but high in other compounds toxic for aerobic organisms. Other homoacetogenic species have not been tested for this application. Isolation of homoacetogens (e.g., strains of the ubiquitous *C. thermoautotrophicum*) from cadmium-polluted anaerobic environments might lead to more tolerant and useful organisms.

18.5   Conclusions

About a dozen patents (including US 4,13084; 4,282,323; 4,506,012; 4,814,273; EP 43,071; 120.369; 120,370; DE 3,705,275; JP 02,261,389; 02,261,391) have been issued during the last decade specifically for the use of homoacetogens. Although no industrial large-scale process presently involves homoacetogenic cultures (except sewage-treatment processes), various members of this physiological group have potentials for advancing to the stage of an "industrial organism." This is especially true for the thermophilic clostridia for the production of Ca-Mg-acetate. One of the present obstacles for a wider acceptance of homoacetogens for industrial processes is the lack of a suitable genetic system for genetic improvements of the organisms in question. So far, only a few genes from *C. thermoaceticum* have been successfully cloned into *Escherichia coli* (Morton et al. 1992). At the present time, it seems unfeasible to genetically convert other organisms which are easier to handle into efficient homoacetogenic bacteria because (i) high acetate production is due to the highly specialized and multifaceted Wood-Ljungdahl pathway and (ii) acetate formation is integrated to the energy metabolism and growth of the cells.

References

Bache, R., and N. Pfenning, 1981. Selective isolation of *Acetobacterium woodii* on methoxylated aromatic acids and determination of growth yields. *Arch. Microbiol.* **130**:255–261.

Baronofsky, J. J., W. J. A. Schreurs, and E.R. Kashket. 1984. Uncoupling by acetic acid limits growth of acetogenesis by *Clostridium thermoaceticum. J. Bacteriol.* **165**:252–257.

Braun, M., S. Schobert, and G. Gottschalk. 1979. Enumeration of bacteria forming acetate from H₂ and CO₂ in anaerobic habitats. *Arch. Microbiol.* **120**:201–204.

Brownell, J. E., and J. P. Nakas. 1991. Bioconversion of acid-hydrolyzed poplar hemicellulose to acetic acid by *Clostridium thermoaceticum. J. Ind. Microbiol.* **7**:1–6.

Bream, P. 1988. Fermentation of single and mixed substances by the parent and an acid-tolerant, mutant strain of *Clostridium thermoaceticum. Biotech. Bioeng.* **32**:444–450.

Bream, P., and R. Datta. 1985. Production of organic acids by an improved fermentation process. US Patent 4,814,273.

Busche, R. M. 1983. Recovering chemical products from dilute fermentation broths. *Biotech. Bioeng. Symp.* **13**:597.

Busche, R. M. 1985. In: *Biotechnology Applications and Research*, P.N. Cheremisinoff and R. P. Oulette (eds.), pp. 88–102. Technicon, Lancaster, PA.

Busche, R. M. 1991. Extractive fermentation of acetic acid. *Appl. Biochem. Biotechnol.* **28/29**:605.

Chollar, B. H., (1984). Federal Highway Administration research on calcium magnesium acetate: An alternative deicer. *Public Roads* **47**:113–118.

Clark, J. E., S. W. Ragsdale, L. G. Ljungdahl, and J. Wiegel. 1982. Levels of enzymes involved in the synthesis of acetate from CO₂ in *Clostridium thermoautotrophicum. J. Bacteriol.* **151**:507–509.

Clyde, R. 1983. Horizontal stainless steel fermentor. U.S. Patent No. 4 407954.

Cunningham, D. P., and L. L. Lundie, Jr. 1993. Precipitation of cadmium by *Clostridium thermoaceticum. Appl. Environ. Microbiol.* **59**:7–14.

Eysmondt, von, J., Vasic-Racki, Dj., Wandrey, Ch. (1990) Acetic acid production by *Acetogenium kivui* in continuous culture-kinetic studies and computer simulations. *Appl. Microbiol. Biotechnol.* **34**:344–349.

Freier, D. 1981. Wechselwirkungen von thermophilen anaeroben Bakterien in cellulolytischen Mischkulturen, M.D. thesis, University of Göttingen, Göttingen, Germany.

Grethlein, A. J., and M. K. Jain. 1992. Bioprocessing of coal-derived synthesis gases by anaerobic bacteria. *TIBTECH* **10**:418–423.

Horner, R., M. Brenner, R. Wagner, and R. Walker. 1991. Environmental evaluation of calcium magnesium acetate. In: *Calcium Magnesium Acetate*, D. L. Wise, Y. A. Lavendis, and M. Metghalchi (eds.), pp. 57–102. Elsevier Science Publisher, Amsterdam.

Hsu, T., S. L. Daniel, M. F. Lux, and H. L. Drake. 1990a. Biotransformations of carboxylated aromatic compounds by the acetogen *Clostridium thermoaceticum*: generation of growth-supportive CO₂-equivalents under CO₂-limited conditions. *J. Bacteriol.* **172**:212–217.

Hsu, T., M. F. Lux, and H. L. Drake. 1990b. Expression of an aromatic-dependent decarboxylase which provides growth-essential CO_2 equivalents for the acetogenic (Wood) pathway of *Clostridium thermoaceticum*. *J. Bacteriol*. **172**:5901–5907.

Ibba, M., and G. H. Fynn. 1991. Two stage methanogenesis of glucose by *Acetogenium kivui* and acetoclastic methanogenic sp. *Biotechnol. Lett*. **13**:671–676.

Inoue, K., S. Kageyama, K. Miki, T. Morinaga, Y. Kamagata, K. Nakamura, and E. Mikami. 1992. Vitamin B_{12} production by *Acetobacterium* sp. and its tetrachloro-methane-resistant mutants. *J. Ferment. Bioeng*. **73**:76–78.

Klemps, R., S. M. Schobert, and H. Sahm. 1987. Production of acetic acid by *Acetogenium kivuii*. *Appl. Microbiol. Biotechnol*. **27**:229–234.

Koesnandar, A., N. Nishio, A. Yamamoto, and S. Nagai. 1989. Production of extracellular 5-aminolevulinic acid by *Clostridium thermoaceticum* grown in minimal medium. *Biotechnol. Lett*. **11**:567–572.

Koesnandar, A., N. Nishio, and S. Nagai. 1991a. Enzymatic reduction of cystine into cysteine by cell-free extract of *Clostridium thermoaceticum*. *J. Ferment. Bioeng*. **72**:11–14.

Koesnandar, A., N. Nishio, and S. Nagai. 1991b. Effects of trace metal ions on the growth, homoacetogenesis and corrinoid production by *Clostridium thermoaceticum*. *J. Ferment. Bioeng*. **71**:181–185.

Kuhn, E. P., J. M. Suflita, M. D. Rivera and L. Y. Young. 1989. Influence of alternate electron acceptors on the metabolic fate of hydroxybenzoate isomers in anoxic aquifer slurries. *Appl. Environ. Microbiol*. **55**:590–598.

Kuo, Y., and H. P. Gregor. 1983. Acetic acid extraction by solvent membranes. *Separation Sci. Tech*. **18**:421–440.

Ljungdahl, L. G. 1986. The autotrophic pathway of acetate synthesis in acetogenic bacteria. *Annu. Rev. Microbiol*. **40**:415–450.

Ljungdahl, L. G., H. Carreira, R. Garrison, N. Rabek, and J. Wiegel. 1985. Comparison of three thermophilic homoacetogenic bacteria from Ca-Mg acetate production. *Biotechnol. Bioeng. Symp*. **15**:207–223.

Ljungdahl, L. G., H. Carreira, R. Garrison, N. Rabek, L. F. Gunther, and J. Wiegel. 1986. CMA manufacture (II): Improved bacterial strains for acetate production. RFD# DTFH-61-83-R-00124 U.S. Department of Transportation, Federal Highway Adminis-tration, Washington, D.C.

Lowe, S. E., M. K. Jain, and J. G. Zeikus. 1993. Biology, ecology, and biotechnological applications of anaerobic bacteria adapted to environmental stresses in temperature, pH, salinity, or substrates. *Microbiol. Rev*. **57**:451–509.

Lundie, L. L., Jr., and H. L. Drake. 1984. Development of a minimal defined medium for the acetogen *Clostridium thermoaceticum*. *J. Bacteriol*. **159**:700–703.

Marynowski, C. W., J. L. Jones, R. L. Boughton, D. Tuse, J. H. Corlopassi, and J. E. Gwinn. 1983. Process development for production of calcium magnesium acetate (CMA), FHWA/RD-82/145. Federal Highway Administration, Washington, D.C.

Marynowski, C. W., J. L. Jones, D. Tuse, and R. L. Boughton. 1985. Fermentation as an advantageous route for the production of acetate salt for roadway de-icing. *I&EC Product Dev.* **24**:457–465.

Morton, T. A., C.-F. Chou, and L. G. Ljungdahl. 1992. Cloning, sequencing, and expressions of genes encoding enzymes of the autotrophic acetyl-CoA pathway in the acetogen *Clostridium thermoaceticum*. In: *Genetics and Molecular Biology of Anaerobic Bacteria*, M. Sebald (ed.), pp. 389–406. Springer-Verlag, New York.

Parekh, S., and M. Cheryan, 1990a. Fed batch fermentation of glucose to acetate by an improved strain of *Clostridium thermoaceticum*. *Biotechnol. Lett.* **12**:681–684.

Parekh, S. R., and M. Cheryan. 1990b. Acetate production from glucose by *Clostridium thermoaceticum*. *Process Biochem. Int.* August 1990:117–121.

Reed W. M., and M. E. Bogdan. 1985. Application of cell recycling to continuous fermentative acetic acic production. *Biotech. Bioeng. Symp.* **15**:641–647.

Reed, W. M., F. A. Keller, F. E. Kite, M. E. Bogdan, and J. S. Ganoung. 1987. Development of increased acetic acid tolerance in anaerobic homoacetogens through induced mutagenesis and continuous selection. *Enzyme Microb. Technol.* **9**:117–120.

Savage, M. D., and H. L. Drake. 1986. Adaptation of the acetogen *Clostridium thermoautotrophicum* to minimal medium. *J. Bacteriol.* **165**:315–318.

Schwartz, R. D., and F. A. Jr. Keller. 1982a. Isolation of a strain of *Clostridium thermoaceticum* capable of growth and acetic acid production at pH 4.5. *Appl. Env. Microbiol.* **43**:117–123.

Schwartz, R. D., and F. A. Jr. Keller. 1982b. Acetic acid production by *Clostridium thermoaceticum* in pH-controlled batch fermentation at acidic pH. *Appl. Env. Microbiol.* **43**:1385–1392.

Shimshick, E. J. 1981. Removal of organic acids from aqueous solutions of salts of organic acids by super critical fluids. U.S. Patent 4,250,331.

Siman'kova, M. V., and A. N. Nozhevnikova. 1989. Thermophilic homoacetate fermentation of cellulose in a combined culture of *Clostridium thermocellum* with *Clostridium thermoautotrophicum*. *Mikrobiologiya* **58**:897–902 (Engl. translation, pp. 723–728).

Stupperich, E., H. J. Eisinger, and B. Krutler. 1988. Diversity of corrinoids in acetogenic Bacteria. *Eur. J. Biochem.* **172**:459–464.

Tanner, R. S., L. M. Miller, and D. Yang. 1993. *Clostridium ljungdahlii* sp. nov., an acetogenic species in clostridial rRNA homology group I. *Int. J. Syst. Bacteriol.* **43**:232–236.

Terracciano, J., W. J. A. Schreur, and E. R. Kashket. 1987. Membrane H$^+$ conductance of *Clostridium thermoaceticum* and *Clostridium acetobutylicum*: Evidence for electronic Na$^+$/H$^+$ antiport in *Clostridium thermoaceticum*. *Appl. Environ. Microbiol.* **53**:782–786.

Traunecker, J., A. Preuß, and G. Diekert. 1991. Isolation and characterization of a methyl chloride utilizing, strictly anaerobic bacterium. *Arch. Microbiol.* **156**:416–421.

Unitika Ltd.-Imabori Kazutomo. 1982. Immobilized thermophilic acetate kinase. Japanese Patent JP 80109055.

Wang, G., and D. I. C. Wang. 1983. Production of acetic acid by immobilized whole cells of *Clostridium thermoaceticum*. *Appl. Biochem. Biotechnol.* **8**:491–503.

Wang, G., and D. I. C. Wang. 1984. Elucidation of growth inhibition and acetic acid production by *Clostridium thermoaceticum*. *Appl. Environ. Microbiol.* **47**:294–298.

Wiegel, J. 1990. Temperature spans for growth: A hypothesis and discussion. *FEMS Microbiol. Rev.* **75**:155–170.

Wiegel, J., M. Braun, and G. Gottschalk. 1981. *Clostridium thermoautotrophicum*. specius novum, a thermophile producing acetate from molecular hydrogen and carbon dioxide. *Curr. Microbiol.* **5**:255–260.

Wiegel, J., L. H. Carreira, Ch. P. Mothershed, L. G. Ljungdahl, and J. Puls. 1984. Formation of ethanol and acetate from biomass using thermophilic and extreme thermophilic anaerobic bacteria. In: *Proceedings 7th Int. FPRS Industrial Wood Energy Forum 83, Nashville*. Forest Product Research Society, Madison, WI.

Wiegel, J., and R. Garrison. 1985. Utilization of methanol by *Clostridium thermoaceticum*, Abstr. I 115, Annu. Meet. Am. Soc. Microbiol., Las Vegas, NV 1985.

Wiegel, J., L. H. Carreira, R. Garrison, N. E. Rabek, and L. G. Ljungdahl. 1991. Calcium magnesium acetate (CMA) manufacture from glucose by fermentation with thermophilic homoacetogenic bacteria. In: *Calcium Magnesium Acetate*, D. L. Wise, Y. A. Lavendis, and M. Metghalchi (eds.), pp. 359–418. Elsevier Science Publisher, Amsterdam.

Wise, D. L., Y. A. Levendis, and Metghalchi (eds). 1991. *Calcium Magnesium Acetate*. Elsevier Science Publisher, Amsterdam.

Wu, Z., S. L. Daniel, and H. L. Drake. 1988. Characterization of a CO-dependent O-demethylating enzyme system from the acetogen *Clostridium thermoaceticum*. *J. Bacteriol.* **170**:5705–5708.

Yang, H., and H. L. Drake. 1990. Differential effects of sodium on hydrogen- and glucose-dependent growth of the acetogenic bacterium *Acetogenium kivui*. *Appl. Environ. Microbiol.* **56**:81–86.

Yates, R. A. 1981. US Patent 4,282,323.

Zhang, X., and J. Wiegel. 1994. Distribution of hydroxybenzoate decarboxylase in clostridia and some other obligately anaerobic bacteria. (in preparation).

V

ACETOGENIC PROCESSES IN OTHER BACTERIAL GROUPS

19

Variations of the Acetyl-CoA Pathway in Diversely Related Microorganisms That Are Not Acetogens

Georg Fuchs

19.1 Introduction

The capacity of bacteria to totally synthesize acetate from CO_2 has been known for approximately 50 years (see Chapter 1, section 1.2.1). However, this process has been unnoticed by most biologists and biochemists. This may be partly due to the fact that essential steps of this process became clear only during the last 15 years (Wood and Ljungdahl, 1991; Ragsdale, 1991; Roberts et al., 1992; see Chapters 1–3 for details). In addition, initially we viewed this process somewhat narrowly relative to the organisms that make use of it. However, this pathway and variations thereof are used by diversely related microorganisms which have in common the key catalyst of the pathway, the nickel-containing enzyme CO dehydrogenase/acetyl-CoA synthase. It catalyzes not only CO oxidation to CO_2 according to the reaction $CO + H_2O \rightarrow CO_2 + 2[H]$, but also CO_2 reduction to an enzyme-bound carbonyl [CO] via the reaction $CO_2 + 2[H] \rightarrow$ enzyme-[CO] $+ H_2O$. The carbonyl-level carbon is then combined with a tetrahydrofolate-derived methyl-level carbon (via an enzyme-bound $[CH_3]$) and coenzyme A (CoA) to give acetyl-CoA. The variants of this biochemical principle are widely distributed in anaerobic *Eubacteria* and *Archaebacteria* and fulfill different central metabolic functions. Many different aspects of this pathway, termed the acetyl-CoA "Wood" pathway, have been covered in preceding chapters [see also Fuchs (1986, 1989); Ljungdahl (1986); Ragsdale (1991); Thauer (1988, 1989, 1990); Thauer et al. (1989); Wood and Ljungdahl (1991); Wood et al. (1986a, 1986b, 1986c); Diekert (1992); Drake (1992, 1993)]. In this chapter, five aspects of the acetyl-CoA pathway and its variants (Fig. 19.1) which are of general biological relevance are briefly discussed.

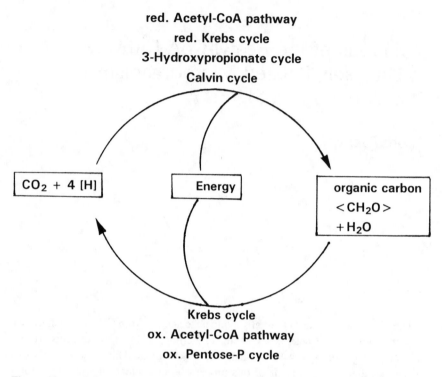

red. Acetyl-CoA pathway

red. Krebs cycle

3-Hydroxypropionate cycle

Calvin cycle

| CO_2 + 4 [H] | | Energy | | organic carbon
<CH_2O>
+H_2O |

Krebs cycle

ox. Acetyl-CoA pathway

ox. Pentose-P cycle

Figure 19.1.   Role of the oxidative and reductive acetyl-CoA pathway in the carbon cycle.

19.2   Diverse Functions of the Acetyl-CoA Pathway

The acetyl-CoA pathway serves six central biological functions, depending on the organism that employs it. These modifications are illustrated in Figure 19.2.

1.   Homoacetate Fermentation. In homoacetogenic bacteria, the reduction of 2 CO_2 with 8[H] (4H_2) to acetate is an energy-yielding metabolic reaction. When grown with CO_2 + H_2, this reductive acetyl-CoA pathway is, in fact, the only energy-yielding process. ATP is synthesized via a chemiosmotic mechanism. This topic is treated in detail in Chapters 2, 4, and 5.

2.   Autotrophic Fixation of Carbon. Many autotrophic, strictly anaerobic bacteria can use the intermediate acetyl-CoA to build up all cell constituents. In these organisms, the acetyl-CoA pathway serves as an autotrophic CO_2 fixation pathway. This applies to most acetogenic and sulfate-reducing *Eubacteria* and

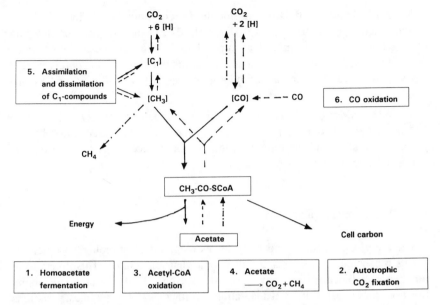

Figure 19.2.   Different metabolic functions of CO dehydrogenase/acetyl-CoA synthase and of the acetyl-CoA pathway in anaerobic bacteria.

all methanogenic and sulfate-reducing *Archaebacteria* studied so far (see Chapter 20).

3.   Acetyl-CoA Oxidation. Some strictly anaerobic bacteria oxidize acetyl-CoA to 2 CO_2 via a reversed, oxidative acetyl-CoA pathway. They can use as an energy source not only acetate but also other organic compounds that can be transformed to acetyl-CoA. In sulfate-reducing bacteria, sulfate serves as the terminal electron acceptor of acetate oxidation. Even in the complete absence of an external electron acceptor, acetate may become oxidized to 2 CO_2 and 4 H_2, provided that the end products CO_2 and H_2 are efficiently removed by syntrophic methanogens (Lee and Zinder, 1988) (see Chapters 12 and 14).

4.   Acetate Disproportionation to CH_4 and CO_2. Methanogenic bacteria that can disproportionate acetate to CO_2 and CH_4 in energy metabolism do so via acetyl-CoA cleavage. During this process, enzyme-bound intermediates are formed ([CO] and [CH_3]) similarly to the bound intermediates formed when the pathway functions in the direction of acetyl-CoA synthesis. Subsequently, the bound [CO] is oxidized to CO_2 + 2[H]. The two reducing equivalents are used to reduce [CH_3] to CH_4. This process is coupled to ATP synthesis via a chemiosmotic mechanism (see Chapter 21).

5. Assimilation and Dissimilation of C₁ compounds. Various C_1 compounds such as HCOOH, CO, CH₃OH, CH₃NH₂, or methylether can either be oxidized or used as biosynthetic precursor by taking advantage of the enzymes of the acetyl-CoA pathway. This pathway needs only peripheral supplementary enzymes to serve as a general assimilatory or degradative pathway for C_1 compounds.

6. CO Oxidation. Some phototrophic bacteria may use CO oxidation to CO_2 as an electron source for carbon fixation and/or energy source. CO dehydrogenase in these organisms resembles the enzyme in acetogens to some extent, although more information for comparison is needed.

19.3 Biochemical Variations

The acetyl-CoA pathway not only serves many different biological functions but its biochemical variations are also somewhat remarkable. This diversity in function and biochemistry may have two reasons. First, different metabolic functions may require different adaptations of enzymes and coenzymes; variations would be the result of evolutionary adaptation. Second, the microorganisms that make use of the acetyl-CoA pathway belong to evolutionarily very distant groups; variations would then reflect large phylogenetic distances. It is still unknown whether the enzymes of the acetyl-CoA pathway in different bacteria are truely homologous. Even if this were the case, it remains unknown whether the enzymes originally functioned as part of this metabolic pathway or whether they have been reorganized in a similar way from independent sources (see Section 19.6).

The largest biochemical differences in the acetyl-CoA pathway exist between the *Eubacteria* and the *Archaebacteria*. The most obvious variations in the individual groups of bacteria are the following:

1. *Pterin coenzymes*. A great variety of tetrahydropterin coenzymes function as C_1 carriers, and the nitrogen atom (N-5 or N-10) to which the C_1 compounds are bound varies.

2. *B₁₂ coenzymes*. A great variety of vitamin B_{12} coenzymes exist which function in methyl transfer reactions in these anaerobes.

3. *Electron transport*. Similarly, the electron carriers in the H_2- or CO-dependent electron transport chains vary.

4. *CO₂ reduction to the formyl level*. The mechanism by which CO_2 is reduced to the formyl level is different in *Eubacteria* and *Archaebacteria*. *Eubacteria* reduce CO_2 to free formate by soluble formate dehydrogenase, and formate becomes activated to formyl-tetrahydropterin in an ATP-dependent reaction by soluble formyltetrahydropterin synthetase.

Archaebacteria do not form free formate from CO_2 but use methano-furan as intermediary CO_2 and formyl carrier and formate is then transferred to a tetrahydropterin. The endergonic reduction of CO_2 to formyl-methanofuran requires energy input in the form of a sodium- or proton-motive force, and, consequently, the CO_2 reductase (formylmethano-furan dehydrogenase) is membrane associated.

19.4 Bioenergetic Considerations

From a physiological rather than mechanistic point of view, the most intriguing question is how acetogens and other organisms gain energy by using the acetyl-CoA pathway. Some of these organisms can live at the expense of an amazingly simple bioenergetic diet, namely, CO_2 plus H_2. Table 19.1 gives the estimated thermodynamic values of the individual reactions required to reduce 2 CO_2 plus 8[H] to acetate in homoacetogens. One ATP is required to activate formate to formyl-tetrahydrofolate (reaction 2), and 1 ATP is reformed by substrate-level phosphorylation during acetyl-CoA conversion to acetate (reactions 7 and 8). The coupling site at which net ATP is being formed is a central issue (see Chapters 2, 4, and 5). Four possibilities for the synthesis of ATP (i.e., conservation of energy) can be considered.

19.4.1 Possible Coupling Sites for Electron Transport Phosphorylation

ATP may be synthesized by electron transport phosphorylation linked to one of the four redox reactions outlined in Table 19.1. Reaction 1, CO_2 reduction to formate, is endergonic and is catalyzed by a soluble (e.g., NADPH-dependent) formate dehydrogenase. There is no evidence that formate dehydrogenase in any of the acetogens is membrane associated or energy driven; yet, the electron donor is not clearly resolved. Reactions 2–4 are catalyzed by soluble enzymes, and NADH or NADPH are the electron donors. Reaction 5 and/or reaction 6 are likely energy-conserving sites. Methylenetetrahydrofolate reductase, catalyzing reaction 5, was found to be a soluble NADH-dependent enzyme in some acetogenic bacteria; in others, the electron donor is not exactly known, and the cellular localization of the reductase has not been unambiguously determined. Therefore, in those acetogens containing a soluble, NADH-dependent methylenetetrahydrofolate reductase, reaction 5 appears to not be a coupling site. In others, this possibility cannot be excluded.

Growth as well as acetogenesis on $H_2 + CO_2$ is Na^+ dependent for several (but not all) acetogens (Heise et al., 1992), and the reduction of formaldehyde to the formal redox level of methanol is the sodium-dependent step; it is not known whether reaction 5 or 6 or both are the Na^+-translocating steps. Reaction

Table 19.1 Estimated thermodynamic values of reactions of the acetyl-CoA pathway in acetogenic bacteria using tetrahydrofolate (FH_4) as a one-carbon carrier. Reaction 6 is considered the sum of partial reactions 6a + 6b. However, note that CO (g) is not the actual intermediate but an enzyme-bound nickel carbonyl.

Reaction	$\Delta G_0'$ if 2[H] = H_2 (kJ/mol)	$\Delta G_0'$ if 2[H] = (NADH) (kJ/mol)	E_0' (mV)
(1) CO_2(g)+2[H] → $HCOO^-$+H^+	+3.4	+21.5	−421
(2) $HCOO^-$+FH_4+ATP → formyl-FH_4+ADP+P_i	−8.4	−8.4	
(3) Formyl-FH_4+H^+ → methenyl-FH_4^++H_2O	+5.9	+5.9	
(4) Methenyl-FH_4^++2[H] → methylene-FH_4+H^+	−23.0	−4.9	−295
(5) Methylene-FH_4+2[H] → methyl-FH_4	−40.2	−22.1	−200
(6) Methyl-FH_4+CO_2(g)+2[H]+CoASH → $CH_3COSCoA$+FH_4+H_2O	−28.7[a]	−10.6[a]	−265[a]
(6a) CO_2(g)+2[H] → CO(g)+H_2O	+20.1	+38.2	−524
(6b) Methyl-FH_4+CO(g)+CoASH → CH_3CO-SCoA+FH_4	−48.8[a]	−48.8[a]	
(7) CH_3CO-SCoA+P_i → CH_3CO-P_i+CoASH	+9.1	+9.1	
(8) CH_3CO-P_i+ADP → CH_3COO^-+ATP	−13.0	−13.0	
(9) $2CO_2$(g)+8[H] → CH_3COO^-+H^++$2H_2O$	−94.9	−22.5	−290
(10) $2CO_2$(g)+8[H]+CoASH → $CH_3COSCoA$+$3H_2O$	−59.2	+13.2	−337

[a] Calculated from the other thermodynamic values. Unless indicated as (g) = gaseous state, all compounds are in the aqueous state.

512

6 is catalyzed by a complex enzyme system that is likely associated with the cytoplasmic membrane. However, CO_2 reduction to the carbonyl level (reaction 6a) is endergonic, and the reverse reaction, $CO + H_2O \rightarrow + 2[H]$, has been shown to be coupled to the translocation of $2H^+/CO$ in methanogens; this probably also applies to the oxidation of the enzyme-bound [CO]. The reverse reaction, therefore, should be driven by a proton- or sodium-motive force.

If the estimated thermodynamic values of the other reactions are correct, reaction 6b is quite exergonic. Part of its energy has to be used to drive reaction 6a, the endergonic CO_2 reduction to the carbonyl level, by an electrochemical coupling mechanism; the simpliest explanation is that reaction 6b drives the export of 3 or 4 H^+/reaction, 2 H^+ of which drive CO_2 reduction. It is not known which catalytic partial step of the complex reaction 6b is the coupling site. The additional 1–2 H^+ translocated may be used by H^+-ATP synthase. Na^+ may substitute for H^+ in some acetogens, and a Na^+-ATP synthase may be operating; a proton- as well as a sodium-motive force are then required to couple CO_2 reduction to acetate to the synthesis of ATP.

19.4.2 Product Export

Export of the fermentation product acetate by a symport mechanism carrying 1 H^+ out per 1 acetate anion is another possible way to contribute to the electrochemical proton potential. Acetate would be used up by other bacteria, e.g., methanogens. However, no evidence is available in support of this possibility.

19.4.3 Methyltransfer

The methyl transfer reaction from CH_3-tetrahydropterin may also vary. In an analogous reaction, methanogens transfer the methyl from CH_3-tetrahydromethanopterin to coenzyme M by a membrane-bound corrinoid protein complex; the free-energy change of that methyl transfer reaction is probably -28 kJ/mol and the methyltransferase is considered to be a coupling site for translocating Na^+ (Becher et al., 1992; Gärtner et al., 1993). One may speculate that the methyl transfer from CH_3-tetrahydrofolate via a corrinoid protein to CO dehydrogenase/acetyl-CoA synthase is coupled similarly to the translocation of Na^+ in *Acetobacterium woodii*. *A. woodii* has a membrane-bound corrinoid, whereas *Clostridium thermoaceticum* does not contain corrinoids in the membrane fraction.

19.4.4 H₂ Oxidation

All reduction reactions of the acetogenic pathway depend on the oxidation of H_2 by hydrogenase(s) when H_2 is the reductant. It is conceivable that H_2 oxidation and concomitant reduction of one of the electron-carrying cofactors is the primary coupling site creating a proton-motive force. This would implicate that hydrogenase oxidizes H_2 to $2 H^+$ at the outer side of the cytoplamic membrane. Electron transport from H_2 to the electron-carrying cofactors still needs to be studied in acetogens. The electron transport chains vary considerably in acetogens; some contain cytochromes and quinones (e.g., *C. thermoaceticum*), whereas others do not (e.g., *A. woodii*).

Hence, bioenergetic mechanisms for the conservation of energy are not likely uniform in acetogens. In this regard, we do not yet have a satisfying answer to the important bioenergetic question: How do these bacteria make their living on CO_2 + H_2? This also applies to the energetic aspects of the variants of the acetyl-CoA pathway in nonacetogenic bacteria.

19.5   Acetyl-CoA Pathway Versus Citric Acid Cycle: A Choice

Another aspect of the acetyl-CoA pathway is the fact that the citric acid cycle may fulfill similar functions. Some anaerobes reduce CO_2 to acetyl-CoA in autotrophic cell carbon fixation from CO_2 by using the reductive acetyl-CoA pathway; others use the reductive citric acid cycle for the same purpose. Vice versa, some bacteria oxidize acetyl-CoA to CO_2 via the oxidative acetyl-CoA pathway, others via an oxidative citric acid cycle that exhibits some remarkable modifications. Hence, there is a choice between the two pathways. What makes the difference? What factors determine which pathway is engaged?

Each of the pathways may operate in the same organism either in the reduction or oxidation direction, depending on whether CO_2 has to be assimilated or acetyl-CoA to be oxidized. Even closely related bacterial species belonging to the same subgroup of proteobacteria use either the citric acid cycle or the acetyl-CoA pathway; for instance, *Desulfobacterium* species use the acetyl-CoA pathway, whereas *Desulfobacter* species use the modified citric acid cycle. What is the rational behind this choice?

19.5.1   Physiology of the Bacteria

The most obvious physiological difference between *Desulfobacter* and *Desulfobacterium* species is the preferential utilization of acetate by *Desulfobacter*, whereas *Desulfobacterium* utilizes preferentially higher fatty acids. *Desulfobacter* species are defined by their ability to grow well on acetate and by their limited (or no) ability to use other organic substrates. *Desulfobacterium* species possess a more versatile nutritional capability, particularly for the utilization formate, fumarate, malate, butyrate, and higher carbon-chain fatty acids. Still, *Desulfobacterium* and *Desulfobacter* species have one problem in common: They need to strictly regulate the direction and flux of the acetyl-CoA pathway and modified citric acid cycle, respectively. It is unknown whether one or two sets of enzymes are responsible for acetate oxidation and CO_2 fixation. This comparison apparently does not give a clear answer to our question.

19.5.2 Utilization of One-Carbon Compounds

It is plausible, however, that the acetyl-CoA pathway is preferred by those anaerobes which can utilize reduced C_1 compounds as a carbon and/or energy source, such as CH_3OH, CH_2O, $HCOOH$, CO, CH_3NH_2, and methyl esters; the pathway may even serve as a sink for CO_2 derived from decarboxylations, e.g., of phenolic acids (Hsu et al., 1990). Here again, however, we do not know why there is selective usage of one process over the other.

19.5.3 Average Redox Potential of Oxidoreductases

Another reason for the choice of one of the two pathways may be the different redox potentials of the oxidoreductase reactions (Thauer, 1988, 1989, 1990; Thauer et al., 1989); for thermodynamic values, see Thauer et al. (1977); Fuchs (1986); Wohlfarth and Diekert (1991). The average redox potentials of the redox reactions of the two pathways have to be compared with the CO_2/acetate couple ($E_0' = -291$ mV). From Table 19.2, it can be seen that the average redox potential of the redox couples of the citric acid cycle (-247 mV) is less negative than that of the acetyl-CoA pathway (-360 mV) and that E_0' of the succinate/fumarate couple ($+32$ mV) is even positive; this requires either reversed electron transport to allow succinate oxidation or a rather positive terminal electron acceptor for acetyl-CoA being oxidized via succinate/fumarate.

19.5.4 Redox Potential for Operation of Key Enzymes

The key enzyme of the acetyl-CoA pathway, CO dehydrogenase/acetyl-CoA synthase, may only function if the redox potential in the cell is very low. This would be prohibited if the terminal electron acceptor is a strong oxidant, e.g., the Fe^{3+}/Fe^{2+} couple ($E_0' = +640$ mV), the Mn^{4+}/Mn^{2+} couple ($E_0' = +650$ mV), or the NO_3^-/NH_4^+ couple ($E_0' = +364$ mV). In this regard, it is interesting that recent results show that nitrate dissimiation to nitrite and ammonium is a preferred electron sink for *C. thermoaceticum* and that in the presence of nitrate, the acetyl-CoA pathway is not functional (Seifritz et al., 1993) (see chapter 10, section 10.2.5).

Table 19.2   Estimated redox potentials of intermediates involved in acetate oxidation to CO_2 [from Thauer (1988), modified]. FH_4 = tetrahydrofolate, MpH_4 = tetrahydromethanopterin, Mf = Methanofuran

Pathway	Redox Couple	E_o'(mV)	Average E_o'(mV)
Citric acid cycle	2-Oxoglutarate+CO_2/ isocitrate	−364	
	Succinyl-CoA+CO_2/ 2-oxoglutarate	−491	−247
	Fumarate/succinate	+32	
	Oxaloacetate/malate	−166	
Acetyl-CoA/CO-dehydrogenase pathway	Methylene-FH_4/ methyl-FH_4	−200	
In acetogens	Methenyl-FH_4/ methylene-FH_4	−295	−360
	CO_2/formate	−421	
	CO_2/CO	−524	
In methanogens	Methylene-MpH_4/ methyl-MpH_4	−323	
	Methenyl-MpH_4/ methylene-MpH_4	−386	−432
	CO_2/formyl-Mf	−496	
	CO_2/CO	−524	
Overall reactions	2CO_2/acetyl-CoA		−337
	2 CO_2/acetate		−290

19.5.5 Net ATP Synthesis

CO_2 reduction to acetate via the reductive citric acid cycle does not allow the net synthesis of ATP. Even if fumarate reduction would be coupled to the synthesis of 1 ATP, this ATP would be required for ATP citrate lyase, catalyzing citrate cleavage to oxaloacetate and acetyl-CoA (acetyl-CoA could, in turn, serve as CoA donor for succinyl-CoA formation from succinate by CoA transfer). The more negative level of the E_0' values of the acetyl-CoA pathway also suggests that only the acetyl-CoA pathway may serve in acetate oxidation to CO_2 + H_2 when the products H_2 and CO_2 are continuously removed in syntrophic bacterial communities. This reaction even allows some organisms to live at the expense of acetate oxidation in the absence of an external electron acceptor (see Chapters 14 and 22).

19.5.6 Oxygen Sensitivity of Key Enzymes

It appears that the acetyl-CoA pathway is confined to strict anaerobes. This may be due to the oxygen sensitivity of some of its enzymes, notably CO dehydrogenase/acetyl-CoA synthase. 2-Oxo acid synthases like pyruvate synthase are also oxygen sensitive; however, they are required for acetyl-CoA assimilation in both pathways. 2-Oxo acid synthases operate even in microaerophilic thermophilic Knallgas bacteria that oxidize H_2 with O_2 and fix CO_2 via the reductive citric acid cycle. This indicates that the enzymes of the modified (reductive or oxidative) citric acid cycle may be less oxygen sensitive than previously thought.

19.6 Historic Relationships and Open Questions

Obviously, both the modified citric acid cycle and the acetyl-CoA pathway may serve two fundamental functions: CO_2 fixation and complete oxidation of organic matter degradable to acetate. Here, only the biosynthetic aspect is considered. The fact that both pathways are widely distributed in anaerobic *Eubacteria* and *Archaebacteria* does not provide an explanation of how a given pathway may have evolved from more primitive "bioorganic" elements and which pathway is more universal. A screening of eucaryotic microorganisms for CO dehydrogenase activity has not yet been undertaken.

19.6.1 Autoorigin of life

The evolution of carbon fixation is of central importance for an autoorigin of life. Wächtershäuser (1992) has made an attempt to retrodict the first autocatalytic carbon fixation cycle. The acetyl-CoA pathway is not a catalytic cycle *sensu strictu*; rather it is a linear pathway making use of cycling of one-carbon-carrying coenzymes. The reductive acetyl-CoA pathway is seen as an anaplerotic pathway for an archaic carbon-fixation cycle which resembles much the reductive citric acid cycle. The citric acid cycle has a central biosynthetic function, in contrast to the acetyl-CoA pathway which—besides using biosynthetic one-carbon units— has no direct link to the origin of cellular constituents. In brief, the acetyl-CoA pathway appears to be a specialized way of living at the expense of C_1 compounds, H_2, and CO_2.

19.6.2 The Acetyl-CoA Pathway, a Universal Pathway?

The distribution of the acetyl-CoA pathway in extant organisms rises another question: Can we consider it a universal pathway? Universality may be either due to antiquity or due to lateral transfer. Vice versa, nonuniversality may be due to lateness of arrival or due to other reasons, such as noneven distribution of pathway takeover, pathway abandonment, pathway reversal, and others. The acetyl-CoA pathway is seen as radiating by pathway abandonments and pathway reversals. This occurs partly for reasons of streamlining and control and partly for reasons of nutritional opportunism. This part of metabolic evolution is closely linked to the parallel evolution of the diversity of coenzymes (Wächtershäuser 1992).

19.6.3 Are the Variants of the Acetyl-CoA Pathway Homologous?

More data are needed to shed light into the important question of whether the variants of the pathway in the different groups are homologous, i.e., have common routes, and what the primordial acetyl-CoA pathway may have looked like. The genes for the enzyme CO dehydrogenase/acetyl-CoA synthase in clostridia and methanogens are only very distantly related, if at all (Morton et al., 1991). It is obvious that the acetyl-CoA pathway used by methanogenic *Archaebacteria* for autotrophic CO_2 fixation, acetate disproportionation, and utilization of reduced C_1 compounds as carbon and energy sources is quite different from that used in the pathway by acetogenic *Eubacteria*.

Nonetheless, the mechanisms for the reduction of CO_2 to acetate by acetogenic bacteria and to methane by methanogenic bacteria clearly have much in common. However, there are two striking dissimilar sets of features which cannot be explained at present. One group of acetogens contains membrane-associated methylene tetrahydropterin reductase and cytochromes, has no corrinoid protein in the membrane, and uses protons as the energetic coupling agent. The other group contains soluble methylene tetrahydropterin reductase and no cytochromes, has corrinoid protein in the membrane, and uses both protons as well as sodium ions as energetic coupling agents.

References

Becher, B., V. Müller, and G. Gottschalk. 1992. The methyltetrahydromethanopterin : coenzyme M methyltransferase of *Methanosarcina* strain Göl is a primary sodium pump. *FEMS Microbiol. Lett.* **91**:239–244.

Diekert, G. 1992. The acetogenic bacteria. In *The Prokaryotes*, 2nd ed., A. Balows, H. G. Trüper, M. Dworkin, W. Harder, K.-H. Schleifer (eds.), pp. 517–533. Springer-Verlag, New York.

Drake, H. L. 1992. Acetogenesis and acetogenic bacteria. In: *Encyclopedia of Microbiology*, Vol. 1, J. Lederberg (ed.), pp. 1–15. Academic Press, San Diego, CA.

Drake, H. L. 1993. CO_2, reductant, and the autrophic acetyl-CoA pathway: alternative origins and destinations. In: *Microbial Growth on C_1 Compounds,* C. Murrell and D. P. Kelly (eds.), pp. 493–507. Intercept Ltd., Andover, England.

Fuchs, G. 1986. CO₂ fixation in acetogenic bacteria: variations on a theme. *FEMS Microbiol. Rev.* **39**:181–213.

Fuchs, G. 1989. Alternative pathways of autotrophic CO_2 fixation. In: *Autotrophic Bacteria*, H.G. Schlegel and B. Bowien, (eds.), pp. 365–382. Science Tech Publishers, Madison, WI.

Gärtner, P., A. Ecker, R. Fischer, D. Linder, G. Fuchs, and R. K. Thauer. 1993. Purification and properties of N^5-methyltetrahydromethanopterin : coenzyme M methyltransferase from *Methanobacterium thermoautotrophicum. Eur. J. Biochem.* **213**:537–545.

Heise, R., V. Müller, and G. Gottschalk. 1992. Presence of a sodium-translocating ATPase in membrane vesicles of the homoacetogenic bacterium *Acetobacterium woodii. Eur. J. Biochem.* **206**:553–557.

Hsu, T., M. F. Lux, and H. L. Drake. 1990. Expression of an aromatic-dependent decarboxylase which provides growth essential CO_2 equivalents for the acetogenic (Wood) pathway of *Clostridium thermoaceticum. J. Bacteriol.* **172**:5901–5907.

Lee, M. J., and S. H. Zinder. 1988. Hydrogen partial pressures in a thermophilic acetate-oxidizing methanogenic coculture. *Appl. Environ. Microbiol.* **54**:1457–1461.

Ljungdahl, L. G. 1986. The autotrophic pathway of acetate synthesis in acetogenic bacteria. *Annu. Rev. Microbiol.* **40**:415–450.

Morton, T. A., J. A. Runquist, S. W. Ragsdale, T. Shanmugasundaram, H. G. Wood, and L. G. Ljungdahl. 1991. The primary structure of the subunits of carbon monoxide dehydrogenase/acetyl-CoA synthase from *Clostridium thermoaceticum. J. Biol. Chem.* **266**:23824–23828.

Roberts, J. R., W.-P. Lu, and S. W. Ragsdale. 1992. Acetyl-coenzyme A synthesis from methyltetrahydrofolate, CO, and coenzyme A by enzymes purified from *Clostridium thermoaceticum*: attainment of in vivo rates and identification of rate-limiting steps. *J. Bacteriol.* **174**:4667–4676.

Ragdsdale, S. W. 1991. Enzymology of the acetyl-CoA pathway of CO_2 fixation. *Crit. Rev. Biochem. Mol. Biol.* **26**:261–300.

Seifritz, C., S. L. Daniel, A. Gößner, and H. L. Drake. 1993. Nitrate as a preferred electron sink for the acetogen *Clostridium thermoaceticum. J. Bacteriol.* **175**:8008–8013.

Thauer, R. K. 1988. Citric acid cycle, 50 years on: Modification and an alternative pathway in anaerobic bacteria. *Eur. J. Biochem.* **176**:497–508.

Thauer, R. K. 1989. Energy metabolism of sulfate reducing bacteria. In: *Autotrophic Bacteria*, H. G. Schlegel (ed.), pp. 397–413. Science Tech Publishers, Madison, WI.

Thauer, R. K. 1990. Energy metabolism of methanogenic bacteria. *Biochim. Biophys. Acta* **1018**:256–259.

Thauer, R. K., K. Jungermann, and K. Decker. 1977. Energy conservation in chemotrophic anaerobic bacteria. *Bacteriol. Rev.* **41**:100–180.

Thauer, R. K., D. Möller-Zinkhan, and A. Spormann. 1989. Biochemistry of acetate catabolism in anaerobic chemotrophic bacteria. *Annu. Rev. Microbiol.* **43**:43–67.

Wächtershäuser, G. 1992. Groundworks for an evolutionary biochemistry; The iron-sulphur world. *Prog. Biophys. Molec. Biol.* **58**:85–201.

Wohlfarth, G. and G. Diekert. 1991. Thermodynamics of methylenetetrahydrofolate reduction to methyltetrahydrofolate and its implications for the energy metabolism of homoacetogenic bacteria. *Arch. Microbiol.* **155**:378–381.

Wood, H. G., and L. G. Ljungdahl. 1991. Autotrophic character of the acetogenic bacteria. In: *Variations in Autotrophic Life*, J. M. Shively and L. L. Barton (eds.), pp. 201–250. Academic Press, New York.

Wood, H. G., S. W. Ragsdale, and E. Pezacka. 1986a. CO_2 fixation into acetyl-CoA: a new pathway of autotrophic growth. *Trends Biochem. Sci.* **11**:14–18.

Wood, H. G., S. W. Ragsdale, and E. Pezacka. 1986b. The acetyl-CoA pathway of autotrophic growth. *FEMS Microbiol. Rev.* **39**:345–362.

Wood, H. G., S. W. Ragsdale, and E. Pezacka. 1986c. The discovery of a new pathway of autotrophic growth using carbon monoxide or carbon dioxide and hydrogen. *Biochem. Int.* **12**:421–440.

20

Autotrophic Acetyl Coenzyme A Biosynthesis in Methanogens

William B. Whitman

20.1 Introduction: Early Studies

Demonstration of the Ljungdahl–Wood pathway of autotrophic acetyl-CoA biosynthesis in methanogens was a difficult task complicated by the unusual coenzymes of methanogenesis and the lack of detailed knowledge for the acetyl-CoA synthase at the time the work was initiated. Because methanogens provided the first evidence for the widespread distribution of the pathway outside the homoacetogenic clostridia, this work was of special importance. Its eventual success was dependent on the description of the C-1 carriers in the pathway of CO_2 reduction to methane and the clostridial acetyl-CoA synthase.

Early attempts to demonstrate the pathway of CO_2 fixation conceptualized the problem as a choice between two alternatives [Fig. 20.1, Taylor et al. (1976)]. The first alternative hypothesized that reduced carbon intermediates of methanogenesis were utilized for cellular carbon biosynthesis. Based on the current biochemical understanding, this model embraced the Ljungdahl–Wood pathway as well as the formaldehyde assimilation pathways of the aerobic methane-oxidizing bacteria. However, early evidence appeared to eliminate both of these pathways.

The Ljungdahl–Wood pathway was counterindicated because cells incorporated formate poorly and the distribution of $H^{14}COOH$ within amino acids and purines implied a limited role for formyltetrahydrofolate in carbon assimilation (Taylor et al., 1976). In addition, autotrophic methanogens contained only very low levels of formyltetrahydrofolate synthetase and methylenetetrahydrofolate dehydrogenase (Ferry et al., 1976). Similarly, the pathways of formaldehyde assimilation were also eliminated. Compared to $^{14}CO_2$, whole cells of *Methano-*

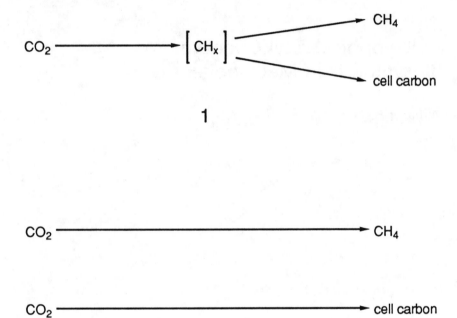

Figure 20.1. Alternatives for autotrophic CO_2 fixation by methane-producing bacteria. In alternative 1, anabolic CO_2 fixation shares intermediates with the pathway of CO_2 reduction to methane. In alternative 2, CO_2 fixation for biosynthesis is an independent process from CO_2 reduction to methane.

bacterium thermoautotrophicum incorporated $^{14}CH_2O$ with specific enrichments of histidine and purines. Thus, free formaldehyde was unlikely to be an early intermediate in CO_2 assimilation (Taylor et al., 1976). Key enzymes of the serine and hexulose phosphate pathways of formaldehyde assimilation in methylotrophic bacteria, hydroxypyruvate reductase and hexulose phosphate synthase, were also undetectable (Taylor et al., 1976; Weimer and Zeikus, 1978). In addition, intermediates of these pathways were not detected among the short-term labeling products for $^{14}CO_2$ fixation in *M. thermoautotrophicum* (Daniels and Zeikus, 1978).

The second alternative hypothesized that CO_2 fixation was a parallel process with methanogenesis and that the two pathways did not share intermediates (Fig.

20.1). Possibilities in this category included the reductive pentose phosphate cycle and the reductive tricarboxylic acid cycle. Several lines of evidence eliminated the reductive pentose phosphate cycle. First, radiolabeled acetate was incorporated into amino acids with a distribution similar to that of $^{14}CO_2$, which suggested that acetate or another C-2 compound was an early intermediate in CO_2 fixation (Taylor et al., 1976). Therefore, a C-3 sugar was an unlikely first product of CO_2 fixation. Likewise, 3-phosphoglycerate, the predominant short-term fixation product of the reductive pentose cycle, was not a detectable early product of CO_2 assimilation in whole cells of *Methanobacterium thermoautotrophicum*, *Methanobrevibacter ruminantium*, and *Methanosarcina barkeri* (Daniels and Zeikus, 1978). The key enzyme of this pathway, ribulose bisphosphate carboxylase, was also not detectable in cell extracts of *M. thermoautotrophicum*, *M. barkeri*, or *Methanococcus maripaludis* (Daniels and Zeikus, 1978; Taylor et al., 1976; Weimer and Zeikus, 1978; Yu and Whitman, unpublished data).

The reductive tricarboxylic acid cycle was also eliminated. Although many of the enzymes required for a reductive tricarboxylic acid cycle were demonstrated in *M. thermoautotrophicum*, isocitrate dehydrogenase and citrate lyase were undetectable (Zeikus et al., 1977). Isocitrate dehydrogenase was also absent from extracts of *M. maripaludis* (Shieh and Whitman, 1987). Likewise, radiolabeled succinate and fumarate were specifically incorporated into glutamate and not alanine and aspartate by whole cells of *M. thermoautotrophicum* (Fuchs and Stupperich, 1978; Fuchs et al., 1978a, 1978b). Therefore, these reduced C-4 compounds were not precursors of pyruvate and oxaloacetate, as expected if the reductive tricarboxylic acid cycle was operative. Instead, *Methanobacterium* utilized an incomplete cycle where oxaloacetate was reduced to 2-ketoglutarate for amino acid biosynthesis.

Although these early studies seemed to exclude all the known pathways of autotrophy, new information required that these conclusions be reevaluated. First, from studies on the reduction of CO_2 to methane, it became clear that methanogens contained an abundant C-1 carrier similar to folate called tetrahydromethanopterin (Fig. 20.2) and that free formate was not an intermediate in the pathway of methanogenesis [Fig. 20.3; for reviews see Wolfe (1985); Rouviere and Wolfe (1988); DiMarco et al. (1990)]. Thus, early experiments which appeared to eliminate the Ljungdahl-Wood pathway were inconclusive. The absence of key folate-dependent enzymes could be explained if the reactions were catalyzed by the analogous tetrahydromethanopterin-dependent enzymes (Fig. 20.3). In addition, the failure of formate to uniformly label autotrophically grown cells did not mean that a C-1 cycle was absent. Second, carbon monoxide dehydrogenase activity was demonstrated in the homoacetogenic clostridia, and this activity was shown to be associated with the highly purified acetyl-CoA synthase (Diekert and Thauer, 1978; Drake et al., 1981). [Because later authors

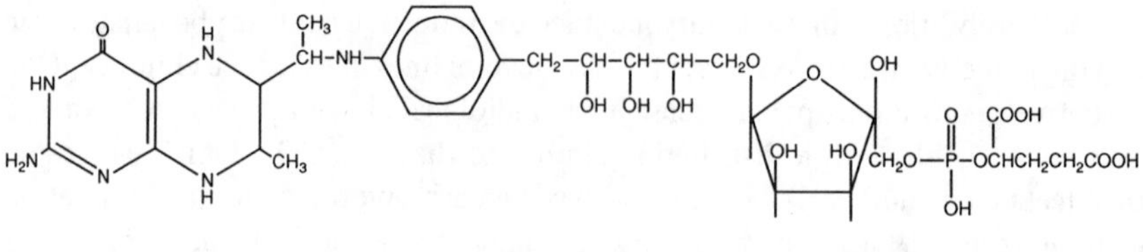

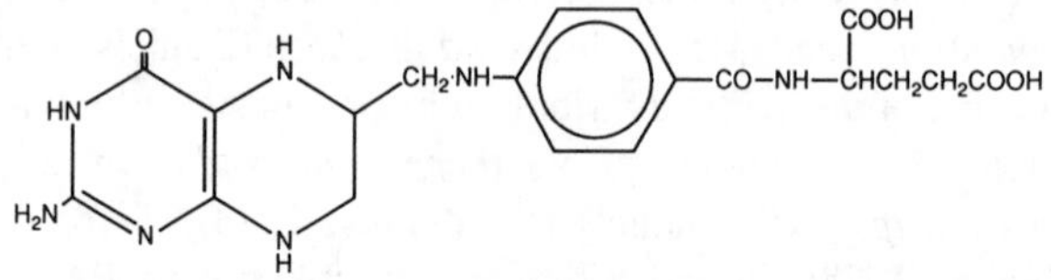

Tetrahydromethanopterin

Tetrahydrofolate glutamate

Figure 20.2.   Comparison of tetrahydromethanopterin with tetrahydrofolate glutamate. Tetrahydromethanopterin is a C-1 carrier in methanogenesis at the formyl, methenyl, methylene, and methyl levels of reduction.

have demonstrated that carbon monoxide dehydrogenase activity is only one reaction catalyzed by the acetyl-CoA synthase, the latter name is used for this enzyme (Ragsdale and Wood, 1985). Although the name "carbon monoxide dehydrogenase" is still in common usage for the same enzyme, it was avoided in preparing this review. Instead, "carbon monoxide dehydrogenase activity" was used to refer to reactions involving CO catalyzed by the acetyl-CoA synthase.] These observations were significant because a similar activity was well known in the methanogens (Kluyver and Schnellen, 1947; Daniels et al., 1977). Although circumstantial, these observations implied that the terminal reactions of the Ljungdahl–Wood pathway and autotrophic CO_2 fixation in methanogens might be the same.

20.2   Demonstration That Acetyl-CoA Was the First Product of CO_2 Fixation

Although early studies implied that acetate or acetyl-CoA was an early product of CO_2 fixation (Taylor et al., 1976), they were not conclusive. Confirmation was provided by examining the long-term incorporation of radiolabeled acetate and pyruvate into *M. thermoautotrophicum* (Fuchs and Stupperich, 1980). In these studies, [3-^{14}C]pyruvate equally radiolabeled alanine, aspartate, and glutamate, as expected from the incomplete reductive tricarboxylic acid cycle pre-

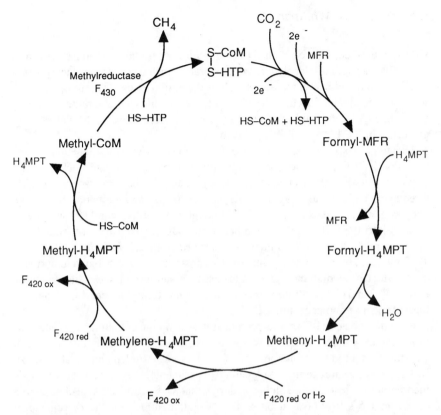

Figure 20.3. Pathway of CO_2 reduction to methane. In the initial reaction, CO_2 is reduced to formylmethanofuran (formyl-MFR). In cell extracts, this reaction requires the heterodisulfide of coenzyme M (HS-CoM) and 7-mercaptoheptanoylthreonine phosphate (HS-HTP). The heterodisulfide is also reduced during the reduction of CO_2. Thus, this step requires four electrons, ultimately derived from H_2. The formyl group is then transfered from methanofuran to tetrahydromethanopterin (H_4MPT) to form formyltetrahydromethanopterin (formyl-H_4 MPT). This reaction is followed by a dehydration to form methenyltetrahydromethanopterin (methenyl-H_4MPT) and a reduction to form methylenetetrahydromethanopterin (methylene-H_4MPT). Two enzyme systems are capable of this reduction. One is H_2 dependent. The other is dependent on the reduced deazaflavin coenzyme F_{420}. In a second coenzyme F_{420}-dependent reaction, methylene-H_4MPT is reduced to methyltetrahydromethanopterin (methyl-H_4MPT). The methyl group is then transfered to coenzyme M to form methyl coenzyme M (methyl-CoM). Methyl coenzyme M is reduced by the methylreductase system, which contains the nickel tetrapyrrole coenzyme F_{430}. HS-HTP is the reductant for this reaction, which regenerates the heterodisulfide. The requirement for the heterodisulfide for the initial reduction of CO_2 is the basis of the stimulation of CO_2 reduction to methane by methyl coenzyme M in cell extracts.

viously identified. However, the N-acetyl moiety of pseudomurein isolated from cell walls was not radiolabeled. In contrast, radiolabeled acetate was incorporated equally into the N-acetyl moiety as well as the amino acids. Therefore, pyruvate was not a precursor of acetate. At this point, the only biosynthetic route remaining for acetate (or acetyl-CoA) was from the condensation of two molecules of CO_2, as occurs in the Ljungdahl–Wood pathway.

This hypothesis was examined specifically. In short-term radiolabeling experiments with $^{14}CO_2$, alanine was a major product (Stupperich and Fuchs, 1981). Previous work had shown that alanine was probably formed from acetyl-CoA via a reductive carboxylation by pyruvate oxidoreductase and amination by alanine dehydrogenase (Zeikus et al., 1977). Although most of the radiolabel was incorporated into the C-1 position, significant radiolabel was also incorporated into the C-2 and C-3 positions, which were derived from acetyl-CoA (Table 20.1). The high rates of incorporation into the C-1 position can be attributed to pyruvate oxidoreductase, which catalyzes an exchange reaction with CO_2. Incorporation into the C-2 and C-3 positions was consistent with autotrophic acetyl-CoA biosynthesis from two molecules of CO_2.

Even though acetyl-CoA was present at very low concentrations in whole cells of *M. thermoautotrophicum*, it was also possible to demonstrate directly that it was the first product of autotrophic CO_2 fixation (Rühlemann et al., 1985). In Calvin-type short-term radiolabeling experiments, the percentage of $^{14}CO_2$ incorporated into ethanol-soluble material that was acetyl-CoA declined from 0.63% at 1.5 s to 0.06% at 60 s. The "negative slope" for the percentage of radiolabel in acetyl-CoA is diagnostic for the first product of CO_2 fixation. The low percentage of radiolabel in acetyl-CoA was explained by the presence of other anabolic carboxylations like the reductive carboxylations of acetyl-CoA and succinyl-CoA to form pyruvate and 2-ketoglutarate, exchange reactions with CO_2, and incorporation into coenzymes in the pathway of methanogenesis.

A comparison with alanine illustrated these points. Alanine typically contained 15–45% of the radiolabeled incorporated in short-term labeling experiments

Table 20.1   Distribution of radiolabel in alanine after short-term labeling of *M. thermoautotrophicum* with $^{14}CO_2$

Length of Time of Radiolabeling(s)	% of ^{14}C in Alanine	DPM ($\times 10^3$) in alanine at carbon position:		
		1	2	3
2	15	9.6	2.3	1.9
8	22	43.6	10.3	7.0
38	24	73.1	16.3	14.7

Source: From Stupperich and Fuchs (1981).

(Daniels and Zeikus, 1978; Stupperich and Fuchs, 1981). The highest radiolabeling was obtained under H_2-limited growth conditions, where the contribution of exchange reactions may have been higher compared to net biosynthesis (Daniels and Zeikus, 1978). Even under conditions of exponential growth, where H_2 concentrations presumably did not limit growth, more than half of the radiolabel incorporated was in the C-1 position (Table 20.1). The sources for this radiolabeling were expected to be pyruvate biosynthesis from acetyl-CoA and CO_2 or an exchange reaction between the C-1 pyruvate and CO_2, both reactions being catalyzed by pyruvate oxidoreductase. Because the pool size for acetyl-CoA was very small and the carbonyl of acetyl-CoA was obtained directly from CO_2 without dilution in an intracellular pool, the radiolabel in the C-1 of alanine obtained from net biosynthesis was expected to be similar to the amount of radiolabel in the C-2 position of alanine. Because it greatly exceeded this amount, a large fraction of the radiolabeled incorporated was due to exchange reactions involving the C-1 position of alanine. In addition, the pool size of alanine was large, 6.4 mM, compared to the pool size for acetyl-CoA, 40–50 μM (Rühlemann et al., 1985). Within the first minute, both pools are fully radiolabeled, which resulted in a 100-fold greater radiolabeling of alanine compared to acetyl-CoA. However, the "slope" of the percent radiolabel in alanine was positive during the initial 20 s, indicating that alanine was not the first product of CO_2 fixation.

20.3 Carbon Monoxide Assimilation and Production by Whole Cells

To substantiate the role of "carbon monoxide dehydrogenase" in *M. thermoautotrophicum*, CO assimilation was examined (Stupperich et al., 1983). Because free CO is incorporated into the carbonyl of acetyl-CoA by the clostridial acetyl-CoA synthase, the incorporation of ^{14}CO into cellular components derived from the C-1 of acetyl-CoA would be strong evidence for the importance of this enzyme in methanogens. During autotrophic growth on $H_2 + CO_2$, ^{14}CO was incorporated into cellular protein, lipid, and nucleic acid, as expected if it were incorporated into an early intermediate of CO_2 fixation. An equal amount of ^{14}CO was also oxidized to $^{14}CO_2$. Because of the large pool size for CO_2, the specific radioactivity for $^{14}CO_2$ was less than 0.5% of the starting ^{14}CO. In contrast, the specific radioactivity of the alanine purified from protein hydrolysates was 15% of that of the initial ^{14}CO, which demonstrated that the radiolabel was not incorporated via CO_2. Importantly, 97% of the radiolabel in alanine was in the C-2 carbon, which was derived from the carbonyl of acetyl-CoA. This observation provided strong circumstantial support for the role of the acetyl-CoA synthase in autotrophy.

This conclusion was further supported upon examination of cyanide inhibition. For the clostridial enzyme, low concentrations of cyanide inhibit the reduction of CO_2 to CO by the carbon monoxide dehydrogenase activity of acetyl-CoA synthase but do not inhibit acetyl-CoA biosynthesis from free CO and a methyl donor (Daniels et al., 1977; Diekert and Thauer, 1978). Cyanide also inhibited autotrophic growth of *M. thermoautotrophicum* but not methanogenesis (Stupperich and Fuchs, 1984b). The inhibition of growth was reversed by carbon monoxide. In the presence of cyanide, the incorporation of ^{14}CO into whole cells was also increased. In the cellular alanine formed, radiolabel was incorporated into the C-2 position at a high specific activity. Therefore, the effects of cyanide were consistent with its inhibition of the carbon monoxide dehydrogenase activity of acetyl-CoA synthase in whole cells. Carbon monoxide bypassed this inhibition by serving as a source of reduced carbon for biosynthesis of the carbonyl of acetyl-CoA.

Further evidence for the importance of the carbon monoxide dehydrogenase activity came from the observation that CO was produced by cultures of *M. thermoautotrophicum* during autotrophic growth on H_2 and CO_2 (Conrad and Thauer, 1983). During the exponential growth phase, the amount produced was about 1/500 of the amount of CO_2 reduced to methane and about 1/40 of the amount of CO_2 fixed into cellular carbon. Although small, the amounts of gaseous CO produced were within the range expected if free CO was in equilibrium with H_2 and CO_2. In contrast, methanogens incapable of autotrophic growth like *Methanobrevibacter ruminantium*, *Methanobrevibacter smithii*, and *Methanococcus voltae* did not produce carbon monoxide during methanogenesis from H_2 + CO_2 (Bott et al., 1985). Similarly, they did not incorporate ^{14}CO into cellular carbon, and cell extracts lacked high levels of carbon monoxide dehydrogenase activity. Thus, CO production and carbon monoxide dehydrogenase activity was correlated with autotrophy. Interestingly, *Methanospirillum hungatei* produced CO and had high levels of carbon monoxide dehydrogenase activity, but it did not assimilate ^{14}CO (Bott et al., 1985). Although initially described as an autotroph, this strain now only grows heterotrophically. Presumably, it contained an inactive acetyl-CoA synthase that retained carbon monoxide dehydrogenase activity.

Additional evidence strongly supported the hypothesis that CO production was a side reaction of the acetyl-CoA synthase during autotrophic CO_2 fixation (Eikmanns et al., 1985). During exponential growth on H_2 + CO_2, propyl iodide stimulated the rate of CO production greater than 10-fold. Alkyl halides inhibit acetyl-CoA biosynthesis, probably because they inactivate the corrinoid Fe-S protein which transfers methyl groups to the synthase, but they have little effect on the CO dehydrogenase activity (Kenealy and Zeikus, 1981; Lu et al., 1990). Therefore, the increase in CO production may be attributed to inhibition of acetyl-CoA biosynthesis, which would be a major sink for CO (Eikmanns et al., 1985). Similarly, resting cells also had an elevated rate of CO production, and CO production was specifically inhibited by cyanide (Eikmanns et al., 1985).

Importantly, resting cells also produced CO concentrations that were higher than equilibrium values, and this CO production was sensitive to uncouplers (Eikmanns et al., 1985). At concentrations that did not inhibit methanogenesis or *in vitro* carbon monoxide dehydrogenase activity, protonophores like TCS (tetrachlorsalicylanilide) abolished CO production. Therefore, CO production may require a membrane potential. Because the reduction of CO_2 to CO by H_2 is endergonic under the growth conditions for methanogens, coupling to the proton- or ion-motive force may be necessary to support high rates of autotrophic acetyl-CoA biosynthesis.

20.4 Enzymology of Autotrophic CO_2 Fixation

Further evidence that autotrophic methanogens assimilate carbon dioxide by a unique variation of the Ljungdahl–Wood pathway relied on the description of enzymatic activities in cell extracts similar to those found in the homoacetogenic clostridia. Whereas these experiments were interpreted to demonstrate the similarities of the clostridial and methanogenic systems, homogeneous enzymes from the autotrophic methanogens have not been available for detailed analysis. Thus, many features of the deduced pathway of autotrophic acetyl-CoA biosynthesis in methanogens is based on analogy with the clostridial pathway (Fig. 20.4). In the clostridial pathway, acetyl-CoA is the first product of CO_2 fixation, and the two carbons of the acetyl group are biosynthesized by different routes (Ljungdahl, 1986). The methyl carbon is derived from CO_2 via the reductions in the tetrahydrofolate pathway. The carbonyl of acetyl-CoA is obtained by the reduction of CO_2 catalyzed by the "carbon monoxide dehydrogenase activity" of acetyl-CoA synthase. Because this enzyme also catalyzes the oxidation of CO to CO_2 in the presence of methyl viologen and other artificial electron acceptors, dye reduction is the basis for the most commonly utilized assay. However, the physiological reaction of the enzyme appears to be the condensation of the methyl group derived from methyltetrahydrofolate, enzyme-bound CO from the reduction of CO_2, and HS-CoA to form acetyl-CoA.

In methanogens, elucidation of the pathway of acetyl-CoA biosynthesis has been hampered by the low specific activities compared to the clostridial system. For instance, on the basis of carbon monoxide dehydrogenase activity, extracts of autotrophically grown methanogens contain about 5% of the activity found in extracts of the homoacetogenic clostridia (Daniels et al., 1977; Ragsdale et al., 1983; Shanmugasundaram et al., 1988). Nevertheless, the evidence strongly supports a model where the methyl group is obtained by the reduction of CO_2 via the tetrahydromethanopterin pathway, which is the central pathway of methane biosynthesis, and the carbonyl group is formed by the "carbon monoxide dehydro-

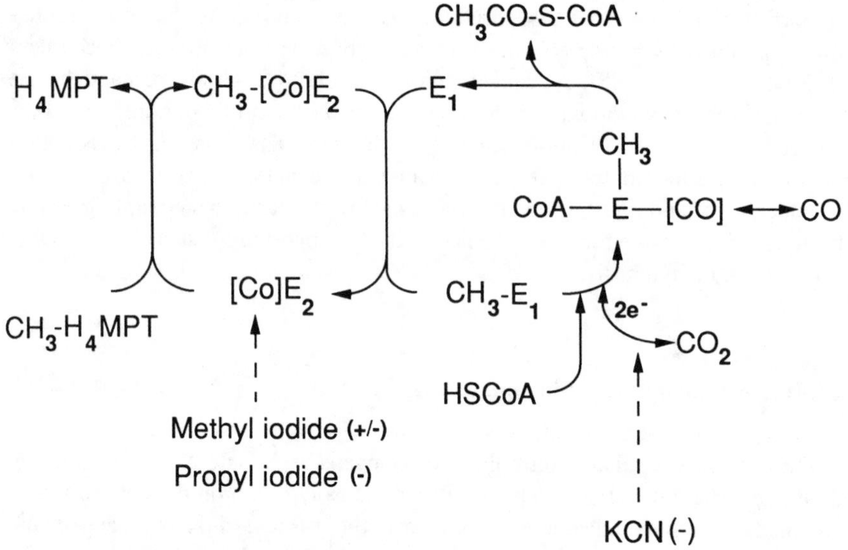

$$CH_3CO\text{-}S\text{-}CoA$$

$$H_4MPT \longleftarrow \curvearrowright CH_3\text{-}[Co]E_2 \curvearrowright E_1 \longleftarrow$$

$$\underset{|}{\overset{CH_3}{CoA\text{---}E\text{---}[CO]}} \longleftrightarrow CO$$

$$CH_3\text{-}H_4MPT \qquad [Co]E_2 \longleftarrow \curvearrowleft CH_3\text{-}E_1 \overset{\nearrow}{\underset{2e^-}{}}$$

$$HSCoA \longrightarrow \qquad \longleftarrow CO_2$$

Methyl iodide (+/-)

Propyl iodide (-)

KCN (-)

Figure 20.4.   The modified Ljungdahl–Wood pathway of autotrophic methanogens. The tetrahydromethanopterin (H_4MPT) pathway of CO_2 reduction to CH_4 is the precursor of the methyl group of acetyl-CoA. Methyltetrahydromethanopterin (CH_3-H_4MPT) is probably the methyl donor to a corrinoid Fe/S protein, $[Co]E_2$. $[Co]E_2$ is inactivated by propyl iodide, and low concentrations of methyl iodide are methyl donors. Therefore, the cobamide, [Co], of E_2 may be the methyl carrier. The methyl group is then probably transferred to E_1, the acetyl-CoA synthase. This enzyme also catalyzes the reversible reduction of CO_2 to the oxidation state of CO, and for this reason it is also called carbon monoxide dehydrogenase. The carbon monoxide dehydrogenase activity is inhibited by cyanide, but cyanide has no effect on its ability to bind free carbon monoxide.

genase" activity of acetyl-CoA synthase in a manner analogous to the clostridial pathway (Fig. 20.4).

The first convincing demonstration of autotrophic acetyl-CoA biosynthesis *in vitro* in methanogens utilized Triton X-100-treated cells of *M. thermoautotrophicum* (Stupperich and Fuchs, 1983, 1984a). Even then, the specific activity was extremely low, less than 0.1–0.2 nmol of acetyl-CoA formed min^{-1} (mg of protein)$^{-1}$. For comparison, *in vivo* rates of acetyl-CoA biosynthesis were about 100 nmol min^{-1} (mg of protein)$^{-1}$ (Eikmanns et al., 1985). In spite of this low activity, this first *in vitro* system had many of the properties predicted from the whole-cell experiments (Stupperich and Fuchs, 1983). Activity was inhibited by low concentrations of cyanide, and this inhibition was reversed by carbon monoxide. Thus, carbon monoxide dehydrogenase appeared to be a component of the methanogen acetyl-CoA synthase. Acetyl-CoA biosynthesis was also dependent on methyl coenzyme M, the terminal C-1 carrier in methanogenesis (Fig. 20.3). In extracts of this methanogen, CO_2 reduction to methane is greatly stimulated by catalytic amounts of methyl coenzyme M (Gunsalus and Wolfe, 1977). By increasing the rate of the initial reduction of CO_2, methyl coenzyme M also increases the levels of C-1 intermediates in methanogenesis. If one of these intermediates was also the methyl donor for acetyl-CoA biosynthesis, the acetyl-CoA synthase would be stimulated. Consistent with this hypothesis, radiolabel from [^{14}C]methyl coenzyme M was not incorporated into acetyl-CoA, both carbons of which were obtained from CO_2 (Stupperich and Fuchs, 1983, 1984a). This *in vitro* system was also insensitive to the folate antagonists methotrexate and aminopterin, as was methanogenesis from CO_2 (Fuchs and Stupperich, 1984a). Similarly, [^{14}C]methyltetrahydrofolate was not a C-1 donor (Fuchs and Stupperich, 1984b). These observations confirmed that folates were not C-1 carriers in the methanogen pathway of acetyl-CoA biosynthesis. Finally, the product of autotrophic CO_2 fixation was trapped with citrate synthase, which confirmed that acetyl-CoA and not acetate was the true product.

The role of tetrahydromethanopterin was confirmed in cell-free extracts. Serine and formaldehyde were found to be C-1 donors for acetyl-CoA biosynthesis (Holder et al., 1985). [3-^{14}C]Serine and $^{14}CH_2O$ specifically radiolabeled the methyl carbon of acetyl-CoA in low amounts. In the case of serine, a serine transhydroxymethylase had been identified in *M. thermoautotrophicum* which utilized tetrahydromethanopterin as the C-1 acceptor (Hoyt et al., 1986). Likewise, formaldehyde had been shown to react chemically with tetrahydromethanopterin to form methylenetetrahydromethanopterin (Escalante-Semerena et al., 1984). Methenyltetrahydromethanopterin was also shown directly to donate the C-1 group to acetyl-CoA (Länge and Fuchs, 1985). In this case, [^{14}C]methenyltetrahydromethanopterin specifically radiolabeled only the methyl carbon of acetyl-CoA.

In the clostridial acetyl-CoA synthase, methyltetrahydrofolate donates the methyl group to a corrinoid iron-sulfur protein that is the proximal methyl donor to the synthase. Evidence for a corrinoid protein in the methanogen system came largely from experiments with alkyl halides on whole cells and extracts. In whole cells, low concentrations of propyl iodide and methyl iodide inhibited autotrophic growth but not methanogenesis (Holder et al., 1985). Reversal of the inhibition by light implicated a corrinoid as the site of action. At slightly inhibitory concentrations, radiolabeled carbon from [^{14}C] methyl iodide was also incorporated into cellular protein, and degradation of the alanine formed demonstrated that the radiolabel was incorporated specifically into the C-3 position, which was derived from the methyl group of acetyl-CoA. Similarly, in cell extracts, acetyl-CoA biosynthesis was inhibited by propyl iodide, and inhibition was reversed by light (Holder et al., 1985). A low concentration of methyl iodide also stimulated acetyl-CoA biosynthesis, and the radiolabel from [^{14}C]methyl iodide was specifically incorporated into the methyl group of acetyl-CoA. Because free cobamides were not methyl donors for acetyl-CoA biosynthesis (Fuchs and Stupperich, 1984b; Länge and Fuchs, 1987), these results were interpreted to imply that a corrinoid protein was a methyl carrier (Holder et al., 1985).

A major criticism of these results with permeabilized whole cells and crude cellular extracts was that the specific activities for acetyl-CoA biosynthesis were extremely low. Thus, their physiological significance was questionable. This criticism was mitigated to some extent by the fact that the *in vitro* results with inhibitors and artificial substrates were consistent with analogous experiments on whole cells (Evans et al., 1986). In addition, some of the results were confirmed with a more highly purified preparation of acetyl-CoA synthase from *M. thermoautotrophicum* (Länge and Fuchs, 1987). This preparation was obtained by ammonium sulfate fractionation of an ultracentrifugation supernatant and had a specific activity of 15 nmol of acetyl-CoA formed min^{-1} (mg of protein)$^{-1}$ (Länge and Fuchs, 1987). Substantially free of endogenous cofactors, acetyl-CoA biosynthesis was dependent on carbon monoxide and tetrahydromethanopterin. Good methyl donors were formaldehyde (+ tetrahydromethanopterin), methylenetetrahydromethanopterin, or methyltetrahydromethanopterin. Methyltetrahydrofolate was not a methyl donor. Although these properties resembled the clostridia enzyme, this preparation was not further characterized.

In summary, both the methanogen and the clostridial enzymes utilized carbon monoxide as a donor for the carbonyl of acetyl-CoA. The methanogen enzyme system also used a form of tetrahydromethanopterin instead of methyltetrahydrofolate as the donor for the methyl carbon. However, the methyl donor in methanogens was not determined with certainty because oxidations and reductions in these cell extracts can rapidly interconvert the methenyl, methylene, and methyl forms of tetrahydromethanopterin (Escalante-Semerena et al., 1984).

20.5   Ljungdahl–Wood pathway in *Methanococcus* and other methanogens

Because the methanogens are phylogenetically diverse, evidence for the Ljungdahl–Wood pathway in other autotrophic methanogens is also important. For instance, on the basis of ribosomal RNA sequence similarity, *Methanobacterium* is about as closely related to representatives of other orders of methanogens like *Methanococcus* or *Methanosarcina* as *Clostridium* is to *Escherichia*. Thus, the discovery of the pathway in many methanogens implies that it may be very widespread and presumably very ancient.

An enzyme with carbon monoxide dehydrogenase activity was purified to homogeneity from autotrophically grown *Methanococcus vannielii*. It was similar in quaternary structure and metal content to the enzyme from the aceticlastic methanogen *M. barkeri* (DeMoll et al., 1987). Because it lacked the acetyl-CoA biosynthetic and exchange activities associated with the acetyl-CoA synthases from *C. thermoaceticum* and *Methanosarcina thermophila*, it was probably only one component of a larger acetyl-CoA synthase complex.

Acetyl-CoA synthase activity has also been demonstrated in cell extracts of the closely related facultative autotroph *Methanococcus maripaludis* (Shieh and Whitman, 1988). This activity was detected by trapping acetyl-CoA via the combined activities of endogenous pyruvate oxidoreductase and added lactate dehydrogenase. Lactate formed was measured spectrophotometrically. The highest rates obtained were 2–3 nmol of lactate formed min^{-1} (mg of protein)$^{-1}$. For comparison, the *in vivo* rate of acetyl-CoA biosynthesis necessary to support autotrophic growth was about 80 nmol min^{-1} (mg of protein)$^{-1}$. Following incubations with $^{14}CO_2$, the percentages of radiolabel in the C-1, C-2, and C-3 positions of lactate were 73, 33, and 11%, respectively. This distribution was expected if the C-1 of lactate (pyruvate) was radiolabeled by $^{14}CO_2$ during the reductive carboxylation of acetyl-CoA and the exchange reaction catalyzed by pyruvate oxidoreductase, and the C-2 and C-3 of lactate were derived from the carbonyl and methyl groups of acetyl-CoA, respectively. The low incorporation of $^{14}CO_2$ into the C-3 position could be explained by dilution from a pool of endogenous C-1 carriers in the tetrahydromethanopterin pathway. Lactate synthesis was stimulated by carbon monoxide + formaldehyde and carbon monoxide + methyl iodide. As expected, ^{14}CO was preferentially incorporated into the C-2 of lactate, which corresponds to the carbonyl of acetyl-CoA. [^{14}C]Formaldehyde, which reacts spontaneously with tetrahydromethanopterin in cell extracts to form [^{14}C]methylenetetrahydromethanopterin, preferentially radiolabeled the C-3 of lactate, which corresponds to the methyl of acetyl-CoA. In contrast to *M. thermoautotrophicum*, methyl coenzyme M inhibited lactate synthesis, and the highest rates were only obtained in the presence of bromoethanesulfonate, an inhibitor, of the methyl coenzyme M methylreductase. These results suggested that C-1 units from the tetrahydromethanopterin pathway were not limiting. In addition, cyanide was only inhibitory at high concentrations. Because CO_2 protected against this inhibition, it was still possible that the carbon monoxide dehydrogenase activity of acetyl-CoA synthase was the site of inhibition. Finally, the activity in extracts of *M. maripaludis* was extremely unstable, and preincubation for 60 min under a N_2 atmosphere reduced activity by 40%. Although these experiments were consistent with the presence of the Ljungdahl–Wood pathway, they provided largely circumstantial evidence and did little to elucidate the biochemical features of the pathway in *Methanococcus*.

More compelling evidence for the Ljungdahl–Wood pathway in *M. maripuladis* was obtained by examination of acetate auxotrophs (Ladapo and Whitman, 1990). These mutants were isolated following ethyl methanesulfonate mutagenesis and enrichment with nucleobase analogs. The six autotrophs obtained were incapable of autotrophic growth with CO_2 as the sole carbon source or growth in rich medium that was deficient only in acetate. Growth in both media with acetate was indistinguishable from the parental strain. All six auxotrophs contained only low levels of carbon monoxide dehydrogenase activity, 1–16% of the wild-type level. In addition, acetyl-CoA synthase, as measured by the lactate trap, was undetectable. In spontaneous revertants, these activities were restored. Therefore, carbon monoxide dehydrogenase activity was essential for autotrophic growth. Interestingly, three of the auxotrophs also contained reduced levels of pyruvate oxidoreductase. Revertants of these auxotrophs regained this activity as well as carbon monoxide dehydrogenase and acetyl-CoA synthase activity. Although the interpretation of this observation was somewhat ambiguous, it could imply a physiological or genetic linkage of these two enzyme systems.

There is also significant evidence for a modified Ljungdahl–Wood pathway in other methanogens. During mixotrophic growth with methanol and H_2, $M.$ $barkeri$ preferentially incorporated [^{14}C]methanol into the C-3 of alanine, which was derived from the methyl of acetyl-CoA (Kenealy and Zeikus, 1982). Methanol oxidation via the tetrahydromethanopterin pathway was probably the source of this enrichment. In extracts of H_2 + CO_2-grown cells, ^{14}CO was also incorporated into acetate, either by an exchange reaction or net synthesis. This activity was dependent on methylcobalamin, which can also serve as a methyl donor to the clostridial enzyme (Kenealy and Zeikus, 1982). Low concentrations of cyanide also inhibited autotrophic growth of $M.$ $barkeri$ on CO_2 or methanol, which implied an important role for carbon monoxide dehydrogenase (Smith et al., 1985).

In conclusion, the evidence for the Ljungdahl–Wood pathway of autotrophic CO_2 assimilation in methanogens is compelling, and the physiological data is much more complete than for any other group of microorganisms outside the homoacetogens. However, the enzymes from the autotrophic methanogens have not been characterized in great detail. For this reason, interpretations of the physiological experiments in whole cells and cell extracts rely heavily on the detailed biochemical studies of enzymes from the homoacetogenic clostridia or aceticlastic methanogens. Thus, the evidence for a corrinoid Fe/S protein is based largely on sensitivity to inhibitors like propyl iodide and the ability of methyl iodide to serve as a methyl donor (Holder et al., 1985; Shieh and Whitman, 1988). Although details on the active site of the acetyl-CoA synthase from autotrophs are unknown, the sensitivity to cyanide and ability of CO to serve as a carbonyl donor suggest strong similarities to the enzymes from other sources.

Acknowledgment

This work was supported in part by a grant DEFG09-91ER20045 from the Department of Energy.

References

Bott, M. H., B. Eikmanns, and R. K. Thauer. 1985. Defective formation and/or utilization of carbon monoxide in H_2/CO_2 fermenting methanogens dependent on acetate as carbon source. *Arch. Microbiol.* **143**:266–269.

Conrad, R., and R. K. Thauer. 1983. Carbon monoxide production by *Methanobacterium thermoautotrophicum*. *FEMS Microbiol. Lett.* **20**:229–232.

Daniels, L., G. Fuchs, R. K. Thauer, and J. G. Zeikus. 1977. Carbon monoxide oxidation by methanogenic bacteria. *J. Bacteriol.* **132**:118–126.

Daniels, L., and J. G. Zeikus. 1978. One-carbon metabolism in methanogenic bacteria: analysis of short-term fixation products of $^{14}CO_2$ and $^{14}CH_3OH$ incorporated into whole cells. *J. Bacteriol.* **136**:75–84.

DeMoll, E., D. A. Grahame, J. M. Harnly, L. Tsai, and T. C. Stadtman. 1987. Purification and properties of carbon monoxide dehydrogenase from *Methanococcus vannielii*. *J. Bacteriol.* **169**:3916–3920.

Diekert, G. B., and R. K. Thauer. 1978. Carbon monoxide oxidation by *Clostridium thermoaceticum* and *Clostridium formicoaceticum*. *J. Bacteriol.* **136**:597–606.

DiMarco, A. A., T. A. Bobik, and R. S. Wolfe. 1990. Unusual coenzymes of methanogenesis. *Annu. Rev. Biochem.* **59**:355–394.

Drake, H. L., S. Hu, and H. G. Wood. 1981. Purification of five components from *Clostridium thermoaceticum* which catalyze synthesis of acetate from pyruvate and methyltetrahydrofolate. *J. Biol. Chem.* **256**:11137–11144.

Eikmanns, B., G. Fuchs, and R. K. Thauer. 1985. Formation of carbon monoxide from CO_2 and H_2 by *Methanobacterium thermoautotrophicum*. *Eur. J. Biochem.* **146**:149–154.

Escalante-Semerena, J. C., K. L. Rinehart, and R. S. Wolfe. 1984. Tetrahydromethanopterin, a carbon carrier in methanogenesis. *J. Biol. Chem.* **259**:9447–9455.

Evans, J. N. S., C. J. Tolman, and M. F. Roberts. 1986. Indirect observation by ^{13}C NMR spectroscopy of a novel CO_2 fixation pathway in methanogens. *Science* **231**:488–491.

Ferry, J. G., D. W. Sherod, H. D. Peck, Jr., and L. G. Ljungdahl. 1976. Levels of formyltetrahydrofolate synthetase and methylenetetrahydrofolate dehydrogenase in methanogenic bacteria. In: *Proceedings of the Symposium "Microbial Production and Utilization of Gases (H_2, CH_4, CO),"* H. G. Schlegel, G. Gottschalk, and N. Pfennig (eds.), pp. 151–155. Goltze, Göttingen.

Fuchs, G. 1986. CO_2 fixation in acetogenic bacteria: variations on a theme. *FEMS Microbiol. Rev.* **39**:181–213.

Fuchs, G., and E. Stupperich. 1978. Evidence for an incomplete reductive carboxylic acid cycle in *Methanobacterium thermoautotrophicum*. *Arch. Microbiol.* **118**:121–125.

Fuchs, G., E. Stupperich, and R. K. Thauer. 1978a. Acetate assimilation and the synthesis of alanine, aspartate, and glutamate in *Methanobacterium thermoautotrophicum*. *Arch. Microbiol.* **117**:61–66.

Fuchs, G., E. Stupperich, and R. K. Thauer. 1978b. Function of fumarate reductase in methanogenic bacteria (*Methanobacterium*). *Arch. Microbiol.* **119**:215–218.

Fuchs, G., and E. Stupperich. 1980. Acetyl CoA, a central intermediate of autotrophic CO_2 fixation in *Methanobacterium thermoautotrophicum*. *Arch. Microbiol.* **127**:267–272.

Fuchs, G., and E. Stupperich. 1986. Carbon assimilation pathways in archaebacteria. *Syst. Appl. Microbiol.* **7**:364–369.

Gunsalus, R. P., and R. S. Wolfe. 1977. Stimulation of CO_2 reduction to methane by methylcoenzyme in extracts of *Methanobacterium*. *Biochem. Biophys. Res. Commun.* **76**:790–795.

Holder, U., D. E. Schmidt, E. Stupperich, and G. Fuchs. 1985. Autotrophic synthesis of activated acetic acid from two CO_2 in *Methanobacterium thermoautotrophicum*. III. Evidence for common one-carbon precursor pool and the role of corrinoid. *Arch. Microbiol.* **141**:229–238.

Hoyt, J. C., A. Oren, J. C. Escalante-Semerena, and R. S. Wolfe. 1986. Tetrahydromethanopterin-dependent serine transhydroxymethylase from *Methanobacterium thermoautotrophicum*. *Arch. Microbiol.* **145**:153–158.

Kenealy, W., and J. G. Zeikus. 1981. Influence of corrinoid antagonists on methanogen metabolism. *J. Bacteriol.* **146**:133–140.

Kenealy, W. R., and J. G. Zeikus. 1982. One-carbon metabolism in methanogens: evidence for synthesis of a two-carbon cellular intermediate and unification of catabolism and anabolism in *Methanosarcina barkeri*. *J. Bacteriol.* **151**:932–941.

Kluyver, A. J., and C. G. Schnellen. 1947. On the fermentation of carbon monoxide by pure cultures of methane bacteria. *Arch. Biochem.* **14**:57–70.

Ladapo, J., and W. B. Whitman. 1990. Method for isolation of auxotrophs in the methanogenic archaebacteria: role of the acetyl-CoA pathway of autotrophic CO_2 fixation in *Methanococcus maripaludis*. *Proc. Natl. Acad. Sci. USA* **87**:5598–5602.

Länge, S., and G. Fuchs. 1985. Tetrahydromethanopterin, a coenzyme involved in autotrophic acetyl coenzyme A synthesis from 2 CO_2 in *Methanobacterium*. *FEBS Lett.* **181**:303–307.

Länge, S., and G. Fuchs. 1987. Autotrophic synthesis of activated acetic acid from CO_2 in *Methanobacterium thermoautotrophicum*. Synthesis from tetrahydromethanopterin-bound C_1 units and carbon monoxide. *Eur. J. Biochem.* **163**:147–154.

Ljungdahl, L. G. 1986. The autotrophic pathway of acetate synthesis in acetogenic bacteria. *Annu. Rev. Microbiol.* **40**:415–450.

Lu, W., S. R. Harder, and S. W. Ragsdale. 1990. Controlled potential enzymology of methyl transfer reactions involved in acetyl-CoA synthesis by CO dehydrogenase and the corrinoid/iron-sulfur protein from *Clostridium thermoaceticum*. *J. Biol. Chem.* **265**:3124–3133.

Ragsdale, S. W., J. E. Clark, L. G. Ljungdahl, L. L. Lundie, and H. L. Drake. 1983. Properties of purified carbon monoxide dehydrogenase from *Clostridium thermoaceticum*, a nickel, iron-sulfur protein. *J. Biol. Chem.* **258**:2364–2369.

Ragsdale, S. W., and H. G. Wood. 1985. Acetate biosynthesis by acetogenic bacteria. Evidence that carbon monoxide dehydrogenase is the condensing enzyme that catalyzes the final steps in the synthesis. *J. Biol. Chem.* **260**:3970–3977.

Rouviere, P. E., and R. S. Wolfe. 1988. Novel biochemistry of methanogenesis. *J. Biol. Chem.* **263**:7913–7916.

Rühlemann, M., K. Ziegler, E. Stupperich, and G. Fuchs. 1985. Detection of acetyl coenzyme A as an early CO_2 assimilation intermediate in *Methanobacterium*. *Arch. Microbiol.* **141**:399–406.

Shanmugasundaram, T., S. W. Ragsdale, and H. G. Wood. 1988. Role of carbon monoxide dehydrogenase in acetate synthesis by the acetogenic bacterium, *Acetobacterium woodii*. *BioFactors* **1**:147–152.

Shieh, J., and W. B. Whitman. 1987. Pathway of acetate assimilation in autotrophic and heterotrophic methanococci. *J. Bacteriol.* **169**:5327–5329.

Shieh, J., and W. B. Whitman. 1988. Autotrophic acetyl coenzyme A biosynthesis in *Methanococcus maripuludis*. *J. Bacteriol.* **170**:3072–3079.

Smith, M. R., J. L. Lequerica, and M. R. Hart. 1985. Inhibition of methanogenesis and carbon metabolism in *Methanosarcina* sp. by cyanide. *J. Bacteriol.* **162**:67–71.

Stupperich, E., and G. Fuchs. 1981. Products of CO_2 fixation and ^{14}C labelling pattern of alanine in *Methanobacterium thermoautotrophicum* pulse-labelled with [14]CO_2. *Arch. Microbiol.* **130**:294–300.

Stupperich, E., and G. Fuchs. 1983. Autotrophic acetyl coenzyme A synthesis in vitro from two CO_2 in *Methanobacterium*. *FEBS Lett.* **156**:345–348.

Stupperich, E., K. E. Hammel, G. Fuchs, and R. K. Thauer. 1983. Carbon monoxide fixation into the carboxyl group of acetyl coenzyme A during autotrophic growth of *Methanobacterium*. *FEBS Lett.* **152**:21–23.

Stupperich, E., and G. Fuchs. 1984a. Autotrophic synthesis of activated acetic acid from two CO_2 in *Methanobacterium thermoautotrophicum*. I. Properties of in vitro system. *Arch. Microbiol.* **139**:8–13.

Stupperich, E., and G. Fuchs. 1984b. Autotrophic synthesis of activated acetic acid from two CO_2 in *Methanobacterium thermoautotrophicum*. II. Evidence for different origins of acetate carbon atoms. *Arch. Microbiol.* **139**:14–20.

Taylor, G. T., D. P. Kelly, and S. J. Pirt. 1976. Intermediary metabolism in methanogenic bacteria (*Methanobacterium*). In: *Proceedings of the Symposium "Microbial Production and Utilization of Gases (H_2, CH_4, CO),"* H. G. Schlegel, G. Gottschalk, and N. Pfenning, (eds.), pp. 173–180. Goltze, Göttingen.

Weimer, P. J., and J. G. Zeikus. 1978. One carbon metabolism in methanogenic bacteria. *Arch. Microbiol.* **119**:49–57.

Wolfe, R. S. 1985. Unusual coenzymes of methanogenesis. *Trends Biochem. Sci.* **10**:396–399.

Zeikus, J. G.,G. Fuchs, W. Kenealy, and R. K. Thauer. 1977. Oxidoreductases involved in cell carbon synthesis of *Methanobacterium thermoautotrophicum. J. Bacteriol.* **132**:604–613.

21

CO Dehydrogenase of Methanogens

James G. Ferry

21.1 Introduction

CO dehydrogenase has a central role in the anaerobic component of the global carbon cycle. The anaerobic decomposition of complex organic matter involves microbial food chains (consortia) in which acetate is the most abundant intermediate. The acetate is primarily metabolized by acetotrophs, terminal organisms of the food chain, utilizing a pathway in which activated acetate (acetyl-CoA) is cleaved by CO dehydrogenase. The enzyme also catalyzes the synthesis of acetyl-CoA in the acetyl-CoA Ljungdahl–Wood pathway of homoacetogens (see Chapters 1 and 3). Although the physiology of these anaerobes are highly variable, the unity of biochemistry predicts that the underlying chemistry of their metabolism is basically the same. This principal of biochemistry is vividly evident in the CO dehydrogenase (acetyl-CoA cleaving) of acetotrohic methanogens from the *Archaea* domain; the enzyme has properties that are surprisingly similar to the CO dehydrogenase (acetyl-CoA synthesizing) of homoacetogens from the *Bacteria* domain. CO dehydrogenases also function in autotrophic methanogens to synthesize acetyl-CoA from CO_2 in a process fundamentally similar to the Ljungdahl–Wood pathway. The extreme phylogenetic diversity between the two domains offer a unique opportunity for studies aimed at understanding the mechanism and evolution of CO dehydrogenases in anaerobes.

21.2 Microbiology of Methanogenesis

The methanogenic decomposition of organic matter is an important component of the global carbon cycle. The process occurs in nearly every conceivable anaerobic environment such as the rumen, the lower intestinal tract of humans, sewage digestors, landfills, rice paddies, and the sediments of lakes and rivers. The strictly aerobic methanotrophs oxidize most of the methane to CO_2, completing the carbon cycle; however, a significant proportion escapes into the upper atmosphere.

At least three interacting metabolic groups of microbes (consortia) are involved in the anaerobic decomposition of complex organic matter to methane (see Chapters 7 and 12). The fermentatives degrade polymers to H_2, CO_2, formate, acetate, and higher volatile carboxylic acids. Syntrophic acetogens oxidize the higher acids to acetate and either H_2 or formate. Methanogens are the final group in the consortia and utilize H_2, formate, and acetate as substrates for growth and methanogenesis. About two-thirds of the methane produced in nature originates from the methyl group of acetate (reaction 21.1), and about one-third from the reduction of CO_2 with electrons derived from the oxidation of H_2 or formate (reactions 21.2 and 21.3).

$$CH_3COO^- + H^+ \rightarrow CH_4 + CO_2, \qquad \Delta G_0' = -36 \text{ kJ/mol}$$
$$\text{(Vogels et al., 1988)} \qquad\qquad (21.1)$$

$$4\,H_2 + CO_2 \rightarrow CH_4 + 2\,H_2O, \qquad \Delta G_0' = -130.4 \text{ kJ/mol } CH_4$$
$$\text{(Vogels et al., 1988)} \qquad\qquad (21.2)$$

$$4HCOO^- + 4\,H^+ \rightarrow CH_4 + 3\,CO_2 + 2\,H_2O, \qquad \Delta G_0' = -119.5 \text{ kJ/mol } CH_4$$
$$\text{(Vogels et al., 1988)} \qquad\qquad (21.3)$$

However, the amount of methane produced from acetate can vary depending on the presence of other metabolic groups of anaerobes and the particular environment. The oxidation of H_2 or formate by homoacetogens increases the amount of acetate and the importance of methanogenic acetotrophs. For example, 80–90% of the methane produced in a rice paddy soil originated from the methyl group of acetate (Thebrath et al., 1992). On the other hand, a nonmethanogenic microbe called AOR (acetate oxidizing rod) has been described which oxidizes acetate to H_2 and CO_2 (Zinder and Koch, 1984). The extent to which acetate oxidizers like the AOR occur in anaerobic environments is unknown; for example, acetate oxidation in a thermophilic anaerobic sewage-sludge disgestor was only of minimal importance when the acetate concentration fell below 1 mM (Petersen and Ahring, 1991). However, the presence of microbes like the AOR would diminish acetate as an intermediate and the relative importance of acetotrophic methanogens. In marine environments, the acetotrophic sulfate-reducing bacteria dominate and the conversion of methylamines becomes the major route of methanogenesis. Acetotrophs grow more slowly than CO_2 reducers, and, therefore, methane from acetate is not likely to predominate when the residence time for organic matter is short.

Methanogens represent the largest and most diverse group within the *Archaea* domain (Woese et al., 1990). Although acetate is a major precursor for methane formation, only two genera of methanogenic acetotrophs (*Methanosarcina* and *Methanothrix*) and a few species have been described. The *Methanosarcina* are the most versatile of all methanogenic species and many have the capability for growth on H_2-CO_2, methanol, methylamines, and acetate. Some *Methanosarcina* species grow weakly on H_2-CO_2 or may require a period of adaptation (Sowers et al., 1984; Zinder et al., 1985; Mukhopadhyay et al., 1991); *Methanothrix,* however, is able to utilize only acetate. The *Methanosarcina* grow on acetate more rapidly than *Methanothrix* species, but the K_m and minimum threshold concentrations for growth are much greater than for the *Methanothrix* (Min and Zinder, 1989; Westermann et al., 1989a; Westermann et al., 1989b; Ohtsubo et al., 1992); as a result, *Methanosarcina* species thrive in environments where acetate concentrations are high, and *Methanothrix* species predominate where acetate is low (Zinder et al., 1984).

21.3   CO Dehydrogenase and the Physiology of Methanogens

21.3.1   The Fermentation of Acetate to Methane

Two independent pathways exist for the conversion of acetate and reduction of CO_2. The pathway of CO_2 reduction to methane has been studied extensively over the past 20 years and is the subject of recent reviews (DiMarco et al., 1990; Ferry, 1992a); however, a fundamental understanding of methanogenesis from acetate (reaction 21.1) has emerged only recently. Figure 21.1 depicts the pathway as it is currently understood in the *Methanosarcina* (Ferry, 1992b). In general, acetate is cleaved and the methyl group is reduced to methane with electrons derived from oxidation of the carbonyl group to CO_2; thus, the pathway is a fermentation. Central to the pathway is the CO dehydrogenase enzyme complex which cleaves activated acetate (acetyl-CoA) synthesized by acetate kinase and phosphotransacetylase. The Ni/Fe-S component of the complex cleaves acetyl-CoA and oxidizes the enzyme-bound carbonyl group to CO_2 to provide an electron pair for reduction of the heterodisulfide CoM-S-S-HTP. The immediate electron acceptor for the CO dehydrogenase is a ferredoxin (Terlesky and Ferry, 1988a; Clements and Ferry, 1992). The electron transport chain from the ferredoxin to the heterodisulfide reductase probably involves membrane-bound electron carriers including *b* cytochromes (Heiden et al., 1993). The requirement for ferredoxin in the pathway of methanogenesis has been confirmed in *M. barkeri* (Fischer and Thauer, 1990b). The CO_2 is converted to carbonic acid by a carbonic anhydrase with a postulated periplasmic location that may assist in proton symport of the acetate anion (Alber and Ferry, 1993). After cleavage by the Ni/Fe-S

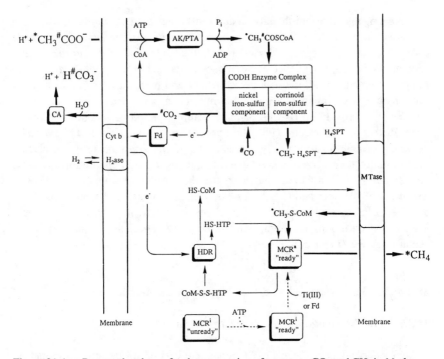

Figure 21.1.   Proposed pathway for the conversion of acetate to CO_2 and CH_4 in *Methanosarcina*. AK, acetate kinase; PTA, phosphotransacetylase; H_4SPT, tetrahydrosarcinapterin; CA, carbonic anhydrase; MCR^i, inactive methylreductase; MCR^a, active methylreductase; HDR, heterodisulfide (CoM-S-S-HTP) reductase; Fd, ferredoxin; CODH, carbon monoxide dehydrogenase; cyt b, cytochrome *b*; H_2ase, hydrogenase; MTase, methyltransferase system. The carbon atoms of acetate are marked with (*) and (#) symbols to distinguish between the carboxyl and methyl groups, respectively.

component, the methyl group is transferred to the corrinoid/iron-sulfur component within the complex and finally to coenzyme M (HS-CoM) involving tetrahydrosarcinapterin and catalyzed by a membrane-bound methyltransferase system (Fischer et al., 1992). The CH_3-S-CoM is reductively demethylated to methane with electrons derived from the sulfur atoms of CH_3-S-CoM and HS-HTP with formation of the heterodisulfide CoM-S-S-HTP. A similar pathway operates in *Methanothrix* (Jetten et al., 1992).

   Most acetotrophic anaerobes from the *Bacteria* domain cleave acetyl-CoA. Unlike methanogens, these microbes oxidize the methyl and carbonyl groups completely to CO_2 reducing a variety of electron acceptors (Thauer et al., 1989; see also Chapter 19); however, acetotrophs from the *Bacteria* and *Archaea* domains that cleave acetyl-CoA have CO dehydrogenase in common which catalyzes the reaction. The extreme phylogenetic difference between the *Archaea* and *Bacteria* offers a unique opportunity for studies aimed at understanding the mechanism and evolution of CO dehydrogenase in anaerobes.

21.3.2 Synthesis of Cell Carbon

CO dehydrogenase also functions in the pathway of CO_2 fixation for cell carbon in autotrophic methanogens where the enzyme catalyzes the synthesis of acetyl-CoA. The methyl group derives from reduction of CO_2 by the same pathway that is operative in the reduction of CO_2 to CH_4 (Ferry, 1992a). CO can supply the carbonyl group or it is generated from the CO dehydrogenase-catalyzed reduction of CO_2 with an electron pair. The reader is referred to Chapter 20 for a more comprehensive treatment of this subject.

21.3.3 Metabolism of CO

In the 1930s and 1940s it was shown that methanogens convert CO to methane with H_2 as reductant (Kluyver and Schnellen, 1947). The process involves a disproportionation of CO as shown in reactions 21.4 and 21.5. It remained until 1977 before growth

$$CO + H_2O \rightarrow CO_2 + H_2, \qquad \Delta G_0' = -20 \, \text{kJ/mol}$$
$$\text{(Thauer et al., 1977)} \qquad\qquad (21.4)$$

$$CO_2 + 4\,H_2 \rightarrow CH_4 + 2\,H_2O, \qquad \Delta G_0' = -131 \, \text{kJ/mol } CH_4$$
$$\text{(Thauer et al., 1977)} \qquad\qquad (21.5)$$

$$CO + 3\,H_2 \rightarrow CH_4 + H_2O, \qquad \Delta G_0' = -151 \, \text{kJ/mol } CH_4 \qquad (21.4+21.5)$$

with CO as the sole energy source (reaction 21.6) was demonstrated with *Methanobacterium thermoautotrohicum* and other CO_2-reducing

$$4\,CO + 2\,H_2O \rightarrow CH_4 + 3\,CO_2, \qquad \Delta G_0' = -211 \, \text{kJ/mol } CH_4$$
$$\text{(Thauer et al., 1977)} \qquad\qquad (21.6)$$

methanogens (Daniels et al., 1977). These results implicate the involvement of CO dehydrogenase in the oxidation of CO. Indeed, extracts of CO-grown *M. thermoautotrophicum* contain CO dehydrogenase activity linked to the reduction of coenzyme F_{420} (Daniels et al., 1977). Whole cells converted CO stoichiometrically to CO_2 and H_2, confirming the involvement of H_2 as an obligatory intermediate in the reduction of CO_2 to CH_4 (O'Brien et al., 1984). However, growth with CO as a sole energy source is slow (O'Brien et al., 1984) and the extent that CO serves as a growth substrate for methanogens in nature is unknown. Potential sources of CO include the combustion of fossil fuels, and as a product of heme degradation by aerobic microbes (Hegeman, 1980). Interestingly, CO is formed by the reductive dehalogenation of CCl_4 and FREONs catalyzed by corrinoids (Krone et al., 1991).

Although purified CO dehydrogenase from *Methanosarcina barkeri* produces H_2 with methyl viologen as an electron carrier (Bhatnagar et al., 1987), it is doubtful that this is the major mechanism of H_2 evolution from CO; rather, it is more likely that an electron transport chain linked to hydrogenase is involved. A CO-oxidizing:H_2-evolving system from *Methanosarcina thermophila* has been reconstituted with purified CO dehydrogenase, ferredoxin, and membranes which contain hydrogenase activity (Terlesky and Ferry, 1988a, 1988b). The ferredoxin mediates electron flow from CO dehydrogenase to the membranes where cytochrome *b* is probably involved in transferring electrons to the hydrogenase. Methanogens could obtain additional energy for growth by converting CO to H_2 (reaction 21.4) in the dismutation of CO to CH_4. Indeed, proton translocation is coupled to the oxidation of CO to CO_2 and H_2 in cell suspensions of acetate-grown *M. barkeri* (Bott et al., 1986; Bott and Thauer, 1989).

It is reported that methanogens produce CO. *M. thermoautotrophicum*, when grown in a fermentor gased with an 80% H_2/20% mixture, contains between 30 ppm and 90 ppm CO in the effluent gas (Conrad and Thauer, 1983). Up to 15,000 ppm CO was present in the gas phase of cell suspensions of several methanogens incubated with CO_2 and H_2 (Bott and Thauer, 1987). The reaction (reaction 21.7) has a positive change in free energy

$$CO_2 + H_2 \rightarrow CO + H_2O, \qquad \Delta G_0' = +20\,\text{kJ/mol}$$
$$\text{(Bott and Thauer, 1987)} \qquad\qquad (21.7)$$

under standard conditions which suggests an energy requirement; indeed, CO is only generated when cell suspensions are producing CH_4 (Bott and Thauer, 1987), and the concentrations in the gas phase of some methanogens greatly surpasses the equilibrium concentration (Eikmanns et al., 1985). Inhibitor studies show that proton-motive-force, and not ATP, drives CO synthesis from CO_2 and H_2 (Bott and Thauer, 1987). CO formation and CO dehydrogenase activity is inhibited by cyanide, which is consistent with an involvement of CO dehydrogenase in the synthesis of CO from CO_2 and H_2 (Eikmanns et al., 1985). Cultures of *Methanothrix* sp. strain CALS-1 growing on acetate reached partial pressures of near 0.6 Pa CO (Zinder and Anguish, 1992). This accumulation of CO during growth on acetate is likely the result of passive equilibration of enzyme-bound CO during cleavage of acetyl-CoA by CO dehydrogenase. These results are in contrast to *M. thermophila* which appears to bind CO tightly to the CO dehydrogenase during acetyl-CoA cleavage (Nelson et al., 1987).

21.4 Biochemistry of CO Dehydrogenases from Methanogens

21.4.1 Acetotrophic methanogens

(a) BACKGROUND

The now familiar similarities in the fundamental chemistry of acetyl-CoA cleavage and synthesis in methanogens and homoacetogens was of major importance in the discovery of the role of CO dehydrogenase in the pathway of acetate fermentation to CH_4. CO dehydrogenases from acetate-grown *Methanosarcina* and *Methanothrix* constitute a major portion of the total cell protein (Krzycki and Zeikus, 1984; Terlesky et al., 1986; Grahame and Stadtman, 1987; Jetten et al., 1989). The prior understanding that CO dehydrogenase catalyzes the synthesis of acetyl-CoA in the Ljungdahl–Wood pathway (see Chapters 1 and 3) led early investigators to hypothesize that the CO dehydrogenase in acetotrophic methanogens is involved in the dissimilation of acetate. The first evidence for the involvement of CO dehydrogenase was obtained when antibodies to the enzyme inhibited methane production from acetate in cell extracts of *M. barkeri* (Krzycki et al., 1985). Finally, the demonstration of an exchange of the carbonyl group of acetyl-CoA with CO (Jetten 1991a, 1991b; Raybuck et al., 1991) provided the first biochemical evidence that supported the hypothesis of an acetyl-CoA cleavage function for CO dehydrogenase of acetotrophic methanogens.

(b) *M. THERMOPHILA*

M. thermophila contains an enzyme complex which catalyzes the cleavage of the C–C and C–S bonds of acetyl-CoA (Raybuck et al., 1991), as well as the synthesis of acetyl-CoA from methyl iodide, CO, and CoA (Abbanat and Ferry, 1990). The complex displays CO dehydrogenase activity and is commonly referred to as the CO dehydrogenase enzyme complex. However, the primary function of the enzyme is cleavage of acetyl-CoA and oxidation of the carbonyl group to CO_2 during growth of the organism on acetate. The five-subunit complex is resolvable into two enzyme components: a 200-kDa CO-oxidizing nickel/iron-sulfur (Ni/Fe-S) component which contains 89-kDa and 19-kDa subunits, and a 100-kDa corrinoid/iron-sulfur (Co/Fe-S) component which contains 60-kDa and 58-kDa subunits (Abbanat and Ferry, 1991). The fifth subunit is a 71-kDa protein that has not been characterized and the function is unknown. The CO-reduced complex has a spin-coupled Ni-Fe-C center with an EPR spectrum nearly identical to the Ni-Fe-C center of the CO-reduced CO dehydrogenase (acetyl-CoA synthase) from the homoacetogenic organism *Clostridium thermoaceticum* (Terlesky et al., 1987). This center is the proposed site for synthesis of the acetyl moiety of acetyl-CoA from CO and a methyl group donated by a methylated corrinoid/

Fe-S protein (Lu et al., 1990; Ragsdale, 1991) associated with the *C. thermoaceticum* CO dehydrogenase. A structure has been proposed for this center with the composition [NiXFe$_{3-4}$S$_4$]-C≡O, where X bridges the Ni and Fe components (Kumar and Ragsdale, 1992; see Chapter 3). The similarities in the EPR spectra suggest a similar structure for the Ni-Fe-C center of the methanogen enzyme; thus, it follows that the center in the Ni/Fe-S component from *M. thermophila* is the proposed site for cleavage of the acetyl moiety of acetyl-CoA (Abbanat and Ferry, 1991).

The Ni/Fe-S component from *M. thermophila* exhibits three low-temperature EPR signals (Lu et al., 1994) with properties similar to those reported for the acetyl-CoA synthase from *C. thermoaceticum* (Ragsdale, 1991). Two of the signals from the *M. thermophila* Ni/Fe-S component have spectra typical of bacterial-like 4Fe-4S centers. The first has a midpoint potential of −444 mV with *g* values of 2.04, 1.93, and 1.89 (g_{av} = 1.95). The second signal is observed at −540 mV and has *g* values of 2.05, 1.95, and 1.90 (g_{av} = 1.97). Based on spectral characteristics, it is likely that the signals of the g_{av} = 1.95 and 1.97 species originate from a single [4Fe-4S]$^{2+/1+}$ cluster which can exist in two conformations. The third low-temperature EPR signal has a midpoint potential of −154 mV and *g* values of 2.02, 1.87, and 1.72 (g_{av} = 1.87) also attributable to a Fe-S center. This atypical Fe-S center has properties similar to the atypical Fe-S center of the *C. thermoaceticum* acetyl-CoA synthase (Ragsdale, 1991). The function of these 4Fe-4S centers in the *M. thermophila* Ni/Fe-S component is unknown, but they may be involved in the transfer of electrons to ferredoxin or the Co/Fe-S component. The CO-reduced Ni/Fe-S component also exhibits the expected spin-coupled Ni-Fe-C EPR spectrum of the enzyme complex (Lu et al., 1994); however, it is not known which, if any, of the Fe atoms from the iron-sulfur centers participate in formation of the Ni-Fe-C center.

Corrinoids are abundant in acetotrophic methanogens (Stupperich and Krautler, 1988; Silveira et al., 1991; Kohler, 1988), and their involvement in methanogenesis from acetate has been suspect for several years (Stadtman, 1967). Initial evidence for the involvement of corrinoids in methyl transfer reactions were obtained from studies with whole cells and cell extracts (Eikmanns and Thauer, 1985; Laufer et al., 1987; van de Wijingaard et al., 1988; Patel and Sprott, 1990). Indications for involvement of a corrinoid protein as the immediate methyl acceptor after cleavage of acetyl-CoA were obtained with studies on the CO dehydrogenase enzyme complex from *M. thermophila* (Abbanat and Ferry, 1991). The Co/Fe-S component of the complex contains factor III (Coα[α-(5-hydroxybenzimidazolyl)]-cobamide), the cobalt atom of which is reduced to the Co^{1+} state with electrons donated directly by the Ni/Fe-S component. The Co^{1+} is a supernucleophile which is proposed to displace the methyl group bound to the Ni/Fe-S component after acetyl-CoA cleavage. EPR spectroscopy of the as-isolated Co/Fe-S component indicates a low-spin Co^{2+} (Jablonski et al., 1993).

There is no superhyperfine splitting from the nitrogen nucleus (I = 1/2) of the 5-hydroxybenzimadazole base in factor III, a result which indicates the absence of a lower axial ligand to the cobalt atom (base-off configuration). The base-off configuration of corrinoids changes the midpoint potential of the $Co^{2+/1+}$ couple from approximately -600 mV (base-on) to a less negative value which allows reduction to the methyl-accepting Co^{1+} redox state by physiological electron donors that have midpoint potentials in the range of -500 mV (Ragsdale, 1991). Accordingly, redox titration of the $Co^{2+/1+}$ couple reveals a midpoint potential of -486 mV for the Co/Fe-S component of the CO dehydrogenase complex from *M. thermophila* (Jablonski et al., 1993). The results are similar to that reported for the *C. thermoaceticum* corrinoid/Fe-S protein that also contains a corrinoid in the base-off configuration (Harder et al., 1989). The Co/Fe-S component of the *M. thermophila* enzyme complex contains a 4Fe-4S center with a midpoint potential of -502 mV which is nearly isopotential with the $Co^{2+/1+}$ couple and is probably involved in electron transfer from the Ni/Fe-S component to the cobalt atom (Jablonski et al., 1993). These properties are similar to the 4Fe-4S center present in the *C. thermoaceticum* corrinoid/Fe-S protein. The genes encoding the two subunits of the Co/Fe-S component from *M. thermophila* have been cloned and sequenced (Maupin and Ferry, 1993); the deduced amino acid sequences share high identity with the sequence deduced from the genes encoding the subunits of the corrinoid/Fe-S protein from *C. thermoaceticum* (Lu et al., 1993). Taken together, the properties of the Co/Fe-S component of *M. thermophila* strongly suggest that it accepts the methyl group of acetyl-CoA after cleavage by the Ni/Fe-S component; however, this reaction has yet to be demonstrated.

In summary, the biochemical properties of the enzyme components from the *M. thermophila* enzyme complex have a striking resemblance to properties of the *C. thermoaceticum* acetyl-CoA synthase and corrinoid/Fe-S protein. The CO dehydrogenases from both organisms are two-subunit, nickel-containing, iron-sulfur proteins with CO-oxidizing activity that reduce ferredoxin. Each of the CO dehydrogenases are associated with a unique two-subunit corrinoid/iron-sulfur protein. Both CO dehydrogenases transfer electrons to the respective corrinoid/iron-sulfur proteins in the absence of additional electron carriers. The acetyl-CoA synthase from *C. thermoaceticum* catalyzes the synthesis of acetyl-CoA from CH_3I, CO, and CoA with an absolute requirement for the corrinoid/ iron-sulfur protein which is methylated with CH_3I followed by transfer of the methyl group to the synthase (Ragsdale, 1991). The five-subunit enzyme complex from *M. thermophila* also catalyzes the synthesis of acetyl-CoA from CH_3I, CO (or $CO_2 + 2e^-$) and CoA (Abbanat and Ferry, 1990).

The composition and properties of the enzyme components from the CODH complex of *M. thermophila* are consistent with a proposed acetyl-CoA cleavage mechanism that is analogous to a reversal of the mechanism proposed for acetyl-

CoA synthesis in *C. thermoaceticum*. In the proposed mechanism (Fig. 21.2), the Ni/Fe-S component cleaves the C–C and C–S bonds of acetyl-CoA at the Ni-Fe site. After cleavage, the methyl group is transferred to the Co^{1+} atom of the Co/Fe-S component. The *M. thermophila* enzyme complex catalyzes an exchange of CoA with acetyl-CoA at rates five-fold greater than the *C. thermoaceticum* acetyl-CoA synthase (Raybuck et al., 1991). The difference in rates may reflect the acetyl-CoA cleavage function for the *M. thermophila* enzyme, as opposed to the *C. thermoaceticum* enzyme which functions in the biosynthesis of acetyl-CoA. The complex also catalyzes an exchange of CO with the carbonyl group of acetyl-CoA (Raybuck et al., 1991); however, the rate of exchange is an order of magnitude less than the *C. thermoaceticum* enzyme. This low rate of exchange is consistent with a low rate of exchange of CO with the carbonyl group during methanogenesis from acetate (Nelson et al., 1987) and no detectable accumulation or consumption of CO during growth on acetate (Zinder and Anguish, 1992). In addition to the C–C and C–S bond cleavage activity, it is proposed that the Ni/Fe-S component binds the carbonyl group and oxidizes it to CO_2 (Lu et al., 1994). This proposed function is supported by the ability of the Ni/Fe-S component to oxidize CO and reduce a ferredoxin purified from *M. thermophila* (Abbanat and Ferry, 1991). In addition to the synthesis of acetyl-CoA from CoA, CO, and CH_3I, the complex is also able to reduce CO_2 to the carbonyl precursor with electrons from titanium (III) citrate; however, CO is preferentially incorporated into acetyl-CoA without prior oxidation to free CO_2 (Abbanat and Ferry, 1990). CO is not directly incorporated during the synthesis of acetate from CH_3I, CO_2, and CO in cell extracts or whole-cell suspensions of *M. barkeri*; rather, CO is a source of electrons for reduction of CO_2 to the carbonyl precursor (Eikmanns and Thauer, 1984; Laufer et al., 1987; Fischer and Thauer, 1990a). This apparent discrepancy between *M. thermophila* and *M. barkeri* is unexplained but could be a result of rapid oxidation of CO to CO_2 in whole cells of *M. barkeri* before any significant acetate synthesis.

(c) *M. BARKERI*

Immunogold labeling shows that the CO dehydrogenase is primarily located in the cytoplasm of *M. barkeri* grown on acetate, a result which agrees with location in the soluble cell fraction after cell disruption (Gokhale et al., 1993). Initially, the enzyme was isolated in an $\alpha_2\beta_2$ oligomer with subunit M_rs of approximately 90 and 19 kDa (Grahame and Stadtman, 1987; Krzycki et al., 1989), results similar to the CO-oxidizing Ni/Fe-S component of the *M. thermophila* complex. Recently, the *M. barkeri* $\alpha_2\beta_2$ CO dehydrogenase was purified in an enzyme complex which also contains a corrinoid protein (Grahame, 1991). The complex catalyzes cleavage of acetyl-CoA and transfer of the methyl group to tetrahydrosarcinapterin (H_4SPT), a result consistent with studies which implicate

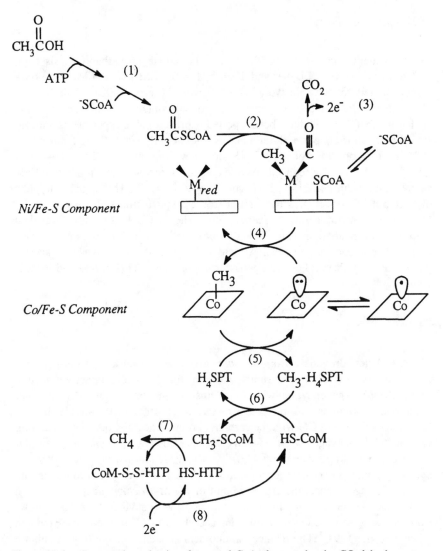

Figure 21.2. Proposed mechanism for acetyl-CoA cleavage by the CO dehydrogenase complex from *M. thermophila*. The M above the rectangle represents the active site metal center of the Ni/Fe-S component; the dots in the orbitals above the rhombus represent the Co^{2+} and Co^{1+} states of the cobamide center of the Co/Fe-S component. The numbering refers to (1) activation of acetate, (2) cleavage by the Ni/Fe-S component, (3) oxidation of the carbonyl group, (4) methyl transfer to the Co/Fe-S component, (5) methyltransfer to tetrahydrosarcinapterin, (6) methyl transfer to coenzyme M, (7) reductive demethylation to methane, and (8) reductive regeneration of cofactors to the corresponding sulfhydryl forms. Adapted from Jablonski et al. (1993).

tetrahydromethanopterin (H_4MPT) as an intermediate in methanogenesis from acetate in *M. barkeri* (Fischer and Thauer, 1989). H_4SPT and H_4MPT are one-carbon carriers found exclusively in the "*Archaea*" (Ferry, 1992a). Although the *M. barkeri* $\alpha_2\beta_2$ CO dehydrogenase component contains Ni and Fe, no EPR signals attributable to Ni have been reported. Core extrusion experiments indicate six [4Fe-4S] clusters per $\alpha_2\beta_2$ tetramer (Krzycki et al., 1989). Low-temperature EPR spectroscopy identifies a 4Fe-4S cluster with *g* values of 2.05, 1.94, and 1.90 ($g_{av} = 1.96$). The reduced enzyme contains a second low-temperature EPR signal with *g* values of 2.005, 1.91, and 1.76 ($g_{av} = 1.89$). This signal is atypical of bacterial 4Fe-4S centers but similar to the atypical iron-sulfur signal in the Ni/Fe-S component from *M. thermophila*. The atypical iron-sulfur signal is perturbed on incubation with CO and is observed in whole cells of *M. barkeri* during methanogenesis, a result which indicates that the cleavage of acetate yields a moiety that CO dehydrogenase recognizes as CO (Krzycki and Prince, 1990).

(d) METHANOTHRIX

The CO dehydrogenase from *M. soehngenii* can be purified in the presence of air but requires strictly anaerobic conditions for CO-oxidizing activity (Jetten et al., 1989). The enzyme, which contains Ni and Fe, is an $\alpha_2\beta_2$ oligomer composed of subunits with molecular masses of 79,400 and 19,400 Da. It catalyzes the exchange of CO with the carbonyl group of acetyl-CoA, demonstrating C–C and C–S cleavage activity (Jetten et al., 1991a). Thus, the enzyme is similar to the CO dehydrogenases from *M. barkeri* and *M. thermophila* except the *M. soehngenii* enzyme is not purified in association with a corrinoid-containing protein. The gene encoding the α subunit of the *M. soehngenii* has a deduced amino acid sequence which has some homology with acyl-CoA oxidases; however, no consensus sequences involved in acetyl-CoA or CoA binding are evident (Eggen et al., 1991). The deduced amino acid sequence of the largest subunit also contains eight cysteine residues with spacings that could accommodate Fe-S centers; accordingly, the anaerobically purified (reduced) enzyme exhibits two low-temperature EPR signals (Jetten et al., 1991a, 1991b). One of these signals has *g* values of 2.05, 1.93, and 1.87 ($g_{av} = 1.95$) and is attributable to a 4Fe-4S center. The other low-temperature signal has *g* values of 2.005, 1.89, and 1.73 ($g_{av} = 1.87$) similar to the atypical Fe-S signals present in the CO dehydrogenases from the *Methanosarcina* and *C. thermoaceticum*. The atypical signal is substoichiometric in intensity and partially disappears when the enzyme is incubated with CO. It is postulated that this signal may arise from a 6Fe-6S prismane-like center. No EPR signals attributable to a Ni center have been reported for the *M. soehngenii* enzyme.

21.4.2 Autotrophic CO₂-Reducing Methanogens

The first evidence for involvement of nickel in a CO dehydrogenase of any methanogen was obtained with the nonacetotrophic CO_2 reducer *Methanobrevibacter arboriphilicus* (Hammel et al., 1984). The synthesis of an active CO dehydrogenase in growing cultures of this organism required nickel in the growth medium. Moreover, the 21-fold purified enzyme from [63]Ni-labeled cells comigrated with [63]Ni during gel filtration.

CO dehydrogenase was purified to homogeneity from *Methanococcus vanielii* grown with formate as the sole carbon source (DeMoll et al., 1987). The enzyme has subunits of 89- and 21-kDa in a $\alpha_2\beta_2$ configuration and contains two nickel atoms. Thus, the enzyme is similar to the CO dehydrogenases from acetotrophic methanogens except for a higher pH optimum of 10.5 and a significantly higher net positive charge. Although the enzyme contains nickel, no Ni EPR signal could be detected at 77°K and the enzyme was incompetent in the exchange of CO with the carbonyl carbon of acetyl-CoA.

Acknowledgments

Research in the author's laboratory is supported by the Department of Energy, Gas Research Institute, National Institutes of Health, National Science Foundation, and Office of Naval Research.

References

Abbanat, D. R., and J. G. Ferry. 1990. Synthesis of acetyl-CoA by the carbon monoxide dehydrogenase complex from acetate-grown *Methanosarcina thermophila*. *J. Bacteriol.* **172**:7145–7150.

Abbanat, D. R., and J. G. Ferry. 1991. Resolution of component proteins in an enzyme complex from *Methanosarcina thermophila* catalyzing the synthesis or cleavage of acetyl-CoA. *Proc. Natl. Acad. Sci. USA* **88**:3272–3276.

Alber, B. E., and J. G. Ferry. 1994. A novel carbonic anhydrase from the archaeon *Methanosarcina thermophila*. *Proc. Natl. Acad. Sci. USA* (in press).

Bhatnagar, L., J. A. Krzycki, and J. G. Zeikus. 1987. Analysis of hydrogen metabolism in *Methanosarcina barkeri:* regulation of hydrogenase and role of CO-dehydrogenase in H₂ production. *FEMS Microbiol. Lett.* **41**:337–343.

Bott, M., B. Eikmanns, and R. K. Thauer. 1986. Coupling of carbon monoxide oxidation to CO₂ and H₂ with the phosphorylation of ADP in acetate-grown *Methanosarcina barkeri*. *Eur. J. Biochem.* **159**:393–398.

Bott, M., and R. K. Thauer. 1987. Proton-motive-force-driven formation of CO from CO₂ and H₂ in methanogenic bacteria. *Eur. J. Biochem.* **168**:407–412.

Bott, M., and R. K. Thauer. 1989. Proton translocation coupled to the oxidation of carbon monoxide to CO_2 and H_2 in *Methanosarcina barkeri*. *Eur. J. Biochem.* **179**:469–472.

Clements, A. P., and J. G. Ferry. 1992. Cloning, nucleotide sequence, and transcriptional analyses of the gene encoding a ferredoxin from *Methanosarcina thermophila*. *J. Bacteriol.* **174**:5244–5250.

Conrad, R., and R. K. Thauer. 1983. Carbon monoxide production by *Methanobacterium thermoautotrophicum*. *FEMS Microbiol. Lett.* **20**:229–232.

Daniels, L., G. Fuchs, R. K. Thauer, and J. G. Zeikus. 1977. Carbon monoxide oxidation by methanogenic bacteria. *J. Bacteriol.* **132**:118–126.

DeMoll, E., D. A. Grahame, J. M. Harnly, L. Tsai, and T. C. Stadtman. 1987. Purification and properties of carbon monoxide dehydrogenase from *Methanococcus vannielii*. *J. Bacteriol.* **169**:3916–3920.

DiMarco, A. A., T. A. Bobik, and R. S. Wolfe. 1990. Unusual coenzymes of methanogenesis. *Annu. Rev. Biochem.* **59**:355–394.

Eggen, R. I. L., A. C. M. Geerling, M. S. M. Jetten, and W. M. Devos. 1991. Cloning, expression, and sequence analysis of the genes for carbon monoxide dehydrogenase of *Methanothrix soehngenii*. *J. Biol. Chem.* **266**:6883–6887.

Eikmanns, B., and R. K. Thauer. 1984. Catalysis of an isotopic exchange between CO_2 and the carboxyl group of acetate by *Methanosarcina barkeri* grown on acetate. *Arch. Microbiol.* **138**:365–370.

Eikmanns, B., and R. K. Thauer. 1985. Evidence for the involvement and role of a corrinoid enzyme in methane formation from acetate in *Methanosarcina barkeri*. *Arch. Microbiol.* **142**:175–179.

Eikmanns, B., G. Fuchs, and R. K. Thauer. 1985. Formation of carbon monoxide from CO_2 and H_2 by *Methanobacterium thermoautotrophicum*. *Eur. J. Biochem.* **146**:149–154.

Ferry, J. G. 1992a. Biochemistry of methanogenesis. *Crit. Rev. Biochem. Mol. Biol.* **27**:473–503.

Ferry, J. G. 1992b. Methane from acetate. *J. Bacteriol.* **174**:5489–5495.

Fischer, R., and R. K. Thauer. 1989. Methyltetrahydromethanopterin as an intermediate in methanogenesis from acetate in *Methanosarcina barkeri*. *Arch. Microbiol.* **151**:459–465.

Fischer, R., and R. K. Thauer. 1990a. Methanogenesis from acetate in cell extracts of *Methanosarcina barkeri:* isotope exchange between CO_2 and the carbonyl group of acetyl-CoA, and the role of H_2. *Arch. Microbiol.* **153**:156–162.

Fischer, R., and R. K. Thauer. 1990b. Ferredoxin-dependent methane formation from acetate in cell extracts of *Methanosarcina barkeri* (strain MS). *FEBS Lett.* **269**:368–372.

Fischer, R., P. Gartner, A. Yeliseev, and R. K. Thauer. 1992. N-5-methyltetrahydromethanopterin:coenzyme M methyltransferase in methanogenic archaebacteria is a membrane protein. *Arch. Microbiol.* **158**:208–217.

Gokhale, J. U., H. C. Aldrich, L. Bhatnagar, and J. G. Zeikus. 1993. Localization of carbon monoxide dehydrogenase in acetate-adapted *Methanosarcina barkeri*. *Can. J. Microbiol.* **39**:223–226.

Grahame, D. A. 1991. Catalysis of acetyl-CoA cleavage and tetrahydrosarcinapterin methylation by a carbon monoxide dehydrogenase-corrinoid enzyme complex. *J. Biol. Chem.* **266**:22227–22233.

Grahame, D. A., and T. C. Stadtman. 1987. Carbon monoxide dehydrogenase from *Methanosarcina barkeri*. Disaggregation, purification, and physicochemical properties of the enzyme. *J. Biol. Chem.* **262**:3706–3712.

Hammel, K. E., K. L. Cornwell, G. B. Diekert, and R. K. Thauer. 1984. Evidence for a nickel-containing carbon monoxide dehydrogenase in *Methanobrevibacter arboriphilicus*. *J. Bacteriol.* **157**:975–978.

Harder, S. R., W. P. Lu, B. A. Feinberg, and S. W. Ragsdale. 1989. Spectroelectrochemical studies of the corrinoid iron-sulfur protein involved in acetyl coenzyme-A synthesis by *Clostridium thermoaceticum*. *Biochemistry* **28**:9080–9087.

Hegeman, G. 1980. Oxidation of carbon monoxide by bacteria. *Trends Biochem. Sci.* **5**:1–3.

Heiden, S., R. Hedderich, E. Setzke, and R. K. Thauer. 1993. Purification of a cytochrome-b containing H_2 : heterodisulfide oxidoreductase complex from membranes of *Methanosarcina barkeri*. *Eur. J. Biochem.* **213**:529–535.

Jablonski, P. E., W. P. Lu, S. W. Ragsdale, and J. G. Ferry. 1993. Characterization of the metal centers of the corrinoid/iron-sulfur component of the CO dehydrogenase enzyme complex from *Methanosarcina thermophila* by EPR spectroscopy and spectroelectrochemistry. *J. Biol. Chem.* **268**:325–329.

Jetten, M. S. M., A. J. M. Stams, and A. J. B. Zehnder. 1989. Purification and characterization of an oxygen-stable carbon monoxide dehydrogenase of *Methanothrix soehngenii*. *FEBS Lett.* **181**:437–441.

Jetten, M. S. M., W. R. Hagen, A. J. Pierik, A. J. M. Stams, and A. J. B. Zehnder. 1991a. Paramagnetic centers and acetyl-coenzyme A/CO exchange activity of carbon monoxide dehydrogenase from *Methanothrix soehngenii*. *Eur. J. Biochem.* **195**:385–391.

Jetten, M. S. M., A. J. Pierik, and W. R. Hagen. 1991b. EPR characterization of a high-spin system in carbon monoxide dehydrogenase from *Methanothrix soehngenii*. *Eur. J. Biochem.* **202**:1291–1297.

Jetten, M. S. M., A. J. M. Stams, and A. J. B. Zehnder. 1992. Methanogenesis from acetate. A comparison of the acetate metabolism in *Methanothrix soehngenii* and *Methanosarcina* spp. *FEMS Microbiol. Rev.* **88**:181–198.

Kluyver, A. J., and C. G. T. P. Schnellen. 1947. On the fermentation of carbon monoxide by pure cultures of methane bacteria. *Arch. Biochem.* **14**:57–70.

Kohler, H. P. E. 1988. Isolation of cobamides from *Methanothrix soehngenii*: 5-methylbenzimidazole as the x-ligand of the predominant cobamide. *Arch. Microbiol.* **150**:219–223.

Krone, U. E., R. K. Thauer, H. P. C. Hogenkamp, and K. Steinbach. 1991. Reductive formation of carbon monoxide from CC14 and FREON-11, FREON-12, and FREON-13 catalyzed by corrinoids. *Biochemistry* **30**:2713–2719.

Krzycki, J. A., and J. G. Zeikus. 1984. Characterization and purification of carbon monoxide dehydrogenase from *Methanosarcina barkeri*. *J. Bacteriol.* **158**:231–237.

Krzycki, J. A., L. J. Lehman, and J. G. Zeikus. 1985. Acetate catabolism by *Methanosarcina barkeri*: evidence for involvement of carbon monoxide dehydrogenase, methyl coenzyme M, and methylreductase. *J. Bacteriol.* **163**:1000–1006.

Krzycki, J. A., L. E. Mortenson, and R. C. Prince. 1989. Paramagnetic centers of carbon monoxide dehydrogenase from aceticlastic *Methanosarcina barkeri*. *J. Biol. Chem.* **264**:7217–7221.

Krzycki, J. A., and R. C. Prince. 1990. EPR observation of carbon monoxide dehydrogenase, methylreductase and corrinoid in intact *Methanosarcina barkeri* during methanogenesis from acetate. *Biochim. Biophys. Acta* **1015**:53–60.

Kumar, M., and S. W. Ragsdale. 1992. Characterization of the CO binding site of carbon monoxide dehydrogenase from *Clostridium thermoaceticum* by infrared spectroscopy. *J. Am. Chem. Soc.* **114**:8713–8715.

Laufer, K., B. Eikmanns, U. Frimmer, and R. K. Thauer. 1987. Methanogenesis from acetate by *Methanosarcina barkeri*: catalysis of acetate formation from methyliodide, CO_2, and H_2 by the enzyme system involved. *Z. Naturforsch.* **42c**:360–372.

Lu, W. P., S. R. Harder, and S. W. Ragsdale. 1990. Controlled potential enzymology of methyl transfer reactions involved in acetyl-CoA synthesis by CO dehydrogenase and the corrinoid/iron-sulfur protein from *Clostridium thermoaceticum*. *J. Biol. Chem.* **265**:3124–3133.

Lu, W. P., I. Schiau, J. R. Cunningham, and S. W. Ragsdale. 1993. Sequence and expression of the gene encoding the corrinoid/iron-sulfur protein from *Clostridium thermoaceticum* and reconstitution of the recombinant protein to full activity. *J. Biol. Chem.* **268**:5605–5614.

Lu, W. P., P. E. Jablonski, M. E. Rasche, J. G. Ferry, and S. W. Ragsdale. 1994. Characterization of the metal centers of the Ni/Fe-S component of the carbon-monoxide dehydrogenase enzyme complex from *Methanosarcina thermophila*. *J. Biol. Chem.* **269**:9736–9742.

Maupin, J. A., and J. G. Ferry. 1993. Corrinoid-containing cobalt/iron-sulfur component of the CO dehydrogenase complex from *Methanosarcina thermophila* strain TM-1: cloning, sequencing, and overexpression in *Escherichia coli*. Abstr. I13, p. 242. *93rd General Meet. Am. Soc. Microbiol.* 1993.

Min, H., and S. H. Zinder. 1989. Kinetics of acetate utilization by 2 thermophilic acetotrophic methanogens: *Methanosarcina* sp strain CALS-1 and *Methanothrix* sp strain CALS-1. *Appl. Environ. Microbiol.* **55**:488–491.

Mukhopadhyay, B., E. Purwantini, E. C. Demacario, and L. Daniels. 1991. Characterization of a *Methanosarcina* strain isolated from goat feces, and that grows on H_2-CO_2 only after adaptation. *Curr. Microbiol.* **23**:165–173.

Nelson, M. J. K., K. C. Terlesky, and J. G. Ferry. 1987. Recent developments on the biochemistry of methanogenesis from acetate. In: *Microbial Growth on C-1 Compounds,* H. W. van Verseveld and J. A. Duine (eds.), pp. 70–76. Martinus Nijhoff, Dordrecht.

O'Brien, J. M., R. H. Wolkin, T. T. Moench, J. B. Morgan, and J. G. Zeikus. 1984. Association of hydrogen metabolism with unitrophic or mixotrophic growth of *Methanosarcina barkeri* on carbon monoxide. *J. Bacteriol.* **158:**373–375.

Ohtsubo, S., K. Demizu, S. Kohno, I. Miura, T. Ogawa, and H. Fukuda. 1992. Comparison of acetate utilization among strains of an aceticlastic methanogen, *Methanothrix soehngenii Appl. Environ. Microbiol.* **58:**703–705.

Patel, G. B., and G. D. Sprott. 1990. *Methanosaeta concilii* Gen-Nov, Sp-Nov (*Methanothrix concilii*) and *Methanosaeta thermoacetophila* Nom-Rev, Comb-Nov. *Int. J. Syst. Bacteriol.* **40:**79–82.

Petersen, S. P., and B. K. Ahring. 1991. Acetate oxidation in a thermophilic anaerobic sewage-sludge digestor. The importance of non-aceticlastic methanogenesis from acetate. *FEMS Microbiol. Ecol.* **86:**149–157.

Ragsdale, S. W. 1991. Enzymology of the acetyl-CoA pathway of CO_2 fixation. *Crit. Rev. Biochem. Mol. Biol.* **26:**261–300.

Raybuck, S. A., S. E. Ramer, D. R. Abbanat, J. W. Peters, W. H. Orme-Johnson, J. G. Ferry, and C. T. Walsh. 1991. Demonstration of carbon-carbon bond cleavage of acetyl coenzyme A by using isotopic exchange catalyzed by the CO dehydrogenase complex from acetate-grown *Methanosarcina thermophila. J. Bacteriol.* **173:**929–932.

Silveira, R. G., N. Nishio, and S. Nagai. 1991. Growth characteristics and corrinoid production of *Methanosarcina barkeri* on methanol-acetate medium. *J. Ferment. Bioengineer.* **71:**28–34.

Sowers, K. R., S. F. Baron, and J. G. Ferry. 1984. *Methanosarcina acetivorans* sp. nov., an acetotrophic methane-producing bacterium isolated from marine sediments. *Appl. Environ. Microbiol.* **47:**971–978.

Stadtman, T. C. 1967. Methane fermentation. *Annu. Rev. Microbiol.* **21:**121–142.

Stupperich, E., and B. Krautler. 1988. Pseudo vitamine B_{12} or 5-hydroxybenzimidazolylcobamide are the corrinoids found in methanogenic bacteria. *Arch. Microbiol.* **149:**268–271.

Terlesky, K. C., M. J. K. Nelson, and J. G. Ferry. 1986. Isolation of an enzyme complex with carbon monoxide dehydrogenase activity containing a corrinoid and nickel from acetate-grown *Methanosarcina thermophila. J. Bacteriol.* **168:**1053–1058.

Terlesky, K. C., M. J. Barber, D. J. Aceti, and J. G. Ferry. 1987. EPR properties of the Ni-Fe-C center in an enzyme complex with carbon monoxide dehydrogenase activity from acetate-grown *Methanosarcina thermophila.* Evidence that acetyl-CoA is a physiological substrate. *J. Biol. Chem.* **262:**15392–15395.

Terlesky, K. C., and J. G. Ferry. 1988a. Ferredoxin requirement for electron transport from the carbon monoxide dehydrogenase complex to a membrane-bound hydrogenase in acetate-grown *Methanosarcina thermophila. J. Biol. Chem.* **263:**4075–4079.

Terlesky, K. C., and J. G. Ferry. 1988b. Purification and characterization of a ferredoxin from acetate-grown *Methanosarcina thermophila*. *J. Biol. Chem.* **263**:4080–4082.

Thauer, R. K., K. Jungermann, and K. Decker. 1977. Energy conservation in chemothrophic anaerobic bacteria. *Bacteriol. Rev.* **41**:100–180.

Thauer, R. K., D. Moller-Zinkhan, and A. M. Spormann. 1989. Biochemistry of acetate catabolism in anaerobic chemotrophic bacteria. *Annu. Rev. Microbiol.* **43**:43–67.

Thebrath, B., H. P. Mayer, and R. Conrad. 1992. Bicarbonate-dependent production and methanogenic consumption of acetate in anoxic paddy soil. *FEMS Microbiol. Ecol.* **86**:295–302.

van de Wijingaard, W. M. H., C. van der Drift, and G. D. Vogels. 1988. Involvement of a corrinoid enzyme in methanogenesis from acetate in *Methanosarcina barkeri*. *FEMS Microbiol. Lett.* **52**:165–172.

Vogels, G. D., J. T. Keltjens, and C. van der Drift. 1988. Biochemistry of methane production. In: *Biology of Anaerobic Microorganisms*, A. J. B. Zehnder (eds.), pp. 707–770. Wiley, New York.

Westermann, P., B. K. Ahring, and R. A. Mah. 1989a. Temperature compensation in *Methanosarcina barkeri* by modulation of hydrogen and acetate affinity. *Appl. Environ. Microbiol.* **55**:1262–1266.

Westermann, P., B. K. Ahring, and R. A. Mah. 1989b. Threshold acetate concentrations for acetate catabolism by aceticlastic methanogenic bacteria. *Appl. Environ. Microbiol.* **55**:514–515.

Woese, C. R., O. Kandler, and M. L. Wheelis. 1990. Towards a natural system of organisms. Proposal for the domains archaea, bacteria, and eucarya. *Proc. Natl. Acad. Sci. USA* **87**:4576–4579.

Zinder, S. H., S. C. Cardwell, T. Anguish, M. Lee, and M. Koch. 1984. Methanogenesis in a thermophilic (58°C) anaerobic digestor: *Methanothrix* sp. as an important aceticlastic methanogen. *Appl. Environ. Microbiol.* **47**:796–807.

Zinder, S. H., and M. Koch. 1984. Non-aceticlastic methanogenesis from acetate: acetate oxidation by a thermophilic syntrophic coculture. *Arch. Microbiol.* **138**:263–272.

Zinder, S. H., K. R. Sowers, and J. G. Ferry. 1985. *Methanosarcina thermophila* sp. nov., a thermophilic, acetotrophic, methane-producing bacterium. *Int. J. Syst. Bacteriol.* **35**:522–523.

Zinder, S. H., and T. Anguish. 1992. Carbon monoxide, hydrogen, and formate metabolism during methanogenesis from acetate by thermophilic cultures of *Methanosarcina* and *Methanothrix* strains. *Appl. Environ. Microbiol.* **58**:3323–3329.

22

Occurrence and Function of the Acetyl-CoA Cleavage Pathway in a Syntrophic Propionate-Oxidizing Bacterium

Alfons J. M. Stams and Caroline M. Plugge

22.1 Degradation of Organic Matter in Methanogenic Ecosystems

In methanogenic ecosystems, mineralization of organic matter proceeds via consortia of different physiological types of bacteria. In sequences of oxidation–reduction reactions, complex organic matter is converted to the most reduced (CH_4) and the most oxidized form (CO_2) of carbon (Zehnder, 1978; Gujer and Zehnder, 1983; Boone, 1982). Methanogens do not use compounds like carbohydrates or amino acids for growth. Therefore, they depend on fermentative and acetogenic bacteria for their substrate supply. The substrates which can be used by methanogens include acetate, formate, H_2/CO_2, methanol, methylated amines, ethanol, and isopropanol (Whitman et al., 1992; Widdel, 1986). However, quantitatively, acetate and hydrogen are the most important substrates for methanogens. During the degradation of complex organic matter about two-thirds of the methane is formed by the cleavage of acetate, whereas the remainder is formed by the reduction of CO_2 with H_2 (Mah et al., 1977; Gujer and Zehnder, 1983).

22.2 Interspecies Electron Transfer

A phenomenon which is associated with methanogenic ecosystems is obligate syntrophic degradation caused by interspecies electron transfer (Boone, 1982; Dolfing, 1988; Schink, 1992). In addition to products that can be used by methanogens, fermentative organisms also produce reduced products like propionate, butyrate, and other fatty acids. Such compounds are anaerobically oxidized to acetate, carbon dioxide, and hydrogen (Dolfing, 1988, Schink, 1992). These acetogenic conversions are energetically very unfavorable; the changes in the Gibbs free energy ($\Delta G_0'$) under standard conditions (1 M for solutes, 1 atm for gases, pH 7, and 298°K) are positive (Thauer et al., 1977). Therefore, they can proceed only if the methanogens keep the concentration of hydrogen and, to some extent, the concentration of acetate low. As a consequence of this, proton-reducing acetogenic bacteria only can grow together with methanogens or other hydrogen-consuming anaerobes. Several types of mesophilic proton-reducing acetogenic bacteria have been described growing on alcohols (Bryant et al., 1967, 1977; Schink, 1984), fatty acids (Boone and Bryant, 1980; McInerney et al., 1979, 1981; Roy et al., 1986; Stieb and Schink, 1985, 1986), aromatic compounds like benzoate (Mounfort and Bryant, 1982; Szewzyk and Schink, 1989), and amino acids (Stams and Hansen, 1984, Zindel et al., 1988).

22.3 Syntrophic Propionate Oxidation

The syntrophic degradation of propionate was extensively studied at our department. Reactions involved in the complete mineralization of propionate are

$$\text{Propionate}^- + 3\,H_2O \rightarrow \text{acetate}^- + HCO_3^- + 3\,H_2 + H^+, \quad \Delta G_0' = +76.1\,kJ$$
$$3\,H_2 + (3/4)HCO_3^- + (3/4)H^+ \rightarrow (3/4)CH_4 + 3\,H_2O, \quad \Delta G_0' = -101.7\,kJ$$
$$\text{Acetate}^- + H_2O \rightarrow CH_4 + HCO_3^-, \quad \Delta G_0' = -31.0\,kJ$$

These reactions are catalyzed by three different types of anaerobic bacteria: one proton-reducing acetogenic bacterium, a hydrogen-consuming methanogen, and an aceticlastic methanogen. Propionate oxidation is obligately coupled to the consumption of hydrogen by methanogens. It can be calculated that at an equal concentration of acetate and propionate, a bicarbonate concentration of 25 mM, and a partial pressure of hydrogen (P_{H_2}) of 10^{-5} atm, the $\Delta G'$ value of the oxidation of propionate to acetate becomes -20 kJ/mol.

It should be mentioned that up to now no direct evidence exists that the degradation of propionate is driven by interspecies hydrogen transfer. It is also possible that formate transfer is the mechanism or an additional mechanism by which electrons are transferred from the acetogen to the methanogen. A possible role of formate transfer in syntrophic degradation was proposed by Thiele and Zeikus (1988) and Boone et al. (1989).

Up to now only a few syntrophic propionate-oxidizing cultures have been described (Boone and Bryant, 1980; Koch et al., 1983; Mucha et al., 1988; Dörner, 1992; Stams et al., 1992, 1993). *Syntrophobacter wolinii* is the only syntrophic propionate-oxidizing bacterium that was obtained in a defined coculture with a *Desulfovibrio* strain (Boone and Bryant, 1980) and with *Methanospirillum hungateii* (Dörner, 1992). Syntrophic propionate-oxidizing bacteria were thought to be highly specialized. However, recently it was shown that *S. wolinii* is able to grow fermentatively on pyruvate and fumarate and is able to grow by the anaerobic oxidation of propionate to acetate with sulfate as electron acceptor (Dörner, 1992).

22.3.1   Biochemical Pathway of Propionate Oxidation

Results from experiments with ^{13}C- and ^{14}C-labeled propionate and results from enzyme measurements in cell-free extracts have indicated that the methylmalonyl-CoA pathway as depicted in Figure 22.1 is involved in propionate oxidation by *S. wolinii* and some other propionate-oxidizing cultures (Mah et al., 1990; Houwen et al., 1987, 1990a, 1990b, 1991; Koch et al., 1983, Robbins, 1988).

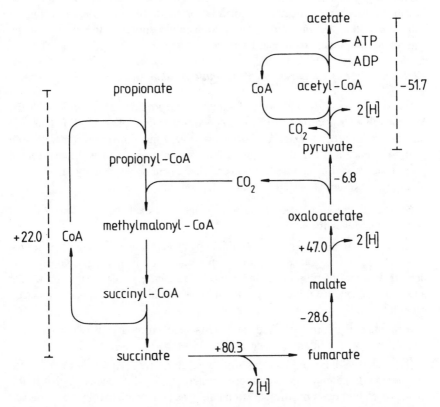

Figure 22.1.   Pathway of propionate oxidation by *Syntrophobacter wolinii*. The numbers indicate the $\Delta G_0'$ values of the reactions. Reproduced from Houwen et al. (1988) with permission.

This pathway is common in propionate-forming anaerobes and is also involved in the anaerobic oxidation of propionate by *Desulfobulbus propionicus* (Stams et al., 1984; Kremer and Hansen, 1988). Typical for the pathway as present in both *Desulfobulbus* and syntrophic cocultures is the coupled carboxylation–decarboxylation of propionyl-CoA to methylmalonyl-CoA and oxaloacetate to pyruvate. The $\Delta G_0'$ of the different reactions of syntrophic propionate oxidation are also given in Figure 22.1. Reducing equivalents are formed in the oxidations of succinate to fumarate, malate to oxaloacetate, and pyruvate to acetyl-CoA. These reducing equivalents are used to form molecular hydrogen from protons. Energetically, succinate oxidation to fumarate is the most difficult reaction in propionate oxidation. It can be calculated that succinate oxidation to fumarate and hydrogen ($\Delta G_0' = + 86.2$ kJ) is only possible if three requirements are fulfilled at the same time: (i) the hydrogen partial pressure has to be 10^{-5} atm, (ii) the succinate : fumarate ratio in the cell has to be 10^4, and (iii) the conversion has to be driven by a membrane potential of about 180 mV. Based on this, it can be hypothesized that fumarate addition might stimulate the growth of the syntrophic propionate-oxidizing bacteria.

22.3.2   Effect of Fumarate on Propionate Oxidation

We started to study the effect of fumarate on propionate oxidation with the *Syntrophobacter wolinii–Desulfovibrio* coculture (Stams et al., 1993). This was not successful because *Desulfovibrio* showed very fast growth on fumarate both in the presence and the absence of sulfate. When we investigated a syntrophic propionate-degrading culture that was enriched at our department by Jan Dolfing, a number of observations were made (Stams et al., 1993):

(i)   In the presence of fumarate or L-malate, the methane yield was lowered and succinate was formed.

(ii)   When methanogens were inhibited by bromoethanesulfonate (BES), propionate was not degraded, but in the presence of fumarate, propionate was degraded; 1 mol of propionate and 3 mol of fumarate yielded 1 mol of acetate and 3 mol of succinate, indicating that fumarate could act as an apparent electron acceptor in the oxidation of propionate to acetate.

(iii)   In the presence of fumarate ^{13}C-labeled propionate was converted to labeled succinate while unlabeled acetate was formed.

(iv)   In BES-inhibited cultures, fumarate alone was fermented to succinate and carbon dioxide in a ratio of approximately 7 : 6 : 4, respectively, but in the presence of hydrogen or formate, fumarate was stoichiometrically converted to succinate.

By a repeated transfer in fumarate-media, the propionate-oxidizing bacterium present in the coculture was obtained free of methanogens. This bacterium degraded propionate to acetate in the presence of *Methanospirillum hungateii*. In the absence of the methanogen, the bacteria could grow by the following reactions:

$$H_2 + \text{fumarate}^{2-} \rightarrow \text{succinate}^{2-}, \quad \Delta G_0' = -86.2 \text{ kJ}$$

$$\text{Formate}^- + \text{fumarate}^{2-} + H_2O \rightarrow \text{succinate}^{2-} + HCO_3^-,$$
$$\Delta G_0' = -84.9 \text{ kJ}$$

$$\text{Fumarate}^{2-} + (1/3) \text{ propionate}^- + 3 H_2O \rightarrow \text{succinate}^{2-} + (1/3) \text{ acetate}^- +$$
$$(1/3) HCO_3^- + (1/3) H^+, \quad \Delta G_0' = -60.8 \text{ kJ}$$

$$\text{Fumarate}^{2-} + (8/7) H_2O \rightarrow (6/7) \text{ succinate}^{2-} + (4/7) HCO_3^- + (2/7) H^+,$$
$$\Delta G_0' = -63.3 \text{ kJ}$$

22.3.3 Involvement of the Acetyl-CoA Cleavage Pathway in Fumarate Oxidation to CO_2

We investigated why the propionate-oxidizing bacterium (MPOB) formed acetate as the oxidized product in the presence of propionate, whereas in the absence of propionate, HCO_3^- was the sole oxidized product (Plugge et al., 1993). Two other anaerobic bacteria, *Proteus rettgeri* and *Malonomonas rubra*, have been reported to grow by fermentation of fumarate to succinate and bicarbonate. These organisms oxidize fumarate via the citric acid cycle to bicarbonate (Kröger, 1974; Dehning and Schink, 1989). Following this pathway (Fig. 22.2A), two fumarate molecules are first converted via malate to two molecules of oxaloacetate. One oxaloacetate molecule is transferred to acetyl-CoA in two decarboxylation steps, and then condensed with the other oxaloacetate molecule to citrate. Citrate is oxidized further to succinate via isocitrate and α-ketoglutarate. In this manner, 2 fumarate are transformed into 1 succinate, 4 CO_2, and reducing equivalents (12[H]). Evidence for this pathway can be obtained by labeling experiments and by enzyme measurements. During growth on fumarate, the MPOB incorporated label from acetate into succinate while no net consumption or formation of acetate took place. Irrespective of whether the C_1 or the C_2 of acetate was labeled, label in succinate was always present at the C_2 or the C_3 atom but never at the C_1 or the C_4 position (Plugge et al., 1993). Incorporation of label at the C_2 and the C_3 positions of succinate can be explained by exchange reactions via the "malate branch" of the pathway. If the citric acid cycle is involved in CO_2 formation from fumarate, the C_1 and the C_4 positions should have been labeled as well. These findings suggested that the citric acid cycle was not involved in fumarate oxidation to CO_2.

The conclusion drawn by the labeling studies could be confirmed by enzyme measurements (Plugge et al., 1993). Some of the key enzymes of the citric acid cycle including α-ketoglutarate dehydrogenase and isocitrate dehydrogenase were not detected. However, when we assayed enzymes of the acetyl-CoA cleavage pathway it appeared that all these enzymes were present in activities comparable with those found in extracts of two sulfate reducers that degrade acetate via the

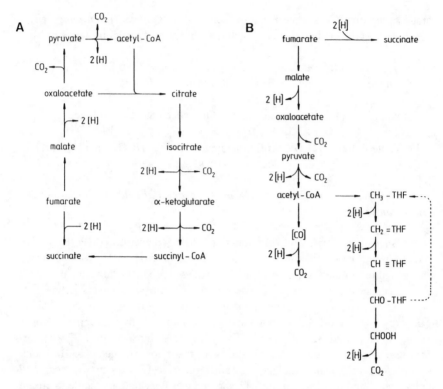

Figure 22.2.   The citric acid cycle (A) and the acetyl-CoA cleavage pathway (B) as possible pathways for the oxidation of fumarate to CO_2.

acetyl-CoA cleavage pathway, *Desulfobacterium autotrophicum* and *Desulfoto-maculum acetoxidans* (Schauder et al., 1989; Spormann and Thauer, 1988). These findings strongly suggested that the propionate degrading bacterium (MPOB) oxidized fumarate to CO_2 as depicted in Figure 22.2B.

22.3.4 Function of the Acetyl-CoA Cleavage Pathway during Growth on Propionate

From the experiments described earlier, it was clear that acetyl-CoA can be oxidized to CO_2, but it remained unclear why acetate was formed as the oxidized product during growth of the MPOB on propionate. It appeared that in cell-free extracts neither propionate and acetate kinase nor propionyl-CoA and acetyl-CoA synthetase were present. However, reasonably high activities of an propionate : acetyl-CoA HS-CoA transferase were detected. This suggested that this enzyme is involved in the activation of propionate. As a consequence, acetate has to be formed during growth on propionate and cannot be formed during growth on fumarate alone. However, the involvement of a HS-CoA transferase in the activation of propionate implies that exactly one acetate molecule has to be formed for each propionate molecule that is activated. Therefore, the organism may have problems with the withdrawal of intermediates from the pathway for cell synthesis. In that case, an anaplerotic CO_2-fixing pathway has to be present for the replenishment of intermediates used for cell synthesis. Most likely, the MPOB possesses the acetyl-CoA cleavage pathway for this purpose during growth on propionate. Recently, a newly enriched propionate-oxidizing bacterium (KoProp) was described, which seems to activate propionate in a similar fashion as our bacterium (Dörner, 1992). However, it is not known which anaplerotic route is present in this organism.

22.3.5 Energetics of Syntrophic Propionate Oxidation

It is intriguing how syntrophic propionate-degrading bacteria conserve energy during growth on propionate. The MPOB possesses a HS-CoA transferase and a transcarboxylase. These enzymes transfer metabolic energy from one compound to another compound with a 100% efficiency. A combination of kinases and phosphotransferases or propionyl-CoA/acetyl-CoA synthetases and (de)carboxylases most likely are less efficient in this respect. In our propionate-degrading bacterium, synthesis of ATP is possible by the cleavage of the thioester bond of succinyl-CoA. However, ATP equivalents, most likely in the form of a membrane potential, have to be invested to pull the unfavorable succinate oxidation reaction. If the latter requires less than 1 ATP, there is a net synthesis of ATP. However, it is difficult to conceive how this is possible because the $\Delta G_0'$ of the conversion of succinyl-CoA to fumarate, HS-CoA, and H_2 is about +51 kJ (Thauer et al., 1977). At a P_{H_2} of 10^{-5} atm, this value still would be about +22 kJ. It is rather unlikely that by differences in concentrations of succinyl-CoA, fumarate, and HS-CoA, the $\Delta G'$ value would become negative enough to allow energy conservation. If the other possibilities are compared, only the oxidation of pyruvate to acetyl-CoA, CO_2, and H_2 is sufficiently exergonic to allow energy conservation. The $\Delta G_0'$ of this reaction is about -12 kJ, whereas at a P_{H_2} of 10^{-5} atm, the $\Delta G'$ is about -41 kJ. A mechanism by which this energy could be conserved is not known yet, especially because it is generally assumed that conversion of pyruvate to acetate yields only 1 ATP. One might speculate that the reduction of protons with reducing equivalents formed in the oxidation of pyruvate takes place at the cytoplasmic membrane and leads to the build up of a proton gradient. The mechanism might be comparable to the extrusion of protons by the transhydrogenase reaction. This extrusion is driven by the conversion of NADH + $NADP^+$ to NAD^+ + NADPH (Hoek and Rydström 1988).

22.4   Comparison with Other Anaerobic Bacteria

Recently, some evidence was obtained that the acetyl-CoA cleavage pathway may also be present in *S. wolinii*. During growth on pyruvate, reducing equivalents formed in the oxidation of pyruvate to acetate were partly used to reduce CO_2 to acetate (Dörner, 1992). It is unknown whether in this organism the acetyl-CoA cleavage pathway is involved during fermentative growth on fumarate.

The involvement of HS-CoA transferases in the activation of substrates is common in anaerobic bacteria. As far as we know, there are no other examples in which this leads to a need for an anaplerotic pathway, such as the acetyl-CoA cleavage pathway. However, an effect similar to that found in this study with respect to the formation of either acetate or CO_2 as the oxidized product was also found with *Desulfobacterium autotrophicum* (Schauder et al., 1986). This organism is able to grow on butyrate and pyruvate; however, it grows poorly on acetate. Butyrate is activated by a butyrate : acetyl-CoA HS-CoA transferase. As a consequence of this, acetate is formed during growth on butyrate, whereas during growth on pyruvate, only CO_2 is formed. In the acetate-degrading *Desulfobacter postgatei* and *Desulfuromonas acetoxidans*, which use the citric acid cycle for the oxidation of acetate to CO_2, acetate is activated by means of an acetate : succinyl-CoA HS-CoA transferase. In these bacteria, low activities of an acetyl-CoA synthetase were detected as anaplerotic enzyme to supply intermediates that have been withdrawn for cell synthesis. However, one might speculate that even sulfate reducers which degrade acetate via the citric acid cycle could have the acetyl-CoA cleavage pathway as an anaplerotic route to overcome the problems of the involvement of the HS-CoA transferase.

References

Boone, D. R. 1982. Terminal reactions in anaerobic digestion of animal waste. *Appl. Environ. Microbiol.* **43**:57–64.

Boone, D. R., and M. P. Bryant. 1980. Propionate-degrading bacterium, *Syntrophobacter wolinii* sp. nov. gen. nov., from methanogenic ecosystems. *Appl. Environ. Microbiol.* **40**:626–632.

Boone, D. R., R. L. Johnson, and Y. Liu. 1989. Diffusion of the interspecies electron carriers H_2 and formate in methanogenic ecosystems, and applications in the measurement of K_m for H_2 and formate uptake. *Appl. Environ. Microbiol.* **55**:1735–1741.

Bryant, M. P., E. A. Wolin, M. J. Wolin, and R. S. Wolfe. 1967. *Methanobacillus omelianksii*, a symbiotic association of two species of bacteria. *Arch. Mikrobiol.* **59**:20–31.

Bryant, M. P., L. L. Campbell, C. A. Reddy, and M. R. Crabill. 1977. Growth of *Desulfovibrio* in lactate or ethanol media low in sulfate in association with H_2-utilizing methanogenic bacteria. *Appl. Environ. Microbiol.* **33**:1162–1169.

Dehning, I., and B. Schink. 1989. *Malonomonas rubra* gen. nov. sp. nov., a microaerotolerant anaerobic bacterium growing by decarboxylation of malonate. *Arch. Microbiol.* **151**:427–433.

Dolfing, J. 1988. Acetogenesis. In: *Biology of Anaerobic Microorganisms,* A. J. B. Zehnder (ed.), pp. 417–468. Wiley, New York.

Dörner, C. (1992). Biochemie und Energetik der Wasserstofffreisetzung in der syntrophen vergärung von Fettsaüren und Benzoat, Ph.D. thesis, University of Tübingen, Germany.

Gujer, W., and A. J. B. Zehnder. 1983. Conversion processes in anaerobic digestion. *Water Sci. Technol.* **15**:127–167.

Hoek, J. B., and J. T. Rydström. 1988. Physiological roles of nicotinamide nucleotide transhydrogenase. *Biochem. J.* **254**:1–10.

Houwen, F. P., C. Dijkema, C. H. H. Schoenmakers, A. J. M. Stams, and A. J. B. Zehnder. 1987. ^{13}C-NMR study of propionate degradation by a methanogenic coculture. *FEMS Microbiol. Lett.* **41**:269–274.

Houwen, F. P., G. Cheng, G. E. Folkers, W. M. J. G. Van de Heuvel, and C. Dijkema. 1988. Pyruvate and fumarate conversion by a methanogenic propionate-oxidizing coculture. In: *Granular Anaerobic Sludge; Microbiology and Technology,* G. Lettinga, A. J. B. Zehnder, J. T. C. Grotenhuis, and L. W. Hulshoff Pol (eds.), pp. 55–70. Pudoc, Wageningen.

Houwen, F. P., J. Plokker, C. Dijkema, and A. J. M. Stams. 1990a. Syntrophic propionate oxidation. In: *Microbiology and Biochemistry of Strict Anaerobes Involved in Interspecies Hydrogen transfer,* J. P. Belaich, M. Bruschi, and J.-L. Garcia (eds.), pp. 281–289. Plenum Press, New York.

Houwen, F. P., J. Plokker, A. J. M. Stams, and A. J. B. Zehnder. 1990b. Enzymatic evidence for involvement of the methylmalonyl-CoA pathway in propionate oxidation by *Syntrophobacter wolinii. Arch. Microbiol.* **155**:52–55.

Houwen, F. P., C. Dijkema, A. J. M. Stams, and A. J. B. Zehnder. 1991. Propionate metabolism in anaerobic bacteria; determination of carboxylation reactions with ^{13}C-NMR spectroscopy. *Biochim. Biophys. Acta* **1056**:126–132.

Koch, M. E., J. Dolfing, K. Wuhrmann, and A. J. B. Zehnder. 1983. Pathways of propionate degradation by enriched methanogenic cultures. *Appl. Environ. Microbiol.* **45**:1411–1414.

Kremer, D. R., and T. A. Hansen. 1988. Pathway of propionate degradation in *Desulfobulbus propionicus. FEMS Microbiol. Lett.* **49**:273–277.

Kröger, A. 1974. Electron-transport phosphorylation coupled to fumarate reduction in anaerobically grown *Proteus rettgeri. Biochim. Biophys. Acta* **347**:273–289.

Mah, R. A., D. M. Ward, L. Baresi, and T. L. Glass. 1977. Biogenesis of methane. *Annu. Rev. Microbiol.* **31**:309–341.

Mah, R. A., L. Y. Xun, D. R. Boone, B. Ahring, P. H. Smith, and A. Wilkie. 1990. Methanogenesis from propionate in sludge and enrichment systems. In: *Microbiology and Biochemistry of Strict Anaerobes Involved in Interspecies Hydrogen Transfer,* J.-P. Belaich, M. Bruschi, and J.-L. Garcia (eds.), pp. 99–111. Plenum Press, New York.

McInerney, M. J., M. P. Bryant, and N. Pfennig. 1979. Anaerobic bacterium that degrades fatty acids in syntrophic association with methanogens. *Arch. Microbiol.* **122**:129–135.

McInerney, M. J., M. P. Bryant, R. B. Hespell, and J. W. Costerton. 1981. *Syntrophomonas wolfei* gen.nov. sp.nov, an anaerobic syntrophic, fatty acid-oxidizing bacterium. *Appl. Environ. Microbiol.* **41**:1029–1039.

Mounfort, D. O., and M. P. Bryant. 1982. Isolation and characterization of an anaerobic benzoate-degrading bacterium from sewage sludge. *Arch. Microbiol.* **133**:249–256.

Mucha, H., F. Lingens, and W. Trösch. 1988. Conversion of propionate to acetate and methane by syntrophic consortia. *Appl. Microbiol. Biotechnol.* **27**:581–586.

Plugge, C. M., C. Dijkema, and A. J. M. Stams. 1993. Acetyl-CoA cleavage pathway in a syntrophic propionate-oxidizing bacterium growing on fumarate in the absence of methanogens. *FEMS Microbiol. Lett.* **110**:71–76.

Robbins, J. E. 1988. A proposed pathway for catabolism of propionate in methanogenic cocultures. *Appl. Environ. Microbiol.* **54**:1300–1301.

Roy, F., E. Samain, H. C. Dubourguier, and G. Albagnac. 1986. *Syntrophomonas sapovorans* sp.nov. a new obligately proton-reducing anaerobe oxidizing saturated and unsaturated long chain fatty acids. *Arch. Microbiol.* **145**:142–147.

Schauder, R., B. Eikmanns, R. K. Thauer, F. Widdel, and G. Fuchs. 1986. Acetate oxidation in CO_2 in anaerobic bacteria via a novel pathway not involving reactions of the citric acid cycle. *Arch. Microbiol.* **145**:162–172.

Schauder, R., A. Preuss, M. Jetten, and G. Fuchs. 1989. Oxidative and reductive acetyl CoA/carbon monoxide pathway in *Desulfobacterium autotrophicum*. 2. Demonstration of the enzymes of the pathway and comparison of CO dehydrogenase. *Arch. Microbiol.* **151**:84–89.

Schink, B. 1984. Fermentation of 2,3-butanediol by *Pelobacter carbinolicus* sp.nov. and *Pelobacter propionicus* sp. nov., and evidence for propionate formation from C_2 compounds. *Arch. Microbiol.* **137**:33–41.

Schink, B. 1992. Syntrophism among prokaryotes. In: *The Prokaryotes,* 2nd ed., A. Balows, H. G. Trüper, M. Dworkin, W. Harder, and K.-H. Schleifer (eds.), pp. 276–299. Springer-Verlag, New York.

Spormann, A. M., and R. K. Thauer. 1988. Anaerobic acetate oxidation to CO_2 by *Desulfotomaculum acetoxidans*. Demonstration of the enzymes required for the operation of an oxidative acetyl-CoA/carbon monoxide dehydrogenase pathway. *Arch. Microbiol.* **150**:374–380.

Stams, A. J. M., and T. A. Hansen. 1984. Fermentation of glutamate and other compounds by *Acidaminobacter hydrogenoformans* gen. nov. sp.nov, an obligate anaerobe isolated from black mud. Studies with pure cultures and mixed cultures with sulfate-reducing and methanogenic bacteria. *Arch. Microbiol.* **137**:329–337.

Stams, A. J. M., D. R. Kremer, K. Nicolay, G. H. Weenk, and T. A. Hansen. 1984. Pathway of propionate formation in *Desulfobulbus propionicus*. *Arch. Microbiol.* **139**:167–173.

Stams, A. J. M., J. B. van Dijk, C. Dijkema, and C. M. Plugge. 1993. Growth of syntrophic propionate-oxidizing bacteria with fumarate in the absence of methanogenic bacteria. *Appl. Environ. Microbiol.* **59**:1114–1119.

Stams, A. J. M., K. C. F. Grolle, C. T. M. J. Frijters, and J. B. Van Lier. 1992. Enrichment of a thermophilic propionate-oxidizing acetogenic bacterium in coculture with *Methanobacterium thermoautotrophicum* or *Methanobacterium thermoformicicum*. *Appl. Environ. Microbiol.* **58**:346–352.

Stieb, M., and B. Schink. 1985. Anaerobic degradation of fatty acids by *Clostridium bryantii* sp. nov., a sporeforming obigately syntrophic bacterium. *Arch. Microbiol.* **140**:387–390.

Stieb, M., and B. Schink. 1986. Anaerobic degradation of isovalerate by a defined methanogenic coculture. *Arch. Microbiol.* **144**:291–295.

Szewzyk, U., and B. Schink. 1989. Degradation of hydroquinonen, gentisate, and benzoate by a fermenting bacterium in pure or defined mixed cultures. *Arch. Microbiol.* **151**:541–545.

Thauer, R. K., K. Jungermann, and K. Decker. 1977. Energy conservation in chemotrophic anaerobic bacteria. *Bacteriol. Rev.* **41**:100–180.

Thiele, J. H., and J. G. Zeikus. 1988. Interactions between hydrogen- and formate-producing bacteria and methanogens during anaerobic digestion, In: *Handbook on Anaerobic Fermentations*. L. E. Erickson and D. Fung (eds.), pp. 537–595. Marcel Dekker, New York.

Whitman, W. B., T. L. Bowen, and D. R. Boone. 1992. The methanogenic bacteria. In: *The Prokaryotes*, 2nd ed., H. G. Trüper, M. Dworkin, W. Harder, and K.-H. Schleifer (eds.), pp. 719–768. Springer-Verlag, New York.

Widdel, F. 1986. Growth of methanogenic bacteria in pure culture with 2-propanol and other alcohols as hydrogen donors. *Appl. Environ. Microbiol.* **51**:1056–1062.

Zehnder, A. J. B. 1978. Ecology of methane formation. In: *Water Pollution Microbiology*, R. Mitchell (ed.), Vol. 2, pp. 349–376. Wiley, New York.

Zindel, U., W. Freudenberg, M. Reith, J. R. Andreesen, J. Schnell, and F. Widdel. 1988. *Eubacterium acidaminophilum* sp.nov., a versatile amino acid degrading anaerobe producing or utilizing H_2 or formate. Description and enzymatic studies. *Arch. Microbiol.* **150**:254–266.

23

Acetate via Glycine: A Different Form of Acetogenesis

Jan R. Andreesen

23.1 Introduction: The Relatedness of Both Acetogenic Pathways

In contrast to acetogenesis via the acetyl-CoA pathway, acetogenesis via glycine has never been reviewed in detail. The carbon flow of the glycine-dependent process was earlier depicted mostly as the "serine bypass" modification rather than the "direct reduction" process (Fuchs, 1986; Ljungdahl, 1984, 1986; Ljungdahl and Wood, 1982, Waber and Wood, 1979; Wood et al., 1986; Ragsdale, 1991; Wood and Ljungdahl, 1991). The glycine reductase modification was dealt with briefly as one of the two possibilities by Wood (1989), although it had been earlier proposed as the likely candidate if the cells are supplemented with trace elements necessary to express enzymes such as formate dehydrogenase/CO_2 reductase and glycine reductase (Dürre and Andreesen, 1982d, 1983; Dürre et al., 1983).

During acetate synthesis via glycine, the first reactions that reduce CO_2 to methylenetetrahydrofolate (methylene-THF) are essentially identical to those of the Wood pathway (Fig. 23.1; see Fig. 1.4 for comparison). In both pathways, ATP has to be expended to activate formate during the synthesis of N-10-formyl-THF. In addition, energy is conserved by substrate-level phosphorylation in the last step of acetate formation in both processes: as acetyl-CoA by the CO dehydrogenase (i.e., acetyl-CoA synthase) in the Wood pathway or as acetyl phosphate in the glycine reductase reaction. The reaction catalyzed by CO dehydrogenase is reversible as shown by the acetate metabolism of some methanogenic and sulfate-reducing bacteria (Thauer, 1988; Thauer et al., 1989; Ferry, 1992), whereas the glycine reductase reaction seems to be irreversible (Arkowitz and

Abeles, 1989). As indicated in Fig. 23.1, both pathways seem to break even relative to energetics.

In contrast to organisms employing the Wood pathway, no autotrophic organism is known that uses the glycine pathway for acetate synthesis. Therefore, some important differences must reside in the intermediate steps. In the glycine pathway, methylene-THF is reductively aminated and carboxylated to glycine by the readily soluble enzyme complex glycine decarboxylase/synthase. Its function as a key enzyme was first shown by Waber and Wood (1979) using *Clostridium acidiurici*. The glycine formed is then reduced by the enzyme complex glycine reductase to acetyl phosphate using the thioredoxin system as natural electron donor (Dietrichs et al., 1991b). Although some of the protein components involved are associated with the cytoplasmic membrane (Freudenberg et al., 1989b; Dietrichs et al., 1991a), there are so far no known integral membrane proteins that might be involved in an energy-conserving step. For some bacteria employing the Wood pathway, this extra energy conservation might be associated with the generation of a sodium gradient (Heise et al., 1992; Becher et al., 1992), reduction of corrinoids (Stupperich et al., 1992), or the presence of menaquinones and cytochromes (Gottwald et al., 1975; Das et al., 1989; Hugenholtz and Ljungdahl, 1989). Clearly, something like this has to occur, as these organisms must live from that extra amount of energy conserved when they grow autotrophically on H_2 plus CO_2.

Sporomusa species are quite interesting acetogens for they use the Wood pathway when they grow autotrophically, on formate or on methanol. An involvement of cytochromes is strongly indicated by mutational analysis (Kamlage and Blaut, personal communication). However, these acetogens prefer to reduce sarcosine (N-methyl glycine) via a sarcosine reductase (quite similar to glycine reductase) to acetyl phosphate, methylamine, and CO_2 (Möller et al., 1986; Oßmer, 1983) (Fig. 23.1). The N-methyl group might be transferred to a corrinoid or tetrahydrofolate (THF) and then be oxidized to CO_2 or reduced to acetate, whereas the remaining glycine moiety might be reduced by glycine reductase or oxidized by glycine decarboxylase to methylene-THF and CO_2. Thus, these organisms seem to integrate both pathways into their physiology. At least the reductase systems seem to have a chance when competing directly with the Wood pathway.

23.2 Conditions for Acetogenesis via Glycine

23.2.1 Redox State of the Substrates

The glycine pathway seems to be preferentially (or even only) used in cases where glycine is an intermediate (Tables 23.1 and 23.2). The redox states of

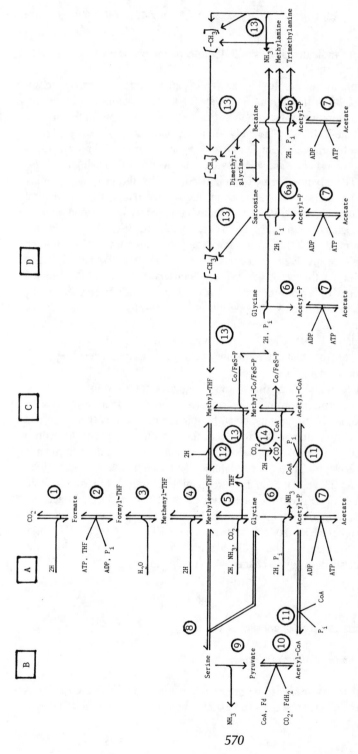

Figure 23.1. Correlation of the two modifications of the glycine pathway (A, B) with the Wood pathway (C) of CO$_2$ reduction to acetate. Pathways: A, glycine reductase pathway; B, glycine-serine-pyruvate-acetate bypass; C, acetyl-CoA (Wood) pathway; D, metabolism of N-methylated glycine derivatives involving parts of both pathways. Components: THF, tetrahydrofolate; Co/FeS-P, corrinoid iron-sulfur protein; 1, CO$_2$ reductase/formate dehydrogenase; 2, N-10 formyl-THF synthetase; 3, methenyl-THF cyclohydrolase; 4, methylene-THF dehydrogenase; 5, glycine decarboxylase/synthase; 6, glycine (a, sarcosine; b, betaine) reductase; 7, acetate kinase; 8, serine hydroxymethyltransferase; 9, serine dehydratase; 10, pyruvate : ferredoxin oxidoreductase; 11, phosphotransacetylase; 12, methylene-THF reductase; 13, methyltransferase(s); 14, CO dehydrogenase/acetyl-CoA synthase.

570

the substrates used by the Wood pathway are generally more reduced than those substrates that yield glycine. Glycine is already more oxidized that acetate. Therefore, no net synthesis from CO_2 has to occur in most cases. The glycine formed as an intermediate can even be partly oxidized to generate extra reducing equivalents for its subsequent reduction to acetyl phosphate. In the case of threonine, glycine with only be reduced. Only in a few cases does a net synthesis of acetate have to take place, e.g., if a purine compound that is more reduced than xanthine (hypoxanthine, adenine, or purine) serves as substrate. Anaerobic purine degradation, e.g., of xanthine, yields via formiminoglycine two redox partners, formate and glycine, that might even be stable products (as formerly observed for *C. cylindrosporum* (Barker and Beck, 1941; Champion and Rabinowitz, 1977). However, under suitable conditions both will be further converted to CO_2 and acetate by formate dehydrogenase and glycine reductase (Schiefer-Ulrich et al., 1984; Hormann and Andreesen, 1989). Thus, only with the unsubstituted purine ring is the net reduction of CO_2 required:

$$\text{Uric acid} \rightarrow 0.75 \text{ acetate} \tag{I}$$
$$\text{Xanthine} \rightarrow \text{formate} + \text{glycine} \rightarrow 1.0 \text{ acetate} \tag{II}$$
$$\text{Hypoxanthine} \rightarrow \text{formate} + \text{acetate} \rightarrow 1.25 \text{ acetate} \tag{III}$$
$$\text{Purine} \rightarrow [2H] + \text{formate} + \text{acetate} \rightarrow 1.5 \text{ acetate} \tag{IV}$$

Only *C. purinolyticum* is able to grow on purine (Dürine et al., 1981). It is likely that these purinolytic organisms partly reroute carbon flow under these special conditions in the reverse direction (i.e., CO_2 to formate and further to glycine) to balance the electrons generated during substrate conversion to xanthine that is split without any redox reactions to formiminoglycine (Vogels and van der Drift, 1976) (Fig. 23.2). With the exception of the first and the last enzymes, formate dehydrogenase and glycine reductase, all other enzymes involved in this sequence seem to work generally in both directions (Fuchs, 1986), also in nonacetogenic bacteria. Thus, such a switch does not create a mechanistic problem. The ability of the formate dehydrogenase present in these purinolytic bacteria to catalyze the reduction of CO_2 has been demonstrated (Thauer, 1973); indeed, about 9% of the acetate synthesized is doubly labeled by $^{13}CO_2$ in the case of *C. acidiurici* and *C. cylindrosporum* (Schulman et al., 1972). In this regard, it is important to note that the fixation of $^{11}CO_2$ and $^{14}CO_2$ into both carbons of acetate had been demonstrated in the classic labeling studies of Barker and coworkers (Barker et al., 1940; Barker and Elsden, 1947; Karlsson and Barker, 1949).

23.2.2 Formation of Formate

As mentioned earlier, the interconversion of CO_2 to formate is rather unique to acetogenic bacteria. Most formate dehydrogenases can only catalyze the oxida-

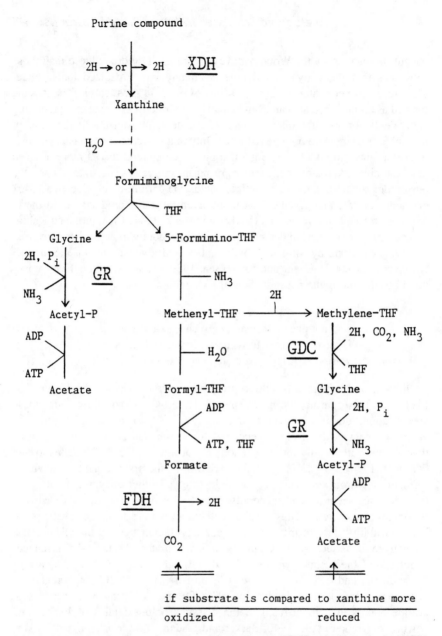

Figure 23.2. Involvement of xanthine dehydrogenase (XDH), glycine reductase (GR), formate dehydrogenase (FDH), and glycine decarboxylase/synthase (GDC) in carbon and electron flow during purine fermentation by acetogenic bacteria that use the glycine pathway.

tion of formate to CO_2. Therefore, the formyl moiety, required as N-10-formyl-THF for biosynthetic purposes (Schirch, 1982; Schirch and Strong, 1989), is generally formed from methylene-THF via methenyl-THF or is formed directly from formate by formyl-THF synthetase, an enzyme present in the cytosol of many organisms (Whitehead et al., 1988). Formate can also be produced by pyruvate-formate lyase, a typical enzyme of enterobacteria; it is also present in some strict anaerobes (Knappe and Sawers, 1990). Breakdown of purines and histidine will give formimino-THF and then methenyl-THF as C_1 products that might be further transformed via N-10-formyl-THF to formate and ATP (Andreesen et al., 1989). Therefore, relatively few reactions are known by which formate can be formed. In all cases, substrates have to be provided that are much more complex or require even formate for their biosynthesis (as in the case with purines). Thus, the reduction of CO_2 to formate might be considered the smartest way to form formate from a simple and abundant source like CO_2.

23.2.3 Serine as an Intermediate

The involvement of serine hydroxymethyltransferase, forming serine from glycine and methylene-THF, was proposed as a key enzyme for acetate synthesis a long time ago (Schulman et al., 1972) (Fig. 23.1B). In earlier experiments, formate and glycine were shown to be condensed to serine (Sager and Beck, 1956; Beck et al., 1957; Hougland and Beck, 1979) that, in turn, was deaminated to pyruvate by a ferrous ion-dependent serine dehydratase (Benziman et al., 1960). A pathway for the oxidation of pyruvate to acetyl-CoA and CO_2 was already outlined in 1961; it included acetate kinase as the final energy-conserving enzyme (Sagers et al., 1961). The involvement of ferredoxin in the pyruvate-dependent formation of NADH was also demonstrated in early studies with *C. acidiurici* (Valentine et al., 1963). The purification of the corresponding enzymes followed in later studies (Uyeda and Rabinowitz, 1968, 1971; Carter and Sagers, 1972).

Although this pathway of glycine conversion to acetate via serine and pyruvate may be of minor relevance *in vivo* (Dürre and Andreesen, 1983), the capacity of the pathway to synthesize acetate from CO_2 has been definitively proven (Schulman et al., 1972; Waber and Wood, 1979). A major breakthrough was the demonstration of the importance of glycine decarboxylase/synthase for acetate synthesis (Fig. 23.1). In the study of Waber and Wood (1979), methylene-THF acts as an intermediate of both carbons of acetate, whereas the C-2 of glycine is only a source of the carboxyl group. All these reactions are well known from one-carbon metabolism (Schirch, 1982, 1984; MacKenzie, 1984). Both serine and glycine provide the THF-bound C_1 compounds required for the formation of thymine, modified RNA bases, methionine, formylmethionine, or the purine ring system (Schirch and Strong, 1989).

Serine is generally formed from 3-phosphoglycerate via a phosphorylated or nonphosphorylated pathway (MacKenzie, 1984). In *E. coli,* an alternative pathway exists for serine biosynthesis (tut-cycle). Threonine is oxidized to 2-amino-3-oxobutyrate and is split to glycine and acetyl-CoA (Ravnikar and Somerville, 1987); two glycine yields one serine by the combined action of glycine decarboxylase and serine hydroxymethyltransferase. The latter transformation is known to occur during the photorespiration in mitochondria (Heldt and Flügge, 1986; Husic et al., 1987). In the peroxisome, glycolate is converted via glyoxylate into glycine; serine is then formed by glycine decarboxylase and serine hydroxymethyltransferase and is transformed via hydroxypyruvate into glycerate. A reaction sequence quite similar to the latter is known as the serine pathway that is employed by some methylotrophic bacteria. Glyoxylate can be generated in the serine pathway by malyl-CoA lyase or, more generally, by isocitrate lyase as part of the glyoxylate bypath of the TCA cycle (Gottschalk, 1986). Thus, it is evident that most of the reactions outlined earlier for C_1 fixation via glycine are, in fact, known from other metabolic capabilities. In this regard, such processes might have simply been rearranged by certain organisms to suit that special catabolic purpose.

23.2.4   Nutritional Factors Required

Earlier studies demonstrated that trace elements play essential catalytic functions for species employing the Wood pathway (Andreesen et al., 1973; Diekert and Thauer, 1978; Diekert et al., 1979) and eventually led to the resolution of many important enzyme systems in these organisms. Similarly, the awareness of the importance of trace elements led to breakthroughs in understanding some of the details of the enzymology of the organisms that form acetate via glycine (Wagner and Andreesen, 1977, 1979; Dürre and Andreesen, 1982a, 1983). The activity of the key enzyme formate dehydrogenase (Fig. 23.2) from *C. acidiurici,* *C. cylindrosporum,* and *C. purinolyticum* is tremendously increased by the presence of selenite and tungstate or molybdate during growth (Schiefer-Ullrich et al., 1984). [185]W-tungsten cochromatographs in all cases with formate dehydrogenase activity, whereas molybdate is a constituent of xanthine dehydrogenase (Wagner et al., 1984; Wagner and Andreesen, 1987). The latter enzyme activity is also positively affected by selenite. Thus, the interconversion of purine compounds to xanthine can be catalyzed quite efficiently by these bacteria (Dürre and Andreesen, 1983). Xanthine is the compound actually split by these anaerobes (Rabinowitz, 1963; Dürre and Andreesen, 1982b). Selenium is also required for growth on glycine (Dürre and Andreesen, 1982a), because it is part of the selenoprotein P_A of glycine reductase (Sliwkowski and Stadtman, 1988b). The presence of these trace elements during enrichment led to the isolation of *C. purinolyticum, Peptostreptococcus barnesae,* and *Eubacterium acidaminophilum*

(Table 23.2) (Dürre et al., 1981; Schiefer-Ullrich and Andreesen, 1985; Zindel et al., 1988).

C. *purinolyticum* metabolizes uric acid by a different pathway when cultured in the absence of selenite (Dürre and Andreesen, 1982c). In the presence of selenite, *P. glycinophilus* can switch pathways for glycine degradation. Under such conditions, the "glycine-serine-pyruvate-acetate" interconversion pathway (Barker et al., 1948) is replaced by the more direct and energetically more favorable reduction of glycine to acetyl phosphate (Fig. 23.1) (Dürre et al., 1983); this metabolic potential is also present in C. *acidiurici* and C. *cylindrosporum* (Barker and Elsden, 1947; Karlsson and Barker, 1949; Dürre and Andreesen, 1983). C. *cylindrosporum* was differentiated from C. *acidiurici* in former studies by producing formate and glycine instead of acetate as main products (Barker and Beck, 1941; Champion and Rabinowitz, 1977). However, selenium-supplemented cells produce only acetate (Schiefer-Ulrich et al., 1984); they thus lose an easily recognizable marker for taxonomic differentiation (Andreesen et al., 1985). Although selenium is a regular contaminant of sulfur compounds, its addition to media as selenite (10^{-7} M) is recommended so that selenium is supplied to the cell in a readily usable form.

These facts demonstrate that experiments prior to 1980 might not reflect the full potential of the organisms studied due to such nutritional insufficiencies. Some species not capable of switching between pathways might have even escaped detection. Indeed, a different selenium content in tap water has a somewhat dramatic effect on the growth and enzyme content of C. *sticklandii* (Turner and Stadtman, 1973). Selenite as well as tungstate are now components of commonly used trace-element solutions (Tschech and Pfennig, 1984).

23.2.5 Sources of Glycine

Many organisms that are known to form acetate via glycine produce this simple amino acid during fermentation of more complex substrates (Table 23.1) such as purines, threonine, or the N-methylated derivatives of glycine (sarcosine, creatine, betaine). Some proteins like gelatin, collagen, spider silk, or certain proteins produced by plants (e.g., species of *Petunia*) are quite rich in glycine (Condit and Meager, 1986; Gosline et al., 1986; Keller et al. 1988). Glycine is also part of conjugates such as hippurate (benzolyglycine) or glycocholate, or of peptides such as glutathione. The hydrolysis of these latter compounds will yield glycine quite readily. Although glycine could be formed from serine by serine hydroxymethyltransferase, this reaction is not involved in the catabolic degradation of serine. In all cases studied, pyruvate is directly formed from serine by serine dehydratase, a unidirectional, pyridoxal phosphate- or iron-sulfur-containing protein (Buckel, 1992).

Table 23.1   Glycine-forming reactions from organic compounds

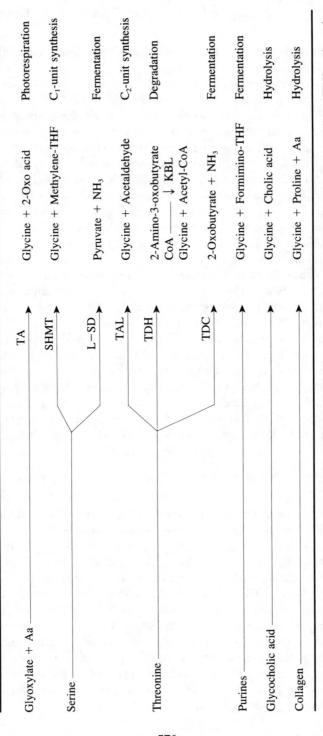

Substrate	Enzyme	Products	Process
Glyoxylate + Aa	TA	Glycine + 2-Oxo acid	Photorespiration
Serine	SHMT	Glycine + Methylene-THF	C$_1$-unit synthesis
Serine	L–SD	Pyruvate + NH$_3$	Fermentation
	TAL	Glycine + Acetaldehyde	C$_2$-unit synthesis
Threonine	TDH	2-Amino-3-oxobutyrate CoA $\longrightarrow$ KBL $\downarrow$ Glycine + Acetyl-CoA	Degradation
Threonine	TDC	2-Oxobutyrate + NH$_3$	Fermentation
Purines		Glycine + Formimino-THF	Fermentation
Glycocholic acid		Glycine + Cholic acid	Hydrolysis
Collagen		Glycine + Proline + Aa	Hydrolysis

Note: TA, transaminase; SHMT, L-serine hydroxymethyltransferase; L-SD, L-serine dehydratase; TAL, L-threonine aldolase; TDH, L-threonine dehydrogenase; KBL, 2-amino 3-ketobutyrate CoA ligase; TDC, L-threonine dehydratase; THF, tetrahydrofolate.

576

23.2.6 Peculiarities of Glycine, N-Methylglycine, and Purine Metabolism

(a) GLYCINE METABOLISM

The organisms dealt with in this chapter should all have the capacity to grow on glycine, for it constitutes the central intermediate. However, some cannot utilize glycine as a substrate or, at least, not under the conditions tested (Table 23.2). Besides the necessity to provide trace elements such as selenium and tungsten, the glycine concentration often used (100 mM) might be too high to achieve growth. Even a specialist such as *E. acidaminophilum* tolerates at first only 25 mM glycine; it must be slowly adapted to 50 mM glycine or higher concentrations (Zindel et al., 1988). The toxic effect of glycine might reside on its interference with the synthesis of the peptidoglycan sacculus, thus creating cell forms sensitive to lysis (Hammes et al., 1973; Heaton et al., 1987; Kämpf et al., 1988).

Bacteria specialized in purine utilization require the addition of a nonreadily utilizable purine compound when glycine serves as a substrate (Barker and Beck, 1941; Dürre and Andreesen, 1982a; Lebertz, 1984; Schiefer-Ullrich and Andreesen, 1985). This requirement can be rationalized, as these specialists normally do not have to synthesize *de novo* the purine ring system (and part of thiamine); indeed, it is their genuine substrate. Because these bacteria must differentiate between purine as a vitamin and purine as a substrate, both the purine concentration and the nature of the substituent groups are important relative to growth on glycine. Although a wide spectrum of variations is observed, all strains of the purinolytic clostridia are able to take up glycine and to degrade it to acetate (Lebertz, 1984). The uptake of glycine and formiminoglycine by *C. acidiurici* requires sodium ions and metabolic energy derived from uric acid or xanthine (Chen, 1975). Growth of *E. acidaminophilum* on glycine also requires sodium ions, and the activity of the glycine decarboxylase from *P. glycinophilus* is stimulated twofold by sodium (Lebertz, 1984).

Regulatory effects might be another reason for observing no or only sporadic degradation of glycine. For *C. sticklandii* and *C. sporogenes,* glycine and proline constitute the most important hydrogen acceptors in the Stickland reaction (Barker, 1981). Although glycine reduction yields acetyl phosphate as a readily usable form of energy, the formation of the necessary enzyme system is repressed by the presence of proline (Venugopalan, 1980; Uhde, 1990). Proline reduction is coupled to ejection of protons in *C. sporogenes* (Lovitt et al., 1986). Betaine (N-trimethylglycine) can only be used as an electron acceptor by *C. sporogenes* if grown in a complex medium (Naumann et al., 1983) but not in a synthetic medium (Lovitt et al., 1987). A similar effect was noted by us for glycine. Although glycine should only be used as an electron acceptor (Stickland, 1984), *C. sporogenes* can also be grown on the Stickland pair glycine/betaine (but

Table 23.2 Selected substrate spectrum of bacteria growing on compounds that usually will be transformed via glycine to acetate

Species	Purines	Glycine	Sarcosine	Betaine	Pyruvate	Serine	Threonine	H_2+CO_2	Formate	Methanol	Methylamines
C. acidiurici	+	(+)[a]	–	–	–	–	–	–	(+)	–	
C. cylindrosporum	+	(+)	–	–	–	–	–	–	(+)	–	
C. purinolyticum	+	+	(+)	(+)	–	–	–	–	(+)	–	
E. angustum	+	(+)	–	–	–	–	–	ND[b]	(+)	ND	
P. barnesae	+	(+)	(+)	(+)	–	–	–	–	(+)	ND	
P. glycinophilus	–	+	(+)	(+)	–	+	–	–	(+)	–	
E. acidaminophilum	–	+	(+)	(+)	(–)[c]	+	–	–	(+)	–	
C. litorale	–	(+)	(+)	–	–	+	+	–	(+)	–	
C. sticklandii	(–)	(+)	–	–	–	+	(+)	–	(+)	–	
C. sporogenes	–	+	+	(+)	+	+	–	+	+	+	+
S. sphenoides	–	–	–	+	+	ND	–	+	+	+	
C. barkeri	–	–	+	–	+	+	–	+	+	+	
E. limosum	+	–	–	+	+	+	+	+	+	–	
P. asaccharolyticus	+	–	–	–	+	+	+	–	–	–	
P. prevotii	+	(+)	(+)	–	+	+	+	–	(+)	–	
P. indolicus	+	–	–	–	+	+	+	–	–	–	–

Note: C., Clostridium; E., Eubacterium; P., Peptostreptococcus; S., Sporomusa. [a] (+), growth in the presence of a cosubstrate. [b] ND, not determined. [c] (–), only substrate under special conditions.

578

not on glycine/proline) (Ludwig and Andreesen, unpublished data). Thus, this organism has a capacity to oxidize glycine (Winter et al., 1987). In *E. coli*, the gcv genes (glycine cleavage) are under the control of the purR repressor (Wilson et al., 1992), as is also the glyA gene (Steiert et al., 1992). Further unknown regulatory systems might also govern the metabolism of glycine.

(b) METABOLISM OF N-METHYLATED GLYCINE DERIVATIVES

The metabolism of betaine provides a good example of where components of the two pathways are merged (Fig. 23.1D). Species of the genera *Sporomusa* and *Acetohalobium* employ the Wood pathway when grown on H_2 plus CO_2, formate, methanol, or methylamines (Breznak, 1992; Zhilina and Zavarsin, 1990). Betaine and the other N-methyl glycine derivatives are successively de-methylated. The methyl group will be oxidized to CO_2; the electrons thus generated are used to reduce betaine, sarcosine, or glycine to acetyl phosphate and the corresponding amines, as observed for *E. acidaminophilum* (Hormann and Andreesen, 1989). Dimethylglycine will be only disproportionated to sarcosine and betaine by *Sporomusa* species (Oßmer, 1983). No organism containing a specific dimethylglycine reductase has been found so far (Ludwig and Andreesen, unpublished results). Sarcosine-grown cells of *S. ovata* contain a low activity of both glycine reductase and sarcosine reductase (Rieth 1987); thus, glycine appears to be formed intracellularly as the demethylated product of sarcosine. Glycine is not utilized as an external substrate by *Sporomusa* species (Breznak, 1992).

Some marine strains of *Desulfobacterium* and *Desulfobacterium autotrophicum* utilize betaine by demethylating and oxidizing just one methyl group to CO_2. They do not reduce betaine but transfer the electrons to sulfate (Heijthuisen and Hansen, 1989a). Marine isolates closely related to the sulfur reducer *Desulfuromonas acetoxidans* reduce betaine; however, they surprisingly oxidize the product acetate to CO_2 instead of using the methyl group as an electron source (Heijthuisen and Hansen, 1989b).

Creatinine and creatine (N-methylguanidinocetate) are anaerobically degraded in most cases via creatinine deiminase to N-methylhydantoin and NH_3 and further via N-carbamoylsarcosine to sarcosine, NH_3, and CO_2 (Möller et al., 1986; Shimizu et al., 1989; Hermann et al., 1992). Both compounds are successively transformed by *E. acidaminophilum* via creatinine amidohydrolase and creatine amidinohydrolase to urea and sarcosine (Hormann and Andreesen, 1989). This organism can use these three glycine derivatives only as electron acceptors for growth (Zindel et al., 1988). In contrast, the creatin(in)e-specialist *E. creatinophilum* (DSM 6911) will grow slowly even in the absence of, e.g., formate, indicating an oxidation of the methyl group. The organism contains reductases for creatine or sarcosine and glycine (Schleicher, 1990). The inability of whole cells to utilize sarcosine or glycine even in the presence of formate might related to transport limitations.

(c) Purine Metabolism

The pathway for purine degradation and the intermediate formation of glycine has only been clearly demonstrated for the three purinolytic clostridia, *C. acidiurici, C. cylindrosporum,* and *C. purinolyticum* (Rabinowitz, 1963; Dürre and Andreesen, 1982b). *E. angustum* and *P. barnesae* have a similar narrow substrate spectrum (Beuscher and Andreesen, 1984; Schiefer-Ullrich and Andreesen, 1985). Added glycine is decomposed, and glycine reductase is present (Lebertz, 1984). Although able to use purines, *C. barkeri* is quite different (Imhoff, 1981) (Table 23.2). Based on its 16S rRNA sequence, murein type, and physiology, it appears to be similar to *E. limosum and Acetobacterium woodii,* acetogenic organisms that grow autotrophically via the Wood pathway (Tanner et al., 1981). *C. barkeri* was isolated for its capacity to degrade nicotinic acid (Stadtman et al., 1972). The nicotinate dehydrogenase is structurally and immunochemically quite related to the xanthine dehydrogenase catalyzing all redox reactions of the purine ring (Reinhöfer, 1985). *C. barkeri* can form in a defined medium more acetate from hypoxanthine than can be accounted for by the carbon skeleton (Imhoff, 1981), thus indicating a net synthesis of acetate. *E. limosum* can also grow on some purines, but neither *C. barkeri* or *E. limosum* are able to grow on glycine, sarcosine, or dimethylglycine (Ludwig and Andreesen, unpublished results). *E. limosum* utilizes betaine only as an electron donor forming dimethylglycine as a product (Müller et al., 1981). Glycine is not taken up by hypoxanthine-grown cells of *C. barkeri,* and glycine reductase activity is not detected in such cell extracts (Lebertz, 1984); in addition, formiminoglycine is not found as an intermediate (Riexinger, 1986). Therefore, it seems unlikely that glycine is involved in the purine degradation pathway of these latter two bacteria.

An alternative pathway was first suggested for purine degradation by *Peptostreptococcus asaccharolyticus* (Micrococcus aerogenes) (Whiteley, 1952). *Peptostreptococcus indolis* and *Peptostreptococcus prevotti* are two other related anaerobic cocci able to degrade purines, in addition to amino acids such as serine, threonine, and arginine (Reece et al., 1976; Spahr, 1982; Riexinger, 1986; Tziaka, 1987). Formiminoglycine could be detected as an intermediate of purine degradation in extracts of these organisms, whereas in selenium-deprived cells, a lumazine derivative was additionally detected (Spahr, 1982), pointing to an alternative mechanism for cleaving the purine-ring system at the imidazole ring, as has been observed for *C. purinolyticum* (Dürre and Andreesen, 1982c). In addition to a rather substrate-unspecific xanthine dehydrogenase, these anaerobic cocci contain a highly specific 2-oxopurine dehydrogenase. Contrary to an earlier report (Woolfolk et al., 1970), the 2-oxopurine dehydrogenase is like xanthine dehydrogenase in that its activity is strongly influenced by the presence of molybdate, tungstate, and selenite during growth (Riexinger, 1986). Therefore, the metabolism of these organisms might be affected likewise by these trace elements.

Collectively, these examples illustrate the fact that strict predictions are difficult to make regarding the mechanisms or pathways by which compounds are anaerobically transformed by these organisms. There are many variations, and each must be evaluated on a case-by-case basis.

23.3 Enzymes Specially Involved in Glycine Metabolism

Glycine decarboxylase/synthase represents a key enzyme in the glycine metabolism of acetogenic bacteria that form acetate via glycine (Waber and Wood, 1979; Wood, 1989). The enzyme catalyzes both the formation of glycine from two C_1 units and also (preferentially) the oxidation of glycine. Specific activities are rather low in extracts of all organisms investigated. Therefore, additional factors might be involved or, alternatively, structural prerequisites necessary, as glycine metabolism *in vivo* occurs at a substantially higher rate (Freudenberg and Andreesen, 1989). A low expression of glycine decarboxylase might be the reason why an organism such as *C. litorale* utilizes glycine slowly as sole substrate; this low expression would be in contrast to when glycine is used in combination with an electron donor such as alanine (Fendrich et al., 1990).

Glycine reductase is the key enzyme required for the direct formation of acetyl phosphate and also for effective energy conservation (Dürre et al., 1983). This is quite obvious when organisms are cultured under conditions when glycine or its N-methylated derivatives sarcosine and betaine act only as electron acceptors. Electron donors can be small compounds like formate or hydrogen gas, amino acids, or certain organic acids, as observed with *E. acidaminophilum, C. litorale, C. histolyticum, C. sporogenes,* and *C. sticklandii* (Costilow, 1977; Fendrich et al., 1990; Hormann and Andreesen, 1989; Lebertz and Andreesen, 1988; Stadtman, 1978; Zindel et al., 1988). As emphasized earlier, the glycine-reductase reaction may be circumvented in cases of selenium deprivation by the "glycine-serine-pyruvate-acetate" interconversion (Dürre et al., 1983).

A bidirectionally acting formate dehydrogenase/CO_2 reductase represents the third important enzyme for acetogenesis via glycine. This enzyme is involved in both the glycine and in the Wood pathway. The following enzymes sequentially catalyze the formation of methylene-THF : formyl-THF synthetase, methenyl-THF cyclohydrolase, and methylene-THF dehydrogenase (Fig. 23.1). In some eukaryotes, the latter three activities are combined into one protein called C_1 synthase that serves a biosynthetic function (Nour and Rabinowitz 1992). Therefore, when used by acetogens, these enzymes are only peculiar in that they are involved in a major catabolic scheme, not only just in biosynthetic reactions.

23.3.1 Glycine Decarboxylase/Synthase

As indicated by its name, this enzyme system both cleaves and synthesizes glycine. It consists of four protein components and involves at one side glycine, and at the other, methylene-THF, NH_3, CO_2, and NAD(P)H (Fig. 23.3). The four proteins are traditionally called P_1, P_2, P_3, and P_4 when isolated from bacteria (Klein and Sagers, 1966a, 1966b, 1967a, 1966b; Kochi and Kikuchi, 1969; Plamann et al., 1983; Freudenberg and Andreesen, 1989) or P-, H-, L-, and T-protein when isolated from eukaryotes (Kikuchi and Hiraga, 1982; Walker and Oliver, 1986; Bourguignon et al., 1988; Fujiwara 1986b, 1987; Hiraga et al. 1988; Kure et al. 1991). The specific names and respective functions of the

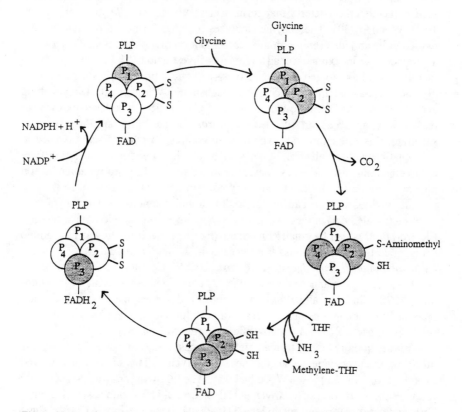

Figure 23.3. Interaction of the four proteins of bacterial glycine decarboxylase (P_1 to P_4) in NADP-dependent glycine oxidation by *Clostridium litorale*. Components: P_1, glycine decarboxylase (= P-protein); P_2, lipoamide-containing electron carrier protein (= H-protein); P_3, dihydrolipoamide dehydrogenase (= L-protein); P_4, transferase protein (= T-protein).

four components are pyridoxal phosphate-containing decarboxylase (P_1 or P), hydrogen carrier protein (P_2 or H), dihydrolipoamide dehydrogenase (P_3 or L), and THF-dependent transferase (P_4 or T). The four proteins work in concert to catalyze the oxidation of glycine and to thus provide the cell with C_1 compounds (at the formaldehyde level) or to synthesize serine in combination with serine hydroxymethyltransferase during photorespiration. In eukaryotes, the enzyme system is exclusively located in the mitochondrial matrix forming up to one-third of the soluble protein (Oliver et al., 1990). The four proteins form a complex consisting of about 2 P-dimers, 27 H-monomers, 1 L-dimer, and 9 T-monomers. However, such a complex is usually difficult to isolate and might already represent a disintegration product (Kikuchi and Hiraga, 1982; Neuburger et al., 1986; Oliver et al., 1990). In phototrophic organisms, the stoichiometry of the four subunits is dependent on a light/dark regime (Tobin et al., 1989; Rogers et al., 1991). Some of the components can form functional associations (P- and H-protein), partly due to their basic or acidic charges (H- and T-protein).

In *E. acidaminophilum*, the content of P_1, P_2, P_4, and of thioredoxin reductase plus thioredoxin as a functional equivalent of P_3, make up more than 10% of the soluble proteins; thus, the glycine decarboxylase complex represents a major enzyme system for that organism, too (Dietrichs et al., 1991b). P_1 and P_2 form a tight complex and are generally required to observe a catalytically active glycine decarboxylase or exchange reaction (Freudenberg et al., 1989b; Hiraga and Kikuchi, 1980b; Kikuchi and Hiraga, 1982; Freudenberg and Andreesen, 1989). If one measures glycine synthesis, the reduced H-protein has to bind first to the T-protein, followed by methylene-THF and ammonia in order to observe a synthesis of the lipoyl(SH)S-aminomethyl group covalently bound to the H-protein (Okamura-Ikeda et al., 1987). Only the P_3/L-protein can be assayed individually, because it is also part of the complex in other 2-oxoacid dehydrogenases (Sokatch and Burns, 1984; Perham, 1991) or of the acetoin cleavage system (Oppermann et al., 1991). The glycine decarboxylase complex of *E. acidaminophilum* represents the only known exception that does not contain a dihydrolipoamide dehydrogenase. Its function is taken over by an unusual thioredoxin reductase (plus thioredoxin) that exhibits some minor dihydrolipoamide dehydrogenase activity and also reacts partly with NAD (Freudenberg et al., 1989a; Meyer et al., 1991). NAD was considered to be the only natural electron carrier for dihydrolipoamide dehydrogenase (P_3- or L-protein) (Williams, 1992). However, quite recently, proteins were purified from anaerobic, glycine-, or purine-utilizing bacteria that couple preferentially, or even only, with NADP (Dietrichs and Andreesen, 1990; Dietrichs et al., 1990; Meyer et al., 1991). As a further exception, the P_3-protein of *C. cylindrosporum* is located at the cell periphery, not in the cytoplasma, as demonstrated by immunocytochemical studies (Dietrichs et al., 1991a).

The enzyme has been studied in detail mainly from eukaryotic sources. The enzyme has also been isolated prokaryotic sources: first from *P. glycinophilus* and *Arthrobacter globiformis,* and later from *C. acidiurici* and *E. acidaminophilum.* In all cases, the selective pressure enforced on the respective bacteria was toward glycine oxidation; however, the reversibility of the reaction (synthase reaction) is always found. The properties of the enzyme complex have been reviewed (Kikuchi and Hiraga, 1982; Schirch, 1984; Husic et al., 1987). Therefore, only the more recent work will be dealt with in the following sections.

(a) P₁ or P-Protein

This protein is a true decarboxylase/carboxylase, although it requires in all cases the presence of the lipoamide-containing P_2- or H-protein to exhibit an activity (Fig. 23.3). Without the latter protein, 0–2% of that activity is obtained in the exchange reaction between bicarbonate and the carboxyl group of glycine. Free lipoic acid stimulates to some degree the decarboxylation and the exchange reactions (Hiraga and Kikuchi, 1980b). The P-proteins from all sources are dimers of 210–280 kDa, except for that of *P. glycinophilus* (125 kDa) (Klein and Sagers, 1966a). The actual subunit size is 97–105 kDa (SDS-PAGE) for the mitochondrial proteins, whereas both prokaryotic proteins from *C. acidiurici* and *E. acidaminophilum* form two different subunits of 58 and 65 kDa and 54 and 60 kDa, respectively (Gariboldi and Drake, 1984; Freudenberg and Andreesen, 1989). These values add up to 123 and 114 kDa for a protomer; thus, they come close to the size of the mitochondrial proteins. Antibodies raised against the P-protein from rat liver demonstrated an immunological relationship to the corresponding proteins from vertebrates ranging from carp to human but not to that of *A. globiformis* (Hayasaka et al., 1980). On the other hand, antibodies raised against the P₁-protein of *E. acidaminophilum* only reacted with extracts of *E. angustum, C. cylindrosporum,* and *C. litorale* but not with extracts from a variety of similar prokaryotic bacteria (Freudenberg and Andreesen 1989).

Mitochondrial P-proteins have recently been sequenced. They all contain a leader sequence of 35–86 amino acids to direct their import into the mitochondria. Based on DNA sequences, the mature proteins are 105 and 108 kDa (pea leaf and chicken liver); thus, they are somewhat larger than as determined by SDS-PAGE (97/98 and 100 kDa, respectively) (Bourguignon et al., 1988; Hiraga and Kikuchi, 1980a; Kume et al., 1991; Walker and Oliver, 1986; Turner et al., 1992a). The protein from rat liver was 105 kDa (Hayasaka et al., 1980). These data indicate a close relationship of the P₁/P-proteins, unless the bacterial subunits represent "split-genes." The amino acid sequence of the mature protein from pea is about 56% identical and 73% similar to those of human and chicken proteins (Turner et al., 1992a); the latter two are 91% similar to each other (Kume et al., 1991).

All P₁- or P-proteins contain noncovalently bound pyridoxal phosphate. The binding region contains defined lysine residues (Lys-738, 754, 792) and a 16 amino acid long glycine-rich sequence in close vicinity to it. This segment is rich in β turns and random coils; thus, it might be well suited for a flexible arm structure required for access to the lipoyl moiety of the P_2/H-protein. The pyridoxal phosphate contributes to an absorption maximum of the protein at about 400 nm in the case of *C. acidiurici* and *E. acidaminophilum* (Gariboldi and Drake, 1984; Freudenberg and Andreesen, 1989) and at about 430 nm in the case of rat liver and *P. glycinophilus* (Hayasaka et al., 1980; Klein and Sagers, 1966b, 1967b). Pyridoxal phosphate and, concomitantly, enzymatic activity are lost during purification, because a substrate activation by Schiff-base formation is no longer possible. A dissociation of the cofactor can be prevented by adding it to the buffers. A reactivation of the coenzyme-free form by addition of pyridoxal phosphate was not achieved in the case of *E. acidaminophilum* (Freudenberg and Andreesen, 1989).

(b) P_2- OR H-PROTEIN

This protein is a small and heat-stable hydrogen carrier protein that contains covalently bound lipoic acid as the redox-active functional group. It serves as a true coenzyme in the glycine decarboxylation reaction catalyzed by P_1/P-protein (Fig. 23.3). The K_m value for the protein isolated from pea mitochondria and *E. acidaminophilum* is about 2 μM; that of rat liver protein is 7 μM (Hayasaka et al., 1980; Walker and Oliver, 1986; Freudenberg and Andreesen, 1989). The K_m value for glycine is generally quite high, 33 mM; however, the affinity increases to about 6 mM by the addition of H-protein in those cases where the P-protein is already active in the absence of H-protein (Hiraga and Kikuchi, 1980a; Fujiwara and Motakawa, 1983). This change coincides with a change in the absorption spectrum and indicates a conformational change in the protein. The peak at 420 nm declines, whereas a new one appears at 330 nm. The addition of bicarbonate prevents this change in absorption because the exchange reaction can now take place (Hiraga and Kikuchi, 1982b). Actually, CO_2, not bicarbonate, is the active species. The pH optimum for the synthesis reaction is 6.5, and for the decarboxylation reaction it is 7.1 in the case of the chicken liver enzyme. The exchange reaction is optimal at pH 6.7 for the rat liver enzyme. The decarboxylation follows a sequential random BiBi mechanism (Fujiwara and Motakawa, 1983; Hayasaka et al., 1980).

The molecular mass of the P_2/H-proteins varies within 12–20 kDa, depending on the source (*A. globiformis, C. acidiurici, E. acidaminophilum, P. glycinophilus, Pseudomonas putida*, mitochondria from chicken, rat, human and pea) (Freudenberg and Andreesen, 1989; Fujiwara et al., 1984; Gariboldi and Drake, 1984; Hiraga et al., 1980a; Kikuchi and Hiraga, 1982; Kochi and Kikuchi, 1974, 1976; Robinson et al., 1973; Sokatch and Burns, 1984; Walker and Oliver,

1986). Antibodies raised against the P_2-protein of *E. acidaminophilum* show only a cross-reaction with extracts of *C. cylindrosporum* and *C. litorale* (Freudenberg and Andreesen, 1989). Eukaryotic proteins are similar according to sequence analysis. The mitochondrial proteins are formed as precursors with a targeting peptide of 34 (pea and Arabidopsis), 39 (chicken), or 48 (bovine) amino acids (Fujiwara et al., 1990; Kim and Oliver, 1990; Koyata and Hiraga, 1991; Macherel et al., 1990, 1992; Srinivasan and Oliver, 1992; Yamamoto et al., 1991). The deduced molecular mass of the 125 amino acid-containing H-protein of human, bovine, and chicken origin is between 13,815 and 13,883; that of the 131 amino acid-containing protein of pea is 13,000. These values are somewhat smaller than those determined by SDS-PAGE (about 15.5 kDa) (Fujiwara et al., 1979; Walker and Oliver, 1986). The H-protein of pea and *Arabidopsis,* exhibit a moderate similarity of 42–46% to the proteins from chicken and bovine origin (Macherel et al., 1990, 1992; Srinivasan and Oliver, 1991), whereas the latter two are highly homologous to the human protein (Koyata and Hiraga, 1991).

The region around the lipoylated lysine residue (Lys-59 or 63) is highly conserved in the four H-proteins sequenced. The protein is lipoylated after its transport into the mitochondrium and its maturation (Fujiwara et al., 1990). The highly conserved amino acids Glu-56 and Gly-70 are necessary for an effective lipoylation (Fujiwara et al., 1991a, 1991b). Some amino acids are conserved in all known lipoylated proteins such as the H-protein and the E_2 components within the complexes of dehydrogenases specific for pyruvate, 2-oxoglutarate, and branched chain 2-oxoacids and of the acetoin-degradation complex (Fujiwara et al., 1991a, Macherel et al., 1990; Priefert et al. 1991). The E_2 component of the pyruvate dehydrogenase complex contains at the N-terminus 3 lipoyl domains that get lipoylated by a lipoate protein ligase (Ali and Guest, 1990; Packman et al., 1991) and might get octanoylated instead of lipoylated in a mutant (Ali et al., 1990). Two proteins of the same M_r, 47 kDa, are present in *E. coli* that use lipoyl-AMP or ATP and lipoate for this reaction (Brookfield et al., 1991). The bovine H-protein can be expressed in *E. coli*. However, most of it is in the unlipoylated, inactive apoform; only about 10% is in the active form (Fujiwara et al., 1992). Bovine liver contains a lipoyl-AMP : N^ε-lysine lipoyltransferase that has no lipoate-activating activity. Thus, the mammalian system requires two proteins for lipoylation.

Investigations on the glycine-cleavage reaction catalyzed by both P- and H-proteins of chicken liver demonstrated that both hydrogen atoms of the C-2 of glycine are retained in the aminomethyl group attached to the lipoamide of the H-protein (Fujiwara et al., 1986a). This excludes a mechanism proposed earlier by Zieske and Davis (1983) in which the formation of a carbanion intermediate necessitates a dissociation of one 2-H_{si} atom of glycine. The methylamine-loaded form of the P_2/H-protein is quite stable and can be isolated (Freudenberg and Andreesen, 1989; Neuburger et al., 1991). A high concentration of glycine is required to saturate the H-protein. The inhibition of the exchange reaction by divalent cations such as Cu^{++}, Zn^{++}, Co^{++} and Ni^{++} is due to their action on the sulfhydryl groups of the H-protein. Therefore, they affect only the carboxylation, but not the decarboxylation, reaction (Hiraga and Kikuchi, 1982a).

(c) P_4 OR T-PROTEIN

The transferase protein reacts directly with the aminomethyl moiety that is bound to the lipoamide of the P_2/H-protein and that is formed from glycine by the combined action of P_1/P- and P_2/H-protein (Fig. 23.3). It reversibly cleaves off the methylene carbon from the aminomethyl group and transfers it to tetrahydrofolate (THF) forming methylene-THF and ammonia (Kikuchi and Hiraga, 1982). This enzyme was partially purified from *P. glycinophilus* and *A. globiformis* (Klein and Sagers, 1967a; Kochi and Kikuchi, 1974). The T-protein from rat liver and the P_4-protein from *E. acidaminophilum* are quite unstable. Their molecular masses are 33 and 42 kDa, respectively (Freudenberg and Andreesen, 1989; Motakawa and Kikuchi, 1974). The monomeric T-protein from pea has a mass of 45 kDa (Okamura-Ikeda et al., 1982; Walker and Oliver, 1986; Bourguignon et al., 1988). During photorespiration, the T-protein seems to operate in a concerted manner with serine hydroxymethyltransferase that removes the methylene-THF formed by the T-protein by condensing it with glycine to form serine. The T-protein from bovine and chicken has a mass of 37–41 kDa (Okamura-Ikeda et al., 1982). DNA sequence analysis indicates a precursor T-protein of 42–43 kDa and a mitochondrial trageting sequence of 22 and 17 amino acids, giving a molecular mass of the mature T-protein of 40,534 and 40,292 Da (375 and 376 amino acids, respectively) (Okamura-Ikeda et al., 1991, 1992).

In vitro, the basic T-protein forms a 1-to-1 complex with the acidic H-protein. The K_m values for the forward reaction are 22 μM for the aminomethyl intermediate bound to the sulfur of the dihydrolipoamide moiety of the H-protein and 50 μM for THF. In the absence of THF, formaldehyde is formed instead of methylene-THF; however, the velocity is at least 2000 times slower, no pH preference is exhibited, and the K_m value is fourfold higher for the intermediate complex (Fujiwara et al., 1984). Serine hydroxymethyltransferase also catalyzes reactions in the absence of THF, again at a much lower rate, presumably by forming a dissociable enzyme-formaldehyde complex (Chen and Schirch, 1973). The methylene-THF formed by the T-protein retains to a high degree the stereospecificity of both hydrogen atoms at C-2 of glycine (Aberhart and Russel, 1984). In the reverse synthase) reaction, methylene-THF will only react with T-protein in the presence of ammonia and the H-protein, indicating the formation of a quarternary complex. The mechanism follows an ordered TerBi mechanism, by which the reduced H-protein first binds the T-protein followed by methylene-THF and ammonia. THF is released as the first product. The K_m values for H-protein, methylene-THF, and ammonia are 0.55 μM, 0.32 mM, and 22 mM, respectively (Okamura-Ikeda et al., 1987). These concentrations are just within the physiological range. Therefore, the synthesis reaction should proceed *in vivo*, although the velocity of the cleavage reaction is about six times faster. The K_m values for methylene-THF in competing reactions such as serine hydroxymethyltransferase, C_1 synthase, and methylene-THF reductase are one to two orders of magnitude lower than for the T-protein (Okamura-Ikeda et al., 1987). Mechanistically, these latter reactions are much simpler.

(d) P_3 OR L-PROTEIN [DIHYDROLIPOAMIDE DEHYDROGENASE (LPD)]

The fourth and final enzyme activity residing in the glycine decarboxylase/synthase complex is catalyzed by the P_3/L-protein, often wrongly called lipoamide dehydrogenase instead of dihydrolipoamide dehydrogenase. Its function is to reoxidize the dithiols of the P_2/H-protein-lipoamide moiety to the corresponding disulfide; this reoxidation is mediated by transferring electrons from the redox-active disulfides and FAD of the LPD to NAD. Some LPDs of anaerobic bacteria are also able to couple with NADP (Fig. 23.3) (Dietrichs et al., 1990; Meyer et al., 1991; Williams, 1992). This reoxidation regenerates the oxidized P_2/H-protein to start a new catalytic cycle in combination with P_1/P-protein. As already mentioned, all lipoylated proteins contain a conserved region around the lipoylated lysine residue. Therefore, many organisms contain only one LPD in order to reoxidize the P_2/H-protein and the E_2 components of 2-oxoacid dehydrogenases. Consequently, only one gene can be found in, for example, plants (Turner et al., 1992b). Glycine metabolism is often strongly affected by the presence of 2-oxoacids (O'Brien, 1978; Kochi et al., 1986). A few bacteria such as *E. coli* and *Alcaligenes eutrophus* contain two distinct LPDs (Richarme, 1989; Pries et al., 1992), and *P. putida* contains three (Burns et al., 1989a; Palmer et al., 1991b). However, the actual function of all of these LPDs is not clear (Burns et al., 1989b). The three LPDs of *P. putida* exhibit an astonishingly low degree of identity according to their sequences (Palmer et al., 1992). In *P. putida*, the P_3-protein of glycine decarboxylase is identical to the E_3-protein (LPD) of pyruvate dehydrogenase (Sokatch and Burns, 1984; Palmer et al., 1991a). In *E. coli*, the LPD forms an operon with the other two proteins (E_1 and E_2) of the pyruvate dehydrogenase (Stephens et al., 1983), whereas in *Pseudomonas fluorescens* and *Azotobacter vinelandii* the LPD gene is associated with 2-oxoglutarate dehydrogenase (Westphal and de Kok, 1988; Benen et al., 1989). Therefore, LPD has to have an additional promoter, for it is also part of the other 2-oxoacid dehydrogenases and—most probably, as in *E. coli* (Steiert et al., 1990b)—of the glycine decarboxylase complex. The latter (gcv) system has only been cloned from *E. coli* and *Salmonella typhimurium* but not sequenced (Stauffer et al., 1986, 1989;

Plamann et al., 1983). The glycine cleavage activity is rather low. An overexpression of the gene seems to be detrimental to *E. coli*.

LPDs are well studied at the biochemical and molecular level (for detailed reviews, see Carothers et al. (1989); Williams, 1992). All LPDs are only enzymatically active as dimers because both subunits have to interact during catalysis. They exhibit a remarkable evolutionary relationship to most of the other FAD-containing pyridine nucleotide-dependent disulfide oxidoreductases. Only the LPDs (P_3/L-proteins) involved in glycine decarboxylase will be dealt with here. By SDS-PAGE analysis, the L-protein from pea mitochondria exhibits a molecular mass of about 60 kDa (Walker and Oliver, 1986; Bourguignon et al., 1988, 1992; Turner et al., 1992b). This is larger than the 50–55-kDa species characterized from other sources (Williams, 1992; Dietrichs et al., 1990); the pea protein has a mitochondrial targeting sequence of 31 amino acids. The mature protein contains 470 amino acids and has an M_r of 50,441 Da (FAD included). The four domaines representing the two characteristic redox-active cysteines (separated by four amino acids), the FAD- and NAD-binding sites, and the interface region for subunit interaction are well conserved (Bourguignon et al., 1992; Turner et al., 1992b). The deduced relationships reflect the general evolutionary distances of the organisms.

A modified type of LPD is present in *C. cylindrosporum* and *C. litorale* that accepts both coenzymes, NAD and NADP, at about equal efficiency (Dietrichs and Andreesen, 1990; Dietrichs et al., 1991a; Meyer et al., 1991). The pyridine nucleotide-binding site might have been modified to allow space for the extra phosphate. The coenzyme specificity of the related enzyme glutathione reductase has been experimentally redesigned from NADP to NAD by protein engineering of the binding domain by changing a few amino acids (Scrutton et al., 1990). Antibodies directed against the LPD of *C. cylindrosporum* cross-react with a corresponding protein of the purinolytic organisms *C. acidiurici*, *P. purinolyticum*, and *E. angustum* but not with the protein of *C. litorale*, *P. barnesae*, *P. micros*, and *Peptostreptococcus* Hare group IX, all of which nonetheless seem to react with both coenzymes (Dietrichs and Andreesen, 1990; Meyer et al., 1991). Because these organisms are quite specialized for purine and glycine fermentation and do not degrade acetoin, their LPDs should only be involved in the glycine decarboxylase. It might be advantageous for these specialists to utilize LPDs that react with both coenzymes.

The LPD of *C. sporogenes* is specific for NADP and has a much lower K_m value for NADP (Dietrichs and Andreesen, 1990). The organism also contains an NADPH-dependent thioredoxin reductase (Dietrichs et al., 1990). Antibodies against the LPD do not react with extracts of a large variety of anaerobic bacteria. *C. sporogenes* represents the classical Stickland-type organism and reduces glycine to acetyl phosphate as does *C. sticklandii* (Costilow, 1977; Venugopalan, 1980; Stadtman, 1989; Schräder and Andreesen, 1992); however, glycine can also be oxidized by *C. sporogenes* (Winter et al. 1987). *C. sticklandii* contains the usual NAD-specific LPD as does *P. glycinophilus* (Stadtman, 1965; Uhde, 1990; Dietrichs and Andreesen, 1990). Both species seem to be unable to utilize acetoin and contain the ferredoxin-dependent enzymes for the metabolism of 2-oxoacids.

E. *acidaminophilum* oxidizes and reduces glycine very efficiently (Zindel et al., 1988). It is exceptional in that it contains no LPD protein. Nonetheless, it exhibits a rather high LPD activity. A protein catalyzing such activity was purified and first called "atypically small lipoamide dehydrogenase" because it was unusually small ($M_r = 35$ kDa), reacted preferentially with NADP, and had a rather low specific activity (Freudenberg et al., 1989a). It was, thus, initially regarded as a member of a new class of LPDs (Dietrichs and Andreesen, 1990). However, the N-terminal amino acid sequence revealed no homology to LPDs. Instead, a considerable homology was observed to the thioredoxin reductase of *E. coli* (Dietrichs et al., 1990). The name was therefore changed to "electron transferring flavoprotein," for no thioredoxin was detected in former studies (Freudenberg et al., 1989a).

Later, an unusual form of thioredoxin was identified by its N-terminal sequence, and both purified proteins together exhibited a high LPD activity as detected in cell extracts (Meyer et al., 1991). DNA sequence analysis of both proteins has proven the relationship of these proteins to thioredoxin reductase and thioredoxin (Lübbers and Andreesen, unpublished results). All other strict NADP-dependent thioredoxin reductases, isolated as homogeneous proteins from *C. litorale, C. sticklandii, C. sporogenes,* and *C. cylindrosporum,* also gain an LPD activity in the presence of their homologous thioredoxins (Meyer and Andreesen, unpublished results). The LPD activity can be further stimulated twofold to threefold by the selenoprotein P_A of the glycine reductase (Dietrichs et al., 1991b). Like a true LPD, the thioredoxin system can function in both directions. Thus, *E. acidaminophilum* likely uses the thioredoxin systems for reoxidizing the P_2-protein of glycine decarboxylase, in addition to its function for the delivery of electrons to glycine reductase (Dietrichs et al., 1991b). Perhaps due to this dual function, the coenzyme specificity is less strict and resembles the LPD protein of *C. litorale* that provides the cell with a balanced supply of both pyridine nucleotides. In *C. litorale,* the LPD protein contributes only 5% of the total LPD activity present in cell extracts, leaving 95% for the thioredoxin system (Meyer et al., 1991). This indicates a high efficiency of the latter system. The thioredoxin system seems not to be involved in proline reduction by *C. sporogenes* (Ludwig and Andreesen, unpublished results). In cases where proline suppresses glycine utilization (Venugopalan, 1980), such bacteria are forced to have both systems. However, the involvement of the thioredoxin system seems to have an advantage for *E. acidaminophilum,* because the formation of reduced pyridine nucleotides can be bypassed, and, thus, the electron flow can be shortened by transferring the electrons directly from the reduced P_2-protein of glycine decarboxylase via the thioredoxin system to the proteins of glycine reductase (Dietrichs et al., 1991b, 1991c).

23.3.2 Glycine Reductase System

Glycine reductase is a key enzyme for acetogenesis via glycine and is only detected in some anaerobic bacteria. The enzyme systems of *C. sticklandii*, *C. purinolyticum*, and *E. acidaminophilum* are examined here in more detail. In all three cases, the glycine reductase system consists of at least three proteins, called P_A, P_B, and P_C (Stadtman, 1978). Sarcosine reductase and betaine reductase are separate enzymes (Hormann and Andreesen, 1989) because they contain different substrate-specific P_B components; nonetheless, they share the same P_A and the P_C components (Fig. 23.4) (Dietrichs et al., 1991b; Schräder and Andreesen, 1992).

P_A forms the bound carboxymethyl selenoether and is redox active. The formation and central activity of the selenoprotein P_A requires the addition of selenite to growth media (Turner and Stadtman, 1973). Under conditions of selenium deprivation, only some bacteria such as *P. glycinophilus* can utilize the energetically less favorable sequence "glycine-serine-pyruvate-acetate" (Dürre et al., 1983); however, other bacteria that depend solely on glycine reductase (e.g., *E. acidaminophilum*) are strictly dependent on selenite (Zindel et al., 1988). Protein P_B is involved in substrate binding (Fig. 23.5). A pyruvyl residue instead of pyridoxal phosphate seems to form a Schiff base with glycine (Arkowitz and Abeles, 1989; Tanaka and Stadtman, 1979; van Poelje and Snell, 1990). The carboxymethylated form of P_A is formed out of this complex and, in turn, interacts with protein P_C to form a P_C-bound acetyl ester that is subsequently phosphorolytically split to acetyl phosphate (Arkowitz and Abeles, 1989, 1990, 1991; Garcia and Stadtman, 1991; Schräder and Andreesen, 1992; Stadtman, 1989; Stadtman and Davis, 1991). Acetate kinase then transforms acetyl phosphate and ADP to the final products acetate and ATP (Stadtman and Elliot, 1956; Stadtman et al., 1958).

(a) SELENOPROTEIN P_A

This protein represents the central key component of the glycine reductase system. It reacts with three different proteins during enzymatic catalysis—thioredoxin, substrate-P_B-complex, and P_C (Fig. 23.4). In comparison to its importance, it is ironically a rather small protein (18.5 kDa by SDS-PAGE); it is an acidic thermotolerant glycoprotein (Dietrichs et al., 1991b; Garcia and Stadtman, 1991; Sliwkowski and Stadtman 1988a, 1988b). The DNA sequence codes for 150 or 158 amino acids (*C. purinolyticum* or *C. sticklandii* and *E. acidaminophilum*,

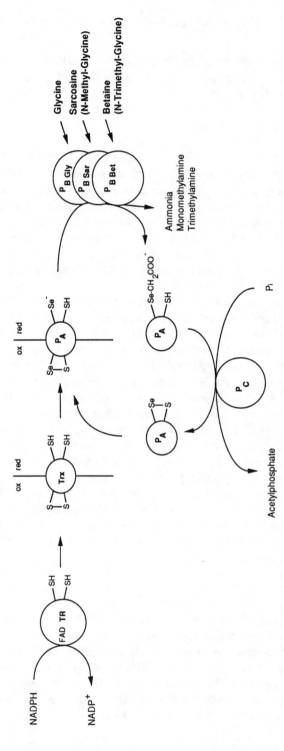

Figure 23.4. Proteins involved in NADPH-dependent reduction of glycine, sarcosine, and betaine by *Eubacterium acidaminophilum*. Components: TR, thioredoxin reductase; Trx, thioredoxin; P_A, selenoprotein P_A; P_B, substrate specific, activating protein P_B; P_C, acetyl-transferring protein P_C. The dithiol–disulfide interactions are indicated for both Trx and P_A by showing both redox states for these proteins.

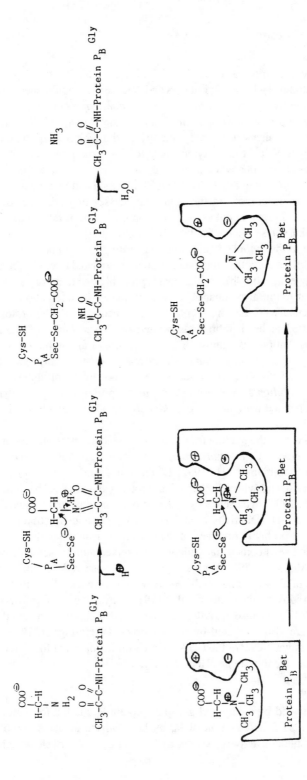

Figure 23.5. Proposed reaction mechanisms for the formation of P_A-bound carboxymethylated selenoether from glycine or betaine by interaction with the respective substrate-specific protein P_B.

respectively), yielding proteins of 15,757, 17,142, and 16,609 Da, respectively (Garcia and Stadtman, 1991, 1992; Lübbers and Andreesen, unpublished results). The amino acid content of the P_A-protein of *E. acidaminophilum* exhibits about 85% similarity to those of *C. purinolyticum* and *C. sticklandii*. The P_A-proteins from these three organisms differ only slightly in the amino acid content and UV spectrum. The UV spectrum is typical for proteins containing no tryptophan, few tyrosines, and many phenylalanines. Only in the P_A of *C. sticklandii* is the N-terminus blocked (Cone et al., 1976, 1977; Slimkowski and Stadtman, 1988a).

In all cases, the differences in the molecular masses of P_A is due to sugar moieties present. Contrary to eukaryotes, glycoproteins are rarely found in the cytoplasm of prokaryotes (Wieland, 1988). Selenoprotein P_A of *E. acidaminophilum* is found within the cytoplasm; however, according to immunocytochemical studies, it also forms a complex with the membrane-associated thioredoxin reductase (Freudenberg et al., 1989b). P_A aggregates probably due to its sugar content. The sugar content might also enhance (help coordinate) the interaction of P_A in an ordered sequence with the three proteins mentioned above. The proteins P_B and P_C are reported to be membrane-associated in *C. sticklandii* (Tanaka and Stadtman, 1979). Antibodies against P_A of *C. sticklandii* also inhibited glycine reductase activity of *C. purinolyticum* (Sliwkowski and Stadtman, 1987; Sliwkowski and Stadtman 1988b; Dietrichs et al. 1991b); however, antibodies against P_A of *E. acidaminophilum* did not cross-react with the protein of *C. purinolyticum* but instead cross-reacted with that of *C. sticklandii* and other anaerobes (Dietrichs et al., 1991b).

A UGA codon within the reading frame codes for the cotranslational incorporation of selenocysteine in *C. purinolyticum, C. sticklandii,* and *E. acidaminophilum* (Garcia and Stadtman, 1991, 1992; Lübbers and Andreesen, unpublished results); this is consistent with all other selenoproteins analyzed at the molecular level (Böck et al., 1991). The selenocysteine-44 of P_A is separated by two amino acids from cysteine-41 (Dietrichs et al., 1991b; Garcia and Stadtman, 1992); cysteine-41 is most probably involved in the reductive cleavage of the P_A-bound carboxymethylated selenoether via oxidation to a mixed sulfide–selenide bridge (Arkowitz and Abeles, 1989). Another cysteine is conserved 15 amino acids further upstream in all 4 proteins analyzed. However, because of the interaction of P_A with thioredoxin (Dietrichs et al., 1991b), cysteine-41 is most likely involved (Table 23.3) because the redox active dithiols in thioredoxin and thioredoxin reductase are also separated by two amino acids (Holmgren, 1989). The mRNA structure downstream of the UGA codon is quite important for its recognition as coding for selenocysteine (Heider et al., 1992).

(b) PROTEIN P_B

Glycine is activated by a carbonyl group of protein P_B via the formation of a Schiff base (Fig. 23.5). Reduction of P_B by borohydride at pH 8 completely inactivates the enzymatic activity, as does hydroxylamine. A very slow exchange

Table 23.3   Redox-active center of thioredoxin reductase, thioredoxin, and selenoprotein P_A

	▼				▼	
Thioredoxin reductase (E.c.)	— A —	C —	A —	T —	C —	D —
Thioredoxin (consensus)	— W —	C —	G —	P —	C —	K —
Thioredoxin reductase (E.a.)	— Y —	C —	A —	T —	C —	D —
Thioredoxin (E.a., C.l.)	— G —	C —	V —	P —	C —	K —
Selenoprotein P_A (E.a., C.p., C.s.)	— E —	C —	F —	V —	SeC —	T —
	▲				▲	

Note: Sequences were obtained for the following: thioredoxin reductase of *E. coli* from Russel and Model (1988), of *E. acidaminophilum* from Lübbers and Andreesen (unpublished); for thioredoxin consensus sequence from Eklund et al. (1991), of *E. acidaminophilum* and *C. litorale* from Meyer et al. (1991); for selenoprotein P_A of *E. acidaminophilum* from Dietrichs et al. (1991b), of *C. purinolyticum* from Garcia and Stadtman (1991), of *C. sticklandii* from Garcia and Stadtman (1992).

reaction of tritium-labeled glycine with water can be used for an independent assay of protein P_B (Tanaka and Stadtman, 1979; Rieth, 1987). The protein from *C. sticklandii* exhibits an M_r of about 240 kDa, being smaller and less hydrophobic than protein P_C. The enzyme has been purified to 90% homogeneity showing one major and three minor protein bands by SDS-PAGE and is said to contain pyruvate as a reactive carbonyl compound (Arkowitz and Abeles, 1989, 1991; van Poelje and Snell, 1990; Tanaka and Stadtman, 1979). In contrast to the betaine-specific protein P_B, the glycine-specific protein P_B of *E. acidaminophilum* of 140 or 250 kDa has so far not been obtained as homogeneous protein. The betaine-specific P_B contains two subunits of 43 and 48 kDa that form an M_r of 160 to 180 kDa by native-gradient PAGE analysis (Meyer and Andreesen, unpublished results). Polyclonal antibodies against betaine-specific P_B also react strongly with one protein of 46 kDa present in both glycine- and sarcosine-grown cells. So far, we cannot conclude if this subunit is identical for both the glycine- and sarcosine-specific protein P_B, because some glycine reductase activity is present in sarcosine-grown cells (Hormann and Andreesen, 1989). We nonetheless speculate that a common antigenic epitop is present.

The P_B specific for betaine exhibits no spectral characteristics and is not inhibited by hydroxylamine. Due to the permethylation of the amino group in betaine to a quarternary amine, no Schiff-base formation is possible for activation of this substrate. Instead, strong ionic interactions can be postulated to bind the zwitter ion betaine to P_B and to polarize further the electron distribution at the carbon–nitrogen bond, thus facilitating a nucleophilic attack by the selenol anion of selenocysteine (Fig. 23.5). One creates a similar polarization of the C–N bond by protonation of the nitrogen of the azomethine group formed from glycine. Thus, the basic chemistry leading from an activated substrate–protein P_B complex to the carboxymethylated selenoether bound to P_A would be rather similar for glycine reductase, sarcosine reductase, and betaine reductase.

(c) PROTEIN P_C

This protein is involved in the final step of glycine reductase activity and catalyzes the transformation of the P_A-bound carboxymethylselenoether to the energy-rich acetyl phosphate ester (Fig. 23.4) (Arkowitz and Abeles, 1989, 1990, 1991; Stadtman, 1989; Stadtman and Davis, 1991; Schräder and Andreesen, 1992). An easy assay for P_C takes advantage of its ability to catalyze an arsenate-dependent hydrolysis of acetyl phosphate (Stadtman, 1989; Arkowitz and Abeles, 1991). P_C from *C. sticklandii* and *E. acidaminophilum* have been purified to homogeneity. Both display an association/dissociation behavior that depends on, e.g., protein concentration or ionic strength. Thus, the native molecular mass differs from 200 to 420 kDa, and a further aggregation is influenced by other proteins, especially one of 63 kDa in the case of *E. acidaminophilum* (Stadtman and Davis, 1991; Schräder and Andreesen, 1992). P_C from *C. sticklandii* and *E. acidaminophilum* are composed of two subunits of 49 and 58 or 48 and 57 kDa, respectively. The N-terminal amino acids exhibit no similarities. Antibodies raised against the 48-kDa subunit from *E. acidaminophilum* cross-react strongly with extracts of *C. litorale, C. purinolyticum,* and *C. sporogenes* but weakly with extracts of *C. sticklandii* and *E. angustum.* Except for *C. sporogenes,* Western immunoblots reveal bands at the same molecular mass (Schräder and Andreesen, 1992).

P_C contains a sulfhydryl group that will easily be alkylated, thereby losing its enzymatic activity (Stadtman and Davis, 1991). A covalently bound acetyl-P_C intermediate can be obtained either by transfer of the carboxymethyl group bound as selenoether to protein P_A or by acetylation of P_C with acetyl phosphate (Arkowitz and Abeles, 1991). The mechanism by which the P_C-bound acetyl thioester is formed from the P_A-bound carboxymethylselenoether is still unresolved. Arkowitz and Abeles (1989, 1990, 1991) propose that the thioester is formed directly during the reductive cleavage without a prior activation of the carboxyl group. This would be rather unusual. Using an abiotic system, the somehow analogous compound thioglycolic acid can get reduced to acetate by a synergism of FeS and H_2S (Blöchl et al., 1992). According to Wächtershäuser (1988, 1990), the carboxylate group and H_2S can be in equilibrium with a thiocarboxylate group, a supposedly prebiotic equivalent of the later CoA thioester. In some enzymes, iron-sulfur cluster remain as a relic from the former pyrite-pulled archaic metabolism (Wächtershäser, 1990). The homogeneous proteins of P_A and P_C do not contain iron-sulfur groups or other readily detectable chromophores and are enzymatically quite active in converting the chemically carboxymethylated selenoether into acetyl phosphate and acetate (Schräder and Andreesen, 1992). However, sulfur and selenium chemistry is prominently involved in glycine reduction by the thioredoxin system (Dietrichs et al., 1991c).

The 57-kDa subunit of protein P_C has been partly sequenced. The high proportion of hydrophobic amino acids agrees with its hydrophobic nature. This suits the hypothetical reaction mechanism proposed by Buckel (1990) by which a ketenoid intermediate is formed via a reductive dehydration reaction (Schräder and Andreesen, 1992). It agrees with the fact that (i) only one ^{18}O of the carboxyl group is retained in the product (Arkowitz and Abeles, 1991) and (ii) both hydrogens of the C-2 glycine become inversed in their configuration (Barnard and Ahktar, 1979). A ketenoid intermediate is quite reactive and would instantly give either an acetyl thioester by reacting with a sulfhydryl group of protein P_C or acetate with water. The former thioester would be further transformed by the smaller 48-kDa subunit in the presence of phosphate to acetyl phosphate as the final product of glycine reductase (Schräder and Andreesen, 1992).

By the reactions outlined above, all proteins of glycine reductase would be regenerated to start a new catalytic cycle except for the selenoprotein P_A that becomes oxidized (Fig. 23.4). Dithiols such as Clelands reagents, but not monothiols, are quite effective *in vitro* as a final reductant (Stadtman, 1978). However, *in vivo* the thioredoxin system reduces the selenoprotein P_A (Dietrichs et al., 1991b). The interaction of thioredoxin reductase, thioredoxin, and selenoprotein P_A is greatly facilitated by the fact that the redox-active cysteines/selenocysteine are just separated by two amino acids in all three proteins (Table 23.3).

(d) THE THIOREDOXIN SYSTEM AS ELECTRON DONOR OF GLYCINE REDUCTASE

The thioredoxin system consists generally of two proteins, an NADPH-dependent thioredoxin reductase and thioredoxin (Holmgren, 1989). Except for its role as electron donor for the three reductive deamination reactions specific for glycine, sarcosine, and betaine (Fig. 23.4) (Dietrichs et al., 1991b; Meyer and Andreesen, unpublished results), this system is involved only in biosynthetic functions. As already mentioned, both proteins together exhibit a dihydrolipoamide dehydrogenase activity when isolated from *E. acidaminophilum, C. litorale, C. cylindrosporum, C. sticklandii,* and *C. sporogenes* (Dietrichs et al., 1991b; Meyer et al., 1991; our laboratory, unpublished results). These five homogeneous thioredoxin reductases cross-react with antibodies raised against the protein of *E. acidaminophilum,* are dimers of 35 kDa, and contain FAD and a vicinal dithiol as redox-active groups (Williams, 1992). The N-terminal amino acid sequences of the thioredoxin reductases of *E. acidaminophilum* and *C. sporogenes* are highly homologous to that of *E. coli* (Dietrichs et al., 1990). The DNA sequence of the thioredoxin reductase gene of *E. acidaminophilum* shows an overall similarity of 64% to that of *E. coli;* a low similarity is noticed especially in the area between the conserved NADP-binding and C-terminal domains (Russel and Model, 1988; Lübbers and Andreesen, unpublished results). In contrast to the enzyme from *E. coli* (Williams, 1992), the thioredoxin reductase from *E. acidaminophilum* can form a semiquinone (Meyer et al., unpublished results).

The thioredoxins isolated from the above-mentioned organisms are rather unique in that they do not bind to DEAE and contain modifications at otherwise strongly conserved amino acid positions such as the tryptophan close to the first redox-active cysteine (Table 23.3) (Meyer et al., 1991). The C-terminus of the thioredoxin from *E. acidaminophilum* contains most of the deviations from the consensus sequence, especially in those amino acids responsible for protein interactions (Eklund et al., 1991; Lübbers and Andreesen, unpublished results). All these thioredoxins are quite heat stable and show an M_r of 12–14.5 kDa (Gleason and Holmgren, 1988). Antibodies raised against the thioredoxin of *E. acidaminophilum* cross-react with corresponding proteins of the other organisms (above) and also those of other anaerobic bacteria but not with thioredoxin of *E. coli*. In contrast to the proteins isolated from the obligate anaerobes, thioredoxin of *E. coli* is completely inactive as an electron donor for glycine reduction (Dietrichs et al., 1991b). The thioredoxin of these anaerobic bacteria might represent an early evolutionary offshoot (Williams, 1992).

23.3.3   *Formate Dehydrogenase/CO₂ Reductase*

The reduction of CO_2 to formate represents a principal reaction by which CO_2 acts as electron acceptor for acetogenic bacteria that use either the Wood or glycine pathways. The half-cell couple formate/CO_2 has a low standard redox potential of -420 mV; this is similar to that of $H_2/2\ H^+$ and of ferredoxin (ox/red) (Thauer et al., 1977). Although ferredoxin seems to be an ideal electron carrier for this reaction, it has been proven only for the formate dehydrogenase of *C. pasteurianum* (Scherer and Thauer, 1978). The autoxidable dye methyl viologen has about the same low redox potential and is known to replace ferredoxin in many reactions. The acetogenic, glycine-utilizing bacteria such as *C. acidiurici*, *C. cylindrosporum*, *C. purinolyticum*, and *E. acidaminophilum* contain a methylviologen-dependent formate dehydrogenase (Kearny and Sagers, 1972; Champion and Rabinowitz, 1977; Schiefer-Ullrich et al., 1984; Zindel et al., 1988).

The purification and study of formate dehydrogenase is hampered by the extreme instability of its activity. Formate dehydrogenase of acetogens contain selenium and tungsten as active components which likely contribute to its instability (Wagner and Andreesen, 1977; Axley et al., 1990; Strobl et al., 1992; Schmitz et al., 1992a, 1992b); this is in contrast to other more simply constructed formate dehydrogenases (Ferry, 1990). The incorporation of tungsten has been examined. [185]W always coelutes with formate dehydrogenase activity of purinolytic clostridia, never with the molybdoenzyme xanthine dehydrogenase (Wagner and

Andreesen, 1987. Some purinolytic clostridia have a separate uptake system for tungstate; thus, it enters the cell at rather low concentrations independently from molybdate (Wagner and Andreesen, 1987). In complex media, the optimal concentrations required for both ions are much higher, indicating an interference with a high-affinity uptake system (Leonhardt and Andreesen, 1977).

For *E. acidaminophilum*, 10^{-9} M tungstate satisfies its tungsten requirement for formate dehydrogenase synthesis and activity (Granderath and Andreesen, unpublished results). Further addition of tungsten labels another protein of about 70 kDa that masks the formate dehydrogenase with a native M_r of about 155 kDa. A preparation of about 25% purity that had already lost substantially in activity still contained about 1000 U/mg of protein. The main contaminating proteins were identified as P_B- and P_C-proteins of betaine or glycine reductase, induced during growth with formate as electron donor (Hormann and Andreesen, 1989). This formate dehydrogenase preparation contained at least 20 times more tungsten than molybdenum and was very rich in Fe/S centers, but it lacked flavins. It had at least one subunit of 95 kDa (Granderath and Andreesen, unpublished results) that exhibited N-terminally no homology to other formate dehydrogenases (Bokranz et al., 1991). Based on the high activity observed with the purified protein, the formate dehydrogenase of *E. acidaminophilum* is probably present in the cell only in low numbers. If true, [185]W-labeling and purification of formate dehydrogenase of *E. acidaminophilum* would be hampered. Formate dehydrogenase *C. thermoaceticum* is also highly active, contains tungsten, selenium, and iron-sulfur, and is composed of two subunits of 96 and 76 kDa (Yamamoto et al., 1983). Although the enzymes reacts with methylviologen, NADPH is the natural electron donor, even for the reduction of CO_2 (Thauer, 1972).

A formate-dependent reduction of a pyridine nucleotide has been observed in cell extracts of *C.acidiurici* and *E. acidaminophilum,* but this activity is lost during purification (Kearny and Sagers, 1972; Wagner et al., 1984; Granderath and Andreesen, unpublished results). Thus, it might be caused by a protein (subunit) catalyzing a transhydrogenase reaction (Strobl et al., 1992). Using reduced methylviologen as donor, a CO_2 reduction to formate has been demonstrated with extracts from *C. acidiurici* and *C. purinolyticum,* and also with a more purified protein preparation from *E. acidaminophilum* (Dürre and Andreesen, 1982d; Thauer, 1973; Granderath and Andreesen, unpublished results). Thus, this important step in acetogenesis appears to be definitely catalyzed by formate dehydrogenase. *E. acidaminophilum* can carry out an interspecies formate or hydrogen transfer, and it can be postulated that its formate dehydrogenase/ CO_2 reductase activity must be high and unaffected by the growth substrate (Zindel et al., 1988).

It was first speculated that the formate dehydrogenase of *E. acidaminophilum* might contribute to a proton gradient if it was localized on the outside of the cytoplasmic membrane (Hormann and Andreesen, 1989). However, no such evidence has been obtained. The point at which electrons from formate will enter the electron transport chain that leads to glycine, sarcosine, or betaine is not resolved. Candidates might be a ferredoxin-thioredoxin oxidoreductase or a ferredoxin-NADP oxidoreductase known to be present in phototrohic and anaerobic bacteria (Hammel et al., 1983; Knaff and Hirasawa, 1991).

Glycine-pathway acetogens do not display carbon monoxide dehydrogenase activity (Diekert and Thauer, 1978; Dürre and Andreesen, 1982d; Zindel et al., 1988). The reader is directed to Chapters 3 and 21 for detailed reviews of this enzyme in Wood-pathway acetogens and methanogens, respectively.

23.3.4   Hydrogenase

Hydrogenase is obviously critical to the H_2-dependent autotrophic growth of acetogenic bacteria. Its involvement in the glycine pathway for acetate formation has been postulated (Dürre and Andreesen, 1982d). However, the production or consumption of hydrogen has not been observed for purinolytic clostridia (Valentine et al., 1963); this is in contrast to glycine-, sarcosine-, and betaine-utilizing bacteria (Fendrich et al., 1990; Zindel et al., 1988).

E. acidaminophilum contains an active viologen-dependent hydrogenase of about 350 kDa (in crude preparations). This organism is unable to consume a large excess of hydrogen and does not respond to nickel starvation (Granderath and Andreesen, unpublished results). Nickel is a typical constituent of the hydrogenase that exhibits conserved sequences for nickel- and iron-sulfur-binding centers (Przybyla et al., 1992). This type of hydrogenase might additionally contain selenocysteine. Methanococcus voltae contains four different gene clusters for hydrogenases (Halboth and Klein, 1992). The iron type of hydrogenase is rarely found and often extremely oxygen sensitive (Adams, 1990). C. pasteurianum contains two Fe hydrogenases that differ significantly in their ability to evolve hydrogen gas.

Relative to acetate synthesis and nonsyntrophic growth, hydrogen evolution seems of no advantage for acetogenic bacteria, as the reduction of CO_2 to acetate by hydrogen is exergonic (Fuchs, 1986). However, the threshold value for hydrogen uptake is much lower for methanogenic and sulfate-reducing bacteria than for acetogens (Conrad and Wetter, 1990; Seitz et al., 1990). E. acidaminophilum produces H_2 during growth on serine or on substrates requiring interspecies hydrogen or interspecies formate transfer in the absence of, e.g., betaine (Zindel et al., 1988). This indicates this organism has an efficient mechanism for the transfer of reductant to hydrogenase. C. sporogenes can oxidize glycine and transfer the electrons to methanogenic bacteria (Winter et al., 1987). Thus, the apparent inability of bacteria employing the glycine reductase pathway to growth autotrophically is not due to their lack of oxidoreductases that can connect the intermediary metabolism with hydrogenase.

23.4 Enzymes Involved in THF-Dependent One-Carbon Metabolism and Acetogenesis

23.4.1 Reactions at the C_1 Level

The C_1 unit bound to THF can undergo redox reactions at the formate, formaldehyde, and methanol levels. At the formaldehyde level, a reversible $C_1 + C_2 \rightleftarrows C_3$ interconversion is possible via serine hydroxymethyltransferase (Schirch and Strong, 1989). In eukaryotes, a trifunctional C_1 synthase catalyses three activities, namely those of formyl-THF synthetase, methenyl-THF cyclohydrolase, and methylene-THF dehydrogenase (Nour and Rabinowitz, 1992). Prokaryotes express these activities usually on separate proteins or might combine the latter two activities (Wood and Ljungdahl, 1991). In *E. acidaminophilum,* the levels of these three activities is coordinatively decreased to about 25% during growth on serine, indicating their involvement under these conditions is less than that during the metabolism of glycine (Zindel et al., 1988).

(a) 10-FORMYL-THF SYNTHETASE

This enzyme catalyzes the reversible activation of formate to N-10-formyl-THF via hydrolysis of ATP to ADP and P_i. In purinolytic anaerobes, the enzyme works mostly in the reverse reaction, thus constituting an important energy conserving step in which ATP is synthesized via substrate-level phosphorylation (Fig. 23.2). The specific activity of formyl-THF synthetase is very high in all purinolytic anaerobes (20–35 U/mg protein) (Ljungdahl, 1984; Dürre and Andreesen, 1982a). Some have been purified easily by heparin Agarose (Staben et al., 1987). The enzyme from *C. cylindrosporum* is a tetramer of 60 kDa and exhibits similar properties to the enzyme of saccharolytic acetogenic species (Ljungdahl, 1984; MacKenzie, 1984). Formyl phosphate can be used by the enzyme as an energy-rich substrate (Smithers et al., 1987). Purinolytic species can be differentiated by antibodies directed against the formyl-THF synthetase (Champion and Rabinowitz, 1977). The nucleotide sequence is known for the enzyme from *C. acidiurici* (Whitehead and Rabinowitz, 1988).

(b) 5,10-METHYLENE-THF CYCLOHYDROLASE

This enzyme catalyzes the reversible cyclization/dehydration of N-10-formyl-THF to N-5,10-methenyl-THF. The forward reaction is chemically favored under acidic conditions (MacKenzie, 1984). Purinolytic acetogens contain additionally a similar activity, an N-5-formimino-THF cyclodeaminase, that forms the identical product by splitting off ammonia (Champion and Rabinowitz, 1977; Dürre and Andreesen, 1982a). Methylene-THF cyclohydrolase from *C. cylindrosporum* is a protein of 38 kDa that contains both activities (Uyeda and Rabinowitz, 1965, 1967a). The N-5 formimino group is formed by glycine formiminotransferase, thus, conserving the energy rich group of formiminoglycine, a typical intermediate of purine degradation (Dürre and Andreesen, 1982a, 1982b).

(c) METHYLENE-THF DEHYDROGENASE

This redox-active enzyme usually couples with NADP(H) in purinolytic and glycine-degrading acetogenic bacteria (Champion and Rabinowitz, 1977; Dürre and Andreesen, 1982a; MacKenzie, 1984; Zindel et al., 1988). From these bacteria, only the protein from *C. cylindrosporum* has been purified; it is similar to those of other origins (Wood and Ljungdahl, 1991; MacKenzie, 1984; Uyeda and Rabinowitz, 1967b). The M_r of the protein is about 70 kDa, the apparent K_m values for NADP and methylene-THF are both about 30 μM, and the equilibrium of the reaction favors the reduced C_1 compound.

(d) REDUCTION OF A FREE CARBOXYLIC ACID

All three of the preceding enzymes are required to activate and reduce formate to the formaldehyde level. Formaldehyde reacts spontaneously with THF to give methylene-THF. Some acetogens are able to reduce free carboxylic acids to the corresponding aldehydes by a tungsten-containing carboxylate reductase (White et al., 1991; Strobl et al., 1992). Unfortunately, only fatty acids longer that acetate were tested as substrates. This enzyme catalyzes preferentially an aldehyde methylviologen oxidoreductase reaction. An acetaldehyde (also formaldehyde) oxidizing activity is present in *E. acidaminophilum* that is stimulated by tungsten (Granderath et al., unpublished results). CO_2 might, thus, be reduced to formaldehyde by the consecutive involvement of the two tungsten-containing enzymes formate dehydrogenase/CO_2 reductase and carboxylate reductase/aldehyde oxidoreductase. However, thermodynamic data (Fuchs, 1986) do not favor this possibility. Methanogenic bacteria reduce CO_2 directly to an activated form of formate, i.e., formylmethanofuran, by a tungsten- or molybdenum-containing formylmethanofuran dehydrogenase, thereby circumventing an expenditure of energy in the form of ATP (see Fig. 20.3). Only the Mo-containing enzyme uses formate as a pseudo-substrate and is inactivated by cyanide (Schmitz et al., 1992a, 1992b). The purine- and glycine-fermenting organisms central to this chapter seem to stick to the THF-dependent enzymes, because they usually oxidize C_1 compounds to CO_2 and, by that, conserve energy at the level of formyl-THF synthetase.

(e) ENZYMES REACTING WITH METHYLENE-THF

Methylene-THF generally represents a major metabolic crossroad leading in several directions. By the enzymes just described, formaldehyde can also be oxidized to the formate level. By 5,10-methylene-THF reductase, formaldehyde can be reversibly transformed to the methanol level. For acetogens that employ the Wood pathway, the latter enzyme is an octomeric flavoprotein; some contain iron-sulfur-zinc (Park et al., 1991). The NADPH-dependent enzyme from *P. productus* contains no iron (Wohlfahrt et al., 1990). The K_m value for methylene-THF is about 0.1 mM. No methylene-THF reductase activity has been reported for purine- and glycine-fermenting anaerobes, although they would require that enzyme for the biosynthesis of, e.g., methionine (Schirch, 1982). Thymidylate synthase uses only some methylene-THF for biosynthetic purposes. The pool of methylene-THF is influenced by glycine decarboxylase/synthase supplying or requiring methylene-THF, NADH, CO_2, and NH_3. Likewise, serine hydroxymethyltransferase adds or drains methylene-THF (Schirch and Strong, 1989).

23.4.2 Enzymes Reacting with Serine

(a) SERINE HYDROXYMETHYLTRANSFERASE

The enzyme catalyzes the reversible interconversion of glycine + 5,10-methylene-THF $\rightleftarrows$ serine + THF (Fig. 23.1). The enzyme is widely distributed in nature because it is the major source of C_1 units required for biosynthesis via splitting serine (Schirch, 1982). The reverse reaction might also be relevant for the biosynthesis of serine and occurs in some purinolytic acetogens such as *C. acidiurici* under selenium-deprived conditions when they use the "glycine-serine-pyruvate-acetate" pathway (Wood, 1989). The purinolytic species contain a high specific activity of that enzyme (0.2–4.4 U/mg of protein), whereas the enzyme is barely detectable in *E. acidaminophilum,* even when grown on serine (Champion and Rabinowitz, 1977; Dürre and Andreesen, 1982a; Zindel et al., 1988). In *E. coli,* the expression of the corresponding gene (glyA) is regulated by the purR gene product (Steiert et al., 1990a, 1992). The partially purified enzyme from *C. cylindrosporum* exhibits K_m vlaues for glycine, 5,10-methylene-THF, serine, and THF of 19, 0.034, 0.22, and 0.057 mM, respectively, and 7.3 μM for the cofactor pyridoxal phosphate (Uyeda and Rabinowitz, 1968). THF is not strictly required for the reaction; however, it stimulates the reaction rate with formaldehyde by more than two orders of magnitude. The reaction of that enzyme (e.g., with threonine which yields glycine and acetaldehyde) proceeds without THF (Schirch, 1984). The affinity for glycine is influenced synergistically by folate compounds. Therefore, the high value obtained for *C. cylindrosporum* (0.25–6 mM) might actually be lower (Schirch et al., 1985). The enzyme of *E.*

coli is a homodimer of 95 kDa. A nonapeptide at the active site is conserved in different organisms. Of the five threonines, only one controls the substrate specificity, histidine-228 (Angelaccio et al., 1992; Stover et al., 1992). The enzyme might also catalyze a hydrolysis of methenyl-THF to 5-formyl-THF (Stover and Schirch, 1990).

(b) SERINE DEHYDRATASE

This enzyme deaminates serine irreversibly to pyruvate. Water is eliminated in the initial step followed by tautomerization and hydrolysis of the imine to ammonia and pyruvate (Buckel, 1992). The enzyme is involved in the serine bypath and present in *C. acidiurici* and *P. glycinophilus* (Klein and Sagers, 1962; Beck et al., 1957). In *E. acidaminophilum*, it is involved in serine degradation (Zindel et al., 1988). The enzyme from *C. acidiurici* can be reactivated by ferrous ions plus dithiothreitol. Pyridoxal phosphate seems to be a firmly bound cofactor (Carter and Sagers, 1972), whereas an iron-sulfur cluster is involved in *P. asaccharolyticus* (Buckel, 1992).

23.4.3   Metabolism of Acetate Esters

Acetyl-CoA is the product or substrate of the reversible pyruvate ferredoxin oxidoreductase (Uyeda and Rabinowitz, 1971), whereas acetyl phosphate is the product of the reductases specific for glycine, sarcosine, and betaine (Figs. 23.1 and 23.4).

(a) ACETATE KINASE

This enzyme transfers the energy-rich phosphate ester to ADP forming ATP and acetate. In anaerobic bacteria, acetate kinase is exclusively involved in energy conservation. This is in contrast to the bidirectional importance of formyl-THF synthetase. Acetate kinase activity has been determined for *C. acidiurici*, *C. cylindrosporum*, *E. acidaminophilum*, and *P. glycinophilus* (Champion and Rabinowitz, 1977; Klein and Sagers, 1962; Zindel et al., 1988). The partially purified enzyme from *C. acidiurici* exhibits K_m values for acetyl phosphate and ADP of 2.1 and 3.2 mM, respectively (Sagers et al., 1961); these are higher than those obtained for *A. woodii* (Winzer, 1992). Due to the generally high K_m value for acetate (20–100 mM), some methanogenic bacteria employ an acetyl-CoA synthetase (AMP forming) for acetate activation (Ferry, 1992); in contrast, other archaea contain the energetically more favorable ADP-forming enzyme (Schäfer et al., 1992). Extracts of *C. cylindrosporum* contain two different kinases for formate: kinase I is associated with formyl-THF synthetase [that also uses formyl phosphate (Smithers et al., 1987)], whereas kinase II might be identical to acetate kinase because it also recognizes acetate (as well as propionate) as substrate (Sly and Stadtman, 1963). The acetate kinase activity of *E. acidaminophilum* is unusually labile (Schräder and Andreesen, 1992); in addition, the enzyme is quite sensitive to sulfhydryl-reactive compounds (Blödorn and Andreesen, unpublished results). A glutamyl residue is phosphorylated in the acetate kinase from *E. coli* (Todhunter and Purich, 1974), and the 43.3-kDa enzyme from this source has been sequenced (Matsuyama et al., 1989).

(b) Phosphotransacetylase

This enzyme converts the energy-rich phosphate ester of acetate into the kinetically more reactive CoA thioester. The phosphotransacetylase activity varies from being high in *C. acidiurici, C. cylindrosporum*, and *P. glycinophilus* (2.9, 4.5, and 1.6 U/mg of protein, respectively) to being low in *E. acidaminophilum* (0.17–0.49 U/mg of protein) (Champion and Rabinowitz, 1977; Klein and Sagers, 1962; Zindel et al., 1988). The enzyme from *C. acidiurici* requires ferrous or manganese ions for activity. In contrast to the 48-kDa subunit of P_C of glycine reductase of *E. acidaminophilum* that catalyzes an arsenate-dependent hydrolysis of acetyl phosphate, phosphotransacetylase from *C. acidiurici* is not affected by 20 mM arsenate (Sagers et al., 1963; Schräder and Andreesen, 1992; Robinson and Sagers, 1972). Thus, the active site of this phosphotransacetylase discriminates between phosphate and arsenate but nonetheless facilitates the transfer of the acetyl group between phosphate and CoA. The enzyme of *C. acidiurici* elutes at an M_r of 63–75 kDa, whereas that of *E. acidaminophilum* elutes at 70 kDa (Robinson and Sagers, 1972; Schräder, 1991); both compare favorably in size with phosphototransacetylase from other sources (Lundie and Ferry, 1989). Monovalent ions generally have a positive effect on the activity of the enzyme.

(c) Pyruvate Ferredoxin Oxidoreductase

This enzyme is of central importance to most anaerobic bacteria because it catalyzes the reversible interconversion from the C_3 to the C_2 plus CO_2 levels (Thauer, 1988). It functions biosynthetically in the direction of C_3 compounds when glycine or betaine are substrates being reduced to acetyl phosphate; in contrast, it functions catabolically during growth on serine (Zindel et al. 1988). Purinolytic clostridia exhibit a high activity of this enzymes (Champion and Rabinowitz 1977). The enzyme of *C. acidiurici* was studied in detail (Uyeda and Rabinowitz, 1971; Rabinowitz, 1972, 1975); more recent studies have focused on the enzyme from *Halobacterium halobium* (Plaga et al., 1992). The DNA sequence from the latter organism codes for two subunits of 68 and 34 kDa, that run on SDS-PAGE as 86- and 42-kDa proteins due to the anomalous behavior of acidic proteins. The enzyme contains per $\alpha_2\beta_2$ tetramer (240–256 kDa) two thiamine diphosphates and two [4Fe-4S] clusters; the latter can only be liganded by the small subunit or by sharing the cysteines of both subunits. This might explain the low iron-sulfur content often determined for the enzyme (Meinecke et al., 1989; Uyeda and Rabinowitz, 1971). The iron-sulfur clusters are involved in intramolecular one-electron redox reactions leading to a catalytically active radical intermediate (Kerscher and Oesterhelt, 1981; Wahl and Orme-Johnson, 1987). A radical mechanism has also been shown for the pyruvate-formate lyase (Wagner et al., 1992). This might also be responsible for the general instability of the pyruvate-ferredoxin oxidoreductase activity.

(d) Ferredoxin

Ferredoxin is for anaerobic bacteria an important, acidic, low-potential iron-sulfur protein that consists mostly of 2 [4Fe-4S] clusters residing on a protein of 6 kDa that exhibits extensive sequence homologies (Bruschi and Guerlesquin, 1988). The ferredoxins from purinolytic bacteria, especially from *C. acidiurici,* were one of the first analyzed (Buchanan et al., 1963; Lovenberg et al. 1963; Rabinowitz, 1972; Champion and Rabinowitz, 1977). The ferredoxin of *E. acidaminophilum* is unusually labile (Claas, 1991). Besides its main involvement in pyruvate-ferredoxin oxidoreductase, the hydrogen evolution reaction by hydrogenase is thermodynamically favored if coupled to ferredoxin (Thauer et al., 1977). Redox reactions involving purines are also at a low redox potential. However, the xanthine dehydrogenase of *Veillonella alcalescens* is the only known xanthine dehydrogenase that reacts with ferredoxin (Smith et al., 1967). The enzyme of *C. acidiurici* and *C. cylindrosporum* efficiently couple only with methylviologen (Bradshaw and Barker, 1960; Wagner et al. 1984), whereas the enzyme from *C. barkeri* couples with NADP (Rienhöfer, 1985).

Transhydrogenases that channel electrons from ferredoxin to pyridine nucleotides have been demonstrated for purinolytic clostridia (Valentine et al., 1963). The regulation of the activity by the redox state and metabolites has been studied in some detail with *C. pasteurianum* and *C. kluyveri* (Thauer et al., 1977).

23.5   Ancillary Enzymes Involved in the Formation of Glycine

23.5.1   Purine Metabolism and Xanthine Dehydrogenase

Purines constitute quantitatively the most important substrates leading to glycine (Fig. 23.2, Table 23.1) (Vogels and van der Drift, 1976). The uptake of purines has not been studied in detail. Preliminary data obtained for *C. acidiurici* indicate a maximum uptake rate of uric acid of 0.44 μmol (mg of dry wt.) min,$^{-1}$

which is about four times higher than that of glycine (Lebertz, 1984). The uptake system for formiminoglycien require metabolic energy and sodium ions and seems to be substrate-specific (Chen, 1975). With the exception of adenine (Dürre and Andreesen, 1983) and guanine (Rakosky et al., 1955), all interconversions at the purine level are catalyzed by the enzyme xanthine dehydrogenase. This enzyme has been studied in detail from *C. acidiurici, C. cylindrosporum,* and *C. barkeri* (Bradshaw and Barker, 1960; Wagner et al., 1984; Rienhöfer, 1985). It is one of the most complex enzymes, is composed of two protomers of 250–300 kDa, and contains molybdenum, iron-sulfur, and FAD as redox-active components (Hille, 1992). Selenium stimulates xanthine dehydrogenase activity in these anaerobic bacteria and is part of the enzyme (Wagner et al., 1984; Rienhöfer, 1985). For the enzyme from *C. barkeri,* involvement of selenium as selenocysteine (Böck et al., 1991) seems unlikely because the metal can be removed by cyanide, resulting in inactivation of the enzyme, but reconstituted by selenide, yielding active enzyme (Rienhöfer, 1985). Selenium might be one of the ligands in the molybdopterin cofactor replacing the cyanolyzable sulfur (Hille, 1992, Rajagopalan and Johnson, 1992), as proposed for the related enzyme nicotinate dehydrogenase (Dilworth, 1982).

By the action of xanthine dehydrogenase, xanthine is formed as the central compound that is split without a redox reaction to formiminoglycine by several enzymes. The intermediates have been detected in *C. cylindrosporum* and *C. purinolyticum* (Rabinowitz, 1963; Dürre and Andreesen, 1982b). Formimino-glycine will be transformed by glycine formiminotransferase to formyl-THF and glycine as described earlier (Uyeda and Rabinowitz, 1965). Only selenium-deprived cells can hydrolyze the lactam of the imidazole ring of uric acid or 8-hydroxypurine to a pyrimidine compound; this process can be envisioned as sort of an escape reaction that overcomes the problems of selenium limitation in central reactions of xanthine dehydrogenase, formate dehydrogenase, and glycine reductase (Dürre and Andreesen, 1982c).

23.5.2 Threonine Metabolism

Threonine is a substrate for many clostridia. *C. histolyticum, C. cadaveris, C. sticklandii,* and *C. subterminale* form only acetate and no propionate from threonine (Elsden and Hilton, 1978). Threonine can be cleaved to glycine and a C_2 unit by two different pathways (Table 23.1) that lead to the same end products (Andreesen et al., 1989). First, glycine and acetaldehyde are directly formed by the action of threonine aldolase [not by an unspecificity of serine hydroxymethyltransferase (Stöcklein and Schmidt, 1985)] (Dainty, 1970). Acet-aldehyde can be oxidized to acetyl-CoA by, e.g., an NAD-dependent, CoA-linked acetaldehyde dehydrogenase as recently found in *A. woodii* (Buschhorn et al., 1992). The reduced pyridine nucleotide could be regenerated by reducing glycine to acetyl phosphate. *C. histolyticum* and *C. sticklandii* contain glycine reductase when grown on threonine (Lebertz and Andreesen, 1988; Schräder and Andreesen, 1992). The latter enzyme would, however, also be involved in a somewhat different pathway proposed for *C. sticklandii* (Golovchenko et al., 1983). Catabolism starts with an NAD-dependent oxidation by threonine dehydrogenase to the unstable 2-amino-3-oxobutyrate. Although this compound might also spontaneously decarboxylate to aminoacetone, it will be thiolytically split by CoA and 2-amino-3-oxobutyrate CoA ligase to glycine and acetyl-CoA (Aronson et al., 1988; Mukherjee and Dekker, 1990). Again, the reduced pyridine nucleotide can serve as electron donor for glycine reductase. Thus, 1 mol of threonine will be converted by both pathways to two energy-rich esters of acetate, that will finally lead to 2 mol of acetate and 2 mol of ATP. It remains to be established by which mechanism NADH is fed into the thioredoxin system that normally channels the electrons of NADPH to glycine reductase. This function could be taken over by NAD-dependent dihydrolipoamide dehydrogenase, transferring electrons by interaction with the thioredoxin system (Dietrichs et al., 1991b,c).

23.6   Conclusions

Because the standard redox potential for the redox couple glycine / acetate is quite high (-10 mV) (Thauer et al., 1977), the reduction of glycine should constitute a main driving force for the enzymatic reactions involved in the reversible synthesis of glycine. However, no known organism seems to use the glycine pathway as a device to grow autotrophically or as major electron sink starting from CO_2. This observation is in direct conflict with the fact that all of the glycine pathway-utilizing organisms studied to date seem to contain the known prerequisites and enzymatic activities for such functions. In contrast to the acetogenic bacteria employing the Wood pathway, the organisms using the glycine pathway contain only low amounts of corrinoids and no cytochromes (Diekert and Thauer, 1978; Dürre et al., 1981; Dürre and Andreesen, 1982d). Bioenergetically, the glycine pathway neither requires nor conserves energy. However, one might imagine a net gain by proton export due to excretion of acetate (Boenigk et al., 1989). In natural ecosystems, the acetate concentration might be kept quite low due to, e.g., acetoclastic methanogenic bacteria (Ferry, 1992; see also Chapter 21). Methanogenic bacteria also have a much lower threshold value for H_2, whereby they might restrict the glycine-forming acetogens to a more fermentative, nonautotrophic metabolism. However, acetogenic bacteria can nonetheless outcompete methanogens by reducing CO_2 to acetate in the ecosystem of termite guts (see Chapter 11). Therefore, one might, indeed, find the right conditions where the glycine reductase pathway has a functional role in the autotrophic synthesis of acetate from CO_2.

Acknowledgments

The work carried out in the author's laboratory was supported by the Deutsche Forschungsgemeinschaft and Fonds der Chemischen Industrie.

References

Aberhart, D. J., and D. J. Russell. 1984. Steric course of N^5,N^{10}-methylenetetrahydrofolate formation from glycine by the glycine cleavage system in *E. coli. J. Am. Chem. Soc.* **106:**4907–4910.

Adams, M. W. W. 1990. The structure and mechanism of iron-hydrogenases. *Biochim. Biophys. Acta* **1020:**115–145.

Ali, S. T., and J. R. Guest. 1990. Isolation and characterization of lipoylated and unlipoylated domains of the E2p subunit of the pyruvate dehydrogenase complex of *Escherichia coli. Biochem. J.* **271:**139–145.

Ali, S. T., A. I. G. Moir, P. R. Ashton, P. C. Engel, and J. R. Guest. 1990. Octanoylation of the lipoyl domains of the pyruvate dehydrogenase complex in a lipoyl-deficient strain of *Escherichia coli. Mol. Microbiol.* **4:**943–950.

Andreesen, J. R., A. Schaupp. C. Neurauter, A. Brown, and L. G. Ljungdahl. 1973. Fermentation of glucose, fructose, and xylose by *Clostridium thermoaceticum:* effect of metals on growth, yield, enzymes, and the synthesis of acetate from CO_2. *J. Bacteriol.* **114:**743–751.

Andreesen, J. R., U. Zindel, and P. Dürre. 1985. *Clostridium cylindrosporum* (ex Barker and Beck 1942) nom rev. *Int. J. Syst. Bacteriol.* **35:**206–208.

Andreesen, J. R. H. Bahl, and G. Gottschalk. 1989. Introduction to the physiology and biochemistry of the genus *Clostridium*. In *Clostridia,* N. P. Minton, and D. J. Clarke (eds.), New York. pp. 27–62. Plenum Press.

Angelaccio, S., S. Pascarella, E. Fattori, F. Bossa, W. Strong, and L. V. Schirch. 1992. Serine hydroxymethyltransferase. Origin of substrate specificity. *Biochemistry* **31:**155–162.

Arkowitz, R. A., and R. H. Abeles. 1989. Identification of acetyl phosphate as the product of clostridial glycine reductase: evidence for an acyl enzyme intermediate. *Biochemistry* **28:**4639–4644.

Arkowitz, R. A., and R. H. Abeles. 1990. Isolation and characterization of a covalent selenocysteine intermediate in the glycine reductase system. *J. Am. Chem. Soc.* **112:**870–872.

Arkowitz, R. A., and R. H. Abeles. 1991. Mechanism of action of clostridial glycine reductase: isolation and characterization of a covalent acetyl enzyme intermediate. *Biochemistry* **30:**4090–4097.

Aronson, B. D., P. D. Ravnikar, and R. L. Somerville. 1988. Nucleotide sequence of the 2-amino 3-ketobutyrate coenzyme A ligase (kbl) gene of *E. coli*. *Nucl. Acid Res.* **16**:3586.

Axley, M. J., D. A. Grahame, and T. C. Stadtman. 1990. *Escherichia coli* formate-hydrogen lyase. Purification and properties of the selenium-dependent formate dehydrogenase component. *J. Biol. Chem.* **265**:18213–18218.

Barker, H. A. 1981. Amino acid degradation by anaerobic bacteria. *Annu. Rev. Biochem.* **50**:23–40.

Barker, H. A., S. Ruben, and J. V. Beck. 1940. Radioactive carbon as an indicator of carbon dioxide reaction. IV. The synthesis of acetic acid from CO_2 by *Clostridium acidi-urici*. *Proc. Natl. Acad. Sci. USA* **26**:477–482.

Barker, H. A., and J. V. Beck. 1941. The fermentative decomposition of purines by *Clostridium acidi-urici* and *Clostridium cylindrosporum*. J. Biol. Chem. **141**:3–27.

Barker, H. A., and S. R. Elsden. 1947. Carbon dioxide utilization in the formation of glycine and acetic acid. *J. Biol. Chem.* **167**:619–620.

Barker, H. A., B. E. Volcani, and B. P. Cardon. 1948. Tracer experiments on the mechanism of glycine fermentation by *Diplococcus glycinophilus*. *J. Biol. Chem.* **173**:803–804.

Barnard, G. F., and M. Akhtar. 1979. Mechanistic and stereochemical studies on the glycine reductase of *Clostridium sticklandii*. *Eur. J. Biochem.* **99**:593–603.

Becher, B., V. Müller, and G. Gottschalk. 1992. N^5-Methyl-tetrahydromethanopterin coenzyme M methyltransferase of *Methanosarcina* strain Gö1 is an Na^+-translocating membrane protein. *J. Bacteriol.* **174**:7656–7660.

Beck, J. V., R. D. Sagers, and L. R. Morris. 1957. Intermediary metabolism of *Clostridium acidi-urici*. I. Formation of pyruvate from glycine. *J. Bacteriol.* **73**:465–469.

Benen, J. A. E., W. J. H. van Berkel, W. M. A. M. van Dongen, F. Müller, and A. de Kok. 1989. Molecular cloning and sequence determination of the lpd gene endcoding lipoamide dehydrogenase from *Pseudomonas fluorescens*. *J. Gen. Microbiol.* **135**:1787–1797.

Benziman, M., R. D. Sagers, and I. C. Gunsalus. 1960. L-Serine specific dehydrase from *Clostridium acidi-urici*. *J. Bacteriol.* **79**:474–479.

Beuscher, H. U., and J. R. Andreesen. 1984. *Eubacterium angustum* sp.nov., a Gram-positive, anaerobic, non-sporeforming, obligate purine fermenting organism. *Arch. Microbiol.* **140**:2–8.

Blöchl, E., M. Keller, G. Wächtershäuser, and K. D. Stetter. 1992. Reactions depending on iron sulfide and linking geochemistry with biochemistry. *Proc. Natl. Acad. Sci. USA* **89**:8117–8120.

Böck, A., K. Forchhammer, J. Heider, W. Leinfelder, G. Sawers, B. Veprek, and F. Zinoni. 1991. Selenocysteine: the 21st amino acid. *Mol. Microbiol.* **5**:515–520.

Bokranz, M., M. Gutmann, C. Körtner, E. Kojro, F. Fahrenholz, F. Lauterbach, and

A. Kröger. 1991. Cloning and nucleotide sequence of the structural genes encoding the formate dehydrogenase of *Wolinella succinogens*. *Arch. Microbiol.* **156**:119–128.

Boenigk, R., P. Dürre, and G. Gottschalk. 1989. Carrier-mediated acetate transport in *Acetobacterium woodii*. *Arch. Microbiol.* **152**:589–593.

Bourguignon, J., M. Neuburger, and R. Douce. 1988. Resolution and characterization of the glycine-cleavage reaction in pea leaf mitochondria. Properties of the forward reaction catalyzed by glycine decarboxylase and serine hydroxymethyltransferase. *Biochem. J.* **255**:169–178.

Bourguignon, J., D. Macherel, M. Neuburger, and R. Douce. 1992. Isolation, characterization, and sequence analysis of a cDNA clone encoding L-protein, the dihydrolipoamide dehydrogenase component of the glycine cleavage system from pea-leaf mitochondria. *Eur. J. Biochem.* **204**:865–873.

Bradshaw, W. H., and H. A. Barker. 1960. Purification and properties of xanthine dehydrogenase from *Clostridium cylindrosporum*. *J. Biol. Chem.* **235**:3620–3629.

Breznak, J. A. 1992. The genus *Sporomusa*. In: *The Prokaryotes. A Handbook on the Biology of Bacteria: Ecophysiology, Isolation, Identification, Applications*, A. Balows, H. G. Trüper, M. Dworkin, W. Harder, and K. H. Schleifer (eds.), pp. 2014–2021. Vol. 2, Springer-Verlag, New York.

Brookfield, D. E., J. Green, S. T. Ali, R. S. Machado, and J. R. Guest. 1991. Evidence for two protein-lipoylation activities in *Escherichia coli*. *FEBS Lett.* **295**:13–16.

Bruschi, M., and F. Guerlesquin. 1988. Structure, function and evolution of bacterial ferredoxins. *FEMS Microbiol. Rev.* **54**:155–175.

Buchanan, B., W. Lovenberg, and J. C. Rabinowitz. 1963. A comparison of clostridial ferredoxins. *Proc. Natl. Acad. Sci. USA* **49**:345–353.

Buckel, W. 1990. Amino acid fermentation: coenzyme B_{12}-dependent and -independent pathways. In: *The Molecular Basis of Bacterial Metabolism*, G. Hauska and R. Thauer (eds.), pp. 21–30. Springer-Verlag, Berlin.

Buckel, W. 1992. Unusual dehydrations in anaerobic bacteria. *FEMS Microbiol. Rev.* **88**:211–232.

Burns, G., T. Brown, K. Hatter, and J. R. Sokatch. 1989a. Sequence analysis of the lpdV gene for lipoamide dehydrogenase of branched-chain-oxoacid dehydrogenase of *Pseudomonas putida*. *Eur. J. Biochem.* **179**:61–69.

Burns, G., P. J. Sykes, K. Hatter, and J. R. Sokatch. 1989b. Isolation of a third lipoamide dehydrogenase from *Pseudomonas putida*. *J. Bacteriol.* **171**:665–668.

Buschhorn, H., P. Dürre, and G. Gottschalk. 1992. Purification and properties of the coenzyme A-linked acetaldehyde dehydrogenase of *Acetobacterium woodii*. *Arch. Microbiol.* **158**:132–138.

Carothers, D. J., G. Pons, M. S. Patel. 1989. Dihydrolipoamide dehydrogenase: functional similarities and divergent evolution of the pyridine nucleotide-disulfide oxidoreductases. *Arch. Biochem. Biophys.* **268**:409–425.

Carter, J. E., and R. D. Sagers. 1972. Ferrous ion-dependent L-serine dehydratase from *Clostridium acidi-urici*. *J. Bacteriol.* **109**:757–763.

Champion, A. B., and J. C. Rabinowitz. 1977. Ferredoxins and formyltetrahydrofolate synthetase: comparative studies with *Clostridium acidiurici, Clostridium cylindrosporum,* and newly isolated anaerobic uric acid-fermenting strains. *J. Bacteriol.* **132**:1003–1020.

Chen, C. S. 1975. Formiminoglycine transport systems in *Clostridium acidiurici. Bot. Bull. Acad. Sinica (Taipei)* **1**:1–9.

Chen, M. S., and L. V. Schirch. 1973. Serine transhydroxymethylase. Studies on the role of tetrahydrofolate. *J. Biol. Chem.* **248**:7979–7984.

Claas, J. U. 1991. Isolierung und Charakterisierung von Ferredoxin und Thioredoxin und deren Einbindung in den Elektronentransport von *Eubacterium acidaminophilum.* Diplom thesis, University of Göttingen.

Condit, C. M., and R. B. Meagher. 1986. A gene encoding a novel glycine-rich structural protein of *Petunia. Nature* **323**:178–181.

Cone, J. E., R. M. Del Rio, J. N. Davis, and T. C. Stadtman. 1976. Chemical characterization of the selenoprotein component of clostridial glycine reductase: identification of selenocysteine as the organoselenium moiety. *Proc. Natl. Acad. Sci. USA* **73**:2659–2663.

Cone, J. E., R. M. Del Rio, and T. C. Stadtman. 1977. Clostridial glycine reductase complex. Purification and characterization of the selenoprotein component. *Biol. Chem.* **252**:5337–5344.

Conrad, R., and B. Wetter. 1990. Influence of temperature on energetics of hydrogen metabolism in homoacetogenic, methanogenic, and other anaerobic bacteria. *Arch. Microbiol.* **155**:94–98.

Costilow, R. N. 1977. Selenium requirement for the growth of *Clostridium sporogenes* with glycine as the oxidant in Stickland reaction. *J. Bacteriol.* **131**:366–368.

Dainty, R. H. 1970. Purification and properties of threonine aldolase from *Clostridium pasteurianum. Biochem. J.* **117**:585–592.

Das, A., J. Hugenholtz, H. van Halbeek, and L. G. Ljungdahl. 1989. Structure and function of a menaquinone involved in electron transport in membranes of *Clostridium thermoautotrophicum* and *Clostridium thermoaceticum. J. Bacteriol.* **171**:5823–5829.

Diekert, G. B., and R. K. Thauer. 1978. Carbon monoxide oxidation by *Clostridium thermoaceticum* and *Clostridium formicoaceticum. J. Bacteriol.* **136**:597–606.

Diekert, G. B., E. G. Graf, and R. K. Thauer. 1979. Nickel requirement for carbon monoxide dehydrogenase formation in *Clostridium pasteurianum. Arch. Microbiol.* **122**:177–120.

Dietrichs, D., and J. R. Andreesen. 1990. Purification and comparative studies on dihydrolipoamide dehydrogenases from anaerobic glycine utilizing bacteria *Peptostreptococcus glycinophilus, Clostridium cylindrosporum,* and *Clostridium sporogenes. J. Bacteriol.* **172**:243–251.

Dietrichs, D., M. Meyer, B. Schmidt, and J. R. Andreesen. 1990. Purification of NADPH-dependent electron-transferring flavoproteins and N-terminal protein sequence data of dihydrolipoamide dehydrogenases from anaerobic, glycine-utilizing bacteria. *J. Bacteriol.* **172**:2088–2095.

Dietrichs, D., M. Bahnweg, F. Mayer, and J. R. Andreesen. 1991a. Peripheral localization of the dihydrolipoamide dehydrogenase in the purinolytic anaerobe *Clostridium cylindrosporum*. *Arch. Microbiol.* **155**:412–414.

Dietrichs, D., M. Meyer, M. Rieth, and J. R. Andreesen. 1991b. Interaction of selenoprotein Pa and the thioredoxin system, components of the NADPH-dependent reduction of glycine in *Eubacterium acidaminophilum* and *Clostridium litorale*. *J. Bacteriol.* **173**:5983–5991.

Dietrichs, D., M. Meyer, A. Uhde, W. Freudenberg, and J. R. Andreesen. 1991c. Electron-transferring flavoproteins from glycine-metabolizing anaerobic bacteria. In: *Flavins and Flavoproteins*, B. Curti, S. Ronchi, and G. Zanetti (eds.), pp. 611–614. W. de Gruyter, Berlin.

Dilworth, G. L. 1982. Properties of the selenium-containing moiety of nicotinic acid hydroxylase from *Clostridium barkeri*. *Arch. Biochem. Biophys.* **219**:30–38.

Dürre, P., W. Andersch, and J. R. Andreesen. 1981. Isolation and characterization of an adenine-utilizing, anaerobic sporeformer, *Clostridium purinolyticum* sp. nov. *Int. J. Syst. Bacteriol.* **31**:184–194.

Dürre, P., and J. R. Andreesen. 1982a. Selenium-dependent growth and glycine fermentation by *Clostridium purinolyticum*. *J. Gen. Microbiol.* **128**:1457–1466.

Dürre, P., and J. R. Andreesen. 1982b. Separation and quantitation of purines and their anaerobic and aerobic degradation products by high-pressure liquid chromatography. *Anal. Biochem.* **123**:32–40.

Dürre, P., and J. R. Andreesen. 1982c. Anaerobic degradation of uric acid via pyrimidine derivatives by selenium-starved cells of *Clostridium purinolyticum*. *Arch. Microbiol.* **131**:255–260.

Dürre, P., and J. R. Andreesen. 1982d. Pathway of carbon dioxide reduction to acetate without a net energy requirement in *Clostridium purinolyticum*. *FEMS Microbiol. Lett.* **15**:51–56.

Dürre, P., and J. R. Andreesen. 1983. Purine and glycine metabolism by purinolytic clostridia. *J. Bacteriol.* **154**:192–199.

Dürre, P., R. Spahr, and J. R. Andreesen. 1983. Glycine fermentation via glycine reductase in *Peptococcus glycinophilus* and *Peptococcus magnus*. *Arch. Microbiol.* **134**:127–135.

Eklund, H., F. K. Gleason, and A. Holmgren. 1991. Structural and functional relations among thioredoxins of different species. *Proteins: Struct. Funct. Genet.* **11**:13–28.

Elsden, S. R., and M. G. Hilton. 1978. Volatile acid production from threonine, valine, leucine and isoleucine by clostridia. *Arch. Microbiol.* **117**:165–172.

Ferry, J. G. 1990. Formate dehydrogenase. *FEMS Microbiol. Rev.* **87**:377–382.

Ferry, J. G. 1992. Methane from acetate. *J. Bacteriol.* **174**:5489–5495.

Fendrich, C., H. Hippe, and G. Gottschalk. 1990. *Clostridium halophilum* sp. nov. and *C. litorale* sp. nov., an obligate halophilic and marine species degrading betaine in the Stickland reaction. *Arch. Microbiol.* **154**:127–132.

Freudenberg, W., and J. R. Andreesen. 1989. Purification and partial characterization of the glycine decarboxylase multienzyme complex from *Eubacterium acidaminophilum*. *J. Bacteriol.* **171**:2209–2215.

Freudenberg, W., D. Dietrichs, H. Lebertz, and J. R. Andreesen. 1989a. Isolation of an atypically small lipoamide dehydrogenase involved in the glycine decarboxylase complex from *Eubacterium acidaminophilum*. *J. Bacteriol.* **171**:1346–1354.

Freudenberg, W., F. Mayer, and J. R. Andreesen. 1989b. Immunocytochemical localization of proteins P_1, P_2, P_3 of glycine decarboxylase, and of the selenoprotein P_A of glycine reductase, all involved in anaerobic glycine metabolism of *Eubacterium acidaminophilum*. *Arch. Microbiol.* **152**:182–188.

Fuchs, G. 1986. CO_2 fixation in acetogenic bacteria: variations on a theme. *FEMS Microbiol. Rev.* **39**:181–213.

Fujiwara, K., K. Okamura, and Y. Motokawa. 1979. Hydrogen carrier protein from chicken liver: purification, characterization, and role of its prosthetic group, lipoic acid, in the glycine cleavage reaction. *Arch. Biochem. Biophys.* **197**:454–462.

Fujiwara, K., and Y. Motokawa. 1983. Mechanism of the glycine cleavage reaction. Steady state kinetic studies of the P-protein-catalyzed reaction. *J. Biol. Chem.* **258**:8156–8162.

Fujiwara, K., K. Okamura-Ikeda, and Y. Motokawa. 1984. Mechanism of glycine cleavage reaction. Further characterization of the intermediate attached to H-protein and of the reaction catalyzed by T-protein. *J. Biol. Chem.* **259**:10664–10668.

Fujiwara, K., K. Okamura-Ikeda, Y. Ohmura, and Y. Motokawa. 1986a. Mechanism of the glycine cleavage reaction: retention of the C-2 hydrogens of glycine on the intermediate attached to H-protein and evidence for the inability of serine hydroxymethyltransferase to catalyze the glycine decarboxylation. *Arch. Biochem. Biophys.* **251**:121–127.

Fujiwara, K., K. Okamura-Ikeda, and Y. Motokawa. 1986b. Chicken liver H-protein, a component of the glycine cleavage system. Amino acid sequence and identification of the N^Y-lipoylsine residue. *J. Biol. Chem.* **261**:8836–8841.

Fujiwara, K., K. Okamura-Ikeda, and Y. Motokawa. 1987. Amino acid sequence of the phosphopyridoxyl peptide from P-protein of the chicken liver glycine cleavage system. *Biochem. Biophys. Res. Commun.* **149**:621–627.

Fujiwara, K., K. Okamura-Ikeda, and Y. Motokawa. 1990. cDNA sequence, in vitro synthesis and intramitochondrial lipoylation of H-protein of the glycine cleavage system. *J. Biol. Chem.* **265**:17463–17467.

Fujiwara, K., K. Okamura-Ikeda, K. Hayasaka, and Y. Motokawa. 1991a. The primary structure of human H-protein of the glycine cleavage system deduced by cDNA cloning. *Biochem. Biophys. Res. Commun.* **176**:711–716.

Fujiwara, K., K. Okamura-Ikeda, and Y. Motokawa. 1991b. Lipoylation of H-protein of the glycine cleavage system. The effect of site directed mutagenesis of amino acid residues around the lipoyllysine residue of the lipoate attachment. *FEBS Lett.* **293**:115–118.

Fujiawara, K., K. Okamura-Ikeda, and Y. Motokawa. 1992. Expression of mature bovine H-protein of the glycine cleavage system in *Escherichia coli* and in vitro lipoylation of the apoform. *J. Biol. Chem.* **267**:20011–20016.

Garcia, G. E., and T. C. Stadtman. 1991. Selenoprotein A component of the glycine reductase complex from *Clostridium purinolyticum*: nucleotide sequence of the gene shows that selenocysteine is encoded by UGA. *J. Bacteriol.* **173**:2093–2098.

Garcia. G. E., and T. C. Stadtman. 1992. *Clostridium sticklandii* glycine reductase selenoprotein A gene: cloning, sequencing and expression in *Escherichia coli*. *J. Bacteriol.* **174**:7080–7089.

Gariboldi, R. T., and H. L. Drake. 1984. Glycine synthase of the purinolytic bacterium, *Clostridium acidiurici*. Purification of the glycine-CO_2 exchange system. *J. Biol. Chem.* **259**:6085–6089.

Gleason, F. K., and A. Holmgren. 1988. Thioredoxin and related proteins in procaryotes. *FEMS Microbiol. Rev.* **54**:271–297.

Golovchenko, N. P., B. F. Belokopytov, and V. K. Akimenko. 1983. Threonine catabolism in the bacterium *Clostridium sticklandii*. *Biochemistry (USSR)* **47**:969–974.

Gosline, J. M., M. E. DeMont, and M. W. Denny. 1986. The structure and properties of spider silk. *Endeavour, New Series* **10**:37–43.

Gottschalk, G. 1986. *Bacterial Metabolism*. Springer-Verlag, New York.

Gottwald, M., J. R. Andreesen, J. LeGall, and L. G. Ljungdahl. 1975. Presence of cytochrome and menaquinone in *Clostridium formicoaceticum* and *Clostridium thermoaceticum*. *J. Bacteriol.* **122**:325–328.

Halboth, S., and A. Klein. 1992. *Methanococcus voltae* harbors four gene clusters potentially encoding two [NiFe] and two [NiFeSe] hydrogenases, each of the cofactor F_{420}-reducing or F_{420}-non-reducing types. *Mol. Gen. Genet.* **233**:217–224.

Hammel, K. E., K. L. Cornwell, and B. B. Buchanan. 1983. Ferredoxin/flavoprotein-linked pathway for the reduction of thioredoxin. *Proc. Natl. Acad. Sci. USA* **80**:3681–3685.

Hammes, W., K. H. Schleifer, and O. Kandler. 1973. Mode of action of glycine on the biosynthesis of peptidoglycan. *J. Bacteriol.* **116**:1029–1053.

Hayasaka, K., H. Kochi, K. Hiraga, and G. Kikuchi. 1980. Purification and properties of glycine decarboxylase, a component of the glycine cleavage system, from rat liver mitochondria and immunochemical comparison of this enzyme from various sources. *J. Biochem.* **88**:1193–1199.

Heaton, M. P., R. B. Johnston, and T. L. Thompson. 1987. Controlled cell lysis and protoplast formation by enhancement of inhibitors of alanine racemase by glycine. *Biochem. Biophys. Res. Commun.* **149**:576–579.

Heider, J., C. Baron, and A. Böck. 1992. Coding from a distance: dissection of the mRNA determinants required for the incorporation of selenocysteine into proteins. *EMBO J* **11**:3759–3766.

Heijthuijsen, J. H. F. G., and T. A. Hansen. 1989a. Anaerobic degradation of betaine by marine *Desulfobacterium* strains. *Arch. Microbiol.* **152**:393–396.

Heijthiujsen, J. H. F. G., and T. A. Hansen. 1989b. Betaine fermentation and oxidation by marine *Desulfuromonas* strains. *Appl. Environ. Microbiol.* **55**:965–969.

Heise, R., V. Müller, G. Gottschalk. 1992. Presence of a sodium-translocating ATPase in membrane vesicles of the homoacetogenic bacterium *Acetobacterium woodii*. Eur. J. Biochem. **206**:553–557.

Heldt, H. W., and U. I. Flügge. 1986. Intrazellulärer Transport in grünen Pflanzenzellen. *Naturwissenschaften* **73**:1–7.

Hermann, M., H. J. Knerr, N. Mai, A. Groß, and H. Kaltwasser. 1992. Creatinine and N-methylhydantoin degradation in two newly isolated *Clostridium* species. *Arch. Microbiol.* **157**:395–401.

Hille, R. 1992. Xanthine oxidase, xanthine dehydrogenase, and aldehyde oxidase. In: *Chemistry and Biochemistry of Flavoenzymes*, F. Müller (ed.), Vol. 3, pp. 21–68. CRC Press, Boca Raton, FL.

Hiraga, K., and G. Kikuchi. 1980a. The mitochondrial glycine cleavage system. Purification and properties of glycine decarboxylase from chicken liver mitochondria. *J. Biol. Chem.* **255**:11664–11670.

Hiraga, K., and G. Kikuchi. 1980b. The mitochondrial glycine cleavage system. Functional association of glycine decarboxylase and aminomethyl carrier protein. *J. Biol. Chem.* **255**:11671–11676.

Hiraga, K., and G. Kikuchi. 19892a. The mitochondrial glycine cleavage system: differential inhibition by divalent cations of glycine synthesis and glycine decarboxylation in the glycine-CO_2 exchange. *J. Biochem.* **92**:937–944.

Hiraga, K., and G. Kikuchi. 1982b. The mitochondrial glycine cleavage system: inactivation of glycine decarboxylase as a side reaction of the glycine decarboxylation in the presence of aminomethyl carrier protein. *J. Biochem.* **92**:1489–1498.

Hiraga, K., S. Kure, M. Yamamoto, Y. Ishiguro, and T. Suzuki. 1988. Cloning of cDNA encoding human H-protein, a constituent of the glycine cleavage system. *Biochem. Biophys. Res. Commun.* **151**:758–762.

Holmgren, A. 1989. Thioredoxin and glutaredoxin systems. *J. Biol. Chem.* **264**:13963–13966.

Hormann, K., and J. R. Andreesen. 1989. Reductive cleavage of sarcosine and betaine by *Eubacterium acidaminophilum* via enzyme systems different from glycine reductase. *Arch. Microbiol.* **153**:50–59.

Hougland, A. E., and J. V. Beck. 1979. The formation of serine from glycine and formaldehyde by cell free extracts of *Clostridium acidi-urici*. *Microbios* **24**:151–157.

Hugenholtz, J., and L. G. Ljungdahl. 1989. Electron transport and electrochemical proton gradient in membrane vesicles of *Clostridium thermoautotrophicum*. *J. Bacteriol.* **171**:2873–2875.

Husic, D. W., H. D. Husic, and N. E. Tolbert. 1987. The oxidative photosynthetic carbon cycle or C₂ cycle. *CRC Crit. Rev. Plant. Sci.* **5**:45–100.

Imhoff, D. 1981. *Clostridium barkeri:* selenabhängiger Nikotinsäure- und Purinabbau sowie taxonomische Neueinodnung, Ph.D. thesis, University of Göttingen.

Kämpf, C., Y. A. Kim, A. D. Cobb, and D. B. Knaff. 1988. Inhibition of phototrophically growing *Chromatium vinosum* D by glycine. *FEMS Microbiol. Lett.* **51**:199–204.

Karlsson, J. L., and H. A. Barker. 1949. Tracer experiments on the mechanism of uric acid decomposition and acetic acid synthesis by *Clostridium acidi-urici*. *J. Biol. Chem.* **178**:891–902.

Kearny, J. J., and R. D. Sagers. 1972. Formate dehydrogenase from *Clostridium acidiurici*. *J. Bacteriol.* **109**:152–161.

Keller, B., N. Sauer, and C. J. Lamb. 1988. Glycine-rich cell wall proteins in bean: gene structure and association of the protein with the vascular system. *EMBO J.* **7**:3625–3633.

Kerscher, L., and D. Oesterhelt. 1981. The catalytic mechanism of 2-oxoacid:ferredoxin oxidoreductases from *Halobacterium halobium*. One-electron transfer at two distinct steps of the catalytic cycle. *Eur. J. Biochem.* **116**:595–600.

Kikuchi, G., and K. Hiraga. 1982. The mitochondrial glycine cleavage system. Unique features of the glycine decarboxylation. *Mol. Cell Biochem.* **45**:137–149.

Kim, Y. H., and D. J. Oliver. 1990. Molecular cloning, transcriptional characterization, and sequencing of cDNA encoding the H-protein of the mitochondrial glycine decarboxylase complex in peas. *J. Biol. Chem.* **265**:848–853.

Klein, S. M., and R. D. Sagers. 1962. Intermediary metabolism of *Diplococcus glycinophilus*. II. Enzymes of the acetate-generating system. *J. Bacteriol.* **83**:121–126.

Klein, S. M., and R. D. Sagers. 1966a. Glycine metabolism. I. Properties of the system catalyzing the exchange of bicarbonate with the carboxyl group of glycine in *Peptococcus glycinophilus*. *J. Biol. Chem.* **241**:197–207.

Klein, S. M., and R. D. Sagers. 1966b. Glycine metabolism. II. Kinetic and optical studies on the glycine decarboxylase system from *Peptococcus glycinophilus*. *J. Biol. Chem.* **241**:206–209.

Klein, S. M., and R. D. Sagers. 1967a. Glycine metabolism. III. A flavin-linked dehydrogenase associated with the glycine cleavage system in *Peptococcus glycinophilus*. *J. Biol. Chem.* **242**:297–300.

Klein, S. M., and R. D. Sagers. 1967b. Glycine metabolism. IV. Effect of borohydride reduction on the pyridoxal phosphate-containing glycine decarboxylase from *Peptococcus glycinophilus*. *J. Biol. Chem.* **242**:301–305.

Knaff, D. B., and M. Hirasawa. 1991. Ferredoxin-dependent chloroplast enzymes. *Biochim. Biophys. Acta* **1056**:93–125.

Knappe, J., and G. Sawers. 1990. A radical-chemical route to acetyl-CoA: the anaerobically induced pyruvate formate-lyase system of *Escherichia coli*. *FEMS Microbiol. Rev.* **75**:383–398.

Kochi, H., and G. Kikuchi. 1969. Reactions of glycine synthesis and glycine cleavage catalyzed by extracts of *Arthrobacter globiformis* grown on glycine. *Arch. Biochem. Biophys.* **132**:359–369.

Kochi, H., and G. Kikuchi. 1974. Mechanism of the reversible glycine cleavage reaction in *Arthrobacter globiformis* I. Purification and function of protein components required for the reaction. *J. Biochem.* **75**:1113–1127.

Kochi, H., and G. Kikuchi. 1976. Mechanism of reversible glycine cleavage reaction in *Arthrobacter globiformis*. Function of lipoic acid in the cleavage and synthesis of glycine. *Arch. Biochem. Biophys.* **173**:71–81.

Kochi, H., H. Seino, and K. Ono. 1986. Inhibition of glycine oxidation by pyruvate Á-ketoglutarate, and branched chain Á-keto acids in rat liver mitochondria: presence of interaction between the glycine cleavage system and α-keto acid dehydrogenase complexes. *Arch. Biochem. Biophys.* **249**:263–272.

Koyata, H., and K. Hiraga. 1991. The glycine cleavage system-structure of a cDNA encoding human H-protein, and partial characterization of its gene in patients with hyperglycinemias. *Am. J. Human Genet.* **48**:351–361.

Kume, A., H. Koyata, T. Sakakibara, Y. Ishiguro, S. Kure, and K. Hiraga. 1991. The glycine cleavage system. Molecular cloning of the chicken and human glycine decarboxylase cDNAs and some characteristics involved in the deduced protein structures. *J. Biol. Chem.* **266**:3323–3329.

Kure, S., H. Koyota, A. Kume, Y. Ishiguro, and K. Hiraga. 1991. The glycine cleavage system. The coupled expression of the glycine decarboxylase gene and the H-protein gene in the chicken. *J. Biol. Chem.* **266**:3330–3334.

Lebertz, H. 1984. Selenabhängiger Glycin-Stoffwechsel in anaeroben Bakterien und vergleichende Untersuchungen zur Glycin-Reduktase und zur Glycin-Decarboxylase, Ph.D. thesis, University of Göttingen.

Lebertz, H., and J. R. Andreesen. 1988. Glycine fermentation by *Clostridium histolyticum*. *Arch. Microbiol.* **150**:11–14.

Leonhardt, U., and J. R. Andreesen. 1977. Some properties of formate dehydrogenase, accumulation and incorporation of [185]W-tungsten into proteins of *Clostridium formicoaceticum*. *Arch. Microbiol.* **115**:277–224.

Ljungdahl, L. G. 1984. Other functions of folates. In: *Folates and Pterins, Vol. 1. Chemistry and Biochemistry of Folates*, R. L. Blakley, and S. J. Benkovic (eds.), pp. 555–579. Wiley, New York.

Ljungdahl, L. G. 1986. The autotrophic pathway of acetate synthesis in acetogenic bacteria. *Annu. Rev. Microbiol.* **40**:415–450.

Ljungdahl, L. G., and H. G. Wood. 1982. Acetate biosynthesis. In: B_{12}, D. Dolphin (ed.), pp. 165–202. John Wiley and Sons, Inc., New York.

Lovenberg, W., B. B. Buchanan, and J. C. Rabinowitz. 1963. Studies on the chemical nature of clostridial ferredoxin. *J. Biol. Chem.* **238**:3899–3913.

Lovitt, R. W., D. B. Kell, and J. G. Morris. 1986. Proline reduction by *Clostridium sporogenes* is coupled to vectorial proton ejection. *FEMS Microbiol. Lett.* **36**:269–273.

Lovitt, R. W., J. G. Morris, and D. B. Kell. 1987. The growth and nutrition of *Clostridium sporogenes* NCIB 8053 in defined media. *J. Appl. Bacteriol.* **62**:71–80.

Lundie, L. L., and J. G. Ferry. 1989. Activation of acetate by *Methanosarcina thermophila*. Purification and characterization of phosphotransacetylase. *J. Biol. Chem.* **264**:18392–19396.

Macherel, D., M. Lebrun, J. Gagnon, M. Neuburger, and R. Douce. 1990. cDNA cloning, primary structure and gene expression for H-protein, a component of the glycine-cleavage system (glycine decarboxylase) of pea (*Pisum sativum*) leaf mitochondria. *Biochem. J.* **268**:783–789.

Macherel, D., J. Bourguignon, and R. Douce. 1992. Cloning of the gene (gdc H) encoding H-protein, a component of the glycine decarboxylase complex of pea (*Pisum sativum* L-). *J. Biochem.* **286**:627–630.

MacKenzie, R. E. 1984. Biogenesis and interconversion of substituted tetrahydrofolates. In: *Folates and Pterins, Vol. 1. Chemistry and Biochemistry of Folates*, R. L. Blakley and S. J. Benkovic (eds.), pp. 255–306. Wiley, New York.

Matsuyama, A., H. Yamamoto, and E. Nakano. 1989. Cloning, expression and nucleotide sequence of the *Escherichia coli* K-12 ack A gene. *J. Bacteriol.* **171**:577–580.

Meinecke, B., J. Bertram, and G. Gottschalk. 1989. Purification and characterization of the pyruvate-ferredoxin oxidoreductase from *Clostridium acetobutylicum*. *Arch. Microbiol.* **152**:244–250.

Meyer, M., D. Dietrichs, B. Schmidt, and J. R. Andreesen. 1991. Thioredoxin elicits a new dihydrolipoamide dehydrogenase activity by interaction with the electron-transferring flavoprotein in *Clostridium litorale* and *Eubacterium acidaminophilum*. *J. Bacteriol.* **173**:1509–1513.

Möller, B., H. Hippe, and G. Gottschalk. 1986. Degradation of various amine compounds by mesophilic clostridia. *Arch. Microbiol.* **145**:85–90.

Motakawa, Y., and G. Kikuchi. 1974. Glycine metabolism by rat liver mitochondria: reconstitution of the reversible glycine cleavage system with partially purified protein components. *Arch. Biochem. Biophys.* **164**:624–633.

Müller, E., K. Fahlbusch, R. Walther, and G. Gottschalk. 1981. Formation of N,N-dimethylglycine, acetic acid, and butyric acid from betaine by *Eubacterium limosum*. *Appl. Environ. Microbiol.* **42**:439–445.

Mukherjee, J. J., and E. E. Dekker. 1990. 2-Amino-3-ketobutyrate CoA ligase of *Escherichia coli*: stoichiometry of pyridoxal phosphate binding and location of the pyridoxyllysine peptide in the primary structure of the enzyme. *Biochim. Biophys. Acta* **1037**:24–29.

Naumann, E., H. Hippe, and G. Gottschalk. 1983. Betaine: new oxidant in the Stickland reaction and methanogenesis from betaine and L-alanine by a *Clostridium sporogenes—Methanosarcina barkeri* coculture. *Appl. Environ. Microbiol.* **45**:474–483.

Neuburger, M., J. Bourguignon, and R. Douce. 1986. Isolation of a large complex from the matrix of pea leaf mitochondria involved in the rapid transformation of glycine into serine. *FEBS Lett.* **207**:18–22.

Neuburger, M., A. Jourdain, and R. Douce. 1991. Isolation of H-protein loaded with methylamine as a transient species in glycine decarboxylase reactions. *Biochem. J.* **278**:765–769.

Nour, J. M., and J. C. Rabinowitz. 1992. Isolation and sequencing of the cDNA coding for spinach 10-formyltetrahydrofolate synthetase. Comparisons with the yeast, mammalian, and bacterial proteins. *J. Biol. Chem.* **267**:16292–16296.

O'Brien, W. E. 1978. Inhibition of glycine synthase by branched-chain alpha-keto acids. A possible mechanism for abnormal glycine metabolism in ketotic hyperglycinemia. *Arch. Biochem. Biophys.* **189**:291–297.

Okamura-Ikeda, K. Fujiwara, and Y. Motokawa. 1982. Purification and characterization of chicken liver T-protein, a component of the glycine cleavage system. *J. Biol. Chem.* **257**:135–139.

Okamura-Ikeda, K. Fujiwara, and Y. Motokawa. 1987. Mechanism of glycine cleavage reaction. Properties of the reverse reaction catalyzed by T-protein. *J. Biol. Chem.* **262**:6746–6749.

Okamura-Ikeda, K. Fujiwara, M. Yamamoto, K. Hiraga, and Y. Motokawa. 1991. Isolation and sequence determination of cDNA encoding T-protein of the glycine cleavage system. *J. Biol. Chem.* **266**:4917–4921.

Okamura-Ikeda, K. Fujiwara, and Y. Motokawa. 1992. Molecular cloning of a cDNA encoding chicken T-protein of the glycine cleavage system and expression of the functional protein in *Escherichia coli*. Effect of mRNA secondary structure in the translational initiation region on expression. *J. Biol. Chem.* **267**:18284–18290.

Oliver, D. J., M. Neuburger, J. Bourguignon, and R. Douce. 1990. Interaction between the component enzymes of the glycine decarboxylase multienzyme complex. *Plant Physiol.* **94**:833–839.

Oppermann, F. B., B. Schmidt, and A. Steinbüchel. 1991. Purification and characterization of acetoin: 2,6-dichlorophenol-indophenol oxidoreductase, dihydrolipoamide dehydrogenase, and dihydrolipoamide acetyltransferase of the *Pelobacter carbinolicus* acetoin dehydrogenase enzyme system. *J. Bacteriol.* **173**:757–767.

Oßmer, R. 1983. Charakterisierung eines anaeroben. Gram-negativen, sporenbildenden Betainverwerters, Diplom. thesis, University of Göttingen.

Packman, L. C., B. Green, and R. N. Perham. 1991. Lipoylation of the E2 components of the 2-oxo acid dehydrogenase multienzyme complexes of *Escherichia coli*. *Biochem. J.* **277**:153–158.

Palmer, J. A., K. Hatter, and J. R. Sokatch. 1991a. Cloning and sequence analysis of the LPD-glc structural gene of *Pseudomonas putida*. *J. Bacteriol.* **173**:3109–3116.

Palmer, J. A., K. T. Madhusudhan, K. Hatter, and J. R. Sokatch. 1991b. Cloning, sequence and transcriptional analysis of the structural gene for LPD-3, the third lipoamide dehydrogenase of *Pseudomonas putida*. *Eur. J. Biochem.* **202**:231–240.

Palmer, J. A., G. Burns, K. Hatter, and J. R. Sokatch. 1992. Structure, function and evolution of multiple lipoamide dehydrogenases of *Pseudomonas putida*. In: *Pseudomonas. Molecular Biology and Biotechnology*, E. Galli, S. Silver, and B. Witholt (eds.), pp. 239–248. American Society for Microbiology, Washington, D.C.

Park, E. Y., J. E. Clark, D. V. DerVartanian, and L. G. Ljungdahl. 1991. 5,10-Methylenetetrahydrofolate reductases: iron-sulfur-zinc flavoproteins of two acetogenic clostridia. In: *Chemistry and Biochemistry of Flavoenzymes*, F. Müller (ed.), pp. 389–400. Vol. 1, CRC Press, Boca Raton, FL.

Perham, R. N. 1991. Domains, motifs, and linkers in 2-oxo acid dehydrogenase multienzyme complexes. A paradigm in the design of a multifunctional protein. *Biochemistry* **30**:8501–8512.

Plaga, W., F. Lottspeich, and D. Oesterhelt. 1992. Improved purification, crystallization and primary structure of pyruvate: ferredoxin oxidoreductase from *Halobacterium halobrium*. *Eur. J. Biochem.* **205**:391–397.

Plamann, M. D., W. D. Rapp, and G. V. Stauffer. 1983. *Escherichia coli* K12 mutants defective in the glycine cleavage enzyme system. *Mol. Gen. Genet.* **92**:15–20.

Priefert, H., S. Hein, N. Krüger, B. Schmidt, and A. Steinbüchel. 1991. Identification and molecular characterization of the *Alcaligenes eutrophus* H16 aco operon genes involved in acetoin catabolism. *J. Bacteriol.* **173**:4056–4071.

Pries, A., S. Hein, and A. Steinbüchel. 1992. Identification of a lipoamide dehydrogenase gene as second locus affected in poly(3-hydroxybutyric acid)-leaky mutants of *Alcaligenes eutrophus*. *FEMS Microbiol. Lett.* **97**:227–234.

Przybyla, A. E., J. Robbins, N. Menon, and H. D. Peck. 1992. Structure-function relationships among the nickel-containing hydrogenases. *FEMS Microbiol. Rev.* **88**:109–136.

Rabinowitz, J. C. 1963. Intermediates in purine breakdown. *Methods Enzymol.* **6**:703–713.

Rabinowitz, J. C. 1972. Preparation and properties of clostridial ferredoxins. *Methods Enzymol.* **24**:431–446.

Rabinowitz, J. C. 1975. Pyruvate-ferredoxin oxidoreductase from *Clostridium acidiurici*. *Methods Enzymol.* **41B**:334–337.

Ragsdale, S. W. 1991. Enzymology of the acetyl-CoA pathway of CO_2 fixation. *Crit. Rev. Biochem. Mol. Biol.* **26**:261–300.

Rajagopalan, K. V., and J. L. Johnson. 1992. The pterin molybdenum cofactors. *J. Biol. Chem.* **267**:10199–10202.

Rakosky, J., L. N. Zimmermann, and J. V. Beck. 1955. Guanine degradation by *Clostridium acidiurici*. II. Isolation and characterization of guanase. *J. Bacteriol.* **69**:566–570.

Ravnikar, P. D., and R. L. Somerville. 1987. Genetic characterization of a highly efficient alternate pathway of serine biosynthesis in *Escherichia coli*. *J. Bacteriol.* **169**:2611–2617.

Reece, P., D. Toth, and E. A. Dawes. 1976. Fermentation of purines and their effect on the adenylate energy charge and viability of starved *Peptococcus prevotii*. *J. Gen. Microbiol.* **97**:63–71.

Richarme, G. 1989. Purification of a new dihydrolipoamide dehydrogenase from *Escherichia coli*. *J. Bacteriol.* **171**:6580–6585.

Rieth, M. 1987. Untersuchungen zur selenabhängigen Glycinreduktase aus *Eubacterium acidaminophilum*, Ph.D. thesis, University of Göttingen.

Rienhöfer, A. 1985. Strukturelle und immunologische Untersuchungen zur Xanthin-Dehydrogenase aus *Butyribacterium barkeri*, ein Seleno-Molybdo-Eisen-Schwefel-Flavoprotein, Ph.D. thesis, University of Göttingen.

Riexinger, S. 1986. Untersuchungen zu purinhydroxylierenden Enzymen einiger anaerober Kokken, Diplom. thesis, University of Göttingen.

Robinson, J. R., and R. D. Sagers. 1972. Phosphotransacetylase from *Clostridium acidiurici*. *J. Bacteriol.* **112**:465–473.

Robinson, J. R., S. M. Klein, and R. D. Sagers. 1973. Glycine metabolism. Lipoic acid as the prosthetic group in the electron transfer protein P_2 from *Peptococcus glycinophilus*. *Biol. Chem.* **248**:5319–5323.

Rogers, W. J., B. R. Jordan, S. Rawsthorne, and A. K. Tobin. 1991. Changes of the stoichiometry of glycine decarboxylase subunits during wheat (*Triticum aestivum* L) and pea (*Pissum sativum* L) leaf development. *Plant Physiol.* **96**:952–956.

Russel, M., and P. Model. 1988. Sequence of thioredoxin reductase from *Escherichia coli:* relationship to other flavoprotein disulfide oxidoreductases. *J. Biol. Chem.* **263**:9015–9019.

Sagers, R. D., and J. V. Beck. 1956. Studies on the formation of formate, glycine, serine, pyruvate, and acetate from purines by *Clostridium acidiurici*. *J. Bacteriol.* **72**:199–208.

Sagers, R. D., and I. C. Gunsalus. 1961. Intermediary metabolism of *Diplococcus glycinophilus*. I. Glycine cleavage and one-carbon interconversions. *J. Bacteriol.* **81**:541–549.

Sagers, R. D., M. Benziman, and I. C. Gunsalus. 1961. Acetate formation in *Clostridium acidi-urici:* acetokinase. *J. Bacteriol.* **82**:233–238.

Sagers, R. D., M. Benziman, and S. M. Klein. 1963. Failure of arsenate to uncouple the phosphotransacetylase system in *Clostridium acidi-urici*. *J. Bacteriol.* **86**:978–984.

Schäfer, T., M. Selig, and P. Schönheit. 1992. Acetyl-CoA synthetase (ADP forming) in archaea, a novel enzyme involved in acetate formation and ATP synthesis. *Arch. Microbiol.* **159**:72–83.

Scherer, P. A., and R. K. Thauer. 1978. Purification and properties of reduced ferredoxin : CO_2 oxidoreductase from *Clostridium pasteurianum*, a molybdenum iron-sulfur protein. *Eur. J. Biochem.* **85**:125–135.

Schiefer-Ullrich, H., and J. R. Andreesen. 1985. *Peptostreptococcus barnesae* sp. nov., a Gram-positive, anaerobic, obligately purine utilizing coccus from chicken feces. *Arch. Microbiol.* **143**:26–31.

Schiefer-Ullrich, H., R. Wagner, P. Dürre, and J. R. Andreesen. 1984. Comparative studies on physiology and taxonomy of obligately purinolytic clostridia. *Arch. Microbiol.* **138**:345–353.

Schirch, L. V. 1982. Serine hydroxymethyltransferase. *Adv. Enzymol.* **53**:83–112.

Schirch, L. V. 1984. Folates in serine and glycine metabolism. In: *Folates and Pterins, Vol. 1. Chemistry and Biochemistry of Folates,* R. L. Blakley, and S. J. Benkovic (eds.), pp. 399–431. Wiley, New York.

Schirch, V., S. Hopkins, E. Villar, and S. Angelaccio. 1985. Serine hydroxymethyltransferase from *Escherichia coli:* purification and properties. *J. Bacteriol.* **163**:1–7.

Schirch, V., and W. B. Strong. 1989. Interaction of folylpolyglutamates with enzymes in one-carbon metabolism. *Arch. Biochem. Biophys.* **269**:371–380.

Schleicher, A. 1990. Anaerober Abbau von Kreatinin und Kreatin unter Beteiligung von Reduktase-Reaktionen, Diplom. thesis, University of Göttingen.

Schmitz, R. A., M. Richter, D. Linder, and R. K. Thauer. 1992a. A tungsten-containing active formylmethanofuran dehydrogenase in the thermophilic archaeon *Methanobacterium wolfei. Eur. J. Biochem.* **207**:559–565.

Schmitz, R. A., S. P. J. Albrecht, and R. K. Thauer. 1992b. A molybdenum and a tungsten isoenzyme of formylmethanofuran dehydrogenase in the thermophilic archaeon *Methanobacterium wolfei. Eur. J. Biochem.* **209**:1013–1018.

Schräder, T. 1991. Isolierung und Charakterisierung von Protein C der Glycin-Reduktase aus *Eubacterium acidaminophilum,* Diplom. thesis, University of Göttingen.

Schräder, T., and J. R. Andreesen. 1992. Purification and characterization of protein Pc, a component of glycine reductase from *Eubacterium acidaminophilum. Eur. J. Biochem.* **206**:79–85.

Schulman, M., D. Parker, L. G. Ljungdahl, and H. G. Wood. 1972. Total synthesis of acetate from CO_2. V. Determination by mass analysis of the different types of acetate formed from $^{13}CO_2$ by heterotrohic bacteria. *J. Bacteriol.* **109**:633–644.

Scrutton, N. S., A. Berry, and R. B. Perham. 1990. Redesign of the coenzyme specificity of a dehydrogenase by protein engineering. *Nature* **343**:38–43.

Seitz, H. J., B. Schink, N. Pfennig, and R. Conrad. 1990. Energetics of synthrophic ethanol oxidation in defined chemostat cocultures. 1. Energy requirement for H_2 production and H_2 oxidation. *Arch. Microbiol.* **155**:82–88.

Shimizu, S., J. M. Kim, and H. Yamada. 1989. Microbial enzymes for creatinine assay: a review. *Clin. Chim. Acta* **185**:241–252.

Sliwkowski, M. X., and T. C. Stadtman. 1987. Purification and immunological studies of selenoprotein A of the clostridial glycine reductase complex. *J. Biol. Chem.* **262**:4899–4904.

Sliwkowski, M. X., and T. C. Stadtman. 1988a. Selenoprotein A of the clostridial glycine reductase complex: purification and amino acid sequence of the selenocysteine-containing peptide. *Proc. Natl. Acad. Sci.* USA **85**:368–371.

Sliwkowski, M. X., and T. C. Stadtman. 1988b. Selenium-dependent glycine reductase: differences in physicochemical properties and biological activities of selenoprotein A components isolated from *Clostridium sticklandii* and *Clostridium purinolyticum. Biofactor* **1**:293–296.

Sly, W. S., and E. R. Stadtman. 1963. Formate Metabolism. II. Enzymatic synthesis of formyl phosphate and formyl coenzyme A in *Clostridium cylindrosporum. J. Biol. Chem.* **238**:2639–2647.

Smith, S. T., K. V. Rajagopalan, and P. Handler. 1967. Purification and properties of xanthine dehydrogenase from *Micrococcus lactilyticus. J. Biol. Chem.* **242**:4108–4117.

Smithers, G. W., H. Jahansouz, J. L. Kofron, R. H. Himes, and G. H. Reed. 1987. Substrate activity of synthetic formyl phosphate in the reaction catalyzed by formyltetrahydrofolate synthetase. *Biochemistry* **26**:3943–3948.

Sokatch, J. R., and G. Burns. 1984. Oxidation of glycine by *Pseudomonas putida* requires a specific lipoamide dehydrogenase. *Arch Biochem. Biophys.* **228**:660–666.

Spahr, R. 1982. Anaerober Aminosäure- und Purinabbau durch einige Arten der Gattung *Peptococcus,* Diplom. thesis, University of Göttingen.

Srinivasan, R., and D. J. Oliver. 1992. H-protein of the glycine decarboxylase multienzyme complex. Complementary DNA encoding the protein from *Arabidopsis thaliana. Plant Physiol.* **98**:1518–1519.

Staben, C., T. R. Whitehead, and J. C. Rabinowitz. 1987. Heparin-Agarose chromatography for the purification of tetrahydrofolate utilizing enzymes: C$_1$-tetrahydrofolate synthase and 10-formyltetrahydrofolate synthetase. *Anal. Biochem.* **162**:257–264.

Stadtman, T. C. 1965. Electron transport proteins of *Clostridium sticklandii.* In: *Non-Heme-Iron Proteins, Role in Energy Conservation,* A. San Pietro (ed.), pp. 439–445. Antioch Press, Yellow Springs, OH.

Stadtman, T. C. 1978. Selenium-dependent clostridial glycine reductase. *Methods Enzymol.* **53**:373–382.

Stadtman, T. C. 1989. Clostridial glycine reductase Protein C, the acetyl group acceptor catalyzes the arsenate-dependent decomposition of acetyl phosphate. *Proc. Natl. Acad. Sci.* USA **86**:7853–7856.

Stadtman, T. C., and P. Elliott. 1956. A new ATP-forming reaction: the reductive deamination of glycine. *J. Am. Chem. Soc.* **78**:2020–2021.

Stadtman, T. C., P. Elliott, and L. Tiemann. 1958. Studies on the enzymic reduction of amino acids. III. Phosphate esterification coupled with glycine reduction. *J. Biol. Chem.* **231**:961–973.

Stadtman, E. R., T. C. Stadtman, I. Pastan, and L. D. Smith. 1972. Clostridium barkeri sp.n. *J. Bacteriol.* **110**:758–760.

Stadtman, T. C., and J. N. Davis. 1991. Glycine reductase protein C. Properties and characterization of its role in the reductive cleavage of Se-carboxymethyl-selenoprotein A. *J. Biol. Chem.* **266:**22147–22153.

Stauffer, L. T., M. D. Plamann, and G. V. Stauffer. 1986. Cloning and characterization of the glycine-cleavage enzyme system of *Escherichia coli*. *Gene* **44:**219–226.

Stauffer, G. V., L. T. Stauffer, and M. D. Plamann. 1989. The *Salmonella typhimurium* glycine cleavage enzyme. *Mol. Gen. Genet.* **220:**154–156.

Steiert, J. G., R. J. Rolfes, H. Zalkin, and G. V. Stauffer. 1990a. Regulation of the *Escherichia coli* gly A gene by the pur R gene product. *J. Bacteriol.* **172:**3799–3803.

Steiert, P. S., L. T. Stauffer, and G. V. Stauffer. 1990b. The lpd gene product functions as the L protein in the *Escherichia coli* glycine cleavage enzyme system. *J. Bacteriol.* **172:**6142–6144.

Steiert, J. G., Ch Kubu, and G. V. Stauffer. 1992. The PurR binding site in the GlyA promoter region of *Escherichia coli*. *FEMS Microbiol. Lett.* **99:**299–304.

Stephens, P. E., H. M. Lewis, M. G. Darlison, and J. G. Guest. 1983. Nucleotide sequence of the lipoamide dehydrogenase gene of *Escherichia coli* K12. *Eur. J. Biochem.* **135:**519–527.

Stickland, L. H. 1934. The chemical reaction by which *Cl. sporogenes* obtains its energy. *Biochem. J.* **28:**1746–1759.

Stöcklein, W., and H. L. Schmidt. 1985. Evidence for L-threonine cleavage and allo-threonine formation by different enzymes from *Clostridium pasteurianum*: threonine aldolase and serine hydroxymethyltransferase. *Biochem. J.* **232:**621–622.

Stover, P., and V. Schirch. 1990. Serine hydroxymethyltransferase catalyzes the hydrolysis of methenyltetrahydrofolate to 5-formyltetrahydrofolate. *J. Biol. Chem.* **265:**14227–14233.

Stover, P., M. Zamora, K. Shostak, M. Gantam-Basak, and V. Schirch. 1992. *Escherichia coli* serine hydroxymethyltransferase. The role of histidine 228 in determining reaction specificity. *J. Biol. Chem.* **267:**17679–17687.

Strobl, G., R. Feicht, H. White, F. Lottspeich, and H. Simon. 1992. The tungsten-containing aldehyde oxidoreductase from *Clostridium thermoaceticum* and its complex with a viologen-accepting NADPH oxidoreductase. *Biol. Chem. Hoppe-Seyler* **373:**123–132.

Stupperich, E., P. Aulkemeyer, and C. Eckerskorn. 1992. Purification and characterization of a methanol-induced cobamide-containing protein from *Sporomusa ovata*. *Arch. Microbiol.* **158:**370–373.

Tanaka, H., and T. C. Stadtman. 1979. Selenium-dependent clostridial glycine reductase. Purification and characterization of the two membrane-associated protein components. *J. Biol. Chem.* **254:**447–452.

Tanner, R. S., E. Stackebrandt, G. E. Fox, and C. R. Woese. 1981. A phylogenetic analysis of *Acetobacterium woodii*, *Clostridium barkeri*, *Clostridium butyricum*, *Clostridium lituseburense*, *Eubacterium limosum*, and *Eubacterium tenue*. *Curr. Microbiol.* **5:**35–38.

Thauer, R. K. 1972. CO_2-reduction to formate by NADPH. The initial step in the total synthesis of acetate from CO_2 in *Clostridium thermoaceticum*. *FEBS Lett.* **27**:111–115.

Thauer, R. K. 1973. CO_2 reduction to formate in *Clostridium acidi-urici*. *J. Bacteriol.* **114**:443–444.

Thauer, R. K. 1988. Citric-acid cycle, 50 years on. Modifications and an alternative pathway in anaerobic bacteria. *Eur. J. Biochem.* **176**:497–508.

Thauer, R. K., K. Jungermann, and K. Decker. 1977. Energy conservation in chemotrophic anaerobic bacteria. *Bacteriol. Rev.* **41**:100–180.

Thauer, R. K., D. Möller-Zinkhan, and A. M. Spormann. 1989. Biochemistry of acetate catabolism in anaerobic chemotrophic bacteria. *Annu. Rev. Microbiol.* **43**:43–67.

Tobin, A. K., J. R. Thorpe, C. M. Hylton, and S. Rawsthorne. 1989. Spatial and temporal influences on the cell-specific distribution of glycine decarboxylase in leaves of wheat (*Tritium aestivum* L.) and pea (*Pisum sativum* L). *Plant Physiol.* **91**:1219–1225.

Todhunter, J. A., and D. L. Purich. 1974. Evidence for the formation of a phosphorylated glutamyl residue in the *Escherichia coli* acetate kinase. *Biochem. Biophys. Res. Commun.* **60**:273–280.

Tschech, A., and N. Pfennig. 1984. Growth yield increase linked to coffeate reduction in *Acetobacterium woodii*. *Arch. Microbiol.* **137**:163–167.

Turner, D. C., and T. C. Stadtman. 1973. Purification of protein components of the clostridial glycine reductase system and characterization of protein A as a selenoprotein. *Arch. Biochem. Biophys.* **154**:366–381.

Turner, S. R., R. Ireland, and S. Rawsthorne. 1992a. Cloning and characterization of the P subunit of glycine decarboxylase from pea (*Pissum sativum*). *J. Biol. Chem.* **267**:5355–5360.

Turner, S. R., R. Ireland, and S. Rawsthorne. 1992b. Purification and primary amino acid sequence of the L subunit of glycine decarboxylase. Evidence for a single lipoamide dehydrogenase in plant mitochondria. *J. Biol. Chem.* **267**:7745–7750.

Tziaka, C. 1987. Untersuchungen zur Charakterisierung von Stämmen der Gattungen *Peptococcus* und *Peptostreptococcus*, Ph.D. thesis, University of Göttingen.

Uhde, A. 1990. Wachstumsphysiologische Untersuchungen zum Abbau von Aminosäuren und mögliche Funktion eines elektronentransferierenden Flavoproteins bei *Clostridium sticklandii*. Diplom. thesis, University of Göttingen.

Uyeda, K., and J. C. Rabinowitz. 1965. Metabolism of formiminoglycine. Glycine formiminotransferase. *J. Biol. Chem.* **240**:1701–1710.

Uyeda, K., and J. C. Rabinowitz. 1967a. Metabolism of formiminoglycine. Formiminotetrahydrofolate cyclodeaminase. *J. Biol. Chem.* **242**:24–31.

Uyeda, K., and J. C. Rabinowitz. 1967b. Enzyme of clostridial purine fermentation. Methylenetetrahydrofolate dehydrogenase. *J. Biol. Chem.* **242**:4378–4385.

Uyeda, K., and J. C. Rabinowitz. 1968. Enzymes of the clostridial purine fermentation: serine hydroxymethyltransferase. *Arch. Biochem. Biophys.* **123**:271–278.

Uyeda, K., and J. C. Rabinowitz. 1971. Pyruvate: ferredoxin oxidoreductase. III. Purification and properties of the enzyme. *J. Biol. Chem.* **246**:3111–3119.

Valentine, R. C., W. J. Brill, and R. D. Sagers. 1963. Ferredoxin linked DPN reduction by pyruvate in extracts of *Clostridium acidi-urici*. *Biochem. Biophys. Res. Commun.* **12**:315–319.

van Poelje, P. D., and E. E. Snell. 1990. Pyruvoyl dependent enzymes. *Annu. Rev. Biochem.* **59**:29–59.

Venugopalan, V. 1980. Influence of growth conditions on glycine reductase of *Clostridium sporogenes*. *J. Bacteriol.* **141**:386–388.

Vogels, G. D., and C. van der Drift. 1976. Degradation of purines and pyrimidines by microorganisms. *Bacteriol. Rev.* **40**:403–468.

Waber, J. L., and H. G. Wood. 1979. Mechanism of acetate synthesis from CO_2 by *Clostridium acidi-urici*. J. Bacteriol. **140**:468–478.

Wächtershäuser G. 1988. Before enzymes and templates: theory of surface metabolism. *Microbiol. Rev.* **52**:452–484.

Wächtershäuser, G. 1990. Evolution of first metabolic cycles. *Proc. Natl. Acad. Sci. USA* **87**:200–204.

Wahl, R. C., and W. H. Orme-Johnson. 1987. Clostridial pyruvate oxidoreductase and the pyruvate-oxidizing enzyme specific to nitrogen fixation in *Klebsiella pneumoniae* are similar enzymes. *J. Biol. Chem.* **262**:10489–10496.

Wagner, R., and J. R. Andreesen. 1977. Differentiation between *Clostridium acidiurici* and *Clostridium cylindrosporum* on the basis of specific metal requirements for formate dehydrogenase formation. *Arch. Microbiol.* **114**:219–224.

Wagner, R., and J. R. Andreesen. 1979. Selenium requirement for active xanthine dehydrogenase from *Clostridium acidiurici* and *Clostridium cylindrosporum*. *Arch. Microbiol.* **121**:255–260.

Wagner, R., R. Cammack, and J. R. Andreesen. 1984. Purification and characterization of xanthine dehydrogenase from *Clostridium acidiurici* grown in the presence of selenium. *Biochim. Biophys. Acta* **791**:63–74.

Wagner, R., and J. R. Andreesen. 1987. Accumulation and incorporation of [185]W-tungsten into proteins of *Clostridium acidiurici* and *Clostridium cylindrosporum*. *Arch. Microbiol.* **147**:295–299.

Wagner, A. F. V., M. Frey, F. A. Neugebauer, W. Schäfer, and J. Knappe. 1992. The free radical in pyruvate formate-lyase is located on glycine-734. *Proc. Natl. Acad. Sci. USA* **89**:996–1000.

Walker, T. L., and D. J. Oliver 1986. Glycine decarboxylase multienzyme complex. Purification and partial characterization from pea leaf mitochondria. *J. Biol. Chem.* **261**:2214–2221.

Westphal, A. H., A. de Kok. 1988. Lipoamide dehydrogenase from *Azotobacter vinelandii*. Molecular cloning, organization and sequence analysis of the gene. *Eur. J. Biochem.* **172**:299–305.

White, H., R. Feicht, C. Huber, F. Lottspeich, and H. Simon. 1991. Purification and some properties of the tungsten-containing carboxylic acid reductase from *Clostridium formicoaceticum*. *Biol. Chem. Hoppe-Seyler* **372**:999–1005.

Whitehead, T. R., and J. C. Rabinowitz. 1988. Nucleotide sequence of the *Clostridium acidiurici* ("*Clostridium acidi-urici*") gene for 10-formyltetrahydrofolate synthetase shows extensive amino acid homology with the trifunctional enzyme C_1-tetrahydrofolate synthase from *Saccharomyces cerevisiae*. *J. Bacteriol.* **170**:3255–3261.

Whitehead, T. R., M. Park, and J. C. Rabinowitz. 1988. Distribution of 10-formyltetrahydrofolate synthetase in eubacteria. *J. Bacteriol.* **170**:995–997.

Whiteley, H. R. 1952. The fermentation of purines by *Micrococcus aerogenes*. *J. Bacteriol.* **63**:163–175.

Wieland, F. 1988. Structure and biosynthesis of prokaryotic glycoproteins. *Biochemie* **70**:1493–1504.

Williams, C. H. 1992. Lipoamide dehydrogenase, glutathione reductase, thioredoxin reductase, and mercuric ion reductase—a family of flavoenzyme transhydrogenases. In: *Chemistry and Biochemistry of Flavoenzymes,* F. Müller (ed.), Vol. 3, pp. 121–211. CRC Press, Boca Raton, FL.

Wilson, R., M. Urbanowski, and G. Stauffer. 1992. Negative regulation of the *Escherichia coli* glycine cleavage enzyme system by the purR gene product, Abstr. H-306, p. 234. *Gen Meeting ASM*.

Winter, J., F. Schindler, and F. X. Wildenauer. 1987. Fermentation of alanine and glycine by pure and syntrophic cultures of *Clostridium sporogenes*. *FEMS Microbiol. Ecol.* **45**:153–161.

Winzer, K. 1992. Reinigung und Charakterisierung der Acetat-Kinase aus *Acetobacterium woodii*, Diplom. thesis, University of Göttingen.

Wohlfahrt, G., G. Geerligs, and G. Diekert. 1990. Purification and properties of a NADH-dependent 5,10-methylenetetrahydrofolate reductase from *Peptostreptococcus productus*. *Eur. J. Biochem.* **192**:411–417.

Wood, H. G. 1989. Past and present of CO_2 utilization. In: *Autotrophic Bacteria*. H. G. Schlegel and B. Bowien (eds.), pp. 33–52. Springer-Verlag, Berlin.

Wood, H. G., S. W. Ragsdale, and E. Pezacka. 1986. The acetyl-CoA pathway: a newly discovered pathway of autotrophic growth. *TIBS* **11**:14–18.

Wood, H. G., and L. G. Ljungdahl. 1991. Autotrophic character of the acetogenic bacteria. In: *Variations in Autotrophic Life,* J. M. Shively and L. L. Barton (eds.), pp. 201–250. Academic Press, London.

Woolfolk, C. A., B. S. Woolfolk, and H. R. Whiteley. 1970. 2-Oxopurine dehydrogenase from *Micrococcus aerogenes*. *J. Biol. Chem.* **245**:3167–3178.

Yamamoto, J., T. Saiki, S. M. Liu, and L. G. Ljungdahl. 1983. Purification and properties of NADP-dependent formate dehydrogenase from *Clostridium thermoaceticum*, a tungsten-selenium-iron protein. *J. Biol. Chem.* **258**:1826–1832.

Yamamoto, M., H. Koyata, C. Matsui, and K. Hiraga. 1991. The glycine cleavage system. Occurrence of two types of chicken H-protein mRNAs presumably formed by the alternative use of the polyadenylation consensus sequences in a single exon. *J. Biol. Chem.* **266**:3317–3322.

Yoch, D. C., and R. P. Carithers. 1979. Bacterial iron-sulfur proteins. *Microbiol. Rev.* **43**:384–421.

Zieske, L. R., and L. Davis. 1983. Decarboxylation of glycine by serine hydroxymethyl-transferase in the presence of lipoic acid. *J. Biol. Chem.* **258**:10355–10359.

Zindel, U., W. Freudenberg, M. Rieth, J. R. Andreesen, J. Schnell, and F. Widdel. 1988. *Eubacterium acidaminophilum* sp. nov., a versatile amino-acid-degrading anaerobe producing or utilizing H_2 or formate. Description and enzymatic studies. *Arch. Microbiol.* **150**:254–266.

Zhilina, T. N., and G. A. Zavarsin. 1990. Extremely halophilic, methylotrophic, anaerobic bacteria. *FEMS Microbiol. Rev.* **87**:315–321.

Index